A Guide to the
Wildflowers of South Carolina

A Guide to the
Wildflowers of
South Carolina

REVISED AND EXPANDED EDITION

Patrick D. McMillan, Richard D. Porcher Jr.,
Douglas A. Rayner, and David B. White

THE UNIVERSITY OF
SOUTH CAROLINA PRESS

© 2022 University of South Carolina

Published by the University of South Carolina Press
Columbia, South Carolina 29208

www.uscpress.com

Manufactured in Korea

31 30 29 28 27 26 25 24 23 22
10 9 8 7 6 5 4 3 2 1

Library of Congress Cataloging-in-Publication Data
can be found at http://catalog.loc.gov/.

ISBN: 978-1-64336-262-5 (hardcover)
ISBN: 978-1-64336-263-2 (paperback)
ISBN: 978-1-64336-264-9 (ebook)

Frontispiece: *Lilium pyrophilum,* photograph by Will Stuart

We dedicate this book to Dr. Wade Thomas Batson (1912–2015), distinguished professor emeritus in the Biological Sciences Department at the University of South Carolina. Dr. "B," as generations of students affectionately called him, was the major professor of Richard D. Porcher Jr. and Douglas A. Rayner when they were graduate students at the university. Both authors entered graduate school intending on careers in zoology but changed to botany after coming under his influence. Field botany became their life's work and passion. This passion was passed down to their pupils, including Patrick D. McMillan, and he passes the love of the amazing world of plants to his students today.

Undergraduate and graduate students were drawn to Dr. B's charisma and his love of field botany. He opened a world unknown to many. Those who spent a semester enrolled in his spring, fall, or summer flora course will never forget the experience. Dr. B's enthusiasm for the flora of South Carolina was contagious. The woods and fields no longer were a mass of indistinguishable brown and green; they became a garden of wood lilies or windflowers. Long after leaving school, students recalled the scientific names of plants they learned in his course and practiced the environmental ethics he taught. He worked hard to make his courses stimulating and informative and was a friend to every student.

Whatever measure of success this book achieves, much of it is owed to Dr. Batson. We thank him for being our mentor and friend and for beginning our life's journey into the world of botany, and particularly for our appreciation of wildflowers.

CONTENTS

PART 2: Species Descriptions and Color Plates

LIST OF FIGURES

PREFACE

The second edition of this state wildflower book was a collaborative effort that was identified as a needed revision and expansion of the first edition due to the dramatic advances that have been made in the past twenty years in plant taxonomy and ecology. Much has changed. Richard Porcher approached me to organize, edit, and add to the compendium of knowledge contained in the first edition. The changes will be immediately obvious to those familiar with the first edition. The number of species included in the book has expanded, the information aiding in the identification and understanding of each species has expanded, and many of the scientific names have been updated. I wanted to give more information on the ranges of each of the species included and that required the creation of accurate county-level range maps. These range maps are the result of examination of over 200,000 herbarium specimens. Images of the vast majority of specimens collected in South Carolina are now available to be viewed online via the SouthEast Regional Network of Expertise and Collections (SERNEC Portal). The challenges presented with access to physical locations of the many museums and universities where these specimens are housed was made exponentially easier with this tool.

Advances in photography have also allowed a much more thorough illustration of the key traits used in identifying each of the species. The first edition depended on slide film. The advancement in digital technology and file sharing has allowed us to choose to represent plants that we desire to include rather than those we have available. The challenge of providing the best images possible required additional expertise. The help of my long-time videographer, photographer, and visual arts expert, David B. White, added a dedicated and extraordinary talent to our team.

The challenges in preparing any manuscript of this scope and nature are taxing both physically and mentally. The collection of images alone has taken Dr. Porcher across the state innumerable times. I am still amazed by the perseverance and indefatigable manner of this vivacious man as he scales up the side of a rock to photograph a small fern. The expertise and additions offered by Douglas Rayner have contributed greatly to the accuracy and expanded the information provided on the medicinal and ecological characteristics. The opportunity to complete this book was ironically assisted by the current global challenge presented by the COVID-19 pandemic. The opportunity to focus most of our attention toward this one project was a boon to the speed at which it has been completed. It has been the pleasure of a lifetime to be able to work with my mentor, Richard D. Porcher Jr., and one of my botanical heroes, Douglas A. Rayner, on this book. Dr. Porcher stated in the preface to the first edition, "we are proud of this final product," and we all are equally proud of this edition. May it bring students of botany countless years of pleasure.

PATRICK D. MCMILLAN

ACKNOWLEDGMENTS

A book of this scope is never the sole product of the authors. Rather, this book was built on the knowledge and data accumulated by many dedicated and talented botanists. Just as this book has made use of that knowledge, the authors hope that botanists will use the material added in *A Guide to the Wildflowers of South Carolina* when future botanical books on South Carolina are written.

This work would not have been possible without the support of Clemson University. The ability to utilize valuable time to complete this book provided by Clemson University Public Service Agriculture as well as the financial support provided by the Glenn and Heather Hilliard Professorship are gratefully acknowledged. Glenn and Heather have been more than patrons, they have become family. Their passion for preserving and enjoying the natural world is evident in the constant efforts and innumerable successes visible as preserved lands today. The publication costs for this book were made possible by their endowment. The patience and support of the staff at the South Carolina Botanical Garden are also greatly appreciated. The staff were more than considerate in allowing the time to complete this book. The support and guidance provided by Dr. George Askew, vice-president of Public Service and Agriculture, and the Clemson University provost, Dr. Bob Jones, are particularly appreciated. No work which takes this much commitment is possible without the support and patience of family, and Patrick McMillan would like to recognize the constant support and assistance provided by his family, Waynna McMillan, Nicholas McMillan, Isabella Wyatt, and Mattie Wyatt.

A book of this scope is dependent on a critical review for technical, scientific accuracy. We are deeply grateful to Keith Bradley and Edward Pivorun for their review of the manuscript and helpful comments and appropriate questions throughout the project. Keith Bradley provided many hours of assistance in locating some of the more difficult species to locate as well as contributing some of his own photographs and assuring technical accuracy through thorough review and editorial suggestions. Mr. Bradley has been the primary sounding board for the first author. This book would not be possible without his expertise and guidance. Indeed, it is rare to find an individual so dedicated to botany, conservation, and assisting others as Mr. Bradley. We are also deeply indebted to Dr. Alan Weakley of the University of North Carolina at Chapel Hill for his aid, corrections, and tolerance of numerous questions concerning the names and distribution of the flora.

We also wish to thank Aurora Bell and the staff of the University of South Carolina Press for their work in bringing this edition to publication.

Numerous colleagues and friends provided suggestions on the manuscript, helped with field photography, and gave encouragement for the project. J. Drew Lanham, Susan Watts, Isabella Wyatt and Waynna McMillan helped significantly with the formatting, editing, and production of the text.

Richard Porcher thanks his collogues and friends for helping bring this book to fruition. John Brubaker, Joel Gramling, Will Stuart, John Nelson, Keith Bradley, Celie Dailey, Jim Fowler, and Harry Shealy have accompanied him on many trips photographing in the field, carrying his equipment and helping in any way needed. When Dr. Porcher needed a place to stay while photographing in the Upstate, Grace and Allison Wilder, Catherine and Milt Boykin, Eva and Sam Pratt, Dabney Peeples and Art Campbell, and Waynna and Patrick McMillan were "home away from home" families. Whatever success this book gains is due, in part, to these collogues and friends.

We gratefully acknowledge the use of the beautiful photographs of plants and habitats provided by Edward Pivorun, Keith Bradley, Alan Cressler, Jim Fowler, Will Stuart, and Bruce Sorrie.

Dr. Rayner would like to thank Wofford College for providing access to computer infrastructure and IT support, as well as access to an office copier. He also especially thanks his wife, Nancy Riser Rayner, for her patience and good humor. Should a recently retired professor need to spend six hours a day "at work?"

As an acknowledgment of their passion for the profession, the authors donate all royalties to support the conservation and education efforts at the South Carolina Botanical Garden and to support future studies of botany in South Carolina at Wofford College.

Purpose and Scope

People are so accustomed to the domesticated landscape of South Carolina that it is hard to imagine the ecosystems that once existed. From the mountains to the seashore, the land was shaped only by the forces of nature and Native Americans. Pine savannas, with their plethora of fire-adapted herbs, Piedmont prairies and oak savannas teeming with wildlife, mountain coves with showy wildflowers, and maritime forests with majestic live oaks all gave evidence of a bountiful land. One has only to read the accounts of the early naturalists who lived in or passed through the state to understand the scope of our lost heritage. Agricultural conversion, urban sprawl, countless roads, and alien species have all contributed to today's mostly domesticated landscape. Still, unparalleled natural beauty is present in the state. The authors hope *Wildflowers of South Carolina* captures glimpses from this earlier era so that readers can experience our predecessors' views of South Carolina, while at the same time appreciating present wildflower communities and the forces that have shaped and will continue to shape the distribution and diversity of South Carolina's natural communities and wildflowers.

Wildflowers of South Carolina includes color photographs of 997 wildflowers and their natural communities and is intended primarily to help amateur botanists identify many native and naturalized wildflowers that grow in South Carolina, as well as in adjacent states. It provides basic information on botanical natural history that hopefully will stimulate interest, enjoyment, conservation, and management of wildflowers. At the same time, technical information in the book not readily available from other sources will be useful to academicians and educators, as well as staff of governmental and private organizations.

This book emphasizes a habitat approach to wildflower identification; therefore, one section describes the natural communities (habitats) where native wildflowers abound along with locations where they can be found. The South Carolina Department of Natural Resources acquired many of these sites, establishing them as state heritage preserves to protect these natural communities and rare species.

A Guide to the Wildflowers of South Carolina separates the weedy species, many of which are nonnative, into a separate category, the ruderal communities. This provides the reader with a useful message: The balance between native and nonnative species has shifted greatly over the years. While some nonnative plants, such as Callery Pear, from Asia, are quite common, some native plants, such as Oconee Bells, are rare today. People contribute to this ruderal shift by intentionally or unwittingly introducing nonnative plants along roadsides and into gardens, where they escape and become established (that is, naturalized). Today, 24% (893 species) of the total plant diversity found in South Carolina are considered introduced from elsewhere. Grouping the majority of the nonnative species into the ruderal communities emphasizes this situation.

Perhaps the most diverse and useful part of the book is the comments entry that follows each species description. The comments include interesting facts about each species, many of which are not widely known or readily available to readers. Some of the information includes: (1) whether a plant is a source for drugs, (2) whether a native plant makes a good cultivated plant, (3) whether a plant is poisonous, (4) the ecological parameters that are necessary for the survival of a plant, and (5) the origin of a species' common or scientific name.

A Guide to the Wildflowers of South Carolina will also be useful to persons living in other adjacent and nearby states, since many of the wildflowers found in South Carolina are found in these areas. Although several recent books cover the wildflowers of these states as well as the Southeast (see "General References"), this book contains treatments and photographs of many plants not included in other books and thus serves as a valuable supplement.

Another goal of this book is to provide a text that can be used in academic courses; therefore, as much information as possible is included on the general botany of South Carolina. This second edition of *A Guide to the Wildflowers of South Carolina* is also intended to complement the first edition (hereinafter abbreviated as *PR,* for the two authors Porcher and Rayner). The second edition elaborates on the treatment contained in this seminal work and includes new and different sections.

The authors hope that stimulating interest in wildflowers will result in more eyes searching the state for rare plants and unique habitats, which can then be recorded with the South Carolina Department of Natural Resources' Heritage Trust Program. Somewhere along a river bluff is an undiscovered limestone outcrop with Wagner's spleenwort, or a mountain cove harboring an undiscovered population of Oconee Bells, or a Native American shell mound with a unique assemblage of calcicoles (plants that thrive in calcium-rich soils). As botanists who have conducted fieldwork for the majority of three lifetimes, the authors still find unrecorded, unique natural areas and rare species. But the scientific community needs the public's help. Organizations such as The Nature Conservancy of South Carolina help preserve natural areas, but first they must know of their existence. Readers who find sites worth recording are encouraged to contact the South Carolina Heritage Trust Program.

Wildflowers are part of nature's grand picture. Wildflowers in natural communities are part of a complex system where energy flow, nutrient cycling, and life histories of all organisms—from soil microbes to higher plants and animals—interact in a "web of life." To fully appreciate the role of wildflowers in the grand picture of nature, you must become a student of ecology. May this book start your journey!

Our Shared and Threatened Natural Heritage

Three lifetimes of careful observation of wildflowers and their habitats have brought the authors great enjoyment and excitement, but it has also elicited overwhelming concern. Our wildflowers are disappearing quickly. In the twenty years since the publication of the first edition of this book, many of the best natural communities in our state have all but vanished. Today, many formerly common plants have become rare, and rare plants have been lost. We have seen breathtaking open savannas and depression meadows become thickets of hard-woods and duck hunting impoundments. We have seen picturesque hillsides of cove forest wildflowers turned into a churned, muddy wallow by hogs. We have seen herbaceous seepages with tens of thousands of pitcherplants reduced to none. How did we get here? To understand this and what can be done about the problem, we must understand how we all fit into the functioning of our landscape, what the problems are, and what potential solutions we can apply.

Each of us plays a role in supporting the plants that we love. We must find common ground and make choices that result in a state that is as or more vibrant than the one we inherit. We face the future with the challenges of con-version of natural communities into developments, fire suppression, indis-criminate use of herbicides, damage caused by invasive wildlife such as wild pigs, invasive exotic plants and insects, and a changing climate. Beginning to heal the land begins with understanding that we are all in this together and that each of us is equally important in shaping the world we leave behind.

We have all contributed to the landscape we see today. Many of us grew up with the concept of wilderness as espoused by John Muir–nature existing in the absence of humans and our influence. This idealistic view of natural areas is not only unrealistic, it's absurd. The reality is that all of the natural com-munities on this continent have existed with humans since they first walked onto the continent, and the erasure of traditional land management can prove very harmful to our biodiversity. The impact that humans had and continue to have on this landscape is a product of the choices we have made. The earli-est inhabitants of South Carolina, the Native Americans, arrived on this con-tinent at least 13,500 years ago; research from the Topper archeological site in Allendale County supports the claim that they arrived perhaps as early as 20,000 years ago. Europeans and European-descended people largely wrote the history most of us were taught, and they conveniently erased most of the culture and impact of our original inhabitants. Interest in the land manage-ment activities of Native Americans during and prior to the colonial era has been a subject of considerable research in recent years.

The structure and composition of some habitats can be ascertained from old photographs and descriptions. The Longleaf Pine woodlands that once covered at least 60% of the Coastal Plain are an excellent example. Old photographs of stands of huge, original-growth trees with an open, grassy understory tell us the appearance and components of the habitat. Surveys of standing lumber, such as that completed by Moses Ashley Curtis, *The Woods and Timbers of North Carolina,* published in 1883 tell a story of overwhelming dominance of Longleaf Pine over most of the Coastal Plain and outer Piedmont of North Carolina. North Carolina has far less cover by Longleaf Pine today than South Carolina and it can be inferred that the same pattern existed here. Other habitats were only a distant memory by the time photography came into existence. There are plants that have perplexed many of us because they don't seem to have a "natural" habitat. Plants, such as Schweinitz's Sunflower, Smooth Coneflower, Georgia Aster, and Piedmont Buckroot, that have global ranges restricted to the Piedmont of the southeastern United States, exist today only on roadsides, power line rights-of-way, and old graveyards. Where did they grow before such habitats existed? Other plants like Prairie Dock, Gray-headed Coneflower, Stiff-leaf Goldenrod, and Green Comet Milkweed are typically found far to the west in the tallgrass prairie but appear also in the same roadsides and power lines in South Carolina's Piedmont. Is it possible that an open prairie or savanna habitat once sheltered these plants that today are exiled to roadsides? Names on old maps and a renewed interest in colonial era collections and writings are helping us to develop a picture of how the Piedmont looked and how it has changed.

Many old maps from the colonial era show an area roughly surrounding Charlotte, North Carolina, but extending south well into South Carolina, that is often labelled as "savannah." The English-speaking settlers had no word in their lexicon for prairie at the time. "Prairie" was a French word that was adapted to describe the vast grasslands of the Louisiana Territory and was not in use in the English language until the nineteenth century.

The names on maps gives us a hint at what at least some portion of the Piedmont was like, but travelers such as Mark Catesby give us much more insight. Catesby published *Natural History of Carolina, Florida and the Bahama Islands* between 1729 and 1747 (Catesby, 1754). This work contains an account that describes the conditions of Carolina. Mark Catesby arrived in South Carolina in 1722 and made two voyages inland from his base in Charleston, in 1723 and 1724, before leaving for points south, and eventually returning to England. His description of Carolina in the appendix as well as in the text accompanying the acclaimed engravings is based on first-hand knowledge as well as the works of those who came before him, such as John Lawson. His description of the Native American management of the land is quite interesting. He describes their use of fire, though he didn't understand why this would be done. His text accompanying his engraving of an American Bison and Hartweg's Locust is very telling (from vol. 1, p. 20 of the *Appendix*): "I never saw any of these trees but at one place near the Apalatchian mountains, where Bufellos had left their dung; and some of the trees had their branches pulled down, from which I conjecture they had been browsing on the leaves. I visited them again at the proper time to get some seeds, but the ravaging Indians had burn'd the woods many miles round, and totally destroyed them, to my great disappointment. . . ." Once you get past the poor grammar and obvious dose of eighteenth century casual racism, you realize that Catesby is describing American bison in South Carolina! Yes, they were here, and they had to have grassy habitats to survive, and the fire that he describes is exactly the way the Native Americans provided these grassy habitats to their advantage and that of the bison. It is now well-known that the Native Americans were using fire to create prairie, to attract and support grazing animals and other game.

Catesby goes on to describe the bison in the appendix, affirming what we suspect, that there were prairies in the Piedmont. In *An Account of Carolina and the Bahama Islands* found at the end of the second volume he states, "They range in Droves, feeding in the open Savannas Morning and Evening, and in the sultry Time of the Day they retire to shady Rivulets and Streams of clear Water, gliding thro' Thickets of tall Canes, which tho' a hidden Retreat, yet their heavy Bodies, causing a deep Impression of their Feet in moist Land, they are often trac'd and shot by the artful Indians; when wounded they are very furious, which cautions the Indians how they attack them in open Savannas, where no Trees are to screen themselves from their Fury" (p. 27). Here, Catesby provides evidence of a treeless habitat he refers to as savannas and he shows a little more respect for the Native Americans, particularly his Creek guides, without whom he could not have found his way or been fed during his collecting forays through this Piedmont landscape. A Piedmont landscape that would appear to be a very foreign to most of us today. Here, in these Piedmont prairies, is the home of species found nowhere else on earth. The habitat for Schweinitz's Sunflower, Smooth Coneflower, Georgia Aster, and Piedmont Buckroot is now relegated to tiny strips measured in square feet, not square miles, along power lines and roadsides.

What happened to the Piedmont Prairie? The Native Americans were forcibly removed from their land by the European colonists. The Piedmont was settled by those colonists and farmed. Vast plantations of cotton, corn, and upland rice made use of the rich soils that were built by the grasslands, forests, and oak savannas of the region. These soils were not protected and quickly lost their deep, rich topsoil through erosion. Original topsoil depths can still be seen in some tiny pieces of the landscape, including at the South Carolina Botanical Garden. Our reconstructed Piedmont prairie exhibit is built on virgin soils and the deep, dark topsoil extends to a depth of two to three feet. Can you imagine the Piedmont with topsoil? The depleted red clay subsoils we see today are the legacy of the lack of soil conservation. With soils depleted and the arrival of the boll weevil, large-scale farming of the Piedmont gradually declined. Today, forests grow over much of the land that was formerly cotton and, before that, prairie or woodland. The most important lesson in this story in regard to understanding wildflowers and their habitats is that humans drove this process. Humans shaped the landscape by encouraging prairie, by tilling the soil, and today, by abandoning traditional management and excluding fire. The prairie, the cotton fields, the soils, and the dense forests of today all depend at least partly on the choices made by humans. The legacy of the hundreds or thousands of years of management by Native Americans is evident in the prairie plants that still inhabit South Carolina's forest margins.

The contribution of African Americans to the landscape is also a blind spot for many naturalists. We visit places like the ACE Basin to view the awe-inspiring numbers of wading birds, shore birds, and migratory waterfowl. When you look out across the extremely productive tidal freshwater marsh at a flock of Tundra Swan, do you think about how those marshes came to be? These are human engineered habitats. Most of this tidal freshwater marsh system was originally tidal swamp forest with a towering old-growth Bald Cypress canopy. The swamp forest had to be cleared and an immense system of dikes, banks, and rice trunk gates installed to support the crops that these fields produced, rice. Carolina gold rice made fortunes for those who owned the land. A portion of the technology and all of the labor was provided by enslaved people. Many of these people were taken from West Africa, where they had cultivated rice for generations and knew the process and the technology needed to produce the crop in South Carolina. It is important to never forget that the beautiful, diverse and life-filled tidal freshwater marsh we enjoy today is the product of kidnapping, slavery, and pain. All of our land has history; it has connections and relationships between all the natural components, including humans. The oak-hickory forest in your back yard may once have been a prairie tended by Native Americans, cotton or tobacco toiled over by enslaved people of color or exploited poor sharecroppers, and now entrusted to you. The choices that were made continue to have an impact today on our natural communities, our lives, and our nation.

Our choices matter, individual choices matter, and we can see that in one of the most interesting natural communities in South Carolina, our shell hammock forests. The shell middens, mounds and rings that dot the coastline of South Carolina were constructed by Native Americans during the late archaic period, most were built roughly 5,000–4,000 years ago. Archaeologists don't have much information on these people, but they left piles of oyster and clam shells along the coastline. The action of piling oyster shells in the marsh or on the islands changed the chemistry of the soil. The oyster shell is composed largely of calcium carbonate (lime). The calcium in the shells raises the soil pH and allows deciduous trees and herbs, normally found far inland, such as Florida Maple, White Basswood, Woodland Pinkroot, and Mottled Trillium, to grow in the maritime strand. These areas would normally be dominated by evergreen species more tolerant of the typical acidic sandy soil. Plants such as Godfrey's Swamp-privet, Small-flowered Buckthorn, Shell-midden Morning-glory, and Leafless Swallowwort are only found in these human-created habitats in South Carolina and are more typical of the limestone-derived soils of Florida. The actions of people who lived 5,000 years ago are evident in the landscape and the wildflowers we see today. Throwing down a shell changes the world! Imagine how important the choices you make as an individual are to the world of your grandchildren and those living 5,000 years from now. Each of us is important. Each of us is changing the world around us, and all of us should be recognized for our contribution. Only by acknowledging the fullest picture of the past and the importance of our actions today do we stand a chance for conserving and healing the land, our legacy—it is painted in the flowers of the fields and forest.

Physiographic Regions of South Carolina

For ease of organization, but mostly for ecological reasons, this book divides South Carolina into the three physiographic provinces that are accepted by most geographers and botanists: The Blue Ridge Province, the Piedmont Province, and the Coastal Plain Province (figure 1). The Coastal Plain Province is further divided into four regions: Fall-line Sandhills; Inner Coastal Plain; Outer Coastal Plain; and Maritime Strand.

Some of the state's natural communities and species occur in all three provinces, while others occur only in one physiographic province or in just one region. Although no natural communities occur only in the Fall-line Sandhills, some are best developed there. Most of the natural communities in the maritime strand (that portion of the Coastal Plain subject to the influence of wind-borne salt spray and/or water with significant salt concentration) are found only in that region.

The Blue Ridge Province, a belt of mountains at the front wall of the Appalachian Mountains, is the westernmost physiographic province in South Carolina. This is an area of generally high relief, although the mountains of the Blue Ridge are often described as "subdued" because they are old and relatively weathered. Once taller than the much younger Rocky Mountains, today

Figure 1. Physiographic Regions of South Carolina.

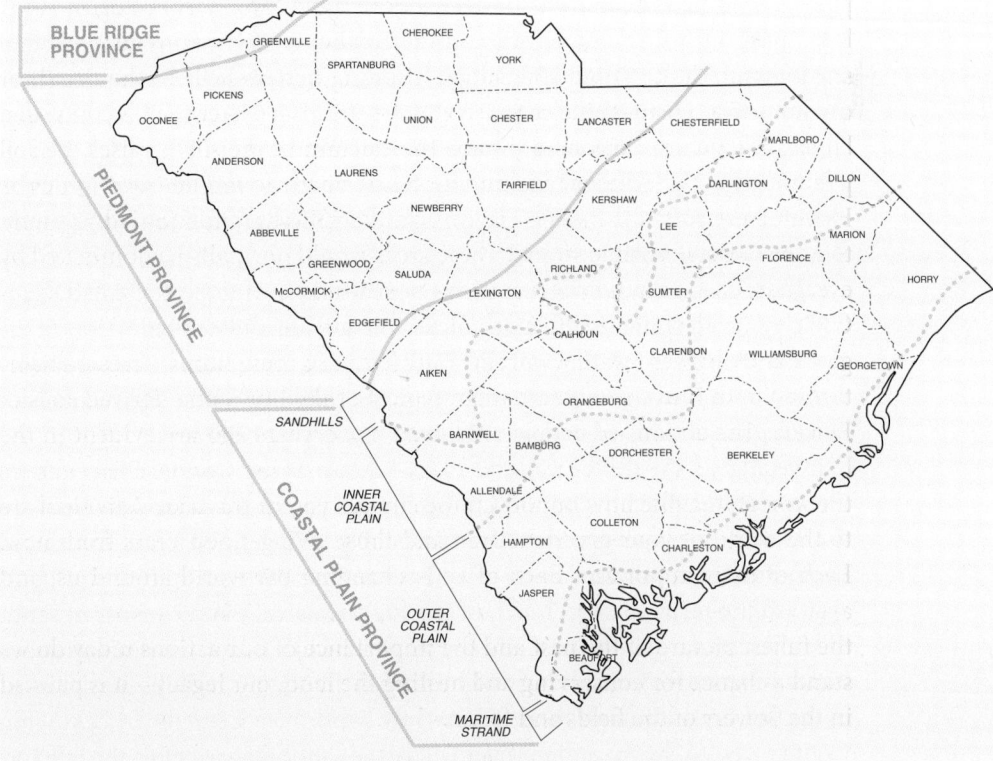

their elevation varies from 1,000 feet to over 3,100 feet. Rocks are mostly granites and gneisses with smaller areas of amphibolite, quartzite, and other rocks. The boundary between the Blue Ridge Province and the Piedmont Province to the east is the Blue Ridge Escarpment, a narrow belt of especially abrupt and steep relief. In a horizontal distance of about 5 to 8 miles, the mountains can rise over 2,000 feet vertically from the Piedmont below. This escarpment is especially noticeable in Oconee and Pickens Counties. The Blue Ridge Province also includes a small part of upper Greenville County.

The Piedmont Province is a broad region of rolling hills that extends from the Blue Ridge Escarpment to the unconsolidated sediments of the Coastal Plain. The term "Piedmont" literally means "foot of the mountain," with the term "foothills" sometimes used as a synonym. The rolling hills of the Piedmont are dissected by numerous small streams; valleys are generally wider than in the mountains, in large part because Piedmont rocks are generally less resistant to erosion than those of the Blue Ridge. Elevation begins at 300 to 400 feet in the east and extends to 1,000 to 1,200 feet in the west. Rocks are mostly gneisses, slates, and schists, with some quartzite and amphibolite. Highly unusual habitats can be found in the isolated dikes and sills of ultramafic rock such as diabase. The Piedmont is also characterized by occasional monadnocks (e.g., Glassy Mountain in Pickens County), isolated hills of resistant rock that arise abruptly above the land.

The boundary between the Piedmont and the Coastal Plain is often described as the "fall line" because of a noticeable change in elevation—that is, a vertical "fall" that was historically recognized as rapids or shoals along rivers that were utilized heavily for trade. Fall-line rapids are most prevalent at the Georgia–South Carolina border, on the Savannah River at Augusta, and at the junction of the Broad and Saluda Rivers, which form the Congaree River at Columbia. Distinguishing the Piedmont from the Coastal Plain is sometimes difficult, especially since some Coastal Plain sandhills are higher in elevation than the adjacent Piedmont. When near the fall line and in doubt, examine the soil; if the soil is clayey, it is likely Piedmont; if the soil is sandy, it is the Coastal Plain.

The Coastal Plain Province is composed mostly of unconsolidated sands and clays, with some sedimentary rocks. The unconsolidated sediments of the Coastal Plain were washed off the Piedmont and mountains. They were deposited in the sea or on land not far from the seashore. These sediments have been repeatedly shifted and sorted by the wandering tidal seas.

Geologists divide the Coastal Plain into an Inner Coastal Plain and an Outer Coastal Plain, using the Orangeburg Scarp as the boundary between the two. The Orangeburg Scarp is a reasonably well-defined terrace produced by an ancient seashore. Lands inland from this scarp are hillier, and lands seaward are relatively flat. Much of the Inner Coastal Plain consists of sandhills that border the fall line. These Fall-line Sandhills form a discontinuous belt, 5 to 20 miles wide, across the middle of the state. Sandhills border most of the fall line, although there is a narrow band from Leesville in Lexington County west to the Savannah River where the sandhills are below the fall line.

The Fall-line Sandhills consist of exposed Cretaceous-age sediments, mostly deep sands (some consolidated into sandstones), which support a distinctive flora. John Barry (1980) in *Natural Vegetation of South Carolina* and Charles Wharton (1978) in *The Natural Environments of Georgia* provide reasonable maps of the extent of the sand hills. The authors agree with these scientists that the deep sands of the Fall-line Sandhills have a significant influence on plant distribution and should be treated as a distinct region of the Coastal Plain Province.

Other "sandhills" (technically sand ridges) are not associated with the fall line that lie in the Coastal Plain, including some that were produced by river-related events (plate 518) and others associated with Carolina bays. These sand ridges produce similar xeric plant communities, but because of the lack of similar relief, they usually don't produce the extensive hillside pocosins, seepages, and Atlantic white-cedar communities that are present in the hilly, Fall-line Sandhills.

The outermost portion of the Outer Coastal Plain is defined as the maritime strand. This area comprises vegetation that is significantly influenced by wind-borne salt spray and/or water with significant concentrations of salt. Although not usually distinguished by geographers or geologists, botanists and ecologists generally agree that this region is distinctive enough to warrant separation from the rest of the Coastal Plain.

In this book, we use the term "Coastal Plain" as a specific physiographic region—the Atlantic Coastal Plain (including the extension into the Gulf Coastal Plain)—and thus capitalize it when referring to it in the ranges of the various species. When a species is endemic (restricted) to the Coastal Plain, we consider the Sandhills region (also capitalized because it is a distinct geographic region) to be a part of the Coastal Plain. When the term "sandhill" is used for a habitat rather than a region, it is not capitalized. Because so many species found in the Sandhills region are either absent in the Inner and Outer Coastal Plain or vice-versa, we report ranges as either Sandhills or Coastal Plain (meant to be exclusive of the Sandhills). If plants occur in both regions, we will report it as Sandhills and Coastal Plain. If a plant is restricted to the Inner Coastal Plain, that is noted. If a plant is restricted to the Outer Coastal Plain, that is also noted. If the species is found in both the Inner Coastal Plain and the Outer Coastal Plain, we state simply Coastal Plain. The term "Piedmont" is also capitalized in this book because we are referring to a specific physiographical region (the Southern Appalachian Piedmont Plateau) of the Appalachian Province. The term "mountains" is not capitalized but in some cases we use the formal term "Southern Blue Ridge Escarpment," which is a specific physiographic region and is capitalized.

Nature of the Flora

Although a small state compared to others in the southeastern United States, South Carolina supports a rich and varied vascular plant flora. At the time of the preparation of this manuscript, as many as 4,050 taxa (species, subspecies, and varieties) have been documented as growing in South Carolina (Weakley, 2020). Native taxa (those historically known from South Carolina at the time of European colonization) number 2,911 species and naturalized, nonnative species number 1,139. Georgia, a state almost twice the size of South Carolina, has 4,378 taxa, 3,370 of which are considered native; and North Carolina, which is 50% larger, has approximately 4,452 taxa, with 3,185 considered native. Three factors explain South Carolina's rich and diverse vascular flora: (1) South Carolina exhibits a wide range of physiographic provinces and regions; (2) South Carolina is the northern range limit for many southern Coastal Plain species but the southern limit for many species of the more northern Coastal Plain flora, which harbors many species found only in the Cape Fear Arch (endemics); and (3) South Carolina's southeast area is slightly influenced by the semitropical climate of Florida, yet its northeast section is influenced more by the mid-Atlantic Coastal Plain climate.

The 1968 publication of the *Manual of the Vascular Flora of the Carolinas* (Radford et al., 1968) was a landmark in our understanding of South Carolina's flora. For the first time, botanists had documented distribution maps of all the known species of vascular plants in the Carolinas. Since then, botanists (and knowledgeable laypersons) have added numerous records of state occurrence and other distribution records to the vascular flora recorded in the *Manual*. With the formation of the South Carolina Heritage Trust Program, a central reservoir of records of significant collections was created, making it easier for botanists to determine collection gaps.

Alan S. Weakley has updated the taxonomy and distribution records of the flora of the Carolinas and Virginia. His *Flora of the Southeastern United States* (latest working draft, 2020) incorporates much of the information from the *Manual of the Vascular Flora of Carolinas* and is augmented by an extensive review of the literature and collection records that were added to herbaria since 1968. His work gives botanists a valuable update of the state's flora. Weakley gives updated ranges for physiographic provinces in South Carolina, and the second edition of *A Guide to the Wildflowers of South Carolina* gives updated, accurate county-level distribution for 997 species that updates and greatly clarifies the extent of documented populations.

What Are Wildflowers?

What is meant by the term "wildflower"? Some authors use it to refer to any flowering plant growing without cultivation. Some restrict the term to native annual or perennial herbs with showy flowers, such as Bloodroot and trillium. However, native trees, such as Bull Bay and Tulip-tree as well as shrubs such as Sweet-shrub and pawpaw, have showy flowers and are more conspicuous than many of the herbaceous species. Many naturalized species such as Japanese Honeysuckle are as showy as the native species and are included in most wildflower books. Native and naturalized species that have small flowers may be abundant and aggressive and are often designated as weeds. When viewed under magnification, their flowers are just as beautiful as the showier species. Showy displays of the flowers of these species, like heliotrope along roadsides or Common Toadflax in fields, give color and character to the land. Some sedges, like White-top Sedge, and grasses, like Seaside Panicum, are as conspicuous as classic wildflowers. And what plants add more beauty to South Carolina than the native woody vines, such as Cow-itch and Coral Honeysuckle? Surely these are also colorful wildflowers.

Wildflowers in this book are defined in a broad sense to include the rich diversity of South Carolina's plant life. Showy native annual and perennial herbs are emphasized, but shrubs, vines, and trees with showy flowers; showy introduced species; and conspicuous grasses, rushes, and sedges are also considered as wildflowers. Species of pteridophytes (ferns and their allies) and gymnosperms (pines, cedars, bald-cypress, and their allies) are included. These plants are a conspicuous and interesting aspect of the native flora. To exclude them would be derelict in representing the state's varied flora. Their inclusion also helps to visually illustrate the composition of the natural communities in which they are found.

Many factors were considered in choosing the species to include in this book. The principal factor was the desire to interest the reader in and educate the reader about native wildflowers and the plant communities in which they grow. Showy herbaceous species that are the most obvious in communities are emphasized, and selected indicator species of each native community are included. For instance, Turkey Oak is featured because it is an indicator species of the xeric sandhills; Pond Cypress identifies the Pond Cypress savannas; and White Oak is pictured because it is a dominant tree of the oak-hickory forests of the Piedmont.

All native species of *Trillium, Rhododendron,* and *Hexastylis* are pictured because these genera have attracted exceptional interest from wildflower lovers.

Certain species were chosen because they represent interesting accounts about the botanical history of the state. Mexican-tea, a weedy introduction, is included because of its use as a folk remedy for "worms"; Sweet Grass is included because of its past and present use in making sweet grass baskets as well as its popularity in the horticultural trade; Poison Oak is noted because of its poisonous nature; Wintergreen is a source of wintergreen oil; and Mistletoe illustrates a fascinating group of plants: the parasitic vascular plants. Many federally listed endangered species such as Swamp Pink, Smooth Coneflower, and American Chaff-seed are included because of the interest in the protection of rare and endangered species.

Another criterion for selecting species was to include many of the species that are not included in the wildflower books that cover part or all of the southeastern US. Obviously, however, overlap occurs.

One final consideration in the choice of plants was recognizing the special interests of our colleagues, our students, and many of our wildflower enthusiast friends. They have made it a pleasure to produce this book. Certain plants, for whatever reason, seem to attract their attention, whether it is a large-flowered herb or inconspicuous sedge. These species appear in appreciation of their love of wildflowers.

Conservation of Native Wildflowers

As natural habitats are altered, native species do not always compete successfully with the weedy introductions that quickly invade disturbed land. Year
after year, more and more native habitats are lost. This escalating loss has
prompted several efforts to conserve native wildflowers. But none of the mul-
tifaceted efforts can be totally successful alone. Successful wildflower conservation requires the mutual work of many individuals and groups.

Most successful conservation of native wildflowers protects them in their
natural habitats. Only by placing large tracts of natural communities under
protection will we ensure that future generations receive the same pleasures
we experience when viewing a Bloodroot or pitcherplant in its natural set-
ting. Many South Carolina organizations are leading the way to conserve
natural communities. The South Carolina Department of Natural Resources,
through its Heritage Trust Program, has purchased more than 160,000 acres
throughout the state, establishing numerous heritage preserves and wildlife
management areas. Numerous nonprofit organizations have augmented the
state's system of protected lands. The Nature Conservancy has developed a
sophisticated and successful system to locate and protect critical habitat for
both plants and animals. Local and regional land trusts have done excellent
work in safeguarding many of the most precious locations. Local land trusts
have been formed as legal vehicles for the acceptance of conservation agree-
ments on private properties. Owners keep legal ownership, but for a tax break,
they give up, in perpetuity, development rights to the property. Much of the
protection success of the ACE Basin (Ashepoo, Combahee, and Edisto basin)
in the lowcountry has been through this method. Despite all this effort, much
work remains to be done if we are to safeguard our biological diversity.

The South Carolina Native Plant Society has done much to bring our
wildflower heritage to the public's attention. Through seminars, workshops,
lectures, and field trips, the society encourages native wildflower conservation and provides information on native wildflowers that was previously unavailable. The reader is encouraged to engage and join this organization for
more involvement with wildflower appreciation and conservation.

Landscape designers have contributed to wildflower conservation by using native plants in their garden designs. If given adequate light, water, and
soil conditions, many native wildflowers respond well to cultivation and are
as beautiful and interesting as the more typical and exotic horticultural species. Although traditional gardens teach little of the habitat and natural history of the plants, gardening is often the first wildflower experience for some,
and it can lead to a lifelong interest in native wildflowers and their habitats.
A brief section on the importance of native flora in our home landscapes and
habitat gardening is provided in this book.

The United States Forest Service has also been active in the conservation
of unique natural areas in South Carolina. In the Francis Marion and Sumter
National Forests, countless acres provide access and protection for our natural

communities. The authors have assisted the Forest Service in identifying areas to be protected and drafting management plans for unique natural areas.

Protection of wildflowers has also benefited from the careful documentation of populations of rare species managed by state and federal agencies. Under the Endangered Species Act of 1973, the United States Fish and Wildlife Service is charged with protecting rare species. Species listed as threatened or endangered are given legal protection. Whether a species is federally designated as threatened or endangered is part of the justification used by the South Carolina Department of Natural Resources for the purchase, or acceptance by donation, of sites that harbor these species. Establishment of preserves concomitantly protects numerous other rare wildflowers, as well as preserving their natural communities.

Conservation of many wildflowers requires a factor that the public is not generally aware of—fire. Coastal Plain communities such as pocosins, pine flatwoods, and pine savannas as well as mountain pine-oak heaths and Piedmont xeric hardpan forests require periodic fire to maintain natural conditions. These communities evolved with fire as a dynamic natural component. Fire suppression can threaten the very existence of countless communities. In Coastal Plain pine savannas and flatwoods, fire suppression ultimately eliminates understory wildflowers because hardwood species invade these communities. One aim of this book, and of those organizations mentioned above, is to raise public awareness of the role of fire in natural communities. The essay "Fire in the South Carolina Landscape" stresses the value of fire in the natural world. Perhaps in the future the public will be more tolerant of smoke generated from fires.

Despite these numerous conservation programs and efforts, much still needs to be done. The areas already protected in South Carolina are not sufficient to ensure the level necessary to conserve all native communities, much less all the species of rare, native South Carolina wildflowers and the threatened and endangered species. Several species are known from only a few sites, and catastrophic events such as Hurricane Hugo could thwart the best conservation efforts. A goal of this book is to make people aware of the rare natural wildflower communities in the hope that more people will join professional botanists and organizations in finding and protecting the sites that support these diverse native communities.

How To Use This Field Guide

Routine, confident wildflower identifications are made by continually studying plants in the field. No one can expect to learn the flora of a state, or even the flora of a region, in a few sessions. Wildflower identification is a lifelong commitment. This book should be a companion on every field trip. As each new species is added to one's understanding of wildflowers, confidence in identification grows.

A Guide to the Wildflowers of South Carolina follows a natural habitat approach to native wildflower identification that was first developed in

Wildflowers of the Carolina Lowcountry and Lower Pee Dee. When you select a plant for identification, first turn to the plant community photographs. Using the descriptions or photographs, choose the one that most resembles the plant's surroundings. Read the description of the community in "South Carolina's Natural Wildflower Communities" to be certain you have made the best selection. Next, turn to the wildflower photographs for that particular community and match the photograph to the plant you have found. If the plant cannot be found in the selected community, select the next best community that resembles the plant's surroundings. Many species occur in more than one natural community, particularly those that are wide-ranging. The species chosen to represent the natural community in this book are generally those that are most obvious or limited to that community. Once you are confident you have made a good selection, the information in the description is used to confirm or reject the identification. Here is an invaluable suggestion: A small, 10x power hand lens will be a great asset in the field for examining flower and plant structures.

Caution: Not every species of wildflower that grows in South Carolina is included in this book. If you check all the photographs for the communities where the plant is most likely to be found and you do not find a match, it may be that the wildflower is not pictured, or it may be pictured in another community.

The habitat approach may not always be the fastest identification method, especially for amateur botanists with little practical field training, but in the long run it will be the most rewarding. Not only will it yield positive identification, but you will also be learning the native communities, something of inestimable value if you seriously journey into the world of native wildflowers.

Ruderal species (weedy species that inhabit disturbed areas) are included in a separate section. Almost everyone is aware of disturbed sites where these species grow, such as lawns, roadsides, and abandoned home sites.

ESSAYS

Topics were chosen to represent a range of the botanical natural history of the state. One suggestion is to read thoroughly, for example, the essay on carnivorous plants, then plan a trip where these occur. Preparation before going to the field will make the experience more meaningful, especially if you are able to locate some of these species.

NATURAL COMMUNITIES

The authors suggest becoming familiar with the natural communities that are included in the book through the descriptions and photographs. Then visit a heritage preserve, state park, or national forest that harbors these natural communities so they can be recognized when they are encountered. One of the best locations for learning about natural communities may be found at the South Carolina Botanical Garden. The South Carolina Botanical Garden is a 295-acre garden and natural area located in Clemson, South Carolina. The garden contains the largest collection of plants native to South Carolina in the world and exhibits them in restored and recreated natural communities. The Natural Heritage Garden trail takes the visitor through most of the natural communities in South Carolina in the span of less than an hour's stroll. The exhibits interpret the intricacies and characteristics of each natural community and are large enough to completely surround the visitor within that community. When you walk through the maritime forest or Longleaf Pine savanna habitats you would never guess you were in Pickens County! Remember that identifying natural communities is a continual learning process.

SPECIES DESCRIPTIONS AND COLOR PLATES

Species are listed within a natural community in which they occur. The species are arranged with the woody plants (trees, shrubs, and lianas) first as they form the visual dominants of many of the habitats. The herbaceous species (including some sub-shrubs) are presented first; last are ferns or fern relatives. The species are arranged according to their period of flowering from earliest spring to latest autumn. The format for each species description contains specific elements that are listed below. Note that not every species description includes all twelve elements. Elements 6 and 9–12 may not be applicable to some species.

1. Color plate number. The color plate number is given in bold type and precedes the common name in each species description; the corresponding photo appears adjacent to its description.

2. Common name(s). At least one common name is given for each species. Common names are currently not standardized, and many species have numerous common names, which would be impracticable to enumerate. When two or more are given, the one thought to be most often used in South Carolina is listed first.

3. Scientific name. The choice of the scientific names used in the book is explained in the section "Origins of Plant Names."

4. Pronunciation guide. A pronunciation guide for each species follows the botanical name. Instructions on how to use the guide are given in the section "Pronunciation Guide to Botanical Names."

5. Family name. The technical family name (which ends in *-aceae*) for each species is included, as well as the common name of the family.

6. Synonymy. These are names that have been applied to the same species but are currently considered not legitimate. Synonymy is provided for species with names that have changed since their publication in *The Manual of the Vascular Flora of the Carolinas* by Radford, Ahles, and Bell (1968; abbreviated as RAB) and the first edition of *A Guide to Wildflowers of South Carolina* by Porcher and Rayner (abbreviated PR). Only synonyms that correspond to these two books are reported. The use of synonyms is explained in the essay "Origins of Plant Names."

7. Description. The plant description is an abbreviated version of that found in standard taxonomic manuals. It should be used to confirm or reject an identification when photographic identification is uncertain. For instance, if the plant in question has opposite leaves with a solitary flower and these features do not agree with the description, identification is suspect. Additional examination is needed.

The simplest and most distinctive characteristics are used in the descriptions. The glossary and the Appendix figure 8, "Illustrations of Plant Structures," contain supplementary information to assist with the terminology used in the descriptions.

Each description contains the flowering and/or fruiting times for the species. Be aware, however, that flowering times vary from year to year and from region to region. Early and warm springs may result in a species flowering in February; the next year a late spring may postpone flowering of the same species until March or later. Similarly, a plant that flowers in April in the Coastal Plain may not flower until May or even later in the mountains. Nevertheless, most wildflowers flower at a particular time in the year. These are the times that are listed in the descriptions.

8. Range-habitat. The range of each species is given, both within and outside the state. The range outside the state is obtained primarily from manuals and data managed by the United States Department of Agriculture. The county level range provided for within the state is based on examination of more than 200,000 images and specimens from herbaria throughout the region and country. Field botany is an ongoing process, and distribution records within states are being continually updated. Terms denoting frequency used in this book (rare, occasional, common, locally abundant) are highly subjective and are especially difficult to apply to widespread species. A plant rare in the state may be common elsewhere, while a plant rare outside the state may be more abundant and even locally abundant in some sites in South Carolina. A plant common in the mountains, such as Downy Rattlesnake Plantain, is rare and scattered in the Coastal Plain. The authors have tried to be consistent in the determinations of frequency and have based these determinations on literature and our own professional observations.

9. Similar species. This section describes how to distinguish between species that look very similar.

10. Taxonomy. This category is used to distinguish between named varieties or subspecies of the particular species presented.

11. Comments. Included here is information on such things as folklore, ecology, and origins of common names, which were taken from the literature as well as from the authors' research and observations. These comments are given to help "personalize" the plant.

12. Conservation status. The definitions of the various federal and state protective statuses that are used in this section are given in the essay "Rarity of Vascular Plants." The protective status given in this book is based on information at the time of publication. As new information is discovered, the status of a particular plant may change. If a plant is federally endangered or threatened, this is noted. South Carolina currently has no official status of endangered and threatened and therefore the rarity of species currently tracked by the Heritage Trust Program is listed in three categories of descending rarity: (1) critically imperiled, (2) imperiled, or (3) vulnerable. The conservation status in South Carolina for species listed here was provided by Keith Bradley, botanist for the South Carolina Department of Natural Resources, and determined with input from all knowledgeable botanists, including the authors.

Origins of Plant Names

SCIENTIFIC NAMES

This taxonomy in this book follows the standard reference for scientific names in the Carolinas, found in the continuously updated electronic draft of the *Flora of the Southeastern United States* by Alan Weakley (2020). The interested wildflower enthusiast or scholar may find the most recent version available for download at http://herbarium.unc.edu/flora.htm.

Manual of the Vascular Flora of the Carolinas by Radford, Ahles, and Bell (1968) was the most widely used and last printed treatment of the flora of the region. The *Manual* remains a landmark scientific work. It updated and generally replaced Small's *Manual of the Southeastern Flora,* published in 1933, for the flora of the Carolinas. The *Manual* is by far the most widely used printed manual in the Carolinas, both for the layperson and the scientist. Because of this, we list in synonymy any scientific name used by the *Manual* that is currently not recognized by Weakley (2020) as legitimate.

Scientific names for plants, unlike their common names, are governed by a system of rules outlined in the International Code of Botanical Nomenclature (ICBN). These rules are reviewed and modified every six years. In this system, every plant no matter how widely distributed has only one scientific name; furthermore, only one plant can have this scientific name. A plant can have only one valid scientific name by which it may be identified at any one point in time. It may have picked up several scientific names, but any name besides the valid one is a synonym. Synonyms, then, are names that have become invalid or discarded by an author for whatever reason.

The scientific or binomial name always consists of two Latin or Latinized words (the binomial system of nomenclature): a genus or generic name and a specific epithet for the species. The specific epithet is followed by the name of the person(s) who first described the plant (the authority) according to the rules of the ICBN. For example, take the following scientific name:

Acer rubrum L.

The generic name is *Acer,* and its specific epithet is *rubrum*. The authority is the Swedish botanist Carl Linnaeus (1707–78); he is honored by the use of the single letter "L." Linnaeus is the scientist who effectively created the modern system of nomenclature. No other botanist is honored with the use of a single letter abbreviation. In this book, all authorities have their names written out.

Scientific names should always be italicized when printed or underlined when typed or written by hand. The generic name is sometimes given as an abbreviation if the context makes its meaning clear: *A. rubrum* L., where *A.* stands for *Acer.*

In using scientific names in text, the genus is always capitalized while the letters of the specific epithet are all in lowercase. Formerly, some botanists used uppercase for the initial letter of the specific epithet if the epithet was honoring a person or serving some other commemorative purpose. For example, in *Gray's Manual of Botany,* the specific epithet *Gronovii* in *Hieracium Gronovii* honors Jan Frederick Gronovius. However, botanists now initiate *all* specific epithets with lowercase. This method is followed throughout the book, and *Hieracium Gronovii* is written as *Hieracium gronovii.*

Changes are made in botanical nomenclature as our biological understanding of a plant improves. For instance, when the generic placement of a species is better understood, it may be shifted from one genus to another. The name of the original author is then placed in parentheses and is followed by the name of the person making this change. The following example illustrates this:

Azalea calendulacea Michaux
changed to
Rhododendron calendulaceum (Michaux) Torrey

Michaux, who originally named and placed the plant in the genus *Azalea,* becomes the parenthetical authority. Torrey reassessed the plant's relationships and moved the species to the genus *Rhododendron,* and he becomes the current authority.

Botanists often recognize variations below the species level in plants that are different yet not sufficiently distinctive to be considered a separate species. These subdivisions are, in descending order of magnitude, subspecies (abbreviated as ssp.), variety (abbreviated as var.), and form. Some authors use the terms subspecies and variety interchangeably. Examples of the first two are given for instructive purposes:

Nuphar luteum ssp. *sagittifolium* (Walter) E. O. Beal
Eryngium aquaticum L. var. *aquaticum*

COMMON NAMES

In this book the common name for a particular species is capitalized (e.g., Shell-midden Morning-glory), while the common name for a group of plants not referring to a particular species is not (e.g., morning-glories).

Common names are not dependable resources for several reasons: (1) the same plant may have many common names (e.g., 140 common names exist for Woolly Mullein); (2) the same common name may apply to several plants, either in the same geographical area or in different geographical areas; (3) common names may be misleading (for example Silkgrass is not a grass, but a composite); and (4) they are not standardized.

Despite these problems, common names remain an important part of plant folklore. They often are appealing and tell us something about the plant. Recall the common name "American Heal-all" for *Prunella lanceolata* in reference to its use in folk medicine as a cure-all, or "Boneset" for *Eupatorium perfoliatum,* from a belief that it could be used to set broken bones. Common names are often the only means of communication for those unfamiliar with the scientific names. It may be easier to remember words such as "Windflower" and "White Oak." Sometimes common names are remarkably descriptive. For example, Dryland Blueberry is a member of the heath family growing in dry uplands with (mostly) blue berries, and Swamp Chestnut Oak refers to a swamp-loving oak with leaves similar to those of American Chestnut.

With these considerations in mind, the authors include at least one common name for each plant. Sometimes there is no published common name; examples include some *Xyris* species in this book. In such cases, the authors, in collaboration with Alan Weakley, author of the *Flora of the Southeastern United States* (Weakley 2020), created common names that could be standardized in both treatments. In some cases, the common name used by resources such as the USDA PLANTS website are misleading, such as Appalachian Mountain Mint used to refer to *Pycnanthemum flexuosum,* which ranges most widely in the Coastal Plain and is extremely rare and local at a handful of actual mountain locations. In such cases, we again consulted with Alan Weakley to use or derive a more acceptable common name. When more than one name is given, the one most frequently used in South Carolina is listed first.

Rarity of Vascular Plants

A rare species is difficult to define since the concept involves two variables: (1) the overall species distribution and (2) the abundance of individuals within that distribution. Figure 2 (from Hardin, 1977) illustrates this relationship. One example of an extreme in rareness in South Carolina, Point A, is Mountain Sweet Pitcherplant (*Sarracenia jonesii,* plate 58), represented by a few individual plants restricted to an extremely limited geographic area (in this case, only a few sites in the mountains and upper Piedmont). Point B is exemplified by Cancer-root (*Orobanche uniflora,* plate 291), which is scattered throughout South Carolina but with only a few specimens at any one site. Point C is exemplified by Pondspice (*Litsea aestivalis,* plate 710), which is locally abundant in Berkeley and Charleston Counties. It is rare since the center of its distribution is limited to these counties. Evaluation of rareness, then, is a function of distribution and abundance and includes all possible intermediates of the three extremes shown in figure 2.

In 1966, the Endangered Species Preservation Act first recognized the intrinsic value of species; it also recognized that some species were so rare that they were near extinction. Unfortunately, the 1966 act recognized only animals. In 1973, the Endangered Species Act broadened the scope to include plants. Since the passage of these acts, numerous organizations have come forward with programs that locate and provide protected habitat for rare species. More important, these groups realized that preservation of our natural heritage depends on preservation of biological diversity. With this concept as its standard, various organizations have protected numerous sites in South

Figure 2. Relative Rareness of Plants as a Function of Distribution and Abundance.

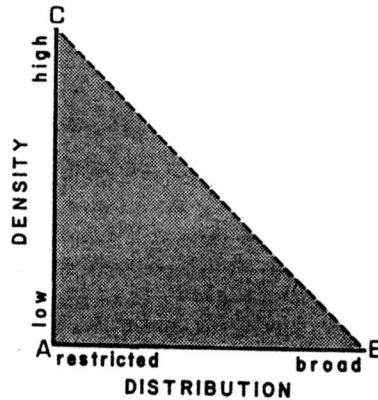

Carolina. At both state and federal levels, programs have been established to document species distributions, to monitor population fluctuations, and to determine appropriate categories for rare species.

Two categories of plant rarity are formally recognized at the federal level and have legal status by the United States Department of Interior under the Endangered Species Act of 1973.

Federal Endangered Species: a species that is in danger of becoming extinct in the near future. In South Carolina, thirteen vascular plants are listed as endangered. One of these, Michaux's Sumac (*Rhus michauxii*) is known historically only from South Carolina.

Federal Threatened Species: a species not now endangered but one that is heading in that direction. Seven species in South Carolina are listed as threatened. One of these, White Fringeless Orchid (*Platanthera integrilabia*), is now possibly extirpated in South Carolina.

At the state level, the Heritage Trust Program keeps a database on all species of vascular plants considered rare in the state. Unlike state law that gives protection to rare animals, there is no state law that provides protection for rare plants. The following categories are used to describe the rarity of plants monitored by the South Carolina Heritage Trust Program. The primary component of the ranking is number of populations, but other factors such as population sizes are also considered. Species may also be classified as extirpated if they are no longer believed to occur in the state or as historical if the species has not been seen recently but could possibly be found.

Critically Imperiled: a species with 1 to 5 documented populations in South Carolina.

Imperiled: a species with 6 to 20 documented populations in South Carolina.

Vulnerable: a species with 21 to 80 populations.

Any species pictured in this book that fits any of the above federal or state categories is so indicated in the protective status section in the species description. If a plant is listed federally, the state status is not given. The status of a plant, however, can change as new information becomes available. The authors expect that over the life of this book, many changes will occur. Interested individuals should obtain updated information from the South Carolina Heritage Trust Program.

Pronunciation Guide to Botanical Names

Popular wildflower books or botanical field guides usually do not include pronunciation guides for botanical names. Many laymen (and some scientists) are hesitant to refer to botanical names because of pronunciation difficulties. With a proper guide, however, the pronunciation of botanical names is not difficult. To assist the reader, the authors provide a pronunciation guide for the botanical name of each plant pictured in this book.

Botanists and gardeners in English-speaking countries today generally use traditional (or standard) English sounds for vowels and consonants, while following the rules of classical Latin for accenting. This book follows this plan and uses the system found in *Gray's Manual of Botany* (8th edition, Fernald, 1950), for which a grave accent (`) denotes a long vowel in the accented syllable and an acute accent (´) denotes a short vowel in the accented syllable. Each botanical name is divided into syllables according to classical Latin. An example:

Yúc-ca glo-ri-ò-sa

If one is curious to know why a vowel is long or short, or why the accent falls where it does, the authors include the following conditions and rules for pronunciation of botanical names. The material is modified from *Vascular Plant Systematics* by Radford et al. (1974), *Botanical Latin* by William T. Stearn (1992), and *Gray's Manual of Botany* by M. L. Fernald (8th edition, 1950).

Commemorative names

Rules of pronunciation cannot be satisfactorily applied to all generic names and specific epithets that commemorate persons. A standard system of pronunciation of commemorative names does not exist because different people use the same letters for different sounds and different letters for the same sounds. The most acceptable method with commemorative names is to pronounce them as nearly as possible like the original name but with a Latin ending. The following examples are illustrative. The English family name James, Latinized as a specific epithet, becomes ja-mè-si-i, since in Latin every vowel must be accented. In this form, the name is unrecognizable. The authors believe it is preferable to pronounce the family name as close as possible as it sounds in English and James becomes jàmes-i-i. In this form the name is recognizable. The English name Hales, used as a generic name, in its Latinized form would be Ha-lè-si-a, which is how it is most commonly pronounced. The authors, however, choose to pronounce it Hàles-i-a to sound as close to the original name as possible. Likewise, the specific epithet *smallii* and the genus *Mitchella* (Latin for Small and Mitchell) are pronounced as smáll-i-i and Mít-chell-a. In these forms, the names more closely represent how they are pronounced in their original English.

Syllabism

Before a word can be accented, it must be divided into syllables, and every Latin word has as many syllables as it has separate vowels or diphthongs.

a. When a single consonant comes between two vowels, it is taken with the vowel that follows: Cà-rex; Mò-rus; Là-mi-um; Smì-lax.

b. When two consonants come between two vowels, one goes with the first vowel and the other with the second vowel: ál-bi-dum; Ver-nòn-i-a; Ver-bás-cum; sep-tem-lò-ba.

c. When two consonants come between two vowels, and the first consonant is b, c, d, g, k, p, or t, and the second is either l or r, both consonants go with the second vowel: glà-bra; Pte-ro-glos-sás-pis.

d. If two consonants that come between two vowels are ch, ph, or th, each pair is counted as one letter and goes with the second vowel: Du-chès-ne-a; Rá-pha-nus; Oe-no-thè-ra.

e. When there are more than two consonants between two vowels, all but the first go with the second vowel: am-bro-si-oì-des; Car-phé-pho-rus.

f. The letter x is always taken with the vowel preceding it: Tax-ò-di-um; Ta-ráx-a-cum

Classical Latin rules of accenting

a. The last syllable of a word is never accented.

b. Words of two syllables are always accented on the first syllable: Cà-rex; Ì-lex.

c. The next to last syllable is called the penult; the second from the last is called the antepenult. The accent is never farther from the end than the antepenult. Thus, the penult is the crucial factor in accenting.

d. In a word of more than two syllables, the accent is on the penult if the penult is long. The penult is long if it ends in a long vowel (cy-mò-sa), a diphthong (gen-ti-a-nòi-des), or a consonant (Pre-nán-thes). In the latter, the vowel is short, but the syllable is accented.

e. When the penult ends in a vowel, one must determine whether that vowel is long or short. This requires a Latin dictionary.

f. If the penult is short, the accent is on the antepenult: gra-mi-ni-fò-li-a. (See 6d below for why the i in the penult is short.)

Consonants

a. When one of the following pairs of consonants begins a word, the first letter is silent: cn, ct, gn, mn, pn, ps, pt, ts, or tm. Examples are Cnidos-colus = ni-dós-co-lus; Ctenium = té-ni-um; Gnaphalium = na-phá-li-um; Psoralea = so-rà-le-a; Pterocaulon = te-ro-caú-lon; Tsù-ga = sù-ga.

b. The consonants c and g have soft sounds of s and j, respectively, when they are followed by e, i, y, ae, or oe. Examples are cy-pe-rì-nus = si-pe-rì-nus and gy-nán-dra = ji-nán-dra.

c. An initial x sounds like z: Xán-thi-um = zán-thi-um; Xỳ-ris = zỳ-rus.

d. When ci, si, and ti follow an accented syllable and are followed by another vowel, they often have the sound of sh or zh: Se-nè-ci-o = se-nè-sh-o; Ar-te-mí-si-a = ar-te-mí-zh-a.

e. Ch, ph, and th are considered one letter each. Ch has the sound of k at the beginning of a word (Chas-mán-thi-um = kas-mán-thi-um; Che-lò-ne = k-lò-ne); ph is sounded as f (phél-los = fél-los); and th as in "thing."

f. When cc is followed by i or y, it sounds like k-si: coc-cí-ne-a = kok-sí-ne-a

Diphthongs

A diphthong consists of two vowels pronounced as a single vowel and classified as a long vowel.

The diphthongs are:

ae: pronounced like long e in "me": laè-vis = leè-vis

oe: pronounced like long e in "me": a-moè-na = a-meè-na

ei: pronounced like long i in "kite": cu-nei-fò-li-a = cu-ni-fò-li-a

eu: pronounced like u in "neuter": mos-cheù-tos = mos-chù-tos

au: pronounced like aw in awful: a-caù-lis = a-càw-lis

oi: Some botanists treat oi as a diphthong, pronounced as in "oil," not as separate vowels according to Latin rules. As an example, a-lo-pe-cu-ro-ì-des becomes a-lo-pe-cu-roì-des. In this book, oi is treated as a diphthong.

Note: In some works, a diacritic mark (diaeresis) is placed over the second of two adjoining vowels when they represent two sounds (e.g., Leucothoë and Isoëtes). Here oe is not treated as a diphthong, and the names are pronounced Leu-cóth-o-ë and I-sò-ë-tes.

Vowels

Following Latin rules, all vowels in words are pronounced, and every word has as many syllables as it has separate vowels or diphthongs.

a. The vowel y is always sounded like i: Ly-co-pó-di-um = Li-co-pó-di-um.

b. Final vowels have the long sound (si-nén-se = si-nén-see), except final a, which is sounded as ah: ca-ro-li-ni-à-na = ca-ro-li-ni-à-nah.

c. Final es is sounded like ease: a-lo-pe-cu-roì-des = a-lo-pe-cu-roì-deez.

d. Two vowels together that do not form a diphthong are always sounded separately, with the first having a short sound: filifolia = fi-li-fò-li-a, where the i is short.

e. Note that the above rule does not hold for words transcribed from Greek. For example, the e in *Staphylea, Centaurea and gigantea* is long because it is a contraction of the Greek diphthong ei. A diphthong is always treated as a long vowel, even if transcribed by a single letter. Since the penult is long, the genera are pronounced as Sta-phy-lè-a, Cen-tau-rè-a, and gi-gan-tè-a.

f. Long penult. As a rule, the penult is long in the following adjective endings and takes the long accent:

-à-ber	-à-re	-è-na	-ì-per	-ò-lus	-ò-vus	-ù-rus
-à-go	-à-ris	-è-nus	-ì-rens	-ò-na	-ù-cus	-ù-sa
-à- gus	-à-rum	-è-tus	-ì-sa	-ò-nis	-ù-bens	-ù-sus
-à-le	-à-ta	-è-um	-ì-sus	-ò-ra	-ù-fus	-ù-ta
-à-li	-à-tes	-ì-cans	-ì-tes	-ò-ris	-ù-go	-ù-tum
-à-lis	-à-tos	-ì-des	-ì-tis	-ò-rum	-ù-lis	-ù-tus
-à-na	-à-tum	-ì-na	-ì-va	-ò-sa	-ù-na	-ỳ-drus
-à-num	-à-tus	-ì-ne	-ì-vus	-ò-sum	-ù-nus	-ỳ-lus
-à-nus	-è-lis	-ì-nis	-ò-bus	-ò-sus	-ù-nes	-ỳ-rum
-à-pa	-è-ma	-ì-nus*	-ò-des	-ò-ta	-ù-ra	

Examples: car-di-nà-lis   bar-bà-tus   fa-ri-nò-sus

g. Short penult. In the following adjective endings, the vowel in the penult is usually short, thus removing the accent to the antepenult:

-ă-cus	-ĕ-na	-ĭ-chus	-ĭ-la	-ĭ-ta	-ŏ-lis	-y-dum
-ă-dos	-ĕ-pis	-ĭ-cum	-ĭ-lis	-ĭ-us	-ŏ-mum	-y-dus
-ă-la	-ĕ-ra	-ĭ-cus	-ĭ-lus	-ĭ-um	-o-pus	-y-la
-ă-lus	-ĕ-ris	-ĭ-da	-ĭ-ma	-ŏ-ba	-o-rus	-y-mus
-ă-nes	-ĕ-rus	-ĭ-dis	-ĭ-mus	-ŏ-da	-ŭ-la	
-ă-ra	-ĕ-us	-ĭ-dum	-ĭ-num	-ŏ-dum	-ŭ-lus	
-ă-ti	-ĭ-as	-ĭ-dus	-ĭ-or	-ŏ-la	-ŭ-us	
-ĕ-ger	-ĭ-ca	-ĭ-ens	-ĭ-pes	-ŏ-lon	-y-ca	

Examples: se-bí-fe-rus   he-pá-ti-cus   cán-di-dus   am-bí-gu-us

h. Compound words ending in the following syllables also have a short vowel in the penult; thus, the accent is moved to the antepenult:

-clă-dus	-fĕ-mur	-gy-nus	-pĭ-lis	-stŏ-mus	-vĭ-rens
-cŏ-lor	-gĕ-rum	-ŏ-lens	-phy-tum	-tŏ-mus	

Examples: di-chó-to-mus   sem-pér-vi-rens

Family names

Family names end in aceae. Since ae is a diphthong and is therefore long, ce (the penult) is short and the accent is on the antepenult. Examples of family names and their pronunciations are As-te-rà-ce-ae, Ro-sà-ce-ae, Li-li-à-ce-ae, Or-chi-dà-ce-ae, and I-ri-dà-ce-ae.

* Exceptions: short in can-ná-bi-nus, gos-sý-pi-nus, se-ró-ti-nus, etc; usually long in cy-pe-rì-nus and sa-li-cì-nus

Part 1

The Nature of South Carolina's Wildflowers

Selected Topics on Natural History and Ecology

Carnivorous Plants

Carnivorous plants are unusual in the plant world because of their unique method of compensating for low mineral availability of their preferred environments. Their leaves have become modified through natural selection to trap insects and other small animals. Carnivorous plants inhabit acidic soil or water. In acidic soils, many minerals, especially nitrogen, are strongly bound to soil particles, and they are unavailable for plants to absorb into their roots. In habitats like these, carnivory is assumed to have arisen as a supplementary means to obtain minerals. After a carnivorous plant traps an animal, usually a small insect, the digestion of animal protein results in the release of minerals that are then absorbed by the plant. As a result of selection pressure, plants that evolved carnivory were able to inhabit and flourish in acidic soils.

Carnivorous plants do not obtain food from their prey, only dissolved minerals. Like all green plants, photosynthesis occurs. To convert carbohydrates made during photosynthesis into other compounds (e.g., lipids, proteins, and nucleic acids), nitrogen is needed. For carnivorous plants, these minerals are supplied from trapped animals.

The leaves of carnivorous plants that have become modified to trap insects are so changed that they only remotely resemble what are normally recognized as leaves. However, their green color indicates that they still function in photosynthesis. There are two major types of traps: passive and active. Passive traps do not employ any type of movement to entrap prey, although ingenious methods have developed to attract prey. Active traps not only attract prey but move to catch, hold, and manipulate the prey for more efficient digestion.

Among the many myths associated with carnivorous plants is that some species are large enough to eat people. The truth is that the largest animals that could be caught by any carnivorous plant, anywhere in the world, are small mammals and reptiles. Slack (1979) reports that species of *Nepenthes* in the Old World tropics do catch small mammals and reptiles, but such catches are probably fortuitous rather than intentional. In the many years of examining carnivorous plants with our students, we have found species such as green anoles, frogs, and even small mice dead in the large traps of Yellow Pitcherplants. Because animals other than insects are often caught as prey, the term "carnivorous" is preferred over "insectivorous."

All carnivorous plants produce fertile flowers and viable seeds. In the case of some pitcherplants, the flowers develop before the leaves, and one often does not equate the two structures as being from the same plant. Other species of pitcherplants and the other four genera of carnivorous plants in South Carolina produce flowers and leaves (traps) at the same time.

Worldwide there are about 450 species of carnivorous plants, representing approximately 18 genera. In South Carolina, there are five genera and 29 species: *Sarracenia*–the pitcherplants (6 species and 2 subspecific varieties, *Drosera*–the sundews (5 species), *Pinguicula*–the butterworts (3 species), *Utricularia*–the bladderworts (14 species), and a single species of *Dionaea*–the Venus' Fly Trap. The reader can refer to the following sources, which are listed in the general reference section, for more information on carnivorous plants: Lloyd (1976); Pietropaola and Pietropaola (1986); Schnell (1976); and Slack (1979).

SARRACENIA (PITCHERPLANTS)

Pitcherplants have perennial rosettes of leaves that are modified to trap insects. The pitchers represent passive traps and are tubular, like elongated funnels, and overtopped by hoods. Usually this hood is supported by a narrow column, which may be reflexed over the pitcher, as exhibited in the Hooded Pitcherplant (*Sarracenia minor*, plate 619), or vertical, as in Southern Purple Pitcherplant (*S. purpurea* var. *venosa*, plate 505). Two mechanisms attract insects: the coloration of the pitchers and the secretions from a nectar gland. The nectar gland is at the rolled-up margin of the hood along the lip opposite the column. In Southern Purple Pitcherplant, once an insect lands on the brim of the tubular opening or on the underside of the hood, downward-pointing stiff hairs drive the insect into the pitcher. Often the insect loses its footing and falls into the pitcher. The upper portion of the pitcher is lined with a slippery, smooth wax that makes it difficult for the insect to crawl out. Ultimately, the insect falls into the watery mixture of enzymes at the base of the pitcher, where digestion takes place and absorption occurs through a wax-free zone. Recent research has shown that most species of pitcherplants include a highly toxic alkaloid, coniine, in their nectar. This may explain a phenomenon that is easy to observe in White-top Pitcherplants (*Sarracenia leucophylla*). Though the white of the pitcher is thought to attract moths, during the day the nectar attracts butterflies. When a butterfly lands and begins to feed they appear to become lethargic and fall deep into the trap.

The leaves of pitcherplants arise from subterranean rhizomes in early spring. The subterranean rhizomes allow the plants to survive the periodic fires that remove the competing vegetation in the savannas and bogs where they grow. These rhizomes also function as a means of asexual reproduction.

Eight taxa of pitcherplants grow in the state: Hooded Pitcherplant (*S. minor*, plate 619), Yellow Pitcherplant (*S. flava*, plate 678), Southern Purple Pitcherplant (*S. purpurea* var. *venosa*, plate 505), Carolina Sweet Pitcherplant (*S. rubra* ssp. *rubra*, plate 732), Georgia Sweet Pitcherplant (*S. rubra* ssp. *viatorum*), Parrot Pitcherplant (*S. psittacina* Michaux), Mountain Purple Pitcherplant (*S. purpurea* var. *montana*, plate 57), and Mountain Sweet Pitcherplant (*S. jonesii*, plate 58). The first six occur in the Coastal Plain and Sandhills regions; the last two are restricted to the mountains and upper Piedmont. One of these, the Parrot Pitcherplant, is only known from a single site on private property in the southeastern corner of South Carolina but has not been observed since the 1990s. The specific habitats of these taxa are provided in the individual species descriptions.

DROSERA (SUNDEWS)

Sundews are herbaceous plants, having a basal rosette of leaves that arise from a fibrous root system. In South Carolina, they are perennials, with the exception of the Dwarf Sundew (*Drosera brevifolia,* plate 600), which is an annual. A fertile stalk, arising from between the leaves, supports from 1 to 25 white flowers in a raceme. The lowest flower opens for about two days and closes, and the flower above it then opens. The flowering process ultimately moves to the top of the inflorescence. Flowering occurs throughout the spring and summer.

The leaves are an active trap to only a limited degree and are often referred to as a flypaper mechanism. The leaf blade is covered with stalked glands, some of which secrete mucilage (which holds the insect), while others secrete digestive enzymes. Small, crawling insects, either attracted to the plant by the leaf coloration and/or by the nectar from the glands or as a result of a chance wandering, become mired in the sticky secretions. As the insect struggles to get free, the motion causes signals to be sent through the leaf blade, by some yet unclear process, to the long-stalked marginal glands, which then bend to the center of the blade and further entangle the insect. In some species, the leaf blade may fold over the insect, which results in a greater leaf surface area contacting the insect, thus increasing the rate of mineral absorption.

The common name of *Drosera,* sundew, comes from the sticky, dewlike tentacles that shine like dew, glittering in the early morning sun.

Four species of sundews currently occur in the state: Round-leaf Sundew (*Drosera rotundifolia,* plate 510), Dwarf Sundew (*D. brevifolia,* plate 600), Pink Sundew (*D. capillaris*), and Water Sundew (*D. intermedia,* plate 509). A fifth, Thread-leaf Sundew (*D. filiformis*), is known only historically from Orangeburg County.

PINGUICULA (BUTTERWORTS)

Butterworts are fibrous-rooted perennials that in summer form a flat rosette of blunt, oblong leaves. The older leaves lie prostrate against the ground, with the younger, nearly flat leaves lying on top of the older leaves. Butterworts retain their leaves over the winter and do not form winter resting buds.

Their leaf surfaces contain two kinds of glands: (1) stalked glands that are important in catching and holding prey and (2) sessile glands that are active in digestion. The trapping mechanism is an active flypaper type. Crawling and flying insects probably come into contact with the leaves by chance or are attracted by the shining sticky surface. Prey become mired in the gland's secretions and are held until digestion and absorption gradually occur. The leaf margins curl over the prey, perhaps preventing the partially digested insect from being washed off by rain, but certainly bringing more leaf surface into contact with the prey and increasing absorption.

The rosette form of the butterworts' leaves is often hidden by other taller plants and thus they are inconspicuous and often overlooked. However, in late spring and early summer, their flowers make butterworts quite apparent.

Three species of butterworts occur in South Carolina, and all grow in the Coastal Plain: Violet Butterwort (*P. caerulea,* plate 616), Yellow Butterwort (*P. lutea,* plate 617 in *WCL*), and Small-flowered Butterwort (*P. pumila,* plate 566).

UTRICULARIA (BLADDERWORTS)

The common name, bladderwort, originates from the modified leaves, or bladders, that function as traps. Species are either aquatic or terrestrial; the terrestrial species grow in moist soil.

During the resting stage of the bladder, much of the fluid inside the bladder is absorbed, creating a higher water pressure outside of the bladder, and the sides of the bladder appear pinched in, or concave. There has thus developed a negative pressure within the bladder, which is now set to be sprung. The traps are activated when an aquatic insect touches one of the sensitive hairs that surround the bladder opening. Functioning like a hinged door, the bladder quickly opens, water rushes into the bladder, sweeping in the insect, then it closes, trapping the prey inside. Digestion takes place inside the bladder. Because of the tiny size of the bladders, only small aquatic insects, such as mosquito larvae, or tiny crustaceans, such as *Daphnia,* are trapped. The most prolific bladder producers are the aquatic species; terrestrial species only produce bladders when they are growing in very moist soil.

One spectacular bladderwort is the Swollen Bladderwort (*Utricularia inflata,* plate 864). This is a Coastal Plain species that forms a remarkable flotation device to support the flowering stalk. Six to ten floats (modified leaves) radiate from the middle of the stalk. Initial development of the stalk and bladders begins under water. As the floats grow, their buoyancy, which is caused by the presence of aerenchyma tissue, makes the entire plant rise to the surface. This fertile stalk then produces from 9 to 14 yellow flowers.

Fourteen species of *Utricularia* occur in South Carolina; only the saline maritime strand has no representatives.

DIONAEA (VENUS' FLY TRAP)

Venus' Fly Trap (*Dionaea muscipula,* plate 622) is a terrestrial carnivorous plant that is endemic to the Coastal Plain of the Carolinas. In South Carolina, it occurs today only in Georgetown and Horry Counties. Populations occur in Lewis Ocean Bay Heritage Preserve and Cartwheel Bay Heritage Preserve. It favors damp, sandy soil with a small portion of peat (like that found around Carolina bays) and open, sunny conditions. It has particularly exacting habitat requirements. The seedling stage is the most sensitive life-history stage because the habitat requirements are very specific at this time. This species is not yet near extinction, but much of its special habitat has been so altered that it must be monitored to ensure its preservation. It is one of many savanna herbs that requires frequent fire to prevent invasion by shrubs and trees, which would cause its elimination. It is also plagued by illegal collecting of wild plants. Perhaps no other species in the Carolinas is so heavily poached.

The trapping mechanism, a modified leaf, is an active, springlike trap. The foliage is a rosette that grows from a perennial rhizome, with each leaf blade consisting of two lobes, and each lobe is attached to the blade's midrib. Prong-like teeth occur along the free margins of the lobes. On the inner surface of the lobes are two types of glands: nectar glands, near the margin to attract insects, and digestive glands on the surface. The latter turn red when exposed to the sun, giving the leaf a reddish coloration, which also attracts insect prey. On the upper surface of each leaf lobe are trigger hairs (generally three or six), which spring the trap when an insect touches them. The lobes close in about half a second, trapping the prey, and then digestion and absorption begin. The initial closure is due to a shape change in the leaf surface from convex to concave. This is followed by a slower closure around the hinge. Springing a trap requires two touches of a single hair or a touch of two separate hairs, occurring in about a 20 second interval. This prevents wind-blown objects from prematurely closing the trap and wasting the plant's energy. Usually, each trap can catch three insects before it dies or no longer responds when stimulated. The continual production of new leaves during the growing season compensates for this loss.

Native Orchids

The orchid family (Orchidaceae) is one of the most fascinating and diverse families of flowering plants in the world. It is a cosmopolitan family that attains its highest development in the mountains of the tropics and sub-tropics of both hemispheres. Orchids are found throughout the world at sites where at least some amount of moisture is present. They are absent only from low, hot deserts and from polar regions, where the ground is permanently frozen. There are at least 30,000 species of orchids recognized worldwide. Within the United States and Canada, botanists recognize more than 210 species and varieties. In South Carolina, 60 species have been identified.

From an aesthetic view, orchids are often accorded first place in nature. The beauty of their flowers has made the orchid family the center of a multimillion-dollar floral industry in the United States and Europe, but the family is otherwise of little economic importance. The most important natural product that comes from an orchid is vanilla flavoring from the seedpods of *Vanilla planifolia,* a climbing orchid native to tropical America and cultivated in many tropical countries.

Despite its wide habitat range and large number of species, the orchid family is often overlooked by botanists and laypersons because the majority of our native orchids have small, inconspicuous flowers. Most people associate orchids with exotic tropical species or with the myriad of large-flowered forms sold for corsages. On close examination, however, our native orchids are just as alluring.

Orchids are specialized perennial herbs that are terrestrial, lithophytic (growing on rocks), or epiphytic (growing on another plant, usually a tree).

Terrestrial orchids are common in temperate areas, such as South Carolina, while epiphytic orchids are more common in tropical areas. With the exception of the Green-fly Orchid (*Epidendrum conopseum,* plate 788), all orchids in South Carolina are terrestrial. North of the sub-tropics, the epiphytic habitat is restricted because of cold temperatures and insufficient moisture. In cold temperatures, exposure to the atmosphere and a lack of insulation from soil kills the roots of epiphytes. Only along the coast of South Carolina, where the ocean moderates the temperatures (often a few degrees above the inland areas in the winter) and where moisture is high, is the epiphytic growth possible for orchids. More commonly, the Green-fly Orchid is found on the branches of tall Bald Cypress and Tupelo Gum trees in swamp forests or on live oaks in upland habitats.

Many orchids, after a normal period of flowering, do not produce aboveground vegetative parts for a year or more. In other words, they "disappear" for a time. Snowy Orchid (*Platanthera nivea,* plate 626) may cover a wet Longleaf Pine savanna one year, then be absent for years, only to return in great numbers another year. Some reports claim that during this hiatus the orchid is building up food supplies in the underground stem that will provide energy for the flowering process.

One of the most unusual specialized orchid lifestyle is mycotropism. Mycotrophic plants often lack chlorophyll and are unable to manufacture their own food. They obtain food from a mostly parasitic association with a fungus that is capable of obtaining carbohydrates from decaying organic matter in the soil. Mycotrophic orchids are terrestrial species that depend on a soil fungus to supply food. The fungus lives in the roots of the orchid and sends its mycelium out into the soil. This mass of mycelium (a large, entangled network of filaments that forms the body of a fungus) greatly increases the surface area that is available for the absorption of water, minerals, and food. The fungus secretes enzymes into the soil to digest organic material, absorbs the digested material and converts it to simpler compounds, and then transfers them to the orchid. Usually saprophytic orchids are confined to humus- and moisture-rich habitats that can support mycelial development.

Two genera of entirely mycotrophic orchids occur in South Carolina: coral roots (*Corallorhiza,* plates 174 and 752), and Crested Coral-root (*Hexalectris spicata,* plate 759). Most if not all species of terrestrial orchid found in South Carolina have a similar association with fungi and can be considered at least partly mycotrophic.

The orchid family is noteworthy among higher plants for three reasons: (1) the diversity of its highly specialized flowers, (2) the large number of extremely minute seeds that lack an endosperm, and (3) the wide diversity of habitats where they are found. The flower is unusual because the male and female parts (stamens and pistil) are fused into one structure called the column. At the apex of the column is the male anther; here pollen grains are aggregated into masses called pollinia. Below the anther is the stigma, the terminal receptive portion of the pistil. The surface of the stigma is sticky, thus allowing pollinia, which are brought from other flowers by insects, to adhere to its surface.

Insects primarily pollinate orchids, with many orchid species apparently having an association with its own insect pollinator. All orchid flowers have three sepals (outer floral whorl) and three petals (inner floral whorl). One of the petals, the central lip, is different from the two lateral ones, often being larger and showier. Usually the flower grows so that the lip is lower than the lateral petals; however, in the genus *Calopogon* (plate 373), the lip is above the lateral petals. Having the lowest petal as the lip is the result of the flower twisting itself upside down prior to opening and is termed a "resupinate flower."

The seed capsule of orchids can require as long as nine months to mature and may contain millions of dustlike seeds. The seeds contain no endosperm and are dependent on external aid for germination and seedling growth. A soil fungus penetrates the seed and establishes a symbiotic relationship with the embryo. The embryo grows by absorbing the externally produced digestion or secretion by-products of the fungus. The fungus will eventually penetrate the orchid's roots, where it remains in a symbiotic mycorrhizal relationship that is permanent and obligate. Thus, all orchids, at least as seedlings, are mycotrophic. The balance of orchid, soil, and soil fungus is so delicate that transplanting orchids from their native habitat is difficult; it is not advised for wildflower gardens.

The diversity of habitats and growth forms attests to the remarkable adaptations of the orchid family. One can find Pink Lady's Slipper (*Cypripedium acaule,* plate 238) in oak-hickory forests and pine-oak heaths; Showy Orchis (*Galearis spectabilis,* plate 120) and Lily-leaved Twayblade (*Liparis lilifolia*) are found in cove forests. In the Coastal Plain, two species of *Spiranthes* have been found in abandoned farm fields and another species in freshwater pools behind coastal dunes. Six species of *Platanthera* occur in the Longleaf Pine savannas, while two species of *Platanthera* grow in swamp forests. Two species grow in the freshwater tidal marshes along the coastal rivers: Water-spider Orchid (*Habenaria repens,* plate 831) and Fragrant Ladies'-tresses (*Spiranthes odorata*). Furthermore, some orchids grow statewide, while others grow only in restricted habitats. For example, Shadow-witch (*Ponthieva racemosa,* plate 763) is confined to the Coastal Plain, growing in calcium-rich sites, while the Small Green Wood-orchid (*Platanthera clavellata,* plate 425) is found in fens, seepages, or swamps throughout the state.

Succession in Natural Communities

Vegetation on most sites is dynamic, not static. Return to an abandoned agricultural field after a number of years, and the field that was once goldenrods and asters has been replaced by a pine forest. Return to an abandoned rice field along a coastal river in the Outer Coastal Plain, and what was once a tidal freshwater marsh is now a swamp forest. A process of ecological succession has occurred—an orderly sequence where one community replaces another over time. If given enough time and if no major disturbance occurs, succession eventually terminates in the climax community: the stable end community that is capable of self-perpetuation under prevailing environmental conditions. The entire sequence of communities that replace one another is called the "sere," and each transitory community in the sere is called a "seral stage."

Two general types of succession may occur. One type is primary succession, in which a community becomes established on a particular substrate for the first time—that is, no living organisms have previously colonized the substrate. Examples are newly formed coastal dunes or sandbars or spits along rivers. Secondary succession occurs where vegetation has occupied the substrate in the past. Some event such as fire, climatic changes, or human intervention has caused the original community to disappear. Examples of secondary succession are abandoned freshwater tidal rice fields and abandoned agricultural lands. In both cases, humans removed the original vegetation. After abandonment, these areas will undergo succession to swamp forests and hardwood forests, respectively.

An understanding of plant succession is critical if one is to fully appreciate the distribution of wildflowers and the composition of natural communities over time. The following two examples illustrate this process of succession.

OLD-FIELD SUCCESSION IN THE PIEDMONT

The pioneer species in an abandoned agricultural field often is crabgrass (*Digitaria* spp.). Crabgrass produces seeds profusely, and its seeds can remain dormant for years. When a field is under cultivation, crabgrass seeds are blown in or carried by birds; they remain dormant in the soil, germinating as soon as the field is abandoned. Crabgrass quickly covers the field. Toward the end of the first year and into the second, horseweed invades the crabgrass. Organic matter that is produced by horseweed inhibits its subsequent establishment and growth. Asters and ragweed quickly invade, and being taller than horseweed, shade out the surviving horseweed (and crabgrass) and assume dominance by the end of the third year. The previously established species have increased the soil's moisture content, which allows broomstraw (*Andropogon* spp.) to invade the field. Once present, broomstraw spreads rapidly and by the fifth year it dominates all other vegetation. Some shrubs may become established, but ultimately, around the second or third year after broomstraw becomes established, seeds of Loblolly Pine (*Pinus taeda*), Virginia Pine (*Pinus virginiana*) or Shortleaf Pine (*Pinus echinata*) blown in from a nearby seed source germinate to produce a dense stand of pines.

When the pines get above the broomstraw, the reduced sunlight quickly eliminates this plant. The pines go on for years in complete charge of the former field. In time, however, the pines go the way of their predecessors. If a source is available, seeds of oaks, sweet gum, hickories, and maple are brought in by wind, birds, and other animals. Seedlings of these species are slightly more shade tolerant and quickly grow in the midst of the developing pine forest. First, the mature hardwoods shade and kill the lower pine branches. When the hardwoods grow to the height of the smaller pines, these trees die out in large numbers. Additionally, many pines are short-lived, like the Virginia Pine, and are merely transient in any one place. This hardwood invasion occurs about 20 to 25 years after the pines are established. After about 150 years, the

OLD-FIELD SUCCESSION

| 0 (fall) | 1 | 2 | 3–5 | 5–25 | 25–150 | 150+ |

crabgrass ➡ horseweed ➡ asters ragweed ➡ broomstraw pine seedlings ➡ pine forest ➡ pine hardwoods ➡ hardwoods

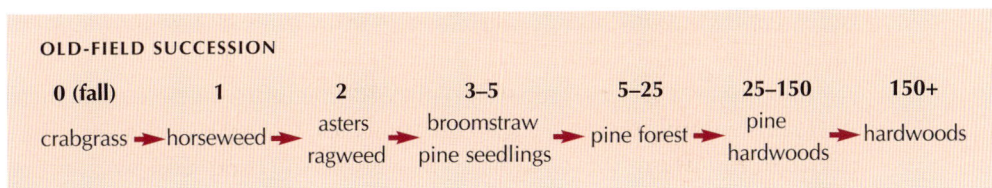

Figure 3. Old-Field Succession

hardwood forest will replace the pine forest. Pine seedlings, which require high light intensities, cannot survive in the shaded forest floor. The pine trees that are large enough to survive the invasion of hardwoods will live out their lives in forest, about 125 to 150 years. When the pines die, a mixed hardwood forest results. Often very shade-tolerant species such as American Beech (*Fagus grandifolia* var. *caroliniana*) become dominant in the understory and can gradually replace many of the oaks and hickories in the canopy. The hardwood forest becomes the climax community since its broadleaf seedlings can trap sufficient sunlight to survive, even in a shaded forest floor. Unless disturbance again removes the hardwood forest, it will perpetuate itself indefinitely. The process of old-field succession is summarized above (figure 3). The numbers refer to the years following abandonment.

FORMER RICE FIELDS OF THE OUTER COASTAL PLAIN

In former tidal rice fields (see the figure 5 in the essay, "Agriculture: Effects on South Carolina's Physical Landscape"), hydrarch succession (succession that begins in water) starts with submerged, anchored hydrophytes such as Brazilian Waterweed (*Egeria densa*), Carolina Fanwort (*Cabomba caroliniana*), pondweeds (*Potamogeton* spp.), and milfoils (*Myriophyllum* spp.). These plants bind the loose soil matrix, trap sediment, and add to the accumulation of organic matter as they die. Next, floating-leaved, anchored hydrophytes such as Common Water-primrose (*Ludwigia hexapetala*) and Alligator-weed (*Alternanthera philoxeroides,* plate 969) become established since their submerged stems can reach to the bottom. Some of the floating aquatics, such as the introduced Water Hyacinth (*Eichhornia crassipes,* plate 866), then become established in the open water of the fields. These plants trap more sediment and continue to add to soil accumulation as they die. When the leaves of the floating aquatics sufficiently block the source of sunlight for the submerged, anchored aquatics, they ultimately die. Finally, the soil level is raised close enough to the surface for emergent, anchored hydrophytes to establish themselves. These include persistent emergents such as cat-tails (*Typha* spp.), Woolly Bulrush (*Scirpus cyperinus*), rushes (*Juncus* spp.), Giant Plume Grass (*Erianthus giganteus,* plate 830), Wild Rice (*Zizania aquatica,* plate 825), Southern Wild Rice (*Zizaniopsis miliacea,* plate 824), and Giant Cordgrass (*Spartina cynosuroides*) and nonpersistent emergents such as Swamp Rose (*Rosa palustris,* plate 833), Swamp Rose-mallow (*Hibiscus moscheutos* ssp. *moscheutos,* plate 839), Pickerelweed (*Pontederia cordata,* 832), Water Hemlock (*Cicuta maculata,* plate 834), Tidal-marsh Obedient-plant (*Physostegia leptophylla,* plate 836),

Water Parsnip (*Sium suave*), and Green Arrow-arum (*Peltandra virginica*, plate 849).

The emergent species ultimately replace the floating species. In turn, these emergent hydrophytes create the conditions for the next seral stage. They reduce soil moisture through transpiration, raise the soil level by trapping more sediment and decomposing, and create a more stable soil due to a mass of interlocking rhizomes. A marsh thicket can then develop as wind, water, and animals bring the seeds of woody plants. These include Swamp Rose, Southern Wax Myrtle, Button-bush (*Cephalanthus occidentalis*), and Common Indigo-bush (*Amorpha fruticosa*). Finally, the soil is raised above the water table, and the wind-borne fruits of Bald Cypress (*Taxodium distichum*, plate 769), Red Maple (*Acer rubrum*, plate 52), willows (*Salix* spp.), Loblolly Pine, cottonwoods (*Populus* spp.), Sweet Gum (*Liquidambar styraciflua*, plate 793), and Swamp Gum (*Nyssa biflora*, plate 707) become established. This new swamp forest will be different in species composition from the original swamp forest because environmental parameters change over the years.

The swamp forests that develop on abandoned rice fields eventually become climax communities. There are enough examples of swamp forests reaching maturity in abandoned rice fields along the coastal rivers to conclude that they are in relative equilibrium with the environment and are self-perpetuating.

Fire in the South Carolina Landscape

Archaeologists tell us that humans first entered North America between 12,000 and 15,000 years ago and that their presence in South Carolina is well documented for the past 12,000 years. Archaeologists also have good evidence that Native Americans changed the composition of forest communities by using wood for fuel and shelters, disturbing large areas of valley bottoms, altering the distribution of some species (e.g., Bald Cypress), and by using fire to increase the proportion of open lands (i.e., meadows, savannas, and savanna woodlands). Reports from early European explorers, such as Jean Ribault, and naturalists William Bartram, Mark Catesby, and John Lawson suggest that both the Coastal Plain and Piedmont of presettlement South Carolina consisted of vast stretches of open savanna or savanna-woodland interspersed with deep forest.

Archaeologists and historians believe that Native Americans regularly used fire for a variety of purposes, including clearing forests for planting crops and driving animals during fall and winter hunts. Early settlers apparently continued the practice of using fire to clear land for agriculture and to improve forage for their free-ranging cattle and pigs. Burning of forest undergrowth in the South has long been an accepted way of life. It was believed to be beneficial, and to this day intentionally set fires still occur in the Coastal Plain, especially in and near large tracts of publicly owned lands such as the Francis Marion National Forest and the Santee Coastal Reserve or on large private tracts managed for quail hunting.

To what extent did fire alter the South Carolina landscape? How important were lightning-set fires compared to fires set by Native Americans and later by settlers? These questions have long been the purview of archaeologists and historians, and ecologists have recently provided new insights. The following observations are particularly pertinent to these questions: (1) the effects of fire differ depending on fire frequency, intensity, and burn season, with fire frequency being the most important; (2) fires seldom run down slopes that have a gradient that is greater than 15%; (3) the amount of acreage that can burn in a single ignition event varies with the variability of the topography; and (4) topographic variability increases from the coast to the inland.

Ecologists believe that nearly 95% of the uplands of the Coastal Plain, and even portions of the adjacent Piedmont, were once dominated by Longleaf Pine, the only tree species in the South with seedlings resistant to fire. The flat Outer Coastal Plain, with natural fire breaks dividing the uplands into fire compartments of 50 to 300 square miles, could easily burn annually, or at least every 1 to 3 years, with just a few ignitions. The hillier Inner Coastal Plain and Sandhills had much smaller natural fire compartments and probably burned naturally every 5 to 15 years. There is convincing evidence that the uplands of the entire Inner and Outer Coastal Plain, excluding the maritime strand, were dominated by fire-adapted vegetation and that lightning alone was adequate for its development and maintenance. Open meadows, savannas, and savanna-woodlands were the dominant plant communities prior to and during Native American occupation.

Numerous reports by early explorers from travels in the Piedmont describe "wide savannas" and "woodlands without undergrowth," as well as "great forests." However, today the Piedmont harbors few fire-maintained communities, so it is unlikely that large parts of the Piedmont were once savannas or savanna-woodlands prior to Native American occupation. In an attempt to assist land managers who are interested in restoring the few parts of the Piedmont that are fire-adapted to prairielike condition, scientists have recently compiled detailed historical information on the prairie landscapes of the Piedmont in the Carolinas. Examination of historical and meteorological evidence suggests that an open landscape was at least as extensive as early reports implied. For example, Barden (1997) has found maps published in 1676 and 1718 that designated much of the South Carolina Piedmont as "Savannae" or "Grand Savannae." Some authors speculate that as much as 95% of the prairies and other open woodlands were the product of Native American burning and agricultural activities, but Barden points out that lightning fires are not uncommon in the Piedmont and that they may have maintained the extensive grasslands. Only in the most extreme upland sites, best represented in parts of Chester and York Counties, where Blackjack Oak is presently a dominant tree, was the natural fire regime adequate to provide for the long-term maintenance of these sites. Brown (1953) cites a report from the time of the American Revolution that describes patriots traveling at night when they went to visit their families in Rock Hill in order to avoid being seen by Tories on the open "plain" during the day. Today, with natural fires long suppressed, nearly a dozen rare plants, usually found only in Midwestern prairies, cling to a perilous existence in canopy gaps, roadside margins, and power-line rights-of-way.

Ecologists are also concerned with this question: What have we lost in the process of altering the natural fire regime? Development has greatly reduced natural fire compartment size, and since about 1950 aggressive efforts have been made to protect forests from wildfires on both public and private lands. Fires still routinely occur in forests, but these fires are almost all prescribed burns and mostly take place in the winter. Available data suggest that Long-leaf Pine-dominated ecosystems in the southeastern United States were once the most significant ecosystem east of the Mississippi River. This extraordinarily diverse ecosystem once occupied from 55 to 85 million acres. Today less than 3% remains in an undegraded condition. One concerned scientist compares this decline to the loss of the North American tall-grass prairie.

With loss of habitat comes a threat to species. An estimated 122 plant species that are associated with this ecosystem are threatened throughout their known range. Nearly fifty of these species occur in South Carolina. Loss of habitat was a significant contributor to the decline of these species, but one of the enduring mysteries in the natural history of the eastern United States, as one researcher put it, is "the spectacular failure of the primeval pine forest to reproduce itself after exploitation." No one knows for sure why this is true, but contributing factors probably include (1) the naturally low plant regeneration, even of the fire-adapted Longleaf Pine, under a heavy fire regime; (2) grazing pressure on pine seedlings by livestock, which until late in the nineteenth century were allowed to graze freely, while croplands were fenced to keep livestock out; and (3) the institution of state fire laws and fire suppression programs. According to the most recent inventory of South Carolina–forest resources, Longleaf Pine continued its long-term decline from 1986 to 1993, declining from 396,000 to 369,000 acres. Longleaf Pine now dominates only 6.7% of the 5.5 million acres of pine-dominated forests in South Carolina.

Carolina Bays of the Coastal Plain

Carolina bays (e.g., Wood Bay, figure 4) are geological formations of unknown origin that occur mainly on the South Atlantic Coastal Plain Province of North and South Carolina and Georgia. They are shallow depressions formed in the sandy coastal soil and vary in depth from a few feet to around twenty feet. They vary in length from a few hundred yards to several miles. The bays occur in three shapes: elliptical, oval, or asymmetrical, with their long axis oriented in a southeast direction. Often the bays overlap each other, as is seen by the outlines of smaller bays in the depressions of larger ones. Furthermore, their distribution is nonrandom. They occur in clusters, and within the clusters they are often aligned along an apparently undetermined physiographic gradient.

Although for years several workers made reference to "certain depressions" in the Coastal Plain, it was not until 1933 when two scientists, Melton and Schriever (1933), viewed aerial photographs of terrain near Myrtle Beach

Figure 4. Aerial View of Woods Bay.

that the true abundance and distinctive characteristics of the bays became known. The photographs also revealed prominent sand ridges on the southeast side of the bays. Melton and Schriever then came to an astonishing conclusion: The depressions were formed by a shower of meteorites that came from the northwest at an angle of 35 to 55 degrees. The impact of the meteorite formed the depression and pushed up the sand ridge.

Some viewed their theory with skepticism while others accepted it as dogma. The scientific community has never accepted an extraterrestrial (or meteoritic) origin of the bays since so much convincing evidence against it has been gathered. At the same time, many laypersons will accept no other theory. As recently as 1982, Henry Savage Jr., in his book *The Mysterious Carolina Bays,* argues for the meteorite theory of origin. His book, a must for students of the bays, is the first to present a complete review of the subject and includes an extensive bibliography.

Of all the theories postulated by scientists to explain the bay's origin, two general themes have been dominant: (1) the catastrophic, which envisioned the sudden shower of meteorites (extraterrestrial) or the sudden formation of artesian springs or sink holes (terrestrial); and (2) the uniformitarianist, which says the origin must be explained by the gradual effects of wind, soil solution, and wave-induced erosion.

With all the theories presented by the scientific community, why is there still such a controversy over the origin of the bays? Simply this: No theory adequately explains all the facts surrounding the bays. Until one does, their origin remains shrouded in uncertainty and mystery.

Whether or not the riddle of their origin is ever solved, the Carolina bays afford a wildflower paradise. Carolina bays act as basins, collecting rainwater from the surrounding uplands, which they hold perched above the normal water table. In the deeper bays, upland swamps develop. If cypress is a component of the swamps, it is always Pond Cypress (*Taxodium ascendens,* plate 675), either in pure stands or mixed with Swamp Gum. In more shallow bays, the beautiful Pond Cypress savannas (plate 672) may occur. For the most part, however, the Carolina bays harbor pocosins (plate 720). Here occur the three bay trees: Loblolly Bay (*Gordonia lasianthus,* plate 729), Sweet Bay (*Magnolia virginiana,* plate 722), and Swamp Bay (*Persea palustris,* plate 721). Did the Carolina bays get their name from the presence of the bay trees? Or did the bay trees get their names because they occur in the bays? (Early on, the depressions were called bays.) The sand ridges are also an important component of the bays. Here the deep sands harbor xeric longleaf pine-turkey oak xeric woodlands (plate 518).

Ecologically the bays are important for several reasons. First, they are wetland habitats, which are an oasis for numerous animals of the surrounding uplands. Several endangered or state rare animals such as the American Black Bear, Flatwoods Salamander, Carolina Gopher Frog, and Pine Barrens Tree Frog, make their homes or breed in the habitats sheltered by the bays. At least 36 plants that are considered rare in South Carolina grow in Carolina bays. Venus' Fly Trap and Carolina Wicky (*Kalmia carolina,* plate 724), two uncommon species, grow on the margins of the pocosins. Add to these the savanna communities, with their floral display of orchids and carnivorous plants, and the xeric sandhills of the rims, and one can see why the Carolina bays are such important habitats.

Unfortunately, the majority of the Carolina bays are highly disturbed. In 1991, Steve H. Bennett and John B. Nelson conducted a study on the distribution and status of Carolina bays. They estimated that approximately 4,000 bays of all size classes occur in South Carolina. Of these, only 400 to 500 were found to be relatively intact; the number of bays remaining in exemplary condition is considerably less. They concluded that there are far fewer bays than had been previously believed, and most bays have been significantly altered. Many have been drained for farmland or timber production; others have been converted to pastures; still others have been used as junk yards to dispose of household goods.

Saving Carolina bays has become a major goal of The Nature Conservancy of South Carolina and the South Carolina Heritage Trust Program. Through the work of these two organizations, plus the United States Forest Service, several areas in the Coastal Plain with Carolina bays have been protected for public use: Lewis Ocean Bay Heritage Preserve and Cartwheel Bay Heritage Preserve in Horry County, Santee Coastal Reserve in Charleston County, and the Francis Marion National Forest in Berkeley and Charleston Counties. Several sites that consist entirely of a Carolina bay have also been protected including: Woods Bay State Park, Savage Bay Heritage Preserve, Cathedral Bay Heritage Preserve, and Bennett's Bay Heritage Preserve.

Agriculture: Effects on South Carolina's Physical Landscape

RICE

The growing of rice, from its introduction in the late 1600s until the end of its cultivation in the early 1900s, forever changed the diversity and structure of the flora (and fauna) of the Outer Coastal Plain (and similar coastal areas in North Carolina and Georgia). Even today its ecological effects are evident. Plants introduced into plantation gardens have escaped and become naturalized; tidal river swamps and inland swamps that were cleared for rice fields and then later abandoned, today support a variety of secondary plant communities; upland woods that were cleared for provision fields and then abandoned, support flora and fauna different from the original woods. Rice culture, however, had the greatest influence on the vegetation of the freshwater inland and river swamps.

Rice was first planted in South Carolina shortly after Charles Town was settled in 1670. The Lords Proprietors were granted a swath of land in the New World by Charles II of England for restoring him to the English Crown. This granted land was the Province of Carolina. The Proprietors, seeking to gain financially from their new land, made plans to settle Carolina. They commissioned three ships, but only one made it to Carolina. The *Carolina* landed at Albermarle Point on a bluff overlooking Old Town Creek, a tributary of the Ashley River. The *Carolina* had stopped at Barbados, and several of the Barbadian planters, at the request of the Proprietors, came to Carolina on board. Land on Barbados was scarce for their sons to begin agriculture because of the vast sugar plantings, so the possibility of new lands in Carolina was attractive. The Barbadian planters settled at Goose Creek, a tributary of the Cooper River. They had brought enslaved people with them, some more than likely first generations from Africa. Securing a source of food would have been paramount. When rice seeds came available (sent by the Proprietors or perhaps brought as food on the voyage from slave ships that began to arrive), the enslaved people, knowing about rice cultivation from Africa, began planting the seeds in the small-stream floodplains of the Cooper River and tributaries after clearing swamp trees. The rice flourished in the damp, fertile soil. Planters, seeing that rice grew well and seeking a crop to profit from, appropriated the enslaved people's early success and knowledge of rice cultivation, and by 1699, hundreds of enslaved people on scores of plantations produced the first significant crops for export. By 1720, rice became the main export crop of the colonies. Rice cultivation ultimately spread from its Cooper River epicenter to North Carolina, Georgia, and Florida, creating the Rice Kingdom.

Rice cultivation first depended on the natural dampness of the soil. Soon, however, planters and enslaved people found that by surrounding the fields with banks, floods could be prevented from destroying the crops and a reservoir was created behind the banks that could hold water when it was needed during a drought. Soon, thousands of inland swamps were banked, some for planting and some for reservoirs. This was the reservoir system of rice cultivation, and it remained the main method of cultivation until after the Revolution.

Although the reservoir system produced great profits and made Charles Town one of the richest cities in the colonies, lack of rain often caused the reservoirs to run dry; there was no water to flood the fields. After the Revolution, planters shifted to a new, more certain supply of water: freshwater tides along the freshwater rivers to flood the fields. This reliable supply of water allowed the tidal system of rice growing to emerge. The freshwater rivers were bordered by fringes of freshwater tidal marshes and wide expanses of low-lying, freshwater tidal swamps. Twice a day, as far as 30 miles inland, the tide ebbed and flowed, draining then flooding the marshes and swamps. An ingenious system was devised to apply this "rhythm of nature" to rice growing, thus ensuring a consistent supply of fresh water. On every river in the low-country above the influence of saltwater and on river sections where at least a three-foot difference in low and high tide occurred, an attempt was made to grow rice. A bank was constructed from the highland through the swamp to the river's edge, along the river's edge, then back through the swamp to the highland. This bank kept the river water out of the banked area during high tide. A series of lower "check-banks" were constructed within the large area, dividing it into smaller fields. Each field was fitted with a trunk-gate system so each could be flooded or drained independently. Next, the task of clearing the swamp began. Enslaved people, using primitive hand tools and oxen, felled, piled and burned the trees. The largest trees were cut at ground level (their stumps can still be seen today in the abandoned fields). Fields were then made ready for planting. Using a trunk-gate system, each field could be flooded when the tide rose or drained at low tide. This dependable supply of fresh water made rice growing profitable. The tidal system replaced the inland system as the principal method of growing commercial rice, and it remained the basic method of commercial rice culture until the end of the industry in the early 1900s.

One hundred and fifty thousand acres of tidal freshwater swamps along the rivers were banked, cleared of trees and planted in rice. All this was done on the backs of enslaved Africans, who literally built the wealth of the early Lowcountry and afterwards. The more certain supply of water vastly increased the profits of the plantations, causing more import of Africans to work the fields. In a short time, an entire ecosystem—the tidal freshwater swamp—was removed from the Lowcountry.

Rice cultivation influenced every aspect of Lowcountry life. A rich plantation system developed, and the planters, accumulating great wealth, sent their sons to Oxford and Cambridge. Fine homes were built on the river plantations and were landscaped with formal gardens. Many of the exotic plants that adorned the gardens escaped over time and have become naturalized. Summer homes were built in cities such as Georgetown and Charleston. Today these homes serve as tourist attractions, especially in Charleston.

Africans, mostly from western Africa were enslaved and forced to work the fields; their descendants are an integral part of the Lowcountry today.

As malaria increased, the planters settled the inland pinelands and seacoast islands to pass the summer months (the malaria season). The pinelands were beyond the flight path of the *Anopheles* mosquitos (the vector of the malaria parasite) that bred in the freshwater swamps. Also, the sandy soils of the high pinelands did not accumulate pools of water that could be used as mosquito breeding. Villages such as Plantersville, Pinopolis, and Cordesville sprang up throughout the Lowcountry. Along the coastal islands, where the adjacent marshes breed the salt marsh mosquito, which does not carry the malaria-causing parasite, settlements developed as retreats for the planters. Pawleys Island in Georgetown County, for example, became the retreat for planters of the Waccamaw area.

The industrial base of the Lowcountry began as a result of the rice industry. With the introduction of the horizontal steam engine in the early 1800s, numerous foundries produced engines and accessory equipment to operate the mills that prepared the rice for market. Many companies that serve the Lowcountry today are descendants of these early foundries.

Both systems of growing rice had major effects on the state's natural history. Thousands of acres of inland swamps were cleared for fields. When the inland system was abandoned, the fields reverted to swamp forests. No records of the species composition of these original inland swamps exist; thus, it is not known how similar the present swamp forests are to the original. But the most pronounced legacy of the inland system is the reservoirs. In many cases the banks are still intact, creating a permanently flooded swamp that provides valuable habitat, especially for wading birds. Washo Reserve on the Santee Coastal Reserve and a reserve at Caw Caw County Park represent just two of the many former inland reservoirs.

The major legacy of the tidal system is approximately 150,000 acres of abandoned tidal fields. Many still have their banks intact and are managed for waterfowl. The ability to control the water in these banked fields allows for a management regime that selects plants that are preferred by waterfowl, such as Redroot (*Lachnanthes caroliniana*) or Wigeon Grass (*Ruppia maritima*). The majority of the abandoned fields, however, have broken banks that allow free exchange with tidal water (figure 5). These fields are undergoing succession to secondary swamp forests. In fact, many of the fields have already reverted to swamp forests. Whether the mature, secondary swamp forests will be similar to the original swamp forests is not known. It is in these abandoned tidal fields that one of the greatest displays of wildflowers occurs: those of the freshwater tidal marshes. One can paddle a small boat through the broken banks and follow a myriad of canals, each revealing a different combination of freshwater marsh plants in spring, summer, and fall.

Rice growing ended around 1911, when two disastrous hurricanes (1910 and 1911) struck, breaking the banks, flooding the fields with saltwater, and destroying the crops. These storms, the loss of slave labor after the Civil War,

Figure 5. Abandoned Rice Fields along the East Branch of the Cooper River.

and the destruction of the mills and associated buildings by Sherman's troops as they swept through the South led to the financial devastation of the rice-growing region after the war and the advent of mechanized rice growing in Arkansas, Texas, and Louisiana around 1880, made it impossible to grow rice profitably.

After the Civil War and the demise of the rice industry, many plantations were bought by wealthy northerners who used them as hunting preserves or business retreats. Even though there was resentment in the South at having had to lose their lands to "outsiders," it was fortuitous since many of these plantations were established as wildlife preserves. The northerners came to love the land so much that rather than see these plantations developed to serve a select few, they preserved them for the people of South Carolina. A prime example: Tom Yawkey, who owned the Boston Red Sox, gave the state 20,000 acres on the North Santee River. Today it is managed by the South Carolina Department of Natural Resources as the Tom Yawkey Wildlife Center. Other such gifts or purchases include the Santee Coastal Reserve and Hobcaw Barony.

COTTON

Cotton farming in South Carolina and the Deep South is a history of unbridled greed, slavery, human suffering, economic bondage to a single-crop agricultural system, and indifference to the ecological consequences of repeated long-term cultivation of agricultural land. Although the effect of the cotton industry on cultivated soils is our major interest here, a brief introduction to the associated social and economic aspects is necessary to understand why cotton was able to maintain its status as a principal money crop in South Carolina for nearly 150 years. In the process, it destroyed the agricultural viability of more than 40% of the Piedmont.

The long-staple, black seed cotton that we now know as sea-island cotton was first grown successfully in South Carolina in about 1790. Sea-island cotton has the long staple needed to make fine cotton garments. Since it could only be grown in a small section of the coastal zone south of Charleston, supply seldom exceeded demand, and prices remained high and profitable. Its demise was brought on by the onslaught of the boll weevil between 1917 and 1922. Because of its long growing season, sea-island cotton was especially susceptible to boll weevil infestation. Much of the acreage devastated by boll weevil was replanted with less-susceptible, short-staple, upland cottons.

The rise of cotton to worldwide prominence, however, involved a number of short-staple, green seed cottons known collectively as upland cotton. Upland cotton could be grown just about anywhere in the South except in the mountains. It soon came into high demand even though its short staple could only be made into relatively coarse cloth. The development of the cotton gin in 1793, at about the same time textile production became mechanized in England, was largely responsible for the rapid rise in the prominence of upland cotton. Robert Mills, in his *Statistics of South Carolina* published in 1826, reported that upland cotton was the major crop throughout the state, except where sea-island cotton and rice were grown. By 1860 cotton production was reported to involve 80% of the South's entire labor pool.

The plantation system, which originated with the cultivation of rice and indigo and involves large land holdings and abundant slave labor, spread from the coast to the lower Piedmont with the spread of cotton farming. Since cotton production is labor intensive, the number of slaves increased rapidly with the increase in cotton cultivation. In fact, cotton has been vilified as the impetus for the eight-fold increase in the slave population in the South between 1784 and 1860. South Carolina was no exception. By 1850, more than 80% of South Carolina counties had slave populations in excess of 50% of the total population. The plantation system soon dominated cotton production, but many small farms were scattered among the plantations.

With the end of the Civil War, loss of slave labor, and nearly all of the best lands already in cultivation, one might have predicted a decline in cotton production. Instead there was an increase in cotton cultivation due to a new labor system: tenancy. By the 1880s, more than 50% of Piedmont crops had liens on them. By 1920, 65% of the farms in South Carolina were run by tenants. This period of economic bondage lasted from the end of Reconstruction in 1876 until after World War II. Even the depredations of the boll weevil, beginning about 1917 and described by some as "a calamity somewhat comparable to the potato blight [in Ireland] of 1840," didn't stop the momentum of cotton farming (Porcher and Rayner, 2001). By reducing production, prices rose and provided encouragement for more planting, and scientific advancements made control of the boll weevil and subsequent crop recovery possible. The Agricultural Adjustment Act of 1933 paid farmers not to grow certain crops, including cotton. It ended overproduction and maintained high prices, but it also heralded the demise of the small-acreage cotton farmer.

In 1860, as much as 77% of the cotton was produced in the lower Piedmont and Inner Coastal Plain, and these regions continued to dominate the state's production until about 1920. After 1920, cotton production shifted to the upper Piedmont, and by 1940 one-third of South Carolina's production was from the upper Piedmont. Evidence suggests that soil erosion and soil infertility in the lower Piedmont and Inner Coastal Plain were responsible for this shift.

Cultivation of upland cotton in South Carolina and the Deep South was typically a very wasteful process, and profits throughout the reign of cotton were based on an exploitation of the land, which has left a legacy of reduced soil fertility and a severely eroded landscape.

The usual practice of cotton cultivation during the plantation era was to clear the land, plant cotton, and when yields declined, plant corn, and eventually abandon the fields. Since land was cheap, and the major cost of production was labor, landowners were seemingly indifferent to the loss of soil fertility that resulted from continual cropping and from the severe erosion that often accompanied cotton cultivation and abandonment to agriculture, especially in hilly regions. This was in sharp contrast to the planters of sea-island cotton, for whom land was a limiting resource. Early on, these planters practiced many of the now-accepted soil conservation practices such as contour plowing, ditching, terracing, and crop rotation; they allowed fields to lie fallow and fertilized with everything from manure and crushed oyster shells to what we now call compost.

As in any economy that is focused on production with no consideration to depletion of resources, many criticized the typical methods of growing upland cotton. Just as there were few who listened to critics of the wanton decimation of the passenger pigeon (now extinct) or the American bison (once reduced to a few thousand), there were few who listened to such critics as Robert Mills, who wrote: "We wish to see them [the farmers] giving back to the soil some portion of the nourishment which they take from it; otherwise the most deplorable results must follow: short crops, and barren fields, the disappearance of the forests, and a desolate country" (Mills 1826).

Post-Reconstruction cotton cultivation differed little from the wasteful practices of the plantation era. They were often worse, since putting new lands in production generally involved clearing and planting marginal, easily eroded lands. Abandoned fields often were again cleared and planted after an extended fallow period of 20 years or so, and sometimes cattle were allowed to graze on the "resting" land. This went on until nearly all the Piedmont had been cleared and cultivated and eventually abandoned at least once. Since the prosperity of landowners and merchants involved in the tenancy system depended on the poverty of their tenant farmers, perhaps it is not surprising that soil conservation practices were not typical of this era. Although some efforts were made to restore land depleted by years of constant cultivation, fertilization tended to be used much more than soil conservation practices.

John Harrington, longtime geology professor at Wofford College and a careful student of the geology and landscape of the South Carolina Piedmont, wondered why Piedmont cotton farmers so often plowed their furrows up and down (perpendicular to) the slope, which even the most casual observer should realize would increase soil erosion. While visiting Ireland, he believed he found an answer: farmers there plowed their furrows perpendicular to the slope with very little erosion occurring due to the hard-to-erode soils. Most of the farmers in the Piedmont were of Scotch-Irish descent.

What is the legacy of 150 years of cotton growing? According to S. W. Trimble (1972), the southern portion of the Piedmont is one of the most severely eroded agricultural areas in the United States, and the South Carolina Piedmont is one of the worst. Based on existing soil profiles and comparisons with expected soil profiles, Trimble estimated that the average depth of erosion in South Carolina was just under 10 inches, and other sources suggest that more than 12 inches were removed from some large areas. Trimble estimated that 40% of the Piedmont in South Carolina was so severely eroded and the topsoil so depleted that it was rendered useless for agriculture. Why so much erosion? First, the soils of the Piedmont are highly erodible because they are easily weathered and friable (easily crumbled). Second, the Piedmont receives abundant rainfall throughout the year, including times when soil is relatively unprotected. Heavy summer thunderstorms may have been particularly damaging. Third, cultivation year after year, with little effort at restoring the soil, and plowing up and down the slope rather than parallel to the slope promoted erosion.

The sediments that eroded off of Piedmont uplands were washed into streams and floodplains, particularly in the Piedmont and Inner Coastal Plain. These areas still carry a heavy burden of sediment. Clarity of all but the smallest streams declined substantially, and most streams turned brown with sediment following modest rain showers, a situation still seen today. Toward the end of the nineteenth century, streambeds in some Piedmont stream valleys had risen as much as 9 to 18 feet, and some bridges were literally buried in sediments. Streambed erosion raises sediment loads downstream and increases the distance that separates the streambed from the adjacent floodplain, which in turn changes the frequency and duration with which the floodplains are flooded. As streambeds become farther and farther separated from their floodplains, floodplains flood less and eventually become so separated that they no longer act as functional floodplain wetlands that store and clean floodwaters and provide temporary fish-foraging habitat. The Piedmont today contains few floodplains that are functional wetlands. Erosion has decreased substantially since the end of World War II, so streambeds are now lowering.

Another legacy of cotton growing is the dominance in the lower Piedmont of plantation forestry. The deeply eroded, infertile soils are suitable for growing pines, especially shortleaf and loblolly, but are too poor for natural regeneration of the oak-hickory forest that once dominated the landscape. It is hard to predict when or if this will change.

The massive alteration of the landscape that is associated with cotton did leave one beneficial legacy: the present system of national forests in South

Carolina. In 1933, with the Great Depression in full swing and an abundance of tax delinquent land, the National Forest Reservation Commission established the Enoree and Long Cane purchase units in the Piedmont, with the express purpose of retiring some marginal farmland and returning it to productive forest. By 1936, nearly 97,000 acres had been purchased, mostly in units of less than 300 acres. The Francis Marion and Sumter National Forests were established by President Franklin Roosevelt in that same year. By 1938, as many as 243,000 acres had been purchased for the Francis Marion National Forest in the Coastal Plain counties of Berkeley and Charleston, most of which had been cut over and regularly burned. By 1941, as many as 253,271 acres had been purchased for the Sumter National Forest, including the Piedmont acreages purchased in 1933 and previously purchased mountain acreages. Purchase of the cut-over and eroding land of the mountain district, the Andrew Pickens District, began soon after the passage of the Weeks Law in 1911, which allowed for federal purchase of lands for watershed protection. The national forest system in South Carolina now totals about 605,000 acres and comprises some of the most heavily forested land and some of the most significant natural areas in the state.

Marshes, Swamps, Peatlands, Bogs, and Fens

Marshes, swamps, peatlands, bogs, and fens are wetlands. No single, accepted term in the scientific community defines a wetland. In general terms, however, wetlands are lands where saturation with water is the dominant factor determining the nature of soil development and the types of plant and animal communities living in the soil and on its surface.

MARSHES

Marshes are wetlands that are inhabited by herbaceous plants that have their roots in the substrate but with their photosynthetic and reproductive organs principally emerged. Marshes have predominantly mineral (nonorganic) soils, even though much organic material may be incorporated. The dominant species are grasses, rushes, and sedges along with numerous broadleaf flowering plants. Marshes in South Carolina can be classified as salt, brackish, or fresh water.

Salt marshes generally occur on a peaty substrate along tidal inlets, behind barrier islands and on spits along the maritime strand. They are regularly flooded and drained by tidal action. The soil is saturated and has a high salt concentration and low oxygen levels. The most extensive salt marsh community in the coastal area is the *Spartina* marsh dominated by Smooth Cordgrass (*Spartina alterniflora,* plate 919). Vast stretches of this marsh system are readily visible as one drives from the mainland to the barrier islands, such as Kiawah, Folly, Hilton Head, and Pawleys.

Brackish marshes are transitional communities between the freshwater and saltwater marshes. They occur in the Outer Coastal Plain where fresh water and saltwater mix; species of both systems occur here. The dominant species are emergent grasses, sedges, and rushes. The brackish marshes occur at varying distances up rivers from the coast. The more freshwater coming down the rivers, the closer the brackish zone occurs to the coast.

Freshwater marshes occur both inland and along the coastal rivers. Inland freshwater marshes occur throughout the state and are a diverse system fed by inflowing water, seepage, or precipitation. The variable water supply results in flooding during high rainfall and drawdown during dry periods, a feature that shapes the structure and composition of the marsh. Inland marshes occur along the edges of lakes and ponds, in canals and roadside ditches, and along the edge of inland swamps.

The dominant freshwater marshes of the state, however, are the tidal freshwater marshes (plate 823) of the Outer Coastal Plain. These marshes occur along the edge of the brownwater and blackwater rivers and are close enough to be affected by daily freshwater tides but far enough to be unaffected by the intrusion of saltwater. Vast expanses of these marshes occur in the abandoned rice fields along rivers. Narrow zones of freshwater marsh also occur where the tidal swamps border the rivers. The low marsh, with its deeper water, is characterized by broadleaf monocots such as Arrow-Arum and pickerelweeds and showy dicots such as Bur-marigold. The high marsh is a mixture of low marsh species plus numerous grasses, rushes, and sedges.

SWAMPS

Woody plants dominate swamp wetlands. The substrate in swamps is flooded for one or more extended periods during each year. Sometimes the flooding is more or less permanent, but usually a swamp is without surface water for at least part of the year. Although much organic material may be incorporated in the soil, swamps have predominantly mineral (nonorganic) soils. South Carolina's swamps are extensive and varied and include the alluvial bottomland swamps (i.e., swamps associated with streams and rivers), tidal freshwater swamps, and nonalluvial swamps.

Bottomland swamps occur on the alluvial floodplains of the brownwater rivers of the lower Piedmont and Coastal Plain, and on the blackwater rivers of the Coastal Plain. Brownwater rivers originate in the mountains and Piedmont and have wide alluvial floodplains, while the blackwater rivers originate in the Coastal Plain and have narrower, less-developed alluvial floodplains. On the regions of the floodplain where the land is almost continually flooded, Bald Cypress-Tupelo Gum swamps (plate 768) occur. Here the trees exhibit typical hydromorphic features in response to growing in water: buttresses, knees (in Bald Cypress), and spongy roots. Hardwood bottoms (plate 792) occur in the Coastal Plain and Piedmont on floodplains that are slightly elevated above the adjoining swamp forests. Here the land is often flooded, but it is dry through much of the year. The dominant vegetation is a mix of water-tolerant, deciduous hardwoods, with oaks predominating and an occasional cypress from the adjacent swamp forests also occurring.

Tidal freshwater swamps occur in the Outer Coastal Plain, from the upper limit of tidal influence to the brackish water line downstream. Tidal freshwater swamps owe their nature to the river tides. Twice a day they are flooded and twice a day they are free of surface water. Both brownwater and black-water rivers support tidal swamp forests. Tidal freshwater swamps today are mainly secondary swamps. The original tidal freshwater swamps were cleared to make rice fields. The fields were abandoned in the early 1900s, and today, secondary tidal swamps are gradually reclaiming the fields. The tidal swamps are similar in composition to the alluvial swamps.

A variety of nonalluvial swamps occur in the state. The main feature distinguishing these swamps is that they are not associated with moving river water. In other words, they are not alluvial or tidal systems. These swamps occur in two types of habitats: (1) upland, isolated sites such as lime sinks, Carolina bays, abandoned rice reservoirs, and irregular depressions; and (2) seepage slopes (see below) associated with river systems along the boundary of floodplains and uplands. The water supply for the bays, sinks, irregular depressions, and seepage slopes is predominately rainwater. One example of an upland, nonalluvial swamp that is covered in this book is the Pond Cypress-Swamp Gum upland swamp (plate 706).

PEATLANDS

Peatlands are simply wetlands whose soils are peat, the partially decomposed remains of dead plants. The dominant peatlands in the state are the pocosins or evergreen shrub bogs. Pocosins (plate 719) are freshwater wetlands found extensively in the Outer Coastal Plain of Virginia, North Carolina, South Carolina, and Georgia. They are fire-adapted systems dominated by a tangled mass of broadleaf evergreen or semi-evergreen shrubs and vines, with scattered Pond Pines and occasional bay trees (Swamp Bay, Sweet Bay, and Loblolly Bay). The most extensive pocosins are found on flat upland areas. Additionally, pocosins may be found in Carolina bays, wet seepage slopes, and in low areas of relict dune fields.

Another type of pocosin is the streamhead pocosin, which is found primarily in the Sandhills. They occur along the headwaters of small streams, streamside flats, and extend up adjacent hillsides. This community is most common in the Fall-line Sandhills, but it does occur in the Inner Coastal Plain.

BOGS

Bogs are ombrotrophic (mineral-poor) wetlands because they receive their water from rain, which contains few minerals. The only true bogs in South Carolina are the sphagnum openings in pocosins. Periodic fires create these openings in the shrub layer where sphagnum develops. If the openings are removed far enough from the boundary of the pocosin and adjacent upland so that they receive no seepage, only rainwater, they are true bogs.

FENS

Fens are wetlands that harbor an open, herb-dominated community and receive water from rain and from seepage. Seepages are not plant communities per se. Seepage refers to a water source, which can allow wetland communities like a fen to develop. Seepage occurs when rainwater penetrates the soil

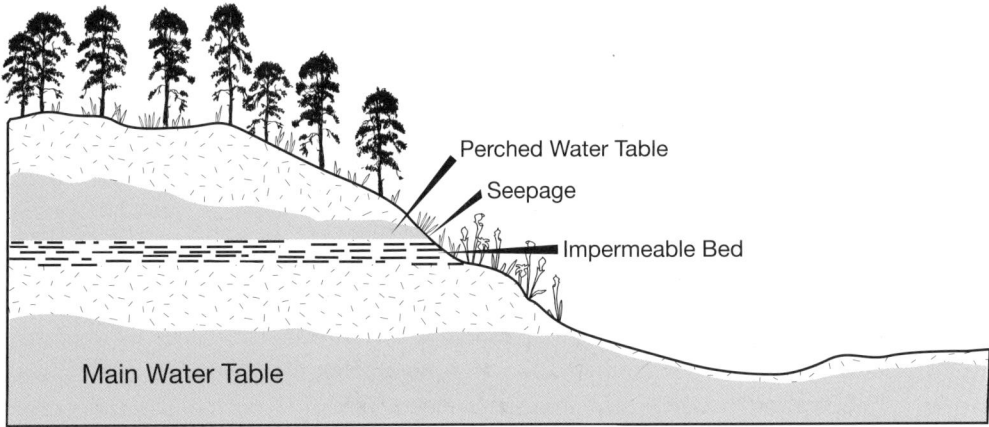

Figure 6. Generalized Seepage System.

to an impermeable layer (such as a clay hardpan). The water then moves laterally through the soil over this impermeable layer. Where the soil slopes so that the impermeable layer is exposed at the surface (outcrops), water seeps out of the soil and flows down the slope. The water picks up dissolved minerals from the soil, which contribute to the mineral supply of the seepage. Figure 6 illustrates a generalized seepage system.

If the seepage water feeding a fen is mineral-poor, it is a poor fen. Often, however, the water seeps through limestone or mafic rock and picks up minerals. These are rich fens because of the high mineral content (minerotrophic).

In the essay "South Carolina's Natural Wildflower Communities," two mountain communities are discussed, the Southern Appalachian fen and the cataract fen. These two communities are not bogs, as defined above, but fens. Their main source of water is seepage, not rainwater. However, the term "bog" has been used for so long for these communities that we recognize that even professional botanists occasionally refer to them as bogs. The reader is now aware of their true nature: they are fens.

South Carolina's Natural Wildflower Communities

The Mountains

MONTANE ROCK OUTCROP COMMUNITIES

Granitic domes, mafic rock outcrops, and shallow soil glades

The typical granitic dome community (plate 1) is found mostly in the mountains but also occurs on monadnocks (mountains of erosion resistant rock) in the Piedmont. It occurs on exposed upper slopes or ridges, usually as steep exposures, except for small gently sloping areas at the top. It is recognized by the smooth (exfoliating) surface of the exposed rock. Granitic domes are sometimes called exfoliation domes because they consist of layers of rock similar to the layers of an onion. These layers were produced during the formation of the rock as a means of relieving pressure. Soils generally are shallow, and trees and shrubs are restricted to the margins or to pockets of deeper soil, which tend to be rare because of the sparsity of fractures or crevices. Vegetation is zoned by soil depth. Only mosses and lichens occupy bare rock, but once a soil layer develops, a variety of vascular plants soon invade. Herbaceous species that are primarily restricted to domes include Rock Harlequin (*Capnoides sempervirens,* plate 12), Cliff Saxifrage (*Micranthes petiolaris* var. *petiolaris,* plate 18), Yarrowleaf Ragwort (*Packera millefolium,* plate 13), and Twisted-hair Spikemoss (*Bryodesma tortipilum,* plate 24). Herbaceous species that are common on granitic domes but also occur in other rocky outcrops include Mountain Dwarf-dandelion (*Krigia montana,* plate 83), Elf Orpine (*Diamorpha smallii,* plate 247), Smallflower Phacelia (*Phacelia dubia*), Appalachian Fameflower (*Phemeranthus teretifolius,* plate 258), Outcrop Rushfoil (*Croton willdenowii,* plate 259), several species of broomsedge (*Andropogon* spp.), Silky Oat-grass (*Danthonia sericea*), Hairy and Woolly Lip-ferns (*Cheilanthes lanosa* and *C. tomentosa,* Plate 263), Pineweed (*Hypericum gentianoides,* plate 260), and Small's Ragwort (*Packera anonyma,* plate 996). Woody species that occur at the edges of the dome or in mats that have deep soil include Fringe-tree (*Chionanthus virginicus,* plate 245), Wafer-ash (*Ptelea trifoliata,* plate 5), Winged Sumac (*Rhus copallina,* plate 948), Eastern Red Cedar (*Juniperus virginiana*), Virginia Pine (*Pinus virginiana*), and, depending on the depth and richness of the soil, a variety of other xerophytic to mesophytic trees. Seepages at the upper edges may harbor interesting species such as Round-leaf Sundew

(*Drosera rotundifolia,* plate 510), Horned Bladderwort (*Utricularia cornuta,* plate 61), Flatrock Pimpernel (*Lindernia monticola,* plate 256), Bog Oat-grass (*Danthonia epilis*), and several species of Meadow-beauty (*Rhexia mariana,* plate 87, and/or *R. virginica,* plate 88). Higher elevation granitic domes may include such species as Sand Myrtle (*Kalmia buxifolia,* plate 440), Prolific St. John's-wort (*Hypericum prolificum,* plate 7), Carolina Hemlock (*Tsuga caroliniana,* plate 4), and Table Mountain Pine (*Pinus pungens,* plate 3).

Granitic domes associated with calcium-producing or magnesium-rich rocks are sometimes described as a calcareous variant of granitic domes. These domes harbor such calcium-loving species as Yellow Honeysuckle (*Lonicera flava,* plate 8), Eastern Prairie Anemone (*Anemone berlanderi,* plate 250), Eastern Columbine (*Aquilegia canadensis*), Smooth Indigo-bush (*Amorpha glabra,* plate 6), Small's Beardtongue (*Penstemon smallii,* plate 16), Hairy Mock-orange (*Philadelphus hirsutus*), and Seneca Snakeroot (*Polygala senega*). An abundant presence of Eastern Red Cedar along the outcrop margins is often a good indicator of this variant. The adjacent woodland often is quite rich, even on the shallow soils of the southern exposures. Additionally, outcrops of pure amphibolite (a mafic, or magnesium-rich rock) are present on some peaks and ridges. These outcrops are referred to as mafic rock outcrops in the habitat section of the species descriptions. Amphibolite outcrops and granite that receives transport of magnesium and other nutrients from amphibolite contain several endemic species found in no other habitat such as the endangered White Irisette (*Sisyrinchium dichotomum,* plate 14) and Biltmore's Sedge (*Carex biltmoreana,* plate 56). Shallow soil found along the margins of such outcrops often has a luxuriant glade community. This natural community, the shallow soil mafic glade, is characterized by widely spaced trees and a dense grass-sedge-herbaceous understory. Most of the species described above extend into this habitat.

Similar to the granitic dome is the acidic cliff. The community is characterized by sheer slopes, exposed rock, and a canopy cover that is less than 25%. The best sites have sheer exposures of fractured, acidic rock. This variant is common in the mountains and rare in the Piedmont. In the driest, most open sites, many species of typical granitic domes or granitic flatrock outcrops are present, especially broomsedges, oat grasses, spikemosses, and ragworts. Because of the fractured rock, crevices are common and rock-loving ferns are present, including lipferns. Two such sites in Pickens County have been discovered to contain four species of ferns that are typical of the dry rocky habitats of the southwestern United States growing hundreds of miles to the east of their typical range. On moister sites, granitic dome species such as Mountain Dwarf-dandelion and Cliff Saxifrage may occur. Shrubs, trees, and other herbs found here are often typical of the surrounding forest community; they may be mesic, xeric, or even boggy species depending on the nature of the adjacent forest community. Because of the open canopy, a variety of weedy species may occur, including fleabanes (*Erigeron* spp.), Butterfly-weed (*Asclepias tuberosa* ssp. *tuberosa,* plate 985), and Woodland Pinkroot (*Spigelia marilandica,* plate 299).

Where to visit the granitic dome community. Excellent examples of granitic domes are easily accessible to the general public on several public lands. Perhaps the most interesting of these is found at Glassy Mountain Heritage Preserve near Pickens, South Carolina. An easy to moderate trail takes you to the edge of a large granitic dome. Glassy Mountain is a Piedmont monadnock and also contains excellent examples of Chestnut Oak forest and shallow soil mafic glade habitats. The granitic dome here is of the calcareous variant and has a much more lush and diverse assemblage of species than other similar outcrops. Yellow Honeysuckle, Yarrowleaf Ragwort, and Smooth Indigobush are all easily observed here. Additionally, a very graceful species of grass that is otherwise unknown from the Piedmont, Wavy Hairgrass (*Avenella flexuosa,* plate 9) is abundant in the glades and margins of the outcrop. A large population of Standing Cypress (*Ipomopsis rubra,* plate 999) is located near the parking area at the termination of Glassy Mountain Church Road and smaller numbers of this startling red-flowered plant occur in the glades adjacent to the rock outcrop. Nearly every characteristic species of the granitic dome habitat can be observed at this remarkable heritage preserve.

Higher elevation granitic domes and acidic cliffs are easily observed at Table Rock State Park along the trail to the summit of Table Rock. One of the species, endemic to South Carolina, Porcher's Ragweed (*Ambrosia porcheri,* plate 23), is found at Governor's Rock and scattered across the summit. Large populations of Sand Myrtle and Prolific St.-John's Wort can be observed near the summit.

Bald Rock Heritage Preserve in Greenville County is also easily accessible and heavily visited. The site has been severely vandalized with graffiti, and much of the area close to the road is heavily trampled. The nearby Eva Russell Chandler Heritage Preserve on Persimmon Ridge Road in Greenville County contains an excellent example of a granitic dome as well as an easily viewed cataract fen.

THE SPRAY CLIFFS AND HUMID GORGE OUTCROPS

Spray cliffs

The spray cliff (plate 30) is a distinctive community that has appeal for the naturalist because it is always associated with waterfalls. Since accessing some of these communities can be dangerous, and the habitat is extremely fragile, care should be taken when exploring the spray cliff community.

Spray cliff communities occur on cliffs, ledges, and gently sloping rock faces that are frequently wetted by the spray or splash from adjacent waterfalls. Because constant spray and/or splash from the waterfall provides high humidity and moisture, mosses and liverworts are often abundant. Vascular plants are found in pockets of shallow soil that collect in tiny crevices and ledges of the rock face. Temperature is moderated in these communities by spray water and because it is sheltered from the sun and wind.

Although the spray cliff community is not high in species diversity, it does harbor a distinct assemblage of plants. The canopy from adjacent forest communities often provides some shade, but there are no trees in the spray cliff community, probably due to the combination of steepness, lack of soil, and wetness. A few dwarfed stems of Canadian Hemlock (*Tsuga canadensis*), Northern Wild-raisin (*Viburnum cassinoides,* plate 225), Appalachian Mock-orange (*Philadelphus inodorus,* plate 273), Great Laurel (*Rhododendron maximum,* plate 158), or Mountain Laurel (*Kalmia latifolia,* plate 226) are sometimes present. Herbs that are typically found here include Thyme-leaf Bluet (*Houstonia serpyllifolia,* plate 33), Mountain Meadowrue (*Thalictrum clavatum,* plate 34), Branch Lettuce (*Micranthes micranthidifolia,* plate 35), Meadow Spikemoss (*Lycopodioides apodum,* plate 50), and Orange Jewelweed (*Impatiens capensis,* plate 37). Rare species are only known from a few spray cliff communities and include American Water-pennywort (*Hydrocotyle americana*).

Humid gorge outcrops

The unique position of our mountains, properly termed the southern Blue Ridge Escarpment, has allowed the development of several unique features. The Appalachian Mountains run essentially north–northwest to south–southeast from their northern terminus in Quebec south to western North Carolina. Near the South Carolina line in North Carolina, the escarpment region (the point where the mountains rise from the Piedmont) turns to become more east to west rather than north to south. This abrupt turn captures moisture and results in the escarpment being one of the wettest places in eastern North America. All of this moisture has carved deep gorges through the escarpment along rivers and creeks such as Eastatoe Creek, the Toxaway River, the Horsepasture River, the Thompson River, the Whitewater River, the Chauga River, and the Chattooga River. The constant rushing water through these shaded and protected gorges raises the humidity and buffers the climate. The rock outcrops found along these gorges support a diverse and unusual assemblage of species that are more typical of higher elevations and even humid tropical forests. Indeed, there is a single outcrop along the Thompson River that contains a population of Mountain Spleenwort (*Asplenium montanum,* plate 45), which is typical of high elevations and more northern latitudes as well as a population of Single-sorus Spleenwort (*Asplenium montanum,* plate 44), which is found here but is typical of tropical forests in Central and South America.

Constantly humid, but never wet, rock surfaces that are protected by ledges above them (plate 31) support plants unique to this deeply shaded habitat (grotto, or cavelike). Plants such as Cave Alumroot (*Heuchera parviflora,* plate 39) are found only here. The leaves are so thin that they are translucent, an adaptation for maximizing the use of the dim light. These plants have extremely short and insignificant root systems and would be swept away if rain washes over them. The constant humidity is vital for their survival, and any disruption of this humidity would lead to their rapid demise. Similarly, this sheltered habitat is critical for the diversity of filmy ferns (family *Hymenophyllaceae*) found in these gorges. The filmy ferns are primarily a group of tropical ferns that are adapted to growing in permanently humid environments and are most diverse in cloud forests (high elevation rainforests) in the tropics. Their leaves, like those of the Cave Alumroot, are very thin, often only a few cell-layers thick and require constant humidity to survive. Our gorges support the highest diversity of filmy ferns in temperate North America, with three species that produce adult, sporophyte plants and several that exist primarily as gametophytes. One of these, Tunbridge Fern (*Hymenophyllum tunbridgense,* plate 43) is found only in Pickens County in temperate North America; the next closest population is in Jamaica.

Where to visit the spray cliff and humid gorge communities. The spray cliff community is easily observed, from a distance, at most of the accessible waterfalls in South Carolina. Station Cove Falls in the Sumter National Forest (accessed through Oconee Station State Park) is one of the easiest to access (plate 30). The trail is easy but requires crossing one small stream with slippery rocks. The trail is one of the premier wildflower hikes in South Carolina and also includes excellent examples of rich cove forest and oak-hickory forest.

Visiting a classic humid gorge natural community requires a bit more effort. One of the finest examples is found near The Narrows of the Eastatoe Gorge in Eastatoe Heritage Preserve. The site is accessed from a point just west of US Highway 178 north of Rocky Bottom in Pickens County. The trail is a little over 4 miles each way and strenuous. The highlight at the end of the trail is an overlook and access to The Narrows, a narrow cut through the rock where the entire volume of Eastatoe Creek flows. The rock outcrops in this area and upstream through the confluence with Rocky Bottom Creek contain many of the unique species restricted to our gorges. All of these species are very rare and fragile and should never be disturbed in the wild.

The easiest way to access humid gorges is to arrange a tour of them, with access provided by boat, on Lake Jocassee. Several tour companies operate tours of the area and Jocassee Lake Tours provides an amazing naturalist-based experience into the gorges. Be prepared for a short (less than 0.25 mile) walk on slippery rocks to see the wonders of the Thompson River gorge or other destinations.

THE SEEPAGE COMMUNITIES

Cataract fens

Cataract fens are one of the most distinctive and aesthetically pleasing wildflower communities (plate 51). These habitats have been called cataract bogs by many authors, including in the first edition of *A Guide to the Wildflowers of South Carolina.* A bog is technically an ombotrophic wetland, meaning that it derives most of its nutrients and water from precipitation. True bogs occur far to the north of our area, primarily in the previously glaciated terrain of the northeastern and midwestern United States and Canada. Fens receive their nutrients and water from seepage. Our cataract fens are correctly termed "poor fens" (if they are mineral poor = oligotrophic) or "rich fens" (if they are mineral rich = minerotrophic). It is continuously moist and shares many of the species typical of Southern Appalachian fens in the mountains and upper Piedmont. The cataract fen is often a very narrow, linear natural community only a couple of feet to yards wide. The best examples of this community are along the margins of small perennial streams that course over rather smooth rock surfaces (granitic domes). Since the water is sliding rather than falling vertically as in a typical waterfall, the term "cataract" is somewhat inaccurate. The community is primarily fed by seepage, although the stream margins sometimes are a significant water source. The community varies in elevation from about 1,200 to 2,400 feet and occurs on middle, upper, and occasionally lower slopes. Fen habitats are best developed where streams slide over a margin of rock outcrops that have a nearly level horizontal component and a slope of 5 to 20 degrees. These habitats are in many ways an ideal habitat for species typical of fens and bogs because (1) light is abundant due to the adjacent rock outcrop, (2) moisture is abundant from seepage, and (3) plant succession is slowed because the fen is on shallow soil that overlays rock.

Cataract fens form a narrow zone immediately adjacent to the associated stream and are shaded in part by trees and shrubs in the adjacent plant communities. Sometimes "fen" trees such as Red Maple (*Acer rubrum*, plate 52) are present. Shrubs include Tag Alder (*Alnus serrulata*), Red Chokeberry (*Aronia arbutifolia*, plate 498), Yellowroot (*Xanthorhiza simplissima*, plate 79), Mountain Laurel, Great Laurel, and Smooth Azalea (*R. arborescens*, plate 81). Herbs tend to be more abundant than grasses and sedges. The distinctiveness of this community is derived from the presence of unusual herbs. Carnivorous plants are sometimes present in abundance, including Mountain Sweet Pitcherplant (*Sarracenia jonesii*, plate 58), Southern Appalachian Purple Pitcherplant (*Sarracenia purpurea* var. *montana*, plate 59), Horned Bladderwort, and Roundleaf Sundew. Other rare or unusual species include Appalachian Small Spreading Pogonia (*Cleistesiopsis bifaria*, plate 241), Bigleaf Grass-of-Parnassus (*Parnassia grandifolia*, plate 67), and in drier margins, Mountain Witch-alder (*Fothergilla major*, plate 54) and Indian paint brush (*Castilleja coccinea*, plate 57). The presence of *Parnassia* suggests a high calcium or magnesium content for at least some of the rocks in these areas. Additional orchids that occur in this community include Small Green Wood-orchid (*Platanthera clavellata*, plate 425), Yellow-fringed Orchid (*P. ciliaris*, plate 214), Rose Pogonia (*Pogonia ophioglossoides*, plate 733), Fragrant Ladies'-tresses (*Spiranthes cernua*), and Common Grass-pink (*Calopogon tuberosus*, plate 625). Other species of interest include Stiff Cowbane (*Oxypolis rigidior*, plate 66), Thymeleaf Bluet, several species of lobelia, and Southern Appalachian Sundrops (*Oenothera tetragona* var. *fraseri*, plate 36).

A unique form of cataract fen is restricted to areas of Pinnacle Mountain in South Carolina. Here the seepage that flows out over the granitic dome habitat is derived from springs farther up slope where amphibolite is prevalent. The seepage water transports minerals downslope onto this outcrop. The elevated magnesium and calcium levels of these fens make this South Carolina's only rich fen, which is termed here a mafic cataract fen. Many species typical of other cataract fens are present but without the pitcherplants. Several plants that are very rare in our mountains are also found here and are diagnostic of this natural community including Ninebark (*Physocarpus opulifolius*, plate 55), Small-leaved Meadowrue (*Thalictrum macrostylum*),

Biltmore Sedge, Short-leaved Sneezeweed (*Helenium brevifolium,* plate 63), large colonies of Bigleaf Grass-of-Parnassus, and the stunning Red Canada Lily (*Lilium canadense,* plate 60).

Southern Appalachian fens

Southern Appalachian fens, sometimes called "upland bogs" or "Southern Appalachian bogs," are in fact poor fens that receive minerals and water from seepage, as discussed under the section on cataract fens above. They are distinguished from other wetland or bottomland forests in the mountains by the presence of sphagnum-dominated openings. These open areas are depressions or seepage channels in streamside flats. In addition to being dominated by a variety of species of sphagnum mosses, they have poor drainage; acidic substrate; and many grasses, sedges, ferns, and broadleaf herbs. The best examples of this community have boggy openings of more than one acre, but the only good example in South Carolina has long linear openings along seepage channels bordering a small stream.

The canopy of the forested portions of this community may be sparse or heavy, and the shrub layer, as the term "forested thicket" suggests, is dense and difficult to traverse. Blasphme-vine (*Smilax laurifolia,* plate 730) is often abundant at the margins of the openings and in the adjacent shrub thickets, making movement difficult anywhere but in the openings.

Since there is only one good example of a Southern Appalachian fen in South Carolina, Matthews Creek fen, refer to the description of this fen for an account of species that dominate the herb, shrub, and tree strata. Southern Appalachian fens are rare and harbor rare species. The most spectacular of the rarities are Swamp Pink (*Helonias bullata* plate 72) and, at least formerly, Bog Rose Orchid (*Arethusa bulbosa*).

Fens are rare in our mountains because of the scarcity of flat, wet sites and because they are successional plant communities that require some sort of disturbance to maintain the early successional status of the sphagnous openings. In the absence of disturbance, such as periodic fire, the boggy openings become filled in by shrubs and trees, and bog plants disappear.

Matthews Creek fen, located in Watson-Cooper Heritage Preserve in Greenville County, was once a good example of a Southern Appalachian fen, but natural succession has degraded the site since 1979. The invasion of shrubs, mostly Great Laurel, has reduced the sphagnum-dominated areas by more than one-half. The population of Swamp Pink has declined and two rare orchids, Appalachian Twayblade (*Neottia smallii*) and Bog Rose Orchid, have disappeared. Shrub diversity also appears to have declined, while species such as male-berry (*Lyonia ligustrina*) and Clammy Azalea (*Rhododendron viscosum,* plate 727) seem to have increased. The herb layer also has decreased in diversity, while American Climbing Fern (*Lygodium palmatum,* plate 75), Northern Long Sedge (*Carex folliculata*), and Galax (*Galax urceolata,* plate 189) appear to have increased. Although the absence of fire is the likely culprit, the exact cause of this rapid natural succession is unknown. If appropriate active management is not initiated soon, this site may not be recoverable.

Where to visit the seepage communities. The best and easiest cataract fen to view in South Carolina is located at Eva Chandler Heritage Preserve along Persimmon Ridge Road in Greenville County. A short, easy-to-moderate trail leads to the granitic dome where the cataract bog is located. The diverse, small, and extremely fragile cataract fen can be viewed from the opposite side of the stream that rapidly flows over the rock here. Most of the characteristic plants of cataract fens are visible from this vantage point. If you bring binoculars you will get a much better look at protected and highly endangered rarities like the Mountain Sweet Pitcherplants. Do not attempt to cross the slippery waterslide to enter the narrow cataract fen: doing so will likely put your life in danger, and if you are successful a simple walk through the fragile habitat could be disastrous for the plants that make their home there. This incredibly rare habitat should be admired from a distance.

South Carolina's only Southern Appalachian fen (located in Watson-Cooper Heritage Preserve) should not be accessed without arranging a visit with the South Carolina Department of Natural Resources Heritage Trust Program. There is no trail leading to the site, and the habitat is extremely fragile. To visit easily accessible examples of this habitat, one needs only visit the Highlands Biological Station and Botanical Garden in Highlands, North Carolina. An easy walk on a boardwalk leads the visitor along the edge of a pond with a broad and open Southern Appalachian fen. Nearly all the characteristic species can be seen at close distance from the boardwalk including Swamp Pink.

CANEBRAKES

This natural community exists only as tiny fragments of nearly monotypical colonies of its dominant species, River Cane (*Arundinaria gigantea,* plate 77). These habitats were essentially native bamboo thickets. At the time of European colonization, the bottomlands in the valleys of the Appalachians and upper Piedmont contained vast amounts of this now all but extinct natural community. They were described by colonial-era travelers such as William Bartram as "an endless wilderness of cane" (Platt and Brantley, 1997). The canebrake community was essentially a monoculture of the single species with other species found in open, moist soils invading here and there. Their maintenance was apparently dependent on fire. Platt and Brantley (1997) describe the conditions that led to their development and maintenance. In their view, the vast canebrakes were the result of reduced population densities of Native Americans following the epidemics brought by the first Europeans. The cane invaded formerly cleared agricultural lands and then was maintained by the Native American practice of burning the landscape to promote the cane. Several species are thought to have depended on these canebrakes in the past, including a species of pitcherplant that is limited to remnants of canebrakes in Alabama. The Bachman's Warbler was apparently restricted to such habitats and to canebrakes of the smaller and more coastal Switch Cane (*Arundinaria tecta,* plate 819). European colonists regarded the canebrakes as a sign of rich soil, and they were quickly cleared to make room for agriculture. The removal of management fires from the landscape combined with conversion quickly led to their near disappearance from the landscape. River Cane still lines the streams of agricultural valleys such as Eastatoe or Brasstown, but they are only a shadow of their former abundance. The many place names, such as the numerous Cane Creeks in South Carolina, North Carolina, Georgia, and Alabama, are indications of the presence of formerly vast canebrakes.

THE ROCKY STREAMSIDE COMMUNITY

The rocky streamside community (plate 78) is in and adjacent to streams and rivers and consists of gravel bars or exposed bedrock and/or boulders with sandy/gravelly soil. It is common in the mountains and much less common in the Piedmont. Plant composition is extremely variable from site to site because of differences in stream size, amount and degree of bedrock exposure, boulders, gravel, and extent and duration of flooding. Trees are usually not present in abundance because these habitats are too rocky, too wet, or too severely scoured by intermittent floods; however, Black Willow (*Salix nigra*) and American Sycamore (*Platanus occidentalis,* plate 815) are usually present, at least as dwarfs in sites that have wide, rocky/gravelly streamsides. Shrub cover is generally sparse to moderate, with occasional dense thickets. A wide variety of shrub species may be present, with the best sites displaying Tag Alder, Sweetspire (*Itea virginica,* plate 773), Smooth Southern Bush-honeysuckle (*Diervilla sessilifolia,* plate 82), and several species of azaleas, including Smooth Azalea and Clammy Azalea. Herbs include mosses and liverworts, as well as grasses, sedges, and rushes. A great diversity of broadleaf herbs may also be present. Sites with variable microtopography may have a mix of dry, mesic, and wetland species as well as many weedy species. Herbs that occur in this community include Brook-saxifrage (*Boykinia aconitifolia,* plate 84), Carolina Tassel-rue (*Trautvetteria caroliniensis,* plate 85), Common Bluet (*Houstonia caerulea,* plate 201), Hollow-stem Joe-pye-weed (*Eutrochium fistulosum,* plate 90), and Cutleaf Coneflower (*Rudbeckia laciniata,* plate 89). Thyme-leaf Bluet and Mountain Meadowrue, species that are more typical of seepages, are often present. Ash-leaf Golden-banner (*Thermopsis fraxinifolia*), a rare species that is more typical of dry ridges, is sometimes also present. Mountain Dwarf-dandelion may be abundant here in dry rock crevices that are above the flood zone. A variety of composites also are present, including many goldenrods, asters, and coreopsis.

Because this is a wildflower community occurring in nearly full sun, species bloom all year, but a visit in early July allows one to see most of the characteristic species in bloom. The associated clear water stream makes this a pleasant plant community to visit any time of the year. Sometimes the shallow soils that are frequently flood-scoured support a prairie like natural community that is referred to as a scour prairie. Dominant species in these sites include grasses normally found in tallgrass prairies such as Big Bluestem (*Andropogon gerardi*), Yellow Indiangrass (*Sorghastrum nutans*), and Common Little Bluestem (*Schizachyrium scoparium*). In such sites, rare species such as

Western Sampson's-snakeroot (*Orbexilum pedunculatum,* plate 86) are found that are more typical of grasslands to the west of SC.

Where to visit the rocky streamside and scour prairie communities. The rocky streamside community can be seen along all of the major rivers that emerge from the southern Blue Ridge Escarpment. The easiest examples to visit are along the Chattooga and Chauga Rivers in Oconee County. Chau-Ram County Park is an excellent example where both the rocky streamside and scour prairie habitats are well developed along the banks of the Chauga River. Other excellent examples can be seen at Burrells Ford on the Chattooga River and near Bulls Sluice on the Chattooga at the US 76 bridge.

THE DECIDUOUS FOREST COMMUNITIES

Rich cove forests

Cove forests are the best known of the mountain communities and are appreciated by botanists and nature lovers for their diversity of plants in all strata, especially the herb layer. It is a mesic community, so it is always found in sheltered locations on the lower slopes of broad ravines or on broad flats adjacent to streams. It occasionally occupies midslope locations on north or northeast-facing slopes, especially where the soil is rich. The definition that fits the historical use of the term "cove" by early mountain residents is a broad convex landform, typically with a stream running down the middle. This is what those of us who grew up in Appalachia refer to as a "holler" (hollow). Soils on these sites are typically deep and rich, and in South Carolina at least some are almost always associated with amphibolite, a metamorphic rock with a high content of the magnesium-rich mineral hornblende. The pH of the soil in rich cove forests is always above 6.0. Habitats that are best described as rich cove forests do occur in the upper Piedmont, but they are rare and associated with the foothills or monadnocks that occur just below the escarpment.

The dense canopy includes a variety of mesophytic trees. Typical trees include Mountain Basswood (*Tilia americana,* plate 97), Yellow Buckeye (*Aesculus flava,* plate 93), and Northern Red Oak (*Quercus rubra*). Other common trees include Tulip-tree (*Liriodendron tulipifera,* plate 95), American Ash (*Fraxinus americana*), and American Beech (*Fagus grandifolia* var. *caroliniana,* plate 311). Few oak species are present other than Northern Red Oak. Trees that are limited or nearly limited to this habitat in South Carolina include Yellow-wood (*Cladrastis kentuckea,* plate 96) and Butternut (*Juglans cinerea,* plate 94).

The understory is usually sparse and includes such species as Flowering Dogwood (*Benthamidia florida*), Hop Hornbeam (*Ostrya virginiana,* plate 314), Ironwood (*Carpinus caroliniana,* plate 794), Common Silverbell (*Halesia tetraptera,* plate 312), and Fraser Magnolia (*Magnolia fraseri,* plate 153).

Typical shrub species include Wild Allspice (*Lindera benzoin,* plate 426), Sweet-shrub (*Calycanthus floridus,* plate 80), American Hazelnut (*Corylus americana,* plate 316), Pagoda Dogwood (*Swida alternifolia),* Silverleaf Hydrangea (*Hydrangea radiata,* plate 199), and Appalachian Mock-orange. Evergreen shrub in the family Ericaceae (*Rhododendron, Kalmia, Leucothoë, Vaccinium*) are rare or absent in this habitat.

The dense and diverse herb layer is characteristic of cove forests in the mountains, just as it is for basic-mesic forests in the Piedmont. A great diversity of herbs may be present, but the species that are rare and distinctive to cove forests include Blue Cohosh (*Caulophyllum thalictroides,* plate 105), Alleghany Spurge (*Pachysandra procumbens,* plate 102), Sweet White Trillium (*Trillium simile,* plate 110), Large-flowered Trillium (*Trillium grandiflorum,* plate 111), Fernleaf Phacelia (*Phacelia bipinatifida,* plate 119), Canada Waterleaf (*Hydrophyllum canadense,* plate 122), Ozark Bunchflower (*Melanthium woodii,* plate 136), Canada Enchanters'-nightshade (*Circaea canadensis,* plate 134), Large Yellow Lady's Slipper (*Cypripedium parviflorum* var. *pubescens,* plate 124), and Walking Fern (*Asplenium rhizophyllum,* plate 147). Many species found in this natural community are also common in other mesic forests with rich soil throughout the mountains and Piedmont. Many of our rich cove forests contain an exceptionally high diversity of *Trillium* species, often four, five, or more species in a single cove.

Unlike many of the communities of the dry ridges and upper slopes in the mountains and Piedmont, cove forests are not fire dependent. For the maintenance of diversity, they apparently do require disturbance in the form of canopy gaps, particularly in the tree and herb layers. High winds that topple tall trees are the main source for the creation of canopy gaps.

Where to visit the rich cove forest community. Rich cove forests can be seen at several locations in the Blue Ridge Escarpment. Peach Orchard Branch, in Pickens County, is accessed from Roy F. Jones Highway (SC-143), 0.7 miles north of the intersection with the Cherokee Foothills Scenic Highway (SC-11). There is a dirt parking area on the right and an orange gate blocking an old logging road that is the access trail. This property is part of the Jim Timmerman Natural Resource Area at Jocassee Gorges, owned and managed by the South Carolina Department of Natural Resources. A short walk (0.3 miles) takes you to the edge of the rich cove forests bordering Eastatoe Creek. The forest here contains most of the typical species found in our rich coves and is an easy, level walk.

Other exceptional locations include Station Cove, accessed through Oconee Station State Park and the Sumter National Forest. This cove is among the first places in our escarpment region to flower in the spring. Peak time for viewing the exceptional spring wildflower diversity here is late March–early April, but the area is beautiful at any time of the year. Tamassee Knob, also in Oconee County, contains good examples of rich cove forest as well as mafic oak-hickory forests with an herb layer nearly as diverse as the rich cove forests. In Greenville County, Chestnut Ridge Heritage Preserve offers easy access to this beautiful natural community.

Acidic cove forests

This is one of the most abundant and easily recognized mesic communities in the mountains. It is found in only a few sites in the upper Piedmont. This community (plate 150) is always on acidic soils and borders nearly every cove and ravine in the mountains that has a small stream running through it. It is also found on broad streamside flats, on north-facing slopes, and in bands of varying width above rich cove forests. The forest is characteristically dense with a high dominance of various oak species including Northern Red Oak, Scarlet Oak (*Quercus coccinea*, plate 223), Chestnut Oak (*Quercus montana*, plate 178), Sweet Birch (*Betula lenta*, plate 151), Eastern White Pine (*Pinus strobus*), and before the invasion by Hemlock Adelgid, formerly Canadian Hemlock (*Tsuga canadensis*). Great Laurel usually forms a dense undergrowth, often mixed with Mountain Laurel. Mountain Doghobble (*Leucothoë fontanesiana*, plate 157) sometimes forms an expansive, dense low shrub layer. In places where the understory is not dense, a variety of shrubs may be present, including Buffalo-nut (*Pyrularia pubera*, plate 182), Witch-hazel (*Hamamelis virginiana*, plate 155), Strawberry-bush (*Euonymus americanus*, plate 317), Mountain White-alder (*Clethra acuminata*, plate 159), and the rare Mountain Camellia (*Stewartia ovata*, plate 154). Herbaceous plant diversity is low, and typical species include Partridge Berry (*Mitchella repens*, plate 331), Indian Cucumber-root (*Medeola virginiana*, plate 172), Foamflower (*Tiarella cordifolia*, plate 106), Round-leaf Yellow Violet (*Viola rotundifolia*), and Sweet White Violet (*Viola blanda*). Ferns sometimes form a dense ground cover. The most common are New York Fern (*Parathelypteris novaboracensis*, plate 196), Hay-scented Fern (*Dennstaedtia punctilobula*, plate 195), and Christmas fern (*Polystichum acrostichoides*, plate 338). At high elevations, Painted Trillium (*Trillidium undulatum*, plate 168), Long-spurred Violet (*V. rostrata*, plate 164), French Broad Heartleaf (*Hexastylis rhombiformis*, plate 71), and Appalachian Twayblade (*Neottia smallii*) can occasionally be found. Where this community occurs on broad streamside flats, the understory is usually quite sparse and Three-birds Orchid (*Triphora trianthophoros*) and Persistent Trillium (*Trillium persistens*, plate 165) are very rare but delightful finds. The most famous inhabitant of this natural community in South Carolina is the Oconee Bells (*Shortia galacifolia*, plate 160), which is nearly restricted as a native species to the drainages that empty into Lake Jocassee and the upper portion of Lake Keowee. Populations found in other areas such as along the Chattooga River have been introduced.

Since this community is often bounded by more diverse communities (i.e., rich cove forests or oak-hickory forest communities), some of the species typical of these communities may be present.

Where to visit the acidic cove forest community. The acidic cove forest is easily observed at many locations in our mountains; it is among the most common habitat types. Some of the most interesting examples, filled with Oconee Bells, are seen along the Foothills Trail in the remote regions of the Horsepasture and Toxaway Rivers. Good examples are found throughout the Sumter National Forest, including near the fish hatchery along the upper Chattooga River. Old-growth examples of this habitat are also present along the trail to Coon Branch, accessed through the Bad Creek project area where the Lower Whitewater River trailhead is located.

Chestnut Oak forests

Chestnut Oak may dominate forests in the Inner Coastal Plain, the Piedmont, and the mountains, but we restrict our definition of Chestnut Oak forests (plate 176) to the definition accepted by most ecologists: low to moderate elevation communities of the mountains and upper Piedmont, where American Chestnut (*Castanea dentata,* plate 179) once dominated. Other Chestnut Oak-dominated communities are included with the broadly defined oak-hickory forest. Chestnut oak forests in the mountains are found on low ridge tops and the upper slopes on any exposure, although they are more common on east- and west-facing slopes. In the upper Piedmont, this forest type is rare and is only found at relatively high elevations, often on upper, north-facing slopes, frequently in association with monadnocks.

This community was the preferred community for American Chestnut, which was once the most valuable tree in eastern North America. Fallen chestnut logs and old stumps are still abundant. Following devastation by chestnut blight in the early 1900s, American Chestnut has been replaced primarily by Chestnut Oak (*Quercus montana,* plate 178) and Scarlet Oak, which now dominate the closed canopy of this sub-xeric community. Chestnut Oak, in particular, is fire-tolerant, and the number of trees probably increases following fire and timber harvest. Additional canopy species include many of the dry to dry-mesic oaks and hickories, especially White Oak (*Quercus alba,* plate 340), Pignut (*Carya glabra*), and Mockernut Hickory (*C. tomentosa*), as well as Black Gum (*Nyssa sylvatica*), various pines, and occasionally, where chestnut oak forests border rocky ledges, Carolina Hemlock. Sprouts of American Chestnut are still found the subcanopy and shrub layer. Destruction by chestnut blight fungus at or just after the first flowering precludes this species ascension to canopy dominance. Additional understory trees include Sourwood (*Oxydendrum arboreum,* plate 347), Sassafras (*Sassafras albidum,* plate 343), Red Maple, and serviceberries (*Amelanchier spp.*).

The shrub layer may be dominated by tall heaths such as Mountain Laurel and Great Laurel or low heaths such as Bear Huckleberry (*Gaylussacia ursina,* plate 230) and Deerberry (*Vaccinium stamineum,* plate 352) or by a mix of tall and low heaths. Occasionally the shrub layer is poorly developed, in which case herbaceous plant diversity increases. Gorge Rhododendron (*Rhododendron minus,* plate 183) and Flame Azalea (*R. calendulaceum,* plate 184) sometimes produce spectacular displays. The rare Mountain Witch-alder is sometimes found here. This community is not known for its spring wildflower diversity, and although herb cover is sparse, some interesting species are present, including Galax, Trailing Arbutus (*Epigaea repens,* plate 232), Solomon's-seal (*Polygonatum biflorum,* plate 361), and Indian Cucumber-root.

Because earlier logging operations took only quality hardwoods, leaving behind the poor-quality trees, and because Chestnut Oak seldom grows straight and true as is desired by lumberman, most of the existing stands of chestnut oak forests are dominated by large, inferior Chestnut Oaks. Restoring these stands to their original high quality will be a challenge.

Where to visit the Chestnut Oak forest community. Since Chestnut Oak forest is the most abundant oak-dominated forest in the mountains, good examples can be found in any of our large mountain parks in Greenville County, the Jim Timmerman Natural Resource Area at Jocassee Gorges in Pickens and Oconee Counties, and the Andrew Pickens District of the Sumter National Forest in Oconee County. One of the easiest to find and access is in Caesars Head State Park along the Raven's Cliff Falls Trail or along mid-elevation trails at Table Rock State Park, once you get up and out of the cove forests.

Montane oak-hickory forests

Montane oak-hickory forests are a high elevation variant of the oak-hickory forests that are common in the Piedmont and mountains. These forests are generally found at elevations above 2,500 feet, either on ridge tops or on upper slopes that slope gently and are exposed but not rocky. The canopy is dominated by oaks and hickories, with White Oak, Northern Red Oak, Chestnut Oak, Pignut Hickory, and Red Hickory (*C. ovalis*) most common. Tulip tree is usually present, and it may be abundant in sites subjected to past large-scale disturbance. Logs and sprouts of American Chestnut may also be abundant. Shrub cover is generally moderate, with Mountain Laurel, Witch-hazel, and Flame Azalea usually present. Common Silverbell and Cucumber Magnolia (*Magnolia acuminata*) are usually present, either as understory trees or tall shrubs. The herbaceous layer is generally more diverse than the Chestnut Oak forest, but many of the species can be found in both natural communities. This community merges downslope with the chestnut oak forest or pine-oak heath community.

Where to visit the montane oak-hickory forest community. Few examples of montane oak-hickory forests are easy to get to. Pinnacle Mountain harbors a good example. The Pinnacle Ridge Trail in Table Rock State Park goes from Panther Gap to the top of Pinnacle Mountain (elevation 3,415′) and back down to the Table Rock Nature Center. It goes in and out of montane oak-hickory forest and Chestnut Oak forests. The best montane oak-hickory forest is just below the top of Pinnacle Mountain. A somewhat shorter trail to Pinnacle Mountain is 4.2 miles one-way. A map of the trails in Table Rock State Park is available from the Visitor Center or the Table Rock Nature Center, which is at the head of all trails in the park. Pinnacle Mountain can also be reached by the Foothills Trail.

Forest margins

Forest margin communities include herbs with high light requirements that are not typical of ruderal environments. Designating such a community would not be necessary if fire had not been suppressed for so long. Several very rare species are largely restricted to this community, including Smooth Coneflower (*Echinacea laevigata,* plate 386), Blue Ridge Bindweed (*Convolvulus sericatus* House, plate 207), Whiteleaf Sunflower (*Helianthus glaucophyllus,* plate 191), and Fraser's Loosestrife (*Lysimachia fraseri,* plate 213). These rare species are nearly confined to the woodland margins along roads, usually old logging roads. All respond well to fire. If natural pine-oak woodlands were subjected to periodic fire, these species probably would occur there. In fact, at Pine Mountain in the Andrew Pickens District of the Sumter National Forest, where fire is now being employed as a management tool, populations of both Blue Ridge Bindweed and Smooth Coneflower have increased dramatically. Efforts to use fire to restore the pine-oak heath community at the Buzzards Roost Heritage Preserve also have increased the abundance and flowering frequency of Blue Ridge Bindweed.

Other dry woodland species, such as Hairy Angelica (*Angelica venenosa,* plate 211) and Indian-tobacco (*Lobelia inflata,* plate 215), are most abundant in openings or in woodland margins along roads. Species such as Robin's-plantain (*Erigeron pulchellus,* plate 202) and Fire-pink (*Silene virginica,* plate 205) are most obvious in the moist to dry woodland margins.

Where to see forest margins. Though this habitat exists quite literally along every road running through our mountains, there are several places where the display is particularly beautiful. A great drive to enjoy wildflowers is along the margins of SC-107, north of Walhalla, and most of the adjacent USFS roads in the Andrew Pickens District of the Sumter National Forest. Additionally, the roadsides through the National Forest along Cassidy Bridge Road and FS-744 (Rich Mountain Road) are among the most beautiful in the area, particularly in the early autumn.

Pine-oak heaths

Pine-oak heaths (plate 221) are reminiscent of the pine barrens of New England and New Jersey. This community occurs on poor, highly acidic soils of narrow ridges, steep south slopes, and the entire tops of small mountains in the upper Piedmont and mountains. A variety of pines and dry-site oaks dominate the often-stunted canopy. Pitch Pine (*Pinus rigida,* plate 222) and Virginia pine (*P. virginiana*) are the most common pines; Table Mountain Pine occurs on very dry soils in high-elevation sites, Eastern White Pine may dominate on more mesic high-elevation sites (especially sites that were previously cleared), and Shortleaf Pine (*P. echinata*) may be present on low-elevation sites, especially in the Piedmont. Oaks in this community typically include Scarlet Oak, Black Oak (*Quercus velutina*), Chestnut Oak, and Blackjack Oak (*Q. marilandica,* plate 375). Understory species include Black Gum, Sourwood, Sassafras, the stump sprouts of American Chestnut, and (rarely) Carolina Hemlock. The shrub layer is usually dense, consisting of evergreen and deciduous member of the heath family, and may be either "tall" or "low." Tall heaths include scattered to dense Mountain Laurel and/or Gorge Rhododendron, interspersed with a dense cover of Bear and Black Huckleberries. Horse Sugar (*Symplocos tinctoria,* plate 224) is one of the few non-heath shrubs typically found here. Low heaths are dominated by a moderate to dense cover of the low-growing Dryland Blueberry (*Vaccinium pallidum,* plate 228).

Herbaceous plant diversity is low, and typical species include Trailing Arbutus, Appalachian Bellwort (*Uvularia puberula,* plate 235), Virginia Goat's-rue (*Tephrosia virginiana,* plate 581), Galax, and Spotted Wintergreen (*Chimaphila maculata,* plate 333). Bracken ferns (*Pteridium* spp.) may form large stands. Species that are characteristic of this community, but are rarely seen, include Turkeybeard (*Xerophyllum asphodeloides,* plate 239), Wintergreen (*Gaultheria procumbens,* plate 234), Blue Ridge Golden-banner (*Thermopsis mollis,* plate 237), Sweet-fern (*Comptonia peregrina,* plate 227), and Mountain Witch-alder. The attractive Appalachian Small Spreading Pogonia is an uncommon orchid found in this community and is seldom seen because it blooms in the heat of June and July. Knowledgeable naturalists can let their nose lead them to a colony of the seldom-seen Sweet Pinesap (*Monotropsis odorata,* plate 231), whose cinnamon-nutmeg and anise-scented flowers make the task of discovery at least possible, if not easy.

Like the pine barrens of New Jersey, this community depends on periodic fires for its maintenance. Lightening-ignited fires probably were important in the past but were almost certainly supplemented with fires set by Native Americans. Today, management with prescribed fire is returning to many areas managed by state and federal agencies.

Where to visit pine-oak heaths. Some of the best pine-oak heaths that are currently managed with fire are easily accessible in Oconee County. Buzzards Roost Heritage Preserve, located off Rich Mountain Road, south of Cassidy Bridge Road and just west of SC-28 north of Walhalla, is one of the best examples. Virtually every species typical of this habitat can be seen at this location.

The Piedmont

THE GRANITIC FLATROCKS COMMUNITY

Granitic flatrocks (plate 244) may be the most spectacular ecosystem in the Piedmont. They are restricted to the Piedmont and are best developed in Georgia and South Carolina. They consist of small to expansive exposures of granite or gneissic rocks that have a smooth (exfoliating) surface. Flatrock outcrops are level or slope gently and are distinguished from granitic domes and cliffs by the small elevation change from the top to the base of the outcrop.

Granitic flatrocks provide a unique set of challenges for plants due to their sunny, shallow-soiled, and droughty nature. These areas are islandlike habitats surrounded by forested and woodland communities. Each granitic flatrock is separated, often by a long distance from the next nearest granitic flatrock. Such natural communities are often referred to as ecological islands. Ecological islands, like true islands, often have plants and animals that are restricted to them, and our granitic flatrocks are no exception. Perhaps no other plant community in the Piedmont has so many endemic species (i.e., species limited to just this plant community). Some large flatrocks such as Forty Acre Rock in Lancaster County have plant species that appear to be limited to a single location.

Because of the shallow soil and excessive heat during the summer, plants that grow in this habitat have many adaptations that enable them to either endure drought or avoid drought. Annual species are particularly abundant in this habitat. Most of the annuals germinated during the late autumn or winter when soil moisture is high and grow throughout the cool season and flower very early in the spring. By summer most of these plants have died and the species survives as seeds capable of waiting for moist conditions to return. Plants with such adaptations are called "drought avoiding species." Other plants have adapted to store water in their stem or leaf tissue such as prickly-pear cacti (*Opuntia* spp.) and fameflowers (*Phemeranthus* spp.). These succulents store water to endure the long, hot season. Other plants have leaves covered with light-colored trichomes such as Woolly Ragwort (*Packera tomentosa,* plate 252) that keep the leaves cool and increase the humidity near the surface of the leaf. Still others, such as the Pineweed (*Hypericum gentianoides,* plate 260), have green stems and greatly reduced, scalelike leaves to reduce water loss. The myriad of adaptations of plants growing on the flatrocks to drought is astounding.

Granitic flatrocks harbor a mosaic of nonforested plant communities, each occupying a habitat with a well-defined soil depth and duration of soil moisture. Habitats include exposed rock surfaces, natural depressions with soil, rock crevices, and outcrop margins that often have open, prairielike vegetation that is called a "shallow-soil glade." Exposed rock surfaces do not provide habitat for any species of vascular plants, but they do contain a distinctive assemblage of mosses and lichens, including the distinctive fruticose Carolina Reindeer Lichen (*Cladonia caroliniana*).

Natural depressions are the most distinctive outcrop habitats and harbor many of the most unusual flatrock plants. These depressions are most abundant at or near the crest of the flatrock and are sometimes called "solution" pools because they were formed as water slowly dissolved the rock. Natural depressions may have intact rims or an eroded downslope rim, but regardless of the rim condition, the soil depth and the length of time that standing water remains in the depression determines the plant species present.

In South Carolina, depressions with one or more eroded rims usually have shallow soil and are typically occupied by Elf-orpine (*Diamorpha smallii,* plate 247), Single-flower Sandwort (*Mononeuria uniflora,* plate 253), and various species of *Cladonia* (reindeer lichen). Large mats of the Elf-Orpine, with its succulent red leaves and its bright white flowers, make this a striking outcrop habitat.

Natural depressions with intact rims, shallow soils, and water one to three inches deep for weeks at a time in the early spring are sometimes called "vernal pools." This is the habitat for several rare aquatic plants that are endemic to flatrock outcrops, including the federally threatened Pool Sprite (*Gratiola amphiantha,* plate 249) and several species of quillwort (*Isoëtes* spp.).

The annual-perennial herb community is the most diverse of the natural depression communities. A zone of haircap mosses (*Polytrichum* spp.) usually delineates this community from the annual-dominated shallow soil community. Visually dominant species in the spring include Woolly Ragwort, Small's Ragwort (*Packera anonyma,* plate 996), and (rarely in South Carolina)

Sunnybell (*Schoenolirion croceum*). This is also a community where a variety of weedy species occur, like Common Toadflax (*Linaria canadensis,* plate 962), Sourgrass (*Rumex hastatulus,* plate 965), Pineweed, and Outcrop Rushfoil.

The margins of flatrock outcrops provide habitat for a number of endemic or near endemic flatrock species. Shallow soils that lay beneath red cedar provide habitat for Puck's Orpine (*Sedum pusillum* Michaux, plate 248). It is only found under red cedar, presumably, because of the high calcium content of fallen cedar leaves. The rare Missouri Rockcress (*Boechera missouriensis*) is found in the deeper soils of some flatrock margins, and the diminutive Rock-loving Draba (*Draba aprica*) is sometime found on open, gravelly flatrock margins.

Seepages at outcrop margins may contain wetland species such as quill-worts (*Isoëtes spp.*), Flatrock Pimpernel (*Lindernia monticola,* plate 256), and bladderworts (*Utricularia* spp.).

Where to visit the granitic flatrock community. The premier site to view the granitic flatrock is undoubtedly Forty Acre Rock at the Forty Acre Rock Heritage Preserve in Lancaster County. This parking area is located at the termination of Conservancy Road just west of US-601 and south of Taxahaw Road (SC-29–123) and southwest of Pageland. This heritage preserve is located at the transition from Coastal Plain to Piedmont. A mostly level trail takes the visitor through sandy oak and Longleaf Pine sandhills, which in the spring are host to a good display of Northern Sundial Lupine (*Lupinus perennis* ssp. *Perennis,* plate 540). This takes the visitor to the edge of the expansive granitic flatrock, which contains nearly every species typical of the habitat. The large vernal pools are exceptionally colorful in April.

THE ROCKY SHOALS COMMUNITY

Rocky shoals (plate 264) are restricted to the Piedmont. The community is characterized by the abundance of small boulders or bedrock near the surface (shoals) during periods of low-water flow. The shoals are mostly submerged during periods of high-water flow. Rocky shoals are most extensive near the fall line that separates the Piedmont from the Coastal Plain on major rivers such as the Saluda, Savannah, Broad, and Congaree.

The rocky islands that are mostly submerged during high water are occupied by emergent aquatic or semiaquatic plants. Three species characterize this habitat: Rocky-shoals Spiderlily (*Hymenocallis coronaria,* plate 265), American Waterwillow (*Justicia americana,* plate 266), and Riverweed (*Podostemum ceratophyllum,* plate 40). Riverweed, with its tiny dissected leaves and fleshy discs, which are used to attach the plant to submerged rocks, resembles an alga more than a vascular plant. Its flowers have no showy parts, and the plant must be viewed with a magnifying lens to fully appreciate its unusual nature.

Rocky-shoals Spiderlily, often present in great abundance, with its five-foot stalks and large whitish flowers, is spectacular. In June 1773, William Bartram, while at the shoals on the Savannah River at Augusta, stated, "nothing in vegetable nature was more pleasing that the odoriferous Pancratium fluitans (*Hymenocallis coronaria* of today's taxonomy), which almost alone possess the little rocky islets which just appear above the water" (Bartram, 1791).

A wide variety of rooted and emergent aquatics, species typical of marshy habitats or weeds typical of wet disturbed habitats, may be associated with the rocky shoals. These include Mermaid-weeds (*Myriophyllum* spp.); Brazilian Elodea (*Egeria densa*); Pickerelweed (*Pontederia cordata,* plate 832); Water Hemlock (*Cicuta maculata,* plate 834); Broadleaf Arrowhead (*Sagittaria latifolia*); Cardinal Flower (*Lobelia cardinalis,* plate 843); Red Tooth-cup (*Ammannia coccinea*); Winged Monkey-flower (*Mimulus alatus,* plate 856); and a variety of rushes, sedges, spikerushes, and bulrushes.

Where to visit the rocky shoals community. Landsford Canal State Park located in Chester County is one of the most dramatic and easily accessible locations to see this visually stunning habitat. A trail along the Catawba River offers great views of the rocky shoals habitat, which in late May through early June are filled with the tall stems of Rocky-Shoals Spiderlily. Other examples of the rocky shoals community may be seen along the Savannah River below the I-20 bridge at Savannah River Bluffs Heritage Preserve and along the Broad River near downtown Columbia along the Three Rivers Greenway.

THE DECIDUOUS FOREST COMMUNITIES

Basic-mesic forests

Basic-mesic forests (plate 267) are one of the most spectacular wildflower communities in South Carolina. They rival the rich cove forests of the mountains and the Longleaf Pine savannas of the Coastal Plain in terms of plant diversity, especially herbaceous plant diversity. In fact, it is the extremely high density and diversity of herbaceous plants that makes this community easily recognizable. As described in this book, basic-mesic forests are restricted to the Piedmont. Coastal Plain sites with basic and mesic soils are included in the beech forest, calcareous bluff, or wet, flat calcareous forest communities.

The adjectives used in the name of this community (i.e., basic and mesic) aptly describe the conditions necessary for its development. Technically, a basic soil is one that has a pH above 7.0. A pH of 7.0 is neutral; a pH below 7.0 is acidic. As used here, and in many nontechnical works in plant ecology, the term "basic" refers to soils that are high in bases, including minerals (especially calcium) that increase soil pH. Since there are very few forest soils in South Carolina that have a pH greater than 7.0, the basic soils here are more technically described as circumneutral (i.e., approximating 7.0 in pH).

Basic-mesic forests are mesic in terms of soil moisture availability—that is, they have good soil moisture for most of the growing season. This requirement restricts basic-mesic forests to "sheltered" topographic positions that reduce total sunlight and minimize loss of soil moisture. Typical locations for this community type are north-facing slopes and lower slopes of small, deep ravines.

Although variable among sites, the canopy is distinctive. Trees that are typical of mesic sites in the Piedmont may be present, but American Beech, the tree common in most mesic Piedmont sites, is never dominant, although it is often present. Also, trees typical of adjacent bottomlands often move upslope into this community type. In a study of the basic-mesic forests at the

Stevens Creek Heritage Preserve in McCormick County, which is the most outstanding example of this community in the Piedmont, Dr. Al Radford (1959) recorded the following canopy dominants in order of their importance: Bitternut Hickory (*Carya cordiformis*), Southern Sugar Maple (*Acer floridanum*, plate 743), Swamp Chestnut Oak (*Quercus michauxii*), Northern Red Oak, Slippery Elm (*Ulmus rubra*), Sugarberry (*Celtis laevigata*), Shagbark Hickory (*C. ovata*, plate 269), Black Walnut (*Juglans nigra*, plate 745), Tulip-tree, and White Oak. Swamp Chestnut Oak, Sugarberry, and Walnut are all bottomland species, and Slippery Elm is well known as a calcium-loving tree. Less mesic sites tend to be dominated by White Oak, with Southern Sugar Maple and bottomland species as important associates. American Beech is usually present but not an important contributor to the canopy.

The understory in basic-mesic forests always includes such trees as Flowering Dogwood, Sourwood, Chalk Maple, and American Holly, but the abundance of Ironwood (*Carpinus caroliniana*), a species more typical of bottomlands, and Hop Hornbeam (*Ostrya virginiana*) distinctively characterize this community. Yellowwood is a rare mountain disjunct found in one example of basic-mesic forest in Aiken County.

Shrub diversity is high, and typical species include Wild Allspice, Bladder-nut (*Staphylea trifolia*, plate 428), Painted Buckeye (*Aesculus sylvatica*, plate 313), Appalachian Mock-orange (*Philadelphus inodorus*, plate 273), Common Pawpaw (*Asimina triloba*, plate 427), and a host of additional shrubs typical of beech forests and oak-hickory forests in the Piedmont. Several very rare shrub species are present in some locations, including Bottlebrush Buckeye (*Aesculus parviflora*, plate 274), Upland Swamp Privet (*Forestiera ligustrina*), and Wahoo (*Euonymus atropurpureus*). Seldom do more than two of these rarities occur at any one site. The presence of Wahoo on upland sites is usually a good indicator of this community type. Stevens Creek Heritage Preserve harbors one of only three populations in the world of Spiny Gooseberry (*Ribes echinellum*, plate 271).

The herbaceous layer is a distinctive characteristic of basic-mesic forests. Some herbaceous plants are disjunct from the mountains, some are typical of rich bottomlands, and some are found only in this community. Rare mountain disjuncts include Doll's-eyes (*Actaea pachypoda*, plate 114), Dutchman's Breeches (*Dicentra cucullaria*, plate 276), Ginseng (*Panax quinquefolius*, plate 301), Lowland Bladder Fern (*Cystopteris protrusa*, plate 307), Cancer-root (*Aphyllon uniflorum*, plate 291), Blue Cohosh, Large Yellow Lady's Slipper, and Perfoliate Tinker's-weed (*Triosteum perfoliatum*). Species that are distinctive to basic-mesic forests include Shooting Star (*Primula meadia*, plate 283), Green Violet (*Cubelium concolor*, plate 297), and Southern Stoneseed (*Lithospermum tuberosum*). Species that are more typical of rich bottomlands include Reflexed Wild Ginger (*Asarum reflexum*, plate 287), Yellow Fumewort (*Corydalis flavula*, plate 281), and Moonseed (*Menispermum canadense*). Additional rare herbaceous species, often restricted to just a few sites, include Lanceleaf Trillium (*Trillium lancifolium*, plate 286), Pale Yellow Trillium (*T. discolor*, plate 109), Relict Trillium (*T. reliquum*, plate 285), Roundleaf Ragwort (*Packera obovata*, plate 290), and Spring Coral-root (*Corallorhiza wisteriana*, plate 752). Additional interesting species include Hairy Spiderwort (*Tradescantia hirsuticaulis*, plate 254), Lopseed (*Phryma leptostachya*), Spring Beauty (*Claytonia virginica*, plate 429), and Dimpled Trout Lily (*Erythronium umbilicatum* ssp. *umbillicatum*, plate 275).

In order to see all the unusual plants in bloom, the visitor to a basic-mesic forest site should plan on visiting a site several times between mid-March and late June.

Where to visit the basic-mesic forest community. The most beautiful and expansive basic-mesic forest in South Carolina is easily accessed at Stevens Creek Heritage Preserve in McCormick County. The site is located along SR-33–88, just east of Clarks Hill. The moderate loop trail takes the visitor through a pine and oak-hickory forest and down a slope into the heart of this spectacular basic-mesic forest habitat. Most of the characteristic species of basic-mesic forest species will be seen along this trail. Highlights include Shooting Stars, Lanceleaf Trillium, Pale Yellow Trillium, Southern Nodding Trillium (*Trillium rugelii*, plate 288), Dutchman's Breeches, and the federally endangered Spiny Gooseberry. The site is typically at its best from late March through mid-April.

Other locations where this habitat may be seen include Savannah River Bluffs Heritage Preserve in Aiken County and Forty Acre Rock Heritage Preserve in Lancaster County.

Beech forests

Beech forests (plate 310), also known as mesic mixed hardwood forests, are restricted to the Piedmont and Coastal Plain. Steep, north-facing river bluffs and sheltered ravines are the locations in which most beech forest communities occur. But they are also found in the Coastal Plain on the upland flats or on islands surrounded by swamp. Associated rocks and soils in the Piedmont are acidic. In the Coastal Plain, they may be either acidic or circumneutral (i.e., calcium-rich). This wildflower community is easily recognized by the abundance of American Beech (*Fagus grandifolia* var. *caroliniana*, plate 311). Canopy dominants also include Tulip tree, Red Maple, Shagbark Hickory, and Northern Red Oak. The general rarity of other oak species is one of the defining characteristics of this community. Southern Sugar Maple, Sweet Gum (*Liquidambar styraciflua*, plate 793), and sometimes Bull Bay (*Magnolia grandiflora*, plate 905) may also be important canopy trees in the Coastal Plain. Many of the same understory trees that are present in beech forests also occur in oak-hickory forests, including Flowering Dogwood, Sourwood, and American Holly, but in the Piedmont, Hop Hornbeam often is abundant. Swamp Bay (*Persea palustris*, plate 721) is often present in beech forests in the Coastal Plain.

Common shrubs include Strawberry-bush and species of blueberry (*Vaccinium* spp.). Small amounts of Mountain Laurel are present in the Piedmont, and Witch-hazel and Horse Sugar are usually present in the Coastal Plain. The herbaceous layer is sparse to moderate in density and diversity. In the Piedmont, herb density and diversity in beech forests are intermediate between basic-mesic forests and oak-hickory forests. In the Coastal Plain, beech forests are the richest upland community. Abundant and attractive species include Bloodroot (*Sanguinaria canadensis*, plate 320), May-apple (*Podophyllum peltatum*, plate 324), Perfoliate Bellwort (*Uvularia perfoliata*, plate 292), various trilliums (Mottled Trillium, *Trillium maculatum*, plate 749) in the Coastal Plain and Catesby's Trillium (*T. catesbaei*, plate 186) in the Piedmont), Wild Geranium (*Geranium maculatum*, plate 329), and Green-and-gold (*Chrysogonum virginianum* var. *brevistolon*, plate 328). Green Adder's-mouth (*Malaxis unifolia*) and Southern Twayblade (*Neottia bifolia*, plate 778) are rarely seen orchids that are commonly found in beech forests in the Coastal Plain.

Where to visit the beech forest community. Beech forests are widespread in the Piedmont and common along slopes, especially along major rivers and larger creeks. In the Coastal Plain, where beech forests are less abundant and higher in diversity and density of wildflowers, an excellent example can be found along the south side of Huger Creek, along SC-402 in the Francis Marion National Forest. This site contains nearly every species that is typical of this habitat. The site was badly damaged by Hurricane Hugo in 1989 but has recovered significantly. The gentle slopes running down toward Huger Creek contain a canopy composed largely of American Beech, Spruce Pine, and Shagbark Hickory, and the wildflowers include nice displays of Bloodroot, May-apple, and Mottled Trillium. The best time to visit is in the early spring.

Oak-hickory forests

This is a complex community that is found in the lower mountains, Piedmont, and Coastal Plain. It is difficult to characterize, with much variation among sites. Oak-hickory forests (plate 339) may be dominated by a surprising variety of canopy species, especially oaks. Considerable variation also exists in the understory and shrub and especially in the herbaceous layer. Despite the site-to-site variation between physiographic provinces, there are unifying characteristics.

Oak-hickory forests are always associated with acid soils and are neither the most mesic nor the driest community in any region. Depending on elevation, aspect, and topography, an oak-hickory forest may occur on ridge tops, upper slopes, or mid-slopes. Occasionally, it occurs on upland flats. Although canopy dominants can vary considerably, white oak is found on most sites and is either dominant or codominant. Mockernut Hickory (*Carya tomentosa*) or Pignut Hickory (*C. glabra*) is common. Southern Red Oak (*Quercus falcata*) is found on drier sites, and Tulip-tree is usually present on mesic sites. Sourwood, Black Gum (*Nyssa sylvatica*), Flowering Dogwood, and Red Maple are present in the subcanopy, with the former two species more abundant on drier sites and the latter two more abundant on more mesic sites. American Holly and Eastern Redbud (*Cercis canadensis*, plate 342) also are abundant subcanopy trees on more mesic sites. On all but the driest sites, Red Maple becomes more abundant when there has been site disturbance.

Shrubs common throughout this community include various species of blueberry, including Dryland Blueberry and Common Deerberry. Vines include Muscadine (*Muscadinia rotundifolia,* plate 357), Coral Honeysuckle (*Lonicera sempervirens,* plate 356), and Poison Ivy (*Toxicodendron radicans,* plate 878). In drier sites, New Jersey Tea (*Ceanothus americanus,* plate 200), Carolina Jessamine (*Gelsemium sempervirens,* plate 355), and Sparkleberry (*Vaccinium arboreum,* plate 351) are common. In more mesic sites, Strawberry-bush and various buckeyes are abundant, including Red Buckeye (*Aesculus pavia,* plate 742) in the Coastal Plain and Painted Buckeye in the Piedmont and mountains.

Herbaceous plant cover is generally sparse and diversity is limited. Herbaceous species common throughout this community include Flowering Spurge (*Euphorbia corollata*), Carolina Wild Petunia (*Ruellia caroliniensis*), Hairy Skullcap (*Scutellaria elliptica*), Veiny Hawkweed (*Hieracium venosum*), Spotted Wintergreen (*Chimaphila maculata,* plate 333), Little Brown Jugs (*Hexastylis arifolia,* plate 321), and Appalachian Oak-leach (*Aureolaria laevigata,* plate 371). Drier sites usually harbor Goat's rue (*Tephrosia virginiana,* plate 581) and Whorled-leaf Tickseed (*Coreopsis major,* plate 398). More mesic sites inevitably harbor Small Solomon's-seal (*Polygonatum biflorum,* plate 361), Catesby's Trillium, and Pale Indian-plantain (*Arnoglossum atriplicifolium,* plate 368).

Oak-hickory forests, as discussed in the section on the shared natural heritage, are unusual habitats because they have been so altered by management activities through the centuries. Many sites were once savannas or prairielike and were converted to agricultural uses. Most of the sites today were once agricultural fields and though oak-hickory forests have always been a part of the landscape, they exist as mostly highly modified habitats. These are the most frequent forest types today throughout the Piedmont and in many sections of the Coastal Plain that have also been subjected to fire suppression.

Where to visit the oak-hickory forest community. Because these forest types are common and widespread, they may be easily seen throughout most of the state. Great examples of these forests may be seen in the Clemson Experimental Forest, particularly along Lake Issaquena Road, located on the north side of Lake Issaqueena and accessed via Old Six Mile Highway, north of Clemson and near Daniel High School. Several parking areas are located on this dirt road, which also involves fording a small stream (no bridge). Exceptional examples of oak-hickory forests on highly acidic to circumneutral soils are found all along this route and miles of hiking trails provide opportunities to explore some of the most diverse oak-hickory forests in South Carolina.

THE EARLY SUCCESSIONAL COMMUNITIES

Piedmont prairie

It is difficult to imagine today that large swaths of the Piedmont supported treeless, grass, and forb-dominated habitats. The evidence suggests this is the case. A well-documented area of prairie habitat existed in the region of York and Lancaster counties, and according to the writings of Mark Catesby, also along the Savannah River drainage of South Carolina's Piedmont. The extent of the Piedmont prairie can only be hypothesized today but it certainly was a natural community and, though extinct as a widespread habitat, it does exist as tiny parcels interspersed into other landscapes. Many of the rarest species in South Carolina today are thought to have been common members of this habitat when it was more widespread. The federally endangered Schweinitz's Sunflower (*Helianthus schweinitzii,* plate 410) as well as Smooth Coneflower, Prairie Dock (*Silphium terebinthinaceum,* plate 396), Southern Obedient-plant (*Physostegia virginiana* ssp. *praemorsa,* plate 397), Gray-headed Coneflower (*Ratibida pinnata,* plate 385), Piedmont Buckroot (*Pediomelum piedmontanum,* plate 383), Low Wild-petunia (*Ruellia humilis,* plate 381), and Stiff-leaf Goldenrod (*Oligoneuron rigidum,* plate 411), which are all imperiled in South Carolina, are all thought to have been limited to this community and the oak savanna. The composition of the original landscape can be reimagined by examining colonial-era collections and visiting the remaining remnants, particularly those found near and in Rock Hill Blackjacks Heritage Preserve.

Where to visit Piedmont prairie. Pockets of habitat approaching what Piedmont prairie was like can best be seen at Rock Hill Blackjacks Heritage Preserve in Rock Hill. Trails through the preserve are all level and easy. The trails take you through good examples of xeric hardpan forest and along the edges of powerlines and other rights-of-way that still maintain a diverse assemblage of prairie species. Most of the plants limited to these habitats can be seen within the preserve. The reason for their persistence here is the presence of a nearly impenetrable hardpan of shrink-swell clay soils that make it difficult for many trees to easily grow and the establishment of the various rights-of-way, which have traditionally been mown has allowed the habitat to persist and even in some cases to expand.

Perhaps the best place to envision what the Piedmont prairies must have been like in precolonial times is along the Natural Heritage Garden trail in the South Carolina Botanical Garden in Clemson. The re-introduced Piedmont prairie here is over nine acres and contains most of the species that are restricted to this habitat. The site was established on virgin soils, those that have never been plowed, so the deep rich soils have allowed the prairie community here to reach full development.

Oak savanna

Oak savannas were once a common and widespread natural community across the Piedmont and portions of the Coastal Plain. Today, only tiny remnants remain. Descriptions of this natural community by colonial era naturalists indicate their abundance in a landscape that was managed by Native Americans with fire. The widely spaced canopy in these habitats is composed primarily of Post Oak (*Quercus stellata,* plate 376) and Blackjack Oak (*Q. marilandica,* plate 375), with other oak species present but not dominant. Shortleaf Pine is often also a component. All three of these species are tolerant of low-intensity fires and require open, sunny locations for establishment. The understory has a low shrub cover and saplings are primarily composed of the dominant canopy species. The herbaceous layer is composed of common tallgrass prairie grass species such as Little Bluestem, Yellow Indiangrass, and Big Bluestem combined with Silky Oatgrass, Black Needlegrass (*Piptochaetium avena-ceum*), multiple species of witchgrass (*Dichanthelium* spp.) and in some cases Slender Woodoats (*Chasmanthium laxum*). The herbaceous wildflowers in this habitat include many of the species we think of today as roadside plants including a vast diversity of tickseeds (*Coreopsis* spp.), sunflowers (*Helianthus* spp.), blazing stars (*Liatris* spp.), Virginia Goat's-rue, and many asters (*Symphyotrichum* spp., *Eurybia* spp., *Doellingeria* spp.) and goldenrods (*Solidago* spp.). A few of our rarest plants, including Georgia Aster (*Symphyotrichum georgianum*, plate 417) and Piedmont Wand Goldenrod (*Solidago austrina*, plate 413), are found in this habitat. Historically, this was probably the primary habitat for the now federally threatened Smooth Coneflower. Fire is a necessity for the management of the oak savanna community. Where examples still exist, fire is applied to the landscape every few years.

Where to visit the oak savanna community. Though this community (plate 374) is nearly extinct in South Carolina, there is one exceptional example on public lands in the Long Cane district of the Sumter National Forest. This site is known as the Post Oak Savanna Natural Area, and directions to the location may be found on the Sumter National Forest website. The mature and widely spaced Post Oaks in this site and diverse herbaceous layer are the closest any of us can get to experiencing the landscape that existed during the time of Mark Catesby in the early 1700s. The site is expansive enough that it allows a full immersion experience that transports us back to a distant time when Native Americans were managing and promoting the biodiversity found in these systems.

Recent management activities in other areas of the Long Cane district and on the Andrew Pickens District of the Sumter National Forest have also recently started to transform formerly dense second-growth oak-hickory forests into more savannalike habitat. In several sites, including near Buzzard's Roost heritage preserve, where Smooth Coneflower populations are growing, the management has resulted in a shift in the community to resemble more closely the precolonial oak savannas.

Piedmont xeric hardpan forests

This community (plate 268) has an impenetrable hardpan of shrink-swell clay near the surface that inhibits infiltration of water and plant roots. This hardpan may also result from rock near the surface, but regardless of its origin, habitat is produced in the upland flats and gentle slopes that are flooded during wet times and are parched during droughts. The rock associated with this community type often is present as boulders that are usually gabbro, which is high in calcium-rich feldspars and weathers to produce a soil with a circumneutral pH.

Stunted trees and a relatively open canopy are characteristic of this community. In undisturbed, old-growth sites, Post Oak and Blackjack Oak, often in association with Carolina Shagbark Hickory (*Carya carolinae-septentrionalis*, plate 377), dominate the canopy. Eastern Red Cedar and Eastern Redbud dominate the understory. On many sites it is often difficult to distinguish the canopy from the understory or subcanopy. Virginia Pine (*Pinus virginiana*) and Shortleaf Pine may be important components of disturbed sites. Additional canopy species include White Oak (*Q. alba*), American Ash (*Fraxinus americana*), a variety of other oaks and Pignut Hickory (*Carya glabra*). Additional understory trees include Winged Elm (*Ulmus alata*), Fringe-tree (*Chionanthus virginicus*), Northern Hackberry (*Celtis occidentalis*) and Persimmon (*Diospyros virginiana,* plate 346). Shrubs are variously abundant; also present are Sparkleberry, Black Haw (*Viburnum prunifolium*), Carolina Buckthorn (*Frangula caroliniana,* plate 931), New Jersey Tea, Aromatic Sumac (*Rhus aromatica*), and one or more hawthorns (*Crataegus* spp.).

The stunted and spaced trees of the canopy provide no hint of the wonderful diversity of unusual plants in the herbaceous layer. Many of the herbs have affinities with midwestern prairies. (Many of these species have already been discussed under the Piedmont prairie community above.)

Where to visit the xeric hardpan forest community. Good quality examples of this forest type are very rare. Only one site, Rock Hill Blackjacks Heritage Preserve, has been partially protected to date and is described in the section on the Piedmont prairie community above.

THE PIEDMONT SPRINGHEAD SEEPAGE FOREST COMMUNITY

Well-developed examples of springhead seepage forests (plate 418) are restricted to the upper Piedmont. The best sites are all in the vicinity of Traveler's Rest in Greenville County. Springhead forests begin at seepages at the base of slopes and may extend downslope for just a few feet or for hundreds or thousands of feet. They are characterized by the presence of seepage channels that have year-round, slow-moving, cool groundwater. Seepage channels often divide downslope from the seephead and may become so braided that water eventually does not flow continuously, which usually determines the lower extent of the springhead forest. All known springhead forests in South Carolina are bounded by Pacolet sandy loam soils, which are apparently ideally suited to the uptake and storage of rainwater and its slow release as seepage.

Good examples of this community typically have a closed canopy composed of Red Maple, Black Gum, and Tulip-tree. Shrubs are sparse to dense, with dense shrub cover confined to canopy openings. A diversity of shrubs that are typical of boggy habitats may be present, especially Tag Alder, Wild Raisin (*Viburnum nudum*), and Red Chokeberry. Additional shrubs present in this community include American Storax (*Styrax americana,* plate 772), Virginia Sweetspire, Male-berry (*Lyonia ligustrina*), Common Winterberry (*Ilex verticillata*), and Poison Sumac (*Toxicodendron vernix,* plate 419). Important vines include Climbing Hydrangea (*Hydrangea barbara,* plate 32) and Blaspheme-vine (*Smilax laurifolia,* plate 730), with the latter species sometimes forming impenetrable tangles.

The herbaceous layer varies from sparse to dense, and grasses and sedges are only abundant in canopy openings. Sphagnum moss is generally not abundant. Important herbs include Cinnamon Fern (*Osmundastrum cinnamomeum*), Royal Fern (*O. regalis* var. *spectabilis*), Stiff Cowbane, Netted Chain-fern (*Lorinseria areolata,* plate 811), and Partridgeberry on drier hummocks. The edges of the seepages have at least some plants of the Small Green Wood-orchid

(*Platanthera clavellata,* plate 425), while the seepages typically harbor Virginia Hedge-hyssop (*Gratiola virginiana*), and in the best sites, the federally endangered Bunched Arrowhead (*Sagittaria fasciculata,* plate 423). Some seepages are being taken over by the nonnative weed Mud-Annie (*Murdannia keisak,* plate 424). Hummocks between the seepages may harbor the rare Dwarf-flowered Heartleaf (*Hexastylis naniflora,* plate 421). Additional rare species that are found only at a few sites include Shortleaf Sneezeweed (*Helenium brevifolium,* plate 63) and Greene's Sedge (*Carex bullata* var. *greenei*).

In the best sites, seepage channels are over solid sands, and the forest is not especially boggy. As silt buildup occurs, which is accelerated by upslope disturbance, this community becomes boggier. The seepage channels change from being clear and constantly flowing to being stagnant and filled with a reddish scum that indicates the activity of anaerobic fungi. It is unclear what management is needed to maintain this community long term.

Where to visit the Piedmont springhead seepage forest community. The best example with public access is Bunched Arrowhead Heritage Preserve near Travelers Rest in Greenville County. This site is home to its namesake, and much of the description above was taken directly from studies of this location.

THE BOTTOMLAND FOREST COMMUNITIES

The Piedmont is home to three bottomland forest communities: the Bald Cypress-Tupelo Gum swamp forests, the hardwood bottoms, and the levee forests. These three communities also occur in the Coastal Plain. In the Piedmont, however, the floodplains have a shorter flooding duration and a lower flooding depth, and the acreage in the Piedmont is considerably less than in the Coastal Plain. In the Piedmont, these communities are a minor component of the vegetation, whereas in the Coastal Plain they are a major component. The detailed description of the bottomland forest communities is given under the Coastal Plain section.

The following wildflowers are found only in the Piedmont bottomland forests: shrubs include Painted Buckeye, Bladdernut, and Silky Dogwood (*Swida amomum,* plate 420); herbs include Golden Ragwort (*Packera aurea*), Virgin's Bower (*Clematis virginiana,* plate 217), Yellow Fumewort, and Honewort (*Cryptotaenia canadensis*). Piedmont bottomland forests are frequently overwhelmed with invasive exotic species such as Japanese Stiltgrass (*Microstegium vimineum*), Japanese Honeysuckle (*Lonicera japonica,* plate 951) and Chinese Privet (*Ligustrum sinense,* plate 944). The overwhelming infestation of invasive exotics has greatly diminished the quality of this habitat for many of our native plants.

Where to visit bottomland forests in the Piedmont. Exceptional and diverse bottomland forests are found at Stevens Creek Heritage Preserve in McCormick County. The site is described above in the section on basic-mesic forests. The loop trail extends along the floodplain of Stevens Creek and is an exceptional example of a Piedmont bottomland forests. This site contains several surprising species for the Piedmont including nice specimens of Bald Cypress along the edges of the creek.

Coastal Plain: The Fall-Line Sandhills

THE XERIC COMMUNITIES

Longleaf Pine-Scrub Oak sandhills

The name of this community (plate 434) references three of its most important components. It is dominated by an open canopy of Longleaf Pine, and a variety of scrub oaks dominate the open to dense subcanopy. It differs from the Longleaf Pine-Turkey Oak community in being less xeric and more fertile, which is probably a result of a shallower sand layer and more organic matter in the soil. This community harbors a greater diversity of species. It is found on middle and lower slopes in the Sandhills region but can be found on sand deposits throughout the rest of the Coastal Plain. Some authors suggest that this community has a clay layer near the surface, which is true of sites with an abundance of Sand Myrtle (*Kalmia buxifolia*, plate 292). It is bounded by Longleaf Pine-Turkey Oak scrub upslope and dry oak-hickory forests or streamhead pocosins downslope. Fire suppression has significantly affected this community.

The Longleaf Pines that form the open canopy become flat-topped with age, creating a distinct and aesthetically pleasing appearance from a distance. The subcanopy is dominated by Blackjack Oak, Turkey Oak (*Quercus laevis*, plate 437), Bluejack Oak (*Q. incana*, plate 438), Sand Post Oak (*Q. margarettiae*), and often Mockernut Hickory (*Carya tomentosa*). Sassafras and Persimmon usually are present, and occasionally Flowering Dogwood (*Benthamidia florida*). The shrub layer is dominated by ericads, especially Sparkleberry, Southern Blueberry (*Vaccinium tenellum*, plate 557), and Southern Dwarf Huckleberry (*Gaylussacia dumosa*, plate 442), Poison Oak (*Toxicodendron pubescens*, plate 535), several species of hawthorn (*Crataegus* spp.), and Carolina Jessamine. Rare to uncommon shrubs include Dwarf Bristly Locust (*Robinia nana*) and Nestronia (*Nestronia umbellula*, plate 353).

The herbaceous layer in areas with frequent fire is well developed and dominated by wiregrass (*Aristida stricta*, plate 445, in the north and *A. beyrichiana*, plate 445, in the south), and Little Bluestem or Creeping Little Bluestem (*Schizachyrium scoparium* var. *stoloniferum*). A wide variety of broadleaf herbs may be present, including Senna Seymeria (*Seymeria cassioides*, plate 490), Carolina Puccoon (*Lithospermum caroliniense*, plate 451), Southern Jointweed (*Polygonum americanum*, plate 479), Narrowleaf Dawnflower (*Stylisma angustifolia*, plate 461), Eastern Green-eyes (*Berlandiera pumila*, plate 471), Sandhills Chaffhead (*Carphephorus bellidifolius*, plate 484), Carolina Sandhill Ironweed (*Vernonia angustifolia*, plate 481) and Sandhills Thistle (*Cirsium repandum*, plate 522).

Species that are more typical of dry oak-hickory forests or dry pine-oak woodlands and are found here include Virginia Goat's-rue and Sweet Goldenrod (*Solidago odora*, plate 401). Species that are more typical of the longleaf pine-turkey oak scrub community include Tread-softly (*Cnidosculus stimulosus*, plate 449), Carolina Ipecac (*Euphorbia ipecacuanhae*, plate 448), and Sandhills St. John's-wort (*Hypericum lloydii*, plate 476). Many additional species characteristic of dry and/or dry and disturbed areas may be present, so don't be surprised if you can't identify everything.

Longleaf Pine-Turkey Oak sandhills

The Longleaf Pine-Turkey Oak sandhills community is the most xeric and least fertile of all the pineland communities in South Carolina. It occurs on deep, coarse sands, most typically wind-blown sands on ridge tops of the Fall-line Sandhills. A similar community also is found on fluvial (water deposited) and aeolian (wind-sorted) sand ridges that parallel and are east of the major Coastal Plain blackwater rivers and on rims of Carolina bays. In the maritime strand, they are even found on old beach dunes. All of these systems are characterized by an open canopy of Longleaf Pine (*Pinus palustris*, plate 436), and all have the subcanopy dominated or co-dominated by Turkey Oak. These systems also are characterized today by large patches of open sands and an abundance of ground lichens.

The Longleaf Pine-Turkey Oak sandhills community is distinguished from Longleaf Pine-Scrub Oak sandhills by its abundance of Turkey Oak and the general lack of other scrub oaks, other than the occasional Sand Post Oak or Bluejack Oak. Wiregrass is present in both communities but is more abundant in the latter. The Longleaf Pine-Turkey Oak scrub community grades downslope into a Longleaf Pine-Scrub Oak sandhills or the streamhead pocosin community.

No one who has ever stood in a xeric sandhill in the middle of a sunny July day has any doubt that this is a harsh habitat, but a number of species have adapted to withstand the extremes. Shrubs are as abundant as herbs, and both may be sparse to moderately dense. Characteristic shrubs include Southern Dwarf Huckleberry, Florida Rosemary (*Ceratiola ericoides*, plate 520), Poison Oak, and the very attractive low shrub Sandhills St. John's-wort. Herbs include Sandhills Milkweed (*Asclepias humistrata*, plate 521), Coastal Plain Wire-plant (*Stipulicida setacea*, plate 460), Carolina Sandwort (*Mononeuria caroliniana*, plate 450), Hairy False Foxglove (*Aureolaria pectinata*, plate 475), Sandhill Wild-buckwheat (*Eriogonum tomentosum*, plate 492), Tread-softly, Carolina Ipecac, and Grassleaf Roseling (*Cuthbertia graminea*, plate 459). Unusual plants found at only a few sites include Pickering's Dawnflower (*Stylisma pickeringii* var. *pickeringii*, plate 477), Northern Golden-heather (*Hudsonia ericoides*, plate 441), Sandhills Pyxie-moss (*Pyxidanthera brevifolia*, plate 446), and Woolly-white (*Hymenopappus scabiosaeus*, plate 463).

Where to visit xeric sandhill communities. Excellent examples of this habitat can be easily visited in Chesterfield County and include the Carolina Sandhills National Wildlife Refuge, which provides a wildlife drive. A stop at the Lake Bee picnic area provides easy access to many acres of this habitat. The nearby Sandhills State Forest contains many additional acres of this habitat. A very unusual site that is not characteristic but is one of the most interesting botanical sites in the area is Sugarloaf Mountain Recreation Area. The hills in this area have outcroppings of sandstone, some of which contain small populations of the extremely rare Bradley's Spleenwort (*Asplenium bradleyi*). Sandhills Pyxie-moss is locally abundant along ridges here and is in flower very early in the spring. These hills apparently have perched water because of the shallow soil underlain by sandstone. These conditions have allowed many typically wetland species like Creeping Blueberry (*Vaccinium crassifolium*, plate 558) and Titi (*Cyrilla racemiflora*, plate 728) to extend their colonies to the very summit of hilltops.

Peachtree Rock Heritage Preserve in Lexington County is another great place to see sandhill habitats. Rarities found here include Rayner's Blueberry (*Vaccinium sempervirens*, plate 516) and Sandhill Goldenaster (*Pityopsis pinifolia*, plate 489). In Aiken County, two excellent locations to view sandhill habitats are found at Hitchcock Woods in Aiken proper and Aiken Gopher Tortoise Heritage Preserve in the southern portion of the county. Gopher tortoises, which are only found in Aiken and Jasper Counties, are one of the highlights of visiting this sandhill habitat.

THE SANDHILLS SEEPAGE COMMUNITIES

Streamhead pocosins

Streamhead pocosins (plate 495) are found along the headwaters of small streams, streamside flats, and extending up hillsides adjacent to streams. This community is most common in the Fall-line Sandhills, but it does occur in the Inner Coastal Plain. Its presence is dependent on flowing water or seepage from the adjacent uplands. Although these communities tend to be long and narrow, fire seldom burns through the entire community because it tends to be wet. Fire often does burn into its upland margins, and it is in these frequently burned ecotones that most of the unusual species are found.

As in pocosins in general, the canopy of this community is dominated by Pond Pine (*Pinus serotina*, plate 720). However, hardwood species such as Tulip-tree, Swamp Gum (*Nyssa biflora*, plate 707), Red Maple, and even Sweet Gum may be abundant depending on the prevalence of fire. Sweet Bay and Swamp Bay sometimes form a distinct subcanopy layer, and occasionally a dense tall shrub layer of Mountain Laurel is present. An alternative name for pocosin is "evergreen shrub bog," so it is not surprising that a dense layer of evergreen shrubs is present including Shining Fetterbush (*Lyonia lucida*, 725), Sweet Gallberry (*Ilex coriacea*), and Inkberry (*Ilex glabra*, plate 563). Species especially abundant at the upland ecotone include deciduous species such as Titi, Sweet Pepperbush (*Clethra alnifolia*, plate 500), and Common Dangleberry (*Gaylussacia frondosa*, plate 561). Unusual shrubs, generally restricted to the frequently burned ecotones, include White Wicky (*Kalmia cuneata*, plate 501) and Coastal Witch-alder (*Fothergilla gardenii*, Murray). Herbs are sparse, except in rare boggy openings. The dominant herb is usually Cinnamon Fern (*Osmundastrum cinnamomeum*), along with a variety of uncommon species such as pitcherplants, Tawny Cottonsedge (*Eriophorum virginicum*, plate 511), Sandhill Heartleaf (*Hexastylis sorriei*, plate 503), and a variety of grasses, sedges, and rushes in boggy areas with an open canopy. The upper margins of streamhead pocosin may transition to herbaceous seepage slopes in areas with frequent fire.

Ecotones maintained by regular fires are delightful places to botanize. However, an unburned ecotone and the wetter portions of the community are almost impenetrable because of the dense shrubs and the tangle of Blaspheme-vine.

Herbaceous seepage slopes

The herbaceous seepage slope community (plate 504) develops where streamheads occur in frequently burned areas. Frequent fires remove the woody vegetation and allow herbaceous species to dominate. Sedges and wetland grasses are abundant and include several species that are dominant here and rarely found elsewhere such as Coastal Plain Beaksedge (*Rhynchospora stenophylla*) and Featherbristle Beaksedge (*R. oligantha*). This natural community has been drastically reduced in the Sandhills because of fire suppression and is one of our most imperiled plant communities. With the absence of fire, shrubs and trees move into the community and it is transformed into a streamhead pocosin. The best examples of this community occur on Fort Jackson and in the Carolina Sandhills National Wildlife Refuge.

This community harbors some of the showiest wildflowers in the Sandhills. Look for three species of pitcherplants and their hybrids: Carolina Sweet Pitcherplant (*Sarracenia rubra* ssp. *rubra*, plate 732), Southern Purple Pitcherplant (*S. purpurea* var. *venosa*, plate 505), and Yellow Pitcherplant (*S. flava*, plate 678). There are also several species of sundews and orchids, several species of bladderworts (*Utricularia* spp.), and Yellow Hatpins (*Syngonanthus flavidulus*, plate 506). Several species endemic or nearly endemic to this community can be seen, which is a testimony to its natural and historical occurrence in the Sandhills and include Scabrous-leaved Yellow-eyed Grass (*Xyris scabrifolia*), Chapman's Yellow-eyed Grass (*Xyris chapmanii*), Texas Hatpins (*Eriocaulon texense*, plate 62), and the stunning Sandhills Bog Lily (*Lilium pyrophilum*, plate 508). Also present are several species that are disjunct between the mountains and Coastal Plain, including Roundleaf Sundew (*Drosera rotundifolia*, plate 510).

Very similar communities are found in the Outer Coastal Plain where there is enough relief to provide seepage. Such areas are known from Horry, Berkeley, Georgetown, Charleston, and Jasper Counties. Many of the dominant species are the same, but the Sandhills endemics are lacking.

Atlantic White-cedar forests

The Atlantic White-cedar forests (plate 514) share many of the same species as pocosin and streamhead pocosin communities, but the thick canopy dominated by Atlantic White-cedar (*Chamaecyparis thyoides*, plate 515) easily distinguishes it from these closely related communities. Atlantic White-cedar forests may occur in Carolina bays or other depressions in the Coastal Plain, but they are best developed in the Fall-line Sandhills at sites with an abundance of flowing water or seepage. If a site has enough seepage and moist soils to retard the regular invasion of fires that originate in the adjacent Sandhills and if a source of Atlantic White-cedar seeds is readily available, then an Atlantic White-cedar forest forms. A high-water table, peaty, acid soils, abundant *Sphagnum* moss, and long fire cycles (50–150 years) are additional distinguishing characteristics of the community.

The canopy of this natural community is typically dense and dominated by Atlantic White-cedar, with Pond Pine, Red Maple, Swamp Gum, and Tulip-tree present. Hardwoods become more abundant as the community ages. The three species of bays, Sweet Bay, Swamp Bay, and Loblolly Bay (*Gordonia lasianthus*, plate 729), dominate the subcanopy layer. The shrub layer is dense and dominated by evergreen species typical of pocosins, especially Shining Fetterbush, Titi, two species of wax-myrtles (*Morella cerifera*, plate 909, and *M. caroliniensis*), and Inkberry and Sweet Gallberry. Rare shrubs to look for include Mosier's Huckleberry (*Gaylussacia mosieri*), Bog Spicebush (*Lindera subcoriacea*, plate 497), and Rayner's Blueberry (*Vaccinium sempervirens*, plate 516). Poison Sumac is usually present.

Where to visit the Sandhills seepage communities. Streamhead pocosin habitat is widespread in the Sandhills, though most of the sites have suffered from fire suppression. The best examples of well-burned streamhead pocosins, herbaceous seepage slopes and good, though young examples of Atlantic white-cedar forests, are found in Chesterfield County in the Carolina Sandhills National Wildlife Refuge. A nice example of a young Atlantic white-cedar forest is located along the margins of Lake Bee, at the beginning of the Wildlife Drive in the National Wildlife Refuge. There is a small fringe of herbaceous seepage slope located at this site as well. The most extensive sandhills herbaceous seepage slope in South Carolina is located along the slopes above Oxpen Branch in the National Wildlife Refuge. This habitat is extremely fragile and filled with critically imperiled species. The authors discourage visitors from walking through the fragile habitat, which is already threatened by damage from hogs and woody plant invasion.

Shealy's Pond Heritage Preserve in Lexington County is the best example of an Atlantic white-cedar forest accessible to the public. This exceptional habitat is one of only a few locations for the critically imperiled Rayner's Blueberry. The edge of the pond also features a narrow zone of sandhills herbaceous seepage slope habitat.

Coastal Plain: The Inner and Outer Coastal Plain

THE XERIC COMMUNITIES

Longleaf Pine-Turkey Oak xeric ridges

The xeric Longleaf Pine-Turkey Oak communities (plate 518) of the Coastal Plain occur on sandy ridges. The ridges have three different origins. Although they share many common species, their distinguishing floristic composition is based on the origin of the sand. All three types of ridges have Longleaf Pine and Turkey Oak as the dominant trees. Some of the classic wildflowers occurring on all three types of xeric ridges in the Coastal Plain include Sandhills Gerardia (*Agalinis setacea,* plate 528), Sandhills Milkweed (*Asclepias humistrata,* plate 521), Sandhills Thistle (*Cirsium repandum,* plate 522), Tread-softly (*Cnidoscolus stimulosus,* plate 449), and Coastal Plain Wire-plant (*Stipulicida setacea,* plate 460).

Brownwater sand ridges. Sand ridges along brownwater rivers, such as the Savannah River, which originates in the mountains, are fluvial (water deposited) in nature and come from erosion in the Piedmont and mountains. They are nutrient-rich, unlike the Carolina bay ridges and blackwater ridges, and harbor several species not found on the ridges associated with Carolina bays and blackwater rivers. These species include Harper's Scrub-balm (*Dicerandra odoratissima,* plate 529), Gopher-apple (*Geobalanus oblongifolius,* plate 531), Soft-haired Coneflower (*Rudbeckia mollis,* plate 526), and Carolina Warea (*Warea cuneifolia,* plate 527).

Blackwater sand ridges. The sands of the blackwater sand ridges are marine in origin since they come from the Coastal Plain (plate 518). They develop along the blackwater, Coastal Plain rivers such as the Little Pee Dee and Edisto and are deposited by wind and water. They are ancient in origin, deposited as the last glaciers retreated from the fall line. Their soils are less fertile than the brownwater sand ridges, and the sands are bright white and are considered to be wind-sorted (aeolian).

Four wildflowers that are characteristic of the blackwater and Carolina bay ridges, but usually absent from the brownwater ridges, are Florida Rosemary (*Ceratiola ericoides,* plate 520), Large-fruited Beaksedge (*Rhynchospora megalocarpa,* plate 524), Carolina Ipecac, and Southern Bogbuttons (*Lachnocaulon beyrichianum,* plate 523). Southern Bogbuttons is usually found along the ecotone of the sand ridge and the adjacent pocosin.

Bay ridges. Bay ridges (often called bay rims) occur most frequently along the southeastern side of Carolina bays. The ridges were deposited by winds blowing during the period when the bays originated. The wildflowers and vegetation of the bay ridges are similar to the blackwater ridges.

Where to visit the Longleaf Pine-Turkey Oak xeric ridge community. Tillman Sand Ridge Heritage Preserve in Jasper County, along the Savannah River, is an excellent example of a brownwater sand ridge. This site is also home to a population of Gopher Tortoise. All of the species limited to this habitat can be seen within the boundaries of the preserve and along Sandhills Road (SC-119) in the areas around the preserve.

Two good sites to see blackwater sand ridges are Little Pee Dee State Park in Dillon County and Little Pee Dee River Heritage Preserve along Marsh Lake Road north of SC-917 in Horry County. Excellent examples of Carolina bay ridges that support typical xeric species occur in Horry County in Cartwheel Bay Heritage Preserve and Lewis Ocean Bay Heritage Preserve, as well as Woods Bay State Park in Sumter County.

Sandy, dry, open woodlands

The sandy, dry, open woodland community (plate 533) is probably a successional community that develops on sand ridges following agricultural abandonment or after disturbance of a dry oak-hickory forests or dry pinelands. This habitat is widespread throughout the Coastal Plain and increasing. It is primarily though perhaps not entirely anthropogenic as fire-maintained dry woodland communities become fire-suppressed and more disturbance occurs in upland habitats. Many of the species occurring here can also be found in Longleaf Pine sandhill and woodland habitats.

Canopy trees are scattered and consist of "dry" oaks and hickories with occasional loblolly pines. Shrubs are sparse. The herbaceous plant layer is open and best defines the community. Invariably some of the following herbaceous wildflowers occur in the community: Common Elegant Blazing Star (*Liatris elegans,* plate 554), Eastern Horsemint (*Monarda punctata,* plate 553), Sandhill Bluestar (*Amsonia ciliata,* plate 541), three species of lupine (*Lupinus perennis,* plate 540, *L. villosus,* plate 539, and *L. diffusus,* plate 538), South Carolina Wild Pink (*Silene caroliniana* var. *caroliniana,* plate 542), Carolina Piriqueta (*Piriqueta caroliniana,* plate 544), Queen's-delight (*Stillingia sylvatica,* plate 545), Southern Dawnflower (*Stylisma humistrata,* plate 550), Vasevine (*Clematis reticulata,* plate 546), and Canada Sunrose (*Crocanthemum canadense*). Several of the above herbs also occur in the xeric communities.

THE MESIC PINE WOODLAND COMMUNITIES

Forests are distinguished by having closed and often overlapping canopies. Woodlands, by contrast, have more widely spaced trees that allow ample sunlight to filter down to the herbaceous and shrub layer. In South Carolina's Coastal Plain, woodlands are maintained by fire or a combination of fire, soil, and hydrological conditions. Another term that is often used in ecology for many types of woodland is "savanna." In South Carolina, "savanna" most often refers to mesic or wet woodlands with grasses, sedges, and forbs dominating beneath the trees. "Flatwoods" is a term that is used to describe savanna habitats with a higher shrub dominance that occupy mostly drier sites.

Longleaf Pine flatwoods

The typical Longleaf Pine flatwoods (plate 555) are dominated by a canopy of tall, Longleaf Pines. The terrain is flat to gently rolling with a sandy soil and a high water table. Although Longleaf Pine characterizes the community, Loblolly and Slash Pine may occur.

In flatwoods where fire is infrequent, a well-developed shrub layer and understory may develop. Under high fire frequency, the shrubs and understory species are kept in check. Because of the site-to-site variation in the understory and shrub layers, pine flatwoods are difficult to characterize. Common understory trees include Sweet Gum, Blackjack Oak, and Black Gum. Common shrubs include Dwarf Wax-myrtle (*Morella pumila,* plate 562), Inkberry, Common Dangleberry, Running Oak (*Quercus elliottii,* plate 566), Downy

Sweet Pepperbush (*Clethra tomentosa,* plate 564), and Sand Post Oak (*Quercus margarettiae*).

The herbs of frequently burned pine flatwoods include grasses, legumes, and composites but few of the showy species of the savannas. Grasses include broomstraws (*Andropogon* spp.), while the legumes including Zornia (*Zornia bracteata,* plate 586), beggar's lice (*Desmodium* spp.), lespedezas (*Lespedeza* spp.), Dwarf Indigo-bush (*Amorpha herbacea,* plate 575), and Virginia Goat's-rue. The composites include Black-root (*Pterocaulon pycnostachyum,* plate 579), asters (*Symphyotrichum walteri,* plate 668, *S. concolor,* plate 494, *Sericocarpus tortifolius,* and *Ionactis linariifolia,* plate 493), and goldenrods (*Solidago* spp., including Sweet Goldenrod, *S. odora,* plate 401). The ubiquitous Southern Bracken (*Pteridium pseudocaudatum*) becomes the dominant herb in the spring after annual winter fires. The rare American Chaff-seed (*Schwalbea americana,* plate 580) occurs in openings of the herb layer.

The pine flatwoods grade into the Longleaf Pine savanna community, and distinguishing between the two can sometimes be difficult. Slight depressions that have a high clay hardpan occur within the pine flatwoods and harbor the typical species that characterize the savannas. At other sites, savannas cover extensive areas that are easily distinguishable from the adjacent pine flatwoods. Two species can be used to identify the two habitats: Hooded Pitcherplant (*Sarracenia minor,* plate 619) and Toothache Grass (*Ctenium aromaticum,* plate 611). Both species require more open and moist conditions than is normally found in the pine flatwoods; their presence indicates pine savannas. Pocosins, cypress savannas, and upland swamps also occur scattered within pine flatwoods. The pine flatwoods are a fire subclimax; prolonged absence of fire will lead to hardwood forests.

Where to visit the Longleaf Pine flatwoods community. Excellent examples of Longleaf Pine flatwoods are easily visited at the James Webb Wildlife Center and Game Management Area in Hampton County and in the Francis Marion National Forest. The flatwoods bordering Steed Creek and Halfway Creek Road are typical of the central portion of the South Carolina Coastal Plain.

Pine/Saw Palmetto flatwoods

The Pine/Saw Palmetto flatwood community (plate 592) is found only in Jasper and Beaufort Counties, its northern limit. Elements of this habitat extend up the Savannah River drainage into Barnwell County and up the coast to Charleston County. It is more extensive in Florida and Georgia. The canopy consists of Longleaf Pine on the ridges and Slash Pine (*P. elliottii,* plate 593) and/or Pond Pine in depressions. A subcanopy of oaks is usually sparse. Saw Palmetto (*Serenoa repens,* plate 597) dominates the shrub layer. Other common shrubs include Hairy Wicky (*Kalmia hirsuta,* plate 599), Rusty Lyonia (*Lyonia ferruginea,* plate 595), and Southern Evergreen Blueberry (*Vaccinium myrsinites,* plate 598). Typical pocosin and Longleaf Pine flatwood species such as Sweet Bay, Swamp Bay, Inkberry, Sweet Gallberry, Shining Fetterbush, Downy Sweet Pepperbush, and Honey-cups are also part of the shrub layer. The sparse herbaceous layer is a mixture of Longleaf Pine flatwood species and those restricted to this community and includes Elliott's Milk Pea (*Galactia elliottii*, plate 606), Vanilla Plant (*Trilisa odoratissima*, plate 587), Pine-barren Aster (*Oclemena reticulata*, plate 602), and Walter's Milkweed (*Asclepias cinerea*, plate 603). Periodic fires promote herbs and saw palmettos. An absence of fire leads to more of a dominance of shrubs.

Where to visit the Pine/Saw Palmetto flatwoods community. The best example of this community available to the public is Victoria Bluff Heritage Preserve in Beaufort County. This site has been fire-suppressed for many years, and burning is difficult because it is surrounded by rapid development in this part of Beaufort County. Many of the characteristic species are still abundant at the site, and some areas remain somewhat open. Rusty Fetterbush and Saw Palmetto are abundant in many areas.

Longleaf Pine savannas

For showy wildflowers, no natural community equals the Longleaf Pine savannas (plate 610). From early spring through late fall, a progression of herbaceous wildflowers graces the Coastal Plain with a mix of colors. Orchids, carnivorous plants, lilies, showy composites, plus many other groups all find a home in the sunny, moist Longleaf Pine savannas.

It would seem a paradox that fire is responsible for maintenance of the savannas. Native Americans burned the savannas to drive game and to clear the ground around settlements. Natural fires, started by lightning, swept through the pinelands, mostly during July and August. Trees and shrubs, with their growing tips at fire level, were killed. Herbaceous species, with their stems (rhizomes) underground, are protected. Shortly after a fire, these herbaceous species put up new growth, and what first appeared as a scene of utter desolation quickly becomes a wildflower garden again.

On the other hand, savannas, protected from fire, quickly succeed to a shrub community, then to a tree-dominated forest. Under a forest canopy, the savanna herbs, which require high light intensity, cannot survive.

One tree that is able to survive the frequent burning of the savannas is Sweet Bay. It does so because portions of its stem are buried in the soil. After a fire, it puts up a cluster of new shoots, giving the appearance of a shrub. If the root-stem system is dug up, one finds a single, enlarged rootstock, which is a testimony that the "tree" may be many years old even though its aboveground stems represent one or two years of growth.

Two general types of Longleaf Pine savannas can be recognized: one dominated by Toothache Grass (*Ctenium aromaticum*, plate 611), Savanna Muhly (*Muhlenbergia expansa*), and bluestem species (*Andropogon* spp.) and the other dominated by wiregrasses (*Aristida stricta* in the northeast and *A. beyrichiana* in the southeast). Both types of savannas develop where the combination of fairly level topography and nondraining subsoil causes a high (or perched) water table in the rainy season. Loblolly Pine, Slash Pine, and Pond Pine, as well as Pond Cypress, may occur with the Longleaf Pine. Wiregrass, however, favors coarse soil where there is some slope to allow lateral drainage, resulting in slightly drier conditions. Toothache Grass and the showy herb species prefer wetter conditions. The greatest display of showy herbs occurs in the Toothache Grass Longleaf Pine savannas.

Most savannas today occur in national forests or on large hunt clubs as a result of prescribed burning. Pine savannas are threatened due to fire suppression, the construction of houses or commercial properties, and drainage. Adequate management with fire has become difficult even on extensive public lands. Drainage canals lower the water table, allowing for the invasion of less moisture-tolerant species. Recent data suggests that more than 95% of Longleaf Pine savannas have been lost. Unless protective measures are taken, in the next few decades this community may be insignificant or lost.

Where to visit the Longleaf Pine savanna community. Beautiful examples of classic Longleaf Pine savannas can still be seen in the Francis Marion National Forest. Two excellent examples occur along Ballfield Road near Awendaw (Awendaw Savanna) in Charleston County and on the west side of Wardfield Road (FS-212), east of SC-45 in Berkeley County (Wardfield Savanna). Additionally, Lynchburg Savanna Heritage Prerserve in Lee County and Lewis Ocean Bay Heritage Preserve in Horry County harbor this beautiful wildflower community. These sites, like all savanna habitats, are most beautiful during years where management fires have been conducted, as many of the plants flower only or more profusely during years with fire.

THE DEPRESSION POND COMMUNITIES

Pond Cypress savannas

Pond Cypress savannas (plate 672) occupy flat, acidic, poorly drained lands within the Longleaf Pine woodlands. They have a slightly longer hydroperiod than that associated with Longleaf Pine flatwoods and Longleaf Pine savannas and are dominated by Pond Cypress, which can tolerate a longer hydroperiod than pines. The cypress canopy is open. Red Maple and Swamp Gum may also be present. Pond Cypress savannas occur scattered throughout the Coastal Plain with many of the best examples occurring within Carolina bays and limesink depressions. Draining and ditching, along with absence of fire, have dramatically reduced the number of Pond Cypress savannas.

Few shrubs occur except for some woody St. John's-worts, including the often abundant Peelbark St. John's-wort (*Hypericum fasciculatum,* plate 677) and Myrtle-leaved Holly (*Ilex myrtifolia,* plate 713). The herbaceous flora, blooming through the spring, summer, and fall, is rich. Some of the showy herbaceous species of the adjacent pine savannas, including many of the carnivorous plants, can also be found in this community. Certain herbs, however, appear to be more common or confined to the Pond Cypress savannas. These include Southeastern Sneezeweed (*Helenium pinnatifidum,* plate 679), Bay Blue-flag Iris (*Iris tridentata,* plate 686), Tall Milkwort (*Polygala cymosa,* plate 684), Awned Meadow-beauty (*Rhexia aristosa,* plate 690), Flax-leaf Agalinis (*Agalinis linifolia,* plate 703), Boykin's Lobelia (*Lobelia boykinii,* plate 687), and the federally endangered Canby's Dropwort (*Tiedemannia canbyi,* plate 705). Other wildflowers of note in the Pond Cypress savannas include Blue Sedge (*Carex glaucescens,* plate 697), Pool Coreopsis (*Coreopsis falcata,* plate 681), Broadleaf Whitetop Sedge (*Rhynchospora latifolia,* plate 682), Ten-angled Pipewort (*Eriocaulon decangulare,* plate 695), Skyflower (*Hydrolea corymbosa,* plate 700), Savanna Obedient-plant (*Physostegia purpurea,* plate 683), Fringed Yellow-eyed-grass (*Xyris fimbriata,* plate 704), Lace-lip Ladies'-tresses (*Spiranthes laciniata,* plate 685), and Carolina grass-of-Parnassus (*Parnassia caroliniana,* plate 667).

Pond Cypress-Swamp Gum upland swamps

Pond Cypress-Swamp Gum upland swamp forests (plate 706) are dominated by Pond Cypress or Pond Cypress and Swamp Gum, with Pond Pine often present as an associate. These swamps occur in upland depressions where some water is on the surface for at least three months. The water is acidic because there is no drainage to remove accumulated acids. The depressions may be limesinks, irregular depressions, or Carolina bays. Generally, the interior is open water, with shrubs and herbs confined to the margin.

The herbaceous flora of this swamp is sparse compared to the riverine and other upland swamps. One rare herb is Violet Burmannia (*Burmannia biflora,* plate 718), which occurs in the drawdown zone. Numerous shrubs occur along the margins of the depressions. In the Francis Marion National Forest, several populations of Pondberry (*Lindera melissifolia,* plate 711), a federally endangered species, occur in this community in the Honey Hill and Cainhoy areas. Another member of the laurel family, Pondspice (*Litsea aestivalis,* plate 710), which is rare throughout its range, is common in many of these swamps in the Francis Marion National Forest and elsewhere in the Coastal Plain. Other shrubs in this upland swamp forest include Titi, Button-bush, Cassena (*Ilex cassine,* plate 712), and Myrtle-leaved holly (*Ilex myrtifolia,* plate 713). The rare Climbing Fetterbush (*Pieris phillyreifolia,* plate 709) is known in South Carolina only from these Pond Cypress-swamp gum swamps. The open water is also habitat for the freshwater aquatics. Floating Bladderwort (*Utricularia inflata,* plate 864) is especially common.

Where to visit the Pond Cypress savanna community. Pond Cypress savannas are still plentiful in the Francis Marion National Forest. Due to the high number of rare species and fragility of the habitat, the authors are reluctant to give directions to any one site in particular for fear of "loving them to death." A ride down Steed Creek or Hoover or Halfway Creek roads in the Francis Marion National Forest will take you past many depressions, easily observed nestled in the pinelands. Any number of these can yield a tremendous view of many of the beautiful species found in this community. The Santee Coastal Reserve, located near South Santee, in Charleston County, is another site with numerous depression ponds. Savage Bay Heritage Preserve in Kershaw County harbors a beautiful pond cypress savanna.

Depression meadows

Depression meadows (plate 673) are temporally flooded herbaceous communities that have saturated soil. Since they are isolated wetlands, these meadows generally do not have a source of water other than rainwater. During times of drought, fires often occur, preventing a woody flora from becoming established. Depression meadows differ from freshwater marshes in that the latter are more or less permanently flooded. The lack of fire in marshes allows woody species to become established and sometimes to ultimately transform the community into a swamp forest.

Depression meadows are among South Carolina's rarest and most threatened natural communities. Only a handful of examples exist and appear to be distinguished from other depression habitats by their highly fluctuating water levels. More than four years of inundation will kill most tree species and extended dry periods help to keep trees like cypress that can tolerate inundation from achieving a foothold.

Classification of depression meadows is unclear, and much fieldwork needs to be done to fully understand these communities. One type of depression meadow being studied occurs in clay-based Carolina bays. These seem to share a basic floristic similarity, with numerous species in common, especially rare ones. The clay soil prevents water from percolating down into the soil, and during times of rain, the bay is flooded. Since the bays are different depths, water level varies. The deeper bays often have open water that supports species of the floating aquatic community such as Fragrant Waterlily and Water-shield. The more shallow bays are dominated by sedges, rushes, and grasses.

Some of the wildflower species of note are Boykin's Lobelia, Shrubby Seedbox (*Ludwigia suffruticosa*, plate 689), Awned Meadow-beauty, Tracy's Beaksedge (*Rhynchospora tracyi*, plate 693), Sclerolepis (*Sclerolepis uniflora*, plate 688), and the endangered Canby's Dropwort.

Where to visit depression meadow community. Excellent examples are found only on private or government property without easy or with restricted public access. One depression meadow, Red Bluff Bay, can be visited; it requires quite a hike from the nearest road. This bay is located south of Echaw Road (FS-204) in the Francis Marion National Forest.

THE PEATLAND COMMUNITY

Peatlands are wetlands where the soils are peat. Peat is the partially decomposed remains of dead plants and, to a lesser extent, animals. Peatlands develop where there is a net gain in organic matter over time; therefore, decomposition cannot exceed production if peatlands are to form. Water slows decomposition in the Coastal Plain. In water, dead plants and animals decompose at a much slower rate than when they are exposed to both air and moisture. Most bacteria and fungi that decompose organic matter need oxygen for respiration. In saturated conditions, no atmospheric oxygen is available. The accumulation of peat is also helped if the water moves slowly since slow-moving water does not carry away organic matter. Acids that build up as by-products of respiration further inhibit decomposition in bogs because bacteria and fungi do not work as effectively under acidic conditions. Bog waters develop a dark color from the acids that are not washed away by moving water.

Pocosins

The dominant peatland community in the Coastal Plain are the pocosins (or evergreen shrub bogs). Pocosins (plate 719) are found on the southeastern Atlantic Coastal Plain of the United States and are botanical treasures. As treated here, the classic pocosin differs from the streamhead pocosins found in the Sandhills and Inner Coastal Plain regions in that they are not predominantly seepage habitats. Pocosins grow on waterlogged, acid, nutrient-poor, sandy, or peaty soils that are located on broad, flat topographic plateaus or in Carolina bays. They are usually removed from large streams and are subject to periodic burning. Natural drainage is poor. Pocosins are dominated by a dense mix of evergreen shrubs, vines, and scattered trees. In some pocosins, the vegetation is so thick and laced with Blaspheme-vine (*Smilax laurifolia*, plate 730) that traversing in this community is difficult. One has to go down "on all fours" to get below the dense tangle of vegetation. Each step in the soft peat is uncertain. The reward is a pocosin community full of wildflowers.

The vegetation of pocosins is a dense growth of shrubs that are associated with scattered trees. Diversity is not great since few species can adapt to the mineral-poor, acid soils and long hydroperiods. The dominant trees are Pond Pine and Loblolly Bay, with Swamp Bay and Sweet Bay occurring as associates. The most frequently found shrubs are the evergreens Shinning Fetterbush, Inkberry, and Sweet Gallberry, combined with deciduous shrubs such as Coastal Plain Serviceberry (*Amelanchier obovalis*, plate 496), Titi, and Honey-cups (*Zenobia pulverulenta*, plate 726). All these species grow with Blaspheme-vine. Because of the evergreen shrubs, some people call the pocosins "evergreen shrub bogs." The shrubs in some sites are short (2–3 feet tall) with scattered Pond Pine and are called "low pocosins." In other sites, both shrubs and trees are taller, and the sites are called "high pocosins." In some pocosins, tall zones exist around the margin with the low pocosin growth in the center. Poison Sumac often grows along the margins of pocosins. Two uncommon evergreen heaths, Leather-leaf (*Chamaedaphne calyculata*, plate 723) and Carolina Wicky, are found along the edges of pocosins.

Fires increase the habitat diversity of pocosins by unevenly burning the peat and creating boggy depressions below the water table. These depressions fill with sphagnum and herbaceous species such as Yellow Pitcherplant and Spoonflower (*Peltandra sagittifolia*, plate 735).

For thousands of years, peat has been building up in these pocosins to a depth of 10 feet or more. During droughts, peat is susceptible to fires, and most pocosins burn about every 10 to 30 years. Pocosin plants have adapted to this fire cycle. After a fire, they vigorously sprout new growth from their rhizomes, which were protected from fire because they are buried in the deeper, wet zone of the peat. Pocosins are also mineral-poor since they receive water mostly from rainfall or drainage through coarse sands.

Pocosins today are being studied intensely as natural reservoirs for water and as habitat for plants and animals. During drought, lower layers of peat hold water that is available to animals or that can be slowly released into the surrounding areas. Pocosins have had a long history of human use; timbering, drainage for agricultural use and timber plantations, peat mining, and urban development all have greatly reduced pocosin habitat. Fortunately, concerned organizations are taking steps to preserve pocosins, especially in Carolina bays.

Where to visit the pocosin community. Lewis Ocean Bay Heritage Preserve in Horry County, Bennett's Bay Heritage Preserve in Clarendon County, and Woods Bay State Park in Sumter and Clarendon Counties offer exceptional pocosin habitats with fairly easy public access. Ocean Bay in the Francis Marion National Forest near the intersection of Steed Creek and Halfway Creek Roads is an exceptional example that burns fairly frequently and has Spoonflower (*Peltandra sagittifolia,* plate 735).

THE CALCAREOUS FOREST COMMUNITIES

The calcareous forest communities do not occupy significant acreage in the Coastal Plain; however, they do contribute considerable diversity to the flora of the Coastal Plain because numerous rare and uncommon species are found there. Most of the undisturbed sites of this community are found on private lands that are not publicly available.

The calcareous forests occur on bluffs, slopes, or moist flats that overlay calcareous substrates. The substrate is either marl or limestone that was laid down as marine deposit when the ocean covered the Coastal Plain. The calcium from the underlying substrate is a major factor shaping the diversity and composition of the vegetation. Certain species of plants, referred to as "calcicoles," thrive in a basic to circumneutral soil that results from the presence of calcium ions. These species generally are mixed with the flora of the surrounding community to form a diverse community. Classification of the various calcareous communities is not well developed; however, two well-developed types are identifiable.

Calcareous bluff forests

Calcareous bluff forests occur on mesic sites that overlay shallowly buried or exposed marl or limestone formations. These forests occur along rivers and creeks where erosion has exposed or brought the marl or limestone formation close to the surface.

Trees that characterize the calcareous bluff forests include Swamp Chestnut Oak (*Quercus michauxii*), Bluff Oak (*Q. austrina,* plate 740), Hop Hornbeam, American Basswood (*Tilia americana*), Slippery Elm (*Ulmus rubra*), Carolina Buckthorn (*Frangula caroliniana,* plate 931), Black Walnut (*Juglans nigra,* plate 745), and Southern Sugar Maple. Elements of the other deciduous forest communities occur and include Eastern Redbud, Flowering Dogwood, and Eastern Red Cedar. Herbaceous species are common and include many that are found in other deciduous communities of the Piedmont and Coastal Plain. Several species, however, are either confined to or are more common in this community. They include Crested Coral-root (*Hexalectris spicata,* plate 759), Shadowwitch (*Ponthieva racemosa,* plate 763), Mottled Trillium (*Trillium maculatum,* plate 749), and Thimbleweed (*Anemone virginiana,* plate 332).

Often, where the calcareous substrate has been exposed, it hardens. Rainwater then erodes the substrate, forming recesses in which rare ferns are able to become established by spore dispersal: Blackstem Spleenwort (*Asplenium resiliens,* plate 767), Wagner's Spleenwort (*A. heteroresiliens,* plate 767), and Venus'-hair Fern (*Adiantum capillus-veneris,* plate 765).

Where to visit the calcareous bluff community. Three sites that harbor this community are Old Santee Canal Park in Berkeley County, Givhans Ferry State Park in Dorchester County, and Santee State Park in Orangeburg County. In these parks, a short walk down shaded trails will take the visitors into the heart of the calcareous bluff community. The sheer walls of limestone that are exposed at Givhans Ferry require a kayak or canoe to view because they lie directly above the Edisto River.

Wet, flat, calcareous forests

This calcareous community occupies low, wet flats adjacent to river systems. The underlying marl formation is not exposed, but it is close enough to the surface to influence plant composition. It is not a common community in the Coastal Plain. In fact, it has been studied only along the western side of the Cooper River in Berkeley County and in several sites in and around Huger Creek in the Francis Marion National Forest. It is not known to what extent it occurs throughout the Coastal Plain, but it appears to be restricted essentially to Berkeley County.

Recent studies on Mulberry and Lewisfield Plantations reveal the distinct flora of the community. The calcicoles that are present include Mottled Trillium, Crested Coral-root, Shadow-witch, and American Alumroot. These species also occur in the calcareous bluff forests. Two rare woody species are Nutmeg Hickory (*Carya myristicaeformis,* plate 744) and Prickly-ash (*Zanthoxylum americana,* plate 433).

Where to visit the wet, flat, calcareous forests. The only location where wet, flat calcareous forest can be visited is along Huger Creek along SC-402. Here the habitat is located just below the beech forest, which also occurs at the site. The habitat transitions into a calcium-rich bottomland forest that is also an excellent example.

THE BOTTOMLAND FOREST COMMUNITIES

Bottomland forests occupy the floodplains above the upper limit of the tidal influence that flanks the Coastal Plain river systems. These forests and their associated fauna comprise remarkably productive riverine communities that are adapted to fluctuating water levels. In the face of intensive land use of the adjacent uplands, the bottomland forests today serve as refuges for floodplain species and upland wildlife species.

Two major types of rivers traverse the Coastal Plain: brownwater and blackwater. Brownwater rivers originate in the mountains and Piedmont areas and have broad and fertile floodplains due to the great quantity of nutrient-rich alluvium that is deposited when the rivers overflow their banks. The brown color of the water comes from the silt and clay that erode from the Piedmont and mountains and are suspended in the water. Most brownwater rivers, such as the Santee and Savannah Rivers, have periods of sustained high flow that result from the cumulative effects of many tributaries and distant rainfall.

Blackwater rivers and tributary streams originate in the Coastal Plain and receive most of their water from local rain. They have narrower, less well-developed floodplains than the brownwater rivers since little alluvium is deposited. Unlike the brownwater rivers, the blackwater rivers may have dry periods when discharge is low. The term "blackwater" comes from the relatively sediment-free but highly colored water that results from the presence of organic acids derived from decaying leaves. Examples of blackwater rivers are the Cooper, Ashley, Combahee, Ashepoo, New, Four Holes, Waccamaw, and Black.

Three major bottomland communities occur on brownwater and blackwater river floodplains above the zone of tidal influence: Bald Cypress-Tupelo Gum swamp forests, levee forests, and bottomland hardwoods.

Bald Cypress-Tupelo Gum swamp forests

The bottomland Bald Cypress-Tupelo Gum swamps (plate 441) represent the forested community least disturbed in the Coastal Plain. Despite this, only a few original growth stands remain.

In swamps, where the land is flooded almost continuously, Bald Cypress (*Taxodium distichum,* plate 769) and Tupelo Gum (*Nyssa aquatica*) may coexist, or each may occur separately in pure stands. Pure tupelo stands, however, often become established following the clear-cutting of cypress-tupelo stands. Knee formations of cypress, reaching 6 feet or more, and buttress formations of cypress and Tupelo Gum are more pronounced in deep sloughs. Shrubs and herbs are sparse because of the flooded conditions and dense canopy. Herbs growing on floating logs and stumps are a distinct swamp microhabitat. Species characteristic of this microhabitat are Mad Dog Skullcap (*Scutellaria lateriflora*), Walter's St. John's-wort (*Hypericum walteri*), False Nettle (*Boehmeria cylindrica,* plate 809), and Clearweed (*Pilea pumila*).

The epiphytes Green-fly Orchid (*Epidendrum conopseum,* plate 788), Spanish Moss (*Tillandsia usneoides,* plate 912), and Resurrection Fern (*Pleopeltis michauxiana,* plate 915) grow on branches of trees. Vines such as Cross-vine (*Bignonia capreolata,* plate 818), Coral Greenbrier (*Smilax walteri,* plate 776), Supplejack (*Berchemia scandens,* plate 800), and Poison Ivy (*Toxicodendron radicans,* plate 878) exhibit pronounced growth, especially at the margin of the swamps. Ladies'-eardrops (*Brunnichia ovata,* plate 785), a rare semi-woody vine, can be found on the margin of the swamp forests near lakes or ponds. Often the swamps have lakes within them (remnants of old streams) where members of the freshwater aquatic and tidal freshwater marsh communities occur.

As the depth and duration of flooding decreases, mesic trees such as Red Maple, Water Ash, Swamp Gum, and Cottonwood form a subcanopy. As the wet soil becomes more exposed, shrubs become common, including Sweetspire (*Itea virginica,* plate 773), Swamp Dogwood (*Cornus foemina*), and Coastal Fetterbush (*Eubotrys racemosus,* plate 770). Southern Rein-orchid (*Platanthera flava,* plate 790) occurs on the edges of muddy sloughs. Where favorable conditions exist, Lizard's-tail (*Saururus cernuus,* plate 781), Southern Blue Flag Iris (*Iris virginica,* plate 779), Butterweed (*Packera glabella,* plate 821), and Goldenclub (*Orontium aquaticum,* plate 848) flourish, adding color to the swamp.

A great diversity of species occurs from swamp to swamp. Age of the forest, past timbering activities, degree of flooding, soil composition, and freedom from disturbance all contribute to the composition of today's swamps. Only Francis Beidler Forest in Four Holes Swamp harbors a significant stand of original growth forest in a blackwater system. Congaree National Park offers a view of original growth forest in a brownwater system.

Hardwood bottom forests

Hardwood bottom forests (plate 792) occur on floodplains that are somewhat elevated above the adjoining cypress-gum swamp forest community. Although flooded for a considerable period, the surface is dry through much of the year. Hardwood bottoms exhibit extreme floral diversity, and the floral composition is variable from one site to another. The vegetation of the hardwood bottoms is dense, and in the more undisturbed sites, trees grow over 3 feet in diameter. Small trees and shrubs are frequent and woody vines luxuriant. In the drier sites, a rich, herbaceous flora flourishes. In areas that have been logged, trees are smaller but still dense. Often hardwood bottom forests are narrow strips with adjacent uplands on one side and the swamp forests on the other. At other times, the forests may be a broad expanse.

Characteristic trees of this community include Sweet Gum, Loblolly Pine, Overcup Oak (*Quercus lyrata*), Water Oak (*Q. nigra*), Willow Oak (*Q. phellos*), Swamp Chestnut Oak (*Q. michauxii*), Laurel Oak (*Q. laurifolia*), Cherrybark Oak (*Q. pagoda*), ash (*Fraxinus* spp.), American Sycamore (*Platanus occidentalis*), American Holly, American Elm (*Ulmus americana*), and Sugarberry (*Celtis laevigata*), among others.

A subcanopy of young canopy species is present, including Ironwood. Among the numerous shrubs that characterize the community are Swamp Dogwood, Arrowwood (*Viburnum dentatum*), Piedmont Azalea (*Rhododendron canescens,* plate 795), Elderberry (*Sambucus canadensis*), and Possumhaw (*Ilex decidua,* plate 797). These species also may occur in the adjacent swamp forests.

Woody vines are especially prominent and include Cow-itch (*Campsis radicans,* plate 952), Poison Ivy, Supplejack, Climbing Hydrangea (*Hydrangea barbara,* plate 32), and Muscadine (*Muscadinia rotundifolia,* plate 357).

Grasses, rushes, sedges, and wildflowers form a rich herbaceous layer in drier sites. Grasses and sedges often form a dense ground layer and separate the hardwood bottoms from the adjacent swamp forests. Many wildflowers from the swamp forests also grace the hardwood bottoms, but several additional species may occur, including Common Atamasco-lily (*Zephyranthes atamasco,* plate 806), Carolina Least Trillium (*Trillium pusillum* var. *pusillum,* plate 804), Common Jack-in-the-pulpit (*Arisaema triphyllum,* plate 803), and Virginia Dayflower (*Commelina virginica,* plate 810).

An interesting microhabitat in the hardwood bottoms is the "windthrow" community. The bottomland trees have shallow but broad root systems. When these trees are blown down, the uplifted soil clings to the roots, which is now in a sunlit area because the fallen tree created a gap in the canopy. This allows for the establishment of numerous weedy herbs. Virtually every "windthrow" harbors Pokeweed (*Phytolacca americana*), its seeds carried by birds from the adjacent uplands.

Levee forests

Natural levees or "fronts" occur along the river edge. Levees are created as vegetation along the river's edge slow floodwaters and allow for the deposit of silt. These levees are slightly higher than the adjacent floodplain, which harbor the hardwood bottom and/or cypress-gum communities. The soil is very fertile, especially along brownwater rivers. Blackwater rivers have less-developed levees due to lower amounts of alluvium. Levee forests can be found throughout the Coastal Plain but also extend into the Piedmont along major river systems.

The levees are occupied by pioneer species, with American Sycamore, River Birch (*Betula nigra,* plate 814), Laurel Oak (*Quercus laurifolia*), Planer Tree (*Planera aquatica*), and willows (*Salix* spp.) being the major trees. Other trees include ash (*Fraxinus* spp.), Tulip-tree, Silver Maple (*Acer saccharinum*), American Holly, and Sweet Gum.

Three common shrubs are Pawpaw (*Asimina triloba,* plate 427), Wild All-spice (*Lindera benzoin,* plate 426), and Switch-cane (*Arundinaria gigantea,* plate 819). "Cane bottoms" or Switch Cane "canebrakes" in the Piedmont and Coastal Plain are associated with large levees. In the Coastal Plain, these extensive areas of cane are now gone. Isolated patches are still found in the Piedmont.

Herbs vary from site to site but generally include River Oats (*Chasmanthium latifolium,* plate 820), Butterweed, and False Nettle.

Where to visit the bottomland forest communities. Two easily visited sites represent not only superb examples of bottomland forests in South Carolina but probably the best examples of old growth systems in the world. Congaree National Park provides a glimpse into the primeval bottomland hardwood forest community. A 2.6-mile handicap accessible trail and boardwalk takes the visitor under some of the tallest trees in the southeastern United States as it winds its way through typical bottomland forest and into an example of a brownwater Bald Cypress-Tupelo Gum swamp forest community. Six national and 25 state champion trees grow here; these represent the largest examples known nationally and in South Carolina, respectively.

Francis Beidler Forest, a National Audubon Society Sanctuary near Harleyville in Dorchester County, is a 3,600-acre preserve in Four Holes Swamp and harbors the largest remaining stand of original growth blackwater Bald Cypress-Tupelo Gum swamp forest in the world. It is easily reached via SC-28 off US 178. A 1.5-mile boardwalk takes visitors through a portion of the ancient groves of Bald Cypress and Tupelo Gum that tower over clear pools and blackwater sloughs. Canoe trips are also offered by reservation. Green-fly Orchid,

Resurrection Fern and Spanish Moss drape the branches of the trees, and common herbs include Cardinal Flower, False Nettle, Butterweed, Swamp Milkweed (*Asclepias perennis*), and Lizard's-tail. Mad Dog Skullcap and Common Marsh St. John's-wort (*Hypericum virginicum*) may be seen growing on floating logs or stumps. The boardwalk also passes through characteristic examples of mature bottomland hardwood forest. The rare Carolina Least Trillium, which flowers in the early spring, also grows along the boardwalk, one of the last locations known in South Carolina.

Levee forests are well-developed along many of our free-flowing river systems. Well-developed levee forests along coastal rivers can be seen north of Echaw Road in the Francis Marion National Forest along the edges of the Santee River. Excellent examples of bottomland hardwood forests and brown-water Bald Cypress-Tupelo Gum swamp forests can also be seen at Guillard Lake Scenic Area in the National Forest. These can be accessed by Forest Service roads, like McConnells Landing Road, leading north to the river. Levee forests extend into the Piedmont, and a good example with an easy hiking trail is found at Landsford Canal State Park in Chester County.

THE FRESHWATER MARSH COMMUNITIES

Marshes are wetlands that are inhabited by herbaceous plants that have their roots in the substrate but with their photosynthetic and reproductive organs principally emergent. Freshwater marshes are more or less permanently flooded. The dominant species are grasses, rushes, and sedges along with numerous broadleaf flowering plants. Freshwater marshes occur along tidal rivers and inland along pond and lake margins, in beaver ponds, in canals and ditches, and in managed impoundments. Of the two types of freshwater marshes, the inland and tidal, the latter covers by far the greater area in the Coastal Plain. Although both inland and tidal freshwater marshes have essentially the same flora, the tidal marshes do have some species that do not occur farther inland. For example, Wild Rice (*Zizania aquatica*, plate 825), Southern Wild Rice (*Zizaniopsis miliacea*, plate 824), and Saw-grass (*Cladium jamaicense*, plate 827) generally tend to be absent from the inland marshes.

Tidal freshwater marshes

Tidal freshwater marshes (plate 823) are much more diverse ecologically and floristically than either salt marshes or brackish marshes. Indeed, they are among the most diverse wetland plant communities in the continental United States. In the freshwater tidal marshes along the Cooper River in Berkeley County, more than 100 species of vascular plants have been identified. A similar diversity occurs in the marshes of the other rivers in the coastal area. The floristic composition varies from site to site within a river as well as between rivers. Zonation may exist within a site, but it is not repeated consistently from site to site.

In the pre-settlement rivers of the Coastal Plain, tidal freshwater marshes occurred as fringes along the rivers where tidal freshwater swamps bordered the rivers. Today, however, the majority of tidal freshwater marshes occur in abandoned rice fields (figure 5). The rice fields, in turn, were originally tidal cypress-gum freshwater swamps that occurred along every tidal river along the Carolina coast. Where there was a swamp that had at least a three-foot difference in tidal amplitude, the swamp was ultimately converted into rice fields. When rice growing ended in the early 1900s, the fields followed two fates: (1) either the banks around the fields were maintained and water control structures were used to select a water regime that encouraged the growth of plant species that attracted waterfowl, which were usually hunted, or (2) the fields were abandoned, allowing nature to take its course. These abandoned rice fields now support the greatest acreage of tidal freshwater marshes.

The species that characterize this marsh community are those with their leaf-bearing stems or leaves extended above the water. They include various rushes, sedges, cat-tails, and broadleaf flowering species. These flowering species, most of which do not occur in brackish or salt marshes, make this one of the greatest wildflower communities. Although the flowering species do not dominate the system, they are sufficiently common to add a distinctive beauty and color to the marsh-scape. The best way to view this community is by boat.

Some of the more conspicuous wildflowers of the tidal marshes are Cardinal Flower, Coastal Carolina Spiderlily (*Hymenocallis crassifolia*, plate 835), Marsh Eryngo (*Eryngium aquaticum*, plate 841), Swamp Rose, Groundnut (*Apios americana*, plate 838), Water Hemlock (*Cicuta maculata*, plate 834), Swamp Rose-mallow (*Hibiscus moscheutos* ssp. *moscheutos*, plate 839), Seashore Mallow (*Kosteletzkya pentacarpos*, plate 842), Pickerelweed, Water-spider Orchid (*Habenaria repens*, plate 831), and Halberd-leaf Tearthumb (*Persicaria arifolia*, plate 844).

Inland freshwater marshes

Inland freshwater marshes occur in a variety of natural and human-made habitats: ditches and canals, lake and pond margins, beaver ponds, and managed impoundments. These marshes contain a wide mix of wetland species. Many of the species of the tidal marshes occur in the inland marshes. Some species that appear to be more common in these inland marshes include showy plants like Carolina Water-hyssop (*Bacopa caroliniana*, plate 851), Golden Canna (*Canna flaccida*, plate 852), Creeping Burhead (*Echinodorus cordifolius*, plate 854), Long Beach Primrose-willow (*Ludwigia brevipes*, plate 855), Winged Monkey-flower (*Mimulus alatus*, plate 856), and Grass-leaf Arrowhead (*Sagittaria graminea* var. *graminea*).

Where to visit the freshwater marsh communities. Tidal freshwater marshes occur along all the major river and creek systems in South Carolina where tidal amplitude allowed the cultivation of rice in the past. Large expanses may be seen in the numerous preserved lands in the ACE Basin. Large areas of tidal freshwater marshes are also found along the Waccamaw and Pee Dee Rivers in Georgetown County. The Samworth Wildlife Management Area located off SC-52 provides a boat ramp for access to the Great Pee Dee River and various creeks with expansive tidal freshwater marshes with most, if not all, of the species found in this habitat.

Tidal freshwater marshes are best observed by boat, kayak, or canoe, and Wambaw Creek in Berkeley County in the Francis Marion National Forest is a great place to see it. One of the best examples of tidal freshwater marsh can be seen by taking a boat trip from Still Landing (at the end of FS 211-B) to the boat ramp on FS 204. Wambaw Creek and the adjoining swamp is a nationally designated Wilderness Area (Wambaw Creek Wilderness Area). Coastal Plain Spiderlily is particularly abundant along the banks of the creek. The upper reaches of the Ashepoo and Combahee Rivers, the Waccamaw River, and both branches of the Cooper River in Berkeley County harbor extensive acreage of tidal freshwater marshes. Best viewing for these rivers requires a boat, but many public boat landings offer easy access.

Good examples of the inland marsh community are observed easily along the shore of Lake Moultrie. Another excellent example of a former rice field that was created by damming a creek is found at Washo Reserve in the Santee Coastal Reserve near South Santee in Charleston County. The habitat may be easily observed from a vehicle or from parking areas along Wildlife Drive in Savannah National Wildlife Refuge in Jasper County and trails through the ACE Basin National Wildlife Refuge.

THE OPEN WATER COMMUNITY

The term "open water community" is used to describe those plant communities forming in the channels of rivers, streams, impoundments, and natural lakes where emergent vegetation is not dominant. The plants that form this community are termed "hydrophytes." Hydrophytes include aquatic plants that normally grow in water or inhabit soils that contain more water than is optimal for the average plant. One of the outstanding structural features shared by most hydrophytes is aerenchyma tissue. This tissue is formed by the disintegration of groups of tissue cells or the separation of cells, which creates enlarged, intercellular cavities (lacunae) that become filled with gases. These air-filled cavities allow hydrophytes to float.

One group of hydrophytes is the freshwater floating aquatics, the true aquatic plants. They consist of three types: (1) the submerged, anchored; (2) the floating-leaf, anchored; and (3) the floating (no soil contact). These aquatics are best developed in lakes, ponds, freshwater sounds, canals, abandoned rice reservoirs, and sluggish streams. They generally occur in the Sandhills, Coastal Plain, and maritime strand but can be found in impoundments throughout the state. Today, much of the habitat for the aquatics is human created. Reservoirs, such as Lake Moultrie and Lake Murray, harbor along their shores a rich, aquatic growth. Inland swamps that were dammed to create water reservoirs for growing rice now support aquatic populations, and many multipurpose canals that were dug for various reasons provide habitat for the aquatics. As was previously mentioned, the abandoned tidal rice fields support aquatics on the fringe of the marsh vegetation. The aquatics are also found in deepwater pockets in inland swamp forests where dams maintain the water level year-round and in ox-bow lakes that form along the major rivers.

A number of submerged, anchored aquatics occur in waterways throughout the state. Often their flowers project above the water's surface. Two species with conspicuous flowers are Carolina Fanwort (*Cabomba caroliniana*) and Brazilian Waterweed (*Egeria densa*). The floating aquatics are represented by Water Hyacinth (*Eichhornia crassipes,* plate 866) and four genera of the duckweed family (Lemnaceae): duckmeats (*Spirodela*), duckweeds (*Lemna*), water-meals (*Wolffia*), and bog-mats (*Wolffiella*). The floating Mosquito Fern (*Azolla caroliniana,* plate 861) is also common.

The most spectacular of the aquatics are the floating-leaf, anchored species, with their showy flowers borne above or on the water's surface. Most prominent of these is Fragrant Water-lily (*Nymphaea odorata,* plate 869). Other species are Broadleaf Pond-lily (*Nuphar advena,* plate 862), American Frog's-bit (*Limnobium spongia,* plate 867), Water-shield (*Brasenia schreberi,* plate 865), Big floating-heart (*Nymphoides aquatica,* plate 863), and Floating Bladderwort (*Utricularia inflata,* plate 864).

Where to view the open water community. Bonneau Ferry Wildlife Management Area in Berkeley County features two former inland rice reservoirs, the Upper Reserve and the Lower Reserve, which harbor the open water community.

Coastal Plain: The Maritime Strand

THE MARITIME COMMUNITIES

Coastal beaches

Along the South Carolina coast, beaches (plate 872) form where ocean currents and waves deposit sand that was picked up by the waters from offshore coastal sites or brought down the rivers from inland areas. No vascular plants grow along the beach below the high-tide line because it is a dynamic zone, constantly changing because of the action of the wind and tides. Above the high-tide zone, a zone of detritus (driftline) develops, which is deposited by the tides. The dominant component of the detritus is the remains of Smooth Cordgrass from the nearby salt marshes, which washed down the tidal creeks. Seeds of two hardy species, Southern Sea Rocket (*Cakile harperi,* plate 874) and Carolina saltwort (*Salsola kali* var. *caroliniana,* plate 875), are generally the first species to become established in this harsh environment. A rare beach plant is the federally threatened Seabeach Amaranth (*Amaranthus pumilus,* plate 876), which occurs from upper Charleston County to North Carolina and also on Long Island in New York.

Coastal dunes and maritime grasslands

Landward from the driftline is the berm, a zone of fairly level, loose sand that is subject to heavy salt spray and is located above all but the highest spring tides. The coastal dunes form here (plate 877). Dunes are mounds of unconsolidated sand formed in the berm area by winds blowing across the beach. Whenever a plant (or object) reduces the force of the wind, wind-borne sand is deposited and ultimately forms a dune. The plant most associated with building coastal dunes in the southeastern United States is Sea Oats (*Uniola paniculata,* plate 880). Young seedlings act as windbreaks, slowing the wind. As the wind slows, the sand it carries accumulates behind and against the leaves. As the sand is added, the plant grows, keeping its leaves above the rising sand. In a few years a dune builds many feet high.

Coastal dunes front the barrier beaches and barrier islands of the Carolina coast. They are interrupted where inlets and sounds allow the ocean to surge through, with the ocean's energy being dissipated inward to the tidal flats and marshes. Dunes are also the first line of defense against oceanic forces, especially hurricanes and winter storms. Just as winds can build dunes, they can destroy what they have created. Where the vegetation is killed, the wind erodes dunes. Preservation-minded coastal residents have long fought to protect dunes from excess disturbance, such as trails made by people or dune buggies. The South Carolina law against disturbing Sea Oats on public property has done much to protect coastal dunes.

Other grasses that help build the dunes are Bitter Seaside Panic-grass (*Panicum amarum* var. *amarum,* plate 882) and Dune Sandspur (*Cenchrus tribuloides,* plate 881). Once the dunes become fairly stable, various forbs become established to help further stabilize the dunes, including Beach Pea (*Strophostyles helvola,* plate 894), Horseweed (*Erigeron canadensis*), Beach Evening-primrose (*Oenothera drummondii,* plate 886), Dune Evening-primrose (*Oenothera humifusa*), Camphorweed (*Heterotheca subaxillaris*), Dune Pennywort (*Hydrocotyle bonariensis,* plate 885), and several species of cacti including Southern Pricklypear (*Opuntia mesacantha* ssp. *lata,* plate 892) and Dune Devil-joint (*O. drummondii,* plate 893).

Swales develop between the dunes. Swales are low-lying areas that are protected from the salt-laden winds and where a fresh to brackish system may develop because rainwater collects and floats atop the heavier saltwater. In this microhabitat, Annual Sea-pink (*Sabatia stellaris,* plate 896) is abundant.

In some areas, such as at Cedar Island in Georgetown County and on Seabrook Island in Charleston County, a more expansive and slightly less dynamic rolling to flat area of dunes has developed. These areas contain most of the typical dune vegetation but have a more continuous cover of grass species and are probably best described as maritime grasslands (plate 877).

The dune communities may be replaced by maritime shrub thickets on barrier islands or on barrier beaches where accretion is occurring and the dunes are building seaward, which protects the inner dunes from the salt-laden winds, or where the shoreline has been stabilized for years. Maritime forests may in turn replace the maritime shrub thickets.

Where to see the coastal dune and maritime grassland communities. A good place to view these habitats in close proximity to Charleston is at the Accreted Beach at Sullivan's Island. The beach and dunes are owned by the town of Sullivan's Island, and the Lowcountry Open Land Trust holds a protective easement. It is open to the public and reached from either Station 18 or Station 16. Limited parking is located at the end of each road. Trails lead directly to the beach. Sea Rocket and Carolina Saltwort are found in the beach community and the dunes harbor Dune Devil-joint, Beach Evening Primrose, Beach Blanket-flower (*Gaillardia pulchella* var. *drummondii*, plate 884), Silver-leaf Croton (*Croton punctatus*, plate 891), Large Sea Purslane (*Sesuvium portulacastrum*, plate 895), Northern Seaside Spurge (*Euphorbia polygonifolia*, plate 890), Beach Pea, Sea Oats, Dune Sandspur, Bitter Seaside Panic-grass and Beach Morning-glory (*Ipomoea imperati*, plate 898). The habitats of the beach and dune communities are familiar to many who visit the beach for vacation and can be found throughout our coastline. Other great examples exist at several state parks, including Huntington Beach State Park in Georgetown County, Edisto Beach State Park in Colleton County, and Hunting Island State Park in Beaufort County. The impact of beach erosion (caused at least in part to sea level rise and alteration of near-shore currents by groins to protect developments) is particularly evident at Hunting Island State Park, where many acres have been lost in recent years.

Maritime forests

Maritime forests (plate 901) occupy the barrier islands and barrier shores of the coast. The characteristic species are a variety of salt-tolerant, evergreen trees and shrubs. There is little herbaceous plant cover except in exposed sites, due to natural breaks in the canopy, and in disturbed areas.

On the ocean side, maritime forests are shaped by the effects of salt spray. The wind blowing from the sea carries salt, depositing it on the windward branches and leaves. These leaves and branches die from the salt, while those on the leeward side, protected from the salt spray, continue growing. The result is a "shearing effect" of the trees and shrubs. Inland, away from the effects of salt spray, the trees assume a more typical appearance.

The characteristic evergreen trees of maritime forests are Live Oak (*Quercus virginiana*, plate 903), Southern Magnolia (*Magnolia grandiflora*, plate 905), Loblolly Pine, and Sand Laurel Oak (*Q. hemisphaerica*). The subcanopy includes the evergreen trees American Holly, Red Bay (*Persea borbonia*, plate 906), Hercules'-club (*Zanthoxylum clava-herculis*, plate 904), Southern Red Cedar (*Juniperus silicicola*, plate 902) and our only tree-sized palm, Cabbage Palmetto (*Sabal palmetto*, plate 907). Shrubs may be abundant, particularly in more open sites and include Wax Myrtle, Common Groundsel Tree (*Baccharis halimifolia*, plate 918), Yaupon (*Ilex vomitoria* Aiton, plate 908) and the semi-shrubby Coral Bean (*Erythrina herbacea*, plate 913). Herbaceous wildflowers are sparse because of the dense canopy, but some can be found in more open sites including the charming little Trailing Bluet (*Houstonia procumbens*, plate 911).

The maritime forests along the Atlantic Coast have been extensively timbered since colonial times. No original growth stands exist along the South Carolina coast, with the possible exception of the interior of St. Phillips Island in Beaufort County, where low swales prevented timbering. Islands such as Capers Island in Charleston County and Daufuskie Island in Beaufort County have had extensive agricultural fields in the interior, and the present forests are secondary growth.

Timbering began on a large scale in the maritime forests in the 1700s, mainly for Live Oak to build wooden sailing vessels. After the War of 1812, "Live Oak mania" began. Expeditions were sent from the northern shipyards to the Atlantic and Gulf coasts to harvest Live Oak. After the invention of iron and steel ships, Live Oak was given a reprieve. Timber companies then began to harvest the pines of the maritime forests. Maritime forests of the Lowcountry today are secondary forests. The existing large Live Oaks are either ones that were left, for whatever reasons, or that came from seedlings. Live Oak is a fast-growing tree on good sites and can reach an impressive size in 50 years. The large Live Oaks that line plantation avenues were planted starting in the 1700s (note their large size today).

Where to visit the maritime forest community. The South Carolina coast from North Island in Georgetown County to Capers Island in Charleston County harbors a string of protected (state and federal) barrier islands that support mature maritime forests. These islands include North Island, South Island, Cedar Island, Murphy Island, Bull Island, and Capers Island. All have to be reached by boat, but the reward is worth the effort. Directions and access are readily obtained from the internet. In addition, Hunting Island State Park in Beaufort County, Edisto Beach State Park in Colleton County, and Huntington Beach State Park in Georgetown provide access by roads. All three harbor at least some acreage of maritime forest and the beach communities.

Salt marshes

Salt marshes (plate 916) occur on regularly flooded substrates along tidal inlets and behind barrier islands and spits. This community is species-poor, often supporting pure stands of Smooth Cordgrass (*Spartina alterniflora*, plate 919); however, it is one of the most productive habitats in the world. It is the dominant wetland in the coastal zone of South Carolina, comprising approximately 150,000 acres. As cordgrass dies, it is washed into the inlets and the ocean, where it forms the basis of the estuarine food chain.

Three forms of Smooth Cordgrass occur: tall, medium, and short. The tall *Spartina* grows next to the tidal creeks in relatively deep water, where it receives an energy subsidy from the tidewater. Away from the creek, the tall cordgrass grades into the medium. The medium then grades into the short marsh. Short forms of Smooth Cordgrass occur at the highest elevation where it is flooded daily, but only to a depth of a few inches to one foot.

In the short zone, termed a "high marsh," two species that also occur in the adjacent salt shrub thickets intermix with the short *Spartina*: Carolina Sea Lavender (*Limonium carolinianum*, plate 923) and Saltmarsh Aster (*Symphyotrichum tenuifolium*, plate 922), adding a touch of color to the otherwise drab salt marshes.

Salt flats

Salt flats (plate 925) are formed where tidal waters drain incompletely, and the soil becomes very salty. As the water evaporates, it leaves behind the salt, which often forms a white crust on the soil. Even the most salt-tolerant species cannot survive in these hypersaline areas, and the center of flats is often barren. However, as salinity decreases toward the margins, a variety of fleshy halophytes (salt-tolerant), grasses, and other herbs appear. Closest to the center are Perennial Glasswort (*Salicornia ambigua,* plate 926) and Saltwort (*Batis maritima,* plate 927), which are obligate halophytes that tolerate high salinity. Salt flats grade into either salt marshes or salt shrub thickets. Intermixed with Saltwort and Perennial Glasswort are diminutive forms of species associated with the salt marshes and salt shrub thickets: Sea Ox-eye (*Borrichia frutescens,* plate 917), Southern Maritime Marsh-elder (*Iva frutescens*), Smooth Cordgrass, Sea Lavender, Saltmarsh Aster, and Southern Seaside Goldenrod (*Solidago mexicana,* plate 924).

Where to visit the salt marsh and salt flat communities. Most highways that traverse the maritime strand and lead to the coast pass through extensive salt marches. Salt flat communities are especially visible along US 21 between US 17 and the city of Beaufort and along SC-174 to Edisto Island. One can exit vehicles and easily access the salt flats. Extensive salts flats abound in and around Mt. Pleasant Memorial Waterfront Park, easily accessible from the park.

Maritime shell forests

Natural and Native American shell deposits are scattered along the coast in salt marshes and on the tips of landmasses within estuaries. The natural shell deposits are not unique floristically. They generally support species of the adjacent maritime communities. The Native American deposits, however, are floristically unique and have only recently been studied along the Carolina coast. The key environmental parameter determining the floristic composition of the Native American shell deposits is the presence of calcium from the shells. Several calcium-loving species, called calcicoles, occur and are mixed with species of the adjacent maritime forests and salt shrub thickets.

Native American shell deposits are of three types: shell rings, shell mounds, and shell hummocks, which occur on marsh islands or on the mainland. Figure 7 shows a shell ring with an associated shell mound. The shell ring is a circular deposit of shells that rises several feet above the marsh. The shell mound is a mass of shells that are raised considerably higher than the shell ring (from 10 to 30 feet high). Shell hummocks are shells that were deposited in piles or in sinuous formations, probably as part of a campsite, on the mainland or on marsh islands. Considerable debate exists about the origin of these Native American shell deposits, but most of the extensive sites are known to be quite old, ranging from approximately 2,000 to 5,000 years before present.

Shell rings are either isolated in the marsh or attached to a landmass. The isolated shell rings harbor few calcicoles, with only Tough Bumelia (*Sideroxylon tenax,* plate 930) and Small-flowered Buckthorn (*Sageretia minutiflora*)

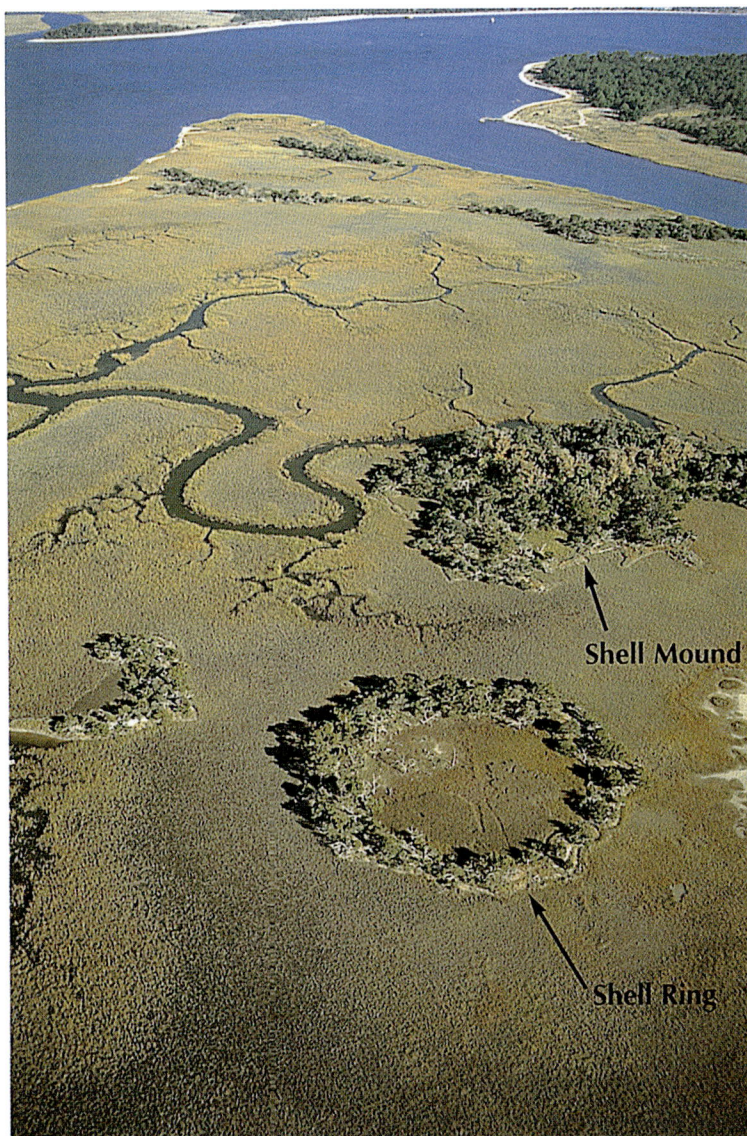

Figure 7. Native American Shell Ring and Shell Mound.

being common. The shell rings attached to an adjacent landmass, however, harbor many of the rare calcicoles that are found on the high shell mounds and shell hummocks.

The shell mounds and shell hummocks harbor a maritime shell forest (plate 928). This is a rare forest community along the coast. Maritime shell forests harbor a variety of trees, shrubs, and herbs that thrive in high-calcium soils. Trees and shrubs include basswoods (*Tilia* spp.), Red Buckeye, Sugarberry (*Celtis laevigata*), Carolina Buckthorn, Small-flowered Buckthorn, Eastern Roughleaf Dogwood (*Swida asperifolia,* plate 747), Southern Sugar Maple (*Acer floridanum,* plate 743), and Godfrey's Forestiera (*Forestiera godfreyi,* plate 929). Rare herbs include Mottled Trillium, Mellichamp's Skullcap (*Scutellaria mellichampii,* plate 934), Leafless Swallowwort (*Orthosia scoparia*),

Shell-midden Morning-glory (*Ipomoea macrorhiza,* plate 933), and Crested Coral-root (*Hexalectris spicata,* plate 759).

Sites with Southern Sugar Maple are the most spectacular. In the fall, their leaves turn a distinctive orangish-red, and the community can be spotted from the air. Maritime shell forests share many species in common with the inland calcareous forests since both are influenced by calcium in the soil.

Where to visit the maritime shell forest community. The most spectacular shell rings and hummocks are on private land and are not accessible except with permission. Pig Island Complex on Botany Bay Plantation (figure 7) on Edisto Island, composed of a shell midden, shell ring, and shell mound, is owned by Department of Natural Resources but is off limits to the public without a DNR guide. An excellent and easy to access maritime shell forest is found in Charleston County at the Sewee Shell Rings. The trailhead is reached by taking FR-243, near Awendaw. A parking area is clearly marked, and the trail winds through maritime forest that was badly damaged during Hurricane Hugo and emerges on the edge of the salt marsh where it continues to a shell-ring composed mostly of oyster shell and another ring composed mostly of clam shells. Many of the typical species of this habitat are visible here, including the Small-flowered Buckthorn, Appalachian Basswood, Tough Bumelia, Eastern Roughleaf Dogwood, and Carolina Buckthorn. Herbs present at the site include Woodland Pinkroot (*Spigelia marilandica,* plate 299), an indication of the connections between the high calcium habitats of the maritime shell forests and the high calcium habitats far inland. On Hilton Head Island, Green's Shell Enclosure Heritage Preserve off Squire Pope Road is home to a mix of calcicoles, including Carolina Buckthorn (*Frangula caroliniana,* plate 931).

Part 2

Species Descriptions and Color Plates

The Mountains

1. Granitic dome

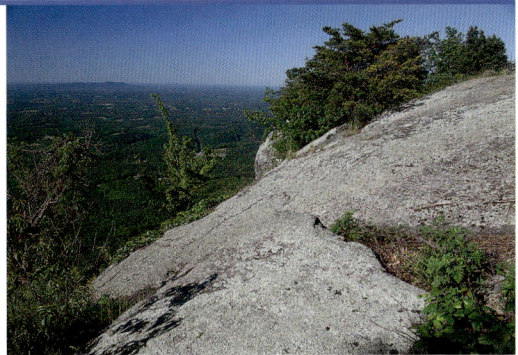

2. Shallow soil glade on mafic rock outcrop

3. Table Mountain Pine; Bur Pine

   Pinus pungens Lambert
   Pì-nus pún-gens
   Pinaceae (Pine Family)

DESCRIPTION: Small, broad-crowned pine, typically less than 30′ tall, but rarely to 70′+; needles in bundles of 2, stout, usually less than 2.5″ long, twisted; cones sessile, ovate, 2–3.5″ long, with stout, sharp upward-curved prickles; cones mature September–October; most cones are closed upon maturation and serotinous (opening only with heat from fires).

RANGE-HABITAT: Endemic to the central and southern Appalachian region; in SC, common in the mountains and rare in the upper Piedmont; shallow, acidic soils of the margins of granitic domes, pine-oak heath and other exposed ridgetop communities.

COMMENTS: Table Mountain Pine is dependent on fire for regeneration and persistence. The cones may remain closed on the tree for years. Some large trees will still

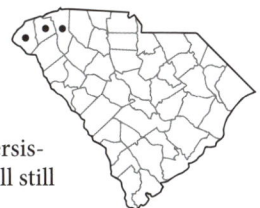

have cones attached toward the base of the tree that were produced decades ago. The heat of fire opens the cones and releases seed. Buzzards Roost Heritage Preserve and some areas in the Sumter National Forest have recently introduced management with fire to help conserve this unique pine.

4. Carolina Hemlock

Tsuga caroliniana Engelmann
Tsù-ga ca-ro-li-ni-à-na
Pinaceae (Pine Family)

DESCRIPTION: Evergreen tree to 90′ tall and 2′ in diameter; leaves flat, linear, 0.4–0.75″ long, smooth on the margins and notched at the tip, with 2 white lines of pores (stomata) below; leaves extending in all directions from the twig; cones about 1.25″ long; cones shed seeds August–October.

RANGE-HABITAT: Endemic to the southern Appalachian region, ranging from TN south to GA; in SC, infrequent in the mountains, rare in the upper Piedmont; open forests on ridge tops, rocky bluffs or gorge walls.

COMMENTS: Easily distinguished from Canadian Hemlock (*Tsuga canadensis* (L.) Carrière) by the leaf arrangement (a flat spray in Canadian Hemlock). Carolina Hemlock is prized as an ornamental. Today it is threatened by the onslaught of the Hemlock Woolly Adelgid (*Adelges tsugae*), although it appears to be slightly more resistant than Canadian Hemlock to this invasive insect pest.

CONSERVATION STATUS: Imperiled

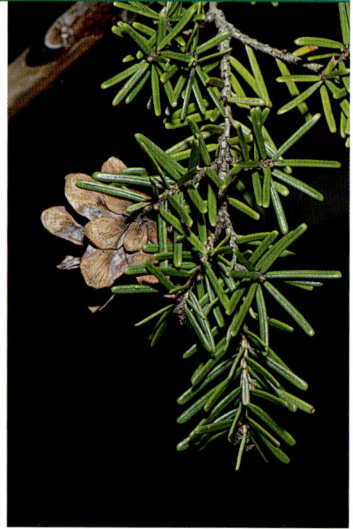

5. Hop-tree; Wafer-ash

Ptelea trifoliata L. var. *trifoliata*
Ptè-le-a tri-fo-li-à-ta
Rutaceae (Citrus Family)

DESCRIPTION: Deciduous shrub or small tree to 20′ tall; leaves with 3, stalkless leaflets, variable in size, usually less than 5″ long; flowers small, greenish-white, in terminal flat-topped clusters; fruits flat, broadly winged; flowers April–June; fruits mature June–August.

RANGE-HABITAT: New England south to FL and west to WI and TX; in SC, occasional throughout; found on rocky bluffs, stream terraces, granitic domes, and mafic outcrops. The species is most abundant in circumneutral soils.

TAXONOMY: Two varieties are recognized. The typical variety, pictured here, has sparse pubescence on the petiole and lower leaf surface and is widespread. Hairy Wafer-ash (*P. trifoliata* var. *mollis* Torrey & Gray) has petioles and lower leaf surfaces pubescent throughout and is found in shallow soils near mafic outcrops and in open, rocky, oak-hickory forests over mafic substrates in the mountains and Piedmont.

COMMENTS: The entire plant is aromatic with a citruslike odor; the flowers are foul-smelling. This species, like other members of the citrus family, is a host for the caterpillars of the Giant Swallowtail, our largest native butterfly. Hop-tree was once thought to be a good substitute for hops in making beer.

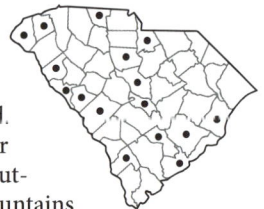

6. Smooth Wild Indigo

Amorpha glabra Desfontaines
    ex Poiret
    A-mór-pha glà-bra
    Fabaceae (Bean Family)

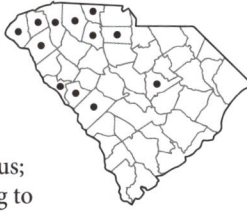

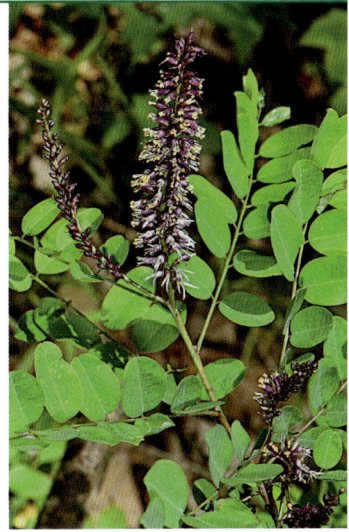

DESCRIPTION: Shrub to 6′ tall; stems glabrous; leaves pinnately compound; leaflets oblong to elliptic, 0.5–1.25″ long, glabrous or minutely pubescent on the veins below; flowers in racemes; petals royal purple; fruit a legume 0.25–0.50″ long and 0.10–0.20″ wide; flowers April–June; fruits mature June–October.

RANGE-HABITAT: Primarily found at low to moderate elevation in the Appalachian region from TN south to AL and GA; in SC, infrequent in the mountains and very rare in the Piedmont and Coastal Plain; margins of granitic domes and other ridgetop communities, particularly where soil is high in magnesium; in the Piedmont and Coastal Plain it is found in forests on rich bluffs along rivers.

SIMILAR SPECIES: Dark Indigo Bush (*Amorpha nitens* Boynton), which is rare in the inner Coastal Plain (Allendale County), is similar, but it has leaves shiny above and minutely puberulent below. It is found in dry, sandy, forest margins and bluffs along the Savannah River in the inner Coastal Plain.

COMMENTS: Smooth Wild Indigo is often confused for a sapling of Black Locust (*Robinia pseudoacacia* L.) but lacks prickles along the stems. The bright purple racemes are very attractive, and plants in cultivation are considerably showier than those seen in the wild.

7. Shrubby St.-John's-wort

Hypericum prolificum L.
    Hy-pé-ri-cum pro-lí-fi-cum
    Hypericaceae (St.-John's-wort
    Family)

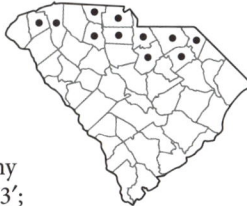

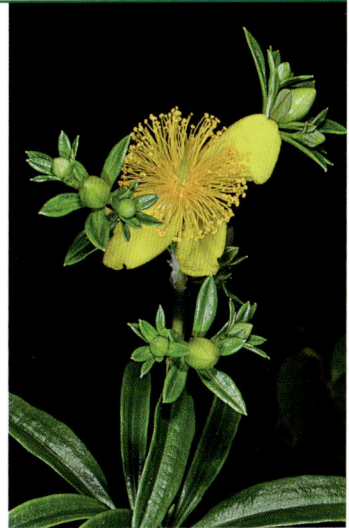

DESCRIPTION: Erect or reclining shrub, many branched, to 6′ tall, but typically less than 3′; leaves glossy, elliptic, oblong or linear, to 1.5″ long, 0.25–0.60″ wide; flowers large and showy, bright yellow, 1–7 per cluster at the tips of branches and/or upper leaf axils; flowers June–September.

RANGE-HABITAT: Widespread in eastern North America from NY west to MN and south to LA and GA; in SC, common in the mountains, rare in the Piedmont and Coastal Plain; granitic domes, stream banks, fens, and marshes, particularly in the mountains.

SIMILAR SPECIES: Dense-flowered St.-John's-wort (*H. densiflorum* Pursh, plate 69) is similar but typically more upright, more abundant in bogs, fens, and swampy forests and has more than 7 flowers per cluster.

COMMENTS: Large colonies of this species may be observed at Table Rock State Park. This species makes a good native ornamental shrub in partial shade to sun throughout SC. In cultivation, it is tolerant of many soil conditions.

8. Yellow Honeysuckle

Lonicera flava Sims
Lo-níc-er-a flà-va
Caprifoliaceae (Honeysuckle Family)

DESCRIPTION: Weak, trailing, or twining vine to 15′ long; stem smooth; leaves opposite; on flowering stems, the terminal 2 (3) pairs of leaves fused at the base; leaves whitish (glaucous) beneath; flowers in opposite terminal clusters of 3 flowers; petals yellow or yellow-orange, reddish with age; flowers April–May.

RANGE-HABITAT: KY west to MO and south to GA and AR; in SC, uncommon, in the mountains and Piedmont where restricted to granitic domes and shallow soils of amphibolite outcrops; restricted to higher pH substrates and often in glade or open woodland communities on the periphery of outcrops.

COMMENTS: This is one of the best indicator species of the calcareous variant of the granitic dome, as it is conspicuous and restricted to higher pH soils. It is common at Glassy Mountain Heritage Preserve in Pickens, SC.

CONSERVATION STATUS: SC-Imperiled

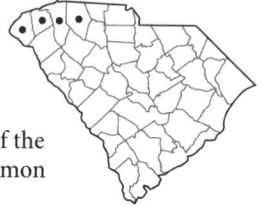

9. Wavy Hairgrass

Avenella flexuosa (L.) Drejer
A-vi-nél-la flex-u-ò-sa
Poaceae (Grass Family)

SYNONYM: *Deschampsia flexuosa* (L.) Trinius—RAB

DESCRIPTION: Tightly clumping grass; flowering culms 1–2′ tall; leaves to 1.5′ long, extremely narrow (capillary or hairlike); spikelets in open panicles, pale straw-colored at maturity; flowers April–May.

RANGE-HABITAT: Circumboreal (throughout the northern hemisphere), south to MN, OH, and GA; in SC, rare, restricted to granitic domes and cliffs in the Blue Ridge Escarpment and one location in the upper Piedmont.

COMMENTS: This graceful, narrow-leaved, clump-forming grass is typical of grassy balds at high elevations farther north. In our area, it forms graceful clumps along the margins of outcrops. Glassy Mountain Heritage Preserve in Pickens County has a large population that is easily observed from the trail.

CONSERVATION STATUS: SC-Critically Imperiled

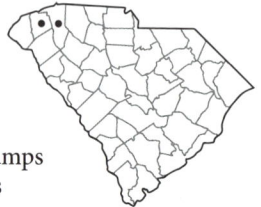

10. Appalachian Sandwort

Mononeuria glabra (Michaux) Dillenberger & Kadereit
Mo-no-neù-ri-a glà-bra
Caryophyllaceae (Pink Family)

SYNONYMS: *Arenaria groenlandica* Retzius var. *glabra* (Michaux) Fernald—RAB; *Minuartia glabra* (Michaux) Mattfeld—PR

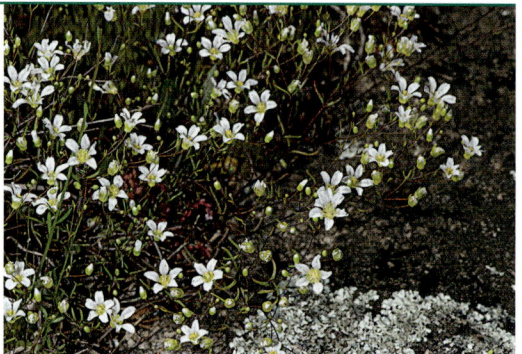

DESCRIPTION: Tufted annual; stems smooth, erect, 3–10″ tall; basal rosette absent at flowering; larger stem leaves 0.4–1.2″ long; 1–50 bright white flowers per plant; flowers March–May.

RANGE-HABITAT: ME and NH south to GA and AL; in SC, common in shallow soil mats and margins of glades on granitic domes, granitic flatrocks, and other outcrops in the mountains and Piedmont.

SIMILAR SPECIES: One-flower Sandwort (*M. uniflora* (Walter) Dillenberger & Kadereit, plate 253) is similar but is a smaller plant with tiny stem leaves (less than 0.2″ long) and smaller flowers. It is restricted to granitic outcrops in the Piedmont.

COMMENTS: During years of good rainfall, Appalachian Sandwort can make huge drifts of white across its favored habitats. It is easily observed at Forty Acre Rock Heritage Preserve in Lancaster County.

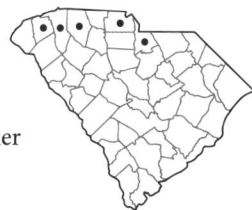

---

11. Shealy's Mountain Lettuce

Micranthes petiolaris
(Rafinesque) Small var. *shealyi*
McMillan & Cushman
Mic-rán-thes pe-ti-o-là-ris var.
shèa-ly-i
Saxifragaceae (Saxifrage Family)

DESCRIPTION: Diminutive fibrous-rooted annual; stems pubescent, 2–6″ tall; leaves pubescent, coarsely toothed, 0.3–2″ long and up to 0.8″ wide; inflorescence a panicle; flowers weakly zygomorphic to actinomorphic; fruit a capsule; flowers February–May.

RANGE-HABITAT: Known definitively from a single location in the upper Piedmont of SC on an extensive granitic dome and granitic flatrock complex at the edge of the Blue Ridge Escarpment in Pickens County; populations in NC are currently under investigation.

COMMENTS: This highly unusual diminutive annual is very similar to, and no doubt closely related to, the widespread perennial Cliff Saxifrage (*M. petiolaris* (Rafinesque) Small var. *petiolaris,* plate 18), but has adapted to the volatile climatic conditions found on the shallow soil of the low elevation granitic outcrops where it is found. This plant has shifted its flowering time to correspond to a narrower window of favorable moisture and climate. It completes its lifecycle by the time temperatures reach lethal levels (June). The plant was discovered by Patrick McMillan in 2002 and described by Laary Cushman and McMillan in 2020 (Cushman, Richards and McMillan, 2020). The specific epithet honors Harry E. Shealy Jr., botanist at the University of South Carolina, Aiken.

CONSERVATION STATUS: SC-Critically Imperiled

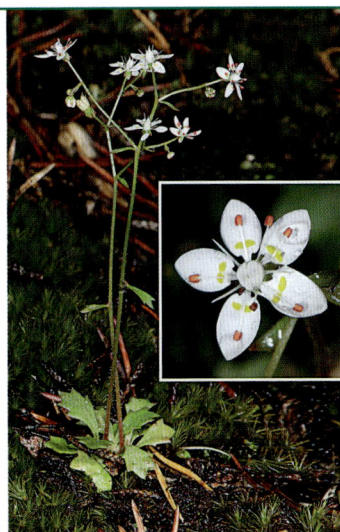

---

12. Rock Harlequin

Capnoides sempervirens (L.)
Borkhausen
Cap-noì-des sem-pér-vi-rens
Fumariaceae (Fumewort Family)

SYNONYM: *Corydalis sempervirens* (L.)
Persoon—RAB

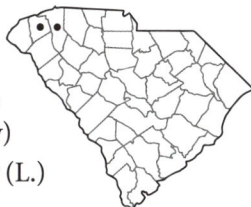

DESCRIPTION: Biennial fleshy herb to 2′ tall with bluish-white, highly divided leaves; inflorescence a raceme or panicle; flowers pink, with yellow tips and a prominent spur, unusually shaped (forming ½ of a heart); fruits are erect capsules; flowers April–June.

RANGE-HABITAT: Across the boreal regions of North America and extending south in the Appalachians to GA; in SC, rare, restricted to granitic domes and other rock outcrops in the mountains.

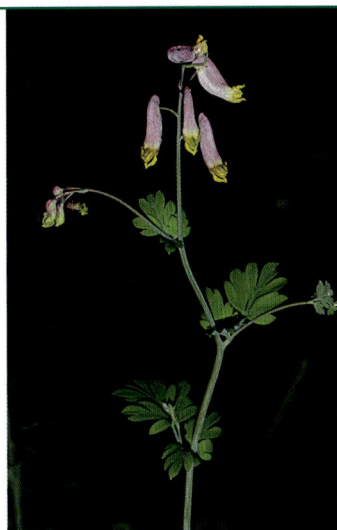

COMMENTS: Plants are similar to a small bleeding heart (*Dicentra*) but with only a half-heart-shaped flower. They may be absent during prolonged droughts as they require two seasons to reach flowering. It persists as dormant seeds in shallow soil on the periphery of outcrops, waiting for optimal conditions. Rock Harlequin may be observed along the trail to the summit of Table Rock in Pickens County.
CONSERVATION STATUS: SC-Critically Imperiled

13. Yarrowleaf Ragwort; Piedmont Ragwort

Packera millefolium (Torrey & Gray) Weber & Löve
Páck-er-a mil-le-fò-li-um
Asteraceae (Aster Family)

SYNONYM: *Senecio millefolium* Torrey & Gray—RAB

DESCRIPTION: Tufted perennial; stems 1–2.5′ tall, smooth or with long, white pubescence near the base when young; leaves in a basal cluster, bipinnately to tripinnately dissected with very narrow leaflets; flowers arranged into involucrate heads of yellow ray and disk flowers; 20 or more heads in a flat-topped arrangement; flowers April–June.
RANGE-HABITAT: Endemic to nw. SC, sw. NC, and ne. GA and disjunct to sw. VA; globally rare but locally common in the mountains and upper Piedmont along the margins of granitic domes, granitic flatrocks, and amphibolite outcrops; indicative of higher pH variants of granitic domes.
COMMENTS: This distinctive species is easy to recognize because of the finely divided leaves. It is very local, rare, and is threatened with genetic swamping through hybridization with the related, native, but weedy, Small's Ragwort (*Packera anonyma* (Wood) Weber & Löve, plate 994). Small's Ragwort has invaded formerly isolated outcrop habitats because of road construction and other disturbances that have allowed it to spread into this habitat along disturbance corridors. Hybrids now appear to outnumber genetically pure plants at Glassy Mountain Heritage Preserve in Pickens, but genetically pure populations still exist at more isolated locations. The genus *Packera* honors Canadian botanist John G. Packer (1929–2019).
CONSERVATION STATUS: SC-Imperiled

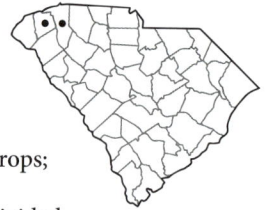

14. White Irisette; Isothermal Irisette

Sisyrinchium dichotomum E. P. Bicknell
Si-sy-rín-chi-um di-chó-to-mum
Iridaceae (Iris Family)

DESCRIPTION: Tufted perennial with dichotomously (forked) branching stems with 2–5 nodes; flowers white with 6 tepals, recurved (bent backwards) at maturity; flowers May–June.
RANGE-HABITAT: A southern Blue Ridge Escarpment endemic that is restricted to sw. NC and Greenville County, SC, where it grows in shallow soil over amphibolite outcrops and in open woodlands (glades).
COMMENTS: This distinctive species is related to the myriad of other "blue-eyed grass" (*Sisyrinchium*) species that are common in SC. It is easily distinguished by the combination of white, recurved tepals, branching habit and unique habitat. The species shares a similar distribution to other narrow endemics such as Yarrowleaf Ragwort and Biltmore Sedge.
CONSERVATION STATUS: Federally Endangered

15. Bastard Toadflax

Comandra umbellata (L.) Nuttall var.
umbellata
Com-mán-dra um-bel-là-ta var.
um-bel-là-ta
Santalaceae (Sandalwood Family)

DESCRIPTION: Perennial herb, forming large colonies via rhizomes; stems to 2′ tall; leaves alternate, elliptic, to 1.5″ long; flowers in alternate cymules (small flat-topped arrangements of 3–5 flowers); sepals whitish, tubular with 5 free lobes; petals absent; fruits are rounded drupes; flowers April–June; fruits mature in July.

RANGE-HABITAT: This variety ranges from ME west to MI and south to GA and AL; other varieties found throughout the northern hemisphere; in SC, uncommon, found throughout the state in shallow soil of open woodlands, grassy margins of granitic domes and other rock outcrops, forest margins along roadsides, pine-oak heaths, and Longleaf Pine sandhills; always in high light situations.

COMMENTS: This species, like all other members of the Sandalwood family, is parasitic, in this case hemi-parasitic; it is green and produces some of its own food and obtains some of its nutrients from a range of hosts, including woody plants, herbs, and grasses. The term "bastard" is used in traditional common names to indicate "false."

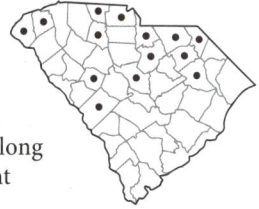

16. Small's Beard-tongue

Penstemon smallii Heller
Pen-stè-mon smáll-i-i
Plantaginaceae (Plantain Family)

DESCRIPTION: Perennial herb with an obvious minty scent; stems to 30″ tall; internodes of the middle stem without trichomes (hairs); flowers rose-purple, two-lipped, with an open, inflated throat and a heavily bearded (yellow-haired) sterile stamen (tongue); flowers in a terminal, paniclelike cluster; lower bracts leaflike; flowers May–June.

RANGE-HABITAT: Endemic to the southern Appalachian region, found in TN, NC, SC, AL, and GA; in SC, rare in the mountains and upper Piedmont where it occurs in woodland margins, rocky stream banks, granitic domes, and amphibolite outcrops.

SIMILAR SPECIES: All other *Penstemon* species in SC have smaller, paler pink or white flowers. Small's Beard-tongue makes an exceptional landscape plant in the Piedmont and mountains, where it thrives in rich, moist soil in partially shaded locations.

COMMENTS: This is the largest-flowered, most attractive beard-tongue in SC. The specific epithet honors John K. Small (1869–1938), distinguished American botanist.

CONSERVATION STATUS: SC-Vulnerable

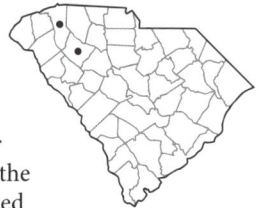

17. Erect Dayflower

Commelina erecta L.
   Com-me-lìn-a e-réc-ta
Commelinaceae (Dayflower
   Family)

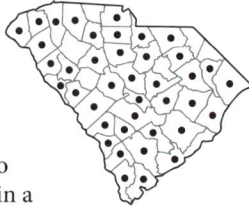

DESCRIPTION: Tufted perennial with erect to reclining stems 1–2′ tall; leaves sheathing in a manner similar to a grass, somewhat succulent, lanceolate to elliptic, 2–6″ long; flowers appear from greenish spathes with 2 bright blue lateral petals and a central whitish petal; flowers June–October.

RANGE-HABITAT: PA south to FL and west to KS and TX; in SC, common throughout on shallow soil of granitic domes; granitic flatrocks; other outcrops; open, dry woodlands; rocky stream sides; and sandy, dry forests.

SIMILAR SPECIES: The Common Dayflower (*Commelina communis* L.), which is introduced from Eurasia, is a common weed and waif in moist, nutrient-rich areas of disturbed soil such as compost piles. It has trailing, nontufted stems.

COMMENTS: A very narrow-leaved variety (*C. erecta* var. *angustifolia* (Michaux) Fernald) is found in Longleaf Pine sandhills and dry, sandy ridges in the Sandhills and Coastal Plain. The ranges of this variety and the typical variety are not distinguished in the range map, and they appear to intergrade in many parts of SC. Dayflowers receive their common name from the fact that each flower is open for only a few hours in the morning. Linnaeus named the genus, with his usual humor, for the brothers Commelin. The two larger petals represent the two brothers who became well-known botanists, and the smaller petal represents the third brother, who died without any botanical achievements. Dayflowers attract pollinators with nutrient-rich pollen produced on the sterile stamens (staminodes).

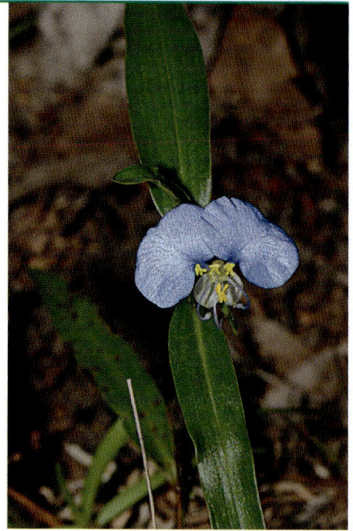

18. Cliff Saxifrage

Micranthes petiolaris
   (Rafinesque) Bush var.
   petiolaris
      Mic-rán-thes pe-ti-o-là-ris var.
      pe-ti-o-là-ris
Saxifragaceae (Saxifrage Family)

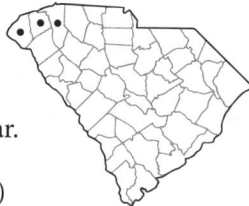

SYNONYM: *Saxifraga michauxii* Britton—RAB, PR

DESCRIPTION: Tufted, perennial herb with flowering stems 6–20″ tall; basal leaves oblanceolate, pubescent, with coarsely toothed margins; flowers in an open panicle; corolla zygomorphic; petals 5, white, the upper 3 each with 2 yellow spots; anthers orange; flowers May–August.

RANGE-HABITAT: Endemic to the southern Appalachian region; in SC, common in the mountains where it occurs in rock crevices and shallow soil on granitic domes and other outcrops, in dry to moist conditions.

COMMENTS: This species is unmistakable in flower because of the irregular symmetry of the petals and the three upper petals, each of which has two yellow spots.

19. Weakleaf Yucca

Yucca flaccida Haworth
Yúc-ca flac-cì-da
Agavaceae (Agave Family)

SYNONYM: *Yucca filamentosa* L. var.
smalliana (Fernald) Ahles—RAB

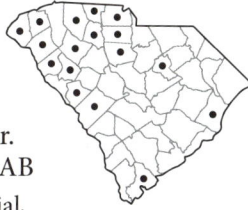

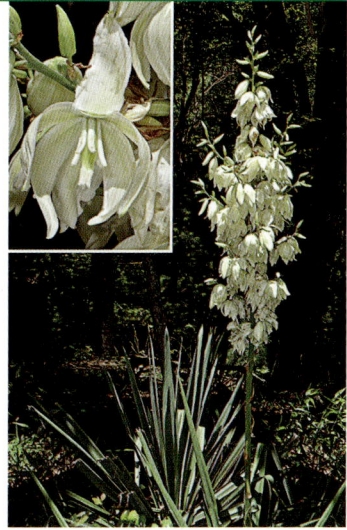

DESCRIPTION: Evergreen, clumping perennial, sometimes with a short trunk; flowering stems to 8′ tall; leaves clustered at base, linear, to 2″ wide, pliable (easily bent without breaking), tapering to a long acuminate tip; margins of leaves fraying into "threads" less than 2″ long; flowers white, composed of 6 tepals, to 2″ long; fruit a capsule; flowering April–June; fruits mature September–October.

RANGE-HABITAT: NC south to FL and west to TN and AL; in SC, common in shallow soils along the margins of granitic domes, granitic flatrocks, and other outcrops in the mountains and upper Piedmont; also in various locations throughout the state where persistent from cultivation and sometimes appearing native.

COMMENTS: This is the common yucca of outcrops in the mountains and Piedmont of SC. The taxonomy of this group is controversial and unclear, with perhaps several undescribed, cryptic species. Plants around isolated rock outcrops in our area are undoubtedly native, but it is not uncommon to come upon this species, or the closely related Spoonleaf Yucca (*Y. filamentosa* L., plate 886), growing in strange colonies far inside moist forests. These unusual populations are the result of cultivation and often indicate the presence of a long-abandoned homesite. Yucca have been cultivated for centuries for their use as fiber, as well as their roots, which can be pounded to produce lather (soap). The flower buds are edible and tasty when fried.

20. Mountain Blazing Star

Liatris spicata (L.) Willdenow
var. *spicata*
Li-à-tris spi-cà-ta var. spi-cà-ta
Asteraceae (Aster Family)

DESCRIPTION: Perennial herb from a swollen, knotty base; stems smooth to sparsely pubescent, 3–6′ tall, unbranched, often with fibrous remains of old leaf stalks at the base; largest leaves up to 0.8″ wide; flowers produced in involucrate heads composed of disk flowers only, with fewer than 14 flowers per head; heads produced in a dense, spikelike arrangement; corolla lavender, smooth within; flowers late June–August.

RANGE-HABITAT: NY, MI, and MO, south to FL and LA; in SC, locally common in the mountains and rare in the Piedmont; moist soils around granitic domes and Southern Appalachian fens, and on roadsides and powerline clearings.

SIMILAR SPECIES: Grass-leaved Blazing Star (*Liatris pilosa* (Aiton) Willdenow, plate 409) is similar but has much thinner leaves, heads arranged in a much less congested, spikelike structure, and is much smaller.

TAXONOMY: A second variety of this species, Savanna Blazing Star (*L. spicata* var. *resinosa* (Nuttall) Gaiser, plate 659), occurs in moist Longleaf Pine savannas of the Coastal Plain. It is distinguished by having less densely packed heads and more copious glandular pubescence along the rachis. Savanna Blazing Star flowers later in the season (August–October). It may be worthy of elevation to distinct species status.

COMMENTS: Mountain Blazing Star is a popular and commonly cultivated species. It is attractive to a wide variety of pollinators, particularly butterflies. It can be easily seen growing along the roadsides through Keowee-Toxaway State Park area in Pickens County or along US-76, in the Sumter National Forest in Oconee County.

21. Creeping Aster

Eurybia surculosa (Michaux)
   G. L. Nesom
      Eu-rý-bi-a sur-cu-lò-sa
Asteraceae (Aster Family)

SYNONYM: *Aster surculosus*
   Michaux—RAB

DESCRIPTION: Perennial from short, creeping rhizomes, forming large mats; leaves less than 0.4″ wide, predominantly basal, smooth; sterile plants outnumbering flowering individuals; flowering stems 1–2′ tall, with 8–14 stem leaves; flowers in involucrate heads, produced in a somewhat flat-topped arrangement; heads 0.5–1.5″ wide, bearing 15–35 blue ray flowers and yellow disk flowers; heads in a somewhat flat-topped arrangement; flowers August–October.

RANGE-HABITAT: Endemic to the southern Appalachian region; in SC, occasional in the mountains and upper Piedmont along the margins of granitic domes; other outcrops on shallow soil and on dry road banks.

SIMILAR SPECIES: The similar Alexander's Rock Aster (*Eurybia avita* (Alexander) G. L. Nesom, plate 261) has significantly paler, light-blue to white ray flowers and much more narrow leaves (0.1–0.35″ wide). It is restricted to granitic flatrocks in the upper Piedmont.

COMMENTS: The large genus *Aster* was recently divided into numerous segregate genera based on their genetic relationships. The reason for this split is that the North American "asters" are not as closely related to the Old-World genus *Aster* as they are to other genera. Many of our "asters" turned out not to be closely related to one another. Creeping Aster is a very attractive species and well-suited to cultivation. Reports of this species from the Fall-line Sandhills and Coastal Plain are based on misidentifications of Slender Aster (*Eurybia compacta* G.L. Nesom), which has leaves wider than 0.6″ and generally has fewer than 14 ray flowers per head.

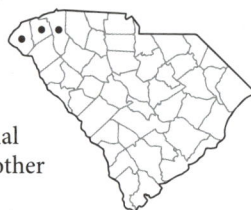

22. Southern Harebell

Campanula divaricata Michaux
      Cam-pá-nu-la di-va-ri-cà-ta
Campanulaceae (Bellflower Family)

DESCRIPTION: Tufted, branching perennial to 2′ tall; leaves lanceolate, with serrate margins, flowers light blue, tiny, 0.25–0.35″ long, bell-shaped (campanulate); flowers July–October.

RANGE-HABITAT: MD and KY south to AL and GA, generally in the Appalachian region; in SC, common in the mountains, uncommon in the Piedmont, and rare in the fall-line region of the Coastal Plain; margins of granitic domes, cliffs, boulders, thin-soil of rocky woodlands, and river bluffs.

COMMENTS: The cheerful light blue to white flowers appear to dance in the air on slender stalks.

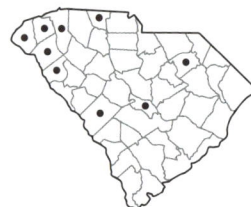

23. Porcher's Ragweed

Ambrosia porcheri
  McMillan & Prevost
  Am-brò-si-a por-chér-i
Asteraceae (Aster Family)

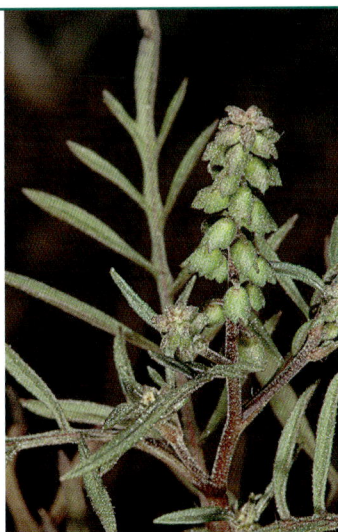

DESCRIPTION: Annual, single-stemmed herb to 2.5′ tall, usually much shorter; leaves heavily divided into very narrow leaf segments; leaves sticky to the touch; entire plant aromatic with the scent of jasmine; flowers inconspicuous, greenish, reduced to anthers and stigmas; flowers arranged in separate male and female involucrate heads; flowers August–October.

RANGE-HABITAT: Restricted to three known locations in Pickens County and one site in Greenville County, where locally common; granitic domes and granitic flatrocks.

COMMENTS: This very distinctive species is easily separated from the weedy Common Ragweed (*Ambrosia artemisiifolia* L., plate 1010) by its very narrow, sticky leaf divisions that have a distinct odor, reminiscent of jasmine or gardenia. The plant produces large amounts of the chemical beta-phellandrene, which may be responsible for the pleasant odor. The larger achenes, more narrow leaves, and short habit of Porcher's Ragweed are all adaptations to the extremely dry, hostile conditions of granitic outcrops. This unique species was first discovered by Patrick McMillan in 2002 and was the focus of a master's thesis by Luanna Prevost and named in honor of Richard Dwight Porcher Jr., botanist and the first author of the first edition of this book.

CONSERVATION STATUS: SC-Critically Imperiled

24. Twisted-hair Spikemoss (A)

Bryodesma tortipilum
  (A. Braún) J. Soják
  Bry-o-dés-ma tor-tí-pi-lum
SYNONYM: *Selaginella tortipila* A. Braún—RAB

Rock Spikemoss (B)

Bryodesma rupestre (L.) J. Soják
  Bry-o-dés-ma ru-pés-tre
Selaginellaceae (Spikemoss Family)

SYNONYM: *Selaginella rupestre* L.—RAB

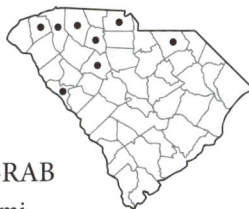

DESCRIPTION: Rock-dwelling, mosslike, primitive vascular plants, forming large, compact, gray-green mounds; *B. tortipilum* with stems 1–6″ tall; *B. rupestre* with stems 0.6–1.5″ tall; in both species, leaves linear, in spirals; apical bristle contorted (twisted) in *B. tortipilum,* and straight in *B. rupestre;* in both species, fertile leaves forming a 4-angled cone about 0.2″ long.

RANGE-HABITAT: *B. tortipilum* is endemic to the southern Appalachian region, found in TN, NC, SC, and GA; in SC, common on granitic domes, cliffs, and granitic flatrocks in the mountains and upper Piedmont; *B. rupestre* is widely distributed across northern North America, extending southward to AL; in SC, infrequent in the mountains and common in the Piedmont on granitic domes, granitic flatrocks, and sandstone outcrops in Longleaf Pine sandhills in Chesterfield County.

COMMENTS: Easily mistaken for a moss, the tiny true roots found scattered along the length of the stem distinguish it as a true vascular plant. Large colonies of these plants are highly susceptible to disturbance through trampling of the habitat, which has occurred on many sites that are accessible to the public. Though *B. rupestre* is a more northern-ranging species, it is more frequently encountered in the Piedmont and is nearly replaced in the mountains by the larger *B. tortipilum*.

25.   Wright's Cliff-brake

Pellaea wrightiana Hooker
Pel-laè-a wright-i-à-na
Pteridaceae (Maidenhair Fern Family)

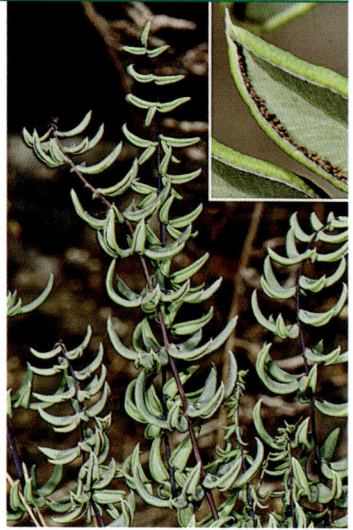

DESCRIPTION: Perennial, evergreen, rock-dwelling fern, forming large clumps on cliff faces; leaves (fronds) stiff, leathery, bluish-green, to 1′ long; leaf stalks purplish brown, smooth, with a single groove; leaf blades 2-pinnate, the upper leaflets arranged in 3s (ternate) almost to the tip of the leaf.

RANGE-HABITAT: This fern is typical of the desert Southwest, ranging from CO to OK south to AZ, TX, and Mexico; in the eastern US known only from Pickens County, SC and Alexander and Stanly Counties, NC; found on southwest and south-facing granitic cliff faces and granitic domes.

SIMILAR SPECIES: Arizona Cliff-brake (*Pellaea ternifolia* (Cavanilles) Link ssp. *arizonica* Windham, plate 26) is similar but with the upper portions of the leaf with single leaflets (not ternate). Purple Cliff-brake (*Pellaea atropurpurea* (L.) Link, plate 49) has stems that are slightly pubescent and not grooved and occurs only on high pH substrates.

COMMENTS: This fern was discovered in SC by Kay Wade and first positively documented in December 2017 (McMillan et al., 2018). The populations in NC are extremely small, but the SC site contains many thousands of large clumps. The SC location is home to two other species that are amazingly disjunct from the desert Southwest. The species is named in honor of Charles Wright (1811–85), who collected plants in an army expedition across Texas in 1849.

CONSERVATION STATUS: SC-Critically Imperiled

26.   Arizona Cliff-brake

Pellaea ternifolia (Cavanilles) Link ssp. *arizonica* Windham
Pel-laè-a ter-ni-fò-li-a ssp. a-ri-zò-ni-ca
Pteridaceae (Maidenhair Fern Family)

DESCRIPTION: Perennial, evergreen, rock-dwelling fern forming small clumps; leaves (fronds) stiff, to 1.5′ long, leathery, bluish-green; leaf stalks purplish brown, smooth with a single groove; leaf blades 2-pinnate on lower portion of the leaf blade, the upper leaflets simply pinnate.

RANGE-HABITAT: This fern is typical of the desert Southwest, ranging from TX to AZ and south into Mexico; disjunct more than 1,200 miles to Pickens County, SC, the only known location in the eastern US; boulders and crevices of south or southwest-facing granitic domes at the base of the escarpment.

COMMENTS: This fern was first discovered for SC by Steve Platt and John Townsend but was mistaken for the similar *Pellaea wrightiana* (Platt and Townsend, 1996). It was later correctly identified by Kerry Heafner (2001). This is likely the most endangered fern in SC. A recent survey by the authors confirmed only a single clump

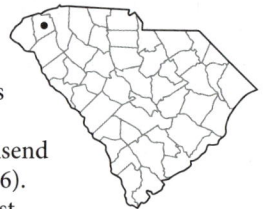

at its only known locality, which has been badly damaged by heavy equipment. It is included here in the hopes that citizen scientists may uncover additional locations for this remarkable and beautiful cliff-brake. Any additional locations should be reported to the South Carolina Heritage Trust Program.
CONSERVATION STATUS: SC-Critically Imperiled

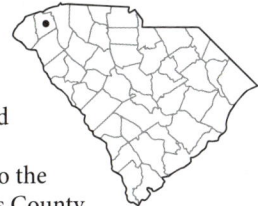

27. Wavy Cloak-fern

Astrolepis sinuata (Lagasca ex Swartz) D. M. Benham & Windham ssp. *sinuata*
As-tró-le-pis si-nu-à-ta ssp. si-nu-à-ta
Pteridaceae (Maidenhair Fern Family)

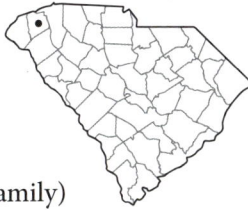

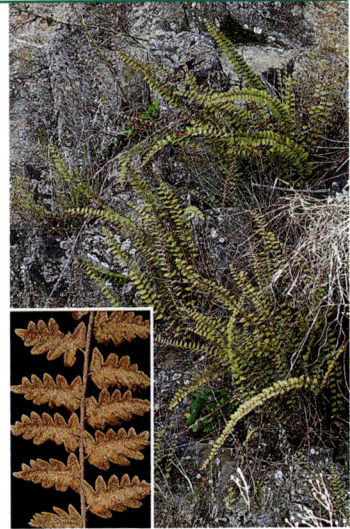

DESCRIPTION: Perennial, evergreen, rock-dwelling fern forming large clumps; leaves (fronds) stiff, to 2.5′ long, leathery, green above, golden-fuzzy below; leaf blades pinnate, the pinnae with wavy margins.

RANGE-HABITAT: This fern is typical of the desert Southwest, ranging from OK and TX west to AZ and south into Central and South America; in SC, found on extremely dry, southwest-facing granitic cliffs in Pickens County.

COMMENTS: A large population of several hundred clumps of this plant was first documented for SC in December 2017. The species has previously been reported from a single site in GA, where possibly introduced, and from one site where it grows on a bridge piling in Louisiana. The SC location includes three species of ferns typical of the desert Southwest growing together in large masses and is a remarkable example of the long-range dispersal capabilities of ferns via long-lived spores. The golden-rusty backs to the pinnae (leaflets) are distinctive. It cannot possibly be confused with any other SC species. The genus name *Astrolepis* refers to the stellate/star-shaped (Astro) scales (lepis).
CONSERVATION STATUS: SC-Critically Imperiled

28. Copper Fern

Bommeria hispida (Mettenius ex Kuhn) Underwood
Bom-mér-i-a hís-pi-da
Pteridaceae (Maidenhair Fern Family)

DESCRIPTION: Small, perennial, evergreen, rock-dwelling fern forming small colonies; leaves to 8″ long, dark green with evident trichomes (hairs); leaf blades pentagonal in outline.

RANGE-HABITAT: This fern is typical of the desert Southwest and ranges from TX to AZ and south into Mexico; it is disjunct over 1,200 miles to the Pickens County, SC location where it is found on south and southwest-facing granitic cliffs.

COMMENTS: This fern is remarkably disjunct to the SC location where it is found with Wright's Cliff-brake and Wavy Cloak-fern. It is extremely rare and easily overlooked. It is included here in the hopes that citizen scientists may uncover other locations in SC. Any additional locations should be reported to the South Carolina Heritage Trust Program. The discovery of this fern in Pickens County in 2017 marked the first known location for eastern North America (McMillan et al., 2018). The addition of the newly discovered populations of disjunct ferns in Pickens County makes this county the most biodiverse county in the United States for ferns and fern relatives. The genus is named in honor of Jean-Édouard Bommer (1829–95), a Belgian botanist who specialized in ferns.
CONSERVATION STATUS: SC-Critically Imperiled

29. Common Woodsia;
Bluntlobe Cliff Fern

Woodsia obtusa (Sprengel)
Torrey ssp. *obtusa*
Woòds-i-a ob-tù-sa ssp. ob-
tù-sa
Woodsiaceae (Woodsia Family)

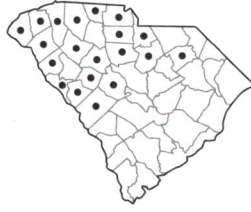

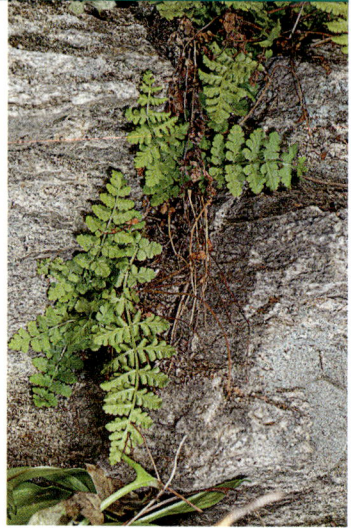

DESCRIPTION: Small, perennial, semi-evergreen, rock-dwelling fern
from short rhizomes, forming small clumps; leaves to 2′ long, usually smaller; petioles light brown to yellowish; blades pale green,
rather thin, twice pinnate; pinnae lobes blunt.

RANGE-HABITAT: Quebec to MN and south to FL and TX; in SC,
common in the mountains, infrequent in the Piedmont, and rare
in the Fall-line Sandhills; found on boulders, crevices in granitic
cliffs and domes, and sandstones in the Sandhills; sometimes
found in shallow soil adjacent to rock outcrops.

SIMILAR SPECIES: Lowland Bladder Fern (*Cystopteris protusa*
(Weatherby) Blasdell, plate 307) is similar but does not have persistent
leaf bases of old leaves clustered around the base. It occurs on shallow soil and outcrops of mafic or
calcareous rocks as well as in deep soil of rich cove forests and basic mesic forests. It typically forms
large colonies, in comparison to the solitary plants or small clumps of Common Woodsia.

COMMENTS: Common Woodsia is the most commonly encountered small "rock fern" in SC.
It is found at most extensive rock outcrops of igneous or metamorphic origin. The genus is
named in honor of Joseph Woods (1776–1864), English architect, botanist, and geologist.

30. Spray cliff

31. Humid gorge outcrops

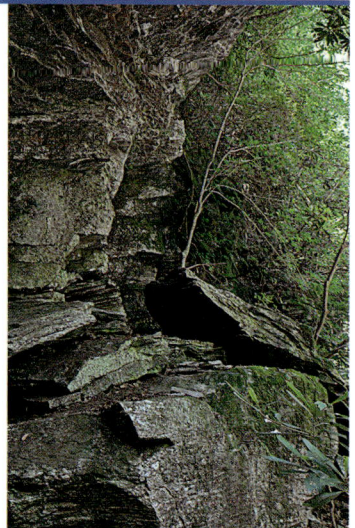

32. Climbing Hydrangea

Hydrangea barbara (L.) B. Schultz
    Hy-drán-ge-a bár-ba-ra
    Hydrangeaceae (Hydrangea Family)

SYNONYM: *Decumaria barbara* L.—RAB

DESCRIPTION: High-climbing woody vine (liana), capable of reaching the tops of the tallest trees; stems with adventitious aerial roots holding the stem firmly in place on trees or rock surfaces; leaves opposite, ovate, shiny dark green and smooth above, pubescent below; flowers white, in corymbs, without sterile florets; fruits produced in small, longitudinally ribbed capsules; flowers May–June.

RANGE-HABITAT: VA south to FL and west to TX; in SC, abundant throughout, in swamp forests, bottomland hardwood forests, floodplain forests, Piedmont seepage forests, humid gorges, cove forests, and stream banks.

COMMENTS: The stem grasps the host tree with fibrous adventitious roots, like the climbing habit of Poison Ivy (*Toxicodendron radicans* (L.) Kuntze, plate 878); it is sometimes mistaken and removed because of this similarity. It produces a juvenile growth form with very small leaves held tight against the growing surface and only produces the larger, adult leaves when it reaches sunlight at the margins of streams or the tops of trees or outcrops. Climbing Hydrangea makes a superb landscape plant that is underutilized in cultivation but suited to all regions of the state.

    Recent molecular evidence indicates that this species is better included within a broad concept of the genus *Hydrangea,* rather than in the traditional genus *Decumaria.* This is predominantly a species of moist forests in the Coastal Plain but is not uncommon in the mountains and Piedmont in appropriate habitats. It is extremely common in the Jocassee Gorges region.

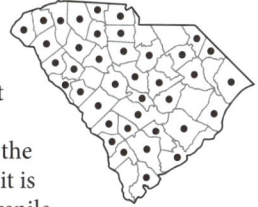

33. Thyme-leaf Bluet; Appalachian Bluet

Houstonia serpyllifolia Michaux
    Hous-tòn-i-a ser-pyl-li-fò-li-a
    Rubiaceae (Madder Family)

DESCRIPTION: Tiny, perennial herb with creeping, much-branched stems; leaves opposite, tiny; flowers deep blue with a yellow or white center (eye), terminal, on long stalks; flowers April–June.

RANGE-HABITAT: Endemic to the Appalachian region found from PA south to SC and GA; in SC, common in the mountains on spray cliffs, in cataract fens, rocky stream sides, seepages, moist woods, and moist road banks; rare in the Piedmont.

SIMILAR SPECIES: Common Bluet (*Houstonia caerulea* L., plate 201) has nearly identical flowers, but forms discrete, small, tightly clumping tufts on exposed road banks, lawns, and margins of outcrops in the mountains and Piedmont.

COMMENTS: Thyme-leaf Bluet is one of the most common and typical species of high elevation stream sides and seeps. The genus honors William Houston (1695–1733), Scottish surgeon and botanist.

34. Mountain Meadowrue; Lady-rue

Thalictrum clavatum
    A. P. de Candolle
    Tha-líc-trum cla-và-tum
    Ranunculaceae (Buttercup
    Family)

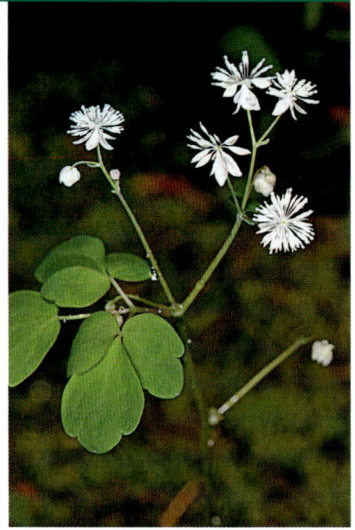

DESCRIPTION: Delicate herbaceous perennial, forming clumps; stems to 2′ tall; leaves biternately compound (compound twice, ending in 3s), segments similar to a maidenhair fern in appearance; flowers perfect, sepals small, white; petals absent; stamens showy, white; flowers April–July.
RANGE-HABITAT: Endemic to the southern Appalachian region found from WV and KY south to SC and GA; in SC, common on spray cliffs, seepages over rock, and small stream margins in cove forests; almost always growing on shallow mats over wet rocks.
SIMILAR SPECIES: Rue-anemone (*Thalictrum thalictroides* (L.) A. J. Eames & B. Boivin, plate 103), is similar but grows in deeper soil, flowers a month earlier, and has showy white sepals.
COMMENTS: This common species is one of the delights of the spray cliff community, gracing the dark rock with the white of the showy stamens in spring. The foliage persists through the growing season; it is like a fern in appearance and very attractive.

35. Branch Lettuce

Micranthes micranthidifolia
    (Haworth) Small
    Mic-rán-thes mic-ran-thi-di-
    fò-li-a
    Saxifragaceae (Saxifrage Family)

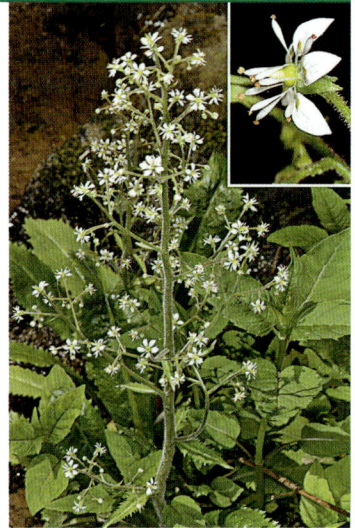

SYNONYM: *Saxifraga micranthidifolia*
    (Haworth) Steudel—RAB, PR

DESCRIPTION: Perennial herb with leaves clustered at the base; basal leaves lanceolate to oblanceolate, to 1′ long; flowering stem leafless, to 30″ tall; corolla, white, actinomorphic; flowers May–June.
RANGE-HABITAT: Endemic to the Appalachian region from PA and WV, south to SC and GA; in SC, common in the mountains on seepages, stream margins, and spray cliffs.
COMMENTS: The young basal leaves are traditionally consumed as a lettuce substitute in Appalachia, often with ramps and drizzled with hot bacon grease. It is a ubiquitous member of the spray cliff habitat. The only other large *Micranthes* in our area, Cliff Saxifrage (*M. petiolaris* (Rafinesque) Small, plate 18), has distinctively zygomorphic corollas.

36. Appalachian Sundrops

Oenothera tetragona Roth var. *fraseri*
    (Pursh) Munz
    Oe-no-thè-ra te-tra-gò-na var. frà-
    ser-i
    Onagraceae (Evening-primrose
    Family)

DESCRIPTION: Perennial herb to 3′ tall; stem pubescent to smooth, branched above; leaves alternate, more than 0.25″ wide; petals yellow, to

1.4″ long; capsule angled, widest at the middle, smooth or with mostly glandular hairs; flowers May–August.

RANGE-HABITAT: Endemic to the Appalachian region from NY and PA south to SC and GA; in SC, common in the mountains on seepages, margins of spray cliffs, and moist forest margins.

SIMILAR SPECIES: Another variety, *O. tetragona* Roth var. *tetragona,* has linear-lanceolate leaves and smaller flowers and is found in dry forests and forest margins throughout SC. A third variety, *O. tetragona* var. *brevistipata* (Pennell) Munz, has smaller flowers and fruits that are distinctively broadest at the tip and club-shaped (clavate). It is found in dry woodlands in the mountains and Piedmont. The taxonomy of SC sundrops is problematic and requires further study.

COMMENTS: The large yellow flowers and smooth leaves are distinctive. The variety is named in honor of Scottish botanist John Fraser (1750–1811). This species makes a good garden plant in moist soil and partial shade.

---

37. Orange Jewelweed; Orange Touch-me-not

Impatiens capensis Meerburgh
Im-pà-ti-ens ca-pén-sis
Balsaminaceae (Touch-me-not Family)

DESCRIPTION: Fleshy annual to 6′ tall with hollow stems; leaves alternate; one sepal forms a prominent sac that ends in a curled spur at the base of the flower; flowers May–frost.

RANGE-HABITAT: Newfoundland south to FL and west to British Columbia and TX; in SC, common throughout; moist forests, spray cliffs, cove forests, alluvial swamps, and tidal freshwater marshes.

SIMILAR SPECIES: Pale Touch-me-not (*Impatiens pallida* Nuttall) is very similar but has pale yellow flowers. It occurs on circumneutral soils in rich cove forests and is very rare in the SC mountains, although it is common and widespread northward.

COMMENTS: The first common name refers to the water-repelling nature of the leaves. On horizontal surfaces, water often rolls into beads, giving the appearance of jewels shining in the light. The second common name refers to the fruits; they are elastically coiled into five sections and, at maturity, explode when touched, expelling the seeds. The stem juice has fungicidal properties and has been used to treat athlete's foot. The stem juice is also a well-known treatment for Poison Ivy and Stinging Nettle, though it must be applied directly after contact with Poison Ivy to counteract the toxin. Duke (1997) recommends it as a remedy for hives.

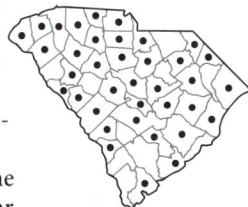

---

38. Rock Alumroot; Crag-jangle

Heuchera villosa Michaux var. *villosa*
Heù-cher-a vil-lò-sa var. vil-lò-sa
Saxifragaceae (Saxifrage Family)

DESCRIPTION: Densely clumping perennial herb with flowering stems to 30″ tall and far exceeding the basal leaves; leaves to 7″ long with a sharp acute tip; petioles often covered with long hairs (villous); flowers small and bell-like; petals white to pinkish; flowers June–September.

RANGE-HABITAT: Endemic to the southern Appalachian region from WV and VA south to AL and

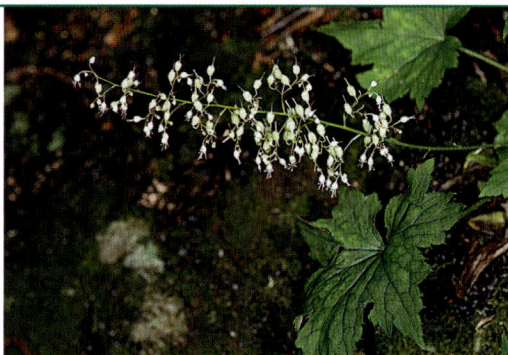

GA; in SC, common in the mountains on boulders, rock crevices, and shallow soil over rock in shaded gorges and coves.

SIMILAR SPECIES: This species is easily distinguished from all other alumroot species in our area by the sharp-pointed leaves and shaggy-pubescent petioles. It is the most abundant alumroot in moist shaded rock crevices of our cove forests and gorges.

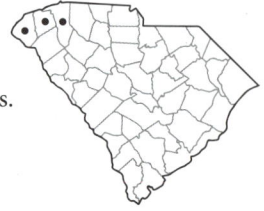

39. Cave Alumroot; Grotto Alumroot

Heuchera parviflora Bartling var. *parviflora*
>   Heù-cher-a par-vi-flò-ra var. par-vi-flò-ra
> Saxifragaceae (Saxifrage Family)

DESCRIPTION: Densely clumping perennial herb with flowering stems to 20″ tall, far exceeding the basal leaves; leaves extremely thin, purplish green to pale green and semi-transparent, round (orbicular) to 6″ long with rounded lobes; flowers small and bell-like; petals white, often recurved and shaped like tiny spoons (spatulate); flowers September–December.

RANGE-HABITAT: Primarily a southern Appalachian species found from OH, WV, and KY, south to GA and SC; in SC, uncommon in the mountains in deeply shaded recesses of overhangs, and grottos in gorges where protected from precipitation and receiving moisture from seepage through the rock, wind-blown spray during rainstorms, or mist and high humidity provided by proximity to rushing water.

COMMENTS: This species is easily distinguished from all other alumroots in our area by the extremely thin leaves. The thin leaves are an adaptation to the low light levels in its preferred habitat. In fact, it is often the only flowering plant capable of growing in the twilight regions of these grottos. If the leaves were thicker, light could not penetrate into the lower cell-layers of the leaf. The plant has very small, surficial roots that barely hold it in place. It is easily washed away if subjected to direct downpours of water. It is also easily dislodged by careless visitors to its habitat. Cave Alumroot often grows in close proximity to filmy fern species (*Hymenophyllaceae*) and in association with mosses and liverworts. Because of its specialized habitat, it is rare throughout its range. It is probably most abundant in the southern Blue Ridge Escarpment region. This plant should never be collected since it is rare and impossible to cultivate. Care should be taken when visiting the sensitive habitat where it occurs.

CONSERVATION STATUS: SC-Imperiled

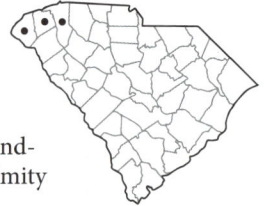

40. Riverweed

Podostemum ceratophyllum Michaux
> Po-dos-tè-mon ce-ra-to-phýl-lum
> Podostemaceae (Riverweed Family)

DESCRIPTION: A highly reduced vascular plant that looks more like a robust algae than a flowering species; plants dark green to brownish, attached to rocks by a disk that is nearly impossible to remove; leaves are divided into long filiform segments; flowers reduced, with 3 scalelike tepals, 1 pistil, and 2 stamens; flowers May–July.

RANGE-HABITAT: Widespread in eastern North America; in SC, common throughout, but most abundant in the mountains and Piedmont where it may be found on spray cliffs and growing on rocks in rapidly flowing, shallow water of small streams to large rivers.

COMMENTS: This unique species is superbly adapted to life in fast-moving water. The rubbery texture of the branches accommodates the movement of the water. It often forms mats adjacent to waterfalls in the wettest areas of spray-cliff habitats but can also be found growing on rocks in any cool, shallow, rapidly flowing body of water.

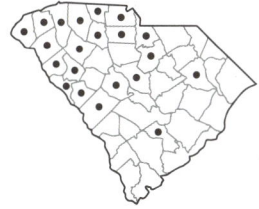

41. Peter's Filmy Fern; Dwarf Filmy Fern

Didymoglossum petersii (A. Gray) Copeland
Di-dy-mo-glós-sum pe-térs-i-i
Hymenophyllaceae (Filmy Fern Family)

SYNONYM: *Trichomanes petersii* A. Gray—RAB

DESCRIPTION: Tiny fern similar to a thalloid liverwort in appearance, forming large, dense patches; rhizomes hairlike covered with tiny black trichomes; leaves (fronds) simple, evergreen, less than 1″ long, thin and translucent green, straplike, often with wavy margins or irregularly lobed; spores produced in sori on an elongate bristlelike extension (receptacle) that extends from a cuplike "involucre" located at the tip of a leaf's midrib.

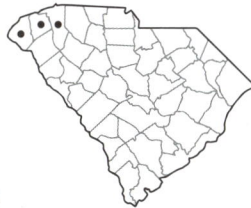

RANGE-HABITAT: Most abundant in the Savannah River drainage in the Blue Ridge Escarpment of SC and NC; also TN, AR, in the Gulf Coastal Plain from FL to LA, and in tropical southern Mexico and Guatemala; vertical faces of rocks in deep recesses of overhangs; most abundant where sheltered from rain and direct sunlight but continually humid from the high humidity of deep gorges along rivers; occasionally found at the bases of trees in forests with dense evergreen *Rhododendron* cover or in more exposed rock outcroppings in deep gorges, always in dim light; in swamps and humid tropical forests in other portions of its range.

COMMENTS: This is the smallest fern species in SC, though it may form mats over 4′ wide on suitable substrate. It hardly looks like a fern and is most often mistaken for a liverwort. Though it is quite rare, it is the most abundant of the three species in this family that commonly exist as sporophytes in SC. Like all filmy ferns, the leaves are only one or two cell layers thick and semi-transparent. Because of their thin leaves, they are extremely susceptible to drought and must remain hydrated continuously. Permanently bathed in high-humidity air, they have the peculiar trait of preferring habitats that are not wet.

CONSERVATION STATUS: SC-Imperiled

42. Appalachian Filmy Fern

Vandenboschia boschiana (Sturm) Ebihara & K. Iwatsuki
Van-den-bósch-i-a. bos-chi-à-na
Hymenophyllaceae (Filmy Fern Family)

SYNONYM: *Trichomanes boschianum* Sturm—RAB

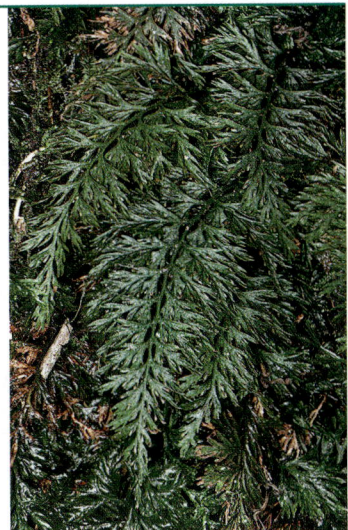

DESCRIPTION: Rhizomatous evergreen fern, forming large, dense patches; rhizomes hairlike, covered with tiny reddish-black hairs (trichomes); leaves (fronds) pinnately divided, ovate-lanceolate, often drooping, thin, and translucent green, to 8″ long; spores produced in clusters (sori) on elongate bristlelike extensions (receptacles) that extend from cuplike "involucres" located in the axils of pinnae lobes.

RANGE-HABITAT: Most abundant in the Savannah River drainage of the Blue Ridge Escarpment of SC and NC, but found in small, highly disjunct occurrences from s. IL and IN south to MS, and northward from SC through the southern Appalachians to WV; also found in AR and Chihuahua, Mexico; vertical faces of rocks, commonly in deep recesses of rock overhangs ("rock shelters"); most abundant in close proximity to rapid water where spray from waterfalls or churning water bathes the plants in humid air and keeps them moist.

COMMENTS: This is the largest of the three filmy ferns that are found as sporophytes in SC. It is often found in moister sites than Dwarf Filmy Fern. Like all filmy ferns, the leaves are one or two cell layers thick and semi-transparent, which serves as an adaptation to low light levels. Because of their thin leaves, they are extremely susceptible to drought and must remain hydrated continuously. This species has been declining rapidly because of prolonged droughts during the past 20 years and may be an excellent barometer for assessing the impacts of climate change on our native species. Populations that are not adjacent to highly turbulent water rely on seepage through the rock surface for much of their moisture, and these populations have suffered greatly from the recent prolonged droughts. The genus is named in honor of Roelof Benjamin van den Bosch (1810–62), a Dutch botanist who specialized in ferns.

CONSERVATION STATUS: SC-Critically Imperiled

43. Tunbridge Fern;
Gorge Filmy Fern

Hymenophyllum tunbridgense
(L.) J. E. Smith
Hy-me-no-phýl-lum tun-brid-
gén-se
Hymenophyllaceae (Filmy Fern
Family)

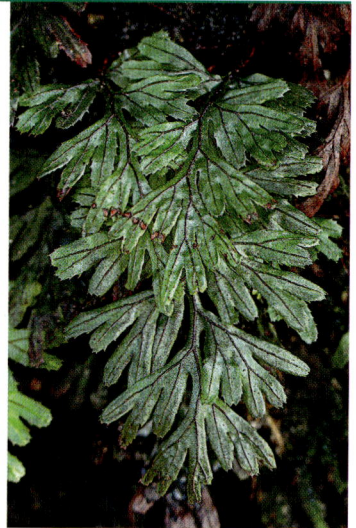

DESCRIPTION: Rhizomatous evergreen fern forming dense patches; rhizomes dark, hairlike, and smooth; leaves (fronds) pinnately divided and lobed, ovate-lanceolate, often drooping, thin and translucent green, 1.0–1.5″ long; spores produced in clusters (sori) contained within a pouchlike "involucre" located along the axils of the pinnae lobes.

RANGE-HABITAT: In the US, confined to a single deep, humid gorge in Pickens County, SC; otherwise known from tropical America (particularly the Caribbean), Ireland, England, western Europe, Africa, Australia, and New Zealand. In SC, found on vertical rock faces of boulders bordering swiftly flowing water and cascades where they are bathed in high humidity air year-round.

COMMENTS: This fern was first discovered in the United States by Mary S. Taylor in the late 1930s during her investigations of the Jocassee Gorges region (Taylor, 1938). Since then, no populations have been found outside of this single stream drainage. Overcollection and drought are threats to this extremely rare fern. The unique characteristics of the gorges area with their high rainfall, tempered climate, and microclimates provided by the deep gorges themselves have allowed this fragile, tiny fern to exist far away from its tropical haunts. The tiny spores of ferns are easily transported on air currents. Theoretically, they established here from spores carried by tropical winds (e.g., hurricanes) from the Caribbean. This may also explain their occurrence in Ireland and England, which often also experience the tropical storms carried along the route of the Gulf Stream. The residency of these plants in the gorges here in SC is a testament to the resiliency of this landscape over a long history. The specific epithet refers to Tunbridge Wells in England, a well-known site for this fern. The genus translates directly to "filmy leaf."

CONSERVATION STATUS: SC-Critically Imperiled

44. Single-sorus Spleenwort

Asplenium monanthes L.
　　As-plè-ni-um mo-nán-thes
Aspleniaceae (Spleenwort
　　Family)

DESCRIPTION: Small, clumping, evergreen fern with dark blackish-brown petioles and very narrow, elongate, pinnately compound leaves (fronds) to 1′ long; pinnae opposite, oblong, and with dentate margins toward the tips; spores produced in a single line or up to 3 lines, on the under surface of the pinnae.

RANGE-HABITAT: Most abundant in the Savannah River drainage of the Blue Ridge Escarpment in SC and NC; found on limestone in FL and AL; wide-ranging in the Neotropics, Hawaii, Azores, Madeira, Philippines, Madagascar, and South Africa.

SIMILAR SPECIES: The species is easily segregated from all other spleenwort species by the distinctive sori arrangement. All other spleenwort species have more than a single line of sori on the pinnae.

COMMENTS: This is primarily a fern of the tropics. It is possibly the only native species SC shares in common with Hawaii. The occurrence of Single-sorus Spleenwort in SC was first reported by Hugo Blomquist in 1948 from just below Whitewater Falls. Since that time, it has been found in small numbers in nearly every drainage of the Jocassee Gorges. Spleenworts found in humid gorges should be examined to see if they have the single line of sori typical of this species. Though it is common on slightly calcareous substrate, it is also found on acidic metamorphic and igneous rock.

CONSERVATION STATUS: SC-Imperiled

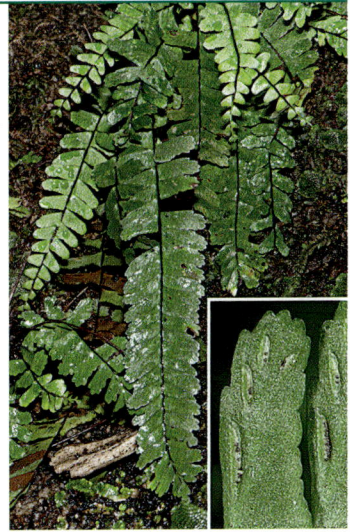

45. Mountain Spleenwort

Asplenium montanum Willdenow
　　As-plè-ni-um mon-tà-num
Aspleniaceae (Spleenwort Family)

DESCRIPTION: A small, clumping, evergreen fern with deltoid-lanceolate, pinnately divided or pinnatifid leaves (fronds) to 7″ long, frequently much smaller; pinnae in 4–7 pairs per leaf, subopposite in arrangement; spores arranged in 2–3 linear groupings (clusters of sori) per pinnae.

RANGE-HABITAT: Primarily in the Appalachian region from MA and VT south to AL and GA; disjunct in MO; in SC, rare in the mountains where restricted to outcrops of metamorphic or igneous rock in humid gorges at lower elevations and on exposed ridges at higher elevations.

COMMENTS: Mountain Spleenwort is not likely to be confused with other spleenworts and is more likely to be mistaken for a small *Dryopteris* or *Cystopteris*. The linear sori and tiny size help to distinguish it from all other ferns in our area. This species is found at some low elevation Piedmont locations farther north, but it is near its southern limit in the gorges of SC. It is most abundant on outcrops at high elevations in NC where it is often surrounded by spruce-fir forest. It is found in the Jocassee Gorges growing in association with otherwise tropical ferns—a unique and most unusual situation that indicates the resiliency of these habitats and their ability to support life from many different climatic types.

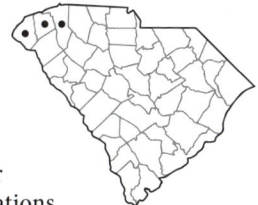

46. Lobed Spleenwort

Asplenium pinnatifidum Nuttall
As-plè-ni-um pin-na-tí-fi-dum
Aspleniaceae (Spleenwort Family)

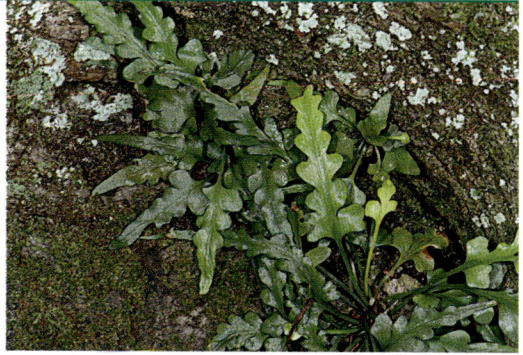

DESCRIPTION: A small, clumping, evergreen fern with ovate-lanceolate, simple but pinnately lobed leaves (fronds) to 10″ long, frequently much smaller; spores arranged in 2–3 linear groupings (clusters of sori) per segment.

RANGE-HABITAT: Primarily found in the Appalachian region from IL to NJ and south to GA and MS; also in AR and OK; in SC, very rare in the Blue Ridge Escarpment region where restricted to outcrops of metamorphic rock at lower elevations near larger streams and rivers, always shaded, but in much drier sites than other rock-dwelling spleenworts; also in Lexington County, on sandstone in the Fall-line Sandhills region.

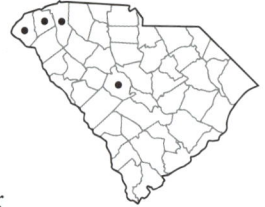

COMMENTS: Lobed Spleenwort is not common in any portion of its range. Like many fern species, it is the result of a polyploid hybridization event between *A. montanum* and *A. rhizophyllum*. This species occupies much drier substrate than either of its parent species. It tends to grow on very large boulders or cliff faces in dry shady sites.

CONSERVATION STATUS: SC-Critically Imperiled

47. Maidenhair Spleenwort

Asplenium trichomanes L. ssp.
trichomanes
As-plè-ni-um tri-chó-ma-nes ssp.
tri-chó-ma-nes
Aspleniaceae (Spleenwort Family)

DESCRIPTION: A small, clumping, evergreen, rock-dwelling fern from a short rhizome; leaf (frond) pinnately compound, oblong, to 10″ long, but generally much shorter; leaflets opposite, with rounded apex and rounded lobes; up to 7 linear groupings (clusters of sori) per leaf segment.

RANGE-HABITAT: Widespread in eastern North America and some western states; in SC, uncommon in the mountains and rare in the Piedmont on moist outcrops of rocks that are high in magnesium or calcium.

COMMENTS: This species is extremely delicate, with leaves that are not as thick as other species in our area. It is a good indicator of high pH substrate.

48. Common Rockcap Fern (A)

Polypodium virginianum L.
Po-ly-pó-di-um vir-gi-
ni-à-num

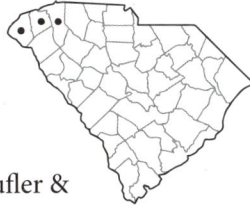

Appalachian Rockcap Fern (B)

Polypodium appalachianum Haufler &
Windham
Po-ly-pó-di-um ap-pa-la-
chi-à-num
Polypodiaceae (Polypody
Family)

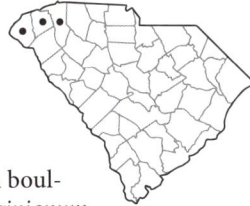

DESCRIPTION: Small, evergreen ferns from creeping rhizomes, forming large mats on boulders ("capping" the rocks); leaves of *P. virginianum* are widest toward the middle, with less tapered tips than leaves of *P. appalachianum,* which are widest toward the base and frequently more robust.

RANGE-HABITAT: *P. virginianum* is widespread in eastern North America; *P. appalachianum* is almost restricted to the Appalachian region from Newfoundland south to AL; in SC, both species are restricted to the mountains, where they are found on boulders and edges of outcrops and cliffs within forested habitats.

SIMILAR SPECIES: Resurrection Fern (*Pleopeltis michauxiana* (Weathreby) Hickey & Sprunt, plate 914) is similar and may grow on rocks in the Piedmont and mountains, particularly on high pH substrates. It may be distinguished by the rusty scales covering the undersides of the leaves and stems.

COMMENTS: Though the two rockcap ferns are very similar, they are distinct species by all definitions. *P. virginianum* is a tetraploid and *P. appalachianum* is a diploid. These two fern species can be difficult for even professional botanists to distinguish. This is made more difficult by the fact that the two species hybridize to produce a sterile triploid that is intermediate in characters. A safe bet is to refer to all rockcap ferns as "*Polypodium virginianum*" complex.

49. Purple Cliff-brake

Pellaea atropurpurea (L.) Link
Pel-laè-a at-ro-pur-pù-re-a
Pteridaceae (Maidenhair Fern
Family)

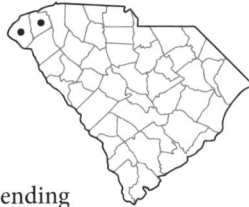

DESCRIPTION: Perennial, evergreen, rock-dwelling fern from a short creeping or ascending rhizome; leaves (fronds) stiff, leathery; leaf stalks purplish-black, round, and sparsely hairy; blades 2x pinnately compound on robust individuals, once compound on small individuals; ultimate segments remote, oblong, sometimes whitened below, but never shiny; separate fertile and sterile leaves.

RANGE-HABITAT: Widespread in eastern North America; in SC, very rare in the mountains and upper Piedmont, where restricted to outcrops and boulders of meta-sedimentary rock such as marble or amphibolite; typical of limestone in other portions of its range.

SIMILAR SPECIES: Two other cliff-brake species (*P. ternifolia* (Cavanilles) Link ssp. *arizonica* Windham, plate 26, and *P. wrightiana* Hooker, plate 25) are known from SC; both have completely smooth leaf stalks that are grooved.

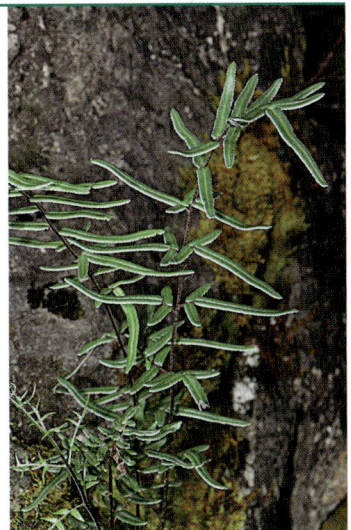

COMMENTS: All species of cliff-brake are highly localized and uncommon in SC. This species is found on amphibolite in the Jocassee Gorges region, where it grows on south and southwest-facing slopes in forests. It is also found at one calcareous variant granitic dome and on marble boulders in dry forests in Oconee County. SC plants are typically smaller than those found north and west of SC, probably due to harsher growing conditions found in SC on the periphery of the species range.

CONSERVATION STATUS: SC-Imperiled

---

50. Meadow Spikemoss

   Lycopodioides apodum
      (L.) Kuntze
      Ly-co-po-di-oì-des á-po-dum
   Selaginellaceae (Spikemoss
      Family)

   SYNONYM: *Selaginella apoda* (L.)
      Spring—RAB

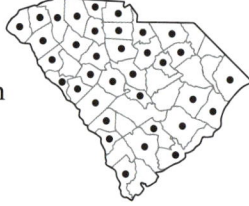

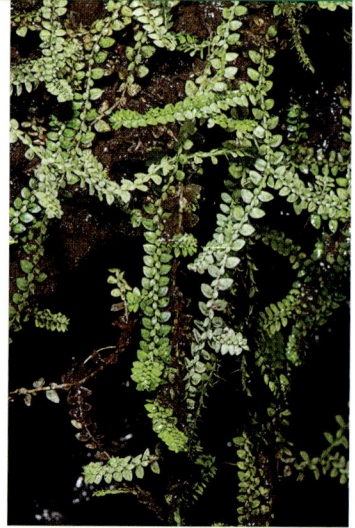

DESCRIPTION: A small, evergreen vascular plant that appears like a true moss; stems creeping and appressed, forming mats; leaves tiny, alternate; spores in inconspicuous clusters in the axils of leaves.

RANGE-HABITAT: Wide-ranging in North America, south to Guatemala; in SC, common throughout, found in continuously moist habitats such as spray cliffs, wet seeps, bogs, fens, wet meadows, and swamp forests.

SIMILAR SPECIES: Meadow Spikemoss looks like a true moss or liverwort but is easily distinguished as a vascular plant by the small, true fibrous roots found sparingly along the creeping stem.

COMMENTS: Though it may look like a moss, this is a vascular plant related to ferns. It is easily overlooked but nearly ubiquitous in the spray cliff community.

---

51. Cataract fen

---

52. Red Maple

   Acer rubrum L. var. *rubrum*
      À-cer rù-brum var. rù-brum
   Aceraceae (Maple Family)

DESCRIPTION: Medium to large, deciduous tree; leaves opposite, usually 3–5-lobed with serrate margins; flowers perfect, or often male and female flowers in separate clusters on the same or different trees; fruit a schizocarp of samaras; flowers January–April; fruit matures March–May.

RANGE-HABITAT: Common throughout eastern North America; common throughout SC; alluvial

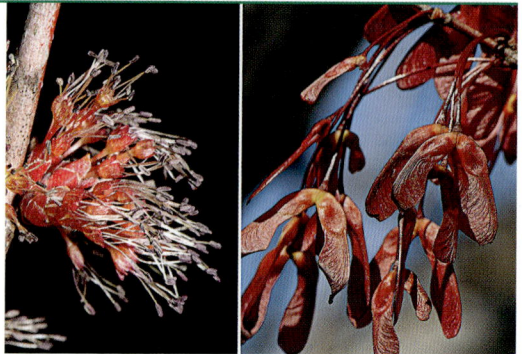

and nonalluvial swamp forests, margins of cataract fens, beech and oak-hickory forests, pine-mixed hardwood forests, pine-oak heath, and ruderal habitats. **SIMILAR SPECIES:** SC is home to seven maple species, but this is the only species with simple leaves that are lobed and serrate. **COMMENTS:** The flowers bloom sometimes as early as January, always long before the leaves appear. Flowers are followed by the conspicuous winged fruits (samaras), bright red to yellow, creating a prominent display of color against the early spring sky. Red Maple is one of the pioneer trees in abandoned rice fields; it also invades cleared, upland sites.

Red maple is often planted as a shade tree. The wood is used for a variety of products and maple syrup can be made from the sap, although the sap contains less sugar than the Sugar Maple.

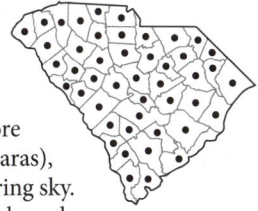

53. Tag Alder; Hazel Alder

Alnus serrulata (Aiton) Willdenow
Ál-nus ser-ru-là-ta
Betulaceae (Birch Family)

DESCRIPTION: Deciduous shrub or small tree to 30′ tall; usually growing in clumps; leaves alternate; flowers in elongate conelike spikes; male spikes (catkins) conspicuous in the spring before leaves appear; female "cones" (catkins) persist through the winter after shedding seeds; flowers February–March.

RANGE-HABITAT: Nova Scotia south to FL and west to OK and TX; common throughout SC; rocky stream sides, riverbanks, fens, freshwater marshes, and in wet places in forests.

COMMENTS: Tag Alder has many uses in folk medicine. Native Americans used bark tea for diarrhea, childbirth pain, coughs, toothaches, and as a "blood purifier"; also as a wash for hives, piles, and Poison Ivy rash. Alders are excellent colonizers of recently eroded or disturbed, moist, mineral soil. Alders can become established in newly exposed, moist-wet habitats as they host symbiotic bacteria that fix atmospheric nitrogen. The flowers shed pollen during the first prolonged warm spells in winter. The catkins are fully formed when the tree sheds its leaves in the autumn and simply wait for the warmth to quickly open.

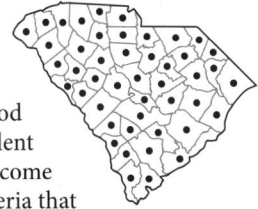

54. Large Witch-alder

Fothergilla major (Sims) Loddiges
Fo-ther-gíll-a mà-jor
Hamamelidaceae (Witch-hazel Family)

DESCRIPTION: Deciduous, multitrunked shrub to 5′ tall; leaves to 5″ long and 4″ wide, ovate, suborbicular or obovate with a toothed margin and cordate to oblique leaf base; flowers in spikes, with inconspicuous, cuplike sepals and no petals; stamens numerous (12–32), elongate, bright white with yellow tips, extremely showy.

RANGE-HABITAT: NC and TN south to GA and AL with disjunct populations in AR; in SC, very rare in the mountains where found along the margins of cataract fens, shallow soils of granitic domes, and dry exposed ridges.

SIMILAR SPECIES: Coastal Witch-alder (*F. gardenii* L., plate 499) and Small-leaved Witch-alder (*F. parvifolia* Karney in Small) are

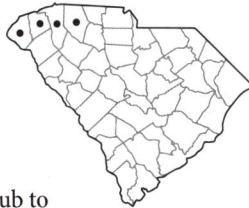

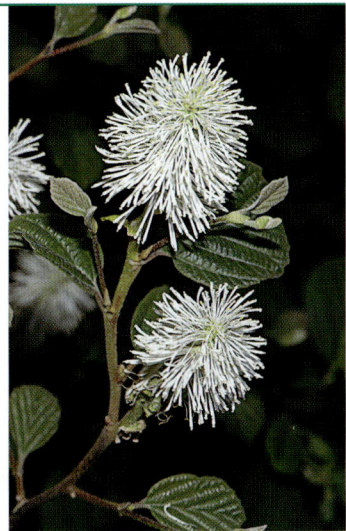

smaller in all respects and occur in the Coastal Plain and Sandhills in pocosins and pocosin margins. Common Witch-hazel (*Hamamelis virginiana* L., plate 155), is more robust and flowers in the autumn through winter. The leaves of Common Witch-hazel have the basal-most lateral veins diverge from the midrib within leaf tissue, while witch-alders have leaves with the basal-most lateral veins beginning along the margin of the leaf blade.

COMMENTS: All witch-alders make excellent landscape plants and are highly adaptable to any portion of SC. The most popular cultivar (*Fothergilla* "Mt. Airy") is a hybrid between *F. major* and *F. gardenii*. The genus is named in honor of the English physician John Fothergill (1712–80), an avid gardener and amateur botanist.

CONSERVATION STATUS: SC-Imperiled

---

55. Eastern Ninebark

Physocarpus opulifolius (L.)
   Maximowicz var. *opulifolius*
   Phy-so-cár-pus o-pu-li-fò-
   li-us var. o-pu-li-fò-li-us
   Rosaceae (Rose Family)

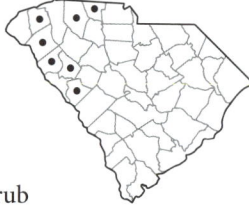

DESCRIPTION: Deciduous, multitrunked shrub to 10′ tall; bark exfoliating into strips; leaves 1.5–4″ long, ovate, three-lobed with three primary veins; flowers in flat-topped arrangements (corymbs), similar to a *Spiraea;* flowers May–July; fruits are aggregates of follicles; fruits mature July–September.

RANGE-HABITAT: Widespread in eastern North America from Quebec to MN and CO and south to FL and AR; in SC, uncommon to rare in the mountains and Piedmont; cataract fens (particularly where seepage is enriched with minerals), margins of rock outcrops (such as amphibolite), rocky stream sides, Piedmont seepage forests, stream banks, and scour prairies.

COMMENTS: Ninebark makes an attractive, fast-growing landscape shrub and is easily cultivated. It is an excellent replacement for exotic *Spiraea* species. When found in the wild, it is an indicator of high magnesium or calcium soils.

---

56. Biltmore Sedge

Carex biltmoreana Mackenzie
   Cà-rex bilt-mor-e-à-na
   Cyperaceae (Sedge Family)

DESCRIPTION: A densely clumping, robust, bluish-green sedge reaching a height of 3′ with dark purple basal sheaths; male and female flowers on separate spikes on the same culm with the male flowers terminal and quickly deciduous; fruit is an achene enclosed by a bluish-green perigynium; flowers April–May.

RANGE-HABITAT: A narrow southern Blue Ridge Escarpment endemic that is restricted to sw. NC, nw. SC, and ne. GA; in SC, rare in the mountains and found only in moist pockets in outcrops, cataract fens, and other seepage over granite or other rock where high magnesium or high calcium levels are present in the substrate or seepage water.

SIMILAR SPECIES: Though all sedges are similar, no other robust sedge with a similar form or coloration is found in this unique habitat.

COMMENTS: This sedge is extremely distinctive due to the powdery bluish foliage and robust nature. In SC, this is an indicator of mafic cataract fens—those with magnesium-rich seepage. The species epithet honors the Biltmore Estate, which sponsored and employed many botanists during the nineteenth century.

CONSERVATION STATUS: SC-Critically Imperiled

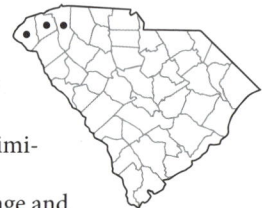

57. Eastern Indian-paintbrush

Castilleja coccinea (L.) Sprengel
Cas-til-lè-ja coc-cí-ne-a
Orobanchaceae (Broomrape
Family)

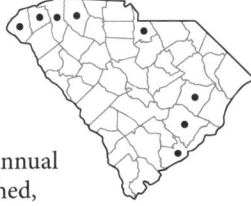

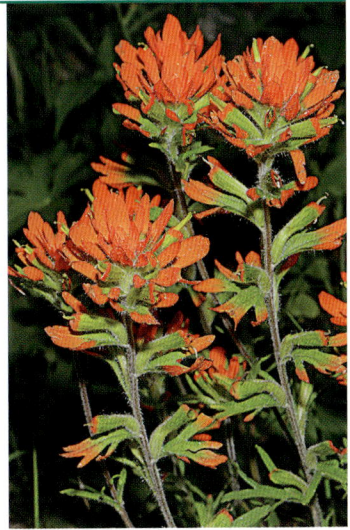

DESCRIPTION: Erect, hairy, hemi-parasitic annual or biennial herb to 28″ tall; stem unbranched, from a basal cluster of leaves; pubescent through-out; flowers yellow to greenish yellow; bracts subtending the flow-ers, vary from the typical scarlet to yellow; flowers April–May.

RANGE-HABITAT: Widespread in eastern North America; woodlands, rock outcrops, meadows, wet pastures, grassy openings, usually over mafic rocks; in SC, rare and primarily a mountain and Pied-mont species, often found along the margins of cataract fens or out-crops with nutrient rich substrate or seepage water; several disjunct populations have been found in the Coastal Plain, where it is very rare in Longleaf Pine savannas on soils with a high clay content.

COMMENTS: The beauty of this plant comes not from the small flow-ers, but from the brightly colored bracts. The common name comes from a Native American legend that tells of a man's discarded brushes being used to paint a brilliant sunset which then grew into flowers. Native Americans used the roots, mixed with iron minerals, to dye deerskins black. The species is partially parasitic on a wide variety of grass and herb species. The genus honors Spanish botanist Domingo Castillejo (d. 1786).

CONSERVATION STATUS: SC-Imperiled

58. Mountain Sweet Pitcherplant

Sarracenia jonesii Wherry
Sar-ra-cèn-i-a jònes-i-i
Sarraceniaceae (Pitcherplant Family)

SYNONYM: Included under a broad concept of *Sarracenia rubra*—RAB

DESCRIPTION: Herbaceous perennial, forming colonies from rhizomes; leaves held upright, 15–30″ tall, modified into hollow tubes (pitchers) that are effective as passive insect traps; the hood ascending and leaving the tube opening exposed; the opening in the tubes with a distinct notch on the outer tip; flower held on a leafless stalk (scape) that is about the same height as the pitcher; petals maroon; flowers May.

RANGE-HABITAT: Endemic to a small region of the southern Blue Ridge, re-stricted to sw. NC and nw. SC; in SC, rare in the mountains and upper Piedmont where restricted to a handful of sites in cataract fens, boggy lake margins, and seepages on granitic flatrocks and granitic domes.

SIMILAR SPECIES: Carolina Sweet Pitcherplant (*Sarracenia rubra* Walter ssp. *rubra*, plate 732) is similar, but is restricted to the Coastal Plain and Sandhills. It has flowers that exceed the leaves and leaves that are generally 12″ tall or less; the opening of the tube does not have as distinc-tive a notch in the opening and the hood is held close to the opening of the pitcher. Georgia Sweet Pitcherplant (*S. rubra* ssp. *viatorum* B. Rice) is also very similar and has similar-sized pitchers, with an ascending hood. It is restricted to White Cedar Swamps and herbaceous seepages in the Sandhills. The leaves of Georgia Sweet Pitcherplant are distinctively pubescent, the flowering scape exceeds the leaves and the hood is weakly cordate in shape.

COMMENTS: Unfortunately, this is probably the most exploited endangered species found in SC moun-tains. Illegal poaching and picking of the unusual plants by uneducated visitors and sophisticated poachers have had a large, negative impact on the easily accessible populations. Entire populations

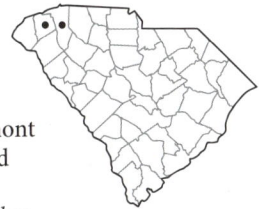

have been eliminated in Pickens County because of poaching. Plants should not be picked under any circumstance. It has sometimes been considered a subspecies or variety of the widespread Coastal Plain species, *Sarracenia rubra*. The genus honors Dr. Michel Sarrasin de l'Étang (1659–1734), physician at the Court of Quebec, who sent a sample of the genus to Europe. The specific epithet honors Dr. F. M. Jones, who studied the insect associates of pitcherplants.

CONSERVATION STATUS: Federally Endangered

59. Southern Appalachian Purple Pitcherplant

Sarracenia purpurea L. var. *montana* Schnell & Determann

Sar-ra-cèn-i-a pur-pú-re-a var. mon-tà-na

Sarraceniaceae (Pitcherplant Family)

SYNONYM: Included under a broad concept of *Sarracenia purpurea* L.—RAB

DESCRIPTION: Rhizomatous perennial, evergreen, carnivorous herb with leaves held horizontally to ascending, near the ground (decumbent), and modified into hollow tubes that are effective as passive insect traps; leaves green, rarely with purplish tints; the lobes of the hood of the tube curving inward and obscuring the opening; flowering stalks 8–16″ tall; flowers May.

RANGE-HABITAT: Endemic to a narrow southern Blue Ridge region, restricted to sw. NC, nw. SC and ne. GA; in SC, very rare and restricted to cataract fens in the mountains.

COMMENTS: This striking plant differs from the widespread Southern Purple Pitcherplant (*Sarracenia purpurea* var. *venosa* (Rafinesque) Fernald, plate 505), by its pitchers having lobes that curve inward and obscuring the opening and the green versus purplish-tinted leaves. It differs from the more northerly Northern Purple Pitcherplant (*Sarracenia purpurea* var. *purpurea*) in having leaves less than 3 times longer than their width and flowers that are red rather than dark maroon. It is likely that with further study this variety may be elevated to a distinct species status. The genus honors Dr. Michel Sarrasin de l'Étang (1659–1734), physician at the Court of Quebec, who sent a sample of the genus to Europe. Hybrids of this species and the Mountain Sweet Pitcherplant, though extremely rare, have been found. These hybrids are intermediate in characters between the two parents.

CONSERVATION STATUS: SC-Critically Imperiled

60. Canada Lily

Lilium canadense L.

Lí-li-um ca-na-dén-se

Liliaceae (Lily Family)

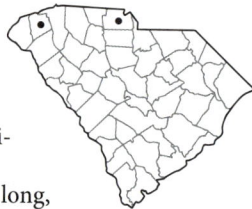

DESCRIPTION: Perennial, deciduous, stoloniferous herb from a scaly bulb; stem stout, upright, to 5′ tall; leaves in whorls, 1.5–6″ long, elliptic to lanceolate, minutely serrulate on the margins; flowers nodding, large and showy; 6 tepals, red or yellow to reddish-orange tepals, heavily spotted toward the base, with tips spreading to slightly recurved outwards; flowers May–June.

RANGE-HABITAT: Widespread in eastern North America from New Brunswick and Ontario south to GA and MS; in SC, rare in the mountains in rich cove forests, oak woodlands over mafic or calcareous substrate, mafic cataract fens, amphibolite outcrops, and forest margins (red-flowered form); also very rare in the Piedmont in xeric hardpan forests and forest margins in York County (yellow-flowered form).

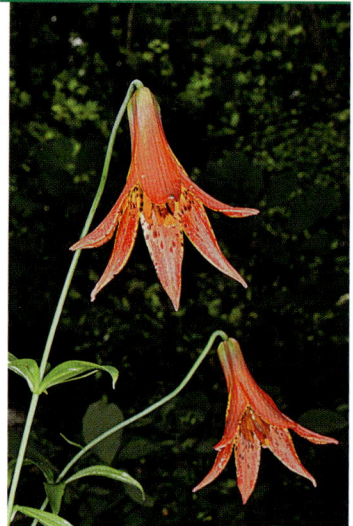

SIMILAR SPECIES: No other lily in SC has similar nodding flowers without heavily recurved sepals. Gray's Lily (*Lilium grayi* S. Watson), which is found in NC, mostly at high elevations, is similar, with wider and more heavily spotted tepals (heavily spotted nearly to the tip), that are not as elongated and are narrowed toward the tip.

COMMENTS: This species is present as sterile plants in large numbers at some rich cove forest sites in the Jocassee Gorges region but rarely flowers; it persists from stoloniferous offshoots. It can be found flowering along roadsides and in sunny cataract fens. The red-flowered form is attractive to hummingbirds and has been referred to as *L. canadense* var. *editorum* Fernald. The yellow-flowered form is more northerly in distribution, and its presence in SC was only recently documented (Schmidt and Barnwell, 2002).

CONSERVATION STATUS: SC-Critically Imperiled

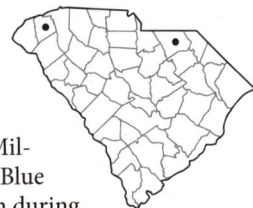

61. Horned Bladderwort

Utricularia cornuta Michaux
U-tri-cu-là-ri-a cor-nù-ta
Lentibulariaceae (Bladderwort Family)

DESCRIPTION: Terrestrial carnivorous herb; subterranean leaves dissected, with small, trapping bladders; surficial leaves, when present, linear and flat; flowering stalk to 1′ tall, leafless, but with small scales and bracts; flowers 2–6 per stem, crowded toward the tip of the stem, yellow, at least 0.6″ long, with a spur 0.3–0.5″ long; flowers May–September.

RANGE-HABITAT: Widespread in eastern North America; also in Cuba and the Bahamas; in SC, rare in cataract fens and seepages on granitic domes in the mountains, seepages on granitic flatrocks in the Piedmont, and margins of depression ponds in the Coastal Plain.

SIMILAR SPECIES: Southern Bladderwort (*Utricularia juncea* Vahl, plate 701) has smaller and more numerous flowers, which are widely spaced on the stem, and nectar spurs less than 0.3″ long. It is common in wet savannas and depression pond margins and herbaceous seepages in the Sandhills and Coastal Plain.

COMMENTS: Horned Bladderwort is the largest flowered of the terrestrial species of bladderworts in SC. *Utricularia* get their name from their distinctive trapping bladders (utriculus = small bladder).

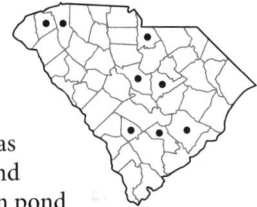

62. Texas Hatpins

Eriocaulon texense Körnicke
E-ri-o-caú-lon tex-én-se
Eriocaulaceae (Pipewort Family)

DESCRIPTION: Perennial herb with flowering stalks to 12″ tall from a tight basal rosette of grasslike leaves to 2.5″ long; flowers white, in dense, flattened heads on the tips of the leafless stalks; heads 0.2–0.4″ in diameter, soft, easily compressed between the fingers; flowers May–June.

RANGE-HABITAT: Primarily on the Coastal Plain from NC south to FL and west to TX; in SC, very rare; found in the upper Piedmont at the edge of the escarpment in cataract fens and seepages on granitic domes and in the Sandhills in herbaceous seepages.

COMMENTS: This diminutive species resembles a very small Soft-headed Pipewort (*E. compressum* Lamarck, plate 680). It was first reported for SC by McMillan et al. (2002). The location in Pickens County is at the transition from the Blue Ridge Escarpment to Piedmont. It is present in large numbers in this location during wet years and nearly absent in years of drought.

CONSERVATION STATUS: SC-Critically Imperiled

63. Shortleaf Sneezeweed

Helenium brevifolium (Nuttall)
      Alphonso Wood
      He-lé-ni-um bre-vi-fò-li-um
Asteraceae (Aster Family)

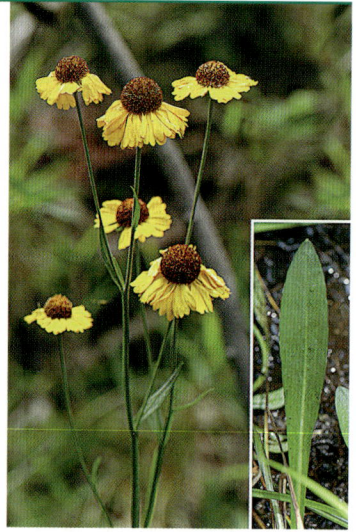

DESCRIPTION: Perennial herb forming clumps with basal rosettes; flowering stems to 32″ tall, with 6–9 nodes below the inflorescence; basal leaves present at flowering; inflorescence composed of 1–4 involucrate heads with yellow sterile ray flowers and reddish fertile 5-lobed disk flowers; flowers May–June.

RANGE-HABITAT: VA and TN south to FL and west to LA; most abundant in the east Gulf Coastal Plain, but in the Carolinas found in the Piedmont and mountains; in SC, rare in the mountains and upper Piedmont in mafic cataract fens, stream banks, and other seepages.

COMMENTS: This species is distinctive among the *Helenium* found in the mountains and Piedmont in that it flowers in the late spring and early summer and has basal leaves present at flowering. It is similar to Savanna Sneezeweed (*Helenium vernale* Walter) and Southeastern Sneezeweed (*H. pinnatifidum* (Nuttall) Rydberg, plate 679), which are found strictly in wetland savannas of the Coastal Plain, produce a single involucrate head per plant, and have 4-lobed disk flowers. Shortleaf Sneezeweed has been erroneously reported for the Coastal Plain based on misidentification with one of these coastal species.

CONSERVATION STATUS: SC-Critically Imperiled

64. Southern Lobelia

Lobelia amoena Michaux
      Lo-bèl-i-a a-moè-na
Campanulaceae (Bellflower
      Family)

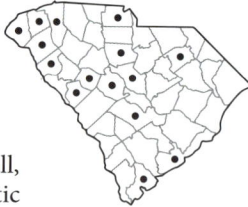

DESCRIPTION: Hebaceous perennial, to 4′ tall, with smooth (glabrous) stems; leaves elliptic to lanceolate, 1.5–6″ long; flowers blue to violet, produced in racemes; flowers July–September.

RANGE-HABITAT: Primarily Appalachian in distribution from NC and TN south to GA and AL; disjunct in the Coastal Plain of SC, GA, and the FL Panhandle; in SC, common in the mountains and upper Piedmont, where found in seepages over granitic outcrops, cataract fens, Southern Appalachian fens, rocky stream sides, streambanks, and other moist habitats; rare in the Coastal Plain in floodplain forests.

SIMILAR SPECIES: Downy Lobelia (*Lobelia puberula* Michaux, plate 218) is similar. It is easily distinguished by the downy pubescent stems versus the smooth stems of Southern Lobelia. The genus honors Matthias de l'Obel (1538–1616), a Flemish herbalist.

65. White Turtlehead

Chelone glabra L
    Che-lò-ne glà-bra
Plantaginaceae (Plantain Family)

DESCRIPTION: Perennial herb with an erect or spreading stem to 3′ tall or long; leaves widest near the middle; leaf stalks less than 0.6″ long; flowers white or tinged with pink or purple toward the tip; staminodes (sterile stamens) with green tips; flowers August–October.

RANGE-HABITAT: Widespread in eastern North America; widely scattered localities throughout SC, but chiefly in the mountains; cataract fens, stream sides, swamps, and other low woodlands.

SIMILAR SPECIES: Purple Turtlehead (*Chelone obliqua* L.) is rare and found in scattered locations throughout SC in swamp forests, stream banks, and rocky stream sides. It is distinguished by having flowers that are pink or purplish throughout and staminodes with white tips.

COMMENTS: Bitter-tasting compounds in the leaves have been used for a variety of ailments, both internal (as a laxative, to expel worms or to treat fever), and external (to treat piles, ulcers, and herpes). The genus *Chelone* is Greek for tortoise or turtle and refers to the flowers, which look like a turtle's head. This is the most widespread and common of the three turtlehead species found in SC.

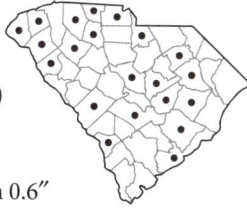

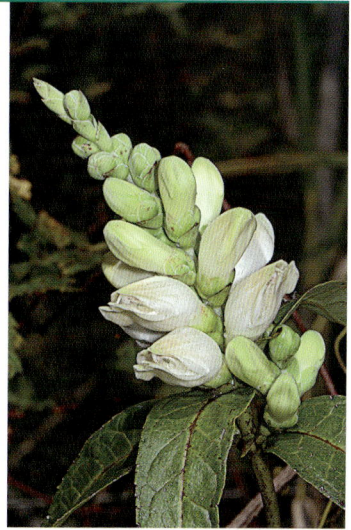

66. Stiff Cowbane; Pig Potato

Oxypolis rigidior (L.) Rafinesque
    Ox-ý-po-lis ri-gí-di-or
Apiaceae (Carrot Family)

DESCRIPTION: Perennial herb to 4.5′ tall, from a cluster of tuberous roots; stems smooth, stout, erect or slender, and arching; leaves with about 7–11 flattened, linear, pinnately arranged leaflets, with few scattered teeth; flowers tiny, white, in a flat-topped compound umbel composed of 20 or more individual umbels; flowers August–October.

RANGE-HABITAT: NY south to FL and west to MN and TX; in SC, occasional in the mountains in cataract fens and stream side wetlands; rare in the Piedmont in swamp forests and wet pastures; occasional in the Coastal Plain in swamp forests, wet savannas, and seepages.

67. Bigleaf Grass-of-Parnassus; Limeseep Grass-of-Parnassus

Parnassia grandifolia A. P. de Candolle
    Par-nás-si-a gran-di-fò-li-a
Parnassiaceae (Grass-of-Parnassus Family)

DESCRIPTION: Perennial herb from a rhizome; leaves slightly longer than broad, on long stalks, in dense basal clusters; flowers white, solitary, on long stalks that bear a single clasping leaf below

the middle; petals not stalked, main veins of each petal 5–9; ovary greenish; flowers late September–October.

RANGE-HABITAT: VA and WV south to FL and west to MO, AR, and TX; in SC, rare in cataract fens and other seepages in the mountains.

SIMILAR SPECIES: Kidney-leaved Grass-of-Parnassus (*Parnassia asarifolia* Ventenat) has leaves that are as wide or wider than long, strongly cordate at the base, and with wider petals. It is very rare in SC and is documented only from Oconee County.

COMMENTS: Bigleaf Grass-of-Parnassus is most frequent where the substrate or seepage water is rich in magnesium or calcium. It is found in more nutrient-rich situations than the Kidney-leaved Grass-of-Parnassus, which prefers highly acidic sites.

CONSERVATION STATUS: SC-Imperiled

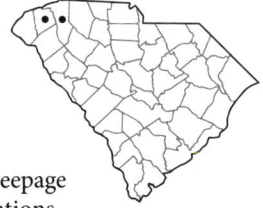

68. Southern Appalachian fen

69. Dense-flowered St.-John's-wort; Mountain Bushy St.-John's-wort

Hypericum densiflorum Pursh
    Hy-pé-ri-cum den-si-flò-rum
    Hypericaceae (St.-John's-wort Family)

DESCRIPTION: Erect multibranched shrub to 6′ tall with smooth, reddish, exfoliating bark; leaves glossy, elliptic, oblong, or linear, to 1.5″ long, less than 0.27″ wide; flowers bright yellow, more than 7 per cluster at the branch tips or upper leaf axils; flowers June–September.

RANGE-HABITAT: PA and NJ south to GA and west to TX; in SC, uncommon in the mountains and Coastal Plain, rare in the Piedmont; Southern Appalachian fens, cataract fens, margins of granitic outcrops, stream sides, floodplain forests, wet meadows, ditches, swamp forests, and other wetlands.

SIMILAR SPECIES: Shrubby St.-John's-wort (*Hypericum prolificum* L., plate 7) is similar but typically shorter. It has leaves that are 0.25–0.60″ wide and 3 flowers per cluster. Shrubby St.-John's-wort has flowers that are larger in all respects than Dense-flowered St.-John's-wort.

COMMENTS: This species achieves its greatest development in open mountain fens and wet meadows in the Appalachian region. The reddish, exfoliating bark, dense growth form, and copious flowering makes this a nice specimen for the native garden.

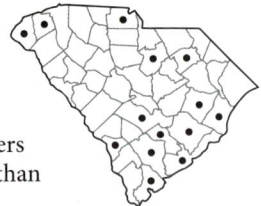

70. Hardhack; Steeplebush

Spiraea tomentosa L.
Spi-raè-a to-men-tò-sa
Rosaceae (Rose Family)

DESCRIPTION: Shrub to 6' tall, sparsely branched; leaves alternate, irregularly toothed, lower surface with a dense covering of white to rusty hairs; flowers pink, reddish, or (rarely) white, in panicles at the ends of branches, pink, reddish, or sometimes white; flowers July–September.

RANGE-HABITAT: From Nova Scotia south to SC and west to MN and AR; in SC, uncommon but scattered throughout; Southern Appalachian fens, wet meadows, Piedmont seepage forests, and swamp margins.

COMMENTS: The name Steeplebush comes from the terminal, dense-flowered panicles that are steeplelike in outline.

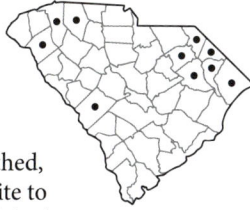

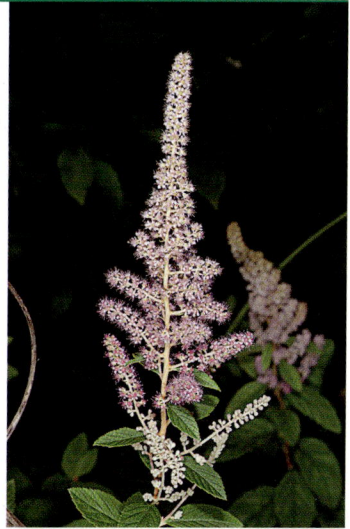

71. French Broad Heartleaf

Hexastylis rhombiformis Gaddy
Hex-ás-ty-lis rhom-bi-fór-mis
Aristolochiaceae (Birthwort Family)

DESCRIPTION: Low evergreen, perennial herb, forming clumps; leaves mostly without mottling, 1.5–2.5" wide and long; flowers solitary in leaf axils; calyx tube rhombic-ovate (widest in the middle), the orifice constricted; calyx lobes very short; inner surface of calyx tube with a close network of reticulate ridges of high relief; flowers late March–June.

RANGE-HABITAT: Restricted to the mountains of Greenville County, SC, and four adjacent counties in NC; in SC, rare, found in acidic cove forests, other acidic forests, and margins of Southern Appalachian fens.

SIMILAR SPECIES: Variable-leaf Heartleaf (*H. heterophylla* (Ashe) Small, plate169) is similar but has large, spotted calyx lobes and a wide orifice at the apex of the calyx tube; Large-flowered Heartleaf (*H. shuttleworthii* (Britten and Baker f.) Small, plate 170) is also similar but has much larger flowers, up to 1.6" long and 1" wide, that are brittle and break easily.

COMMENTS: This species of heartleaf was described by SC botanist L. L. Gaddy (1986). It is unclear how it is related to a species described by W. W. Ashe, Memminger's Heartleaf (*Asarum* (*Hexastylis*) *memmingeri* Ashe); one of the type specimens of this previously described species has been determined to be representative of French Broad Heartleaf.

CONSERVATION STATUS: SC-Critically Imperiled

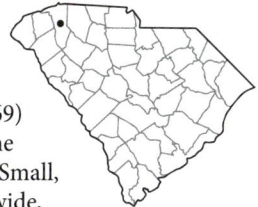

72. Swamp Pink

Helonias bullata L.
He-lò-ni-as bul-là-ta
Heloniadaceae (Swamp Pink Family)

DESCRIPTION: Perennial, evergreen herb to 3′ tall, sometimes forming dense clumps; leaves in a dense basal rosette, flat and glossy, pointed at the tips, and wider toward the tips than toward the base; flowers fragrant, in a dense cluster terminating a leafless stalk, pink with blue stamens; flowers April–early May.

RANGE-HABITAT: Swamp pink ranges from NY and NJ to VA on the Coastal Plain and from VA through NC to SC and GA in the Blue Ridge Mountains; in SC, very rare in the mountains; sphagnum moss-dominated openings in Southern Appalachian fens.

COMMENTS: The only SC population is protected within a South Carolina Heritage Preserve. Though protected, this population is threatened by encroachment by woody plants in its restricted habitat. Plants should never be picked or disturbed in the wild.

CONSERVATION STATUS: Federally Threatened

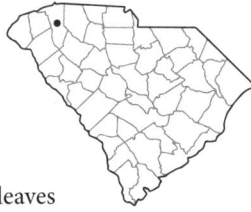

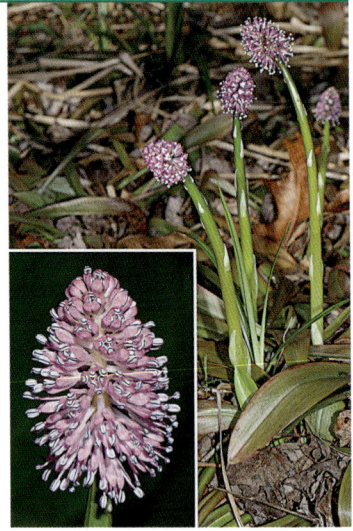

73. Swamp Dewberry

Rubus hispidus L.
Rù-bus hís-pi-dus
Rosaceae (Rose Family)

DESCRIPTION: Evergreen, trailing perennial with glabrous or glandular-pilose stems and scattered prickles; leaves with 3 leaflets on flowering and nonflowering stems; leaflets with crenate margins, dark green, smooth and shiny above, smooth to sparingly pubescent below; flowers white, held on pedicels less than 0.8″ long; fruit an aggregate of drupelets; flowers May–June; fruits mature June–July.

RANGE-HABITAT: Nova Scotia west to WI and south to GA and SC; in SC, rare in the mountains in Southern Appalachian fens and wet, acidic woodlands.

COMMENTS: This delicate little dewberry has small but tasty fruit and is less aggressive and smaller than the other two weedy species found in SC. The small size and short pedicels serve to distinguish it from all other similar *Rubus* species in our state.

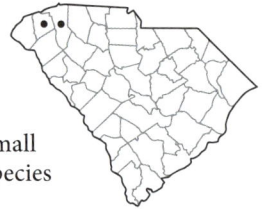

74. Green Fringed Orchid;
Ragged Fringed Orchid

Platanthera lacera (Michaux)
D. Don
Pla-tán-the-ra lá-ce-ra
Orchidaceae (Orchid Family)

SYNONYM: *Habenaria lacera* (Michaux)
Loddiges—RAB

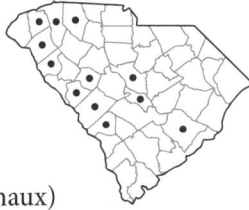

DESCRIPTION: Perennial herb, 8–20″ tall, with tuberlike roots; leaves stiffly erect, to 5″ long; flowers whitish green or yellowish green; lip petal deeply fringed; flowers June–August.
RANGE-HABITAT: Widespread in ne. North America and south to SC, GA, AL, AR, and OK; in SC, rare, scattered throughout; bogs, openings in wet swamps, Longleaf Pine savannas in the Coastal Plain, and in wet meadows.
COMMENTS: No one habitat predictably harbors this species. It is more abundant in the Appalachian region of NC. Like many orchid species, it may be present one year and difficult to find or absent in others.
CONSERVATION STATUS: SC-Imperiled

75. American Climbing Fern;
Hartford Fern

Lygodium palmatum Swartz
Ly-gò-di-um pal-mà-tum
Lygodiaceae (Climbing Fern Family)

DESCRIPTION: Perennial from a long creeping rhizome; leaf (frond) vinelike, coiling, and climbing; pinnae (leaflets) deeply and palmately 4–8 lobed.
RANGE-HABITAT: Widespread in eastern North America; in SC, rare throughout; Southern Appalachian fens, moist forests, and roadsides in strongly acid soil.
COMMENTS: The climbing habit is distinctive to *Lygodium* in SC. The only species easily confused with it is the invasive-exotic, Japanese Climbing Fern (*Lygodium japonicum* (Thunberg) Swartz), which has leaflets pinnately divided into saw-toothed segments and is much more robust. The fronds, not stems, of climbing ferns continue to expand from the tips to form the "vines."

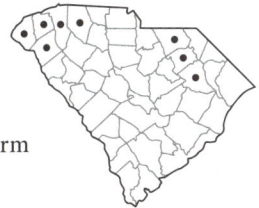

76. Canebrake

77. River Cane

Arundinaria gigantea (Walter) Walter
ex Muhlenberg
A-run-di-nà-ri-a gi-gan-tè-a
Poaceae (Grass Family)

DESCRIPTION: Rhizomatous, perennial, evergreen grass (bamboo) forming extensive colonies (canebrakes); stems 6–30′ tall and to 1.2″ wide, with culm internodes usually grooved; plants flower only every 40–50 years; flowering April–May.

RANGE-HABITAT: VA south to FL and west to MO, OK, and TX; in SC, common in the mountains and upper Piedmont, rare elsewhere; habitats include low lying, moist to wet places such as low woodlands, riverbanks, and stream banks, often abundant along streams in mountain valleys.

SIMILAR SPECIES: The genus *Arundinaria* has three species, all of which occur in SC. Switch Cane (*Arundinaria tecta* (Walter) Muhlenberg, plate 819) is a smaller, evergreen plant, without the dense branching common to River Cane. It is confined to the lower Piedmont and Coastal Plain and has round stem internodes (not grooved) and flowers every 3–4 years. Hill Cane (*Arundinaria appalachiana* Triplett, Weakley & L. G. Clark, plate 185) is a small species with deciduous leaves and round stems (not grooved). It is common on hillsides in a variety of forested habitats in the mountains and upper Piedmont.

COMMENTS: *Arundinaria* is the only genus of the bamboo tribe in the grass family (Poaceae) native to the United States. Numerous other species of the bamboo tribe are cultivated.

Stems of large canes were traditionally used by Native Americans and colonial settlers for many purposes including fishing poles, mats, baskets, blowguns, and furniture. The stems are split and made into chair bottoms. The tender shoots are edible, and the large grains can also be used for food. However, Foster and Duke (1990) warn that ergot, a highly toxic fungus, occasionally replaces the large seeds of River Cane, making them poisonous. Stock browse on the young leaves and stems.

The term "canebrake" comes from this plant. Canebrakes are large areas dominated by dense stands of cane. In colonial times, extensive canebrakes were common in the mountains and Piedmont. Today these brakes are virtually gone or reduced to small stands along streams in pastures or fields. The large canebrakes were managed by Native Americans with fire to encourage their growth and reduce woody competition.

78. Rocky stream side

79. Yellowroot

Xanthorhiza simplicissima
Marshall
Xan-tho-rhì-za sim-pli-cís-si-ma
Ranunculaceae (Buttercup Family)

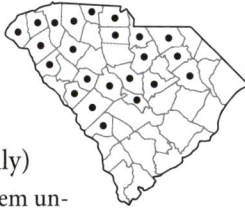

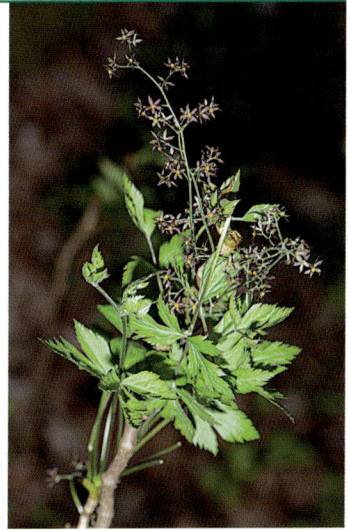

DESCRIPTION: Deciduous shrub to 3′ tall; stem unbranched; leaves in a dense cluster at the tip, each with 3–5 leaflets, margins toothed and cleft; flowers brownish purple, on long stalks in terminal drooping panicles of racemes, appearing before or as the leaves expand; flowers March–May.
RANGE-HABITAT: NY and PA south to FL and AL; in SC, common in the mountains and Piedmont, rare in the Sandhills and inner Coastal Plain; acid soils of shady rocky stream sides and stream margins.
COMMENTS: The inner bark and wood of both stems and roots contains a bright yellow, bitter-tasting alkaloid, which has been used for a dye and medicine. It contains berberine, known to have many useful properties, including the stimulation of bile and bilirubin. It is also a natural antimicrobial and boosts the immune system. It may be useful in lowering tyramine for cirrhosis of the liver but is potentially toxic in large amounts. Duke (1997) recommends Yellowroot as an herbal remedy for athlete's foot and ulcers. A gargle made from the stems is effective in treating mouth ulcers and sore-throat and is still widely used in Appalachia.

80. Sweet-shrub; Sweet Bubby-bush

Calycanthus floridus L.
Ca-ly-cán-thus fló-ri-dus
Calycanthaceae (Sweet-shrub Family)

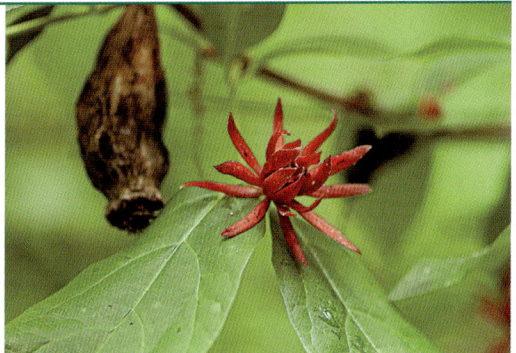

DESCRIPTION: Deciduous shrub, 3–6′ tall; older branches purplish, enlarged and flattened at the nodes, with distinct dichotomous arrangement; leaves opposite, with smooth margins; flowers solitary and terminating axillary branches, purplish-brown or greenish-purple, scented like concord grapes, fetid like rotting flesh, or fungal; all vegetative parts aromatic (like concord grapes or allspice); flowers March–June; fruits mature September–October.
RANGE-HABITAT: PA and OH, south to FL and MS; in SC, common throughout in mixed deciduous forests and rocky stream sides.

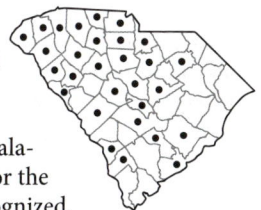

COMMENTS: Sweet-shrub is planted as an ornamental because of the spicy fragrance of its flowers and tidy appearance. It is extremely popular among Appalachian families, where it is called "Sweet-bubby," and planted near the home for the spicy odor of the flowers. Though historically several varieties have been recognized, these appear to not warrant taxonomic recognition.

81. Sweet Azalea; Smooth Azalea

Rhododendron arborescens (Pursh)
 Torrey
  Rho-do-dén-dron ar-bo-rés-cens
 Ericaceae (Heath Family)

DESCRIPTION: Nonclonal deciduous shrub or small tree to 20′ tall; young stems, leaves, and buds smooth and with a whitish hue on the lower leaf surface; flowers white or pink; flowers late May–early July.

RANGE-HABITAT: From PA and KY, south to GA and AL, mostly in the Appalachians; in SC, common in the mountains and upper Piedmont where it occurs on rocky stream sides, cataract fens, and moist soils adjacent to rock outcrops.

COMMENTS: This is the only SC azalea with sweetly scented white-pink flowers and smooth stems and leaves. Both common names highlight distinctive features. It is one of the distinguishing species of the rocky stream side habitat and is common along the Chattooga and Chauga Rivers.

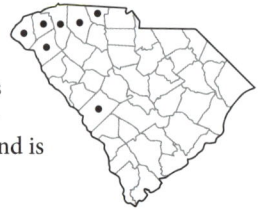

82. Smooth Southern Bush-honeysuckle

Diervilla sessilifolia Buckley
 Di-er-víll-a ses-si-li-fò-li-a
 Diervillaceae (Bush-honeysuckle
 Family)

DESCRIPTION: Deciduous shrub to 6′ tall; leaves opposite, sessile or with very short petioles; plants without hairs except for twig angles; yellow flowers in clusters at branch tips or in leaf axils; flowers June–August.

RANGE-HABITAT: Endemic to the southern Appalachian region in TN, NC, SC, GA, and AL; in SC, occasional in the mountains; rocky stream sides, outcrops, and ridges at moderate to high elevations.

COMMENTS: Flowers resemble those of honeysuckle but are only 0.8″ long, light yellow with a darker center, and turn reddish with age. This is the only species of bush-honeysuckle in SC. The genus honors late seventeenth- and early eighteenth-century botanist, Dr. Marin Dièreville, a French surgeon, who discovered the genus while in Canada and introduced it into cultivation in Europe.

CONSERVATION STATUS: SC-Vulnerable

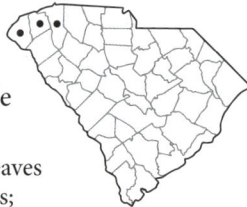

83. Mountain Dwarf-dandelion

Krigia montana (Michaux) Nuttall
 Kríg-i-a mon-tà-na
 Asteraceae (Aster Family)

DESCRIPTION: Perennial herb with milky juice, from a nontuberous rootstalk; leaves linear-lanceolate, in a basal rosette and on a branched stem; flowers yellow, arranged in involucrate heads of ray flowers only; flowers May–September.

RANGE-HABITAT: Endemic to the southern Appalachian region in TN, NC, GA, and SC; in SC, occasional, restricted to the mountains where it grows in moist crevices in cliffs, granitic domes, and rocky stream sides.

COMMENTS: Mountain Dwarf-dandelion often forms clusters of many stems, which, along with its perennial habit and nontuberous rootstalk, help to distinguish it from its weedy relatives. The genus honors David Krig (1667–1713), a German physician and botanist, who was among the first to collect plants in Maryland.

CONSERVATION STATUS: SC-Vulnerable

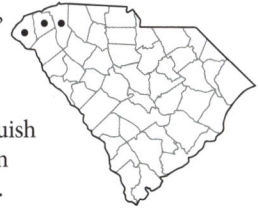

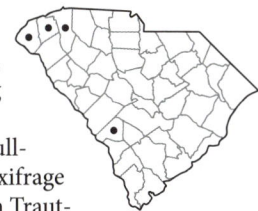

84. Brook-saxifrage; Eastern Boykinia

Boykinia aconitifolia Nuttall
  Boy-kín-i-a a-co-ni-ti-fò-li-a
Saxifragaceae (Saxifrage Family)

DESCRIPTION: Perennial herb to 20″ tall; basal and stem leaves deeply 5–7-lobed and toothed, wider than long, to 4″ wide; inflorescence is a roughly pyramidal cyme of small white flowers; fruits are capsules; flowers June–mid July; fruits mature August–September.

RANGE-HABITAT: Endemic to the southern Appalachian region from WV south to GA and AL; in SC, rare in the mountains on rocky stream sides and small seeps.

COMMENTS: Leaves superficially resemble Thimbleweed (*Anemone virginiana* L., plate 332) or Carolina Tassel-rue (*Trautvetteria caroliniensis* (Walter) Vail, plate 85). The epithet *aconitifolia* translates literally as "with leaves of *Aconitum*." Brook-saxifrage's leaves are much less similar to common Blue Monkshood (*Aconitum uncinatum* L., plate 142) than they are to thimbleweed or tassel-rue. The glandular hairs on the rachis of all stems of the flower cluster separate this species from any of the above. The genus honors Samuel Boykin (1786–1848), a Georgia botanist.

CONSERVATION STATUS: SC-Imperiled

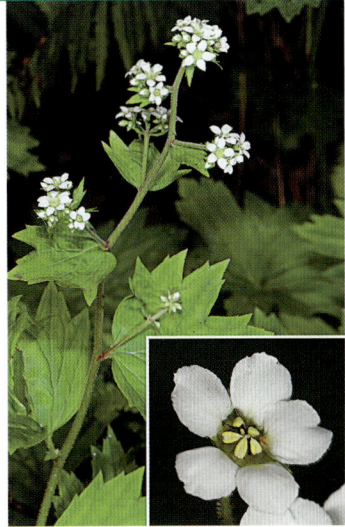

85. Carolina Tassel-rue; False Bugbane

Trautvetteria caroliniensis (Walter) Vail
  Traut-vet-tèr-i-a ca-ro-li-ni-én-sis
Ranunculaceae (Buttercup Family)

DESCRIPTION: Perennial herb to 4′ tall; with basal and stem leaves, to 12″ wide, reduced upward on the stem, palmately 3–11-lobed; inflorescence a terminal cyme (flat-topped); flowers without petals; the numerous long, whitish filaments of the stamens are the showy part of the flowers; fruits are aggregates of utricles (bladderlike and 1-seeded); flowers June–early July; fruits mature August–September.

RANGE-HABITAT: PA and KY, south to GA; in SC, occasional in the mountains, very rare in the Coastal Plain; stream sides, cataract fens, and seepages along mountain streams.

COMMENTS: Tassel-rue is distinctive when in flower. Sterile plants when not full-sized resemble Thimbleweed (*Anemone virginiana* L., plate 332) or Brook-saxifrage (*Boykinia aconitifolia* Nuttall, plate 84). The genus honors Ernst Rudolph von Trautvetter (1809–89), a distinguished Russian botanist.

CONSERVATION STATUS: SC-Vulnerable

86. Western Sampson's-snakeroot

Orbexilum pedunculatum (P. Miller) Rydberg
Or-béx-i-lum pe-dun-cu-là-tum
Fabaceae (Bean Family)

SYNONYM: *Psoralea psoralioides* (Walter) Cory var. *eglandulosa* (Elliott) F. L. Freeman—RAB

DESCRIPTION: Perennial herb, 1–3′ tall, arising from a cigar-shaped taproot; leaves trifoliolate; leaflet surfaces, bracts, calyces, and pods without glands or very sparsely glandular; flowers blue-purple, in racemes (almost spikelike); flowers May–July; legume matures July–September.

RANGE-HABITAT: Primarily a midwestern plant; scattered and uncommon in and east of the Blue Ridge; in SC, rare in the mountains and upper Piedmont; rocky stream sides, scour prairies, xeric hardpan forests, open woodlands, fields, and clearings.

SIMILAR SPECIES: Eastern Sampson's-snakeroot (*Orbexilum psoralioides* (Walter) Vincent, plate 574) has leaflet surfaces, bracts, calyces, and pods conspicuously glandular; it is found in the lower Piedmont, Sandhills, and Coastal Plain of SC where it is common.

COMMENTS: One of many plants that was believed to act against snake venom. Western Sampson's-snakeroot can be locally abundant in rocky stream side, prairielike habitats (termed "scour prairies"). These tiny prairie pockets are important for the preservation of many species. This plant may be seen growing along the Chauga River at Chau-Ram County Park in Oconee County.

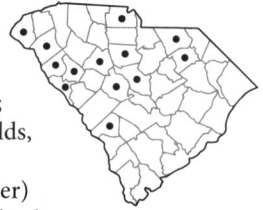

87. Maryland Meadow-beauty

Rhexia mariana L. var. *mariana*
Rhéx-i-a ma-ri-à-na var. ma-ri-à-na
Melastomataceae (Melastome Family)

DESCRIPTION: Aggressively colonial rhizomatous perennial herb; stems unbranched or branched, hirsute (hairy), unequally angled or grooved, to 30″ tall; leaves opposite, lanceolate to ovate, three-nerved; flowers with four, large, pale-pink petals, subtended by an urn-shaped hypanthium that is sparsely covered with glandular hairs; flowers May–October.

RANGE-HABITAT: MA south to FL and west to IL and TX; in SC, common throughout the state in open, moist situations such as rocky stream sides, ditches, wetland savannas, and moist road banks.

SIMILAR SPECIES: Virginia Meadow-beauty (*Rhexia virginica* L., plate 88) has stems with four equal faces that are sharply angled or winged. Nash's Meadow-beauty (*Rhexia nashii* Small, plate 631) is more robust, with larger, brighter pink flowers and hypanthia that are smooth (glabrous or glabrate) with a few long hairs at the tips, between the sepals.

COMMENTS: Like all meadow-beauties, this one has flowers that are open for only a few hours, in the morning through early afternoon. Maryland Meadow-beauty is the most abundant *Rhexia* in SC. It has been collected in every county in the state.

88. Virginia Meadow-beauty

Rhexia virginica L.
    Rhéx-i-a vir-gí-ni-ca
    Melastomataceae (Melastome
        Family)

DESCRIPTION: Perennial herb with tuberous roots, forming colonies; stems 1–3′ tall; branched, hirsute (hairy), sharply-angled with four, relatively equal sides and winged corners; leaves opposite, lanceolate to ovate, three-nerved; flowers with four large bright pink petals, subtended by an urn-shaped hypanthium with glandular hairs; flowers May–October.

RANGE-HABITAT: Widespread in eastern North America; in SC, common in the mountains and Coastal Plain, less common in the Piedmont; rocky stream sides, Southern Appalachian fens, wet meadows, ditches, Longleaf Pine savannas and flatwoods, pocosin margins, upper edges of depression ponds, and ditches and other open, wet areas.

SIMILAR SPECIES: See comments under *R. mariana* L., above. Swollen Meadow-beauty (*R. ventricosa* Fernald & Griscom) is very similar but lacks the tuberous roots and is restricted to the Coastal Plain in SC.

COMMENTS: This is the most widespread species of meadow-beauty in the eastern United States. It is generally not as abundant as other *Rhexia* species in the Coastal Plain but is common in the mountains and upper Piedmont. Like all meadow-beauties, the flowers are open only for a few hours in the morning, and the petals fall in the afternoon.

89. Cutleaf Coneflower; Sochan

Rudbeckia laciniata L.
    Rud-béck-i-a la-ci-ni-à-ta
    Asteraceae (Aster Family)

DESCRIPTION: Perennial herb; stem smooth, 3–8′ tall, branched; leaves alternate, long-stalked, reduced in size and number of divisions from base upward; lower leaves pinnately dissected, with segments coarsely toothed; flowers in involucrate heads of yellow ray flowers and yellow disk flowers; disk flowers are green in bud and yellow when open; flowers July–October.

RANGE-HABITAT: Widespread in eastern North America; in SC, common throughout in moist soils of levee forests, bottomland hardwoods, rocky stream sides, and moist meadows.

COMMENTS: The Cherokee refer to this plant as "Sochan," and it is utilized as a nutritious source of greens as a potherb. *Rudbeckia* species in general are reported to contain compounds that stimulate the immune system and may be useful in treating HIV. Several varieties of this variable taxon have been named. The genus name honors professors Rudbeck (Olaf, 1630–1702, and Olaf, 1660–1740, his son), predecessors of Linnaeus at Uppsala University in Sweden.

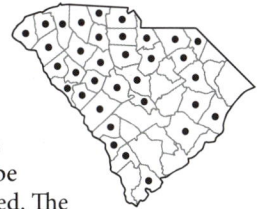

90. Hollow-stem Joe-pye-weed

Eutrochium fistulosum (Barratt)
    E.E. Lamont
    Eu-trò-chi-um fis-tu-lò-sum
Asteraceae (Aster Family)

SYNONYM: *Eupatorium fistulosum*
    Barratt—RAB, PR

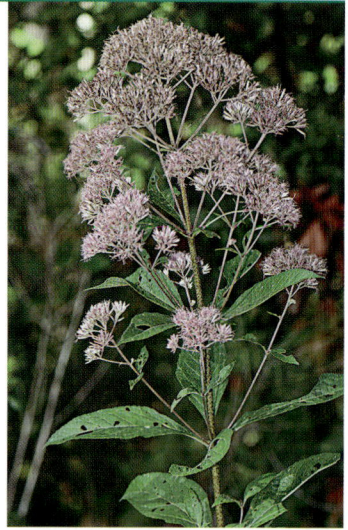

DESCRIPTION: Perennial herb; stems to 10′ tall, hollow, green with whitish and purple spots; leaves in whorls of 3–7; flowers arranged in involucrate heads of pinkish disk flowers only; flowers, 4–7 per head; flowers July–October.

RANGE-HABITAT: Widespread in the eastern and central US; in SC, common throughout in marshes, wet ditches, moist meadows, rocky stream sides, forest margins, and Southern Appalachian fens.

SIMILAR SPECIES: Purple Joe-pye-weed (*Eutrochium purpureum* (L.) E. E. Lamont) is similar to Hollow-stem Joe-pye-weed, but it has solid stems, seldom attains such a lofty stature, and occurs in drier, mostly forested habitats.

COMMENTS: This is one of the most attractive species to butterflies, particularly swallowtails. It is extremely useful as an addition to the home landscape to attract and support pollinators, but it often grows so tall that it falls over. It can be staked or caged to keep the stems upright.

91. Tall Scouring-rush

Equisetum hyemale L. ssp. *affine*
    (Englemann) Calder &
    R. L. Taylor
    Eq-ui-sè-tum hye-mà-le ssp.
    af-fi-ne
Equisetaceae (Horsetail Family)

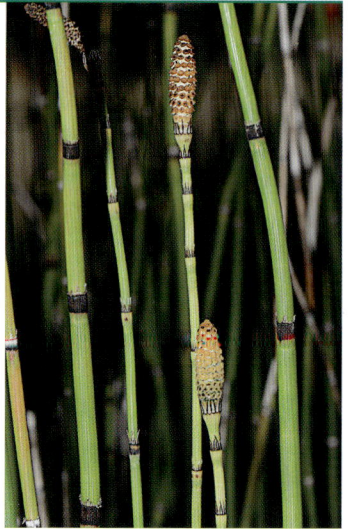

DESCRIPTION: Evergreen, perennial, nonflowering, vascular plant from a creeping underground stem; aerial stems stiffly erect, unbranched, with ribbed internodes; leaves scalelike, in whorls that are fused at the base into a sheath; sporangia clustered into a terminal conelike structure (strobilus).

RANGE-HABITAT: Scattered throughout North America and in Mexico and Guatemala. In SC, occasional, chiefly in the mountains and Piedmont; scattered and rare in the Coastal Plain; rocky or sandy stream sides, railroad banks, and roadsides.

COMMENTS: Silicon dioxide (sand) in the cell walls gives the stems a rough texture. The stems were once used to scour pots and to give wood a fine finish. It is poisonous to livestock and alters vitamin (thiamine) metabolism. The German Commission E in Germany has approved and Duke (1997) recommends its use as an herbal remedy for gallstones and kidney stones.

92. Rich cove forest

93. Yellow Buckeye

Aesculus flava Solander
Aès-cu-lus flà-va
Hippocastanaceae (Horse-chestnut Family)

SYNONYM: *Aesculus octandra* Marshall—RAB

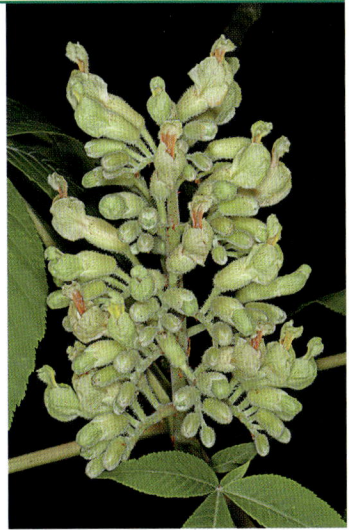

DESCRIPTION: Deciduous tree to 100′+ tall and 6′ in diameter; bark shaggy, forming large slabs, gray; leaves deciduous, palmately compound with 5–7 leaflets; calyx and pedicels of flowers with glandular hairs; petals yellow or rarely salmon; capsule leathery, with 1–3 large, dark-brown, shiny seeds; flowers March–April; fruits mature July–August.

RANGE-HABITAT: PA south to GA and west to IL and AL, primarily in the Appalachian region; in SC, restricted to the mountains and upper Piedmont where often abundant in rich cove forests, along streams and in basic-mesic forests.

COMMENTS: This is our only buckeye that becomes a large tree. It is abundant in rich cove forests, although it also grows in more acidic situations. The leaves and seeds of all buckeyes are poisonous, and the crushed fruits were once used to stupefy fish for easy collection. All buckeye species produce their leaves very early in the spring and drop their leaves early in the fall. They make excellent landscaping plants. Some populations in the mountains of SC have salmon-pink flowers, particularly in Eastatoe Valley. This unusual coloration may be due to hybridization with Red Buckeye (*A. pavia* L., plate 742) in the distant past. The genus is named in honor of the Greek playwright Aeschylus (525–456 B.C.).

94. Butternut; White Walnut

Juglans cinerea L.
Jùg-lans ci-nè-re-a
Juglandaceae (Walnut Family)

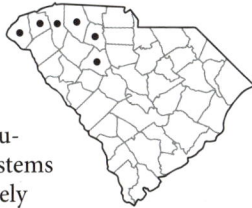

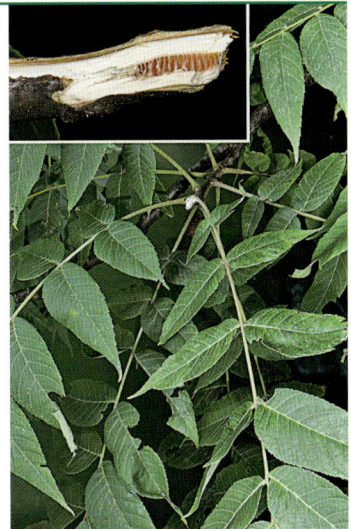

DESCRIPTION: Deciduous tree to 90′ tall, usually much smaller; bark pale gray; pith of stems chambered and dark brown; leaves pinnately compound with 7–17 leaflets, resembling Black Walnut but with a terminal leaflet; inflorescence of long catkins; flowers April–May; fruits large, ellipsoid, densely pubescent; fruits mature September–October.

RANGE-HABITAT: New Brunswick to MN, south to n. GA and AR; in SC, uncommon and restricted to rich cove forests, basic-mesic forests, and rich bottomlands along streams and rivers in the mountains and Piedmont.

COMMENTS: Butternut is a good indicator of high pH soils and is often associated with other rare species. Butternut produces a good quality, tasteful nut, though it is nearly as hard to extract as Black Walnut. It has declined dramatically throughout its range because of Butternut Canker Disease, caused by an invasive, exotic fungus (*Sirococcus clavigignenti-juglandacearum*). This fungus was first detected in Wisconsin in 1967 and is thought to have originated in Asia. Although never a common tree in our area, Butternut is threatened with extinction because of the Butternut canker disease.

CONSERVATION STATUS: SC-Imperiled

95. Tulip-tree; Yellow Poplar

Liriodendron tulipifera L.
Li-ri-o-dén-dron tu-li-pí-fe-ra
Magnoliaceae (Magnolia Family)

DESCRIPTION: Large deciduous tree, 98–165′ tall, with a straight trunk; leaves and flower unmistakable; winter buds flattened and enclosed by two nonoverlapping scales; flowers April–June.

RANGE-HABITAT: Widespread in eastern North America; in SC, common throughout in moist, rich soils of cove forests and virtually all other forest types; easily moves into drier sites such as abandoned fields and timbered uplands; in the Coastal Plain, it also grows in acidic soils with species such as *Pinus serotina, Nyssa biflora,* and *Acer rubrum* in streamhead pocosins.

COMMENTS: Tulip-trees of the original forest grew to 165′ tall. The SC state champion tree measured 135′ tall in 1984. Trees grown in forest conditions had 80–100′ tall straight trunks, a feature that made it a superior lumber tree. Tulip-tree wood was historically used for lumber more than any other hardwood. Today, its wood has a myriad of uses such as interior finishes, furniture, general construction, and plywood. The Charleston furniture maker Thomas Elf (1719–75) employed it in his works. It is planted as a shade tree. The seeds are an important food for wildlife. The common and scientific names come from the flowers resembling cultivated tulips.

Plants growing in blackwater swamps and streamhead pocosins in the Fall-line Sandhills apparently represent an undescribed variety and may be separated by their rounded lobes (Weakley 2020). Tulip Trees are one of the hosts of the Tiger Swallowtail, SC's state butterfly as well as the huge Tulip Tree Silkmoth.

96. Yellow-wood

Cladrastis kentuckea
(Dumont de Courset) Rudd
Cla-drás-tis ken-túc-ke-a
Fabaceae (Bean Family)

SYNONYM: *Cladrastis lutea* (Michaux f.)
K. Koch—RAB

DESCRIPTION: Deciduous tree to 60′ tall with smooth, dark bark; freshly cut wood is bright yellow; leaves pinnately compound with leaflets arranged alternately along the rachis; inflorescence to 1.5′ long composed of white, fragrant, beanlike flowers; flowers April–May; fruit a legume, mature July–August.

RANGE-HABITAT: OH south to GA and west to IN and OK; the range is primarily in the Appalachians and Ozarks; in SC, rare in the mountains where it is found in rich cove forests; very rare in the Piedmont and known from a single location in a basic-mesic forest in Aiken County.

COMMENTS: Yellow-wood is restricted to the high pH soils of extremely rich, mesic forests. For many years it was known in SC only from a single location in Aiken County along the Savannah River. Large populations of mature trees are found in some areas of the Jocassee Gorges and also on Pinnacle Mountain, in Table Rock State Park. This species makes a very handsome landscape tree. Beautiful horticultural specimens can be observed at Fort Hill Estate, on the Clemson University campus.

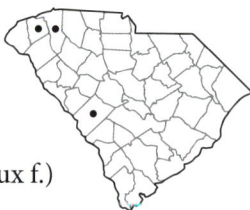

97. Mountain Basswood; White Basswood

Tilia americana L.
   Tí-li-a a-me-ri-cà-na
Malvaceae (Mallow Family)

DESCRIPTION: Deciduous tree to 100′ tall; often multitrunked; young twigs smooth; leaves simple, alternate, broadly ovate and oblique or cordate at the base; leaf surface pale beneath and densely pubescent with microscopic, star-shaped trichomes; flowers perfect, produced in a cyme that is fused to a distinctive leaflike bract; flowers small, whitish, fragrant; flowers May–June; fruits dry, brownish, and nutlike; fruits mature July–August.

RANGE-HABITAT: PA south to GA and sporadically to the FL Panhandle; also in the Ozarks; in SC, common in the mountains where characteristic of rich cove forests; less abundant, with disjunct populations in the Piedmont in basic-mesic forests, and Coastal Plain in marl forests, shell hammock forests, and river bluffs.

COMMENTS: The taxonomy of basswoods is very complex and poorly understood. Currently, a single species is found in the eastern United States with three varieties, two of which, White Basswood (*T. ameriana* var. *heterophylla* (Ventenat) Loudon), and Carolina Basswood (*T. americana* var. *caroliniana* (P. Miller) Castiglioni), are found in SC. Carolina Basswood is a smaller tree with grayish or brownish undersurfaces to the leaves and is found in marl forests and shell hammocks in the Coastal Plain. The distinction between the two can be extremely difficult at times and the general range of *Tilia americana* is mapped here. All basswoods are excellent landscape trees and provide nectar and pollen for insects; the highly fragrant flowers are very attractive to honeybees.

98. South Carolina Sedge

Carex austrocaroliniana
   L. H. Bailey
      Cà-rex aus-tro-ca-ro-li-ni-à-na
Cyperaceae (Sedge Family)

DESCRIPTION: Densely tufted, evergreen perennial sedge; leaves primarily basal, grasslike, less than 0.25 inch wide; flowering stems (culms) to 18″ long, often flopping over or reclined, dark purplish basally; stem leaves mostly reduced and bractlike; female flower spikes held on long, weak, extremely narrow, and hairlike peduncles that dangle from the culm; flowers March–May; fruits are achenes enclosed by a green perigynium; fruits mature May–June.

RANGE-HABITAT: Endemic to the southern Appalachian region from sw. NC and se. KY south to GA; in SC, common in the mountains, where found in rich cove forests and other nutrient-rich forests along streams and bases of slopes.

COMMENTS: Sedges can be extremely frustrating to identify for those who are not professional sedge taxonomists. They are also one of the most diverse groups of plants in SC and are often excellent indicators of special habitats or ecosystem health. South Carolina Sedge is easy to identify because of the narrow brilliant green, tight, attractive clumps of leaves; the purplish leaf bases and bracts; and the delightful dangling spikes of female flowers. It is present in nearly every rich cove forest in SC mountains, and the abundance in mountains makes it worthy of the common name. The specific epithet translates as South Carolina.

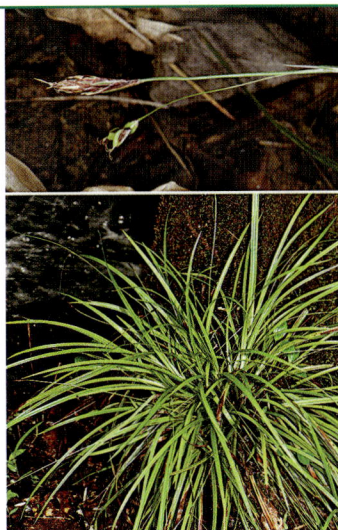

99. Seersucker Sedge; Plantainleaf Sedge

Carex plantaginea Lamarck
Cà-rex plan-ta-gí-ne-a
Cyperaceae (Sedge Family)

DESCRIPTION: Densely tufted, evergreen perennial sedge; leaves primarily basal, to 1.25″ wide, appearing puckered; flowering stems (culms) to 20″ long, often flopping over or reclined, dark purplish basally; stem leaves mostly reduced and bractlike with purplish bases; female flower spikes held on very short peduncles that do not dangle from the culm; flowers March–May; fruits an achene enclosed by a green perigynium; fruits mature May–June.

RANGE-HABITAT: New Brunswick south to KY and west to MN and GA; in SC, restricted to the mountains where it is common and characteristic of rich cove forests and other nutrient-rich forests along streams or bases of slopes.

COMMENTS: The extremely wide, "puckered" leaves may appear like seersucker, hence the common name. This distinctive sedge is easy to identify and also a beautiful addition to the woodland garden where it provides year-round interest. It is available from many reputable nurseries and should not be dug from the wild.

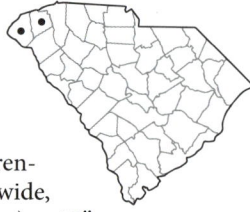

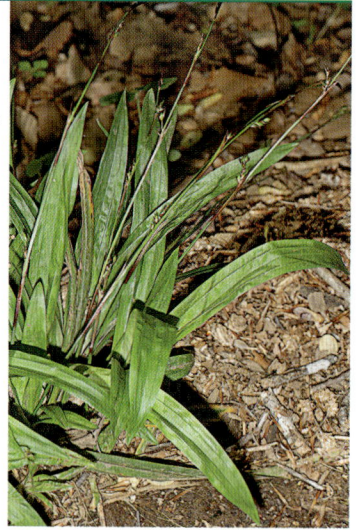

100. Radford's Sedge

Carex radfordii Gaddy
Cà-rex rad-fòrd-i-i
Cyperaceae (Sedge Famliy)

DESCRIPTION: Robust, densely tufted, evergreen perennial sedge; leaves powdery-blue; primarily basal, to 1″ wide; flowering stems (culms) to 30″ tall; base of stem green, white, and brownish-striped; female flower spikes held on very short peduncles that do not dangle from the culm; flowers April–May; fruits and achene enclosed by a perigynium with a long, curved apex; fruits mature May–June.

RANGE-HABITAT: Endemic to the southern Blue Ridge Escarpment in sw. NC, at Whitewater Falls, and from Pickens County, SC west through Rabun County, GA; rich cove forests over high magnesium (mafic rock) such as amphibolite or other high pH substrates.

COMMENTS: This sedge was named in honor of Albert E. Radford (1918–2006), professor of botany at the University of North Carolina, Chapel Hill in 1995 by SC botanist L. L. Gaddy in honor of. It is endemic to a very narrow area with most of the populations found within the Jocassee Gorges region of Pickens County.

CONSERVATION STATUS: SC-Vulnerable

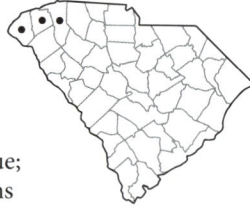

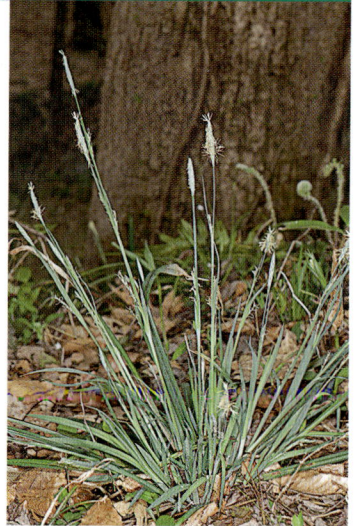

101. Sharp-lobed Liverleaf; Sharp-lobed Hepatica

Anemone acutiloba (Augustin de Candolle) G. Lawson
A-ne-mò-ne a-cu-tí-lo-ba
Ranunculaceae (Buttercup Family)

SYNONYM: *Hepatica acutiloba* Augustin de Candolle—RAB, PR

DESCRIPTION: Herbaceous, evergreen perennial from a short underground stem; leaves basal,

deeply cleft into 3 acute-tipped lobes; flowers solitary on elongated, hairy stalks; petals absent; sepals petal-like and whitish, sometimes bluish or pinkish; flowers March–April.

RANGE-HABITAT: Widespread in eastern North America; in SC, common in the mountains, extending barely into the Piedmont along the edge of the escarpment; rich cove forests and other mesic forests, often associated with circumneutral soils.

SIMILAR SPECIES: Round-lobed Liverleaf (*A. americana* (A.P. de Candolle) Ker-Gawler, plate 319) is similar but smaller with blunt-lobes at the leaf tips. It is common in the Piedmont and uncommon in the Coastal Plain where found in cool, shaded slopes and bluffs, usually on acidic soils. Though the two species barely overlap in range, both may be found growing together at Nine Times Preserve in Pickens County.

COMMENTS: Old leaves are present at early flowering; new leaves develop well after flowering. The liver-leafs were formerly placed in the genus *Hepatica,* which refers to the similarity between the leaves that turn brownish in winter to the human liver (hepatic organ). Historically, it was used to treat issues that were thought to be linked to the liver, such as cowardice and freckles. The traditional use of plants to treat portions of the body they resemble in appearance is termed the "Doctrine of Signatures."

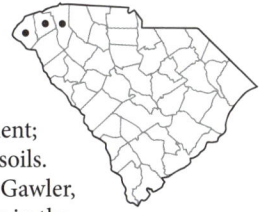

102. Allegheny Spurge

Pachysandra procumbens
    Michaux
        Pa-chy-sán-dra pro-cúm-bens
    Buxaceae (Boxwood Family)

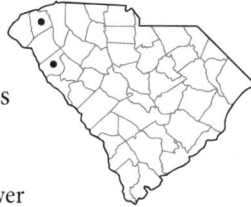

DESCRIPTION: Perennial herb from a semi-woody base; stems 6–9″ long, with the lower portion on the ground and the upper portion erect; the few, alternate leaves are mottled with lighter green and located near the tip of the erect stem; the ill-scented flowers consist of 4–5 sepals produced on 1–several elongated stalks that originate from the lower portion of the stem; flowers March–April.

RANGE-HABITAT: KY south to FL; in SC, rare, in rich cove forests and basic-mesic forests in a single mountain county and, at least formerly, in the upper Piedmont.

COMMENTS: The majority of the distribution of this species is west of the Blue Ridge Mountains. It is still locally common at a few locations along the Eastatoe River in Pickens County, SC, and was formerly known from rich forests in Abbeville County, at a location that was flooded during the construction of Lake Russell. It should be sought in basic-mesic forest habitats in other regions of the Piedmont. It is more common in cultivation, among wildflower enthusiasts, than it is in the wild in SC. Though it is similar to the popular horticultural plant, Japanese Pachysandra (*Pachysandra terminalis* Siebold & Zuccarini), it does not form dense patches readily and is better utilized as a specimen than a groundcover.

CONSERVATION STATUS: SC-Imperiled

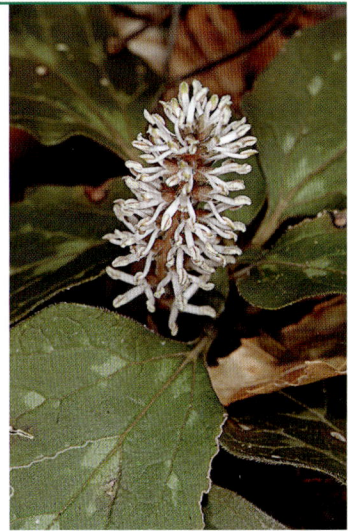

103. Windflower; Rue-anemone

Thalictrum thalictroides (L.)
    A. J. Eames & B. Boivin
        Tha-líc-trum tha-lic-troì-des
    Ranunculaceae (Buttercup Family)

DESCRIPTION: Perennial herb with tuberous roots; stem to 10″ tall; leaves compound, with leaf-lets in 3's and 3-lobed at the apex; flowers in a single, terminal umbel subtended by opposite leaves; sepals 5, white; petals not present; flowers March–May.

RANGE-HABITAT: Widespread in eastern North America; in SC, common in cove forests in the mountains and basic-mesic and beech forests in the Piedmont; rare in beech forests in the Coastal Plain.

SIMILAR SPECIES: False Rue-anemone (*Enemion biternatum* Rafinesque, plate 277) is very similar but is found on high pH soils of basic-mesic forests in the Piedmont. These two look-alikes can be distinguished by the opposite leaves subtending the inflorescence in Windflower versus the alternate leaves of False Rue-anemone. False Rue-anemone also does not produce umbels of flowers but has axillary and terminal flowers.

COMMENTS: Some authors report that the tuberous roots are edible, but according to recent reports, they are potentially toxic.

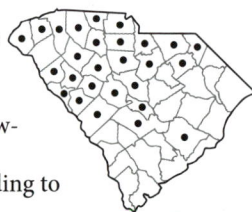

104. Sessile-leaf Bellwort; Straw-lily

Uvularia sessilifolia L.
U-vu-là-ri-a ses-si-li-fò-li-a
Colchicaceae (Meadow Saffron Family)

DESCRIPTION: Perennial herb forming colonies from underground stolons; stems to 18″ tall, 1-branched when flowering, upper portion without hairs; leaves sessile, dull green, usually glaucous (whitish) below; flowers usually 1 per stem, light yellow to cream and nodding, bell-like; flowers late March–early May.

RANGE-HABITAT: Widespread in eastern North America; in SC, common in the mountains and Piedmont, rare in the Coastal Plain; moist forests, especially in rich cove forests and along streams or bases of slopes where nutrients accumulate in acidic cove forests; basic-mesic forests, beech forests and hardwood bottoms in the Piedmont, and hardwood bottoms in the Coastal Plain.

SIMILAR SPECIES: Appalachian Bellwort (*U. puberula* Michaux, plate 235) is similar. It is distinguished by its lack of stolons, its minutely pubescent upper stems and backs of leaves, its deep green leaves, and its occurrence in much drier habitats.

COMMENTS: Sometimes called "wild oats" due to the clasping and upward-pointing leaves of the young stems, which is reminiscent of oats. Large colonies of Sessile-leaf Bellwort may be seen very near the parking lot for the visitors' center at Jones Gap State Park in Greenville County.

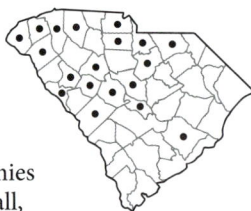

105. Blue Cohosh

Caulophyllum thalictroides (L.) Michaux
Cau-lo-phýl-lum tha-lic-troì-des
Berberidaceae (Barberry Family)

DESCRIPTION: Smooth, erect perennial herb to 3′ tall, from a knotty rootstalk; all parts with a bluish-white cast; the stem bears a single leaf with 3 long stalks, each bearing many leaflets that are 2–5-lobed above the middle; leaf appearing to be 3 leaves because of the obscure common leaf stalk; 1–3 clusters of yellowish green or greenish purple flowers terminate the stem; flowers to 0.4″ in diameter; flowers April–May; fruits are blue when mature, July–August.

RANGE-HABITAT: Widespread in eastern North America; in SC, common in the mountains; rare in the Piedmont; rich cove forests and basic-mesic forests.

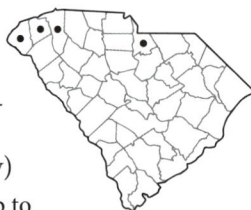

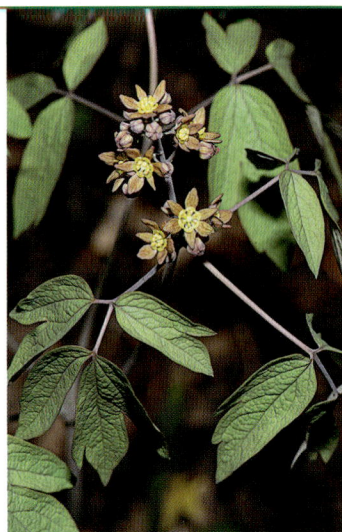

COMMENTS: The developing seeds rupture the ovary soon after fertilization, so the dark blue seeds, which resemble drupes, ripen fully exposed. Native Americans used a root extract to induce labor and reduce menstrual cramps. Duke (1997) recommends it as an herbal remedy for inducing labor. The flowers have a distinctive odor, like cardboard. They may just remind you of playing in a homemade fort made from a large cardboard box as a child.

106. Foamflower

Tiarella cordifolia L.
  Ti-a-rél-la cor-di-fò-li-a
Saxifragaceae (Saxifrage Family)

DESCRIPTION: Perennial herb, 4–20″ tall; basal leaves with a heart-shaped base and long leaf stalk; flowering stalk leafless; petals white and stalked; fruit a two-parted capsule; flowers April–June.

RANGE-HABITAT: From Nova Scotia south to AL and west to Ontario and MO; in SC, common in the mountains and Piedmont; rich cove forest, acidic cove forests, basic mesic forests, and beech forests.

COMMENTS: Foamflower is easily confused with Two-leaved Miterwort (*Mitella diphylla* L.) when not in fruit or flower; Two-leaved Miterwort usually has just two basal leaves, oppositely arranged, whereas foamflower has many basal leaves. *Tiarella* means "little tiara," apparently in reference to the appearance of the fruit. Two forms of Foamflower occur in SC: one is tightly clumping and the other is long-rhizomatous. These two forms have previously been considered to be separate varieties.

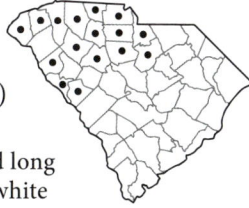

107. Golden-seal

Hydrastis canadensis L.
  Hy-drás-tis ca-na-dén-sis
Hydrastidaceae (Goldenseal Family)

DESCRIPTION: Perennial herb, 6–18″ tall; sterile plants with 1 leaf, fertile plants with 2 alternate leaves, both near the apex of the flowering stem; leaves pubescent, with cordate bases and with 3–7 lobes, serrate; flowers without petals and consisting of numerous showy stamens and pistils; fruits an aggregate of red berries; flowers March–April; fruits mature May–June.

RANGE-HABITAT: VT south to GA and west to MN and AR; in SC, very rare and restricted to rich cove forests in the mountains.

COMMENTS: Goldenseal, like American Ginseng, has been dramatically exploited for traditional medicinal uses and is threatened or has disappeared in many parts of its range due to overcollection. The bright yellow rhizome contains berberine and is thought to boost the immune system. It is also used as a cold remedy, treatment for allergies, and as a digestive aid, among many other uses. Berberine is toxic in large quantities and can lead to cardiac arrhythmia and death. The same chemical is found in other much more abundant species, such as Yellowroot, and its collection from the wild is unwarranted. Though its occurrence in SC has been ignored by most sources, Francis P. Porcher, in 1869, reports that "it grows among the mountains of South Carolina." The species was verified for SC by Kathy Kegley and Patrick McMillan in 2002. Goldenseal grows in the highest pH moist soils of cove forests over amphibolite in the Jocassee Gorges region but should be sought in other sites in Oconee and Greenville Counties.

CONSERVATION STATUS: SC-Critically Imperiled

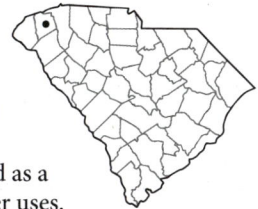

108. Little Sweet Betsy; Purple Toadshade

Trillium cuneatum Rafinesque
Tríl-li-um cu-ne-à-tum
Trilliaceae (Trillium Family)

DESCRIPTION: Perennial herb from a short rhizome; stem erect, to 15″ tall; leaves sometimes so broad that the margins of the 3 whorled leaves overlap; flowers sessile; petals purple, green, or yellow, usually less than 4x as long as they are wide; the two whorls of stamens are alike; flowers March–April.

RANGE-HABITAT: NC south to GA and west to AL, KY, and TN; common in the mountains and portions of the upper Piedmont; usually found in rich cove forests, basic mesic forests, beech forests, and floodplains.

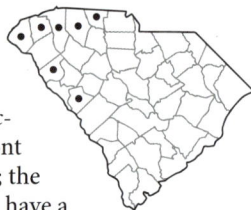

SIMILAR SPECIES: This species closely resembles Mottled Trillium (*Trillium maculatum* Rafinesque, plate 749). Mottled Trillium is found in the lower Piedmont and Coastal Plain and has petals more than 4.5 times as long as they are wide; the petals are typically bright red, sometimes yellow or brownish, and the flowers have a spicy, bananalike odor.

COMMENTS: Little Sweet Betsy is either strongly fetid (like rotting meat or fungal) or like sweet concord grapes. Plants with sweet odors are common in Greenville County and to the north and east. Plants with fetid odor are abundant in Oconee County. Huge colonies may be observed at Station Cove Falls in the Sumter National Forest or at Chestnut Ridge Heritage Preserve in Greenville County. Yellow-flowered forms have been confused with Yellow Trillium (*T. luteum* (Muhlenberg) Harbison), which does not occur in SC.

109. Pale Yellow Trillium; Faded Trillium

Trillium discolor Wray ex Hooker
Tríl-li-um. dís-co-lor
Trilliaceae (Trillium Family)

DESCRIPTION: Herbaceous perennial from a short, thick rhizome; stem erect, to 12″ tall; leaves in a single whorl of 3, stalkless, mottled with 2–3 shades of green, sometimes almost circular in outline; flowers sessile; petals spoon-shaped, incurved, pale yellow with a greenish or purplish stalklike base; at least one petal terminated by a distinct point; ovary distinctly 6-angled; flowers late March–early May.

RANGE-HABITAT: Restricted to the mountains and Piedmont of the Savannah River drainage of NC, SC, and GA; common; rich cove forests and basic-mesic forests.

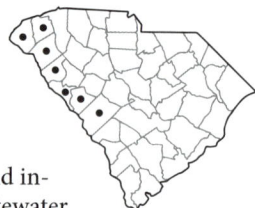

COMMENTS: This is the most abundant *Trillium* in the upper Savannah River drainage, although it has a very small range. The pale spoon-shaped petals and intense clovelike odor are distinctive. The range barely extends into NC at Whitewater Falls, and the vast majority of populations are found in the Jocassee Gorges region.

110. Sweet White Trillium

Trillium simile Gleason
Tríl-li-um sí-mi-le
Trilliaceae (Trillium Family)

DESCRIPTION: Perennial herb from a thick rhizome; stem erect, usually 10–24″ tall; leaves in a single whorl of 3, uniformly green, wider than long; flower stalk long and held nearly erect; petals white, broadly ovate, overlapping at the base and producing a funnel-shaped flower when young, spreading with age; sepals curled inward and boatlike at the tip; anther sacs yellow; ovary dark purple, strongly angled; flowers late March–early July.

RANGE-HABITAT: Endemic to the southern Appalachian region in NC, SC, TN, and GA; in SC, local and uncommon in the mountains; rich cove forests or other forests with circumneutral soils.

COMMENTS: The epithet *simile,* meaning similar, is appropriate since it resembles several members of the *T. erectum* species group. Its wide petals and boatlike sepals resemble Vasey's Trillium (*Trillium vaseyi* Harbison, plate 112), but the flowers are held erect on long stalks. The flowers are said to have the odor of green apples.

CONSERVATION STATUS: SC-Vulnerable

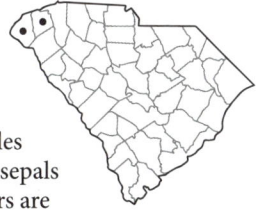

111. Large-flowered Trillium

Trillium grandiflorum
(Michaux) Salisbury
Tríl-li-um gran-di-flò-rum
Trilliaceae (Trillium Family)

DESCRIPTION: Herbaceous perennial from a short, thick rhizome; stem erect, to 12″ tall; leaves in a single whorl of 3, uniform green, sometimes with a reddish tinge in pink-flowered forms, without a leaf stalk; flower borne slightly below the vertical, on a long stalk (to 3″); petals white, rarely light or dark pink, and often fading to pink with age, overlapping at the base to form a strongly funnel-shaped flower, flared at the tip; anthers white or greenish white between the yellow anther sacs; flowers April–May.

RANGE-HABITAT: New England and WI, south to AL, GA, and SC; in SC, rare in the mountains; rich cove forests.

COMMENTS: Four-leaved forms occasionally occur; forms with double, 2, or 4 petals also occur. Infection with mycoplasma may result in unusual forms without petals or with green or green-and-white petals or "knots" for sepals. Though one of the most abundant species of *Trillium* in the US, it is very rare in SC and restricted to a handful of sites very close to the NC border. This species uses pollen to attract potential pollinators; the majority of our *Trillium* lure pollinators with scent but offer no reward. Bees and small flies are the primary pollinators of Large-flowered Trillium.

CONSERVATION STATUS: SC-Critically Imperiled

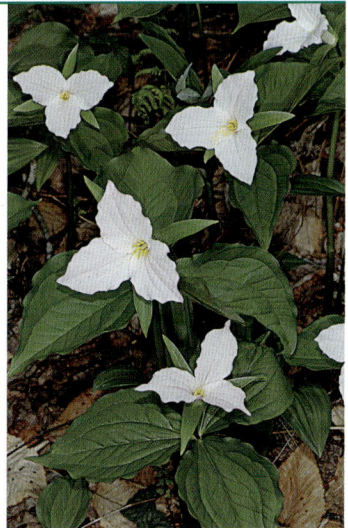

112. Vasey's Trillium; Sweet Beth

Trillium vaseyi Harbison
Tríl-li-um và-sey-i
Trilliaceae (Trillium Family)

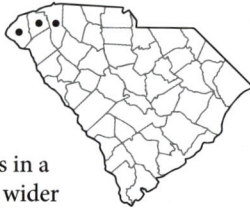

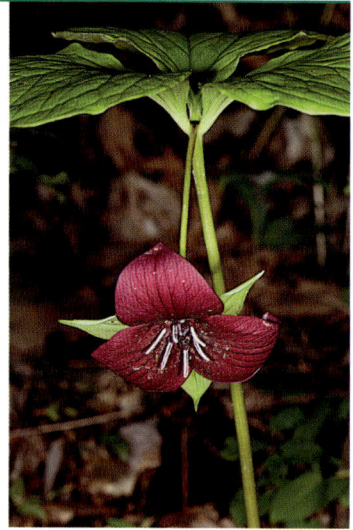

DESCRIPTION: Perennial herb from a short, thick rhizome; stem erect, to 20″ tall; leaves in a single whorl of 3, uniformly green, usually wider than long; flowers on a long stalk, declined below the leaves or rarely nearly horizontal; petals maroon or rarely white, margins not wavy, bent backward between the sepals; stamens extending well beyond the pistil; filaments partly maroon; anthers yellow to maroon; ovary small and dark purple; flowers April–May.
RANGE-HABITAT: Endemic to the southern Appalachian region in TN, NC, SC, GA, and AL; in SC, occasional in the mountains and extending into the edge of the upper Piedmont; rich cove forests and bases of slopes and along streams in acidic cove forests, where minerals have accumulated and soils are richer.
COMMENTS: This species may produce incredibly large flowers on robust individuals. The authors have observed flowers more than 9″ wide from petal tip to petal tip. Flower odor is foul. The specific epithet honors George R. Vasey (1822–93), curator of botany at the Smithsonian Institution and an expert on the grass family.

113. Canada Violet

Viola canadensis L.
Vì-o-la ca-na-dén-sis
Violaceae (Violet Family)

DESCRIPTION: Perennial herb from a rhizome; erect stem bears both leaves and flowers; leaves to 4″ long, gradually narrowed to a sharp apex, base heart-shaped; stipules long-triangular, with a sharp tip and smooth margins; petals white with a yellow center; spur shorter than 0.4″ long; flowers April–July.
RANGE-HABITAT: Widespread in eastern and central North America; in SC, common in the mountains and uncommon in the Piedmont; rich cove forests and basic-mesic forests.

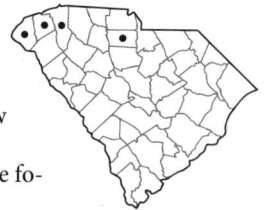

SIMILAR SPECIES: Cream Violet (*V. striata* Aiton) is similar but has creamy colored flowers, not bright white and grows in floodplains and other rich, low forests in the mountains and Piedmont.
COMMENTS: This is one of the most characteristic cove forest species in SC. The foliage and flowers have a faint odor and taste of wintergreen.

114. Doll's-eyes; White Baneberry

Actaea pachypoda Elliott
Ac-taè-a pa-chý-po-da
Ranunculaceae (Buttercup Family)

DESCRIPTION: Erect perennial herb to 24″ tall; leaves large, 2–3-times ternately compound (in 3s); inflorescence a short raceme; petals white, 3–7, appearing like modified stamens; fruit is a berry, white and capped by a broad, red stigma; flowers April–mid-May; fruits August–October.

RANGE-HABITAT: Widespread in eastern North America; in SC, occasional in the mountains in rich cove forests; rare in the Piedmont in basic mesic forests and beech forests.

COMMENTS: Doll's-eyes refers to the mature fruit. The brilliant white berries are toxic. The leaves are distinctive and readily distinguishable by the swellings at the tips of the leaflets.

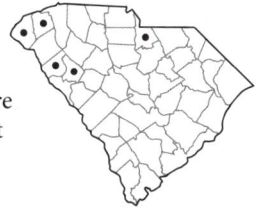

---

115. Dwarf Crested Iris

Iris cristata Aiton
    Ì-ris cris-tà-ta
Iridaceae (Iris Famly)

DESCRIPTION: Perennial herb usually less than 12″ tall in flower, forming colonies from elongate rhizomes that are alternately tapered and widened; leaves broadly linear, 4–16″ long, 1″ wide, conduplicate (folded together lengthwise); flowers with showy sepals, bluish to violet, with 2 crested ridges in the middle of a white or yellow central band; flowers April–May.

RANGE-HABITAT: MD south to GA and west to OK; in SC, common in the mountains, occasional in the Piedmont; cove forests, nutrient-rich oak-hickory forests, beech forests, and basic-mesic forests.

COMMENTS: Readily distinguished from the similar Dwarf Iris (*I. verna* L., plate 359) by habitat, the crested sepals, and wider leaves.

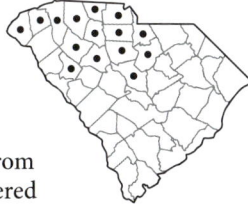

---

116. Early Meadowrue

Thalictrum dioicum L.
    Tha-líc-trum di-oì-cum
Ranunculaceae (Buttercup
    Family)

DESCRIPTION: Perennial herb to 2′ tall; leaves ternately compound; male and female flowers on separate plants (dioecious); sepals very small, purplish; anthers dangling, showy, with yellow filaments; stigmas of female plants purple; fruits are an aggregate of follicles; flowers March–April; fruits mature May–June.

RANGE-HABITAT: Wide-ranging in the eastern United States from Quebec south to SC and west to MN and MO; in SC, common in the mountains where found in rich cove forests and other nutrient-rich forests.

COMMENTS: The specific epithet refers to the dioecious nature of the plant. It is the earliest flowering of the tall species of meadow-rues and is already in fruit by the time other species come into flower.

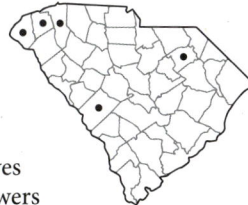

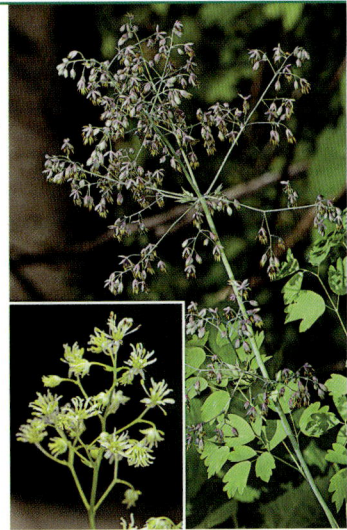

117. False Solomon's-seal;
Solomon's-plume

Maianthemum racemosum (L.)
Link ssp. *racemosum*
Mai-án-the-mum ra-ce-mò-
sum ssp. ra-ce-mò-sum
Ruscaceae (Ruscus Family)

SYNONYM: *Smilacina racemosa* (L.)
Desfontaines—RAB, PR

DESCRIPTION: Perennial herb from a long rhizome; stem arching, unbranched, to 2′ long; leaves alternate, 2-ranked, to 6″ long and 2″ wide, somewhat leathery, with raised veins above and with fine hairs below; flowers in a terminal panicle; sepals and petals white to green; flowers mid-April–June.

RANGE-HABITAT: Widespread in North America; in SC, common in the mountains, Piedmont, and inner Coastal Plain; rare in the outer Coastal Plain; found in a wide variety of moist, rich forests.

COMMENTS: False Solomon's-seal is vegetatively similar to true Solomon's-seals (*Polygonatum* spp., plate 361), from which it is easily separated when not in flower by its brighter green leaves and somewhat zig-zagged stem. Native Americans used the root smoke as a treatment for insanity and to quiet sobbing children.

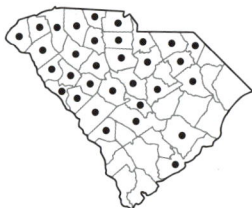

118. Whorled Horsebalm

Collinsonia verticillata Baldwin
Col-lin-sòn-i-a ver-ti-cil-là-ta
Lamiaceae (Mint Family)

DESCRIPTION: Pubescent, perennial herb from a thick, tuberous rootstalk; stem 8–30″ tall and unbranched; leaves obovate, margins coarsely toothed, well-developed leaves in 2–3 pairs, with very short internodes, opposite but appearing whorled; flowers in a narrow, terminal, panicle-like cluster; petals lavender to pink or salmon, the middle lobe of the lower lip long and fringed; flowers late April–May.

RANGE-HABITAT: TN, OH, VA, NC, SC, GA, and MS; in SC, common in the mountains and upper Piedmont; rich cove forests, basic-mesic forests, and oak-hickory forests; always on circumneutral or high magnesium soils.

COMMENTS: The pairs of opposite leaves, appearing whorled, are distinctive to this species. This is the only *Collinsonia* to flower in the spring in SC. The range is very odd, being found mostly west of the Blue Ridge with populations in the Piedmont of GA, SC, and Polk County, NC, and also occurring in coastal VA. The genus honors Peter Collinson (1694–1768), British horticulturalist and correspondent of John Bartram.

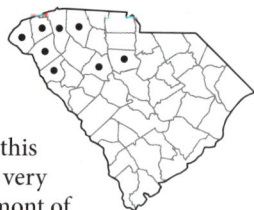

119. Fernleaf Phacelia

Phacelia bipinnatifida Michaux
   Pha-cè-li-a bi-pin-na-tí-fi-da
Hydrophyllaceae (Waterleaf
   Family)

DESCRIPTION: Biennial to 18″ tall from a taproot; stem and leaf stalks with glandular hairs; stem and basal leaves present; stem leaves pinnately dissected and incised, appearing twice compound; petals deep bluish-purple; flowers April–May.

RANGE-HABITAT: WV south to AL, and IL and AR; in SC, very rare in the mountains in rich cove forests.

COMMENTS: This is the largest and showiest species of *Phacelia* in SC. It is restricted to small populations in the mountains but is much more abundant farther north. The genus is from the Greek *phákelos*, "bundle," in reference to the flower arrangement.

CONSERVATION STATUS: SC-Critically Imperiled

120. Showy Orchis

Galearis spectabilis (L.) Rafinesque
   Ga-le-à-ris spec-tá-bi-lis
Orchidaceae (Orchid Family)

SYNONYM: *Orchis spectabilis*
   L.—RAB, PR

DESCRIPTION: Perennial, herbaceous, terrestrial orchid to 12″ tall from a fleshy, tuberlike rootstalk; leaves basal, to 4″ long, ovate to nearly round, rounded at the apex; bracts subtending each flower leaflike; the lip petal is uppermost and has a prominent nectar spur; flowers April–May.

RANGE-HABITAT: Widespread in eastern North America; in SC, uncommon in the mountains; grows in rich cove forests and bases of slopes where nutrients accumulate in acidic cove forests.

COMMENTS: The spurred lip has nectar at its base; its length and position assure cross-pollination. The flowers have a strong perfumelike scent. The genus name comes from the Greek *galea*, "helmet," in reference to the odd-shaped, distinctive flowers.

CONSERVATION STATUS: SC-Vulnerable

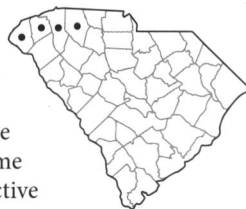

121. Yellow Mandarin; Yellow Fairybells

Prosartes lanuginosa (Michaux)
   D. Don
      Pro-sár-tes la-nu-gi-nò-sa
Liliaceae (Lily Family)

SYNONYM: *Disporum lanuginosum*
   (Michaux) Nicholson—RAB, PR

DESCRIPTION: Perennial herb from a knotty rhizome; stem to 3′ tall, branched, the lower portion brown and wiry; leaves alternate, to

4″ long and 2″ wide, soft hairy below; flowers terminating the branches and opposite the last leaf; tepals greenish and without spots, nodding; flowers April–May.

RANGE-HABITAT: Ontario south to GA and AL; in SC, common in the mountains in rich cove forests.

COMMENTS: The many-branched stem produces an overall look that is dissimilar to any other herbaceous plants in the cove forest of SC.

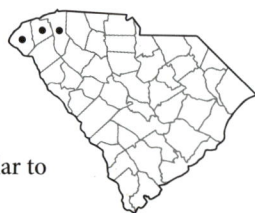

122. Canada Waterleaf; Mapleleaf Waterleaf

Hydrophyllum canadense L.
    Hy-dro-phýl-lum ca-na-dén-se
    Hydrophyllaceae (Waterleaf Family)

DESCRIPTION: Perennial herb; basal leaves pinnately lobed, often variegated with 2–3 shades of green; stem leaves alternate and palmately lobed, broader than long, about 6″ wide; flowers white, produced in coiled clusters, appearing as racemes in fruit; flowers May–June.

RANGE-HABITAT: VT south to AL and west to Ontario, AR, and MO; in SC, rare and local in the mountains where restricted to rich cove forests and other moist forests with circumneutral soils.

COMMENTS: This is the only waterleaf in SC. The size and shape of the stem leaves are distinctive. The common name waterleaf comes from the variegated (water-stained) appearance of the leaves and the genus name translates as hydro (water) phylum (leaf).

CONSERVATION STATUS: SC-Vulnerable

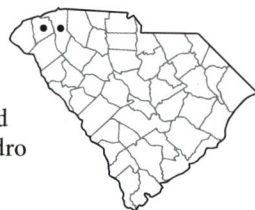

123. Southern Jack-in-the-pulpit

Arisaema quinatum (Nuttall) Schott
    A-ri-saè-ma qui-nà-tum
    Araceae (Arum Family)

DESCRIPTION: Erect, perennial herb 8–30″ tall from a corm; leaves palmately divided with 5, sometimes 3 leaflets; leaflets glaucous (whitish) beneath; flowers on a fleshy spadix, male above, female below; spathe (the pulpit) with a tube and hood that arches over the spadix (jack); spathe hood green; flowers April–May; fruits are red berries attached to the spadix; fruits mature July–August.

RANGE-HABITAT: sw. NC and e. TN south to the FL Panhandle and TX; in SC, uncommon in the mountains and Piedmont; rich cove forests, other nutrient-rich forests along streams and rivers, and basic mesic forests.

COMMENTS: Only a single species of Jack-in-the-pulpit was recognized as growing in the eastern United States by most twentieth century authors. The Common Jack-in-the-pulpit (*Arisaema triphyllum* (L.) Schott) is a tetraploid (4 sets of chromosomes) and is reproductively isolated from the other four species occurring in the eastern United States. Southern Jack-in-the-pulpit is more abundant farther south. It extends into SC in the Savannah River watershed, where it reaches well into the mountains. Five leaflets on the leaves that are glaucous beneath are typical, and this species typically only produces green spathes. The only other species in our area that has leaves that are glaucous beneath is Common Jack-in-the-pulpit, which has three leaflets and green-striped or purple-striped spathes.

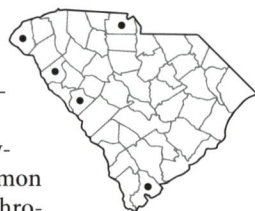

124. Large Yellow Lady's-slipper

Cypripedium parviflorum
Salisbury *var. pubescens*
(Willdenow) Knight
Cy-pri-pè-di-um par-vi-flò-
rum var. pu-bés-cens
Orchidaceae (Orchid Family)

SYNONYM: *Cypripedium calceolus* L. var. *pubescens*
(Willdenow) Correll—RAB

DESCRIPTION: Showy perennial herbaceous terrestrial orchid to
28″ tall; leaves 3–5, alternate, to 8″ long and 4″ wide, with glandular hairs and parallel, raised veins; flowers 1 or 2, terminal; the
2 lateral petals twisted, green streaked with purple; the "slipper"
petal golden yellow and streaked or spotted with purple on the
inside; flowers April–May.

RANGE-HABITAT: Widespread in eastern North America; in SC,
occasional in the mountains and rare in the central Piedmont; rich
cove forests in the mountains and basic-mesic forests in the Piedmont.

COMMENTS: This showy species is becoming less abundant, primarily due to collection by
thoughtless nature "lovers." Just "snatching the top" has long-term consequences because
it significantly reduces energy capture by photosynthesis. *Cypripedium* comes from the
Latin *Cypris* for Venus and *pedilon* for shoe, so lady's slipper could be called "Venus'
slipper." It was heavily collected in the nineteenth century for its root, which has sedative
properties.

CONSERVATION STATUS: SC-Imperiled

125. Puttyroot; Adam-and-Eve

Aplectrum hyemale (Muhlenberg ex
Willdenow) Torrey
A-pléc-trum hye-mà-le
Orchidaceae (Orchid Family)

DESCRIPTION: Perennial herb from a beadlike horizontal rootstalk; solitary basal leaf elliptic, with
a purplish lower surface, and distinctly pleated
with raised whitish veins; leaf produced in the
fall and withering, but not disappearing, before
the appearance of the leafless flowering stem in
spring; flowering stem 12–16″ tall; flowers on
distinct stalks, greenish, yellowish, or whitish, and marked with purple or
violet; flowers April–June.

RANGE-HABITAT: Widespread in the eastern US; in SC, occasional in the mountains and Piedmont; in a variety of moist forests, including cove forests, moist
oak-hickory forests, and upper margins of floodplains.

COMMENTS: Puttyroot is seldom found in colonies of more than a few plants and is
probably more common than records indicate. It is called puttyroot for the putty-
like substance obtained from its roots. A thin connector attaches the current year's leaf and its swollen
underground stem (Eve) to last year's leaf and corm (Adam). It is easily distinguished from Cranefly
Orchid (*Tipularia discolor* (Pursh) Nuttall, plate 370), which has spurred flower and leaves without
prominent raised whitish veins.

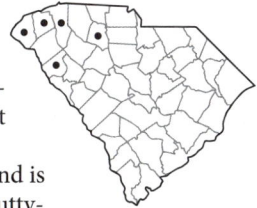

126. Common Black Cohosh

Actaea racemosa L.
Ac-taè-a ra-ce-mò-sa
Ranunculaceae (Buttercup Family)

SYNONYM: *Cimicifuga racemosa* (L.) Nuttall—RAB, PR

DESCRIPTION: Perennial herb to 5′ tall, from a thick, knotted rootstalk; leaves large, 2–3-times ternately compound (in 3s); leaflets with serrate margins; flowers without petals, in long terminal racemes; showy parts are stamens; fruits are follicles, mostly 1 per flower, on a very short stalk; flowers May–July.

RANGE-HABITAT: Fairly widespread in eastern North America; in SC, common in the mountains and Piedmont; rich cove forests, rich oak-hickory forests, basic-mesic forests, and beech forests.

SIMILAR SPECIES: Mountain Black Cohosh (*Actaea podocarpa* A.P. de Candolle) is very similar but flowers later (July–September). It may also be distinguished by the persistent groove on the petiole of basal leaves and the cluster of fruits being held on a stalk (stipe), rather than nearly sessile as in Common Black Cohosh. It is rare in SC. It is known from Greenville and Pickens counties, at high elevations, very near the NC state line.

COMMENTS: Common Black Cohosh is one of several unrelated species of cove forests wildflowers with 2–3-times ternately compound leaves. The vascular bundles of the rootstalk are arranged in a 3-, 4-, or 5-parted star. Native Americans used cohosh for a variety of "female complaints"; research has confirmed sedative, estrogenic, anti-inflammatory, and hypoglycemic activity. Duke (1997) recommends it as an herbal remedy for menopausal symptoms. It has become a popular herbal remedy that is widely available.

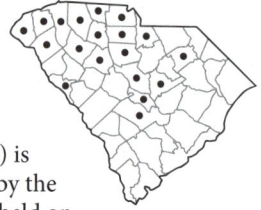

127. Eastern Goat's-beard

Aruncus dioicus (Walter) Fernald
A-rún-cus di-oì-cus
Rosaceae (Rose Family)

DESCRIPTION: Perennial herb to 5′ tall with several stems from the same rootstalk; leaves 6–8 per stem, large, 2–3-times ternately compound (in 3s); leaflets large, with doubly serrate margins; terminal leaflet unlobed; male and female flowers on separate plants, arranged into a large pyramidal, terminal panicle; petals white in female flowers, greenish white on male flowers; fruits are follicles; flowers May–June.

RANGE-HABITAT: PA south to GA and west to IN and AL; in SC, common in the mountains, rare in the Piedmont; rich cove forests, acidic cove forests, rich oak-hickory forests, and forest margins.

SIMILAR SPECIES: Goat's-beard superficially resembles Appalachian False Goat's-beard (*Astilbe biternata* (Ventenat) Britton), which is very rare in similar habitats in the mountains and has glandular hairs on the lower leaf surface, trilobed terminal leaflets, a 2-locular fruit, and does not have separate male and female plants.

COMMENTS: The epithet *dioicus* refers to the presence of separate male and female plants (dioecious). Goat's-beard refers to the dried cluster of male flowers, which resemble a goat's beard.

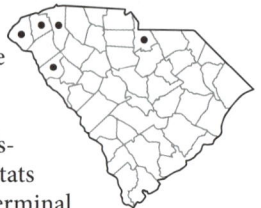

128. Dutchman's-pipe; Pipevine

Isotrema macrophyllum (Lamarck)
   C. F. Reed
   I-so-trè-ma mac-ro-phýl-lum
   Aristolochiaceae (Birthwort Family)

SYNONYM: *Aristolochia macrophylla*
   Lamarck—RAB, PR

DESCRIPTION: Large woody vine that may climb high into trees; leaves large (to 12″ wide) and heart-shaped; flowers borne singly; corolla absent; calyx dull brown-purple and shaped like an ornate tobacco pipe; flowers May–June.

RANGE-HABITAT: Endemic to the Appalachian region from PA south to GA; in SC, locally common in rich cove forests in the mountains.

COMMENTS: This species requires high light to flower and may be found in canopy gaps in communities upslope or downslope from cove forests. It contains aristolochic acid, which is known to have antitumor properties. This, and related species, are the hosts for the caterpillars of Pipevine Swallowtails. Their cultivation is possible in cool climates, but in most of SC the related Wooly Dutchman's-pipe (*Isotrema tomentosum* (Sims) H. Huber, plate 817) is a better candidate for attracting butterflies.

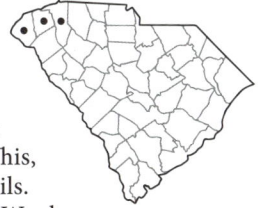

129. Spikenard

Aralia racemosa L.
   A-rà-li-a ra-ce-mò-sa
   Araliaceae (Ginseng Family)

DESCRIPTION: Perennial herb 3–5′ tall from a large aromatic root; leaves large, with 3 main divisions, each with 3–7 leaflets; leaflets with serrate margins; flowers small, white, unisexual and bisexual on the same plant, in large terminal clusters of many umbels; flowers May–July; fruits are fleshy, red to purplish drupes; fruits July–September.

RANGE-HABITAT: Widespread in eastern and central North America; in SC, occasional in the mountains in rich cove forests, at bases of slopes, and along streams through acidic cove forests.

COMMENTS: Well known in the Northeast for its large, aromatic roots. The roots have a licoricelike flavor, are used to flavor soups, and traditionally were used as a substitute for Sarsaparilla (*Smilax ornata* Lemaire) in making root beers. Spikenard was also traditionally used as a treatment for lung disorders. A close relative from the Himalayas was the source of the expensive ointment that many believe Mary poured on Jesus' head in the New Testament parable told in the book of Mark.

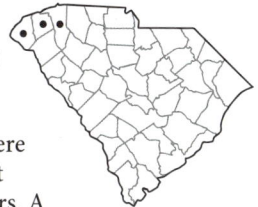

130. Fourleaf Milkweed

Asclepias quadrifolia Jacquin
As-clè-pi-as qua-dri-fò-li-a
Apocynaceae (Dogbane Family)

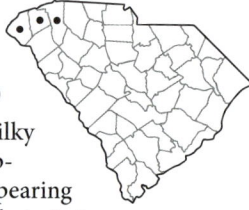

DESCRIPTION: Perennial herb with white milky sap; stems single, to 20″ tall; lowest and uppermost leaves opposite, middle leaves appearing whorled in robust individuals; flowers pink, in 2–4 umbels from the upper leaf axils; fruit an erect follicle, to 4.5″ long; flowers May–June; fruits mature August–September.

RANGE-HABITAT: Widespread in eastern North America; in SC, uncommon in the mountains and Piedmont; rich cove forests, montane oak-hickory forests, and basic-mesic forests; always in circumneutral or high magnesium soil.

COMMENTS: This beautiful, small milkweed occurs on well-drained convex landforms when found in rich cove forests. It is the only milkweed native to the mountains with midstem leaves appearing whorled. The genus *Asclepias* is named in honor of the Greek god of medicine, Asclepius, son of Apollo.

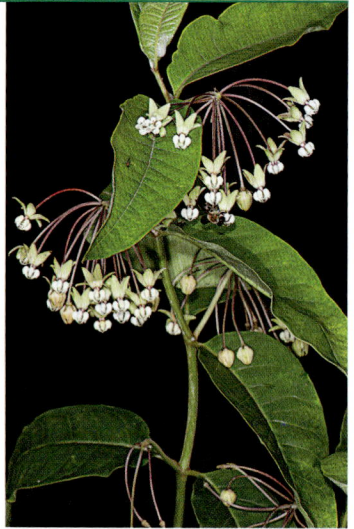

131. Tall Milkweed; Poke Milkweed

Asclepias exaltata L.
As-clè-pi-as ex-al-tà-ta
Apocynaceae (Dogbane Family)

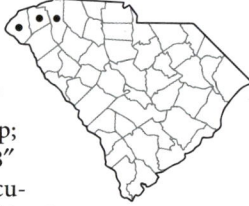

DESCRIPTION: Perennial herb with milky sap; stems single, to 3′ tall; leaves opposite, to 8″ long and 4″ wide, with obvious petioles, acuminate tips and attenuate bases; flowers white, in umbels from the upper leaf axils; fruit a smooth follicle, to 9″ long; flowers June–July; fruits mature August–September.

RANGE-HABITAT: Widespread in eastern North America; in SC, common in the mountains in rich cove forests and forest margins.

COMMENTS: This is SC's only milkweed with stalked, broad, smooth leaves, and smooth, erect follicles. The epithet *exaltata* is literally translated as "very tall." The plant is often much taller in the northern portion of its range. Duke (1997) recommends Tall Milkweed as an effective herbal remedy for warts. The genus *Asclepias* is named in honor of the Greek god of medicine, Asclepius, son of Apollo.

132. Basil Bergamont

Monarda clinopodia L.
Mo-nár-da cli-no-pò-di-a
Lamiaceae (Mint Family)

DESCRIPTION: Perennial herb to 3′ tall; leaves ovate, usually 2x as long as wide; flowers in a terminal headlike cluster, subtended by greenish or whitish, leaflike bracts; flowers white, greenish white, or pinkish, lower lip spotted with purple, upper lip without hairs; flowers late May–September.

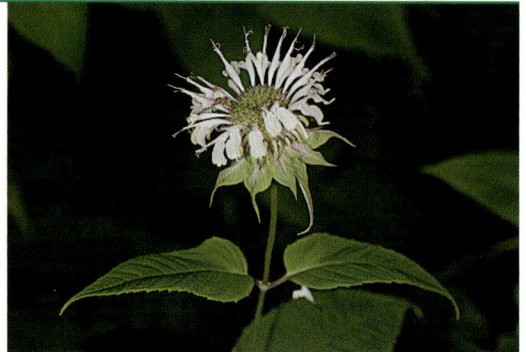

RANGE-HABITAT: NY south to GA and west to IL and AL; in SC, common in the mountains, rare in the Piedmont; mesic forests and moist forest margins; most abundant in rich cove forests.

COMMENTS: Recognizable as a *Monarda* by the terminal, headlike cluster of flowers subtended by leaflike, often light-colored bracts; recognizable as Basil Bergamont by the single, white, flower cluster, the leaves about 2x as wide as long, and the hairless upper lip of the corolla. True Bergamont is a species of orange (*Citrus bergamia* Risso); its rind is the source of an oil used in perfumery.

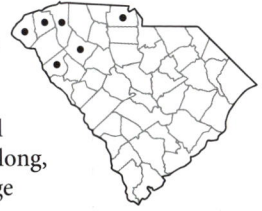

133. Appalachian Mountain-mint

Pycnanthemum montanum
Michaux
Pyc-nán-the-mum.
mon-tà-num
Lamiaceae (Mint Family)

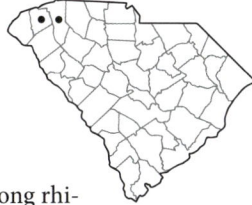

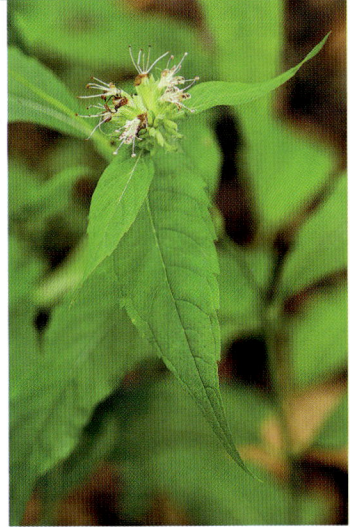

DESCRIPTION: Herbaceous perennial with long rhizomes; stems to 3′ tall; leaves opposite, lanceolate, to 5″ long and 1″ wide; the upper leaves not whitish; flowers in headlike clusters terminating the stem and in the axils of the upper 2 pairs of leaves; bracts subtending flowers have long hairs on the margins; corolla with purple spots on a white-to-yellowish background; calyx lobes with long, spaced hairs; flowers June–August.

RANGE-HABITAT: Endemic to the southern Appalachian region from WV south to SC and GA; in SC, occasional in the mountains in rich cove forests and montane oak-hickory forests, always in circumneutral or high magnesium soils.

SIMILAR SPECIES: All other *Pycnanthemum* species growing in the mountains with broad leaves have whitish, densely pubescent leaves subtending the flowers. While most of our species grow in open habitats, Appalachian Mountain-mint is primarily found under closed canopies.

CONSERVATION STATUS: Vulnerable

134. Canada Enchanters'-nightshade

Circaea canadensis (L.) Hill
Cir-caè-a ca-na-dén-sis
Onagraceae (Evening-primrose Family)

SYNONYM: *Circaea lutetiana* ssp. *canadensis* (L.) Ascherson & Magnus—RAB

DESCRIPTION: Erect perennial to 18″ tall (to 3′ in some parts of its range), with rhizomes; flowers with parts in 2′s, in racemes from the axils of reduced upper leaves; fruits with curved, bristly hairs; flowers June–August.

RANGE-HABITAT: Nova Scotia south to GA and west to LA, NE, and Manitoba; in SC, occasional in the mountains and very rare in the Piedmont; rich cove forests, basic-mesic forests, rich woods of floodplains, and stream sides.

COMMENTS: Although not spectacular, this is a distinctive wildflower with very unusual fruit that is not easily confused with other species.

135. Mountain Bunchflower

Melanthium parviflorum
(Michaux) S. Watson
Me-lán-thi-um par-vi-flò-rum
Melanthiaceae (Bunchflower
Family)

SYNONYM: *Veratrum parviflorum*
Michaux—RAB

DESCRIPTION: Perennial herb with large, pleated, oblanceolate to obovate basal leaves; flowering stem to 3' tall, pubescent, highly branched, with numerous flowers; flowers small, greenish, with 6 tepals; fruit a capsule; flowers June–August; fruits mature September–October.

RANGE-HABITAT: Endemic to the southern Appalachian region from WV and KY south to GA and AL; in SC, infrequent in the mountains where found in rich cove forests and montane oak-hickory forests at higher elevations.

COMMENTS: Mountain Bunchflower is distinctive because of its large, showy leaves that appear in the spring. It is sometimes confused with Ramps (*Allium tricoccum* Aiton). Bunchflower, like many plants in this family, contains toxic alkaloids that lead to cramps, vomiting, heart abnormalities, and even death. The pleated leaves and lack of an "onion" odor serves to distinguish this toxic species from Ramps.

CONSERVATION STATUS: SC-Imperiled

136. Ozark Bunchflower

Melanthium woodii
(J. W. Robbins ex Wood)
Bodkin
Me-lán-thi-um woòd-i-i
Melanthiaceae (Bunchflower
Family)

SYNONYM: *Veratrum woodii* J. W. Robbins ex Wood

DESCRIPTION: Perennial herb with large, pleated, elliptic to oblanceolate basal leaves to 1.5' long; flowering stem to 3' tall, pubescent, highly branched with numerous small, purplish or brownish flowers with 6 tepals; fruit a capsule; flowers July–August; fruits mature September–October.

RANGE-HABITAT: Primarily in the Ozark and Midwest region from Ohio south to TN and west to MO, AR and OK; isolated populations in GA, AL, Panhandle FL, nw. SC, and sw. NC; in SC, rare and restricted to a rich cove/montane oak-hickory forest complex with circumneutral soils over amphibolite.

COMMENTS: Ozark Bunchflower was first documented for SC by Patrick McMillan in 2002. The species is more robust, with longer, pleated leaves that are narrower in outline, and darker green, than Mountain Bunchflower (*M. parviflorum* (Michaux) S. Watson, plate 135). When in flower, the purplish tepals are distinctive. Its distribution east of the Ozarks is highly disjunct. It should be sought in similar habitats throughout the Carolinas.

CONSERVATION STATUS: SC-Critically Imperiled

137. Broadleaf Coreopsis; Broadleaf Tickseed

Coreopsis latifolia Michaux
Co-re-óp-sis la-ti-fò-li-a
Asteraceae (Aster Family)

DESCRIPTION: Smooth, perennial herb with long rhizomes; stems erect, 10–30″ tall; leaves simple, opposite, coarsely serrate, 1.5–4.5″ wide, lance-ovate to lance-elliptic; flowers arranged in involucrate heads of a few yellow ray flowers and yellow disk flowers; flowers August–September.
RANGE-HABITAT: Broadleaf Coreopsis is endemic to the southern Appalachian region from sw. NC and se. TN south into nw. SC and ne. GA; in SC, rare in the mountains in rich cove forests and rich wooded slopes.
COMMENTS: Broadleaf Coreopsis is endemic to high magnesium soils in the Blue Ridge. It is the least *Coreopsis*-like of the tickseeds and is most often misidentified for a sunflower due to its upright stem, lack of basal leaves, and inflorescences with few ray flowers. It often occurs with Whiteleaf Sunflower (*Helianthus glaucophyllus* D.M. Smith, plate 191). It is one of the best indicators of high magnesium soils and is often also associated with many other rare species.
CONSERVATION STATUS: SC-Imperiled

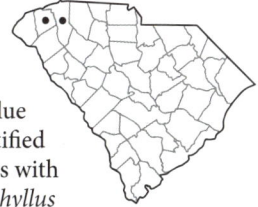

138. Wood-nettle

Laportea canadensis (L.) Weddell
La-pórt-e-a ca-na-dén-sis
Urticaceae (Nettle Family)

DESCRIPTION: Perennial herb 1–4′ tall, with stinging hairs throughout; leaves alternate, coarsely toothed, to 6″ long, long-stalked; flowers in terminal and axillary panicles, inconspicuous, the upper flowers are female, the lower male; no corolla present; flowers late June–August.
RANGE-HABITAT: Nova Scotia south to FL and west to Manitoba and OK; in SC, common in the mountains in rich cove forests, especially near seepages; common in the Piedmont on floodplains and in basic-mesic forests; rare in the Coastal Plain in alluvial woods.
COMMENTS: Like all true nettles, Wood-nettle has urticating hairs that inject histamine, acetylcholine, serotonin, formic acid, as well as oxalic acid, which along with the formic acid, causes immediate and long-lasting pain. The closely related Common Stinging Nettle (*Urtica dioica* L.) has been shown to be, for some users, an effective remedy for rheumatoid arthritis. Duke (1997) recommends wood-nettle for the same use. The plant is wind-pollinated and produces the male flowers below the female flowers to avoid self-pollination. The genus honors François de Laporte (1810–80), a French naturalist who worked in Canada.

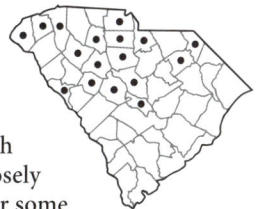

139. Appalachian White Snakeroot

Ageratina roanensis (Small)
   E. E. Lamont
   A-ge-ra-tì-na ro-a-nén-sis
Asteraceae (Aster Family)

SYNONYM: *Eupatorium rugosum*
   Houttuyn—RAB, in part, PR,
   in part

DESCRIPTION: Perennial herb usually 3–4′ tall, often forming dense colonies; leaves opposite; leaf stalks to 2″ long; blade coarsely toothed, distinctly longer than the leaf stalk, not resinous, usually more than 2.5″ long; flower heads in flat-topped clusters terminating the stem and from the upper leaf axils; flowers late July–August.

RANGE-HABITAT: Endemic to the southern Appalachian region from VA south to SC and west to KY and GA; in SC, common in the mountains in rich cove forests and montane oak forests; one known disjunct Piedmont location.

SIMILAR SPECIES: The similar Common White Snakeroot (*Ageratina altissima* var. *altissima*) is more widespread, being found in moist, rich forests throughout the state. The typical variety is a less-robust plant and often does not form the large masses typical of Appalachian White Snakeroot. It also has involucral bracts, which are not drawn out to long acuminate tips.

COMMENTS: Appalachian and Common White Snakeroot caused the notorious "milk sickness" prevalent in earlier times. The plant produces a poison that is transmissible in cow's milk.

140. Appalachian Turtlehead; Lyon's Turtlehead

Chelone lyonii Pursh
   Che-lò-ne. ly-òn-i-i
Plantaginaceae (Plantain Family)

DESCRIPTION: Perennial herb with erect stems, 12–40″ tall; leaves ovate with round bases, on distinct leaf stalks (0.5–1″ long); corolla pink, with yellow hairs on the inside; sterile stamens all white or with a rose tip; flowers July–September.

RANGE-HABITAT: Chiefly in the mountains of NC, SC, and TN; in SC, uncommon in the mountains in rich cove forests on circumneutral soils.

COMMENTS: The genus name means "tortoise or turtle," and the flower of turtlehead simulates the head of a turtle to a remarkable degree; the two lips (a half-open mouth) terminate a large open corolla. The flowers are a perfect fit for large, heavy bumblebees that land on the lower petal and open the flower for access. Appalachian Turtlehead is typical of high-elevation northern hardwood and spruce-fir forests in NC, but in SC it occurs in a very different habitat. The specific epithet honors John Lyon (1765–1814), a Scottish botanist and explorer of the southern Allegheny Mountains.

CONSERVATION STATUS: SC-Imperiled

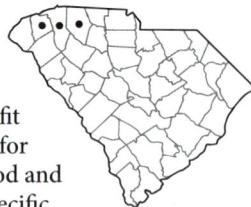

141. Northern Horsebalm; Richweed

Collinsonia canadensis L.
Col-lin-sòn-i-a ca-na-dén-sis
Lamiaceae (Mint Family)

DESCRIPTION: Perennial herb from a thick woody rhizome; stem to 3′ tall, with several pairs of large (10–20″ long), toothed leaves scattered on the stem; flowers yellow, two-lipped, the middle lobe of the lower lip fringed with narrow segments; flowers in an open terminal cluster (panicle); flowers late July–September.

RANGE-HABITAT: Widespread in eastern North America; in SC, common in a wide variety of rich forests in the mountains, especially rich cove forests; common in basic-mesic forests and mesic oak-hickory forests in the Piedmont; rare in beech forests and calcareous bluff forests in the Coastal Plain.

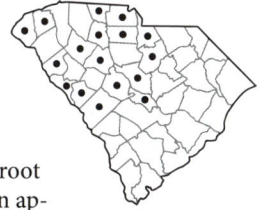

COMMENTS: Horsebalm is a coarse (which is the old traditional meaning implied by "horse") plant similar to balm (*Melissa*). It is sometimes called Stoneroot because of its very hard roots, which were once used to treat kidney stones, an application of the Doctrine of Signatures. The roots contain rosmarinic acid, an antioxidant and preservative. The genus honors Peter Collinson (1694–1768), British horticulturalist and correspondent of John Bartram.

142. Eastern Blue Monk's-hood

Aconitum uncinatum L.
A-co-nì-tum un-ci-nà-tum
Ranunculaceae (Buttercup Family)

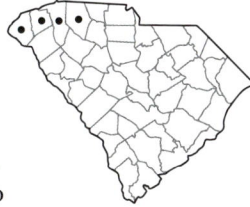

DESCRIPTION: Weak-stemmed perennial to 3′ tall; leaves alternate, deeply divided into 3–5 segments; flowers blue, irregular, the upper-most petal helmetlike; flowers August–October.

RANGE-HABITAT: MD, PA, and IN, south to GA and SC; in SC, rare in rich cove forests, spray cliff seepages, and moist forest margins.

COMMENTS: Eastern Blue Monk's-hood is a highly poisonous relative of the European Wolfsbane (*A. napellus* L.), which is the source of the drug aconite, famous in European legend for its use in poisoning wolves. Some populations are extremely late flowering in SC, such as at Station Cove Falls, where it seldom is in bloom before late September.

CONSERVATION STATUS: SC-Imperiled

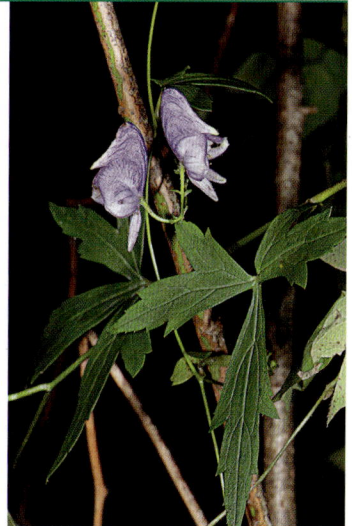

143. Curtis's Goldenrod

Solidago curtisii Torrey & Gray
So-li-dà-go cur-tís-i-i
Asteraceae (Aster Family)

DESCRIPTION: Perennial herb from long rhizomes; stem usually 3–5′ tall, green, angled, and grooved; leaves numerous, the larger to 6″ long, 3–10x as long as wide, sharply toothed, tapered to a fine tip and a broad base; flowers yellow, 5–10 per involucrate head; heads in clusters of 3–15 from the axils of the upper leaves; flowers August–October.

RANGE-HABITAT: Endemic to the southern Appalachian region from WV south to GA and AL; in SC, common in rich forests in the mountains, especially rich oak-hickory forests and rich cove forests.

SIMILAR SPECIES: Appalachian Goldenrod (*S. flaccidifolia* Small) is similar but has wider proportioned leaves that are 1–3x as long as wide, vs. 3–10x as long as wide. It is found in rich cove forests in the mountains and basic-mesic forests and bluffs in the Piedmont. Zigzag Goldenrod (*S. flexicaulis* L.) also has much wider proportioned leaves that are 1–2x as long as wide and distinctly zig-zagged stems. It is found in basic-mesic forests, floodplain forests, and other forests with nutrient rich soil throughout SC. Axillary Goldenrod (*S. caesia* L.) has round stems that are frequently glaucous (with a whitish cast) and is common in deciduous forests throughout the state.

COMMENTS: Goldenrods are difficult to identify, but the most common mountain species has grooved and angled green stems; numerous, large, long stem leaves; and heads in short clusters in the axils of upper leaves. The specific epithet honors its discoverer, Moses Ashley Curtis (1808–72), a clergyman and botanical explorer from Massachusetts who spent most of his life in NC and SC.

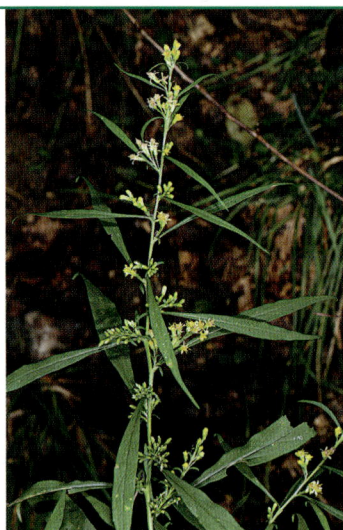

144. Gorge Goldenrod

Solidago faucibus Wieboldt
So-li-dà-go faù-ci-bus
Asteraceae (Aster Family)

DESCRIPTION: Perennial herb with stout, long rhizomes; stem usually 3–8′ tall, green, angled, and grooved; basal leaves numerous, larger leaves to 8″ long, broadly ovate, sharply toothed with a scabrous upper surface; flowers yellow, in involucrate heads of 4–6 ray flowers and 5–7 disk flowers per head; flowers August–October.

RANGE-HABITAT: Endemic to the southern Appalachian region from VA, WV, and KY, with disjunct populations in the mountains and upper Piedmont of SC; uncommon in rich cove forests and montane oak forests with circumneutral, high magnesium soils.

SIMILAR SPECIES: Atlantic Goldenrod (*Solidago arguta* Aiton var. *caroliniana* Gray) is common in mesic to dry woodlands throughout SC and has much smaller basal leaves and is smaller in all regards.

COMMENTS: This species was independently "discovered" by Thomas Wieboldt in VA, KY, and WV and by Patrick McMillan in Pickens County, SC. It was formally described in 2003 (Wieboldt & Semple, 2003). The species contains 10 sets of chromosomes (a decaploid) and rarely flowers. It persists as large colonies of rhizomatous basal rosettes until large light gaps open in the forest habitat, at which time it will flower. The distribution is very strange: the SC locations are hundreds of miles from the remainder of the species range. It is likely that this species arose as a polyploid hybrid within the *Solidago arguta* species complex, and the two populations may represent separate hybridization events leading to the decaploid species.

CONSERVATION STATUS: SC-Imperiled

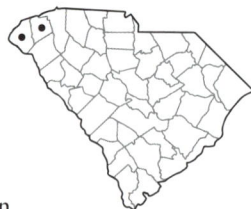

145. Heart-leaved Aster; Common Blue Wood Aster

Symphyotrichum cordifolium (L.)
  G. L. Nesom
   Sym-phy-o-trì-chum cor-di-fò-
   li-um
Asteraceae (Aster Family)

SYNONYM: *Aster cordifolius* L.—RAB, PR

DESCRIPTION: Perennial herb forming clumps; stems 2–4′ tall; basal and lower stem leaves with heart-shaped bases, acuminate tips, and well-developed nonwinged petioles; flowers produced in involucrate heads of blue to lavender ray flowers and yellow to reddish disk flowers; heads arranged in an elongate paniclelike structure; flowers September–October.

RANGE-HABITAT: Nova Scotia south to GA and west to MN and AL; in SC, common throughout; rich cove forests, basic-mesic forests, beech forests, calcareous bluff forests, and forest margins.

SIMILAR SPECIES: White Arrowleaf Aster (*S. urophyllum* (Lindley ex A.P. de Candolle) G. L. Nesom) is similar but has narrower leaves that are truncate to very shallowly cordate at the base. It is common in the mountains and upper Piedmont and rare elsewhere in SC.

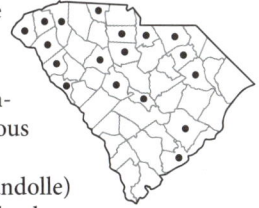

146. Northern Maidenhair Fern

Adiantum pedatum L.
  A-di-án-tum pe-dà-tum
Pteridaceae (Maidenhair Fern
  Family)

DESCRIPTION: Deciduous, rhizomatous fern; fronds horseshoe-shaped in outline; stalk of frond black and shiny, about as long as the frond blade; blade with two equal divisions, each bearing pinnae on just one side; spores produced in clusters (sori) grouped in lines along the outer margins of the pinnae; spores produced June–August.

RANGE-HABITAT: Widespread in eastern North America; in SC, common in the mountains, uncommon in the Piedmont; moist soils of rich cove forests, basic-mesic forests, beech forests, or in seepages associated with cliffs.

COMMENTS: The shape and composition of the leaf is distinctive. The common name Maidenhair is derived from the first species described in the genus, Southern Maidenhair Fern (*Adiantum capillus-veneris* L.). The specific epithet for Southern Maidenhair Fern translates to Venus' hair.

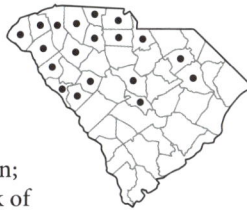

147. Walking Fern

Asplenium rhizophyllum L.
  As-plè-ni-um rhi-zo-phýl-lum
Aspleniaceae (Spleenwort Family)

DESCRIPTION: Small, evergreen fern from a short, erect rhizome; leaves (fronds) in clusters, 2–12″ long, simple and without lobes, the apex very long-pointed and usually producing a small plant at the tip when it comes into contact with the substrate; spores produced May–October.

RANGE-HABITAT: From s. Quebec south to GA and MS and west to MN, AR, and OK; in SC, rare in the mountains where it is restricted to calcium-rich or magnesium-rich rocks such as amphibolite or marble, in rich cove forests.

COMMENTS: This distinctive little fern "walks" as it produces plantlets from successive leaf tips. Many of the most accessible populations in SC have been trampled out of existence by hikers. This fern should not be disturbed if encountered in the wild.

CONSERVATION STATUS: SC-Vulnerable

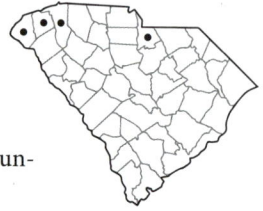

148. Glade Fern

Diplaziopsis pycnocarpa (Sprengel) M. G. Price

Dip-la-zi-óp-sis pyc-no-cár-pa

Diplaziopsidaceae (Glade Fern Family)

SYNONYM: *Athyrium pycnocarpon* (Sprengel) Tidestrom—RAB

DESCRIPTION: Deciduous, rhizomatous fern, resembling a Christmas Fern but with thin, pinnately compound fronds that are held strictly erect; fronds to 43″ long; bright green; spores produced in clusters (sori) grouped in two rows extending outwards from the midrib of the pinnae; spores produced July–September.

RANGE-HABITAT: Widespread in eastern North America from Quebec and Ontario south to GA and LA; in SC, rare in the mountains where it is restricted to rich cove forests with high calcium or high magnesium soils with circumneutral pH.

COMMENTS: This species is found only in high pH soils in moist conditions and could be mistaken for a Christmas Fern (*Polystichum acrostichoides* (Michaux) Schott, plate 338). Glade Fern is easily distinguished by the much less leathery, deciduous leaves that are held strictly upright and are bright spring green vs. the darker green of the evergreen Christmas Fern. It is frequently associated with many other rare species in its preferred habitat.

CONSERVATION STATUS: SC-Imperiled

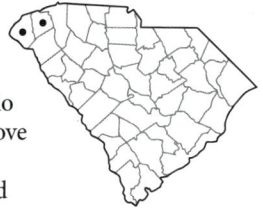

149. Silvery Spleenwort; Silvery Glade Fern

Deparia acrostichoides (Swartz) M. Kato

De-pá-ri-a ac-ros-ti-choì-des

Athyriaceae (Lady Fern Family)

SYNONYM: *Athyrium thelypterioides* (Michaux) Desvaux—RAB

DESCRIPTION: Deciduous, pubescent fern with erect, twice pinnatifid, dull silvery green leaves (fronds), 1 to 3.5′ long; blades gradually tapered toward the tip and at the base, producing a plumelike appearance; spores in clusters (sori) produced in lines on either side of veins on underside of pinnae; spores produced June–September.

RANGE-HABITAT: Widespread in eastern North America; in SC, common in rich cove forests in the mountains.

COMMENTS: This species looks similar to a huge New York Fern (*Parathelypteris noveboracensis* (L.) Cheng, plate 196) but is much larger. It is characteristic of the rich cove forest habitat but is found in very moist areas and is tolerant of more acidic conditions than Glade Fern.

150. Acidic cove forest

151. Sweet Birch; Black Birch

Betula lenta L. var. *lenta*
Bé-tu-la lén-ta var. lén-ta
Betulaceae (Birch Family)

DESCRIPTION: Deciduous tree to 90′ tall and 34″
in diameter; inner bark with odor and taste of
wintergreen; outer bark resembles Black Cherry
(*Prunus serotina* Ehrhardt, plate 345); leaves
alternate, to 4″ long and 2.4″ wide, sharply
pointed and toothed; male and females flowers
on separate trees; fruit a samara (apically winged-
achene), in headlike clusters (catkins); flowers
March–April; fruits June–July.

RANGE-HABITAT: ME and OH, south to GA and AL; in SC, common in the
mountains in acidic cove forests and montane oak-hickory forests.

COMMENTS: The bark of young stems is distinctive (reddish brown with horizontally elongate lenticels), as is the wintergreen taste and odor. It was once the
major source of oil of wintergreen flavoring used in medicines and confections.
Oil of wintergreen is toxic if ingested in quantity, but it takes very little to provide
flavoring (less than 0.04%). The lumber is valuable for cabinetwork and veneers. Wildlife use the twigs,
leaves, buds, flowers, and fruits for food. In early spring, the stems can be tapped for sugar water. Duke
(1997) recommends species of *Betula* as a useful remedy for warts. Though the wood is soft, it was often
referred to as "mahogany" by local mountain folks in the Carolinas. Many place names in NC refer to
this species (e.g., Mahogany Rock).

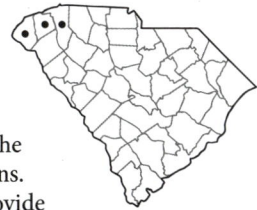

152. Yellow Birch

Betula alleghaniensis Britton
Bé-tu-la al-le-ghe-ni-én-sis
Betulaceae (Birch Family)

SYNONYM: *Betula lutea* Michaux
f.—RAB

DESCRIPTION: Deciduous tree to 100′ tall and 36″
in diameter, though typically much smaller; bark exfoliating in
narrow, gray or golden shaggy strips, dark reddish brown and
lustrous on young branches and stems; inner bark and twigs with
the odor and taste of wintergreen; leaves alternate, ovate to ovate-
oblong, to 4″ long and 2.25″ wide, sharply pointed and doubly-
serrate; male and female flowers on separate trees; fruit a samara,
winged apically, in headlike clusters (catkins); flowers March–April;
fruits June–July.

RANGE-HABITAT: Widespread in the northern and northeastern portions of North America, ranging south to GA in the Appalachians; in SC, rare in the mountains where found in acidic cove forests along streams.

COMMENTS: The shaggy bark of this species is distinctive. It is abundant at higher elevations in NC; in SC it is rare in small, cool microclimates along streams. It may be encountered near the Whitewater River, on the trail to Coon Branch, in Oconee County.

CONSERVATION STATUS: SC-Critically Imperiled

153. Fraser Magnolia; Umbrella Tree

Magnolia fraseri Walter
Mag-nòl-i-a frà-ser-i
Magnoliaceae (Magnolia Family)

DESCRIPTION: Tree to 100′ tall and 36″ in diameter; leaves deciduous, obovate, 16–29″ long, base with ear-shaped lobes; flowers very large, to 8″ across; petals creamy or white; flowers April–May; fruits are an ellipsoid aggregate of follicles, reddish at maturity; fruits mature July–August.

RANGE-HABITAT: Endemic to the southern Appalachians; in SC, common in the mountains, rare in the upper Piedmont; acidic cove forests, montane oak-hickory forests, and other moist forests on acidic soils.

COMMENTS: The genus name honors Pierre Magnol (1638–1715), a professor of botany at Montpellier. Thomas Walter honored his publisher, John Fraser (1750–1811), by naming this species for him. The SC state champion was recorded in 1981 at 86′ tall and 24″ in diameter. Though a common and beautiful native species, it is very difficult to cultivate in home landscapes.

154. Mountain Camellia

Stewartia ovata (Cavanilles) Weatherby
Ste-wárt-i-a o-và-ta
Theaceae (Tea Family)

DESCRIPTION: Small, deciduous tree, seldom over 25′ tall; leaves to 6″ long and 3″ wide, alternate, ovate, thin in texture, with finely serrate and ciliate margins and acuminate tips; flowers large and showy, resembling a white camellia; filaments purple; styles 5, separate; fruit a woody capsule; flowers June–July; fruits mature August–September.

RANGE-HABITAT: VA south to GA and west to KY and AL; primarily in the highlands but also in coastal VA; in SC, rare and restricted to the mountains and upper Piedmont in acidic cove forests and montane oak-hickory forests, especially on steep slopes near water.

COMMENTS: The showy flower is very similar to that of camellia and other members of the tea family. This delightful tree is not common anywhere in its range. It is most abundant on acidic soils in natural openings such as occur along rivers and streams and on thinly wooded slopes. It is easily observed along the Oconee Bells Trail in Devil's Fork State Park where it grows near the beaver pond on dry slopes. The similar Silky Camellia (*Stewartia malacodendron* L., plate 318) is very rare in the upper Piedmont and mountains of SC and has styles that are united; smaller leaves; and smaller, plump seeds. Silky Camellia flowers a month earlier than Mountain Camellia. The genus honors John Stuart (1713–92), Scottish nobleman and botanical enthusiast.

CONSERVATION STATUS: SC-Imperiled

155. Northern Witch-hazel

Hamamelis virginiana L. var. *virginiana*
Ha-ma-mè-lis vir-gi-ni-à-na var.
vir-gi-ni-à-na
Hamamelidaceae (Witch-hazel
Family)

DESCRIPTION: Small tree or shrub, seldom more
than 20′ tall; leaves ovate to obovate to 6″ long
and 4″ wide, the margins crenate to serrate with
oblique bases; flowers in clusters of 3 produced
in the leaf axils; flowers consist of 4 yellow,
orangish, or rusty-colored linear petals up to
3/4″ long, appearing to unroll as they emerge; fruit a pubescent capsule;
flowers October–January; fruits mature the following October–November.
RANGE-HABITAT: Widespread in eastern North America; in SC, common
throughout but most abundant in the mountains; acidic cove forests, montane
oak-hickory forests, and other forests with acidic substrate.
COMMENTS: Though there are numerous accounts of why this plant received
the name "witch," it appears that it refers to the extreme plasticity of the stems
and comes from an old English word "wyche," meaning bendable. This species is distinct while
flowering, which occurs as the leaves are falling in the autumn or just after leaf fall. Plants will
flower into late January in SC. The coastal form of this plant with smaller, more pubescent leaves
has been described as a separate variety: Small-leaved Witch-hazel (*H. virginiana* L. var. *henryae*
Jenne ex C. Lane).

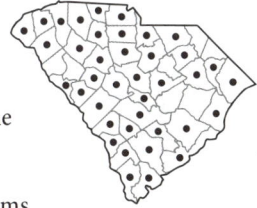

156. Mountain Fetterbush

Eubotrys recurvus (Buckley) Britton
Eù-bo-trys re-cúr-vus
Ericaceae (Heath Family)

SYNONYM: *Leucothoë recurva* (Buckley)
A. Gray—RAB

DESCRIPTION: Deciduous shrub to 12′ tall; twigs
pubescent when young, becoming smooth with
age; leaves elliptic to ovate, pubescent on the
veins below, 2–5″ long, 1–2″ wide, with toothed
margins; flowers white, urn-shaped, produced in
racemes; flowers open before the leaves expand in
spring; flowers March–April; fruits are capsules, dark brown when dry; fruits
mature August–October.
RANGE-HABITAT: Nearly endemic to the southern Appalachians, from VA and
KY south to GA and SC; in SC, common in the mountains on ridges, rocky
Chestnut Oak forests, margins of rock outcrops, pine-oak heaths, and other habitats with acidic soils; rare in the Piedmont on bluffs and in ravines.
COMMENTS: The flowers of Mountain Fetterbush are reminiscent of the more abundant Mountain Doghobble and for many years it was thought to be closely related and placed in the
same genus (*Leucothoë*). Recent molecular evidence indicates it is more closely related to Leatherleaf
(*Chamaedaphne*), and it is now placed in a separate genus, *Eubotrys*. Mountain Fetterbush is very easy
to locate in the early spring when it is adorned with flowers, but the rest of the year it can be mistaken
for a blueberry species. The persistent, dried capsules of previous years are a good way to separate it
from blueberries when not in bloom.

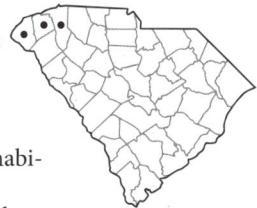

157. Mountain Doghobble

Leucothoë fontanesiana (Steudel) Sleumer
Leu-cóth-o-ë fon-ta-ne-si-à-na
Ericaceae (Heath Family)

SYNONYM: *Leucothoë axillaris* var. *editorum* (Fernald and Schubert) Ahles—RAB

DESCRIPTION: Evergreen shrub, 2–5′ tall, forming dense colonies; branches arching; leaves alternate, shiny, leathery, long tapering at the tip, toothed on the margins; flowers white, urn-shaped, in dense racemes to 4″ long; flowers April–May.

RANGE-HABITAT: Nearly endemic to the southern Appalachians, from TN and VA, south to GA and SC; in SC, common in the mountains and upper Piedmont, rare in the lower Piedmont; moist acid soils, usually along streams; most common in acidic cove forests in the mountains and acidic bluffs in the Piedmont.

COMMENTS: Mountain Doghobble is one of the most easily recognizable plants of ravines in the mountains and upper Piedmont. Dense thickets of these plants apparently "hobble" the progress of hunting dogs. The genus is from the mythical Leucothoë, daughter of Orchamus, King of Babylon, who was buried alive by her father but resurrected by Apollo as an incense shrub. Recent observations by the authors indicate that an unidentified species of caterpillar has decimated many populations in the acidic coves around Lake Jocassee.

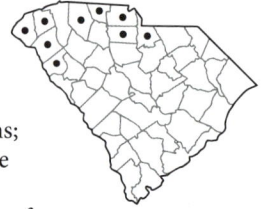

158. Great Laurel; White Rosebay

Rhododendron maximum L.
Rho-do-dén-dron máx-i-mum
Ericaceae (Heath Family)

DESCRIPTION: Evergreen shrub or small tree to 30′ tall; leaves leathery, to 10″ long; leaf usually 3–5x as long as wide, base wedge-shaped, apex acute; petals pale pink to white, the longest lobe spotted with green; fruit an elongated capsule; flowers June–August; fruits mature September–October.

RANGE-HABITAT: New England south to GA and west to Ontario and AL; in SC, common in the mountains in a variety of habitats on acidic soils, especially acidic cove forests; rare in the upper Piedmont on north-facing slopes.

SIMILAR SPECIES: Catawba Rhododendron (*Rhododendron catawbiense* Michaux) has leaves with rounded bases and bright purple flowers and flowers April–May in SC. It is rare and restricted to a couple of locations in the mountains on thin soils near rock outcrops at high elevations but has not been seen in several decades.

COMMENTS: The angle of the leaves below the horizontal correlates highly with temperature: the lower the temperature, the greater the angle of drooping. At 32° F, the leaves droop, below 28° F, they also roll lengthwise into a tube. Great Laurel is an excellent ornamental planted along streams at higher elevations and cool climates in acidic soils.

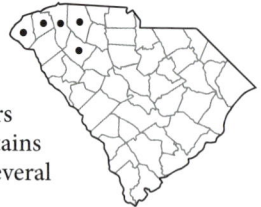

159. Mountain White-alder;
Mountain Sweet-pepperbush

Clethra acuminata Michaux
Clè-thra a-cu-mi-nà-ta
Clethraceae (Clethra Family)

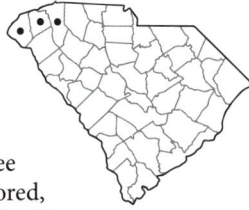

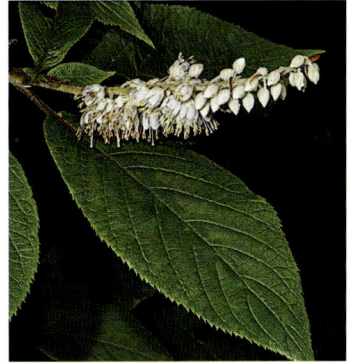

DESCRIPTION: Deciduous shrub or small tree to 15′ tall; bark distinctive, cinnamon-colored, exfoliating in strips; leaves to 7–8″ long, 3″ wide, toothed and long tapered at the tip; flowers in terminal racemes; flowers small, white; sepals and flower stalks densely covered with star-shaped hairs; flowers July–August.

RANGE-HABITAT: Endemic to the Appalachian Mountains, from PA and WV south to GA; in SC, common in the mountains; moist forests with acidic soils, especially acidic cove forests and stream margins.

COMMENTS: The shredding, reddish bark is distinctive (cinnamon-like) and therefore very useful in identification. The flowers have been described as "deliciously fragrant." Three species of *Clethra* are known from SC, but this is the only one with treelike stature and native to the mountain region. This species was first collected by André Michaux in 1787 in the mountains of SC.

---

160. Oconee Bells; Southern Shortia

Shortia galacifolia Torrey & Gray
Shórt-i-a ga-la-ci-fò-li-a
Diapensiaceae (Diapensia Family)

DESCRIPTION: Low rhizomatous, evergreen sub-shrub, forming dense colonies; leaves shiny, circular to broadly elliptic, margins crenate-serrate; leaf tip squared across or slightly indented; the white or pinkish flowers resemble nodding bells; flowers March–mid-April.

RANGE-HABITAT: A narrow endemic of the southern Blue Ridge Escarpment and adjacent upper Piedmont; known naturally in SC only from Oconee and Pickens Counties; acidic cove forests along mountain streams in the Blue Ridge Escarpment, usually under Great Laurel (*Rhododendron maximum* L.) or Mountain Laurel (*Kalmia latifolia* L.).

COMMENTS: Oconee Bells was first collected in fruit in 1787 in Oconee County, SC, by the French botanist and early explorer of eastern North America, André Michaux (1746–1802). Harvard botanist Asa Gray (1810–88) discovered Michaux's collection in Paris in 1830 and named the genus in 1839 in honor of Charles W. Short (1794–1863), a prominent botanist and physician of Louisville, Kentucky. Gray and others made many excursions to the "high mountains of Carolina," as the location of Michaux's original collection was described, with no success—not surprisingly, since the plant occurs at elevations of 600–2100′. In 1877, 17-year-old George Hyams "rediscovered" the genus near Marion, NC. It was not until 1886 that Charles Sprague Sargent (1841–1927) relocated the species in SC, in the region of the Jocassee Gorges, nearly 100 years after the original discovery by Michaux. The construction and filling of Lake Jocassee destroyed at least 60% of the species' population. Recent molecular and morphological work indicates that the McDowell County, NC population represents a distinct species, Northern Shortia (*Shortia brevistyla* (P. A. Davis) Gaddy) (Gaddy et al., 2020). Therefore, all known natural populations of true Oconee Bells are known from the drainages of Lake Jocassee and the upper end of Lake Keowee and some adjacent streams in a small area of Oconee and Pickens Counties and a small region of Transylvania County, NC. The plant has been naturalized along the Chattooga River, northern GA, and several areas in Greenville County and the NC and TN mountains, where it was planted and is not native.

CONSERVATION STATUS: SC-Vulnerable

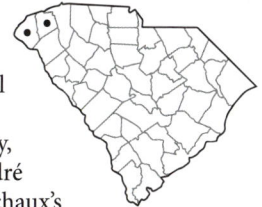

161. Round-leaf Yellow Violet; Early Yellow Violet

Viola rotundifolia Michaux
    Vì-o-la ro-tun-di-fò-li-a
Violaceae (Violet Family)

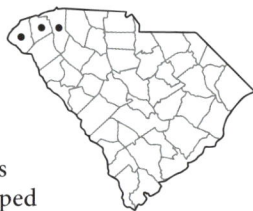

DESCRIPTION: Herbaceous perennial; leaves basal only, nearly round, with a heart-shaped base, hairy and fleshy; flowers yellow, on stalks at least as long as the leaves; flowers March–early April.

RANGE-HABITAT: ME south to SC and west to Ontario, OH, and GA; in SC, common in the mountains in acidic cove forests and moist areas in montane oak-hickory forest.

COMMENTS: This is the only yellow violet with flowers on a leafless stalk; its flowers appear when its leaves are only partly developed.

162. Halberd-leaved Violet; Spearleaf Violet

Viola hastata Michaux
    Vì-o-la has-tà-ta
Violaceae (Violet Family)

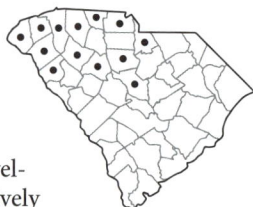

DESCRIPTION: Herbaceous perennial with yellow flowers on leafy stems; leaves distinctively spearpoint-shaped, usually mottled with silvery gray areas, clustered near the stem tip; flowers late February–May.

RANGE-HABITAT: PA south to FL and west to OH and AL; in SC, common in the mountains and upper Piedmont in moist deciduous forests, especially beech forests, acidic cove forests, and moist oak-hickory forests.

COMMENTS: The species was first collected by André Michaux in the vicinity of the Jocassee Gorges in 1787.

163. Sweet White Violet

Viola blanda Willdenow
    Vì-o-la blán-da
Violaceae (Violet Family)

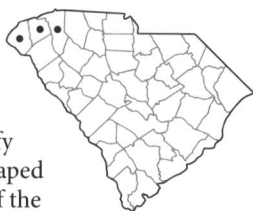

DESCRIPTION: Perennial herb without a leafy stem; leaves nearly round, with a heart-shaped base, hairy, at least on the upper surface of the basal lobes, which often overlap; flowers white; flower stalk tinged with red; flowers April–early June.

RANGE-HABITAT: Widespread in eastern North America; in SC, occasional in the mountains in rich cove forests and acidic cove forests, particularly along the bases of slopes and near streams.

SIMILAR SPECIES: Sweet White Violet is easily confused with Wild White Violet (*V. pallens* (Banks ex de Candolle) Brainerd), which is uncommon in the mountains. Wild White Violet occurs in a different habitat (acidic seepages and banks of small streams and fens) and has smaller flowers and completely smooth leaf blades.

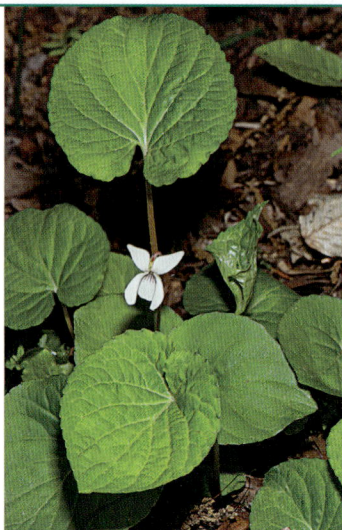

164. Long-spurred Violet

Viola rostrata Pursh
Vì-o-la ros-trà-ta
Violaceae (Violet Family)

DESCRIPTION: Perennial herb with a leafy stem; leaves ovate, with a heart-shaped base, smooth or with a few hairs on the midrib; stipules present; flowers light bluish-violet with a darker throat; nectar spur elongated, 0.4–0.8″ long; flowers April–May.

RANGE-HABITAT: Quebec west to WI and south to GA and AL; in SC, common in the mountains in acidic cove forests, along streams, and in other moist, acidic woodlands.

COMMENTS: This is one of the most easily identified and charming species of SC native violets. The long nectar spur is distinctive. Though it is often cited as growing under hemlocks, it is frequently encountered far away from hemlocks in SC. This species is easily seen in moist forests along the stream at Eva Chandler Heritage Preserve in Greenville County and in nearby Jones Gap State Park. The only similar species is American Dog Violet (*V. labradorica* Schrank), which has a much shorter nectar spur, smaller flowers, and is rare in SC and known only from small stream floodplains in Oconee and Pickens Counties.

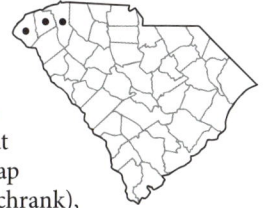

165. Persistent Trillium

Trillium persistens Duncan, Garst & Neece
Tríl-li-um per-sís-tens
Trilliaceae (Trillium Family)

DESCRIPTION: Small perennial herb from a rhizome; leaves in a single whorl of 3, solid green and narrowed to a pointed apex; the single terminal flower is on a stalk that droops slightly but is held above the leaves; sepals not recurved; petals white, turning pink with age; stamens are light yellow and straight; flowers early March–mid-April.

RANGE-HABITAT: A narrow southern Blue Ridge Escarpment endemic, restricted to the Tallulah-Tugaloo River system and known only from Rabun and Habersham counties in GA and Oconee County in SC; acidic cove forests of deep ravines or gorges, often under or near Great Laurel (*Rhododendron maximum* L.) or Gorge Rhododendron (*R. minus* Michaux).

COMMENTS: As with most *Trillium,* this species is ant dispersed, with a protein body on the seed (an elaiosome) that attracts ants. The small, narrow petals, lance-shaped leaves, and pale yellow and straight anthers are characteristics that separate this species. The vegetative portions of the plant persist well into the late summer and autumn, which is the reason for the name *persistens.*

CONSERVATION STATUS: Federally Endangered

166. Jones Gap Trillium

Trillium sp. nov.

Tríl-li-um

Trilliaceae (Trillium Family)

DESCRIPTION: Low perennial herb from a rhizome; leaves in a single whorl of 3, solid green and narrowed to a pointed apex; a single terminal flower is on a stalk that is held erect or slightly drooping; white petals turn pinkish with age; stamens are cream to yellow; flowers mid-March–mid-April.

RANGE-HABITAT: Endemic to the southern Blue Ridge Escarpment, known from sw. NC and adjacent SC; rich cove forests and bases of slopes in acidic cove forests along streams where nutrients have accumulated; often occurs with Catesby's Trillium (*Trillium catesbyi* L., plate 186).

COMMENTS: This species is very similar to the closely related Catesby's Trillium, but it flowers 3–4 weeks earlier and often forms large colonies. Jones Gap Trillium is a smaller plant in all regards with paler stamens and is reproductively isolated from Catesby's Trillium by flowering time. These distinctive populations have been known for some time but thought to be part of the variation within Catesby's Trillium. Joseph Townsend, Pickens County botanist, championed the recognition of this plant and brought it to the authors' attention in the early 2000s. It is easily observed in Jones Gap State Park. As of this writing (spring 2021), there has been no formal description of the species.

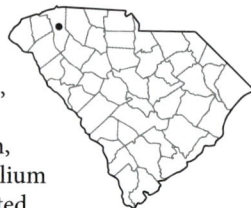

167. Wake Robin; Red Trillium; Stinking Benjamin

Trillium erectum L.

Tríl-li-um e-réc-tum

Trilliaceae (Trillium Family)

DESCRIPTION: Perennial herb from a short, thick rhizome; stem erect, to 18″ tall; leaves in a single whorl of 3, uniformly green, about as wide as long; flower stalk erect and stiff, usually positioning the flower at a near horizontal angle; petals dark purple, pale yellow, or white; sepals flat to weakly boat-shaped at the tip; flowers April–May.

RANGE-HABITAT: Widespread in eastern North America; in SC, rare in the mountains in acidic cove forests.

COMMENTS: The fragrance of the flower is like that of a wet dog. The flowers are pollinated by flies, including one of the carrion flies, the green flesh fly. Wake Robin has long been attributed to SC, but all previous documentation has been found to be based on mis-identifications of other species. The species was finally, and definitively, discovered in SC by Pickens County resident and botanist Joseph Townsend in 2012, along the Chattooga River very near the NC state line.

CONSERVATION STATUS: SC-Critically Imperiled

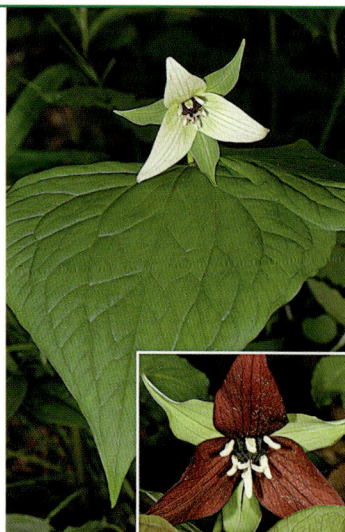

168. Painted Trillium

Trillidium undulatum (Willdenow)
   Floden & E. E. Schilling
   Tril-lí-di-um un-du-là-tum
Trilliaceae (Trillium Family)

SYNONYM: *Trillium undulatum*
   Willdenow—RAB, PR

DESCRIPTION: Herbaceous perennial from a thick rhizome; stem erect, to 8″ tall; leaves in a single whorl of 3, uniformly dark green and with a short but distinct petiole; flower stalk erect, 1–2″ long; petals white, with an inverted red V at the base, margins strongly wavy; anthers lavender to white, opening toward the outside of the flower; flowers April–May.

RANGE-HABITAT: Eastern Canada south to GA and west to KY and TN; in SC, rare in the mountains in relatively high elevation acidic cove forests.

COMMENTS: This is one of SC's rarest members of the trillium family. (It is more abundant in NC.) The inverted V at the base of the petals is distinctive. Recent molecular evidence has indicated that Painted Trillium belongs to a separate genus from the rest of the *Trillium* species. The species is impossible to transplant or cultivate and should never be disturbed in the wild.

CONSERVATION STATUS: SC-Imperiled

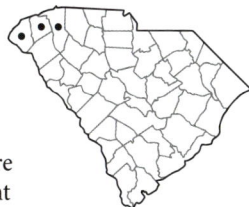

169. Variable-leaf Heartleaf

Hexastylis heterophylla (Ashe) Small
   Hex-ás-ty-lis he-te-ro-phýl-la
Aristolochiaceae (Birthwort Family)

DESCRIPTION: Low evergreen, perennial herb forming small, dense clumps; leaves mottled along the veins or unmottled, rounded, and heart-shaped, to 4″ long and 3.5″ wide; flowers solitary in the axils of leaves; calyx tube urn- or bell-shaped, to 0.60″ long and wide; flowers March–May.

RANGE-HABITAT: Endemic to the southern Appalachian region from WV and KY south to AL and GA; in SC, common in the mountains and the upper Piedmont; acidic cove forests, oak-hickory forests, and near streams, always on acidic soils.

COMMENTS: This is the most abundant heartleaf in the mountains of SC. It is frequently confused with Large-flower Heartleaf (*Hexastylis shuttleworthii* (Britten and Baker) Small, plate 170), which has much larger flowers that are produced later in the season and often forms large patches. It is also confused with Dwarf-flower Heartleaf (*Hexastylis naniflora* Blomquist, plate 421), which has much smaller flowers and ranges to the east of Variable-leaf Heartleaf in the Piedmont.

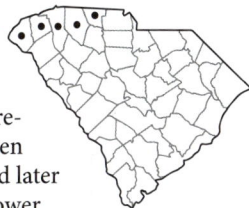

170. Large-flower Heartleaf

Hexastylis shuttleworthii
(Britten & Baker) Small
Hex-ás-ty-lis shut-tle-wórth-i-i
Aristolochiaceae (Birthwort Family)

DESCRIPTION: Low evergreen, perennial herb, often forming large patches; leaves mottled along the veins or unmottled, rounded, and heart-shaped, to 4″ long and 3.5″ wide, from terminal portion of root-stalk only; flowers solitary in the axils of leaves; calyx tube urn- or bell-shaped, to 1.6″ long and 1″ wide; flowers April–June.

RANGE-HABITAT: VA and TN, south to GA and AL; in SC, common in the mountains; usually found in acidic soils, along creeks and under Great Laurel (*Rhododendron maximum* L.) in acidic cove forests.

COMMENTS: The very large, easily broken flowers of this species make it readily identifiable. The specific epithet honors Robert J. Shuttleworth (1810–74), an English plant collector.

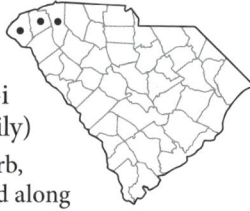

171. Speckled Wood-lily

Clintonia umbellulata (Michaux)
Morong
Clin-tòn-i-a um-bel-lu-là-ta
Liliaceae (Lily Family)

DESCRIPTION: Perennial herb; leaves 2–4, clustered on the lower stem, appearing basal, to 12″ long and 3.5″ wide, with obvious hairs on the margins (ciliate); flowers in a terminal umbel; tepals white and tipped or speckled with purple; fruit a black berry; flowers May–mid-June; fruits mature in August.

RANGE-HABITAT: NY and OH, south to GA; in SC, occasional in the mountains; acidic cove forests and mesic or dry ridges and slopes.

COMMENTS: Named after DeWitt Clinton (1769–1828), governor of New York. The name is said to have annoyed Henry David Thoreau, writer and self-proclaimed naturalist, who was never so honored. The basal leaves are like those of several other species, such as Fairy Wand, Showy Orchis, and Carolina Lily. Speckled Wood Lily can immediately be distinguished from these species by the ciliate margins of the leaves.

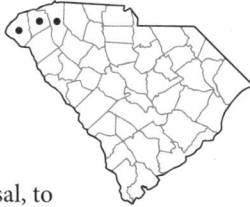

172. Indian Cucumber-root

Medeola virginiana L.
Me-dè-o-la vir-gi-ni-à-na
Liliaceae (Lily Family)

DESCRIPTION: Perennial herb from a white, swollen rhizome; stems 0.8–2.5′ tall; leaves of flowering plants usually in two whorls, the upper whorl generally of 3 leaves, the lower of 6–10 leaves; flowers borne on recurved stalks from the leaf axils and projecting below the top whorl of leaves; sterile plants with a single whorl of leaves; flowers April–June; in fruit, the stalks are ascending or erect; fruits are blackish-blue berries that mature September–October.

RANGE-HABITAT: Quebec south to FL and west to MN and LA; in SC, common in the mountains and Piedmont, uncommon in the Coastal Plain; generally found in moist forests, usually with acidic soils.

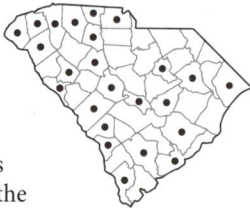

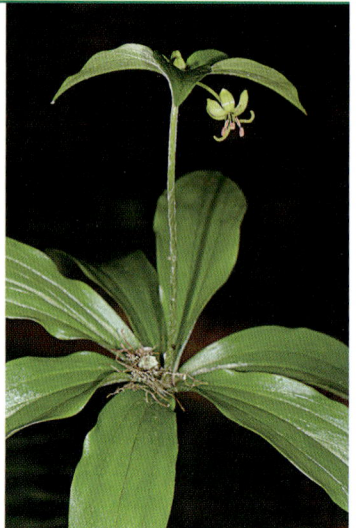

COMMENTS: Native Americans used the rhizomes for food; they are crisp and starchy, with a taste similar to cucumber. They should not be harvested, particularly on public lands, because of their limited numbers. When the fruits are mature, the upper leaves become tinged with scarlet, possibly attracting animals that aid in seed dispersal.

173. Small Whorled Pogonia
Isotria medeoloides (Pursh)
Rafinesque
I-só-tri-a me-de-o-loì-des
Orchidaceae (Orchid Family)

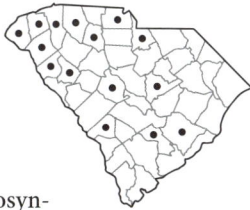

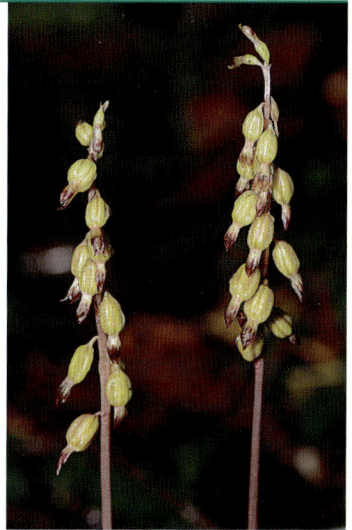

DESCRIPTION: Perennial herb to 10″ tall; stem green or purplish; leaves whorled at the top of the stem, pale green and somewhat glaucous (with a whitish tint); flowers yellowish green, nearly sessile with very narrow linear-oblanceolate tepals and a distinctive lip; flowers May–June.

RANGE-HABITAT: Widespread, but very rare in eastern North America, ranging south to GA; in SC, very rare and localized in acidic cove forests and oak-hickory forests in the mountains, often associated with Eastern White Pine (*Pinus strobus*).

COMMENTS: Vegetatively this closely resembles Indian Cucumber-root. When not in flower it is distinguished by its pale green and somewhat glaucous (with a whitish tint) leaves. Like many orchids, it may be apparent one year and unobservable the next. Foraging wild pigs have decimated the largest SC population. This species is extremely uncommon and most often located by accident. It should never be disturbed in the wild, and all sightings should be reported to the South Carolina Heritage Trust Program.

CONSERVATION STATUS: Federally Threatened

174. Autumn Coral-root
Corallorhiza odontorhiza
(Willdenow) Poiret
Co-ral-lo-rhì-za o-don-to-rhì-za
Orchidaceae (Orchid Family)

DESCRIPTION: Perennial herb without photosynthetic tissues, from a cluster of branched, coral-like roots; without leaves, but with a few sheathing scales toward the base; flowers small, with partly fused sepals and petals, purplish or purplish green, except for the lip petal, which is white, spotted with purple with 2 ridges near the base; flowers August–October.

RANGE-HABITAT: Widespread in the eastern US; in SC, occasional throughout; oak-hickory forests, acidic cove forests, beech forests, and other mesic woodlands.

COMMENTS: Autumn Coral-root depends on saprophytic fungi associated with its roots for all of its nutritional needs, a habit that is now described as mycoparasitism. Its small size distinguishes it from other saprophytic orchids, except for the Spring Coral-root (*C. wisteriana* Conrad, plate 752), which blooms in the spring and is found growing in richer soils. There are two forms of this species: one has flowers that remain closed and are self-pollinating, the other has flowers that are open and are at least potentially cross-pollinated.

175. Fancy Fern; Evergreen Wood-fern

Dryopteris intermedia (Muhlenberg ex Willdenow) Gray
Dry-óp-te-ris in-ter-mè-di-a
Dryopteridaceae (Wood-fern Family)

DESCRIPTION: Large, semi-evergreen, rhizomatous fern with leaves surrounding the crown (growth point); petioles and rhizome with rusty scales; leaves 3-pinnate, appearing very "lacy," to 20″ long and 10″ wide; spores produced in distinct rounded (reniform) clusters (sori) on the back sides of the pinnae and not restricted to the edges of the pinnae; spores produced June–September.

RANGE-HABITAT: Newfoundland west to MN and south to MO and GA; in SC, uncommon and restricted to acidic cove forests and other moist acidic forests near streams in the mountains.

COMMENTS: The fronds of this species are heavily dissected and appear very light and attractive. They are evergreen but often are bent to the ground by the arrival of the winter season and die and decompose in the early spring. The wood ferns are fairly easy to identify because of their habit of forming a circle of leaves around the growth point and the brownish scales that are produced on the lower portions of the petioles.

CONSERVATION STATUS: SC-Vulnerable

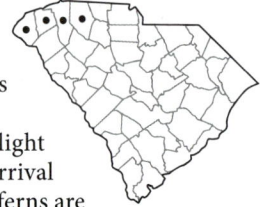

176. Chestnut Oak forest

177. Montane oak-hickory forest

178. Chestnut Oak; Basket Oak

Quercus montana Willdenow
Quér-cus mon-tà-na
Fagaceae (Beech Family)

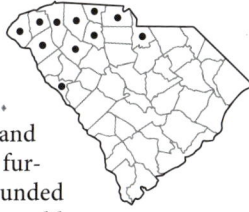

DESCRIPTION: Deciduous tree to 100′+ tall and 6′ in diameter; bark dark gray and deeply furrowed; leaves with 9–16 evenly spaced, rounded teeth on each side; lower leaf surface with starlike clusters of 2–5 hairs; female flowers 1–2 on a short stalk, separate from the male flowers, which are in catkins; acorns mature in September–November of the same year as flowering.

RANGE-HABITAT: ME south to AL and west to IL and MS; in SC, common in the mountains and Piedmont and rare in the Sandhills; found in a variety of dry habitats and mostly on ridges and mid- to upper slopes; often the dominant tree in Chestnut Oak forests, but also common in pine-oak heaths and dry oak-hickory forests.

COMMENTS: Chestnut Oak does not grow well in shade. It has replaced much of the American Chestnut that was lost in the first half of the twentieth century to the blight. It is also called Basket Oak because the wood is easily split into long, tough ribbons that are used in making baskets and barrel staves. The wood was once prized for railroad cross ties. The SC state champion was measured in 1987 at 130′ tall and 4.2′ in diameter in 1987.

179. American Chestnut

Castanea dentata (Marshall) Borkhausen
Cas-tà-ne-a den-tà-ta
Fagaceae (Beech Family)

DESCRIPTION: Medium to tall tree, formerly reaching heights of 115′; leaves alternate, deciduous, 6–11″ long, tapering to a short or long-pointed tip, leaf surfaces smooth, with numerous coarse, sharp-pointed teeth along the margin; fruits are large spiny burs enclosing 1 to 3 nuts; flowers June–July; fruits mature September–October.

RANGE-HABITAT: Throughout much of eastern North America, where it reached its greatest size in the southern Appalachian Mountains; in SC, occasional in the mountains and Piedmont; mesic and xeric forests.

COMMENTS: Today chestnut exists as stump sprouts or small trees. A fungal blight, *Cryphonectria parasitica,* introduced from Asia to New York in 1904, swept through the entire range of the American Chestnut, killing nearly every standing tree. When a stump sprout reaches the size of first fruit production, the blight, which persists in the forests on oaks, infects the tree. Only rarely does a stump sprout grow large enough to produce fruits. One can still see evidence of vast numbers of almost completely decayed fallen chestnut trees and stumps in the forests of the mountains and Piedmont of SC. Interestingly, populations of chestnut in NC are persisting as small flowering trees on amphibolite outcrops and ridges. People and animals prized the chestnut's large, sweet nuts. The flowers are beetle pollinated and visited by large swarms of various species of longhorn and flower tumbler beetles. The reddish-brown wood was lightweight, soft, easy to split, and resistant to decay. Both the bark and wood were rich in tannins, which were once used to tan animal hides to make leather.

180. Smooth Serviceberry

Amelanchier laevis Wiegand
A-me-lán-chi-er laè-vis
Rosaceae (Rose Family)

SYNONYM: *Amelanchier arborea*
(Michaux f.) Fernald var. *laevis*
(Wiegand) Ahles—RAB

DESCRIPTION: Small to medium-sized tree reaching 80′ but typically much smaller (< 40′), with smooth grey bark; the young emerging leaves reddish and smooth; mature leaves alternate, toothed with 6–16 teeth per 0.5″ of margin; flowers white, produced in racemes; fruit a pome that is dark blackish-purplish and sweet when ripe; pedicels of fruits 0.5–1.25″ long; flowers February–April; fruits mature May–July.

RANGE-HABITAT: Widespread in eastern North America from Nova Scotia to MN south to AL and SC; in SC, locally common at the highest elevations in the mountains in acidic cove forests, acidic oak-hickory forests, and margins of rock outcrops.

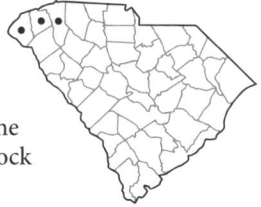

SIMILAR SPECIES: Smooth Serviceberry has formerly been confused with Downy Serviceberry (*Amelanchier arborea* (Michaux f.) Fernald, plate 341). Downy Serviceberry is a shrub or small tree with downy pubescent undersides and whitish to green leaves when young, has shorter pedicels (0.3–0.7″), and has lighter purple or reddish fruits that are sour tasting when ripe. Downy Serviceberry is found throughout SC in a variety of dry wooded habitats.

COMMENTS: Smooth Serviceberry is sometimes called Juneberry in the Appalachian region. The ripe fruit are picked and eaten fresh. The tree provides good wildlife food and is relished by frugivorous birds. The name "serviceberry" is variously interpreted: some think the name refers to the fact that it is often in flower for Easter Sunday church services; others think it refers to the plant flowering when the ground was no longer frozen, and thus those who had died during winter could be buried and services held.

181. Mountain Holly; Mountain Winterberry

Ilex montana Torrey & Gray
Ì-lex mon-tà-na
Aquifoliaceae (Holly Family)

SYNONYM: *Ilex ambigua* var. *montana*
(Torrey & Gray) Ahles—RAB

DESCRIPTION: Deciduous shrub or small tree to 35′ tall; leaves 2.5–6.0″ long, margins sharply fine toothed, with long acuminate tips; leaf stalks (petioles) more than 0.4″ long; flowers on short stalks, white; sepals ciliate; male and female flowers on different plants; drupes red, spherical, with 4 seeds, each seed surrounded by a pit (pyrene); flowers April–June; fruits mature August–September.

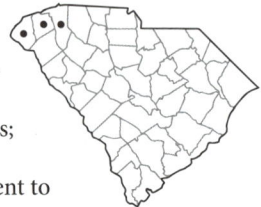

RANGE-HABITAT: MA south to AL, with the range centered on the Appalachians; in SC, common in the mountains and upper Piedmont; acidic cove forests, oak-hickory forests; sometimes found in drier upland forests or in and adjacent to montane fens.

COMMENTS: This is the only treelike deciduous holly in mesic forests in the mountains of SC. The much smaller Carolina Holly (*Ilex ambigua* (Michaux) Torrey) is similar but has smaller, rounder leaves with inconspicuous teeth along the margins and petioles of the leaves less than 0.4″ long. Carolina Holly is uncommon in the mountains and upper Piedmont in similar habitats and common in dry forests of the Coastal Plain.

182. Buffalo-nut; Oil-nut

Pyrularia pubera Michaux
  Py-ru-là-ri-a pù-be-ra
  Santalaceae (Sandalwood Family)

DESCRIPTION: Deciduous shrub to 12′ tall; stem highly branched, arching; overwintering buds bright green; leaves alternate, thin, veins on lower surface raised; plants dioecious (male and female flowers on separate plants); flowers produced in a spike; flowers small, petals absent, sepals 5, greenish-yellow; fruit a pear-shaped drupe, about 1″ long; flowers April–May; fruits mature July–October.

RANGE-HABITAT: PA south to GA; common in the mountains and upper Piedmont in a variety of dry to moist forests on acidic soils.

COMMENTS: The oil is very poisonous. The source of the primary common name requires little imagination. Like all members of the sandalwood family in SC, the species gets some of its nutrition by parasitizing the roots of hardwoods.

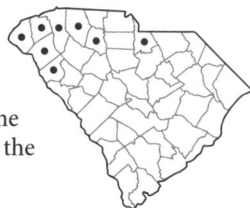

183. Gorge Rhododendron; Punctatum Carolina Rhododendron

Rhododendron minus Michaux
  Rho-do-dén-dron mì-nus

Carolina Rhododendron

Rhododendron carolinianum Rehder
  Rho-do-dén-dron ca-ro-li-
  ni-à-num
  Ericaceae (Heath Family)

DESCRIPTION: *R. minus* is an evergreen shrub to 10′ tall; leaves leathery, to 4″ long and 2″ wide, lower surface dotted with circular brown scales; corolla with the tube longer than the corolla lobes; petals pink to purplish, the largest lobes spotted with green; fruit a 5-parted capsule; flowers late April–June; *R. carolinianum* is vegetatively almost identical to *R. minus;* flowers smaller with corolla tubes about the same length as the corolla lobes; flowers generally lighter pink to white; flowers late March–April.

RANGE-HABITAT: Gorge Rhododendron is found from NC, mostly south of Asheville, to GA and AL; in SC, it is found in the mountains (common), Piedmont (occasional), and Sandhills (rare); in the mountains in rocky woods of acid bluffs from stream side to ridge top, depending on slope and aspect; acidic bluffs along streams in the Piedmont and Sandhills. Carolina Rhododendron is restricted to the mountains where it ranges from Alleghany County, NC and e. TN south to nw. SC and ne. GA. It is found in acidic forests in the mountains of SC, especially Chestnut Oak forests.

COMMENTS: Gorge Rhododendron is the common species over most of the escarpment and flowers from late April through June, peaking in May. The flowers are most commonly pink to purple. Carolina Rhododendron flowers from late March through April, peaking in mid-April, is found in similar habitats but is more abundant in Greenville County. The two species are completely separated by flowering time: Carolina Rhododendron flowers a month earlier, even at high elevations. Both species co-occur in Jones Gap State Park. Recent vagaries in the climate, including extreme droughts and abnormally warm winters, have led to many plants flowering in late fall through mid-winter. The impact of such unseasonal flowering is unknown.

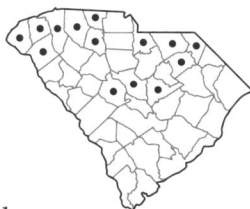

184. Flame Azalea

Rhododendron calendulaceum (Michaux) Torrey

Rho-do-dén-dron ca-len-du-là-ce-um

Ericaceae (Heath Family)

DESCRIPTION: Deciduous, upright shrub to 12′ tall, not stoloniferous; leaves to 3.5″ long with gray hairs, at least along the midrib, to 3.5″ long; flowers open just before or with the emergence of leaves in the spring; petals bright orange, yellow, or red; corolla tube with glandular hairs outside (sticky); flowers May–June.

RANGE-HABITAT: From PA south to GA and west to OH and AL; in SC, common in the mountains and rare in the Piedmont; mid- to upper slopes of dry-to-mesic oak-dominated forests, especially Chestnut Oak forests.

COMMENTS: Darker red-flowered forms are very similar to Oconee Azalea (*R. flammeum* (Michaux) Sargent, plate 349). Their ranges do not overlap, and Oconee Azalea is found in the middle and lower Piedmont, Sandhills, and Coastal Plain. The related Cumberland Azalea (*R. cumberlandense* E. L. Braun) is also similar but has flowers opening well after the leaves have expanded and is stoloniferous. It is a rare disjunct in the Piedmont of SC.

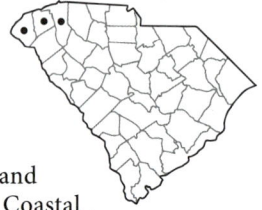

185. Hill Cane

Arundinaria appalachiana Triplett, Weakley, & L. G. Clark

A-run-di-nà-ri-a ap-pa-la-chi-à-na

Poaceae (Grass Family)

DESCRIPTION: Rhizomatous, perennial, bamboo (grass) with deciduous leaves; stems 2–6′ tall, with culm internodes terete (round) in cross-section; plants flower very infrequently; flowering April–May.

RANGE-HABITAT: NC, TN, SC, GA, AL, primarily in the Appalachian region; in SC, common in the mountains and upper Piedmont, in oak-hickory forests and in often dry and other forested communities, most often on slopes (hence the common name).

SIMILAR SPECIES: The genus *Arundinaria* has three species, all of which occur in SC. Switch Cane (*Arundinaria tecta* (Walter) Muhlenberg, plate 819) is a similar-sized, evergreen plant, confined to the lower Piedmont and Coastal Plain and flowers every 3–4 years. River Cane (*Arundinaria gigantea* (Walter) Muhlenberg, plate 77) is a much larger evergreen species, 6–30′ tall, with grooved internodes and copious branching of the stem. It is found in bottomlands, rather than hillsides, in the mountains and Piedmont.

COMMENTS: *Arundinaria* is the only member of the bamboo tribe of the grass family (Poaceae) native to the United States. Numerous other species of the bamboo tribe are cultivated. The stems of this species do not attain the size of River Cane and were less important traditionally for use as building materials. This species does not form the dense thickets typical of River Cane or Switch Cane.

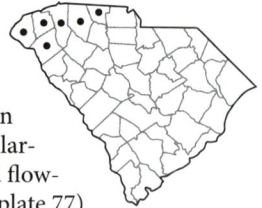

186. Catesby's Trillium; Rose Wake-robin

Trillium catesbyi Elliott
    Trĺl-li-um càtes-by-i
    Trilliaceae (Trillium Family)

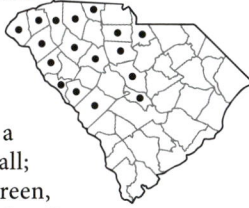

DESCRIPTION: Herbaceous perennial from a short, thick rhizome; stem erect, to 16″ tall; leaves in a single whorl of 3, uniformly green, with purplish tinge in sunny habitats; leaf stalk to 0.5″ long; flower stalk held below leaves; petals white, pink, or rose and all fading to pink with age; sepals sickle-shaped; anthers twisted outward; pollen sacs dark yellow; flowers April–early May.

RANGE-HABITAT: NC and TN south to AL; in SC, common in the mountains, occasional in the Piedmont; acidic soils of mesic woods on slopes and floodplains in the Piedmont; on drier, more exposed slopes in the mountains, usually oak-hickory forests merging into pine-oak heaths or acidic cove forests.

COMMENTS: This is one of many specific epithets honoring British naturalist Mark Catesby (1679–1749), who visited SC from 1722 to 1724 and published *The Natural History of Carolina, Florida and the Bahama Islands* (1729–47).

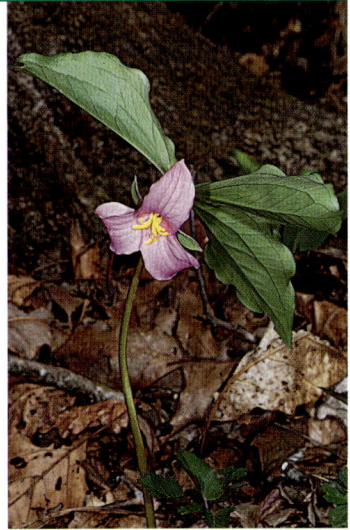

187. Gaywings; Fringed Polygala

Polygaloides paucifolia (Willdenow)
    J. R. Abbott
    Po-ly-ga-loì-des pau-ci-fò-li-a
    Polygalaceae (Milkwort Family)

SYNONYM: *Polygala paucifolia* Willdenow

DESCRIPTION: Perennial herb, forming colonies from slender rhizomes; stem 3–6″ tall; scalelike leaves low on the stem, and 3–6 well-formed leaves clustered above; flowers 1–4 in a terminal raceme, pink to purple (rarely white), wings large, to 0.6–.8″ long; stamens 6; flowers April–May.

RANGE-HABITAT: Widespread in eastern North America; in SC, rare in the mountains; oak forests at moderate to high elevations.

COMMENTS: This is SC's largest-flowered polygala, and the few large flowers above the sparsely clustered leaves are distinctive. Self-pollinating flowers are hidden under leaf litter. This species is most easily seen in SC on Pinnacle Mountain in Table Rock State Park and in Jones Gap/Caesars Head State Parks.

CONSERVATION STATUS: SC-Imperiled

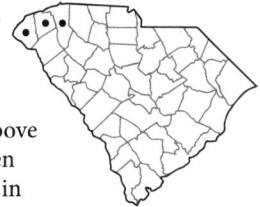

188. Bowman's Root; Mountain Indian Physic

Gillenia trifoliata (L.) Moench
Gil-lèn-i-a tri-fo-li-à-ta
Rosaceae (Rose Family)

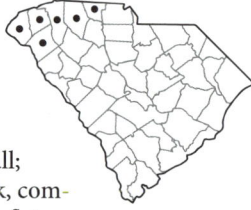

DESCRIPTION: Perennial herb; stems to 3′ tall; leaves alternate, with a very short leaf stalk, comprised of 3 palmately arranged, toothed leaflets; stipules persistent, about 0.25″ long; petals 5, white, unequal in length; flowers April–June.

RANGE-HABITAT: Ontario and New England south to MO and GA; in SC, common in the mountains, rare in the Piedmont; forest margins, oak-hickory forests, and forest margins.

COMMENTS: The five unequal white petals and 3-parted leaves together are distinctive. Native Americans used the powdered, dried root as a laxative and emetic—hence, the common names. The genus is named for German botanist and physician Arnold Gillen (1586–1633).

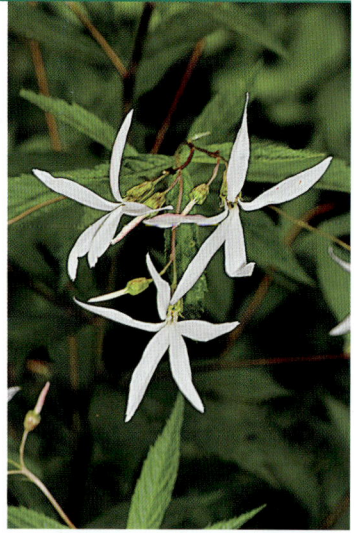

189. Galax

Galax urceolata (Poiret) Brummitt
Gà-lax ur-ce-o-là-ta
Diapensiaceae (Diapensia Family)

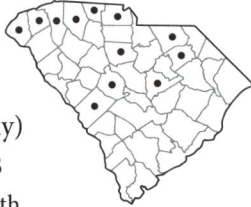

SYNONYM: *Galax aphylla* L.—RAB

DESCRIPTION: Evergreen perennial herb, with leaves clustered on a very short stem; leaves long-stalked, round, sharply dentate or crenate, with cordate bases; flowers produced in a raceme on a long leafless stalk; flowers small, with 5 white petals; flowers May–July.

RANGE-HABITAT: MD, WV, and VA, south to GA and AL; in SC, common in the mountains and upper Piedmont, rare elsewhere; dry to moist woods such as Chestnut Oak forests and oak-hickory forests, often with Mountain Laurel (*Kalmia latifolia*).

COMMENTS: Plants in the mountains of SC have four complements of chromosomes (tetraploid); those elsewhere in the state have two (diploid). This plant produces the distinctive musty smell of many mountain locations. Vigorously brush the leaves with your hand and then smell the leaves; you can pick up the scent. The cause of the scent is still unknown but may be related to long chain, sulphur-containing chemicals.

190. Downy Rattlesnake Plantain

> *Goodyera pubescens*
>   (Willdenow) R. Brown
>   Good-yèr-a pu-bés-cens
> Orchidaceae (Orchid Family)

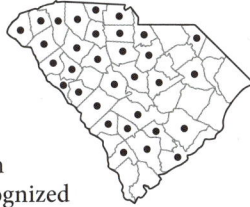

DESCRIPTION: Perennial herbs covered with short hairs, from a creeping rhizome; recognized by its distinctive basal leaves, which are blue-green and variegated with a network of white on the veins; flowers are small, globose, and whitish; they are produced in a raceme and are barely discernable as orchids without close inspection; flowers June–August.

RANGE-HABITAT: New Brunswick to Ontario and MN and south to FL, MS, and AR; in SC, common in the upper Piedmont and mountains, rare in the Coastal Plain; found in a wide variety of moist to dry forests.

COMMENTS: The common name refers to the belief of superstitious country people that the mottled snake-striped leaves, when chewed and applied to a rattlesnake bite, would act as an antidote. After the rosette produces the flowering stalk, it withers and dies. From the rhizome, however, other plantlets sprout, creating a colony that ensures survival. Pollination must be effective since virtually every capsule produces a myriad of ripe seeds in the fall. The genus honors John Goodyer (1592–1664), an English botanist.

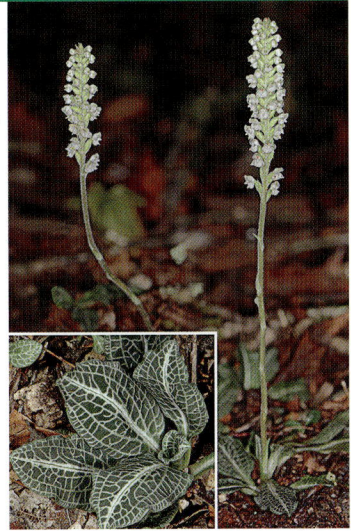

191. Whiteleaf Sunflower

> *Helianthus glaucophyllus* D. M. Smith
>   He-li-án-thus glau-co-phýl-lus
> Asteraceae (Aster Family)

DESCRIPTION: Perennial herbs 3–7′ tall, with long rhizomes; stem leaves opposite, 4–10″ long and 1–3″ wide, rough above, smooth and with a whitish bloom (glaucous) below; flowers produced in involucrate heads of yellow ray and yellow disk flowers; head showy, 2–3″ broad from tip of ray flower to tip of ray flower, with individual ray flowers up to 1.75″ long; flowers July–September.

RANGE-HABITAT: Mostly endemic to the southern Appalachians; w. NC, ne. TN, and SC; in SC, locally common in mafic oak-hickory forests and woodland borders in the mountains; also found in McCormick County in the Piedmont in basic-mesic forests where associated with other typically montane species.

COMMENTS: Whiteleaf Sunflower was initially considered a mountain species found at medium elevations; however, Hill and Horn (1997) reported its occurrence along Stevens Creek in McCormick County, expanding its range into the Piedmont. The leaves, whitened below, separate this species of *Helianthus* from other sunflowers in the same habitats.

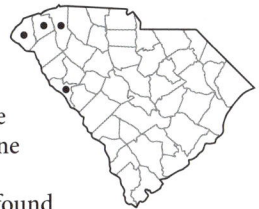

192. Gall-of-the-earth; Lion's-foot

Nabalus serpentarius (Pursh) Hooker
Ná-ba-lus ser-pen-tà-ri-us
Asteraceae (Aster Family)

SYNONYM: *Prenanthes serpentaria* Pursh—RAB

DESCRIPTION: Perennial herb with milky sap, from a thickened rootstock; stems to 3′ tall; leaves very variable in shape and size, lobed or dissected below or unlobed, much reduced upward, becoming less cut or lobed and with shorter leaf stalks upward; flowers arranged into involucrate heads of cream to yellowish ray flowers; involucral bracts with long, ciliate hairs; heads droop and are arranged in an open terminal cluster; flowers August–frost.

RANGE-HABITAT: Widespread in eastern North America; in SC, common throughout; moist oak-hickory forests, bluff forests, beech forests, and other rich, deciduous forests in the Piedmont and Coastal Plain.

COMMENTS: *Nabalus* comes from Greek roots that translate to "drooping blossom." The drooping heads make identification of the plant as a *Nabalus* easy; identification to species requires technical examination of the bracts and pappus. Gall-of-the-earth has highly variable leaves that range from extremely narrowly lobed to triangular and are of little help in identification to species.

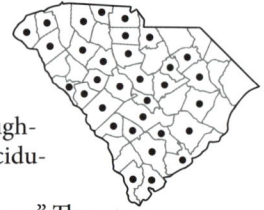

193. Curtis's Aster

Symphyotrichum retroflexum (Lindley ex A. P. de Candolle) G. L. Nesom
Sym-phy-o-trì-chum re-tro-fléx-um
Asteraceae (Aster Family)

SYNONYM: *Aster curtisii* Torrey & Gray—RAB

DESCRIPTION: Perennial herb from short rhizomes, forming clumps; stems erect, smooth; leaves cauline, narrow, smooth, dark green 2.25–6″ long and 0.5–1″ wide; flowers with purplish to lavender ray flowers and yellow disk flowers produced in involucrate heads; the bracts of the involucre leafy, spreading toward the horizontal or sometimes recurved; flowers August–October.

RANGE-HABITAT: Endemic to the southern Appalachian region from NC and TN south to GA and SC; in SC, common in the mountains in Chestnut Oak forests, other montane oak-hickory forests, acidic cove forests, and forest margins.

COMMENTS: This species is abundant in the escarpment region of Oconee County. Though showy, it is seldom cultivated. Numerous other purple and blue-flowered "asters" occur in our area. The bracts of the involucre, the lack of basal leaves, and the smooth stem and leaves help to identify this species. Only Smooth Blue Aster (*Symphyotrichum laeve* (L.) Löve & Löve) and Narrow-leaved Smooth Aster (*Symphyotrichum concinnum* (Willdenow) Mohlenbrock) are similar, but both have involucral bracts that are appressed upward toward the flowers. Though this and many other native North American species were formerly placed in the genus *Aster*, molecular evidence shows that they are not closely related to typical members of the genus *Aster* in Eurasia; they are more closely related to other North American genera and thus are North American species in different genera.

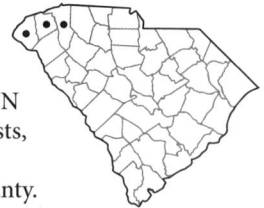

194. Appalachian Gentian

Gentiana decora Pollard
Gen-ti-à-na de-cò-ra

Soapwort Gentian

Gentiana saponaria L.
Gen-ti-à-na sa-po-nà-ri-a
Gentianaceae (Gentian Family)

DESCRIPTION: *G. decora:* Perennial herb 10–20″ tall; stems densely covered with short hairs, unbranched, 1–8 from a root crown; leaves opposite, smooth; flowers blue, white, or violet, striped with blue or violet; corolla with unevenly 2-lobed plaits (appendages) between the 5, short corolla lobes; *G. saponaria:* similar but with smooth stems and solid blue petals; both species flower September–frost.

RANGE-HABITAT: *G. decora* ranges from WV south to SC and GA; in SC, uncommon in the mountains, rare in the Piedmont; in a wide variety of upland, mostly mesic, forest types, especially Chestnut Oak forests and other montane oak-hickory forests; *G. saponaria* ranges from NY and IL south to FL and TX; in SC, occasional throughout in moist forests, stream banks, and boggy places.

COMMENTS: Gentian flowers are often barely open at the top. They are pollinated by bumblebees, which are adapted to working their way down into the flowers. Two other gentians are present in the same habitats in the mountains. Agueweed (*Gentianella quinquefolia* (L.) Small, plate 219) has no plaits between the corolla lobes and produces many, small flowers. Striped Gentian (*G. villosa* L.) has a greenish white corolla sometimes tinged with purple.

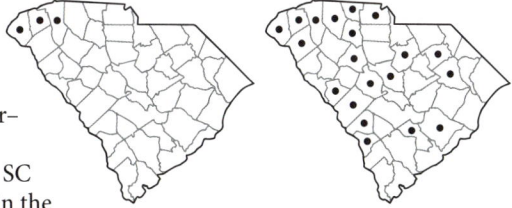

195. Hay-scented Fern

Dennstaedtia punctilobula
(Michaux) T. Moore
Denn-staèdt-i-a punc-ti-ló-bu-la
Dennstaedtiaceae (Bracken Family)

DESCRIPTION: Large, perennial, deciduous fern from long, creeping slender rhizomes, forming dense, large patches; leaves 2–3′ long and up to 1′ wide, twice pinnate, light green, heavily aromatic when crushed (like fresh cut hay); spores produced in round masses (sori) in the sinuses at the margins of the pinnae lobes and covered by recurved margins of tissue.

RANGE-HABITAT: Widespread in eastern North America from Nova Scotia west to MN and south to AL and AR; in SC, common in the mountains in Chestnut Oak forests, montane oak-hickory forests, and acidic cove forests.

COMMENTS: The common name is derived from the strong odor of fresh-cut hay produced when the frond is crushed or dried. This species is aggressively colonial farther north. In the NC mountains, it forms vast patches in open forests. It is not favored browse by white-tailed deer and becomes quite common when other competition for space is removed. In SC it does not attain the abundance it does farther north.

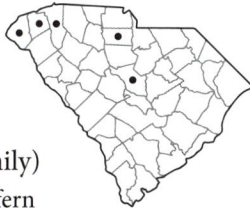

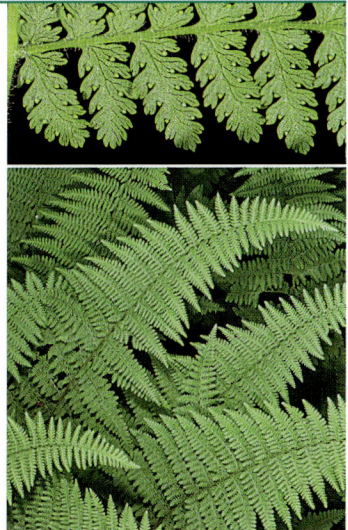

196. New York Fern

> *Parathelypteris noveboracensis*
> (L.) Ching
> Pa-ra-the-lỳp-te-ris no-ve-
> bo-ra-cén-sis
> Thelypteridaceae (Marsh Fern
> Family)

> SYNONYM: *Thelypteris noveboracensis* (L.)
> Nieuwland—RAB

DESCRIPTION: Small, perennial, deciduous fern from slender rhizomes, forming large colonies; leaves (fronds) to 2′ tall; frond blade twice pinnatifid, widest in the middle and tapered to the tip and base, producing a plumelike appearance; spores produced in rounded clusters (sori) along the margins of the pinnae segments.
RANGE-HABITAT: Widespread in eastern North America from Newfoundland west to Wisconsin and south to AR and AL; in SC, common in the mountains and Piedmont, rare in the Coastal Plain; mesic oak-hickory forests, acidic cove forests, and other mesic hardwood forests.
COMMENTS: New York Fern is probably the most abundant fern species in the mountains and upper Piedmont. It can be found in most open hardwood forests with even slightly mesic soils and forms very large patches in some areas. Like Hay-scented Fern, white-tailed deer do not eagerly browse this species, one probable reason for its abundance. It is easily distinguished as the only small fern with upright fronds that is equally tapered both at the tip and at the base—as the saying goes, burning the candle at both ends, like a New Yorker.

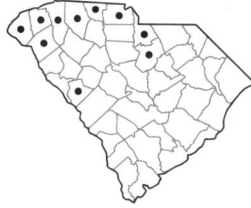

197. Common Running-cedar;
Turkey-foot (A)

> *Diphasiastrum digitatum*
> (Dillenius ex A. Braun) Holub
> Di-pha-si-ás-trum di-gi-tà-
> tum
> SYNONYM: *Lycopodium flabelliforme*
> (Fernald) Blanchard—RAB, PR

Common Ground-pine (B)

> *Dendrolycopodium obscurum*
> (L.) A. Haines
> Den-dro-ly-co-pò-di-um ob-
> scù-rum
> Lycopodiaceae (Club-moss Family)

> SYNONYM: *Lycopodium obscurum* L.—RAB

DESCRIPTION: Evergreen, perennial fern relatives forming large colonies by means of surficial or sub-surficial runners; *D. digitatum:* stems upright, with repeatedly forked branches forming flattened fans in outline; branches flattened and with appressed leaves; spores produced in cones (strobili) in a forked (dichotomously) branched structure like a candelabra; *D. obscurum:* stems upright, branching stems with leaves not appressed, and appearing like a small fir tree; spores produced in cones (strobili) with a central axis and branches not dichotomous; spores produced August–October.
RANGE-HABITAT: Both species widespread in eastern North America; in SC, *D. digitatum* is common in dry oak-pine forests throughout but more abundant in the mountains and Piedmont; *D. obscurum* is common in the mountains and Piedmont; rare in the Sandhills and Coastal Plain; acidic oak-hickory forests, acidic cove forests, and bay forests; *D. obscurum* is also found

in Asia; in SC, common and scattered throughout but also more abundant in the mountains and Piedmont in moist to dry acidic forests; bay forests and streamheads in the Coastal Plain. **COMMENTS:** Common Running-cedar and Common Ground-pine are types of club mosses, a very ancient group of plants. The spores are explosive when ignited and burst quickly with a flash. The spores were historically used in flash photography as well as in magic, when a burst of cool flame was needed. The spores are also used as a natural lubricant for latex. Though they form a blanket over large areas of poor soil in forests, they are not good landscape plants because they do not transplant successfully. This is likely due to complex relationships with mycorrhizae in the soil, as occurs with lady's slipper orchids.

198. Forest margin

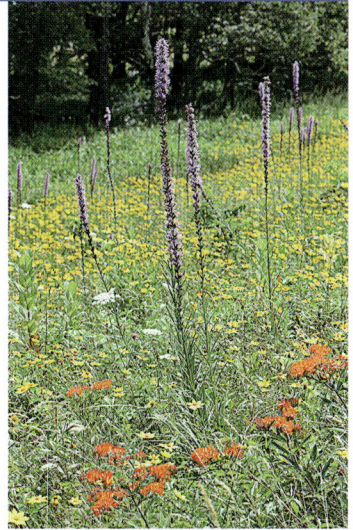

199. Silverleaf Hydrangea; Snowy Hydrangea (A)

Hydrangea radiata Walter
Hy-drán-ge-a ra-di-à-ta
SYNONYM: *Hydrangea arborescens*
L. ssp. *radiata* (Walter)
McClintock—RAB

Northern Wild Hydrangea; Smooth Hydrangea (B)

Hydrangea arborescens L.
Hy-drán-ge-a ar-bo-rés-cens
Hydrangeaceae (Hydrangea Family)

DESCRIPTION: *H. radiata:* shrub to 6′ tall; leaves deciduous, opposite, ovate, densely covered with white hairs below; flowers in corymbs, often with sterile florets with showy bracts toward the periphery of the cluster; similar to *H. arborescens,* but the leaves are green and smooth or minutely pubescent beneath and typically lack the sterile florets along the margin of the corymb; both species flower April–June.
RANGE-HABITAT: *H. radiata* is a southern Appalachian endemic extending a short distance into the Piedmont, from NC and TN, south to GA; in SC, common in the mountains and upper Piedmont; also found at one location on a rich bluff in the inner Coastal Plain; in a variety of moist forests in rich soil, often rocky; most conspicuous on woodland margins along streams, rock outcrops, and roads. *H. arborescens* is widespread in eastern North America; in SC, locally common in the mountains where it is found in a variety of moist forests, roadsides, and stream sides.

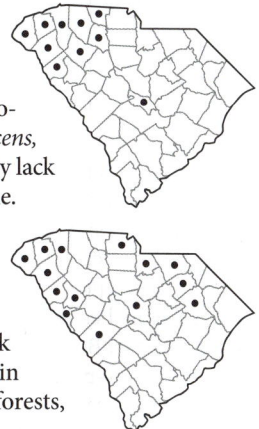

SIMILAR SPECIES: A third shrubby species, Ashy or Southern Wild Hydrangea (*H. cinerea* Small), is distinguished by the pubescence on the lower leaf surface that is not densely matted, does not obscure the lower leaf surface, and is grayish. In SC it is found only in the north-central Piedmont.

COMMENTS: Silverleaf Hydrangea is SC's most common hydrangea, often lining mountain roadsides. Northern Wild Hydrangea is far less abundant in SC but is locally common in portions of Oconee County.

The root and bark of hydrangeas were once used medicinally for both external and internal ailments. They have recently been shown to contain cyanogenic compounds that can result in cyanide poisoning; they also contain a substance known as hydrangin, which can produce bloody diarrhea and painful inflammation of the lining of the intestines and stomach.

200. New Jersey Tea

Ceanothus americanus L.
   Ce-a-nò-thus a-me-ri-cà-nus
Rhamnaceae (Buckthorn
   Family)

DESCRIPTION: Low shrub to 3′ tall; roots dark red; leaves simple, alternate, deciduous, 1–2.5″ long; strongly 3-nerved beneath; inflorescence terminal and axillary in racemes of umbellate cymes; petals white with a narrow base and wide tip (like a hood); flowers May–June.

RANGE-HABITAT: ME south to FL and west to Manitoba and TX; in SC, common throughout; oak-hickory forests, Longleaf Pine flatwoods, gladelike openings, dry pine or oak ridge forests in the mountains, and forest margins.

COMMENTS: The dried leaves were used during the Revolutionary War as a substitute for tea, but it contains no caffeine and does not provide the lift of tea. The root is reported to be a stimulant and a sedative and is used to loosen phlegm. The root is also strongly astringent because of its high tannin content. Nodules on the roots harbor bacteria that fix atmospheric nitrogen.

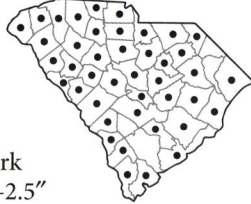

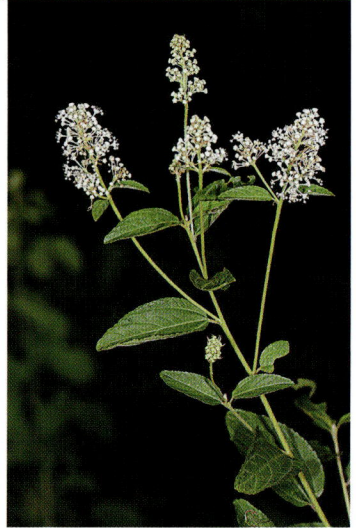

201. Common Bluet

Houstonia caerulea L.
   Hous-tòn-i-a cae-rù-le-a
Rubiaceae (Madder Family)

DESCRIPTION: Diminutive perennial herb, forming small tight clumps; l flowering stems to 6.5″ tall; leaves basal, with only tiny reduced leaves on the flowering stem; flowers with 4 petals, pale blue with a yellow center; flowers March–May.

RANGE-HABITAT: Widespread in eastern North America; in SC, common throughout; woodland margins, road banks, margins of outcrops, and thin oak woodlands.

SIMILAR SPECIES: Thyme-leaf Bluet (*Houstonia serpyllifolia* Michaux, plate 33) has nearly identical flowers but has long, creeping surficial stems and slightly brighter blue flowers and occupies very different habitats.

COMMENTS: Common Bluet is one of the most cheerful of early spring wildflowers. It is much more abundant in the mountains of NC. Bee flies in the genus *Bombylius* are often seen visiting the flowers in spring. The genus honors William Houston (1695–1733), Scottish surgeon, botanist, and explorer.

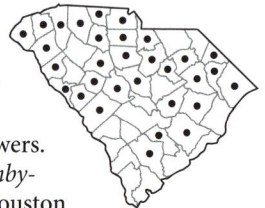

202. Robin's-plantain; Showy Fleabane

Erigeron pulchellus Michaux
  var. *pulchellus*
    E-rí-ge-ron pul-chél-lus var.
    pul-chél-lus
Asteraceae (Aster Family)

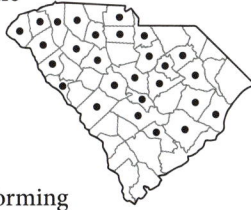

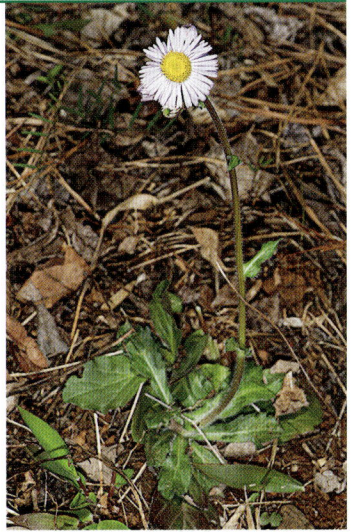

DESCRIPTION: Perennial herb 15–20″ tall, forming clusters from long stolons, with spreading hairs throughout; leaves of the basal rosette usually obovate, large and obvious; stem leaves smaller and reduced upward; flowers produced in involucrate heads of 50–100 white to lavender ray flowers and yellow disk flowers; flowers March–May.

RANGE-HABITAT: ME south to GA and west to MN and TX; in SC, common in the mountains, occasional in the Piedmont, Sandhills, and Coastal Plain; in a variety of mesic habitats but most abundant in forest margins in the mountains.

COMMENTS: Fleabane was used as a stuffing for mattresses in an unsuccessful attempt to reduce bites from fleas (and bed bugs?). Showy Fleabane produces the largest inflorescences of SC's 10 fleabane species. Most other species flower much later in the season.

203. Carolina Vetch

Vicia caroliniana Walter
    Ví-ci-a ca-ro-li-ni-à-na
Fabaceae (Pea or Bean Family)

DESCRIPTION: Perennial herb, sprawling to climbing; leaves compound with 10–18 leaflets; leaflets elliptic to oblong, terminal leaflets modified into simple or branched tendrils; flowers in loose racemes, mostly white, often with some blue or purple on the keel petal; flowers March–May.

RANGE-HABITAT: NY south to GA and west to MN and TX; in SC, common in the mountains and Piedmont, occasional in the Sandhills and Coastal Plain; mesic oak-hickory forests, forest margins, and along stream sides.

COMMENTS: Though similar to many of the introduced vetch species found in SC, this is the only one commonly found in undisturbed woodland situations.

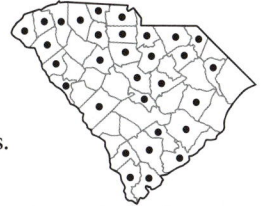

204. Wild Strawberry

Fragaria virginiana P. Miller
    Fra-gà-ri-a vir-gi-ni-à-na
Rosaceae (Rose Family)

DESCRIPTION: Stemless herb with short rhizomes and long surficial stolons; leaves basal with 3 leaflets; flower stalk 3–6″ tall, flowers white; achenes imbedded in the fleshy receptacle (the strawberry); flowers March–June.

RANGE-HABITAT: Labrador and Newfoundland to Alberta and south to GA, AL, and OK; in SC,

common throughout the mountains and Piedmont, occasional in the Coastal Plain; old fields and forest margins.

SIMILAR SPECIES: Wild Strawberry is frequently confused with the weedy Indian-strawberry (*Potentilla indica* (Andrews) T. Wolf, plate 955). Indian-strawberry is introduced from Asia and has yellow flowers and tasteless fruit.

COMMENTS: The generic name refers to the fragrance of the fruit. Wild Strawberry was one of the parents that produced the cultivated table strawberry; the other parent is from South America (*F. chiloensis* var. *ananassa* Bailey). The fruit of Wild Strawberry is the sweetest of the strawberries.

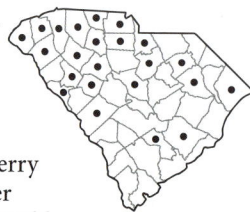

205. Fire-pink

Silene virginica L. var. *virginica*
     Si-lè-ne vir-gí-ni-ca var. vir-gí-ni-ca
     Caryophyllaceae (Pink Family)

DESCRIPTION: Perennial herb to 30″ tall forming small clumps; plants with glandular (sticky) pubescence and stems branching near the base; leaves opposite; entire plant with glandular (sticky) pubescence; flowers with 5 brilliant red petals that are notched at the tip; fruit a capsule with free-central placentation; flowers April–July.

RANGE-HABITAT: NY south to GA and west to Ontario, MI, and OK; in SC, common in the mountains and upper Piedmont, rare elsewhere; moist to dry habitats, usually with ample light (especially forest margins and roadsides), often on rock outcrops; also in Longleaf Pine sandhills.

COMMENTS: The sticky hairs on the calyx serve to reduce theft of pollen and nectar by non-pollinating flower visitors, especially ants. The brilliant crimson flowers with cleft petals cannot be confused with any other plants in SC. The pink in the name does not refer to coloration, but rather the notch in the petals that look like they have been cut with pinking shears ("pinked").

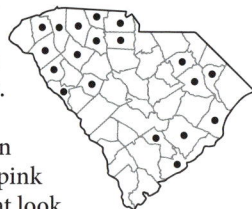

206. Narrow-pod White Wild Indigo

Baptisia albescens Small
     Bap-tí-si-a al-bés-cens
     Fabaceae (Bean Family)

SYNONYM: *Baptisia alba* (L.)
     R. Brown—RAB

DESCRIPTION: Perennial herb from a rhizome; stems to 4′ tall; leaves trifoliolate, glaucous (whitish); flowers in racemes that terminate the stem and often the branches; petals white, to 0.5″ long with short petioles (0.2–0.4″ long); fruit a cylindrical yellowish-brown legume that is 0.75–1.20″ long and 0.25–0.35″ wide; flowers April–June; fruits mature June–October.

RANGE-HABITAT: VA and TN south to FL and AL; in SC, common on forest margins, powerline clearings, open oak woodlands, and other "prairielike" habitats in the mountains and Piedmont; occasional in the Coastal Plain in open woodlands and Longleaf Pine flatwoods on fine-textured sub-mesic soils.

SIMILAR SPECIES: This is our only species with cylindrical fruits. It is similar to Thick-pod White Wild Indigo (*Baptisia alba* (L.) Ventenat), which has much more globular, blackish-colored legumes, and slightly larger flowers with longer petioles and is confined to the lower Piedmont and Coastal Plain in SC.

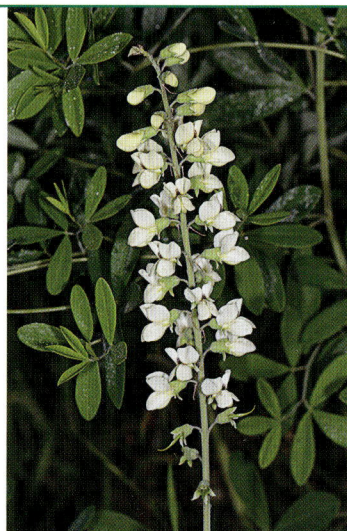

COMMENTS: This is one of the most obvious wildflowers blooming on late spring roadsides throughout the Sumter National Forest of Oconee County. It is found in remnants of former fire-maintained habitats and with the addition of prescribed fire to many areas of the Sumter National Forest; it is becoming more common in the open oak woodland habitat.

207. Blue Ridge Bindweed; Silky Bindweed

Convolvulus sericatus House
    Con-vól-vu-lus se-ri-cà-tus
Convolvulaceae (Morning-glory Family)

SYNONYM: *Calystegia sericata* (House) Bell—RAB, PR

DESCRIPTION: Perennial trailing vine, stem often less than 3′ long, with silky hairs; leaves tapered to a sharp point, the base heart-shaped, with a dense, silky white covering of hairs on the lower surface; flowers showy, white, funnel-shaped, to 2″ long, borne in the axils of middle and upper leaves; flowers May–July.

RANGE-HABITAT: Endemic to the southern Blue Ridge in NC, SC, and GA; in SC, uncommon in the mountains; found in open woodlands or forest margins, usually on high-magnesium or -calcium soils.

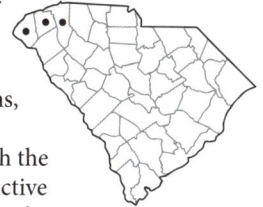

COMMENTS: The soft, velvety, grayish pubescence of the entire plant, along with the large, pure-white, showy flowers, make this one of the most showy and distinctive members of the morning glory family in SC. It is considerably more abundant than once thought and is locally abundant along roadsides and powerline clearings in the Jocassee Gorges area, particularly on sites derived from amphibolite.

CONSERVATION STATUS: SC-Vulnerable

208. Skunk Meadowrue; Waxy Meadowrue

Thalictrum amphibolum Greene
    Tha-líc-trum am-phí-bo-lum
Ranunculaceae (Buttercup Family)

SYNONYM: *Thalictrum revolutum* A.P. de Candolle—RAB

DESCRIPTION: Robust, perennial herb with flowering stems to 5′ tall; leaves ternately divided and leaflets typically 3-lobed and 0.6–1.0″ wide; undersides of leaflets, pedicels, and achenes stipitate glandular pubescent or smooth; male and female flowers on separate plants (dioecious); male flowers with showy, drooping anthers; flowering May–August.

RANGE-HABITAT: Wide-ranging in eastern North America; in SC, occasional in the mountains and Piedmont; margins of outcrops, forest margins, and glade-like openings in forests.

COMMENTS: This is the most abundant of several tall species of meadowrue. The leaves are reminiscent of a robust maidenhair fern.

209. Carolina Phlox

Phlox carolina L.
Phlóx ca-ro-lì-na
Polemoniaceae (Phlox Family)

DESCRIPTION: Perennial rhizomatous herb to 3′ tall with narrow, lanceolate to ovate-lanceolate, opposite, smooth leaves; stems with 6–25 nodes; calyx subcylindric; corolla various shades of pink; flowering May–September.

RANGE-HABITAT: VA south to GA and west to IL and MO; in SC, common throughout in forest margins and open woodlands with fine-textured soils.

COMMENTS: This is the only common tall phlox in SC. Plants in the mountains are more robust with more leaf nodes and have been treated as *P. carolina* ssp. *alta* Wherry by some authors.

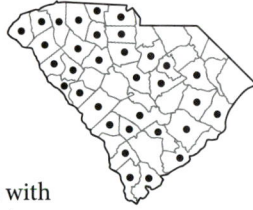

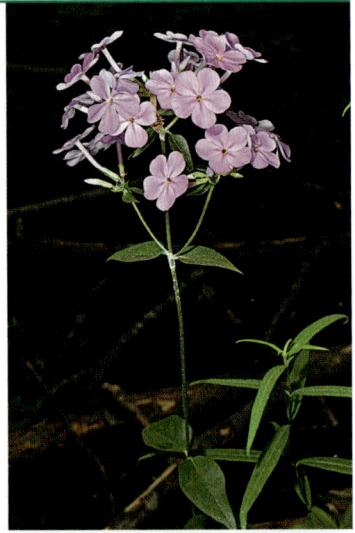

210. Appalachian Milkwort

Polygala curtissii Gray
Po-lý-ga-la cur-tíss-i-i
Polygalaceae (Milkwort Family)

DESCRIPTION: Smooth annual herb; stem solitary but often highly branched; leaves alternate, narrowly elliptic; inflorescence produced at the tips of the stem and branches, spikelike, rounded to cylindric; flowers small, pink, white, green, or a combination of these, subtended by persistent pink bracts, in headlike clusters about 0.4″ in diameter; flowers June–October.

RANGE-HABITAT: DE south to SC and west to OH and MS; in SC, common in the mountains, Piedmont, and Sandhills; forest margins, dry road banks, old fields, margins of outcrops, and other openings with dry rocky or sandy soils.

COMMENTS: The specific epithet honors Allen Hiram Curtiss (1845–1907), one of the first professional botanists in Florida.

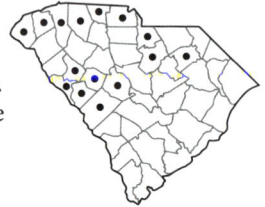

211. Hairy Angelica

Angelica venenosa (Greenway) Fernald
An-gé-li-ca ve-ne-nò-sa
Apiaceae (Carrot Family)

DESCRIPTION: Perennial herb from a taproot; stems to 3′ tall, covered with dense, fine hairs; leaves with long leaf stalks and blades twice compound, produced basally and on the stem, becoming smaller upward; leaflets thick-textured; flowers small, white, arranged in terminal and axillary, compound umbels; flowers June–August.

RANGE-HABITAT: MA south to FL and west to MN, MS, and AR; in SC, common throughout the mountains and Piedmont, rare in the Sandhills and Coastal Plain; sunny dry woodlands, margins of outcrops, forest margins, rocky stream sides, and Longleaf Pine sandhills.

COMMENTS: The European angelica is reportedly useful for a variety of ailments, including heartburn, colic, and psoriasis. Duke (1997) recommends all species of *Angelica* as herbal remedies for these aliments.

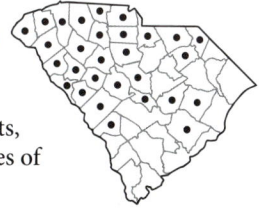

---

212. Southern Mountain-mint

Pycnanthemum pycnanthemoides (Leavenworth) Fernald
  Pyc-nán-the-mum pyc-nan-the-moì-des
Lamiaceae (Mint Family)

SYNONYM: *Pycnanthemum incanum* (L.) Michaux—RAB, in part, PR, in part

DESCRIPTION: Perennial herb with branched stems and rhizomes; lower surface of leaves pubescent and lighter than the upper surface; upper leaves whitened near the 2–3 headlike flower clusters; flowers whitish to yellowish, bilabiate, with petals heavily spotted with purple; sepals narrowly triangular, tips acuminate and bearing clusters of bristlelike hairs; flowers June–August.

RANGE-HABITAT: VA south to SC and west to GA and IL; in SC, common throughout, most abundant in the mountains; forest margins, roadsides, margins of rock outcrops, and other open, sunny habitats.

COMMENTS: When crushed, the leaves have a strong mint odor, though there is much variability in intensity and overall "mintyness." A leaf tea was once used for fevers, colds, and coughs. Many mountain-mints contain pennyroyal oil, which is toxic in large doses and may cause abortions; plants with pennyroyal oil should not be used by women who may be pregnant.

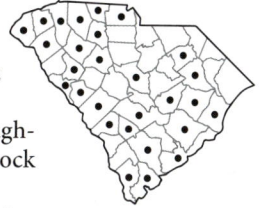

213. Fraser's Loosestrife

Lysimachia fraseri Duby
  Ly-si-má-chi-a frà-ser-i
Primulaceae (Primrose Family)

DESCRIPTION: Perennial herb 2 to 4′ tall, forming colonies from underground stems; leaves in whorls of 3–6; a translucent red band outlines the leaf margins; flowers produced in a terminal panicle; petals yellow; flowers June–July.

RANGE-HABITAT: From w. NC and TN south to n. SC, n. GA, and AL; in SC, rare in the mountains and, at least historically, in Anderson County in the upper Piedmont; moist forest margins, meadows, along roadsides, and sunny rocky slopes.

COMMENTS: The translucent red marginal band on the leaves and the distinctive knob-tipped hairs on the upper stems, stalks of flower clusters, and sepals are probably the best field characters. This plant has a high light requirement and probably was much more abundant in precolonial times when Native Americans routinely used fire as a tool to maintain open, grassy habitats. It generally flowers when exposed to extra sunlight created by a large canopy opening.

CONSERVATION STATUS: SC-Vulnerable

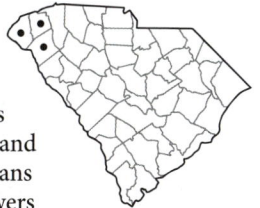

214. Yellow Fringed Orchid

Platanthera ciliaris (L.) Lindley
   Pla-tán-the-ra ci-li-à-ris
Orchidaceae (Orchid Family)

SYNONYM: *Habenaria ciliaris*
   L.—RAB

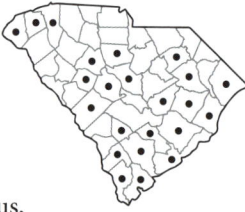

DESCRIPTION: Robust, perennial, herbaceous, terrestrial orchid, 1–2.5′ tall; stem smooth, leafy below, reduced to bracts above; flowers produced in a raceme, corolla bright orange; lip petal fringed; nectar spur 0.8–1.3″ long; flowers June–August.

RANGE-HABITAT: NH south to FL and west to MI and TX; in SC, common in the Coastal Plain, occasional in the upper Piedmont and mountains; Longleaf Pine savannas, herbaceous seepages, pocosin ecotones, cataract fens, forest margins, and roadside ditches.

SIMILAR SPECIES: Crested Fringed Orchid (*P. cristata* (Michaux) Lindley, plate 641) is similar in color but is a smaller species with a nectar spur 0.2–0.4″ long. It is common in Longleaf Pine savannas, moist roadsides, and ditches in the Coastal Plain and Sandhills.

COMMENTS: The common name Yellow Fringed Orchid does not really describe this plant. The coloration of the flowers is brilliant, almost neon, orange. This species is easily observed along roadsides in the Sumter National Forest in Oconee County in late July through early August. It should not be disturbed in the wild.

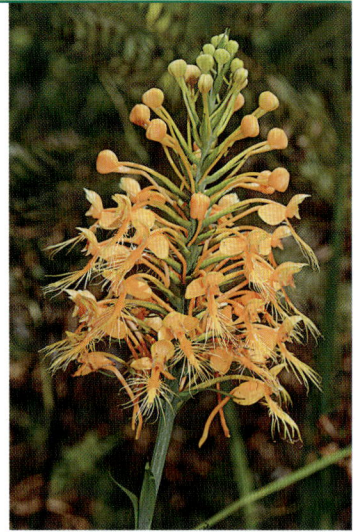

215. Indian-tobacco; Pukeweed

Lobelia inflata L.
   Lo-bèl-i-a in-flà-ta
Campanulaceae (Bellwort Family)

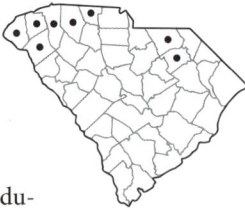

DESCRIPTION: Branched annual to 3′ tall; midstem leaves greater than 0.3″ wide, gradually reduced upward; flowers blue to lavender or sometimes white; most distinctive when the fruit swells (inflates) to become subglobose at maturity; flowers July–frost.

RANGE-HABITAT: ME south to GA and west to MI and MS; in SC, common in the mountains, rare in the Piedmont; open, dry-to-moist forests and forest margins.

COMMENTS: Historically, Indian-tobacco had many uses. Native Americans smoked the leaves to relieve asthma, sore throats, and coughs. Traditionally it was used to induce vomiting and sweating. Lobeline (an alkaloid) is used in commercial "quit-smoking" preparations. Lobeline is like nicotine and binds to the same (nicotinic) nerve receptors. The genus honors Matthias de l'Obel (1538–1616), a Flemish herbalist.

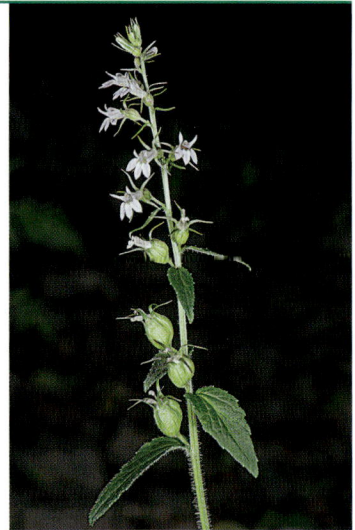

216. Sunfacing Coneflower

Rudbeckia heliopsidis
    Torrey & Gray
        Rud-béck-i-a he-li-óp-si-dis
    Asteraceae (Aster Family)

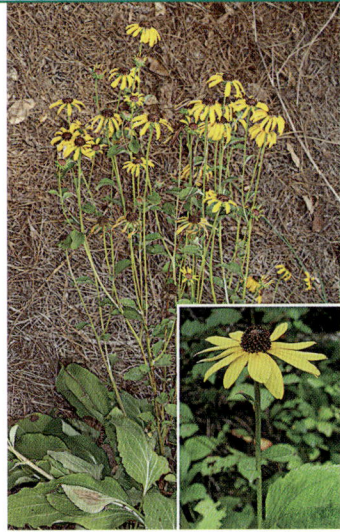

DESCRIPTION: Rhizomatous, perennial herb to 3.5′ tall, forming large colonies; basal leaves ovate, to 5″ long, with long petioles and serrate margins; flowers produced in involucrate heads; heads with yellow ray flowers similar to other *Rudbeckia* but with brownish rather than dark purple or black disk flowers; flowers July–September.

RANGE-HABITAT: VA south to GA and MS; in SC, very rare in forest margins and road banks in the mountains.

COMMENTS: This species is locally abundant at the single extant site in SC in Oconee County, where it occurs over a large area. The range is very discontinuous, being most abundant today in the Cumberland Plateau of AL. It is distinctive in forming large rhizomatous colonies and is a long-lived perennial. It should never be disturbed in the wild.

CONSERVATION STATUS: SC-Critically Imperiled

217. Virgin's-bower

Clematis virginiana L.
        Clé-ma-tis vir-gi-ni-à-na
    Ranunculaceae (Buttercup Family)

DESCRIPTION: Scrambling, climbing, herbaceous vine, climbing via the twisted stems or leaf stalks that wrap around other plants; leaves opposite, compound, with long stalks and 3 nearly symmetrical leaflets; flowers in many-flowered clusters from the leaf axils, sepals white, petal-like; flowers July–September.

RANGE-HABITAT: Widespread in eastern North America; in SC, common in the mountains, uncommon in the Piedmont, and rare in the Coastal Plain; moist woodland margins, on stream sides, and moist forest margins.

COMMENTS: All parts are reported to be highly irritating to the skin and mucous membranes.

218. Downy Lobelia

Lobelia puberula Michaux
Lo-bèl-i-a pu-bé-ru-la
Campanulaceae (Bellflower
Family)

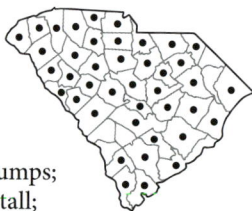

DESCRIPTION: Perennial herb from small clumps; stems with short downy pubescence, to 3′ tall; leaves lanceolate to elliptic; flowers blue, produced in racemes; the fused petals 0.25–0.40″ long; flowers July–October.

RANGE-HABITAT: Widespread in eastern North America; in SC, common throughout in moist to dry soils of forest margins, open oak hickory woodlands, margins of rock outcrops, glades, and other open habitats with fine-textured soils.

SIMILAR SPECIES: Southern Lobelia (*Lobelia amoena* Michaux, plate 64) has stems that are completely smooth and is found in wetter habitats.

COMMENTS: This is the most common blue *Lobelia* in the mountains and Piedmont.

The genus honors Matthias de l'Obel (1538–1616), a Flemish herbalist.

219. Agueweed

Gentianella quinquefolia (L.)
Small var. *quinquefolia*
Gen-ti-a-nél-la
quin-que-fò-li-a
Gentianaceae (Gentian Family)

SYNONYM: *Gentiana quinquefolia* L.—RAB

DESCRIPTION: Annual herb with a stem to 2.5′ tall, frequently much shorter; stem branched nearly to the base of the plant; leaves sessile, opposite, ovate to lanceolate, 1–3″ long; flowers with petals fused, erect, bottlelike, blue to purple produced in 3–5 flowered clusters on the main stem and on branches; flowers September–October.

RANGE-HABITAT: ME and Ontario south to GA, mostly in the Appalachians; in SC, uncommon in the mountains and upper Piedmont in forest margins, particularly moist road banks and other clearings.

COMMENTS: This beautiful, late-flowering annual is immediately recognizable as a gentian because of the bottle-shaped, bluish or purple flowers that appear not to open. It is distinguished from all other gentians in SC by the small flowers that are produced in large numbers, the annual habit, and the heavy branching of the stem.

220. Southern Grapefern

Sceptridium biternatum
(Savigny) Lyon
Scep-trí-di-um bi-ter-nà-tum
Ophioglossaceae (Adder's-tongue
Family)

SYNONYM: *Botrychium biternatum*
(Savigny) Underwood—RAB, PR

DESCRIPTION: Small, herbaceous, perennial fern; plant green (often purplish) through the winter, dormant during the majority of the warm season; leaves (fronds) thick and leathery, 2–3 pinnate; stipe forking into an inclined, sterile blade and an erect, fertile segment, tipped by clustered globular sporangia containing the spores.
RANGE-HABITAT: Widespread in the southeastern US; common throughout SC; moist forests, clearings, old fields, pinelands, and swamps.
COMMENTS: Grapeferns in this genus are unusual in being "wintergreen," rather than evergreen. They emerge in the late summer through autumn and hold their leaves through the winter when the light in the forest is more abundant, but temperatures are not as conducive to photosynthesis. The common name comes from the grapelike clusters of sporangia.

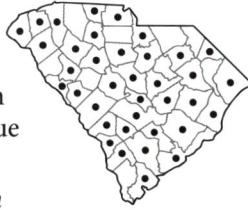

221. Pine-oak heath

222. Pitch Pine

Pinus rigida P. Miller
Pì-nus rí-gi-da
Pinaceae (Pine Family)

DESCRIPTION: Medium to tall pine to 101′ tall, though frequently much shorter; needles 2–5″ long, produced in bundles of 3, rarely in 2 or 4; female cones broadly ovoid, 1–2″ long.
RANGE-HABITAT: Widespread in eastern North America from eastern Canada south to GA; in SC, common in the mountains in pine-oak heaths and other forests with acidic soils, rare in the upper Piedmont.
COMMENTS: One of four pine species found in pine-oak heaths. In the most exposed and acidic habitats it is often found with Table Mountain Pine (*P. pungens* Lambert, plate 3). The bark does not contain obvious resin ducts, which serves as an easy way to distinguish it from Shortleaf Pine (*Pinus echinata* P. Miller). There are only two pines in SC, Pitch Pine and Pond Pine; the latter commonly sprouts from the trunk when burned. This characteristic is referred to as epicormic sprouting. Pitch Pine sprouts vigorously when lightly burned and frequently send out numerous small shoots from the trunk even in the absence of fire.

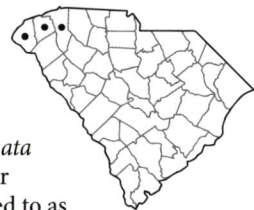

223. Scarlet Oak

Quercus coccinea Muenchhausen
    Quér-cus coc-cí-ne-a
Fagaceae (Beech Family)

DESCRIPTION: Medium to tall deciduous tree to 98′ tall, typically much shorter; bark grayish-brown with scaly ridges; leaves with 5–9 deep lobes with the sinuses extending more than one-half of the distance to the midrib; the lobes and tip of leaf with long awns; leaves turn scarlet-red in autumn; acorn matures the year after flowering; acorn with the cup turbinate and covering one-third to one-half of the brown nut; flowers April; fruit matures in the autumn of the next year.

RANGE-HABITAT: Widespread in eastern North America; in SC, common in the mountains and Piedmont in forests with highly acidic, well-drained soils; also found occasionally in the Sandhills and Coastal Plain, also in acidic oak forests.

COMMENTS: With its bright scarlet leaves, this oak produces one of the most colorful autumn displays with its bright scarlet leaves. It is one of the most abundant tree species in the pine-oak heaths.

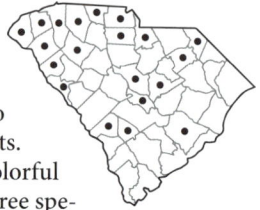

224. Horse Sugar; Sweetleaf; Yellow-wood

Symplocos tinctoria
    (L.) L'Heritier de Brutelle
    Sým-plo-cos tinc-tò-ri-a
Symplocaceae (Sweetleaf Family)

DESCRIPTION: Shrub or small tree to 46′ tall with dark stems and dark grey, smooth bark with longitudinally thickened ridges; leaves simple, alternate, deciduous, or tardily deciduous; flowers fragrant, in dense clusters on naked branches (or almost naked if some leaves persist); flowers March–May.

RANGE-HABITAT: Horse sugar is widespread in the southeastern US, from DE south to FL and west to TX and OK; in SC, common, primarily in the mountains and Coastal Plain, but interestingly, present in the Piedmont mainly near the borders of the adjacent regions (mountains or Sandhills), and in scattered sites in the central Piedmont; pine-oak heaths, oak-hickory and beech forests, Longleaf Pine flatwoods, maritime forests, thickets, ridge top forests, and stream side habitats.

COMMENTS: The bark and leaves yield a bold yellow dye. Dying with Yellow-wood, a practice several centuries old in the South, is still practiced. Cattle and horses greedily consume the leaves, and white-tailed deer browse on the whole plant. Older leaves often show signs of insect damage. The leaves can be chewed as a refreshing trail tidbit. The wood has no commercial value. Where protected from fire, it can grow tall. The SC state champion was measured in 1984 at 46′ tall.

225. Northern Wild-raisin; Withe-rod

Viburnum cassinoides L.
Vi-búr-num cas-si-noì-des
Viburnaceae (Viburnum Family)

DESCRIPTION: Shrub or small tree, usually 6–12′ tall, often forming dense thickets; leaves opposite, elliptic to ovate, to 4.25″ long, with acuminate to acute tips and undulate-crenate margins, fully expanded by flowering; flowers white, produced in a flat-topped compound cyme; flowering May–June; fruit a drupe, maturing in the autumn.

RANGE-HABITAT: Widespread in eastern North America from Ontario south to GA and west to WI and AL; in SC, common in the mountains and rare in the Piedmont; acidic oak-hickory forests, pine-oak-heath, and other moist to dry forests with acidic soils.

COMMENTS: This is the common *Viburnum* growing on dry, highly acidic soils in the mountains. The plant often makes its presence known, particularly in the autumn by the rank odor, resembling dirty socks, or sometimes described as slightly "skunky." This odor is probably the result of the emission of butyric acid. The fruits are edible and fairly sweet tasting when fully ripe; the pits of the drupes are very large.

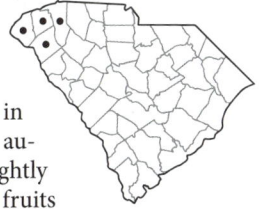

226. Mountain Laurel; Kalmia

Kalmia latifolia L.
Kálm-i-a la-ti-fò-li-a
Ericaceae (Heath Family)

DESCRIPTION: Shrub or small tree, usually 6–12′ tall, often forming dense thickets; leaves leathery and evergreen; flowers white or pink, usually with purple around the 10 pockets in which the anthers are fitted; flowers April–June.

RANGE-HABITAT: Widespread in eastern North America; in SC, found throughout; common in the mountains, Piedmont, Sandhills, and inner Coastal Plain; rare in the outer Coastal Plain; extending down the Savannah River along bluffs to Jasper County; pine-oak heaths, acidic forests, bluffs, Longleaf Pine sandhills, and a wide range of other habitats on acidic soils.

SIMILAR SPECIES: Though in flower Mountain Laurel could hardly be confused with any other plant, vegetatively it is very similar to Gorge Rhododendron (*Rhododendron minus* Michaux, plate 183). It may be distinguished from Gorge Rhododendron by the absence of rusty scales and dots on the petioles and lower midrib of the leaves and the absence of the large ovate floral bud typical of *Rhododendron*.

COMMENTS: The anthers forcefully spray pollen at maturity after a touch releases the tension that holds them in the pockets of the corolla. Pollination requires a pollinator large enough to trip the anthers, such as bumble bees or carpenter bees. The hard root was once used to make pipe bowls. It makes a beautiful shrub on rocky hillsides that are not suitable for other shrubs because of acidic soils. The genus honors Pehr Kalm (1716–79), a pupil of Linnaeus, who traveled and collected in America. In the Aiken area, where it is abundant, it is known by the common name of Kalmia.

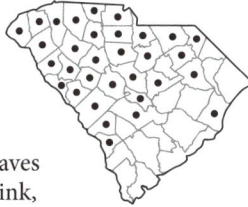

227. Sweet-fern

Comptonia peregrina
  (L.) Coulter
    Comp-tòn-i-a pe-re-grì-na
Myricaceae (Bayberry Family)

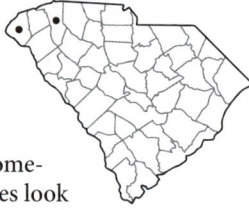

DESCRIPTION: Deciduous shrub to 3′ tall, sometimes forming large colonies; leafy branches look
very much like fern fronds; flowers without sepals
or petals; male and female flowers on same plant; male flower
clusters cylindrical and about 1.25″ long; female clusters burrlike
and below the male; flowers April.

RANGE-HABITAT: Nova Scotia south to SC and west to Manitoba and
GA; in SC, rare in the mountains on dry, open ridges and roadsides, and in fire-maintained pine-oak heath communities on very
poor soils.

COMMENTS: The sweet-smelling foliage can sometimes be detected
from a distance. Native Americans made a leaf tea from Sweet-fern
that was used to treat Poison Ivy rash. The genus honors Henry
Compton (1632–1713), a bishop of London and a patron of botany.

CONSERVATION STATUS: SC-Critically Imperiled

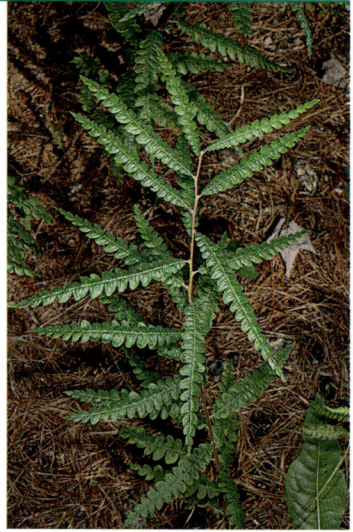

**228. Dryland Blueberry;
Hillside Blueberry**

Vaccinium pallidum Aiton
    Vac-cí-ni-um pál-li-dum
Ericaceae (Heath Family)

SYNONYM: *Vaccinium vacillans* Kalm
  ex Torrey—RAB, PR

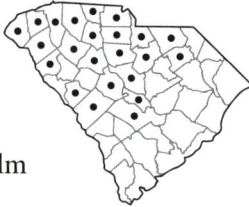

DESCRIPTION: Shrub, 12–24″ tall, forming extensive colonies from
rhizomes; leaves deciduous, pale and whitish below; flowers white
or greenish with a pink tinge, produced in a few short racemes
from the previous year's wood; corolla urn-shaped; fruit a many-
seeded berry, blue with a thin covering of whitish wax that is easily
rubbed off; flowers March–April; fruits June–July.

RANGE-HABITAT: Widespread in the eastern US; in SC, common
in the mountains and Piedmont, occasional in the Sandhills and
Coastal Plain; in a variety of forested slopes that are usually rather
dry with highly acidic soils.

COMMENTS: The small berries are sweet and edible and are an
important food for wildlife. Duke (1997) recommends species of *Vaccinium* as herbal
remedies for multiple sclerosis and bladder infections. It is often one of the most abundant species in pine-oak heaths. This species is often called Lowbush Blueberry by locals,
although this name is more appropriately applied to the more northern *Vaccinium angusti-
folium* Aiton.

229. Black Huckleberry

Gaylussacia baccata
(Wagenhouse) K. Koch
Gay-lus-sàc-i-a bac-cà-ta
Ericaceae (Heath Family)

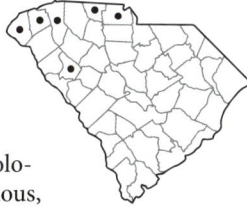

DESCRIPTION: Erect shrub, forming large colonies, usually less than 3′ tall; leaves deciduous, ovate to oblong, with golden, sessile glands on the upper and lower surface; flowers orange to red, in racemes in the axil of leaves; fruits are blackish, juicy berries with 10 seeds; flowers April–May; fruits mature July–August.

RANGE-HABITAT: Widespread in the eastern United States; in SC, locally common in the mountains upper Piedmont in pine-oak heaths, acidic oak-hickory forests and other exposed acidic soil sites in the mountains and upper Piedmont.

COMMENTS: Though not as abundant as Bear Huckleberry (*Gaylussacia ursina* (M.A. Curtis) Torrey, plate 230) in SC, this species is locally common in parts of Pickens and Greenville counties. It is easily separated from Bear Huckleberry by the less pointed and less narrow leaves and the reddish-orange flowers. The genus honors Louis Joseph Gay-Lussac (1778–1850), a French chemist.

230. Bear Huckleberry

Gaylussacia ursina (M. A. Curtis)
Torrey & Gray ex Gray
Gay-lus-sàc-i-a ur-sì-na
Ericaceae (Heath Family)

DESCRIPTION: Erect shrub, usually about 3′ tall; leaves deciduous, with a blunt projection at the tip of the tapering apex; leaves with golden, sessile glands only on the lower surface; flowers white, sometimes tinged with pink, produced on old wood, in racemes; flowers May–June; fruits ripen July–August.

RANGE-HABITAT: Restricted to the mountains of TN, NC, SC, and GA; upper slopes and ridge tops, most common in pine-oak heaths, Chestnut Oak forests, and oak-hickory forests.

COMMENTS: This species is extremely abundant in Oconee County but much less abundant in the rest of the SC mountains. It often forms dense, expansive clonal stands. It is similar in appearance to Black Huckleberry (*G. baccata* (Wangenheim) K. Koch, plate 229), which has golden glands on both leaf surfaces, and to Dangleberry (*G. frondosa* (L.) Torrey & Gray, plate 561), which has leaves that are somewhat whitened on the lower surface. Huckleberry tea once was widely used to treat rheumatism and arthritis. The genus honors Louis Joseph Gay-Lussac (1778–1850), a French chemist.

231. Sweet Pinesap

Monotropsis odorata Schweinitz
ex Elliott
   Mo-no-tróp-sis o-do-rà-ta
Ericaceae (Heath Family)

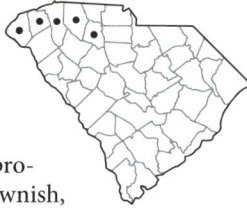

DESCRIPTION: Inconspicuous perennial, saprophytic herb; stems, leaves, and flowers brownish,
purplish, or pinkish; stems to 2.5″ tall, nodding
in flower and erect in fruit; leaves reduced to scales that are
spirally arranged and overlapping; flowers very fragrant, smelling like cloves; flowers February–March and (rarely) September–
November.

RANGE-HABITAT: MD south to SC and west to KY and GA; in SC,
uncommon in the mountains, very rare in the Piedmont; pine-
oak heaths and acidic oak-hickory forests, often under Mountain
Laurel.

COMMENTS: Leaf litter often partially covers Sweet Pinesap, and it
blooms at times when few botanists are in the woods. This prob-
ably explains its reported rarity. The flowers are fragrant and have been compared to smelling
like a mixture of cinnamon, nutmeg, and cloves. Heyward Douglas, a retired Clemson staff
member, has had great success "smelling" for the species in mid-March in pine-oak heath
and other acidic forest communities. He has documented many new populations for SC. It
appears the species reaches its greatest abundance in the SC mountains. It can be seen, or
smelled, along the Oconee Bell Trail in Devil's Fork State Park.

CONSERVATION STATUS: SC-Vulnerable

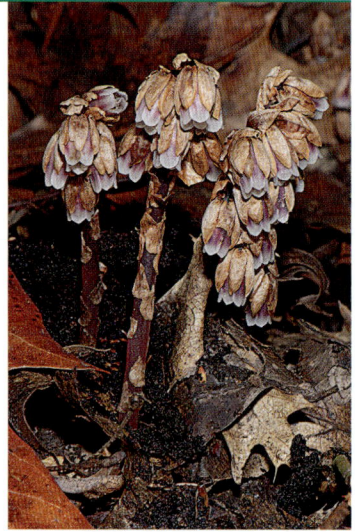

232. Trailing Arbutus; Mayflower

Epigaea repens L.
   E-pi-gaè-a rè-pens
Ericaceae (Heath Family)

DESCRIPTION: Creeping subshrub; stem 8–16″
long, branching, with spreading hairs; leaves
evergreen, leathery, elliptic, to 2.5″ long; flowers
fragrant, in tight terminal and axillary spikes,
white to pink; flowers late February–early May.

RANGE-HABITAT: Widespread in eastern North
America; in SC, common in the mountains,
Piedmont, Sandhills, and inner Coastal Plain and
rare in the outer Coastal Plain; in a wide variety of dry to somewhat moist
habitats with acidic usually sandy and/or rocky soils.

COMMENTS: This is another of SC's many ant-dispersed species, the ant attrac-
tant in this case being the placental tissue around the seeds. This is the Mayflower of the Pilgrims, described as a messenger of hope in Whittier's poem *The
Mayflower*. Finding this plant in the early spring at Plymouth Rock was a hopeful
sign that the cold and apparently barren land of the New World was actually bountiful. Natural populations have been greatly depleted because the sweet-scented blossoms were picked
for bouquets; now it is protected by law in many states. The genus means "upon" (epi) "earth" (Gaea,
Greek goddess of earth), in reference to its habit of growing flat against the ground.

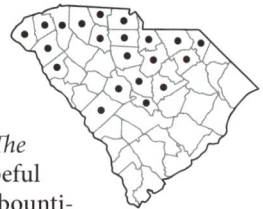

233. Plantain Pussytoes

Antennaria plantaginifolia (L.)
 Richardson
   An-ten-nà-ri-a plan-ta-gi-ni-fò-li-a
Asteraceae (Aster Family)

DESCRIPTION: Perennial, white, woolly herb with long, ascending stolons; leaves in a basal rosette, alternate, reduced and alternate on the stem, white-woolly above; flowers white to purplish and tiny, in involucrate heads of disk flowers only, with several heads terminating the stem; flowers March–early May.

RANGE-HABITAT: Widespread in eastern North America; common throughout SC except in the outer Coastal Plain, where it is rare; dry woods, including pine-oak heaths, oak-hickory forests, and Chestnut Oak forests.

COMMENTS: Soft, downy flower heads are crowded together into a cluster resembling a cat's paw—hence, the common name. A very similar species, Big-head Pussytoes (*Antennaria parlinii* Fernald ssp. *fallax* (Greene) Bayer & Stebbins) is distinguished by having taller flowering heads, leaves that lose their pubescence on the upper surface with age, and stolons that are flat against the ground when young, rather than ascending as in Plantain Pussytoes.

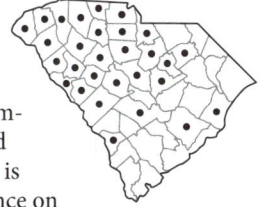

234. Wintergreen; Teaberry

Gaultheria procumbens L.
   Gaul-thèr-i-a pro-cúm-bens
Ericaceae (Heath Family)

DESCRIPTION: Diminutive subshrub, forming colonies from a horizontal rhizome; aerial stem usually 5–6″ tall, with a few leaves crowded toward the tip; leaves evergreen, to 2.5″ long; flowers white or pale pink; fruits are small red berries with the odor and flavor of wintergreen; flowers June–August; fruits mature September–November.

RANGE-HABITAT: Widespread in ne. North America, extending south in the Appalachian Mountains to GA and AL; in SC, rare, known only from a few populations in the mountains—in openings in pine-oak heaths and under Great Laurel or Mountain Laurel in acidic montane oak-hickory forests.

COMMENTS: Leaves contain oil of wintergreen (methyl salicylate), which is used topically to treat muscle pain. Because of the potential for salicylate toxicity, the oil apparently should not be applied after heavy exercise or in hot humid weather. Duke (1997) describes wintergreen as a useful herbal remedy for sore throats, backache, and sciatica. The genus honors Jean-François Gaultier (1708–56), a naturalist and court physician in Quebec.

CONSERVATION STATUS: SC-Vulnerable

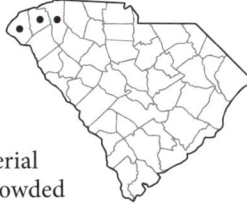

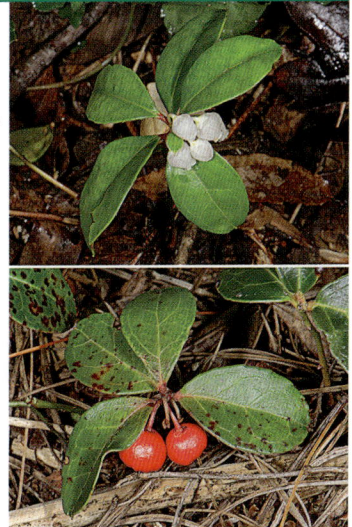

**235. Appalachian Bellwort;
Carolina Bellwort**

Uvularia puberula Michaux
U-vu-là-ri-a pu-bé-ru-la
Colchicaceae (Meadow-saffron
Family)

SYNONYM: *U. pudica* (Walter) Fernald
(in part)—RAB

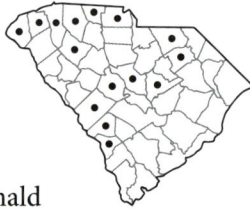

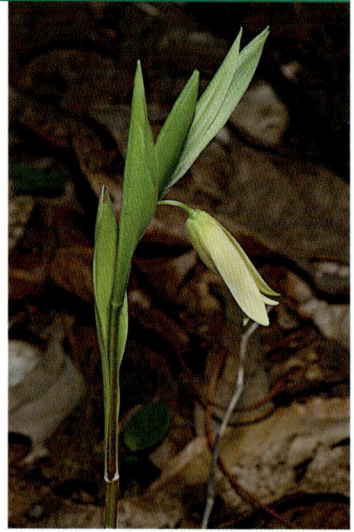

DESCRIPTION: Perennial herb; stems to 16″ tall, from a crown,
usually with 1–2 branches; leaves sessile, leaves shiny, light green;
flowers solitary in upper leaf axils, 1–3 per stem, yellow and nodding, bell-like; flowers April–early May.

RANGE-HABITAT: PA south to GA, mostly in the mountains; in SC,
common in the mountains, uncommon in the Piedmont; dry
to moist forests on acid soils, usually in openings, especially in
pine-oak heaths and acidic montane oak-hickory forests in the
mountains and xeric oak-hickory forests in the Piedmont.

COMMENTS: Readily distinguished from other bellwort species by its preference for dry
acidic habitats and its noncolonial growth.

236. Hairy Yellow Stargrass

Hypoxis hirsuta (L.) Coville
Hy-póx-is hir-sù-ta. var. hir-
sù-ta
Hypoxidaceae (Yellow Stargrass
Family)

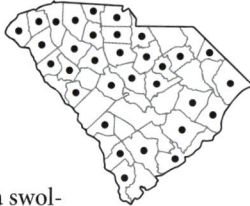

DESCRIPTION: Herbaceous perennial from a swol-
len underground stem; leaves basal only, pubescent, grasslike,
ascending to nearly erect; petals yellow and hairy on the outside;
flowers March–June.

RANGE-HABITAT: From MA to Manitoba, south to FL and TX; in SC,
common throughout except the outer Coastal Plain; open woodlands, meadows, road banks, and pine-oak heaths.

COMMENTS: This is the only species of yellow stargrass in the
mountains and Piedmont of SC. In the Coastal Plain the much
more delicate Fringed Stargrass (*Hypoxis juncea* J. E. Smith) has
filiform, pubescent leaves less than 1 mm wide; Swamp Stargrass
(*Hypoxis curtissii* Rose) has much wider, smooth leaves and is
found in swamp forests.

237. Blue Ridge Golden-banner

Thermopsis mollis
(Michaux) M. A. Curtis
Ther-móp-sis mól-lis
Fabaceae (Pea or Bean Family)

DESCRIPTION: Perennial herb to 2′ tall from long, thin rhizomes, forming large colonies; leaves dark green, trifoliolate, to 2.5″ long and with obvious, persistent stipules; flowers in terminal racemes; corolla pale to deep golden yellow; fruit a legume covered with short, appressed trichomes; flowers March–May.

RANGE-HABITAT: Endemic to the southern Appalachian region at low to middle elevations from VA south to AL; in SC, uncommon, but locally abundant, in the mountains and upper Piedmont; pine-oak heaths and dry oak-hickory forests with acidic soils on ridges.

SIMILAR SPECIES: Ash-leaf Golden-banner (*Thermopsis fraxinifolia* (Nuttall) M. A. Curtis), is also found in open oak woodlands and rocky streamside habitats of the mountains. It is a taller plant, not forming large colonies, as it lacks the spreading rhizomes of Blue Ridge Golden-banner. The two species are also separated by flowering time with the early flowering Blue Ridge Golden-banner completes its flowering two weeks or more before Ash-leaf Golden-banner begins.

CONSERVATION STATUS: SC-Vulnerable

238. Pink Lady's Slipper; Moccasin-flower

Cypripedium acaule Aiton
Cy-pri-pè-di-um a-caù-le
Orchidaceae (Orchid Family)

DESCRIPTION: Showy, perennial, herbaceous, terrestrial orchid to 18″ tall; basal leaves to 9.5″ long and 5.5″ wide, glandular-hairy, with riblike parallel veins; flower solitary, terminating a leafless stalk; 3 sepals and 2 lateral petals are green and often tinged with purple; the "slipper" or "moccasin" petal is pink or rarely white; flowers April–May.

RANGE-HABITAT: Widespread in eastern North America; in SC, common in the mountains, occasional in the upper Piedmont, and rare in the Sandhills; dry acidic woodlands under pines, pine-oak heaths, or under thickets of Great Laurel.

COMMENTS: Thoughtless nature enthusiasts often collect Pink Lady's Slipper and bring it into the home garden, where it seldom survives. As with all wild orchids, its roots (and seeds) require the presence of highly specific soil fungi (mycorrhizae) for proper development and maintenance. Lady's slipper was widely used in the nineteenth century as a sedative for nervous headaches, hysteria, and insomnia. The pollination mechanism is very interesting. A large bee is usually attracted to the pink flower and enters through the lip. The lip has an entrance with inward folded edges that traps the bee when it enters. The only escape is through the top of the lip where it has a "lid" formed by a staminal shield adjacent to balls of pollen. The bee must lift the "lid" to exit and in the process becomes covered in pollen. When it visits another flower, this pollen will be deposited onto the stigma of the trap. This staminal shield and opening are illustrated in the photograph.

239. Turkeybeard; Beargrass

Xerophyllum asphodeloides (L.) Nuttall
Xe-ro-phýl-lum as-pho-de-loì-des
Xerophyllaceae (Beargrass Family)

DESCRIPTION: Perennial herb with a thick cluster of basal leaves; basal leaves grasslike, stiff, to 20″ long, with fine saw-teeth on the margin; stem leaves similar but greatly reduced upward; flowers produced in a terminal raceme a long stem with narrow, reduced leaves; flowers May–June.
RANGE-HABITAT: Two ranges: (1) Coastal Plain of s. NJ and DE; and (2) mountains of w. VA south to e. TN, w. NC, nw. SC, and ne. GA; in SC, rare in the mountains and upper Piedmont; pine-oak heaths and dry pine-oak-hickory forests.
COMMENTS: The basal leaves of Turkeybeard greatly resemble the "grass-stage" of the Longleaf Pine and apparently serve the same function, protecting the apical meristem from the effects of fire. The rarity of this species is probably due to fire suppression in its fire-adapted habitat. In the mountains it is usually associated with Pitch Pine (*P. rigida* Miller, plate 222) or Table Mountain Pine (*P. pungens,* plate 3); in GA it is usually associated with Shortleaf Pine (*P. echinata* Miller) or Virginia Pine (*P. virginiana* Miller).
CONSERVATION STATUS: SC-Imperiled

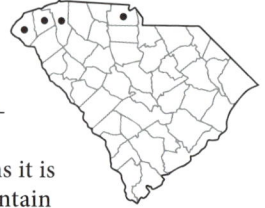

240. Fly Poison

Amianthium muscitoxicum (Walter) Gray
A-mi-án-thi-um mus-ci-tóx-i-cum
Melanthiaceae (Bunchflower Family)

DESCRIPTION: Perennial herb from a bulbous base with a thick cluster of grasslike basal leaves; basal leaves to 2′ long and up to 0.9″ wide, though generally less than 0.4″ wide; stem leaves similar but greatly reduced upward; flowers produced in a terminal raceme; tepals white; fruit a suborbicular capsule; flowers May–July.
RANGE-HABITAT: Widespread in eastern North America from NY south to FL and west to OK and MS; in SC, common in the mountains where it is found in pine-oak heaths and acidic montane oak-hickory forests; rare in the Piedmont and Coastal Plain on hardwood bluffs, Longleaf Pine savannas and sandhills.
SIMILAR SPECIES: Crow Poison (*Stenanthium densum* (Desrousseaux) Zomlefer & Judd, plate 620) is very similar but restricted to the Coastal Plain where it is found in mesic pine savannas. It can be distinguished by having a purplish basal sheath that surrounds the base of the basal leaves and a conical rather than subglobose capsule.
COMMENTS: The common name refers to a characteristic of many members of the family *Melanthiaceae,* namely, that it produces highly toxic alkaloids.

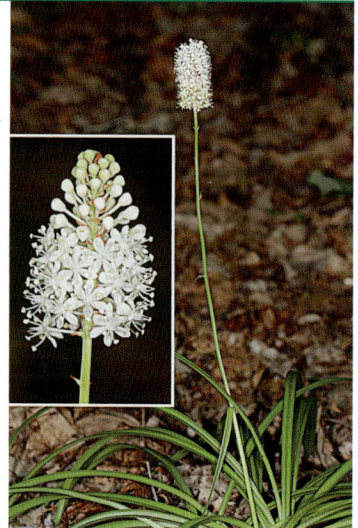

241. Appalachian Small Spreading Pogonia

Cleistesiopsis bifaria (Fernald) Pansa-
rin & F. Barros
Cleis-te-si-óp-sis bi-fà-ri-a
Orchidaceae (Orchid Family)

DESCRIPTION: Perennial, herbaceous, terrestrial orchid with a single, smooth stem to 2′ tall; a single thick, smooth, bluish-green, fleshy leaf produced on flowering plants and a leafy floral bract subtends the flower; the floral bract exceeds the length of the pedicellate flower; flowers solitary; lip and dorsal tepals generally pale pink, and lateral tepals and brownish to maroon lateral tepals; the lip with a central keel containing 5–7 irregular ridges; fresh flowers odorless; flowering May–July.

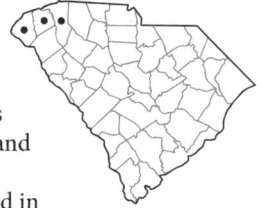

RANGE-HABITAT: WV and KY south to GA and AL; generally, in the mountains and Piedmont; in SC, rare, in the mountains where found in pine-oak heath and dry margins of cataract fens.

COMMENTS: Three species of showy orchids in the genus *Cleistesiopsis* are found in SC. They may be distinguished by range; the other two species are restricted to the Sandhills and Coastal Plain. Appalachian Small Spreading Pogonia is very rare in SC and should never be picked or disturbed. It is probably more abundant than collections suggest; it seldom flowers in fire-suppressed habitats.

CONSERVATION STATUS: SC-Imperiled

242. Carolina Lily; Michaux's Lily

Lilium michauxii Poiret
Lí-li-um mi-chaùx-i-i
Liliaceae (Lily Family)

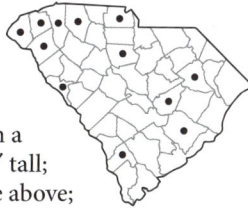

DESCRIPTION: Smooth, perennial herb from a scaly bulb; stem stout, strictly erect, to 3.5′ tall; leaves deciduous, whorled below, alternate above; flowers nodding, in terminal umbels of 1–2, rarely 3–4; flower segments strongly recurved, orange-red and purple-spotted in the lower half; flowers July–August.

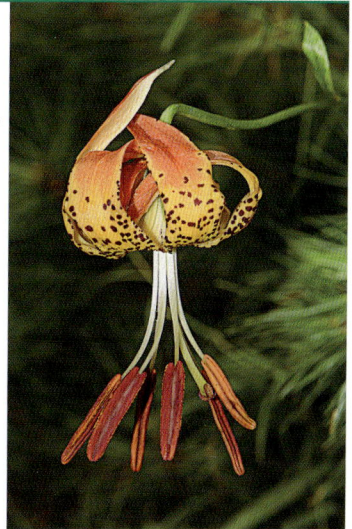

RANGE-HABITAT: VA south to FL and west to TN and LA; in SC, occasional in the mountains and rare in the Piedmont and Coastal Plain; pine-oak heaths, dry margins of rock outcrops, open oak-hickory forests on ridges and upper slopes, and in forest margins.

COMMENTS: Carolina Lily is the only true lily in the Piedmont and mountains that prefers highly acidic soils. Turk's-cap Lily (*Lilium superbum* L.) is similar but much larger, with many more flowers per stem; it has narrower and thinner lanceolate, rather than oblanceolate to obovate leaves. Turk's-cap Lily is very rare in the mountains of Greenville County and has also been located once in Hampton County in the Coastal Plain. The specific epithet honors André Michaux (1746–1802), a French botanist and author of two important works, *The Oaks of North America* (1801) and *Flora Boreali-Septentrionalis* (Flora of North America) (1803, published posthumously by his son).

243. Featherbells

Stenanthium gramineum
(Ker-Gawler) Morong
Ste-nán-thi-um gra-mí-
ne-um
Melanthiaceae (Bunchflower
Family)

DESCRIPTION: Perennial herb with a thick cluster of basal leaves from a slender bulbous base; basal leaves grasslike, stiff, to 2′ long, and less than 0.5″ wide; flowering stems to 4′ tall; stem leaves similar but greatly reduced upward; flowers produced in a large, showy panicle; tepals white; fruit a lanceolate capsule; flowers July–September.

RANGE-HABITAT: Widespread in eastern North America, from PA south to FL and west to IL and TX; in SC, occasional in pine-oak heaths and forest margins in the mountains.

COMMENTS: The showy, large inflorescences are easily observed along roads through the Andrew Pickens District of the Sumter National Forest in Oconee County. Like other members of the family, this species contains toxic alkaloids.

The Piedmont

244. Granitic flatrock

245. Fringe-tree; Old Man's Beard

Chionanthus virginicus L.
    Chi-o-nán-thus vir-gí-ni-cus
Oleaceae (Olive Family)

DESCRIPTION: Fast-growing, short-lived shrub or small tree, occasionally reaching 30′ tall; leaves simple, opposite, deciduous, and entire; each flower has a 4-parted calyx and 4 white straplike petals; flowers April–May; fruits are blue drupes, maturing July–September.

RANGE-HABITAT: NJ south to FL and west to MO, OK, and TX; common throughout SC; a wide variety of habitats, including dry, mesic, or wet forest margins, granitic flatrocks and domes, and pocosins.

COMMENTS: Fringe-tree is a widely cultivated native tree in SC; its attraction is the airy clusters of fragrant, white flowers. A wide variety of wildlife eat the drupes.

246. American Beautyberry; French-mulberry

Callicarpa americana L.
    Cal-li-cár-pa a-me-ri-cà-na
Lamiaceae (Mint Family)

DESCRIPTION: Deciduous shrub up to 8′ tall; stems arching, with star-shaped hairs; leaves opposite, simple, pubescent, with crenate to serrate margins; flowers produced in axillary cymes;

petals light pink to lavender; flowers June–July; fruits are brilliant lavender to purple drupes; fruits mature August–October.

RANGE-HABITAT: Primarily southeastern US; common throughout SC; margins of granitic flatrocks, maritime forests, fencerows, woodland borders, pine-mixed hardwood and oak-hickory forests, and sandy or rocky woodlands and forest margins.

COMMENTS: The berries are edible but not very palatable, and although sweet at first, they are pungent and astringent afterward. The plant makes a very adaptable, drought-tolerant ornamental and the drupes are consumed late in the autumn by birds, particularly catbirds. The leaves have been shown to repel biting insects, and fresh leaves can be wiped over skin as a natural insect repellent.

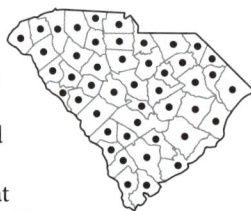

247. Elf-orpine

Diamorpha smallii Britton ex Small
Di-a-mór-pha smáll-i-i
Crassulaceae (Stonecrop Family)

SYNONYM: *Sedum smallii* (Britton ex Small) Ahles—RAB

DESCRIPTION: Annual herb to 4″ tall; leaves fleshy, cylindrical, to 0.25″ long, reddish; fruit opening by a small flap on the undersides of each follicle; flowers March–May.

RANGE-HABITAT: From VA to AL and locally north into se. TN; in SC, common on granitic flatrocks in the Piedmont, rare on sandstone outcrops in the Coastal Plain, and rare on granitic domes in the mountains.

COMMENTS: This species is uncommon because of its restricted habitat. It occurs on isolated flatrocks that are as small as 0.02 acres. Its habitat is shallow soils of rock-rimmed depressions that hold water following a rain and on sunny outcrop margins. It is one of several species that are adapted to avoid the extreme heat and drought of summer by persisting as seeds, a strategy termed "drought avoidance." The specific epithet honors John K. Small (1869–1938), botanist at the New York Botanical Garden, author, and botanical explorer of the southern states.

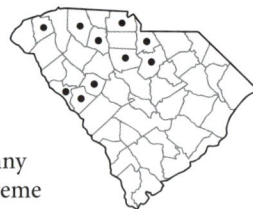

248. Puck's Orpine; Granite Stonecrop

Sedum pusillum Michaux
Sè-dum pu-síl-lum
Crassulaceae (Stonecrop Family)

DESCRIPTION: Diminutive annual herb; stems to 3″ tall, few-branched; leaves fleshy, alternate, in spirals, cylindrical, to 0.5″ long, blue-green or sometimes reddish in full sun; flowers white, in flat-topped clusters; flowers March–April.

RANGE-HABITAT: NC, SC, GA, and AL; in SC, rare, and restricted to the Piedmont; found on granitic flatrocks, in shallow soil in partial shade, almost always under Eastern Red Cedar.

COMMENTS: Sometimes confused with Elf Orpine (*Diamorpha smallii,* plate 247), especially plants with reddish leaves growing in full sun. Puck's Orpine blooms two weeks earlier, and the fruit opens by a lengthwise slit on the upper surface; the fruit of *D. smallii* opens by a small flap on the underside. Additionally, the petals of Puck's Orpine are flat and those of Elf Orpine are often hood shaped, with rolled margins.

CONSERVATION STATUS: SC-Imperiled

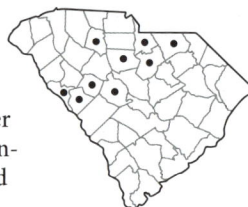

249. Pool Sprite; Snorkelwort

Gratiola amphiantha D. Estes &
   R.I. Small
   Gra-tì-o-la am-phi-án-tha
   Plantaginaceae (Plantain Family)

SYNONYM: *Amphianthus pusillus*
   Torrey—RAB, PR

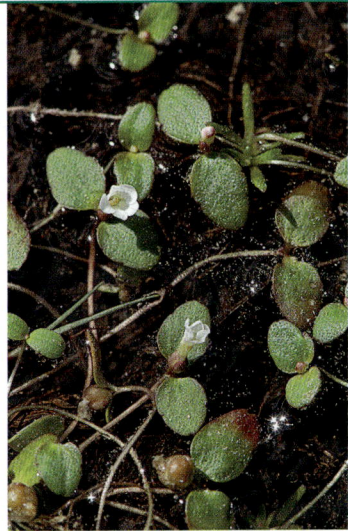

DESCRIPTION: Diminutive annual herb; leaves both floating and
submerged; submerged leaves less than 0.4″ long and in a rosette
at the tip of a short stem; floating leaves in single pairs at the tips
of threadlike stems, ovate, and less than 0.3″ long; flowers tiny, in
clusters in the axils of both floating and submerged leaves, white to
pale violet; flowers March–April.

RANGE-HABITAT: Endemic to granitic flatrocks in the Piedmont of
AL, GA, and SC; very rare in SC; in shallow soil of rock-rimmed
depressions that hold water after a rain; only three populations are
known from SC, one of which is at Flat Creek Heritage Preserve in
Lancaster County.

COMMENTS: The presence of both exposed and submerged flowers provides for both cross- and
self-pollination. Based on the extremely low amount of genetic diversity found both within
and between populations, self-pollination has apparently been the primary mode of sexual
reproduction.

CONSERVATION STATUS: Federally Threatened

250. Eastern Prairie Anemone

Anemone berlandieri Pritzel
   A-ne-mò-ne ber-lan-di-èr-i
   Ranunculaceae (Buttercup
   Family)

SYNONYM: *Anemone caroliniana*
   Walter—RAB, in part

DESCRIPTION: Herbaceous, perennial, spring ephemeral from a vertically oriented bulb; stem to 6″ tall at flowering, pubescent above
and below the leaves; leaves with thin linear segments, located
above the middle of the stem when flowering; petals absent; sepals
white; flowers March–April; pedicel elongates well above the leaves
after flowering to produce a fruit that is an aggregate of achenes
held up to 18″ above the ground; fruits mature in April.

RANGE-HABITAT: From AR and KS south to LA and TX and locally
in isolated locations that are highly disjunct from the core range
in AL, GA, FL, NC, SC, and VA; in SC, very rare in the Piedmont,
where it is found in shallow-soiled glades associated with mafic (high magnesium) rock
outcrops such as amphibolite or calcareous variants of granitic flatrocks, and in rocky, dry,
oak-hickory forests with circumneutral soil.

COMMENTS: This is the earliest flowering of our *Anemone* species. It is very rare in SC where
it has been documented from five sites. The flowers appear well before most species in the
habitat, and it is most often encountered in fruit, with "fluffy" tufts of achenes each with tufts
of hair for wind dispersal. Additional populations of this species should be sought in shallow
soils of rock outcrops throughout the Piedmont of SC and reported to the South Carolina
Heritage Trust Program. A very similar species, Carolina Anemone (*Anemone caroliniana*
Walter), can be distinguished by its flowers, which are produced high above the stem leaves;
stems that are only pubescent above the leaves; and a rhizome rather than a vertical bulb. It is
very rare in thin soils of outcrops and oak-hickory forests in the Piedmont.

CONSERVATION STATUS: SC-Critically Imperiled

251. Flatrock Phacelia

Phacelia maculata Wood
Pha-cè-li-a ma-cu-là-ta
Hydrophyllaceae (Waterleaf Family)

DESCRIPTION: Erect or sometimes sprawling annual herb, 4–16″ tall; stems hairy; leaves pinnately divided; flowers purple, ranging to pale blue or white, spotted with purple; inflorescences are coiled terminal and axillary flower cluster that straighten and elongate as the flowers reach maturity; flowers April.

RANGE-HABITAT: NC south to GA and west to LA; in SC, uncommon on and around granitic flatrocks in the Piedmont and granitic domes in the upper Piedmont and mountains.

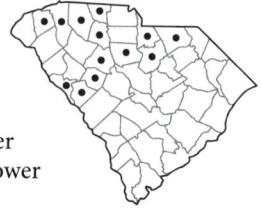

COMMENTS: The more widespread Appalachian Phacelia (*Phacelia dubia* (L.) Trelease), is very similar, but Flatrock Phacelia has longer sepals (to 0.3″ long versus to 0.15″ long) and the sepals have spreading, stiff, marginal hairs longer than 1 mm. Flatrock Phacelia also has an erect versus spreading habit. The flower color is also more intensely purple or blue in Flatrock Phacelia.

252. Woolly Ragwort

Packera tomentosa (Michaux) C. Jeffrey
Páck-er-a to-men-tò-sa
Asteraceae (Aster Family)

SYNONYM: *Senecio tomentosus* Michaux—RAB, PR

DESCRIPTION: Perennial herb; stems 8–30″ tall with dense, white hairs; leaves toothed, basal, and on stem; leaves covered with dense, white hairs, particularly when young and held vertically; heads of ray and disc flowers, both yellow, in open, terminal, flat-topped clusters; flowers April–early June.

RANGE-HABITAT: NJ to FL and west to TX; uncommon in SC; moist, open habitats on or around granitic flatrocks in the Piedmont; rare on moist sands in the Coastal Plain.

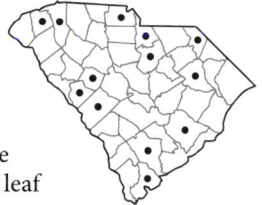

COMMENTS: Many ragworts contain highly poisonous alkaloids. Woolly Ragwort is uniquely adapted to thrive in the intense sun and heat of outcrops. The leaves are held upright, which prevents direct sunlight from beating down on leaf blade and the light-colored hairs reflect light and keep moisture close to the leaves and stem. The result of these adaptations is that the leaf temperature is often several degrees cooler than the air temperature—natural "air conditioning."

253. One-flower Sandwort; Piedmont Sandwort

Mononeuria uniflora (Walter) Dillenberger & Kedereit
Mo-no-neù-ri-a u-ni-flò-ra
Caryophyllaceae (Pink Family)

SYNONYMS: *Arenaria uniflora* (Walter) Muhlenburg—RAB; *Minuartia uniflora* (Walter) Mattfeld—PR

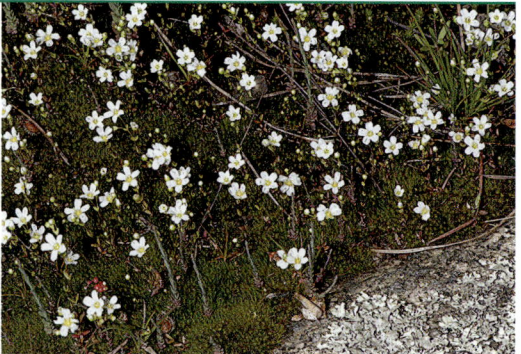

DESCRIPTION: Small, smooth annual with basal rosette and taproot, to 5″ tall; leaves narrow, to 0.20″ long; flowers March–May.

RANGE-HABITAT: From NC south to GA, west to AL; primarily in the Piedmont, but extending into the Coastal Plain of GA and AL; in SC, common on granitic flatrocks of the lower Piedmont.

COMMENTS: Piedmont Sandwort can easily be seen at Forty Acre Rock at Flat Creek Heritage Preserve in Lancaster County. The ecology of this species is similar to many outcrop endemics such as Puck's Orpine. It grows during the cool, moist seasons (winter–spring), flowers, and then dies. It avoids the harsh conditions of the summer as seeds. This is termed "drought avoidance."

CONSERVATION STATUS: SC-Vulnerable

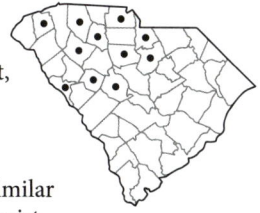

254. Hairy Spiderwort

Tradescantia hirsuticaulis Small
Tra-des-cánt-i-a hir-su-ti-
caù-lis
Commelinaceae (Spiderwort
Family)

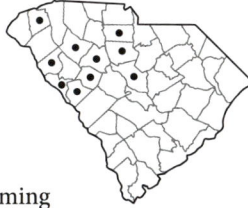

DESCRIPTION: Perennial herb to 15″ tall forming dense tufts; stem densely hairy; leaves densely hairy, the opened leaf sheath as wide as or wider than the leaf; flowers blue to rose, with a mix of glandular and nonglandular hairs; flowers April– June and occasionally again in autumn.

RANGE-HABITAT: NC, SC, GA, AL, AR, and OK; occasional in the Piedmont and rare in the mountains; dry woods adjacent to rock outcrops, especially on calcium-rich soils upslope from basic-mesic forests, and around granitic outcrops or domes.

COMMENTS: The dense hairiness of this spiderwort is distinctive, as are its short stature and habitat. The genus honors John, the elder Tradescant (1570s–1637), gardener to Charles the First of England. This species is an excellent choice for growing in the home landscape because it is short and clumping and produces a vivid display of large flowers.

255. False Garlic; Grace Garlic

Nothoscordum bivalve (L.) Britton
No-tho-scór-dum bi-vál-ve
Alliaceae (Onion Family)

SYNONYM: *Allium bivalve* (L.)
Kuntze—RAB

DESCRIPTION: Perennial herb from a bulb without an onion odor; stems to 18″ tall; leaves straplike and narrow; flowers produced in an umbel; petals white to cream colored with green or purple coloration along the midrib of the petal; fruit is a capsule; flowers March–May; fruits mature September–October.

RANGE-HABITAT: Widespread in the southern US from VA west to KS and south through South America; in SC, common throughout on shallow soil of rock outcrops, dry roadsides, and ruderal habitats.

COMMENTS: This common species is almost always present on granitic outcrop communities, where it is a native member of the flora. It has also spread to many ruderal habitats throughout our area. It can be distinguished from true onions by the lack of any onionlike or garliclike odor—hence the common name.

256. Flatrock Pimpernel

Lindernia monticola
   Muhlenberg ex Nuttall
      Lin-dér-ni-a mon-tí-co-la
Linderniaceae (False-pimpernel
   Family)

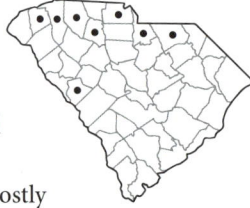

DESCRIPTION: Small perennial herb with mostly
leafless stems, 4–8″ tall, arising from a basal rosette; basal leaves
0.25–0.6″ long, elliptic to obovate; stem leaves reduced to inconspicuous bracts; pedicels elongate, glandular-punctate; corolla
bilabiate, blue to purplish-blue on the outside and pale lavender
within; flowers April–June.

RANGE-HABITAT: NC south to FL and AL; in SC, widely scattered
locations, but locally common in the Piedmont and mountains;
seepages on granitic domes and granitic flatrocks, cataract fens,
and rocky streamsides.

COMMENTS: This species is frequently seen growing from seasonally moist mats and near permanent seepage water over granitic
outcrops. It is plentiful on some outcrops at Forty Acre Rock in Flat Creek Heritage Preserve
in Lancaster County and at Boggs Rock, near Liberty, in Pickens County. No other species
closely resembles this beautiful and delicate wildflower in SC.

257. Eastern Prickly Pear

Opuntia mesacantha
   Rafinesque ssp. *mesacantha*
      O-pún-ti-a me-sa-cán-tha
      ssp. me-sa-cán-tha
Cactaceae (Cactus Family)

SYNONYMS: *Opuntia compressa*
   (Salisbury) J. F. Macbride—RAB; *Opuntia*
   humifusa (Rafinesque) Rafinesque—PR

DESCRIPTION: Creeping succulent with stems modified into
rounded pads; pads spiny; spines stout and present on at least
some of the internodes on each pad, though they may be broken
or fall off with age; flowers with bright yellow or orangish petals; flowers April–June; fruits are fleshy, red berries that mature
June–May, often ripening over winter.

RANGE-HABITAT: NJ south to FL and west to LA; in SC, common
throughout the state on granitic flatrocks, granitic domes, dry
roadsides, Longleaf Pine sandhills, and other dry, sandy woodlands.

COMMENTS: Prickly pear have stems that are produced in pads of several internodes with each
internode containing spines and/or glochids (small, stiff, irritating hairs). Longer spines are
modified leaf petioles; shorter spines are modified stipules. In this primitive cactus genus,
true leaves are reduced to small fleshy structures that fall off when the pads are very young.
This is the most widespread and common of SC's cacti. The rounded pads without a scalloped
margin distinguish it from Southern Prickly Pear (*O. mesacantha* ssp. *lata* (Small) Majure,
plate 892), which is confined to the Coastal Plain. Eastern Prickly Pear is a tetraploid (4 sets
of chromosomes), while the related Southern Prickly Pear is diploid (2 sets of chromosomes).
A valid argument could be made to recognize the two as separate species due to the inability
to produce fertile offspring.

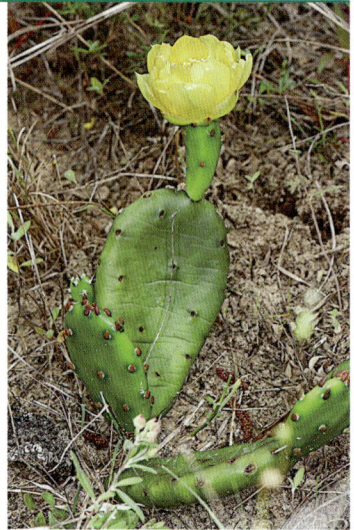

258. Appalachian Fameflower (A)

Phemeranthus teretifolius (Pursh)
  Rafinesque
  Phe-me-rán-thus te-re-ti-fò-li-us
SYNONYM: *Talinum teretifolium*
  Pursh—RAB, PR

Menges' Fameflower (B)

Phemeranthus mengesii (W. Wolf) Kiger
  Phe-me-rán-thus men-gès-i-i
Montiaceae (Montia Family)

DESCRIPTION: Smooth, fleshy, perennial herbs, 4–14″ tall in *P. teretifolius,* 8–24″ tall in *P. mengesii;* leaves cylindrical, arranged low on the stem; stems frequently branching in *P. mengesii;* flowering stems long, with few, pink flowers in an open terminal cluster; flowers bright pink; *P. teretifolus* with 12–30 stamens; *P. mengesii* with 50–80 stamens; flowers June–September.

RANGE-HABITAT: *P. teretifolius:* PA and WV, south to GA and AL; in SC, common in the Piedmont and mountains in shallow soils on granitic flatrocks and granitic domes; *P. mengesii:* TN and SC, south to AL and GA; in SC, very rare and known from two locations in the upper Piedmont where it grows in shallow soil mats of granitic domes and flatrocks.

COMMENTS: Appalachian Fameflower is a tetraploid species (4 sets of chromosomes), with flowers that are open 3:30–7:00 p.m. Menges' Fameflower is a diploid species (2 sets of chromosomes) with flowers open 1:00–3:00 PM. The difference in flowering time may prevent hybridization between diploid and tetraploid plants that would result in a sterile triploid.

CONSERVATION STATUS: *P. mengesii:* SC-Critically Imperiled

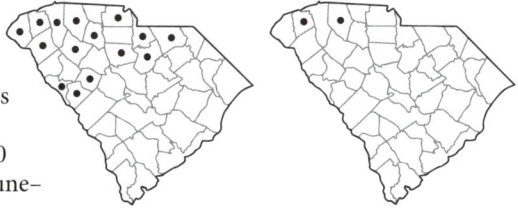

259. Outcrop Rushfoil

Croton willdenowii G. L. Webster
  Crò-ton will-de-nòw-i-i
Euphorbiaceae (Spurge Family)

SYNONYM: *Crotonopsis elliptica*
  Willdenow—RAB, PR

DESCRIPTION: Annual herb to 14″ tall, with dichotomous or trichotomous branching; entire plant, except leaves, with red spots and starlike hairs; leaves alternate; male and female flowers on separate plants; flowers June–October.

RANGE-HABITAT: From CT, PA, IL, and se. KS, south to FL and TX; in SC, common throughout but chiefly in the Piedmont; granitic flatrocks, thin soils around other outcrops, and sandy disturbed soil.

COMMENTS: The specific epithet honors Carl Ludwig Willdenow (1765–1812), botanist, taxonomist, pharmacist, and one of the founders of the study of plant distribution (phytogeography).

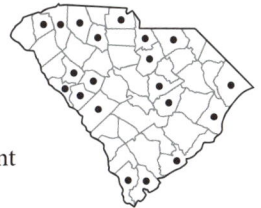

260. Pineweed; Orange-grass

Hypericum gentianoides (L.) Britton,
    Sterns & Poggenburg
    Hy-pé-ri-cum gen-ti-a-noì-des
Hypericaceae (St. John's-wort Family)

DESCRIPTION: Annual herb with wing-angled stems to 15″ tall, and numerous filiform branches toward the tip; leaves appressed, scalelike, and less than 0.04″ wide; flowers with 5 small green sepals and 5 larger yellow to orange petals; flowers July–October.

RANGE-HABITAT: Widespread in the eastern US; in SC, common throughout, growing in and around granitic flatrocks, granitic domes, and in sandy fields and roadsides.

COMMENTS: The bruised stems emit a fragrance that smells like orange juice. In this plant, water loss is reduced because green stems do most of the photosynthesis. Though it may not appear similar on first glance, this is a species of St.-John's-wort and contains the chemical hypericin, which is thought to be an antibiotic, increases dopamine levels, and is used as an herbal antidepressant.

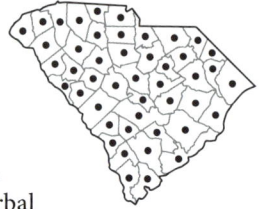

261. Alexander's Rock Aster

Eurybia avita (Alexander) G. L. Nesom
    Eu-rý-bi-a a-vì-ta
Asteraceae (Aster Family)

DESCRIPTION: Perennial herb forming large tufts from short rhizomes; flowering stems to 20″ tall; basal leaves linear to linear-lanceolate, long and grasslike and less than 0.4″ wide; stem leaves reduced upwards; flowers produced in involucrate heads arranged in a narrow flat-topped arrangement (corymbose); ray flowers bluish white to white; disk flowers yellow; flowers August–September.

RANGE-HABITAT: Limited to a handful of sites in central GA and one site in Pickens County, SC. In SC, very rare but locally abundant, and dominant on shallow-soil glades along the margins of a single granitic flatrock, Boggs Rock, in Pickens County.

COMMENTS: Alexander's Rock Aster is a distinctive and attractive species that should be sought at other locations in the Piedmont of SC. It is extremely rare and localized and should not be disturbed or picked. Any suspected additional populations should be reported to the South Carolina Department of Resources Natural Heritage Program.

SIMILAR SPECIES: Creeping Aster (*Eurybia surculosa* (Michaux) G. L. Nesom, plate 21) is similar but has longer rhizomes and forms looser colonies. It has lavender or blue ray flowers and less prominent basal leaves that may wither before flowering. Creeping Aster is also often found along the margins of rock outcrops in shallow soil in the mountains and upper Piedmont.

CONSERVATION STATUS: SC-Critically Imperiled

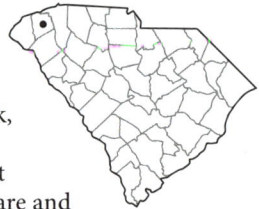

262. Common Blue Curls

Trichostema dichotomum L.
Tri-chos-tè-ma di-chó-to-mum
Lamiaceae (Mint Family)

DESCRIPTION: Annual, upright, branching herb from a tap root; stems to 2′ tall; stem leaves elliptic to elliptic-lanceolate, entire or slightly toothed, 1.0–2.75″ long and up to 0.9″ wide; flowers produced in a panicle of helicoid cymes; petals arranged in a bilabiate flower, blue to purple; stamens long and curled between the lobes and bent toward the lip; flowers August–November.

RANGE-HABITAT: Widespread through most of the eastern US; in SC, common throughout in thin soil of outcrops, dry road banks, and other dry often sandy or rocky habitats.

SIMILAR SPECIES: Narrowleaf Blue Curls (*T. setaceum* Houttuyn) is similar but has leaves that are 5–15 times as long as wide and very short pubescence along the stem versus the longer pubescence along the stem leaves that are 2.5–4.0 times as long as wide for Common Blue Curls. Narrowleaf Blue Curls is found in similar habitats but is not common. Dune Blue Curls (*T. nesophilum* K.S. McClelland & Weakley, plate 899) is a perennial, with a bushy appearance that is limited to the maritime strand.

COMMENTS: The common name is appropriate as the curled stamens dominate the lateral view of the flower.

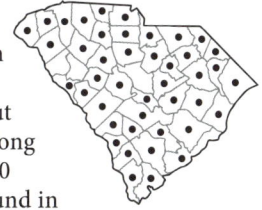

---

263. Woolly Lip-fern (A)

Myriopteris tomentosa (Link) Fée
My-ri-òp-te-ris to-men-tò-sa
SYNONYM: *Cheilanthes tomentosa*
Link—RAB, PR

Hairy Lip-fern (B)

Myriopteris lanosa (Michaux) Grusz &
Windham
My-ri-òp-te-ris la-nò-sa
SYNONYM: *Cheilanthes lanosa* (Michaux)
D. C. Eaton—RAB

Pteridaceae (Maidenhair Fern Family)

DESCRIPTION: *M. tomentosa*: creeping or ascending evergreen ferns forming tight clumps; blades 3-pinnate, bluish green in color on the upper surface; petioles and blades densely hairy, the petioles with a mixture of hairs and thin scales; the spores are partially covered by the rolled margins of the leaf blade, which are

modified into a pale flap of protective tissue; *M. lanosa* is similar but forms loose clumps; leaves dark green on upper surface; petioles and blades densely hairy, the petioles without scales; the margin of the leaves rolled under, but without modified protective flaps around the spores.

RANGE-HABITAT: *M. tomentosa*: primarily Appalachian, from PA south to KY, GA, LA, and scattered localities west; in SC, occasional in the mountains and Piedmont; granitic domes, granitic flatrocks, and shallow soil of other rock outcrops, in exposed, dry, sunny locations; *M. lanosa*: widespread in the eastern US from NY and IL south to FL and TX; in SC, locally common in the mountains and Piedmont

where found on granitic domes, flatrocks, and shallow soil of other rock outcrops, in exposed, dry, sunny locations.

COMMENTS: Two SC species of lip-fern often grow in close proximity. Woolly Lip-fern is easily distinguished, even at a distance, due to the bluish coloration and more robust, tightly clumping growth form. The genus is mostly found in the western United States but extends in dry, desertlike conditions to the east. Though we may not think of ferns growing in dry habitats, they are uniquely adapted to endure these conditions as true xerophytes, capable of losing most of their water and springing back to life with moisture, much like the Resurrection Fern.

264. Rocky shoals

265. Rocky-shoals Spiderlily

Hymenocallis coronaria (Leconte) Kunth

Hy-me-no-cál-lis co-ro-nà-ri-a

Amaryllidaceae (Amaryllis Family)

DESCRIPTION: Perennial herb from a large bulb; bulb usually wedged in rock crevices; leaves basal, to 1.6″ wide and 32″ long; flowering stalk 1–3 per plant, to more than 40″ tall, each with an umbel of 6–9 flowers; flowers about 6″ across with a staminal cup greater than 1.75″ long; flowers mid-May–mid-July.

RANGE-HABITAT: SC, GA, and AL; rare in the Piedmont; found in rocky shoals of major rivers and streams.

SIMILAR SPECIES: Hammock Spiderlily (*Hymenocallis occidentalis* (LeConte) Kunth) is similar but flowers later in the season (July–August), occurs in rich floodplain forests or other rich soil situations, and has a staminal cup less than 1.75″ long. Hammock Spiderlily is extremely rare in SC; it is found only at a few locations in the upper Piedmont in the Savannah River drainage.

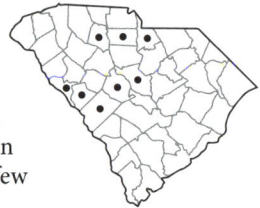

COMMENTS: Rocky-shoals Spiderlily is more robust than any of the other spiderlily species in SC, and its habitat is distinctive. Plants utilize contractile roots to grab the substrate and pull them downward into crevices between the rocks, thus achieving anchorage in their tumultuous habitat. Plants are easily observed at Landsford Canal State Park in Chester County and in the Rocky Shoals of the Broad River in Columbia. A large population is also now managed by the South Carolina Native Plant Society along Stevens Creek in McCormick County.

CONSERVATION STATUS: SC-Imperiled

266. American Waterwillow

Justicia americana (L.) Vahl
Jus-tíc-i-a a-me-ri-cà-na
Acanthaceae (Acanthus Family)

DESCRIPTION: Perennial herb 9–30″ tall; leaves opposite, narrow; flowers produced in headlike clusters at the tip of long stalks from the upper leaf axils; petals purple to white, with purple at the base of the lower lip; flowers June–October.

RANGE-HABITAT: Quebec south to GA and west to WI, KS, and TX; in SC, common in rocky shoals in the Piedmont, occasional along rocky streamsides in the mountains and along sandy streambeds and lake shores in the Coastal Plain.

COMMENTS: The unusual shape of the flowers is distinctive. The genus honors James Justice (1698–1763), a Scottish horticulturist and botanist.

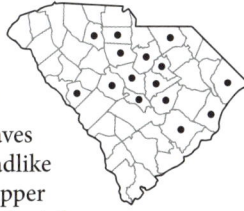

267. Basic-mesic forests

268. Chalk Maple

Acer leucoderme Small
À-cer leu-co-dér-me
Aceraceae (Maple Family)

DESCRIPTION: Small, often multitrunked deciduous tree with pale chalky gray bark; leaves lobed, entire, similar to Southern Sugar Maple in outline, but smaller, typically with two additional lobes and green and pubescent on the underside; leaves turn a rich salmon color in the autumn; some of the dried leaves remain attached to the tree until spring (marcescence); flowers March–April; fruits are schizocarps of samaras that mature May–September.

RANGE-HABITAT: Southeastern US from NC south to FL and west to OK and TX; in SC, common in the mountains and Piedmont on rich soils with elevated magnesium, basic mesic forests, margins of granitic outcrops, mafic outcrops, mafic oak-hickory forests, rich cove forests, beech forests, and river bluffs; rare in the Coastal Plain where found in marl forests and river bluff forests.

SIMILAR SPECIES: Southern Sugar Maple (*Acer floridanum* (Chapman) Pax, plate 743) has a single trunk, leaves whitish beneath, and the terminal lobes of some leaves are broader toward the apex than toward the base, with lobe tips acute to rounded. Chalk Maple usually has multiple trunks, leaves greenish yellow beneath, and the terminal lobe is narrower toward the apex than the base, with lobe tips pointed and often drooping.

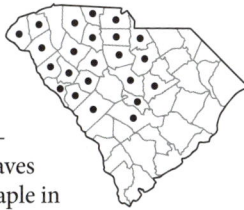

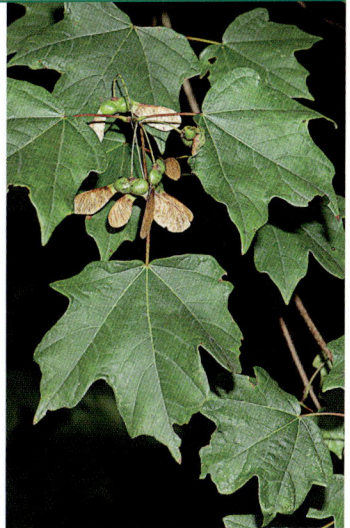

COMMENTS: Chalk Maple is a very attractive small landscape tree with a growth form similar to Japanese Maples. It is prized for its bold autumn color. In the wild, this species is an indicator of rich soils, typically associated with amphibolite, marble or coquina limestone (marl). If a stand of Chalk Maple is located, there is a good chance that the location will also contain rare and unusual species restricted to such habitats.

269. Common Shagbark Hickory

Carya ovata (P. Miller) K. Koch
Cà-ry-a o-và-ta
Juglandaceae (Walnut Family)

DESCRIPTION: Large, deciduous tree with grayish bark shredding in long strips on large trees; leaves pinnately compound, typically with 5 (7) leaflets, the terminal leaflet much larger than the lateral; leaflets often pubescent on the undersurface; fruits spherical in outline; flowers March–April; fruits mature in autumn.

RANGE-HABITAT: Widespread in eastern North America from southern Canada south to GA and west to TX; in SC, restricted to the Piedmont and Coastal Plain; basic-mesic forests, levee forests, marl forests, and other forests with calcium or magnesium-rich soil; locally common in the Piedmont, rare in the Coastal Plain.

COMMENTS: Mature trees are easily identified by the straplike shredding bark. Young trees and saplings may often be distinguished at a distance by the oversized leaflets. The nut is sweet and edible and was heavily utilized by Native Americans. The populations in Berkeley County along both the Santee and Cooper River drainages in marl forests are surprisingly disjunct from the center of the range in the Piedmont.

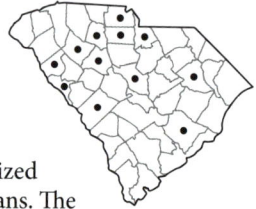

270. Bigleaf Magnolia

Magnolia macrophylla Michaux
Mag-nòl-i-a mac-ro-phýl-la
Magnoliaceae (Magnolia Family)

DESCRIPTION: Medium-sized deciduous tree generally less than 50′ tall; young twigs and buds pubescent; leaves huge to 43″ long and over 12″ wide (but typically 24–36″ long); leaf blades obovate with cordate or auriculate bases, pubescent and glaucous on the undersides; flowers up to 17″ wide; tepals cream-white with a purple spot toward the inner base of inner tepals; flowers April–May.

RANGE-HABITAT: VA and KY south to GA and LA, the range discontinuous; in SC, rare as a native in basic-mesic forests and mafic oak-hickory forests of the Piedmont; naturalized in the Coastal Plain.

COMMENTS: This species produces the largest simple leaves and largest flower of any native plant. It is uncommon throughout its range but is often planted as a novelty. It has thoroughly naturalized much of the natural area in the South Carolina Botanical Garden, and the state champion is located there. A showy display can be seen along the Bigleaf Magnolia Trail at the South Carolina Botanical Garden.

CONSERVATION STATUS: SC-Critically Imperiled

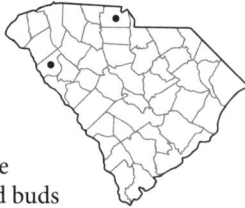

271. Spiny Gooseberry; Miccosukee Gooseberry

Ribes echinellum (Coville) Rehder
Rì-bes e-chi-nél-lum
Grossulariaceae (Currant Family)

DESCRIPTION: Low spiny shrub with palmately lobed and veined leaves borne alternately on new wood and in clusters on short lateral branches; stiff spines at the nodes; flowers greenish-yellow; ovary and fruit are densely covered in long, gland-tipped bristles; flowers March–April.

RANGE-HABITAT: Known only from two small regions, one near Lake Miccosukee, Jefferson County, FL, and a second at Stevens Creek, McCormick and Edgefield counties SC; very rare but locally abundant in basic-mesic forest.

COMMENTS: In SC, this species drops its leaves in August or September and puts on new, small leaves in December, apparently as an adaptation to obtain food from photosynthesis at a time when the generally closed canopy is leafless. Gooseberries are mostly found in cold, northern climates but two southern species exist, Spiny Gooseberry, and Granite Gooseberry (*Ribes curvatum* Small); the latter is found to the west of SC. Both have leaves in the winter and have adapted to life in cool microclimates in the Southeast. It is thought that Spiny Gooseberry represents relict populations from a wider distribution during cooler climates. Though abundant at Stevens Creek, this species is threatened by continued invasion of its habitat by Chinese Privet. Recent attempts to control privet at Stevens Creek Heritage Preserve appear to have increased the vigor and population of Spiny Gooseberry. It is also susceptible to extended drought, so climate change is probably the greatest threat to this species. The SC population has declined by at least 50% since 1977.

CONSERVATION STATUS: Federal Threatened

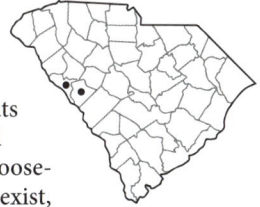

272. Leatherwood

Dirca palustris L.
Dír-ca pa-lús-tris
Thymelaeaceae (Mezereum Family)

DESCRIPTION: Small, deciduous shrub to 6′ tall, taller in other portions of the range; stems dark brown to gray, very flexible, with swollen nodes; leaves opposite, obovate to ovate with a wedge-shaped base; glabrous on both surfaces; flowers yellowish with exerted showy stamens; flowers February–March; fruit a green drupe, turning red at maturity; fruit matures July–August.

RANGE-HABITAT: Nova Scotia west to Ontario and MN, south to FL, LA, and OK; in SC, restricted to the Piedmont where found in basic-mesic forests, margins of floodplains, and other forests with circumneutral soils.

COMMENTS: Leatherwood is unmistakable with its low growth form and incredibly flexible branches and swollen nodes. The tough bark was used by Native Americans yo make rope, fabric, and baskets. Southern populations of this species are diploid, while those occurring north of the extent of the last glacial advance are tetraploid. These northern populations are more robust and found in varying habitats.

CONSERVATION STATUS: SC-Imperiled

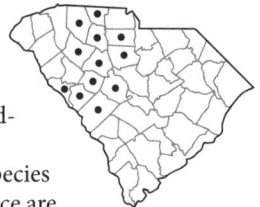

273. Appalachian Mock-orange

Philadelphus inodorus L.
Phi-la-dél-phus i-no-dò-rus
Hydrangeaceae (Hydrangea Family)

DESCRIPTION: Shrub usually 4–8′ tall; stems reddish brown and smooth; leaves opposite, smooth below; flowers white, showy; petals 4, styles separate above; sepals erect in fruit; flowers April–May.

RANGE-HABITAT: VA and TN, south to GA and AL; in SC, locally common in the mountains and Piedmont, rare in the Coastal Plain; rich cove forests, streamsides, spray cliffs, mafic oak-hickory forests, basic-mesic forests, and rocky bluffs, especially over calcium-rich or magnesium-rich (mafic) soils.

SIMILAR SPECIES: The rare Hairy Mock-orange (*P. hirsutus* Nuttall), which is restricted to the mountains, is found in drier, often rocky habitats and is distinguished by its dense, soft hairs on the lower leaf surface, pubescent young twigs, united styles, smaller flowers, and nonerect sepals in fruit.

COMMENTS: Appalachian Mock-orange is a great indicator of circumneutral soils. It is often found along the bases of slopes and along creeks in otherwise acidic habitats. The soils here receive extra nutrients through transport by the streams (alluvial transport) and downhill movement (colluvial transport).

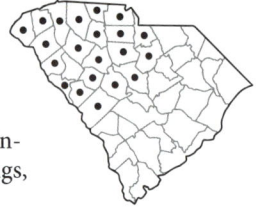

274. Bottlebrush Buckeye

Aesculus parviflora Walter
Aès-cu-lus par-vi-flò-ra
Hippocastanaceae (Buckeye Family)

DESCRIPTION: Shrub or occasionally trees to 16′ tall; leaves deciduous, opposite, palmately compound with 5–7 leaflets; inflorescence a narrow raceme; flowers white with stamens 3–4x the length of the petals; flowers June–July; fruits are capsules with 1–2 seeds, maturing October–November.

RANGE-HABITAT: Throughout Alabama, extending into GA with a single known location in Aiken County, SC; basic-mesic forest on fall-line river bluffs with calcium-rich soils.

SIMILAR SPECIES: Bottlebrush buckeye is easily separated in flower from Painted Buckeye (*Aesculus sylvatica* Bartram, plate 313) by its long-protruding stamens; in the latter, the stamens are included or are slightly longer than the petals.

COMMENTS: Bottlebrush Buckeye is found in the Savannah River Bluffs Heritage Preserve in Aiken County. This buckeye makes an excellent ornamental. "Bottlebrush" refers to the distinctive cluster of flowers with long stamens. The SC population is highly disjunct from the rest of the range of the species. The Aiken County location is probably the type location—where it was first discovered, as it appears on a list of exports in a shipping manifest of plants collected by Andre Michaux that were sent to Europe (McMillan, 2007). This population occurs at the crossing for the Savannah River that would have been taken by Michaux and John Fraser when travelling from Augusta, GA to SC, and was likely given to Walter by Fraser from this location for Walter's description and inclusion in his flora.

CONSERVATION STATUS: SC-Critically Imperiled

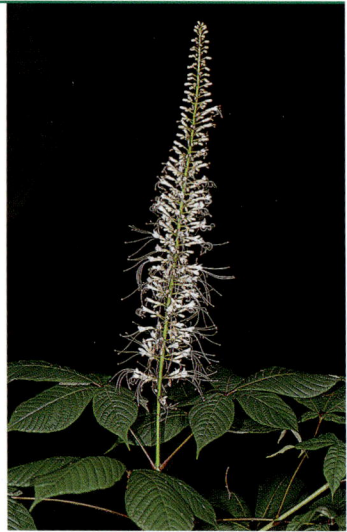

275. Dimpled Trout Lily

Erythronium umbilicatum
  Parks & Hardin ssp.
umbilicatum
    E-ry-thrò-ni-um um-bi-li-cà-
    tum ssp. um-bi-li-cà-tum
  Liliaceae (Lily Family)

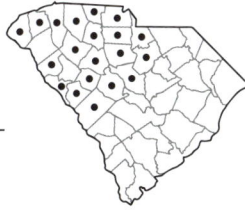

DESCRIPTION: Perennial herb from a bulb; the paired mottled leaves and the single, nodding, yellow flower flecked with purple are distinctive characteristics; this species is distinguished by the absence of an auricle at the base of the petals, the slight flecking with purple, and the presence of an indentation at the apex of the fruit, which is typically held on stems that have bent down to the ground by the time the fruit is mature; flowers January–April.
RANGE-HABITAT: WV, VA, and TN, south to AL and GA; in SC, common in the mountains and Piedmont, rare in the Sandhills and Coastal Plain, and absent in the maritime strand; basic-mesic forests, beech forests, cove forests, and bottomland forests; also sometimes found in shallow-soil mats on granitic domes and granitic flatrocks.
SIMILAR SPECIES: American Trout Lily (*Erythronium americanum* Ker-Gawler) is very rare in SC and restricted to basic-mesic forests in the Piedmont; it is similar but does not have auricles at the base of the petals and has a short beak at the end of the fruit, which is held above the ground.
COMMENTS: A true spring ephemeral, the vegetative development and flowering in late winter and early spring allow Dimpled Trout Lily to take advantage of the sunlight that filters through leafless trees; by May this species is typically dormant. This strategy works well to avoid the drought experienced on granitic outcrops as well. Populations may easily be observed in early spring on the Carrick Creek Trail in Table Rock State Park and growing on a granitic dome at Glassy Mountain Heritage Preserve, both in Pickens County.

276. Dutchman's Breeches

Dicentra cucullaria (L.)
  Bernhardi
    Di-cén-tra cu-cul-là-ri-a
  Fumariaceae (Fumitory Family)

DESCRIPTION: Perennial herb from a bulblet-bearing rootstalk; leaves basal, with long leaf stalks and deeply dissected blades; the unique flowers are diagnostic, resembling a pair of pants hung upside down; flowers February–March.
RANGE-HABITAT: Widespread, from Nova Scotia to GA and AR, and in a few western states; rare in SC and restricted to basic-mesic forests in the Piedmont.
COMMENTS: The resemblance of the flowers to a Dutchman's breeches led believers in the Doctrine of Signatures to think this plant was useful in the treatment of syphilis. A true spring ephemeral, it is typically dormant by May. This species is only known in SC from Stevens Creek Heritage Preserve in McCormick County and Hilton Pond Preserve in York County.
CONSERVATION STATUS: SC-Critically Imperiled

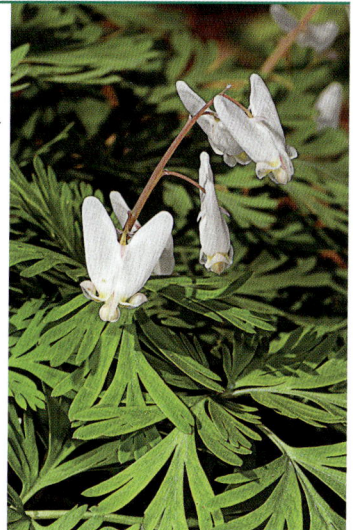

277. False Rue-anemone

Enemion biternatum Rafinesque
E-né-mi-on. bi-ter-nà-tum
Ranunculaceae (Buttercup
Family)

SYNONYM: *Isopyrum biternatum* (Rafinesque) Torrey & Gray—RAB, PR

DESCRIPTION: Herbaceous perennial that mostly withers and dis-
appears by mid-May, sometimes with a few leaves evident until frost;
4–16″ tall; basal leaves on long stalks, 2–3x ternately compound, with
3-lobed leaflets; stem leaves smaller and with only 3 leaflets; flowers
less than 0.75″ in diameter, with 5 white, petal-like sepals; flowers
solitary, terminal, and in few-flowered clusters in the axils of upper
leaves; flowers March–April; fruit a cluster of 4 podlike follicles.
RANGE-HABITAT: Ontario, NY, and MN, south to AR and FL; in SC
known only from the lower Piedmont; rare; basic-mesic forests.
SIMILAR SPECIES: False Rue-anemone is similar in appearance to
Windflower (*Thalictrum thalictroides,* plate 103), which has a terminal umbel of white flowers
and a fruit that is a cluster of achenes. Vegetatively, these two look-alikes can be distinguished
by the single tier of leaves in Windflower versus the multitiered leaves of False Rue-anemone.
CONSERVATION STATUS: SC-Critically Imperiled

278. Lanceleaf Anemone

Anemone lancifolia Pursh
A-ne-mò-ne lan-ci-fò-li-a
Ranunculaceae (Buttercup Family)

DESCRIPTION: Herbaceous perennial, typically
less than 8″ tall; with stem and basal leaves;
stem leaves in a single whorl; leaves dissected
into 3 segments, the outer segments not deeply
cut; leaflets broadest at or above the middle and
serrate throughout; flowers single on erect stalks;
petals absent; sepals white; flowers March–May.
RANGE-HABITAT: From s. PA south to GA and west
to AL; in SC, common throughout the Piedmont, occasional in the mountains and rare in the Coastal Plain; basic-mesic forests, cove forests, and
hardwood bottoms.
SIMILAR SPECIES: Lanceleaf Anemone is similar to Wood Anemone (*A. quinquefolia* L.). Lanceleaf Anemone has leaves with 3 segments, the lateral segments not
deeply cut; *A. quinquefolia* has leaves dissected into 5 segments, or 3 segments and
the lateral segments deeply cut. The leaflets of Wood Anemone are serrate only above
the middle and usually widest above the middle. In SC Wood Anemone is restricted to the
mountains.

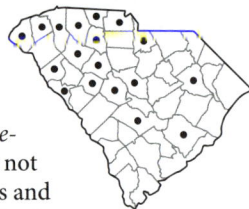

279. Eastern Slender Toothwort (A)

Cardamine angustata O. E. Schulz
Car-dá-mi-ne an-gus-tà-ta

Cutleaf Toothwort (B)

Cardamine concatenata (Michaux)
O. Schwarz
Car-dá-mi-ne con-ca-te-nà-ta
Brassicaceae (Mustard Family)

DESCRIPTION: *C. angustata:* perennial herb with segmented, white rhizome; rhizome segments 0.75–1.5″ long; stems 8–16″ tall; basal leaves usually present, often much wider than stem leaves; stem leaves usually 2, subopposite, palmately divided with lateral leaflets not deeply incised; *C. concatenate:* similar, but has basal leaves of sterile plants, narrow and similar in size to the stem leaves of fertile plants; lateral leaflets deeply incised, thus appearing to have 5 leaflets; flowers March–April.

RANGE-HABITAT: *C. angustata:* NJ and IN south to GA, TN, and MS; in SC, rare in rich cove forests, basic-mesic forests, and alluvial forests in the Piedmont, rare in the mountains. *C. concatenata:* ME south to FL and west to MN and TX; in SC, common in the mountains and rare in the Piedmont and Coastal Plain; rich cove forests, basic-mesic forests, and alluvial forests.

COMMENTS: Eastern Slender Toothwort and Cutleaf Toothwort are both easily seen in the basic-mesic forests at Flat Creek Heritage Preserve in Lancaster County and at Stevens Creek Heritage Preserve in McCormick County. The toothworts are the host for the West Virginia White Butterfly (*Pieris virginiensis*), which has been found feeding on this species in the Eastatoe Valley of Pickens County.

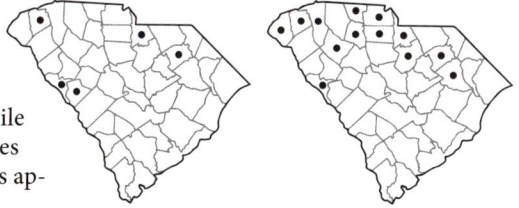

280. Early Saxifrage

Micranthes virginiensis
(Michaux) Small
Mic-rán-thes vir-gi-ni-én-sis
Saxifragaceae (Saxifrage Family)

SYNONYM: *Saxifraga virginiensis*
Michaux—RAB, PR

DESCRIPTION: Perennial herb with basal leaves only; plant 4–20″ tall; leaves scalloped or shallowly toothed; hairy flower stalk has branched clusters of fragrant, white flowers; flowers March–May.

RANGE-HABITAT: New Brunswick west to Manitoba, south to GA, LA, and AR; in SC, common in the mountains, infrequent in the Piedmont, rare in the Coastal Plain; basic-mesic forests, rock outcrops, rich cove forests, and moist alluvial forests and slopes.

SIMILAR SPECIES: Carey's Saxifrage (*Micranthes careyana* (Gray) Small) is similar but has coarse teeth along the margins of the leaves and is known only from Greenville County, very close to the NC state line.

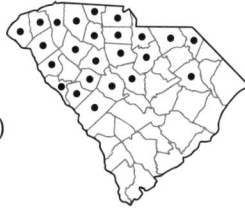

281. Yellow Fumewort

Corydalis flavula (Rafinesque)
A. P. de Candolle
Co-rý-da-lis flà-vu-la
Fumariaceae (Fumitory Family)

DESCRIPTION: Annual herb with a branching stem, erect or reclined to 15″ long; plants usually dormant by May; leaves green to bluish-green, highly divided with linear terminal segments; flowers with a fused yellow corolla, a flared opening, a nectar spur, and appearing as half a heart in shape; fruit is a cylindrical capsule; flowers March–April.

RANGE-HABITAT: NY and Ontario west to SD and south to OK, LA, and FL; in SC, occasional throughout growing in basic-mesic forests, rich floodplain forests, and other nutrient rich forests.

282. Creeping Phlox

Phlox stolonifera Sims
Phlóx sto-lo-ní-fe-ra
Polemoniaceae (Phlox Family)

DESCRIPTION: Creeping perennial herb, forming colonies; flowering stems 4–15″ tall; leaves opposite; creeping stem leaves evergreen, obovate, and larger than the lanceolate to elliptic flowering stem leaves; inflorescence an open, few-flowered cyme; corolla salverform (with a long tube and a whorl of lobes at the tip); petals lavender but ranging from pink to white; flowers March–April.

RANGE-HABITAT: PA and OH south to GA; in SC, uncommon in the mountains and Piedmont where found in basic-mesic forests, bases of slopes along streams, and rivers and small stream floodplains.

COMMENTS: Creeping Phlox is one of the most attractive and useful spring wildflowers for the shaded woodland garden. The dense, low cover of stems provides a perfect living mulch that prevents the growth of weeds while allowing larger seasonal interest plants to grow unobstructed. This species is uncommon in SC but readily available from native plant nurseries. Good populations can be seen at Flat Creek Heritage Preserve in Lancaster County and at the George H. Aull Natural Area in the Clemson Experimental Forest. Plants from SC are slightly smaller than populations from farther north and have consistently pinkish to lavender-colored flowers. Bright pink and white selections, from populations found farther north, are available.

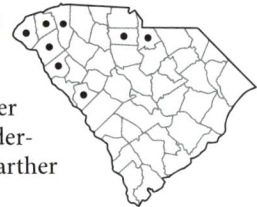

283. Shooting Star

Primula meadia (L.)
   A. R. Mast & Reveal
   Prím-u-la mèad-i-a
Primulaceae (Primrose Family)

SYNONYM: *Dodecatheon meadia*
   L.—RAB, PR

DESCRIPTION: Smooth, perennial herb; leaves basal; flowering stalk leafless, terminated by a cluster of nodding flowers; flowers white or pink, the reflexed sepals and petals and erect stamens give the flower a very distinctive appearance; flowers late March–May.

RANGE-HABITAT: WI and PA south to GA; in SC, uncommon and restricted to the high-calcium soils of basic-mesic forests in the Piedmont, in similar forests in the inner Coastal Plain, and historically in Berkeley County.

COMMENTS: Shooting stars have traditionally been placed in the genus *Dodecatheon,* which in Greek means "twelve gods," which apparently are not easily seen by anyone other than Linnaeus, the person who described the species. Recent molecular work has indicated that they belong in the genus *Primula.* The specific epithet honors Dr. Richard Mead (1673–1754), an English physician and patron of Mark Catesby. The flowers, with their unique reflexed sepals and petals and exerted stamens, are adapted for pollination by sonication, often referred to as "buzz pollination." In this type of pollination, a bumblebee lands on the stamens and buzzes to cause the pollen to fall from the stamens. A similar morphology for "buzz pollinated" flowers can be seen in completely unrelated species such as cranberries and even tomatoes.

CONSERVATION STATUS: SC-Imperiled

284. Eastern Baby Blue-eyes

Nemophila aphylla (L) Brummitt
   Ne-mó-phi-la a-phýl-la
Hydrophyllaceae (Waterleaf Family)

SYNONYM: *Nemophila microcalyx* (Nuttall) Fischer & C. A. Meyer—RAB

DESCRIPTION: Small, annual herb with weak stems lying flat along the forest floor; leaves deeply lobed, resembling a *Phacelia;* flowers tiny, produced singly at nodes opposite the leaves; petals white; flowers March–April.

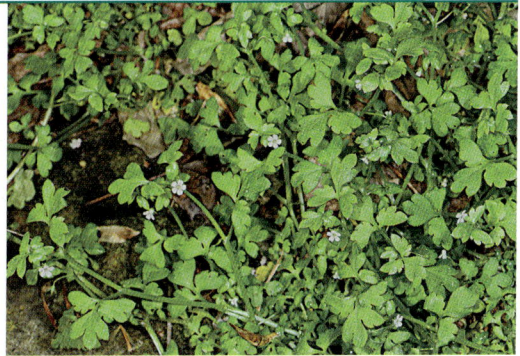

RANGE-HABITAT: MD west to AR and south to the FL Panhandle and TX; in SC, common in basic-mesic forests, floodplains, and other nutrient rich forests in the Piedmont and Coastal Plain.

COMMENTS: This species can be extremely abundant in basic-mesic forests and floodplains. It closely resembles the much showier *Phacelia* but produces tiny flowers.

285. Relict Trillium

Trillium reliquum Freeman
Tríl-li-um re-lí-qu-um
Trilliaceae (Trillium Family)

DESCRIPTION: Low perennial herb; stem 2–8″ long, curved, often positioning the single whorl of leaves just above the ground; leaves stalkless, elliptic, and heavily mottled with blue-green, with a silvery streak down the middle; flowers with a foul odor similar to rotting meat or manure; flowers March–April.

RANGE-HABITAT: Known only from widely separated populations in GA and SC; rare but locally abundant in a narrow band along the Savannah River at the fall line; calcium-rich soils in basic-mesic forests of ravines and adjacent bottomlands.

COMMENTS: The curved stem, silver streak down the middle, and anthers prolonged into a distinct beak are characteristics found in no other trillium in SC. The foul odor of the flowers is typical of many plants that depend on flies for pollination. The only protected populations in SC are threatened by increased herbivory by deer.

CONSERVATION STATUS: Federally Endangered

286. Lanceleaf Trillium

Trillium lancifolium Rafinesque
Tríl-li-um lan-ci-fò-li-um
Trilliaceae (Trillium Family)

SYNONYM: *Trillium lanceolatum* Boykin—RAB

DESCRIPTION: Low perennial herb from a short horizontal rootstalk; leaves in 1 whorl of 3, sessile, very narrow and mottled silver and green; the single sessile flower is terminal and erect, with green, reflexed sepals and maroon to brownish lanceolate petals that point straight up and are often slightly twisted; flowers March–April.

RANGE-HABITAT: Highly scattered locations from TN and SC south to FL Panhandle and west to AL; in SC, very rare and known from only a few populations in the lower Piedmont within the drainage of the Savannah River; basic-mesic forests or bottomland hardwoods on calcium-rich soils.

COMMENTS: This plant has a unique symmetry; the leaves are parallel to the ground, the sepals point straight downward, and the petals point straight up.

CONSERVATION STATUS: SC-Critically Imperiled

287. Wateree River Trillium

Trillium oostingii Gaddy
    Tríl-li-um oos-tíng-i-i
    Trilliaceae (Trillium Family)

DESCRIPTION: Low perennial herb from a short horizontal rootstalk; leaves in 1 whorl of 3, sessile, ovate, drooping and lightly mottled leaves; the single sessile flower is terminal and erect, with green, drooping sepals and clawed (narrow at the base) oblanceolate petals; petals green or yellowish green toward the tips and maroon, brownish, or green at the narrow clawed base; flowers March–April.

RANGE-HABITAT: Endemic to SC; found only in Kershaw and Richland Counties; rare; basic-mesic forests or bottomland hardwoods on calcium-rich soils.

COMMENTS: This distinctive *Trillium* is one of a handful of species currently thought to be endemic to SC. It was first collected by Duke University botanist, Henry J. Oosting (1903–68), in 1937. The species was not recognized as distinct from the related Lanceleaf Trillium until officially described by SC botanist L. L. Gaddy in 2008. The publication of this species led to a mass collection of this rare plant from the few known locations by Trillium enthusiasts and the largest threat to its persistence is illegal collecting activities. It should never be disturbed in the wild.

CONSERVATION STATUS: SC-Critically Imperiled

288. Southern Nodding Trillium

Trillium rugelii Rendle
    Tríl-li-um ru-gél-i-i
    Trilliaceae (Trillium Family)

DESCRIPTION: Perennial herb from a thick rhizome; stem to 18″ tall; leaves in a single whorl of three, uniformly bright green, broader than long; flower stalk strongly recurved, positioning the flower beneath the leaves; petals broadly ovate, thick in texture, white or maroon, and generally reflexed, with the margins straight; stamens at most 1.5x longer than the pistil; filaments shorter than the ovary; anther sacs purple; flowers April–early May.

RANGE-HABITAT: TN, NC, SC, GA, and AL; in SC, occasional in the mountains and Piedmont; rich cove forests and basic-mesic forests.

COMMENTS: Southern Nodding Trillium is included under *T. cernuum* L. in Radford et al. (1968). *T. cernuum* is now considered to be restricted to the northeastern United States, and *T. rugelii* is restricted to the Southeast. The specific epithet honors Ferdinand Rugel (1806–1878). This species can be observed at Flat Creek Heritage Preserve in Lancaster County and at Stevens Creek Heritage Preserve in McCormick County. Some populations in the Blue Ridge Escarpment have intensely beautiful scents resembling a sweet perfume; others may be fetid.

CONSERVATION STATUS: SC-Imperiled

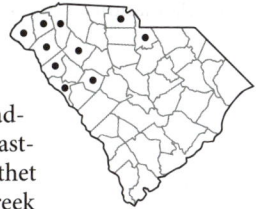

289. Reflexed Wild Ginger

Asarum reflexum E. P. Bicknell
Á-sa-rum re-fléx-um
Aristolochiaceae (Birthwort
Family)

SYNONYM: *Asarum canadense* L.—
RAB, PR, in part (see discussion
below)

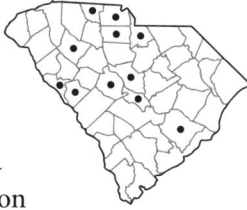

DESCRIPTION: Rhizome-producing, perennial herb, growing at
ground level; leaves opposite, one pair on a stem; flower solitary,
arising between the 2 leaf stalks; flowers often inconspicuous, hidden by the forest litter; corolla absent; calyx brownish or reddish
with reflexed lobes that are acute or acuminate sometimes with a
short, tail-like tip to 0.15″ long; flowers March–April.
RANGE-HABITAT: CT to Manitoba and south to GA and OK; uncommon throughout SC; basic-mesic slopes along rivers and streams
and diabase dikes in the Piedmont, cove forests in the mountains,
and calcareous bluffs and wet marl flats in the Coastal Plain.
SIMILAR SPECIES: Until recently, all deciduous wild ginger in the eastern United States were
considered the same species, Canada Wild Ginger (*Asarum canadense* L.). Recent evidence
suggests that there are three species in the eastern United States. Canada Wild Ginger may be
distinguished by the longer tail-like tips to the calyx lobes, 0.15–0.78″ long, and by the longer
calyx lobes in general. The range of the two species is in need to of clarification in SC. It appears that Reflexed Wild Ginger may be the only representative of this group found in SC.
COMMENTS: Wild ginger is a plant of many uses. The rhizomes have a strong odor similar to
true ginger and can be used as a substitute. Native Americans made candy of the rhizomes
by boiling until tender, then dipping in syrup. Numerous sources suggest, however, that
consumption of excessive amounts of *Asarum* should be avoided because it may have cancer-
causing properties.

290. Roundleaf Ragwort

Packera obovata (Muhlenberg
ex Willdenow) W. A. Weber
& A. Love
Páck-er-a o-bo-và-ta
Asteraceae (Aster Family)

SYNONYM: *Senecio obovatus* Muhlenberg
ex Willdenow—RAB, PR

DESCRIPTION: Perennial with aboveground stolons; the basal cluster
of obovate-shaped leaves is distinctive to this species of *Packera* in
SC; flowers April–June.
RANGE-HABITAT: Widespread in the eastern US; uncommon in the
mountains and Piedmont of SC and restricted to calcium-rich or
magnesium-rich soils of basic-mesic forests, mafic oak-hickory
forests, calcareous variants of granitic domes and rich cove forests;
very rare in the Coastal Plain growing on limestone bluffs.
COMMENTS: An easy-to-recognize species whose occurrence sug-
gests the presence of other interesting species. This species makes an excellent groundcover
in the woodland garden, forming a living mulch that prevents the growth of weeds while
allowing larger plants to survive.

291. Cancer-root; Ghostpipe

Aphyllon uniflorum (L.)
  Torrey & Gray
  A-phýl-lon u-ni-flò-rum
  Orobanchaceae (Broomrape
  Family)

SYNONYM: *Orobanche uniflora*
  L.—RAB, PR

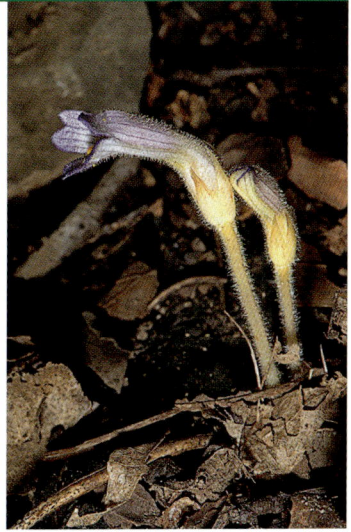

DESCRIPTION: Root parasite lacking chlorophyll; stems often clustered, mostly underground, 0.5–2″ long, with bractlike leaves; flowers white to violet, terminal and solitary on a flower stalk 2–5″ long; flowers April–May.

RANGE-HABITAT: Widespread in the US; in SC, uncommon throughout; basic-mesic forests, levee forests, cove forests, and streamsides.

COMMENTS: This species is a holoparasite that lacks chlorophyll. Host plants in SC may include goldenrods, asters, and sunflowers, among other species. Like many parasites, it may be present in an area one year and seemingly absent the next.

CONSERVATION STATUS: SC-Imperiled

292. Perfoliate Bellwort

Uvularia perfoliata L.
  U-vu-là-ri-a per-fo-li-à-ta
  Colchicaceae (Meadow-saffron
  Family)

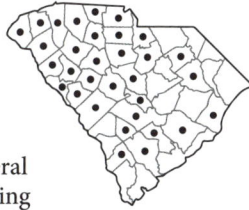

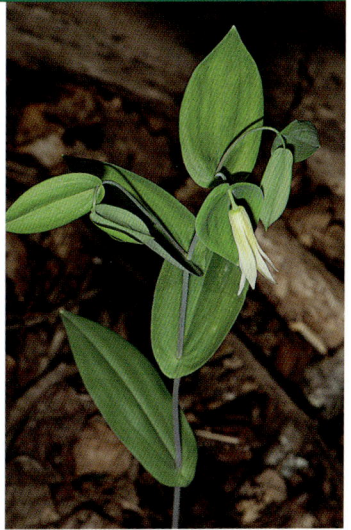

DESCRIPTION: Perennial herb; stems 1–several from a crown, with elongate stolons, forming colonies; usually with 1 branch; stems to 16″ tall; leaves perfoliate (appearing to be pierced by the stem), bluish-green; flower solitary, in upper leaf axil, appearing terminal, yellow and nodding, bell-like; flowers March–early May.

RANGE-HABITAT: Widespread in eastern North America from NH and Ontario west to IL and south to LA and the FL Panhandle; in SC, common in the mountains and Piedmont and rare in the Coastal Plain; basic-mesic forests, mafic oak-hickory forests, rich cove forests, beech forests, river bluff forests, and other rich deciduous forest systems.

COMMENTS: The leaf attachment is distinctive among our bellworts, appearing to be pierced by the stem. The specific epithet is derived from the term for this attachment, perfoliate.

293. Dwarf Stinging Nettle; Ortiguilla

Urtica chamaedryoides Pursh
  Ur-tì-ca cha-mae-dry-oì-des
Urticaceae (Nettle Family)

DESCRIPTION: Herbaceous, rhizomatous annual with stems to 4′ tall, though generally smaller, erect or reclining; leaves and stem with urticaceous (stinging) trichomes; leaves opposite, marginal teeth are blunt; male and female flowers on the same plant, together in globular clusters; male flowers with 5 anthers, shedding pollen explosively; flowers April–May.

RANGE-HABITAT: WV west to OK and south to FL and TX, also in Mexico; in SC, rare, in basic-mesic forests in the Piedmont and rich forests in floodplains in the Coastal Plain.

COMMENTS: This is a native relative of the Eurasian weed, Stinging Nettle (*Urtica dioica* L.). The stinging hairs found on this species are typical of many members of the family. Nettles have been used as an edible cooked green and the stinging hairs as a folk remedy for arthritis. The plant is very rare in SC and should not be picked.

CONSERVATION STATUS: SC-Imperiled

294. Meadow Parsnip (A)

Thaspium barbinode (Michaux) Nuttall
  Thás-pi-um bar-bi-nò-de

Common Golden Alexanders (B)

Zizia aurea (L.) Koch
  Zí-zi-a aù-re-a
Apiaceae (Carrot Family)

DESCRIPTION: Erect perennial herbs; stems 1–1.5′ tall, 1–several from the base; stems and petioles sparsely to moderately short pubescent in *T. barbinode,* and smooth in *Z. aurea.* basal and stem leaves 2-ternately or more compound with serrate margins, more deeply lacerate-serrate in *T. barbinode;* basal and lower stems leaves large, stem leaves gradually reduced upward; flowers creamy-yellow to golden-yellow, produced in compound umbels; each umbel consisting of 8–10 flowers in *T. barbinode;* more than 10 flowers in each umbel in *Z. aurea;* flowers April–May.

RANGE-HABITAT: Both species are widespread in the eastern US; in SC, both species are common in the mountains and Piedmont and uncommon in the Coastal Plain; rich cove forests, basic-mesic forests, wet, flat, calcareous forests, calcareous bluffs, and other rich, deciduous forests.

COMMENTS: Though these two yellow-flowered carrot relatives are in different genera, they are very frequently confused. They are vegetatively very similar. In addition to the characters listed above, the central flower in each umbel is sessile or nearly so in *Zizia* and pedicellate in *Thaspium.* Both species make attractive additions to the shaded woodland garden.

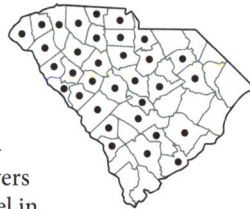

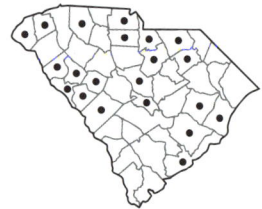

295. Carolina Larkspur

Delphinium carolinianum
   Walter ssp. *carolinianum*
   Del-phí-ni-um ca-ro-li-
   ni-à-num
Ranunculaceae (Buttercup Family)

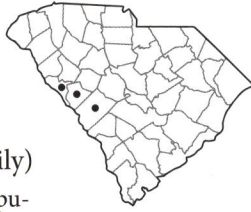

DESCRIPTION: Perennial herb; stems erect, pubescent, to 30″ tall; leaves pubescent, highly dissected, the lobes
linear; flowers with 4 blue or whitish petals, 2 with nectar spurs
projecting into the sepal spur; fruit an aggregate of follicles; flowers March–May.

RANGE-HABITAT: Primarily midwestern with disjunct populations
in SC, GA, and FL; in SC, rare in the lower Piedmont where
restricted to basic-mesic forests and glades on high calcium substrate.

COMMENTS: This is SC's only native larkspur. It is very rare in the
eastern United States but abundant in the tall grass and mixed
grass prairies of the Midwest. The species was described from
plants collected by John Fraser and given to SC botanist Thomas Walter (1740–89), likely
from near the Rocky Shoals of the Savannah River near North Augusta. It is ironic that the
common and specific epithet refers to a portion of the range where it is very rare.

CONSERVATION STATUS: SC-Critically Imperiled

296. Southern Hound's-tongue; Southern
Wild Comfrey

Andersonglossum virginianum (L.)
   J. I. Cohen
   An-der-son-glós-sum vir-gi-ni-à-
   num
Boraginaceae (Borage Family)

SYNONYM: *Cynoglossum virginianum*
   L.—RAB

DESCRIPTION: Harsh, perennial herb 12–30″ tall,
with long, bristly hairs throughout; basal and low
stem leaves elliptic-ovate, 4–8″ long; inflorescence a coiled arrangement that
elongates as flowers mature; petals blue, sometimes white; flowers April–May.

RANGE-HABITAT: CT south to FL and west to OK and LA; in SC, occasional
throughout but most abundant in the Piedmont; basic-mesic forests, rich
floodplains, and other nutrient-rich, deciduous forests.

COMMENTS: The large, harsh leaves of this plant make it conspicuous. The flowers,
small for the size of the plant, are a delicate and pleasant blue. Though this plant
frequently grows at the base of slopes and in small stream floodplains, it grows in
drier portions of the floodplain. The genus is named in honor of American botanist William
Russell Anderson (1942–2013).

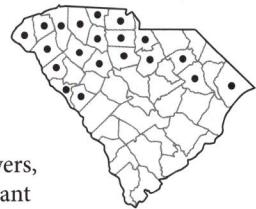

297. Green Violet

Cubelium concolor
(T. F. Forster) Rafinesque ex
Britton & A. Brown
Cu-bé-li-um cón-co-lor
Violaceae (Violet Family)

SYNONYM: *Hybanthus concolor* (Forster)
Sprengel—RAB, PR

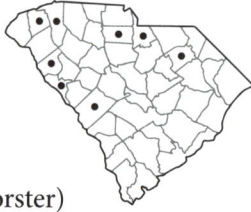

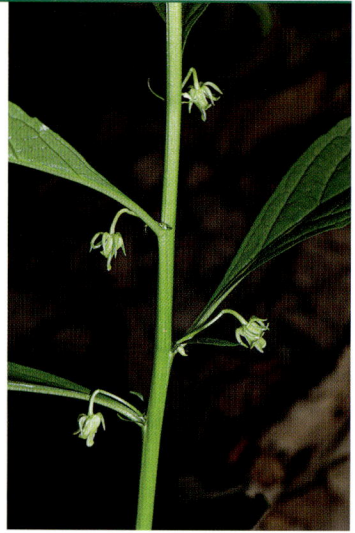

DESCRIPTION: Perennial herb; stems erect, 1.2–3′ tall, 1–several from the base; leaves alternate, widest toward the middle or just past the middle with long tapering tips and with entire or remotely toothed margins; flowers green, irregular, nodding; flowers that never open (cleistogamous) are tiny and solitary in the axils of the uppermost leaves; flowers that open are produced in clusters in the axils of lower leaves; flowers late May–June.

RANGE-HABITAT: Widespread in the eastern US; in SC, occasional in the mountains and Piedmont and very rare in the inner Coastal Plain; basic-mesic forests, rich cove forests, and other moist forests with circum-neutral soils.

COMMENTS: Although not appearing violetlike, it shares many technical features with the genus *Viola*, including the production of some flowers that open and are cross-pollinated (chasmogamous) and other flowers that never open and are self-pollinated (cleistogamous). This species is one of the best indicators of high pH soil and are visible throughout the growing season. If you find Green Violet during the summer, it indicates there may be many other interesting species at the site in the spring.

298. Virginia Marbleseed

Lithospermum virginianum L.
Li-tho-spér-mum vir-gi-ni-
à-num
Boraginaceae (Borage Family)

SYNONYM: *Onosmodium virginianum*
(L.) Alphonse de Candolle—RAB, PR

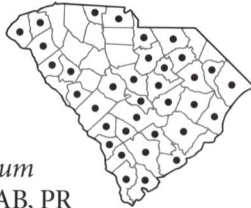

DESCRIPTION: Harsh, strigose, perennial herb 8–32″ tall; stems slender, branched above; corolla light yellow to orange; flowers in a coiled, terminal arrangement that straightens as the flowers mature; flowers April–September.

RANGE-HABITAT: LA to FL, north to NY and MA; occasional throughout SC; barrens, glades, or woodlands over calcareous or mafic (high Mg) rocks in the mountains, basic-mesic forests in the Piedmont, woodlands in the Sandhills, and shell deposits and sandy woodlands in the Coastal Plain.

COMMENTS: Weakley (2001) states that the unifying factor determining the distribution of Virginia Marbleseed may be open woodland conditions maintained by fire and that it generally occurs in small populations.

299. Woodland Pinkroot; Wormgrass

Spigelia marilandica (L.) L.
  Spi-gèl-i-a ma-ri-lán-di-ca
Loganiaceae (Strychnine Family)

DESCRIPTION: Erect perennial herb to 28″ tall; stems with 4–7 pairs of opposite, stalkless leaves; flower scarlet on outside, yellow-green on inside; flowers May–June.

RANGE-HABITAT: SC west to IN and OK, south to FL and TX; common and scattered throughout SC; in the mountains and Piedmont in rich cove forests, mafic oak-hickory forests, basic-mesic forests; in the Sandhills, Coastal Plain, and maritime strand in calcareous bluff forests; wet, flat, calcareous forests; shell hammock forests; and shell rings.

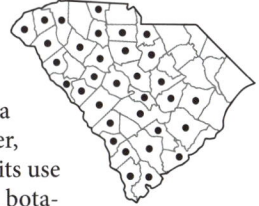

COMMENTS: In the early 1800s, the demand for Woodland Pinkroot for use as a vermifuge (for treating parasitic worms) almost lead to its extinction; however, severe side effects accompanied its use so that by the early twentieth century its use was discontinued, and the plant is again common. The genus honors Flemish botanist and anatomist Adrian Spiegel (1578–1625), who was perhaps the first to give directions for preparing an herbarium. The species makes an exceptional addition to the woodland garden and is pollinated by hummingbirds and butterflies.

300. Columbo

Frasera caroliniensis Walter
  Frà-ser-a ca-ro-li-ni-én-sis
Gentianaceae (Gentian Family)

SYNONYM: *Swertia caroliniensis*
  (Walter) Kuntze—RAB

DESCRIPTION: Robust determinate perennial herb, 3–8′ tall; leaves 4–12″ long, in whorls of 3–9; flowers form an ample inflorescence at the summit; each of the 4 greenish yellow corolla lobes bears a gland surrounded by a fringe of hairs; flowers late May–June.

RANGE-HABITAT: From NY and Ontario, west to IL, MI, MO, and OK, south to SC, GA, and LA; in SC, rare and known only from Fairfield, Newberry, Abbeville, Laurens, and Greenwood Counties; basic-mesic forests adjacent to streams or forests within the stream floodplain.

COMMENTS: Columbo was a drug erroneously believed to have come from Colombo in Ceylon, and *Frasera* provided a substitute. A root tea was used for a variety of folk remedies, including treatment for colic, cramps, dysentery, and diarrhea; also used as a general tonic. Thomas Walter honored his publisher, John Fraser (1750–1811), by naming this genus in his honor. The plant is considered a determinate perennial similar to century plants (*Agave*) that can grow for a number of years until they gather enough stored energy reserves to produce the massive inflorescence. Once the plant has flowered, it dies.

CONSERVATION STATUS: SC-Imperiled

301. American Ginseng; Sang

Panax quinquefolius L.
Pà-nax quin-que-fò-li-us
Araliaceae (Ginseng Family)

DESCRIPTION: Smooth, hairless perennial herb with a fusiform root and single stem, 8–24″ tall; leaves palmately compound with 3–5 leaflets; flowers in terminal, solitary umbels; flowers May–June; drupes mature August–October.

RANGE-HABITAT: Quebec west to MN and SD, south to e. VA, NC, SC, GA, c. AL, LA, and OK; in SC, rare and scattered in the mountains and Piedmont; basic-mesic forests, cove forests, and mesic deciduous forests.

COMMENTS: The Chinese highly prize ginseng root as an aphrodisiac and heart stimulant. In the US it is in demand as a tonic. With the US rediscovery of the healing power of plants, more pressure is being put on native populations of ginseng and other plants. Formally abundant and occurring in large populations, ginseng has been reduced in most of its range to small, scattered populations. Today's method of collection has not helped. Rural people in past days would remove the fruits from the plant and plant them, ensuring future populations. Much of the collection today is done in haste, often before the plants produce mature fruits, and the seeds are not planted. Today in SC, it is hard to find good populations. Duke (1997) recommends ginseng as an herbal remedy for loss of libido and memory loss associated with aging. Because of the enormous number of herbal products said to contain ginseng, there is simply not enough ginseng grown or harvested for all of them to contain significant quantities. Buyer beware.

CONSERVATION STATUS: SC-Vulnerable, because of exploitation

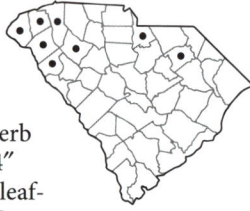

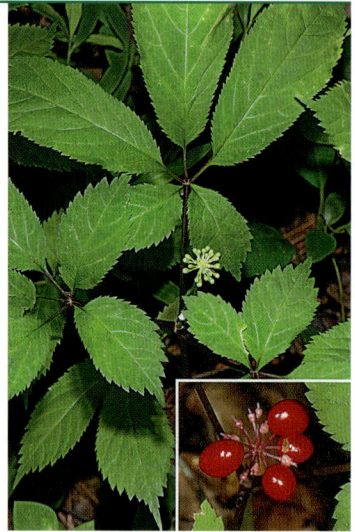

302. American Figwort (A)

Scrophularia lanceolata Pursh
Scro-phu-là-ri-a lan-ce-o-là-ta

Eastern Figwort (B)

Scrophularia marilandica L.
Scro-phu-là-ri-a ma-ri-lán-di-ca
Scrophulariaceae (Figwort Family)

DESCRIPTION: Erect, perennial herbs typically less than 4′ tall; stems square; branched with opposite leaves that are finely serrate in *S. marilandica* and very coarse in *S. lanceolata;* flowers produced in a panicle with small maroon petals, partly fused with the upper two projecting and liplike; American Figwort flowers May–June; Eastern Figwort flowers June–September.

RANGE-HABITAT: Wide-ranging in eastern North America with American Figwort extending south only to SC and Eastern Figwort found essentially throughout the eastern United States; in SC, Eastern Figwort found throughout the state but uncommon, in high-calcium and high-magnesium soils in forests that are dry to mesic; American Figwort is very rare and confined to basic-mesic forests and small stream floodplain forests in Lancaster County (Flat Creek) and perhaps historically in Berkeley County in the Coastal Plain.

COMMENTS: Our two figworts are very similar, but they do not overlap in flowering times. The yellowish staminodes and very coarsely serrate leaves in American Figwort vs. purplish staminodes and finely

serrate leaves in Eastern Figwort help to distinguish them. Though American Figwort was collected by H. W. Ravenel and others in SC, the plant went virtually unknown until the population at Flat Creek was made widely known by John Schmidt. Though the flowers are small, both species are visited by bees, flies, butterflies, and even hummingbirds, making them excellent pollinator plants.

CONSERVATION STATUS: *S. lanceolata:* SC-Critically Imperiled

303. Piedmont Aster

Eurybia mirabilis (Torrey & Gray)
G. L. Nesom
Eu-rỳ-bi-a mi-rá-bi-lis
Asteraceae (Aster Family)

SYNONYM: *Aster comixtus* (Nees) Kuntze—RAB

DESCRIPTION: Erect, perennial herb forming clumps; stems to 2–3′ tall; leaves basal and on the stem; 10–16 leaves per stem, the lower and basal leaves ovate and often cordate (heart-shaped) at the base, reduced upward to elliptic or elliptic to elliptic lanceolate and sessile leaves; flowers produced in a flat-topped arrangement of involucrate heads; ray flowers white, disk flowers yellow; heads subtended by squarrose bracts; flowers July–September.

RANGE-HABITAT: Endemic to NC and SC in the lower Piedmont; in SC, uncommon in basic-mesic forests and floodplain forests on rich soils.

COMMENTS: This distinctive "aster" is one of a handful of Piedmont forest endemics found only in the Carolinas. It is seldom recognized or included in other wildflower treatments. The majority of the range is found in the Piedmont of SC.

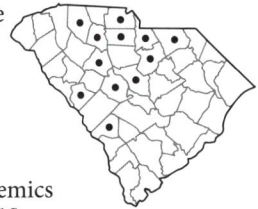

304. Eared Goldenrod

Solidago auriculata Shuttleworth ex
Blake
So-li-dà-go au-ri-cu-là-ta
Asteraceae (Aster Family)

SYNONYM: *Solidago notabilis* Mackenzie—RAB

DESCRIPTION: Erect, short-rhizomatous perennial forming dense clumps; leaves basal and on the stems; leaves ovate to elliptic, pubescent below and rough above; leaf margins serrate; leaves are distinctive due to the broadly winged petiole with auriculate-clasping bases (surrounding the stem like ear lobes); flowers produced in involucrate heads arranged in a narrow paniclelike formation; ray flowers yellow, disk flowers yellow; flowers September–October.

RANGE-HABITAT: SC and TN west to OK and south to the FL Panhandle and TX; in SC, restricted to basic-mesic forests of the Piedmont along the Savannah River drainage.

COMMENTS: This species is apparent during the flowering period at Stevens Creek Heritage Preserve. It is virtually unknown to most wildflower enthusiasts but is one of the easiest of the many confusing goldenrods to identify because of the clasping auriculate leaf bases. It makes an excellent and well-behaved specimen for the woodland garden bringing color to the shade during a time of year when not much is flowering.

CONSERVATION STATUS: SC-Vulnerable

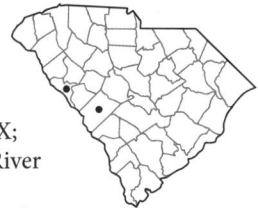

305. Sang-find; Rattlesnake Fern

Botrypus virginianus (L.)
   Michaux
      Bò-try-pus vir-gi-ni-à-nus
   Ophioglossaceae (Adder's-tongue
   Family)

SYNONYM: *Botrychium virginianum* (L.)
   Swartz—RAB, PR

DESCRIPTION: Small, deciduous fern with stems to 18″ tall; fronds triangular, held nearly horizontal, 3-pinnatifid, bright green, produced in the spring; fertile fronds erect, appearing to extend upward from the central base of the sterile frond, extending up to 8″ above the sterile frond, 2-pinnate with spores produced in stalked sporangia that look like tiny clusters of green grapes; spores produced April–May.

RANGE-HABITAT: Nearly throughout North and South America; in SC, common throughout in a variety of rich upland and bottomland forest types.

COMMENTS: Though similar to the other "grape ferns," this species is the only one in our area that produces leaves in the spring. The others (genus *Sceptridium*) produce their leaves in the autumn, and they become dormant in spring. The common name Sang-find is commonly used in Appalachia and is given because this species often shares habitat with the less common American Ginseng and was used to divine the best areas for hunting the rare plant.

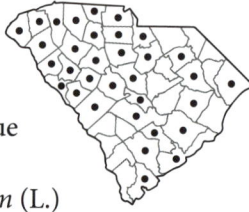

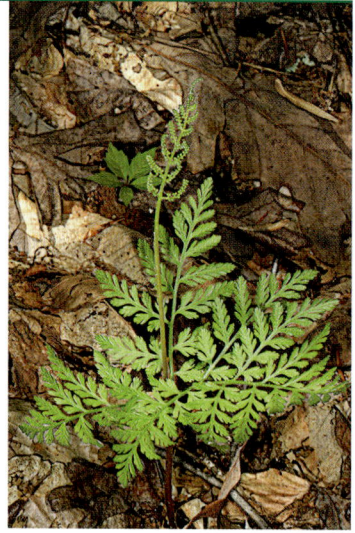

306. Southern Adder's-tongue

Ophioglossum pycnostichum
   (Fernald) A. & D. Löve
      O-phi-o-glós-sum pyc-nós-
      ti-chum
   Ophioglossaceae (Adder's-tongue
   Family)

SYNONYM: *Ophioglossum vulgatum* L. var. *pycnostichum*
   Fernald—RAB

DESCRIPTION: Strange terrestrial, spring ephemeral fern from rhizomes forming large colonies; leaves of sterile individuals a solitary, smooth, ovate blade; leaves of fertile plants with one ovate sterile leaf and a very narrow fertile leaf that is held on a long petiole and is spikelike; spores produced March–June.

RANGE-HABITAT: NJ and IL south to GA and TX; in SC, infrequent throughout where found in floodplain forests, rich cove forests, and basic-mesic forests and oak-hickory forests with circumneutral soils.

COMMENTS: The common and scientific names of this plant refer to the fact that fertile plants look like a snake's head with an extended tongue. This species is restricted to high-calcium or high-magnesium soils and are green during the late winter and early spring, generally withering away by late May. Several other species are found in the Carolinas, all in sunnier conditions such as graveyards and roadsides.

307. Lowland Bladder Fern; Southern Fragile Fern

Cystopteris protusa (Weatherby) Blasdell
    Cys-tóp-te-ris pro-trù-sa
Cystopteridaceae (Brittle Fern Family)

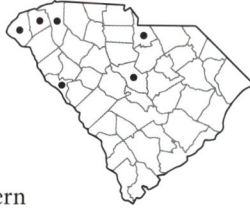

DESCRIPTION: Small, deciduous, creeping fern from rhizomes; fronds thin, to 8″ long, 2-pinnatifid or 2-pinnate with segments ovate to lance-ovate; bright spring green in coloration; spores produced in sori in clusters near the margin of the pinnae; spores produced April–June.

RANGE-HABITAT: Ontario south to FL and west to MN and TX; in SC, common in the mountains and Piedmont in basic-mesic forests, rich cove forests, and shaded outcrops, always on nutrient-rich, high-calcium or high-magnesium soils; rare in moist forests on high-pH soils in the Coastal Plain.

COMMENTS: This fern is characteristic of rich forests of upstate SC. It may be distinguished from the similar genus *Woodsia* by its habitat, being restricted to moist forests, forming large clumps, and the absence of the persistent stems of the dead fronds, which are dark in *Woodsia* and pale in *Cystopteris*.

CONSERVATION STATUS: SC-Vulnerable

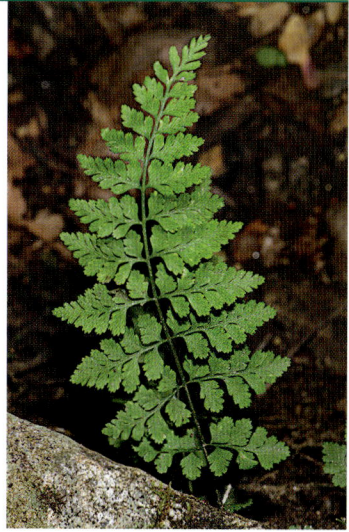

308. Southern Lady Fern

Athyrium asplenioides
    (Michaux) A. A. Eaton
    A-thý-ri-um as-ple-ni-oì-des
Athyriaceae (Lady Fern Family)

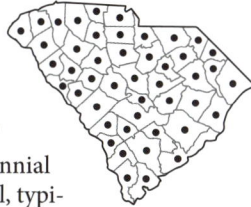

DESCRIPTION: Large, erect, deciduous, perennial fern from a short rhizome; fronds to 3′ tall, typically smaller, with long reddish or greenish petioles about the same length as the blade; blades lanceolate to ovate, 2-pinnate, serrate; spores produced in sori that are arranged in lines or curved lines on the undersides of pinnae; spores produced May–September.

RANGE-HABITAT: MA and IL south to FL and TX; in SC, abundant in moist forests such as floodplain forests, bottomland hardwoods, basic-mesic forest, cove forests, and a number of other moist, shaded habitats throughout the state.

COMMENTS: Though not all plants display red petioles, a lacy, large fern with reddish petioles is likely to be a Southern Lacy Fern. This extremely abundant species can be distinguished from similar wood ferns (*Dryopteris spp.*) by its linear sori vs. sori produced in rounded clusters.

309. Marginal Woodfern

Dryopteris marginalis (L.) Gray
   Dry-óp-te-ris mar-gi-nà-lis
   Dryopteridaceae (Woodfern
   Family)

DESCRIPTION: Large, evergreen, fern from ascending rhizomes, often solitary plants; fronds evergreen, to 30″ long; petioles with dense, cinnamon-brown scales toward the base; blades 2-pinnatifid, leathery, thick, dark green, oblong-ovate; spores produced in sori that are arranged in rounded clusters near the margin of the undersides of the pinnae.

RANGE-HABITAT: Ontario and MI south to GA and TX; in SC, common in the mountains and Piedmont, where it is found in basic-mesic forests, cove forests, oak-hickory forests, and other forested systems, often in shallow, nutrient-rich soils of rocky slopes.

COMMENTS: This is the only large evergreen species with 2-pinnatifid leaves growing in dry situations in SC. Log Fern (*Dryopteris celsa* (W. Palmer) Knowlton, plate 812A) and Southern Woodfern (*Dryopteris ludoviciana* (Kunze) Small, plate 812B) are found in much moister surroundings, such as floodplains and swamp forests.

310. Beech forests

311. American Beech

Fagus grandifolia Ehrhart var. *caroliniana* (Loudon) Fernald & Rehder
   Fà-gus gran-di-fò-li-a var. ca-ro-li-ni-à-na
   Fagaceae (Beech Family)

DESCRIPTION: Large tree with smooth, gray bark; leaves deciduous, but often remaining throughout the winter (marcescent), the leaf margins denticulate; terminal buds long, sharp-pointed, slender, about 1″ long before just breaking open; flowers in March–April; male and female flowers in separate clusters on the same tree; large fruit production occurs every 2–3 years; fruits mature September–October.

RANGE-HABITAT: Common forest tree of lower elevations (mostly below 3,500′) over most of eastern North America; throughout SC; grows best in rich, moist, loam soils with a high humus content.

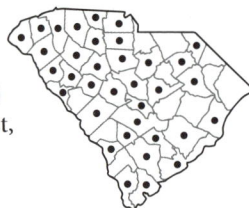

COMMENTS: Beech is a beautiful, long-lived tree and is ideal for cultivation. The oily seeds are an important wildlife food. Beech-nut oil has never been important in America, but in Europe beech oil (from *Fagus sylvatica* L.) is an important commercial product. The nut is one of the sweetest in North America. At one time, New England country groceries sold it. The wood has never been an important commercial product. It did gain short-lived fame when it was made into clothespins because of its elastic nature. Beechwood creosote is used in Japan as an effective anti-diarrheal that can act in as little as 15 minutes. Beech wood contains an abundance of phenol guaiacol, which gives the smoky odor/taste to most whiskeys (Dauncey and Howes, 2020).

A second variety, var. *grandifolia,* is found at higher elevations in NC and points to the north. It has serrate rather than denticulate margins on the leaves and less smooth bark.

312. Common Silverbell

Halesia tetraptera Ellis
Hàles-i-a te-tráp-te-ra
Styracaceae (Storax Family)

DESCRIPTION: Deciduous shrub or tree to 50′ tall; young stems with a distinctive pattern of dark and light vertical stripes; flowers white, bell-shaped, 0.6–0.8″ long, on long stalks and drooping in clusters from the twigs of the previous year; style included or slightly exerted; fruit broadly 4-winged and 1.2–2″ long, broadest near the middle; flowers March–May; fruits mature August–September.

RANGE-HABITAT: From w. VA, WV, s. OH, and s. IL, south to FL and e. TX; throughout SC (but absent in the maritime strand); common in the mountains and upper Piedmont, occasional in the lower Piedmont and Coastal Plain; moist slopes, creek banks, river bottoms, beech woods, and cove forests.

TAXONOMY: There has been considerable confusion over the application of the name *H. tetraptera*. RAB misapplied the name *H. carolina* L. to the Common Silverbell. They applied the name *Halesia parviflora* Michaux to the plant that should be referred to as *H. carolina*. Carolina Silverbell (*H. carolina*), is rare in the Piedmont and has corollas 0.3–0.5″ long, styles strongly protruding, and fruits that are broadest toward the tip, and narrowly winged.

COMMENTS: Silverbells are excellent native small trees for planting in landscapes, either in shade or full sun. The snow-white flowers are spectacular in the spring. They can either be planted in groups or as single plants. Veneer from its wood, when the log is turned against the rotary, has a beautiful bird's-eye figure. The genus honors the English clergyman Stephen Hales (1677–1761), the author of *Vegetable Staticks*.

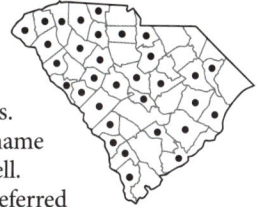

313. Painted Buckeye

Aesculus sylvatica Bartram
Aès-cu-lus syl-vá-ti-ca
Hippocastanaceae (Buckeye Family)

DESCRIPTION: Shrub or small tree, seldom over 20–25′ tall, typically much smaller; leaves deciduous, palmately compound with 5 leaflets; flowers yellow-green to cream, tinged with red, or salmon pink in some upper Piedmont populations; stamens included or barely longer than the petals; flowers April–mid-May; capsule leathery, with 1–3 large, dark brown, shiny seeds; fruits mature July–August.

RANGE-HABITAT: Southeastern US, from sc. VA south through c. NC, c. SC, and nc. GA to nc. AL; common in the low mountains and Piedmont, uncommon in the inner Coastal Plain; found in cove forests, basic-mesic forests, alluvial swamp forests, and riverbanks.

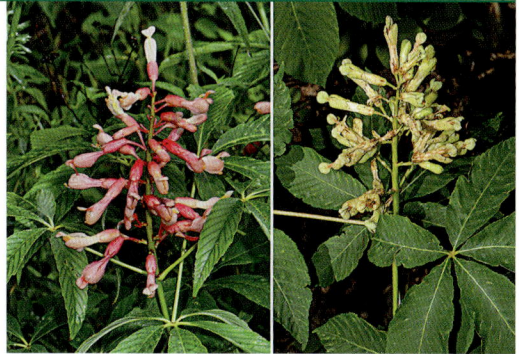

COMMENTS: Painted Buckeye is an excellent ornamental in shady, mesic forest sites along slopes and bluffs. The flowers are pollinated by butterflies and hummingbirds. The crushed fruits were once used to stun fish for easy harvest. Like all buckeyes, the fruits are toxic to humans. Populations in the upper Piedmont of SC have flowers that are a very attractive salmon color. These plants apparently have a few genes that have been transported by pollen from Red Buckeye (*Aesculus pavia* L., plate 742) creating hybrid plants that are genetically mostly Painted Buckeye while maintaining a tinge of color through genes incorporated from Red Buckeye (Thomas et al., 2008). A similar phenomenon can be seen in salmon-colored flowers of large Yellow Buckeyes (*Aesculus flava* Solander, plate 93) in the Eastatoe Valley in Pickens County.

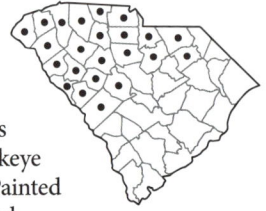

314. Hop Hornbeam

Ostrya virginiana (Miller) K. Koch
Ós-try-a vir-gi-ni-à-na
Betulaceae (Birch Family)

DESCRIPTION: Small, deciduous tree with brown, shredding bark; occasionally reaching 35–50′ tall; male and female catkins produced in April–May; mature female catkin consists of nutlets, each enclosed by membranous inflated sacs, arranged in a spike; fruits mature August–October.

RANGE-HABITAT: Widespread in s. Canada; e., mw, and se. US; and south to Honduras; in SC, common in the Piedmont and mountains in mesic to dry forests, especially on calcium-rich soils; rare in the Coastal Plain on limestone bluffs and wet, marl forests.

COMMENTS: Hop hornbeam is usually an understory species and not large enough to be of commercial importance. The wood is extremely hard and used to make tool handles. The common name comes from the resemblance of the catkins to commercial hops. "Witches brooms," broomlike clusters of short branches, are common on this species.

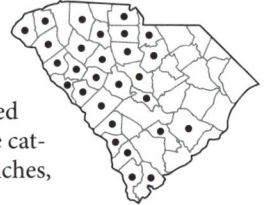

315. Umbrella Magnolia

Magnolia tripetala (L.) L.
Mag-nòl-i-a tri-pé-ta-la
Magnoliaceae (Magnolia Family)

DESCRIPTION: Small deciduous tree to 50′ tall, typically much smaller with smooth gray bark; buds smooth; leaves elliptic-oblong to oblanceolate in whorl-like clusters to 30″ long with a tapered cuneate base; flowers with fetid odor similar to that of a Wake Robin; tepals white with the outer whorl reflexed; flowers April–May; fruit an aggregate of follicles maturing July–October.

RANGE-HABITAT: PA and IN south to Panhandle FL and MS and disjunct in AR and OK; in SC, infrequent in the Piedmont, rare in the Coastal Plain; beech forests, basic-mesic forests, ravines, bluff forests, and bottomlands and oak-hickory forests where soils are rich and moist.

COMMENTS: Umbrella Magnolia is unmistakable; it is distinguished from all other deciduous *Magnolia* with such massive leaves in SC by the lack of lobes at the leaf base. The plant makes a stunning landscape tree in light shade. The flowers of most *Magnolia* have a pleasant odor, while this species is fetid, emitting large amounts of methyl benzoate (Azuma et al., 2005). Most *Magnolia* are pollinated by beetles, and the odors are important for their attraction. The flowers of *Magnolia* have even been shown to produce heat, which encourages pollinator visitation.

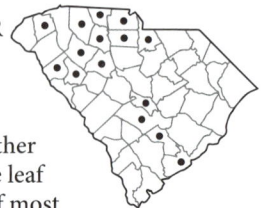

316. American Hazelnut

Corylus americana Walter
Có-ry-lus a-me-ri-cà-na
Betulaceae (Birch Family)

DESCRIPTION: Shrub 4–5′ tall, forming small colonies; leaves broadly ovate, to 5″ long, margins finely doubly serrate; leaf stalk glandular-hairy; flowers produced in catkins, February–March; fruit a nut enclosed in bracts, widest at apex and lobed into teethlike projections; nut matures September–October.

RANGE-HABITAT: Widespread in eastern North America; in SC, rare in the Coastal Plain and lower Piedmont, common in the upper Piedmont and mountains; most common in beech forests and other mesic woods, but also found in rocky woodlands; calcareous bluff forests; and wet, flat, calcareous woodlands and thickets.

COMMENTS: The nuts of this species are every bit as tasty as the commercial hazelnut. Native Americans used the twig hairs to expel intestinal worms. Heavy deer browse decimates this species in many landscapes.

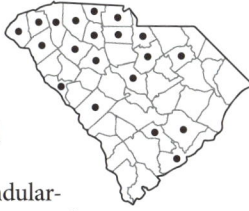

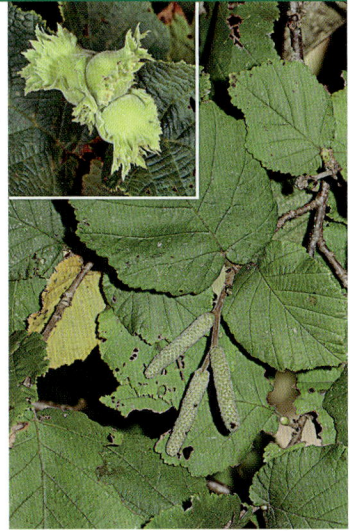

317. Strawberry-bush; Heart's-a-bustin'

Euonymus americanus L.
Eu-ó-ny-mus a-me-ri-cà-nus
Celastraceae (Bittersweet Family)

DESCRIPTION: Straggling or erect deciduous shrub to 6′ tall; stems distinctively dark green; leaves opposite and bright green, on short leaf stalks; flowers greenish, on long stalks from axils of the uppermost leaves; the fruits are distinctive with their warty, red exterior that splits to expose the seeds; flowers May–June; fruits September–October.

RANGE-HABITAT: NY to MO, south to FL and TX; common throughout SC in a variety of upland and lowland forests, including beech forests, oak-hickory forests, cove forests, hardwood bottoms, and levee forests.

COMMENTS: The seeds are a strong laxative, and the fruits are toxic, causing coma if consumed in sufficient quantity. The plant is often cultivated by native plant enthusiasts for the curious fruits. In areas with an overpopulation of deer, this species has been decimated by deer browsing.

318. Silky Camellia; Virginia Stewartia

Stewartia malacodendron L.
Ste-wàrt-i-a ma-la-co-dén-dron
Theaceae (Tea Family)

DESCRIPTION: Deciduous shrub or small tree, seldom over 20′ tall; bark smooth, gray and flaking into strips; leaves thin, alternate, elliptic to elliptic-oblong, with fine hairs along the serrulate margin; flowers large and showy, 2–4.75″ wide; stamen filaments purple; styles united with 5 stigma lobes; flowers May–June.

RANGE-HABITAT: From VA south to FL and west to se. TX; in SC, occasional in the Coastal Plain; rare in the Piedmont and mountains; most often in hummocks in swamps or on beech-dominated bluffs, along streambanks in the Piedmont and mountains.

SIMILAR SPECIES: Mountain Camellia (*S. ovata* (Cavanilles) Weatherby, plate 154) is similar but flowers later in the season (June–July), has five separate styles, larger leaves and petioles that enclose the terminal, and lateral buds at the base. It is uncommon in the mountains. Both species only occur together in Oconee County.

COMMENTS: The flower of Silky Camellia resembles the cultivated camellias and is among the most beautiful of SC's flowering trees. The genus honors John Stuart (1713–92), or as he was often referred to in formal writing, Stewart, third Earl of Bute. Trees are extremely slow-growing and difficult to cultivate but are prized by native plant enthusiasts, who are willing to patiently wait and pamper their finicky specimens.

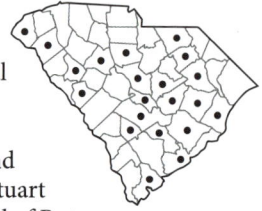

319. Round-lobed Liverleaf

Anemone americana (A. P. de Candolle) H. Hara
A-ne-mò-ne a-me-ri-cà-na
Ranunculaceae (Buttercup Family)

SYNONYM: *Hepatica americana* (A. P. de Candolle) Ker-Gawler—RAB, PR

DESCRIPTION: Herbaceous perennial from a short rhizome; leaves basal, purplish beneath, 3-lobed, the lobes rounded; petals absent; sepals bluish, rarely white or pink; flowers February–April.

RANGE-HABITAT: Widespread in eastern North America; in SC, common in the Piedmont, rare in the mountains and Coastal Plain; moist, rich, deciduous forests.

COMMENTS: Old leaves are present at early flowering; new leaves develop well after flowering. The liverleafs were formerly placed in the genus *Hepatica,* which refers to the similarity between the leaves that turn brownish in winter to the human liver (hepatic organ). A "liver tonic" boom resulted in the consumption of 450,000 pounds of the dried leaves in 1883 alone. No medical evidence exists that indicates liverleaf has any medical benefit in treating liver diseases. Liverleaf often grows at the base of trees, prompting someone to call this plant "squirrel-cups."

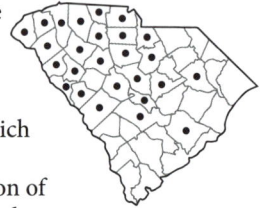

320. Bloodroot

Sanguinaria canadensis L.
San-gui-ná-ri-a ca-na-dén-sis
Papaveraceae (Poppy Family)

DESCRIPTION: Rhizomatous, perennial herb, 4–16″ tall, with no leafy stem; rhizome with blood-colored sap; flower single on a leafless stem, opening only on bright or sunny days, closing at night, and lasting only a short time; the single, lobed leaf continues to enlarge after the petals drop; flowers early March–April.

RANGE-HABITAT: Nova Scotia west to MN and Manitoba, south to FL and OK; in SC, common in the mountains and Piedmont; scattered in the Coastal Plain; in a variety of moist, nutrient-rich forests.

COMMENTS: Both the common name and the genus refer to the blood-colored sap. Native Americans used the red sap from the roots to dye baskets and clothing and to make war paint. The name "puccoon" was applied to this plant and it was used medicinally as an escharotic (causing tissue death) to remove warts, skin cancers, and other unwanted growths. The sap was also used as an insect repellent. Appalachian crafters today still use the red juice to dye baskets and cloth. When it became known that the Native American's use of the root was somewhat successful in treating tumors, interest in the plant increased. The plant should not be used without supervision because all parts can be highly toxic.

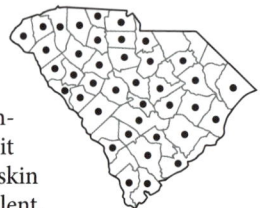

321. Little Brown Jugs; Little Pigs

Hexastylis arifolia (Michaux) Small
Hex-ás-ty-lis a-ri-fò-li-a
Aristolochiaceae (Birthwort Family)

DESCRIPTION: Rhizomatous, evergreen, perennial herb with no aboveground stem; leaves arrowhead-shaped, with mottling between the veins, aromatic, smelling like Sassafras or root beer, evergreen, often in clusters; flowers without petals; calyx with three lobes, fused basally, flask-shaped; flowers often hidden under leaf litter; flowers March–May.

RANGE-HABITAT: From se. VA west to sw. VA, south to n. FL and s. MS; common throughout SC in dry to moist deciduous forests.

COMMENTS: This is the common *Hexastylis* in SC and is the only *Hexastylis* with arrowhead-shaped leaves, though some may display less arrowhead shaped leaves than others. This is also the only *Hexastylis* with mottling between the veins rather than along the veins. Many rural folks refer to them as "little pigs," in reference to the flowers looking like a pig's head before the lobes open. Native Americans used the plant for spice and for candy but the presence of safrole in the leaves may indicate that it is carcinogenic if ingested regularly and in quantity.

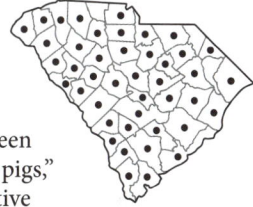

322. Pennywort

Obolaria virginica L.
O-bo-là-ri-a vir-gí-ni-ca
Gentianaceae (Gentian Family)

DESCRIPTION: Fleshy perennial, 3–6″ tall; roots brittle; leaves rounded, opposite; flowers usually in groups of three in the axils of purplish, bract-like leaves; flowers February–April.

RANGE-HABITAT: NJ to IL and south to FL and TX; in SC, common throughout the mountains and Piedmont; rare in the Coastal Plain; moist, nutrient-rich forests such as rich cove forests, basic-mesic forests, and beech forests.

COMMENTS: Pennywort is often overlooked because of its small size and early flowering period. *Obolaria* is a monotypic genus, meaning this is the only species in the genus. The genus name comes from "obolos," a Greek coin, similar to a penny, in reference to its round leaves.

323. Three-part Violet (A)

Viola tripartita Elliott
Vì-o-la tri-par-tì-ta

Southern Yellow Violet (B)

Viola tenuipes Pollard
Vì-o-la te-nù-i-pes
Violaceae (Violet Family)

DESCRIPTION: Erect, herbaceous perennials with pubescent stems (stems very short puberulent in *V. tenuipes*); 3–7 pubescent stem leaves on

flowering plants, divided into three segments in *V. tripartita,* simple and lance-triangular or rhombic-lanceolate, broadest toward the base, in *V. tenuipes;* solitary yellow flowers produced in the axils of the leaves; flowers of *V. tripartita* with dark nectar-guide lines extending more than two-thirds the length of the petal; flowers of *V. tenuipes* with dark nectar-guide lines extending approximately one-half the length of the petal; flowers March–May.

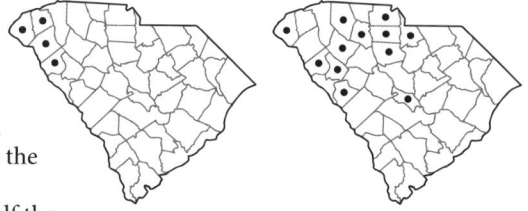

RANGE-HABITAT: *V. tripartita:* NC and TN south to SC and GA; in SC, locally common in the Piedmont and low mountains in mafic, dry, or moist oak forests; beech forests; and basic mesic forests and bottomlands; *V. tenuipes:* NC south to the FL Panhandle; in SC, very rare in basic-mesic forests and oak-hickory forests with circumneutral soils.

COMMENTS: Three-part Violet is unmistakable with its heavily divided leaves. The plant is far more common than once thought, though its range is restricted. Southern Yellow Violet, without divided leaves, was formerly considered a variety of *V. tripartita.* Neither is abundant at any one site; they are usually discovered by accident either as solitary plants or as a few, widely spaced plants.

324. May-apple; American Mandrake

Podophyllum peltatum L.
    Po-do-phý-llum pel-tà-tum
    Berberidaceae (Barberry Family)

DESCRIPTION: Smooth, rhizomatous, perennial herb 12–18″ tall; flower solitary, nodding, from the base of a pair of large, deeply lobed leaves; leaves deciduous; leaf stalk attached in the center on the underside of the leaf (peltate); flowers March–April; berry yellow or red when ripe in May–June.

RANGE-HABITAT: Widespread throughout most of eastern North America; common throughout SC in rich forests, bottomlands, meadows, pastures, and along moist road banks.

COMMENTS: The plant has long been known to contain anti-tumor alkaloids. Native Americans used extracts of the rhizomes as purgatives and for skin disorders and tumorous growths. May-apple contains the ligand podophyllotoxin and a resin called podophyllin, which are toxic to skin cells infected with specific wart-causing viruses and is used to treat ailments such as plantar's wort. The leaves, rhizomes, and seeds are poisonous if eaten in large quantities. Ripe fruits are edible but are said to have the flavor of earthworms and can be made into marmalade. Only two-leaf plants bear flowers.

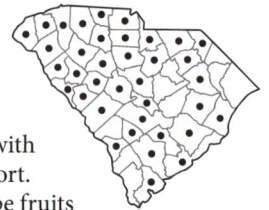

325. Piedmont Barren Strawberry

Waldsteinia lobata (Baldwin ex Elliott)
    Torrey & Gray
    Wald-steìn-i-a lo-bà-ta
    Rosaceae (Rose Family)

DESCRIPTION: Evergreen perennial herb forming strawberrylike clumps through short subsurface runners; leaves in basal rosettes, rounded, with a heart-shaped base and 3–5 shallow lobes, each with irregular teeth; flowers with petals 0.15–.25″ long; flowers March–May; fruits June–July.

RANGE-HABITAT: NC, SC, and GA; known in SC only from the upper Piedmont of Oconee County; stream terraces and adjacent slopes in beech forests and oak-hickory forests on magnesium-rich (mafic) soils.

SIMILAR SPECIES: Southern Barren Strawberry (*Waldsteinia doniana* Trattinick) is similar with more heavily lobed and incised leaves and flowers with petals generally less than 0.1″ long. Southern Barren Strawberry is found in similar habitats, primarily in the Piedmont.

COMMENTS: The leaves turn an attractive burgundy red in the fall and are retained until replaced by new leaves in the spring. The leaves are quite distinctive, and close attention to their appearance should prevent possible misidentifications. The fruit is brown and dry and "barren" in that it is inedible.

CONSERVATION STATUS: SC-Vulnerable

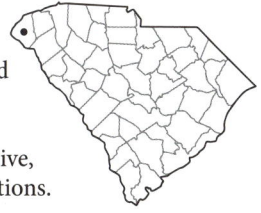

326. Giant Chickweed; Star Chickweed

Stellaria pubera Michaux
Stel-là-ri-a pù-be-ra
Caryophyllaceae (Pink Family)

DESCRIPTION: Perennial herb; stems erect or ascending, 5–12″ tall with hairs in lines or dispersed; leaves opposite, those of sterile shoots much larger than those of the fertile shoots; flowers white, the 5 petals so deeply incised as to appear to be 10; fruits split to the base to expose seeds; flowers April–June.

RANGE-HABITAT: NJ to IN and south to FL and AL; in SC, common in the mountains and upper Piedmont; rare in the Sandhills and Coastal Plain; moist uplands and extending into adjacent hardwood bottoms.

COMMENTS: The Common Chickweed (*S. media* (L.) Villars, plate 954), a weedy species of disturbed habitats, is smaller in all parts and has weak stems. Like common chickweed, Giant Chickweed may contain compounds that make the plant useful as an herbal medicine. Though it is often overlooked as an ornamental, it is very attractive in a garden.

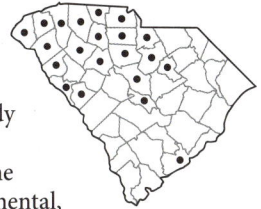

327. Devil's-bit; Fairy-wand

Chamaelirium luteum (L.) Gray
Cha-mae-lí-ri-um lù-te-um
Chiongraphidaceae (Fairy-wand Family)

DESCRIPTION: Perennial herb with a distinct whorl of basal leaves; dioecious—with separate male and female plants; female plants to more than 3′ tall, with erect, short racemes; male plants shorter but have longer, arching, showy racemes; flowers white, turning yellow on drying; flowers March–May.

RANGE-HABITAT: Ontario south to FL and west to AR and LA; in SC, common in the mountains and Piedmont; rare in the Coastal Plain; wide variety of upland and lowland habitats, including beech forests, oak-hickory forests, cove forests, hardwood bottoms, pocosin edges, and wet pine savannas.

COMMENTS: The pale green basal leaves are distinctive, as are the wandlike terminal clusters of flowers.

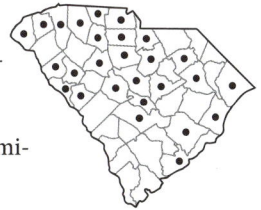

328. Green-and-gold

Chrysogonum virginianum L. var.
    brevistolon G. L. Nesom
        Chry-só-go-num vir-gi-ni-à-num
        var. bre-ví-sto-lon
    Asteraceae (Aster Family)

SYNONYM: *Chrysogonum virginianum*
    L.—RAB, PR

DESCRIPTION: Herbaceous perennial with stolons, forming colonies; flowering early when very small; at first stemless or short-stemmed, but later flowering stems elongating to 4″ tall; leaves opposite; flowers late March–early June.

RANGE-HABITAT: From NC south to FL and west to TN and AL; throughout SC; common; moist to fairly dry forests, woodlands, and forest margins.

TAXONOMY: The similar *C. virginianum* var. *virginianum* is cespitose, without stolons, not mat forming, and with flowering stems 6–12″ tall. Variety *virginianum* occurs in the nc. Coastal Plain of SC and otherwise ranges to the north.

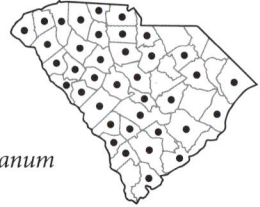

329. Wild Geranium

Geranium maculatum L.
    Ge-rà-ni-um ma-cu-là-tum
    Geraniaceae (Geranium Family)

DESCRIPTION: Herbaceous perennial from a thick rhizome; stems 8–24′ tall; basal leaves with long stalks; flowers pink to purplish-pink; fruit with a long beak; flowers April–June.

RANGE-HABITAT: MA south to GA and west to Manitoba, SD, and AR; in SC, common in the mountains and Piedmont; scattered and rare in the Coastal Plain; cove forests, beech forests, basic-mesic forests, bottomland forests, and other mesic, calcium-rich forests.

COMMENTS: Native Americans made a powder from the dried rhizome and used it as a styptic and astringent to slow bleeding from cuts. The method of seed dispersal is unusual. The fruit of Wild Geranium is actually five slightly fused fruits, and the long beak consists internally of five slender, stiff, append-ages (awns), each of which cups a single seed. At maturity the awns dry, bend, and create tension. The tension is released by touch and results in the catapulting of each seed away from the mother plant.

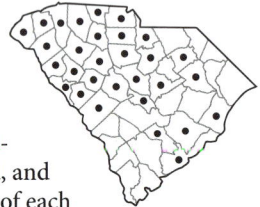

330. American Alumroot

Heuchera americana L.
    Heù-cher-a a-me-ri-cà-na
    Saxifragaceae (Saxifrage Family)

DESCRIPTION: Rhizomatous perennial with long leaf stalks and rounded blades, often with varie-gation between the veins; flowering stems 2–3′ tall; stamens with orange anthers that obviously protrude from the calyx (at least 3 mm); styles extending from the calyx by 2.6 mm or more; flowers April–June.

RANGE-HABITAT: NY and Ontario south to TX and west to IL and GA; in SC, common in the mountains and Piedmont, rare in the Coastal Plain; rich woods and rock outcrops; calcareous bluff forests; and wet, flat, calcareous forests, usually on circumneutral soils,.

SIMILAR SPECIES: Carolina Alumroot (*Heuchera caroliniana* (Rosendahl, Butters & Lakela) E. F. Wells) is almost identical but is distinguished by the stamens barely exceeding the calyx and the styles included in the calyx. It is uncommon in the upper Piedmont in similar habitats, mostly close to the NC line.

COMMENTS: The genus honors Johann Heinrich Heucher (1677–1747), a German botanist. Alumroot species have become incredibly popular landscape plants, and most selections are hybrids or morphological variants with attractive leaves.

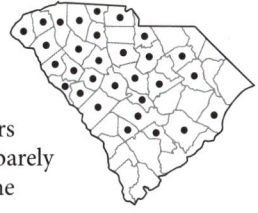

---

331. Partridge Berry

Mitchella repens L.
Mít-chell-a rè-pens
Rubiaceae (Madder Family)

DESCRIPTION: Trailing, creeping, evergreen perennial; leaves opposite with a distinctive pale stripe along the midrib; flowers in pairs; a single berry forms from the fusion of the ripened ovaries of two flowers, leaving two scars on the mature fruit; flowers April–June; fruits mature in June–July but persist throughout the winter.

RANGE-HABITAT: From Nova Scotia south to peninsular FL and west to MN and TX; disjunct in Guatemala; common throughout SC; deciduous and coniferous forests, stream sides, and maritime forests.

COMMENTS: The berries are edible although bland and seedy. It makes an excellent groundcover for home gardens, especially under acid-loving shrubs. The fruits are consumed by Hermit thrush and other birds during the late winter through spring. Native American women made a tea from the leaves as an aid in childbirth. The genus honors Dr. John Mitchell (1676–1768), an early correspondent of Linnaeus and a Virginia botanist.

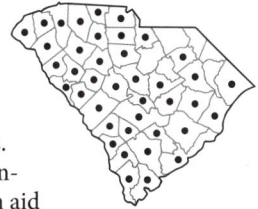

---

332. Thimbleweed

Anemone virginiana L.
A-ne-mò-ne vir-gi-ni-à-na
Ranunculaceae (Buttercup Family)

DESCRIPTION: Herbaceous, hairy perennial, to 3′ tall from a short, thick rhizome; often growing in colonies; basal and stem leaves present, ternately compound and ternately lobed; petals absent; sepals are showy-white; receptacle in the center of the flower is tall and becomes taller as the fruits mature; fruit an aggregate of achenes; flowers May–July.

RANGE-HABITAT: Widespread in eastern North America; in SC, primarily a mountain and Piedmont species; rich forests, woodlands, and forest margins such as roadsides; rare in the Coastal Plain in calcareous wooded communities such as limestone bluffs.

COMMENTS: The thimble-shaped cluster of pistils or fruits is the distinctive feature of this species and accounts for the common name.

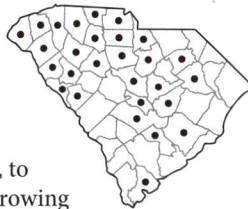

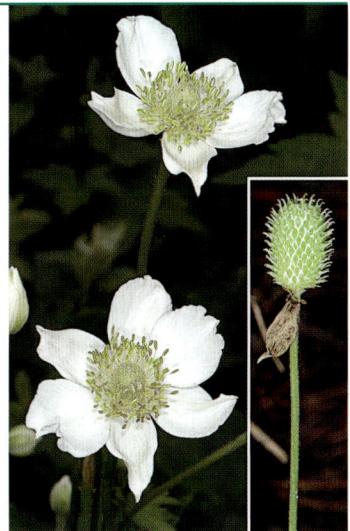

333. Spotted Wintergreen; Pipsissewa

Chimaphila maculata
(L.) Pursh
Chi-má-phi-la ma-cu-là-ta
Ericaceae (Heath Family)

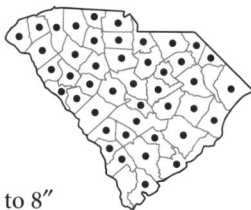

DESCRIPTION: Small rhizomatous subshrub to 8″ tall; leaves evergreen, whorled, and variegated, with white on the midrib and larger veins; flowers nodding, fragrant; flowers May–June.
RANGE-HABITAT: Widespread in eastern North America; common throughout SC; beech, oak-hickory, pine-mixed hardwood forests, pine forests, and plantations.
COMMENTS: Pipsissewa has had a variety of medicinal uses. Native Americans used the plant to treat kidney stones. They used a poultice from the leaves to heal blisters or to reduce swelling in legs or feet. Early settlers used the leaves to make tea that served as a febrifuge to break the fever associated with typhus. A popular refreshing hot tea was made by pouring boiling water over chopped leaves. Until recently, species of this genus were used to flavor commercial root beer, and they are still available for home use.

334. Summer Bluet

Houstonia purpurea L.
Hous-tòn-i-a pur-pù-re-a
Rubiaceae (Madder Family)

DESCRIPTION: Perennial herb with several stems from the base; stems often branching near the tip; leaves widest toward the base; flowers in terminal clusters, dark lavender to almost white; flowers May–July.
RANGE-HABITAT: MD south to GA and west to OH and TX; in SC, common except in the outer Coastal Plain; found along road banks, around rock outcrops, and in a variety of moist to dry upland forests.

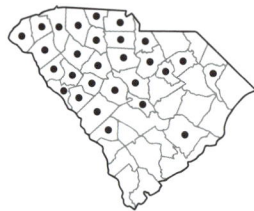

335. Eastern Anglepod

Gonolobus suberosus (L.)
   R. Brown
      Go-nó-lo-bus su-be-ró-sus
Apocynaceae (Dogbane Family)

SYNONYMS: *Gonolobus gonocar-pus* (Walter) Perry—PR; *Matelea gonocarpa* (Walter) Shinners—RAB

DESCRIPTION: Perennial, twining herb with milky sap; leaves opposite, widely ovate-elliptic to elliptic-oblong; corolla smooth within; petals with a lighter colored (greenish or yellowish) margin and tip and dark maroon to brownish base, linear-lanceolate, often reflexed, 3x as long as the sepals; flowers June–July; follicles (fruit) smooth, sharply angled; fruits mature September–November.

RANGE-HABITAT: From VA and KY south to FL, west to MS; in SC, primarily in the Piedmont; occasional; beech forests and thickets.

COMMENTS: This species has formerly been divided into two species (*G. gonocarpus* and *G. suberosus*). Recent evidence suggests that there is only a single species in our area, though it may be divided into two varieties. The plant is best distinguished from the very similar spinypod species (*Matelea* spp.) by the fruits. The fruits of Eastern Anglepod have distinct angles rather than bumps or blunt, spinelike protrusions, and petals with a lighter margin, rather than all one color (as in *Matelea*). If no flowers or fruits are evident, the two similar genera can be told apart by crushing a leaf. The leaves of *Gonolobus* have a strong, rancid, peanut butter odor, and the leaves of *Matelea* do not.

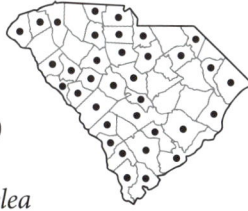

336. Beech-drops

Epifagus virginiana (L.) Barton
      E-pi-fà-gus vir-gi-ni-à-na
Orobanchaceae (Broomrape Family)

DESCRIPTION: Annual holoparasite on the roots of beech trees; annual; plant 4–18″ tall, branched; dried stalks persisting throughout the winter; leaves reduced to scales; upper flowers sterile, lower flowers not opening and self-pollinated, producing abundant seeds; leaves reduced to scales; flowers September–November.

RANGE-HABITAT: Nova Scotia south to FL and west to WI and LA; common throughout SC, wherever beech trees occur.

COMMENTS: The dustlike seeds are carried through the soil by percolating water where they germinate only in the rhizosphere of beech roots.

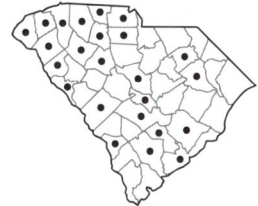

337. Broad Beech Fern

Phegopteris hexagonoptera (Michaux) Fée
Phe-góp-te-ris hex-a-go-nóp-te-ra
Thelypteridaceae (Marsh Fern Family)

SYNONYM: *Thelypteris hexagonoptera*
(Michaux) Weatherby—RAB

DESCRIPTION: Small, deciduous, rhizomatous fern with elongate rhizomes with pale scales; fronds held perpendicular to the ground, pubescent, triangular in outline, 2-pinnatifid, 2.25–8″ long; spores produced in roundish clusters (sori); spores produced April–August.

RANGE-HABITAT: Throughout most of the eastern United States; in SC, common throughout, often in beech forests but also in other mesic hardwood forests on rich soils.

COMMENTS: This is our only common fern with small, triangular blades that are held perpendicular to the ground. Eastern Bracken Fern (*Pteridium latiusculum* (Desvaux) Hieronymus ex Fries) is a much larger plant with leathery leaves found in poor soils and fire-maintained habitats.

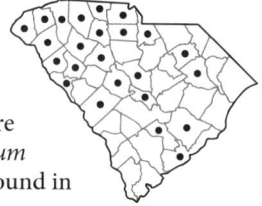

338. Christmas Fern

Polystichum acrostichoides
(Michaux) Schott
Po-lýs-ti-chum ac-ro-sti-choì-des
Dryopteridaceae (Wood-fern Family)

DESCRIPTION: Large, evergreen, clump-forming fern; petiole with many brown scales; fronds thick, leathery, pinnately compound; fertile portions of the frond located toward the tip of the blade, which appears reduced in size because this portion withers soon after the spores mature; spores mature June–September.

RANGE-HABITAT: Throughout the eastern United States and into Mexico; in SC, common throughout in a variety of mesic forests, often with rich soil.

COMMENTS: The evergreen habit gives this fern the common name. It has been traditionally gathered for decoration during the winter season. No other large, evergreen fern in our area has simple pinnate fronds with the exception of the introduced and invasive Asian holly fern (*Cyrtomium* spp.). The individual leaflets are shaped a bit like a Christmas stocking.

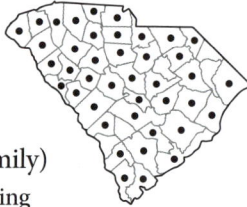

339. Oak-hickory forests

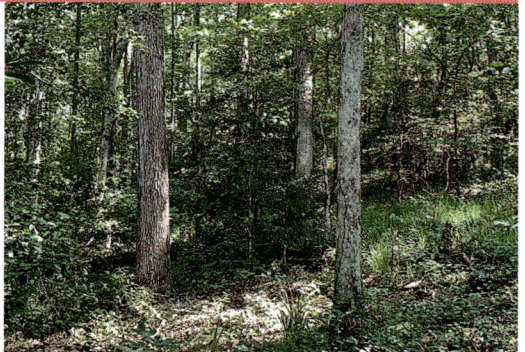

340. White Oak; Stave Oak

Quercus alba L.
  Quér-cus ál-ba
  Fagaceae (Beech Family)

DESCRIPTION: Medium to tall deciduous tree to 115′ tall with pale grayish bark that peels or flakes, sometimes significantly; leaves alternate, divided into 7–10 shallow-to-deep rounded lobes without bristle-tips; female flowers 1–2 on a short stalk, separate from the male flowers, which are in catkins; acorns mature September–November.

RANGE-HABITAT: Widespread in eastern North America; found throughout SC; mesic to xeric forests and woodlands, but best growth is achieved in rich, well-drained loamy soils.

COMMENTS: Historically, White Oak was one of the most valuable lumber trees in eastern North America. Its wood is hard, tough, strong, and close-grained, and although used in many ways, it is best suited for support timbers, furniture, flooring, and interior finishing. White Oak was the mainstay in the construction of North American ships prior to the use of steel. It is often referred to as "stave oak" because the wood is used for whiskey and wine barrels. It is an attractive, long-lived, shade tree, and its acorns are important to wildlife. The inner bark is astringent, and a tea from the bark was once used for diarrhea, dysentery, bleeding, and a gargle for sore throats.

341. Downy Serviceberry; Shadbush

Amelanchier arborea Michaux f.
  A-me-lán-chi-er ar-bò-re-a
  Rosaceae (Rose Family)

DESCRIPTION: Deciduous shrub or rarely a small tree to 20′ tall, rarely taller in SC; leaves alternate, margins toothed, apex pointed or gradually tapering, downy below when unfurling in spring but becoming smooth with age; flowers produced in a drooping inflorescence; petals to 0.75″ long, white or infused with pink, appearing with or before the leaves; flowers March–April.

RANGE-HABITAT: From Nova Scotia south to the FL Panhandle and west to MN and TX; in SC, common in the mountains, Piedmont, and inner Coastal Plain; rare in the outer Coastal Plain; dry to moist habitats including Chestnut Oak forests, moist oak-hickory forests, margins of outcrops, forest margins, and margins of bottomland hardwood forests.

SIMILAR SPECIES: There are five species of serviceberry in SC. Dwarf Serviceberry (*A. spicata* (Lamarck) K. Koch) is very similar and is best separated by the densely pubescent summit to the ovary and the densely tomentose undersides to the leaves when they are young. It is rare in sandy, dry woodlands in the Coastal Plain. Eastern Serviceberry (*Amelanchier canadensis* (L.) Medikus) is also very similar but produces inflorescences that are more erect and has very short petals, less than 0.5″ long. It is found in pocosins, wet flatwoods, and other mesic forests in the Coastal Plain. Smooth Serviceberry (*Amelanchier laevis* Wiegand, plate 180) is a larger tree with larger flowers and leaves that are smooth when unfurling and is confined to the mountains. Coastal Plain Serviceberry (*Amelanchier obovalis* (Michaux) W. W. Ashe, plate 496) is a rhizomatous shrub found in pocosins and wet flatwoods in the Sandhills and Coastal Plain.

COMMENTS: There are several colloquial explanations for the term "service" in the name. The most common indicates that the "service" referred to is Easter, but in Appalachia it is thought that the flowering of this plant indicated the ground had thawed enough to dig graves and hold memorial services for those who had died during the winter. The "shad" referred to in the other common name is a small fish

that usually has its mating runs up major rivers about the time this species is in bloom. It is also sometimes called Juneberry or Mayberry in the Appalachians, because of the season of fruit production. The fruits contain antioxidants that retard spoilage. Native Americans used serviceberries with animal fat to make a type of pemmican. The juicy fruits can be eaten raw or used in jams, pies, and muffins.

342. Eastern Redbud; Judas-tree

Cercis canadensis L.
        Cér-cis ca-na-dén-sis
    Fabaceae (Pea or Bean Family)

DESCRIPTION: Small, short-lived tree with alternate, simple, deciduous, heart-shaped leaves; flowers, pink, purplish or rarely white, borne on old wood, appearing in spring before the leaves; flowers March–April.

RANGE-HABITAT: From s. Ontario south to FL and west to NE and TX; common throughout SC; most often on rich, moist sites; oak-hickory, beech, and pine-mixed hardwood forests; river bottoms and stream sides; forests over calcareous soil or mafic rocks; and maritime shell hammocks.

COMMENTS: Redbud is one of the most popular native trees in cultivation. This species is very common on exposed red clay roadsides and prefers to grow along forest margins. It is unusual in producing flowers on short shoots emerging along the length of the branches, even on the trunk. The edible flowers have the flavor of raw green beans and can be used in salads or pickled. The buds, flowers, and young pods are good when fried in butter or made into fritters. The common name, Judas-tree, is sometimes transferred to the eastern redbud from the related species of the Mediterranean Judas-tree (*Cercis siliquastrum* L.), the tree on which Judas is believed to have hanged himself. Legend states that the flowers were white but turned to red, either with shame or from the drops of blood shed by Jesus.

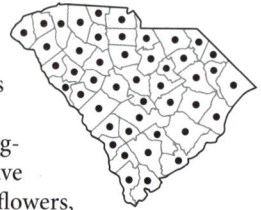

343. Sassafras

Sassafras albidum (Nuttall) Nees
        Sás-sa-fras ál-bi-dum
    Lauraceae (Laurel Family)

DESCRIPTION: Small to medium-sized, fast-growing tree, often forming thickets from lateral root sprouts; male and female flowers on separate plants, appearing in the spring before the leaves; leaves deciduous, alternate and polymorphic (either entire, with a 1-sided lobe, or 3-lobed); twigs, leaves, and roots spicy aromatic; flowers March–April.

RANGE-HABITAT: Throughout the eastern US; common throughout SC; wide variety of forests, along fencerows, abandoned fields, pine-mixed hardwood forests, and woodland borders.

COMMENTS: Sassafras is an early successional tree and, because it is shade intolerant, it is seldom found as an understory tree in a closed canopy forest. It is one of the first trees to invade abandoned fields, often appearing as small groves spreading from lateral root offshoots. Fruit production is sparse, but it does provide an important food for wildlife. Oil of Sassafras is distilled from the bark of the roots and once was a flavoring material of considerable importance. The oil has been used to flavor tobacco, root beer, and other beverages, soaps, perfumes, and gums. A tea made from boiling the bark was once used as a spring tonic to "thin the blood" before the advent of summer; proven ineffective, it quickly fell into disrepute. Young leaves are ground into a fine powder to produce the mucilaginous gumbo of Creole cooking. The US Food and Drug Administration has banned the use of safrole, found in oil of sassafras, because it may be carcinogenic.

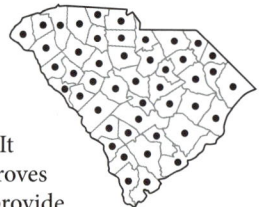

344. Crab-apple

Malus angustifolia (Aiton) Michaux
Mà-lus an-gus-ti-fò-li-a
Rosaceae (Rose Family)

DESCRIPTION: Small tree or thicket-forming shrub from root sprouts; petals white or light-pink and highly fragrant; leaves deciduous, simple, alternate, with margins of some leaves scalloped, toothed, or nearly entire; flowers April–May; fruit a yellowish green pome, very sour, ripe in August–September.

RANGE-HABITAT: MD south to FL and west to AR and TX; throughout SC, but more common in the Coastal Plain; moist soil in oak-hickory forests, woodland borders, thickets along riverbanks, and fencerows.

COMMENTS: Crab-apple is often used as an ornamental because of its showy and fragrant flowers. The fruit is used to make jelly, preserves, and cider. It is an important wildlife food, being consumed by deer, foxes, raccoons, quail, and turkeys. The wood is not commercially important.

345. Black Cherry

Prunus serotina Ehrhart
Prù-nus se-rò-ti-na
Rosaceae (Rose Family)

DESCRIPTION: Medium to large tree reaching 80–90′ tall; bark of small trees with horizontal lenticles, becoming fissured and scaly with age; leaves and inner bark contain the almond-flavored hydrocyanic acid that can readily be detected by breaking a twig and smelling the broken end; leaves alternate, simple, deciduous; flowers produced in racemes, April–May; fruit a drupe, black when mature in July–August.

RANGE-HABITAT: Nova Scotia to MN, south to c. TX, and east to FL; common throughout SC; in a variety of natural and disturbed habitats such as cove forests, oak-hickory and pine-mixed hardwood forests, fencerows, thickets, pastures, Longleaf Pine flatwoods, and bottomland forests.

COMMENTS: Black cherry is one of SC's most versatile native trees. Its reddish brown, close-grained wood takes a beautiful polish and is widely used to make furniture, veneers, and small wooden wares. Its use as a lumber tree is reduced because of the lack of large trees; today it is used mainly as a specialty wood. Trees reach their greatest size in the Appalachian region.

As an astringent, the bark is used to make cough medicines and expectorants to treat sore throats. Duke (1997) recommends Black Cherry as an herbal remedy in treating flu. Pioneers in the Appalachians used the ripe fruits to make a drink called cherry bounce. The juice was pressed from the fruits and infused in brandy or rum to give it the bitter taste desired. Even today the fruit is used to flavor liqueurs.

Though not widely used as a landscape tree, research has shown that no fewer than 456 species of native butterflies and moths, including the Eastern Tiger Swallowtail and Red-Spotted Purple, make use of the tree as a food source for caterpillars. It is the favored food plant for the eastern tent caterpillar, whose tentlike webs in the crotches of branches are clearly visible in early spring. These facts make it ideal for planting to support biodiversity, including the neotropical migrant birds that eat the caterpillars.

All parts of the plant are poisonous (except the pulp of the ripe fruit) because of hydrocyanic acid. Children have been poisoned by sucking the twigs. A wide variety of wildlife eat the fruits. The hydrocyanic acid in the wilted leaves may be harmful to deer and cattle, but whitetail deer can apparently eat the fresh leaves without ill effects.

346. Persimmon

Diospyros virginiana L.
Di-os-pỳ-ros vir-gi-ni-à-na
Ebenaceae (Ebony Family)

DESCRIPTION: Small- to medium-sized tree reaching 70–80′ tall with deeply furrowed and blocky bark; leaves deciduous, simple, alternate, blades ovate or elliptic, upper surface dark green, often with black blemishes, lower surface light pale green to bluish-green; male and female flowers on separate trees, produced April–June; fruit an orange-yellow berry, mature in September–November.

RANGE-HABITAT: CT west to IL, MO, and KS, south to FL and TX; common throughout SC; dry deciduous forests, floodplains, pinelands, xeric Longleaf Pine sandhills, mesic forests, fencerows, and old fields.

COMMENTS: Persimmon wood is extremely hard but is not used as commercial lumber since it yields an inferior grade. Its main use is (was) for golf clubs. The sweet fruit is edible when fully ripe and can be eaten raw or made into a variety of dishes, including bread and pudding. When green, the fruit is strongly astringent. The genus name translates to "divine fruit" or literally "God's fruit." In folk usage, an astringent tea from the inner bark was used as a gargle for sore throat and thrush. A wide variety of wildlife eat the fruits, and it is relished by coyotes and domestic dogs. The SC state champion was measured in 1995 at 132′ tall.

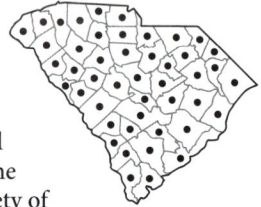

347. Sourwood

Oxydendrum arboreum (L.) A. P. de Candolle
Ox-y-dén-drum ar-bò-re-um
Eriaceae (Heath Family)

DESCRIPTION: Medium-sized tree with deeply ridged bark and often very twisted and contorted branches; tree has sour-tasting sap and leaves; leaves deciduous, alternate, smooth, lanceolate to elliptic-lanceolate; flowers white, erect, in a terminal panicle; flowers June–July.

RANGE-HABITAT: PA south to FL and west to IN and LA; in SC, common throughout, except rare or absent in the outer Coastal Plain; mesic to xeric deciduous forests, especially dry-mesic to xeric oak-hickory, pine-oak heath, and oak-pine forests; in the fall-line Sandhills in Longleaf Pine sandhill-streamhead pocosin ecotones.

COMMENTS: This is the only species in the genus (monotypic), and it has no close relatives in the family. The wood is not important commercially, but the flowers yield the prized sourwood honey. Sourwood has considerable ornamental value because of its drooping racemes of white flowers and early fall color (brilliant red). It is especially valuable along roadsides, where it can grow on the poor soils of road cuts. Native Americans chewed the bark for mouth ulcers and used a leaf tea for asthma, diarrhea, and indigestion.

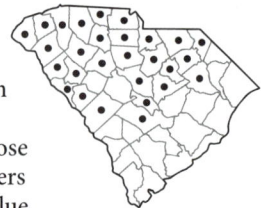

348. Pinxterflower; Common Wild Azalea

Rhododendron periclymenoides
(Michaux) Shinners
Rho-do-dén-dron pe-ri-cly-
me-noì-des
Ericaceae (Heath Family)

SYNONYM: *Rhododendron nudiflorum*
(L.) Torrey—RAB

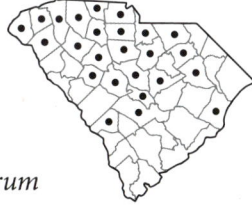

DESCRIPTION: Deciduous shrub to 10′ tall; leaves not present or not fully expanded at flowering, with hairs on the margins and stiff hairs on the main veins below; floral tube pubescent but without sticky glands; flowers pink to white; sepals less than 0.04″ long, without stalked glands; capsules cylindric; flowers late March–May.

RANGE-HABITAT: MA south to GA and west to OH and AL; in SC, common in the mountains and Piedmont, rare in the Coastal Plain; streambanks, beech forests, and a variety of moist to dry oak-hickory forests.

SIMILAR SPECIES: This species is easily confused with Piedmont Azalea (*R. canescens* (Michaux) Sweet, plate 795), which may be distinguished by having bud scales that are densely pubescent, leaves that are densely canescent (soft-short pubescent) below, and floral tubes that are pubescent with scattered sticky glands. Piedmont Azalea occurs in bottomlands, swamp forests and flatwoods in the Coastal Plain and Piedmont, and rarely in the mountains.

COMMENTS: SC native deciduous azaleas make exceptional landscape plants. The flowers are pollinated by butterflies and moths.

349. Oconee Azalea

Rhododendron flammeum (Michaux)
Sargent
Rho-do-dén-dron flàm-me-um
Ericaceae (Heath Family)

DESCRIPTION: Deciduous shrub to 15′ tall; flowers appearing before or with the leaves; petals brilliant, deep orange to orange-red; corolla tube with short hairs on the outside, not glandular and sticky; scales of flower buds with marginal hairs; capsules ovoid; flowers April.

RANGE-HABITAT: GA and SC; rare in the Sandhills and Piedmont and restricted to the Savannah River drainage; dry-mesic oak-hickory forests on slopes and bluff.

COMMENTS: Oconee Azalea could be confused with Flame Azalea, but they have very different blooming periods (April vs. May–June) and technical floral characters. Their ranges also do not overlap. Oconee Azalea is noted for its range of flower color; under cultivation it will tolerate drier soils than any other native azalea. Though Oconee County, SC tries to claim this beautiful azalea, it has never grown as a wild plant in that region of the state. The name is derived from the Oconee River in GA, which is derived from the colonial English name of Native Americans of the Creek and Seminole nations.

CONSERVATION STATUS: SC-Vulnerable

350. Maple-leaved Arrowwood

Viburnum acerifolium L.
Vi-búr-num a-ce-ri-fò-li-um
Viburnaceae (Viburnum Family)

DESCRIPTION: Low deciduous shrub, 3–6′ tall, rarely taller, usually colonial; leaves opposite, simple, 3-lobed (like a maple leaf); flowers white, produced in cymes April–early June; drupes black, mature August–October.

RANGE-HABITAT: Quebec south to GA and west to MN and TN; in SC, common in the mountains and Piedmont, occasional in the inner Coastal Plain and Sandhills; deciduous forests communities.

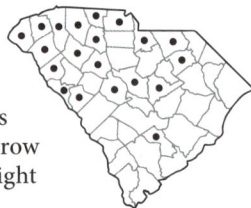

COMMENTS: Arrowwood makes a good shade-tolerant ornamental since the foliage is especially colorful in the fall; allow it to develop into large, loose colonies. The drupes are used by wildlife. A number of different species of arrowwoods and dogwoods were used by Native Americans to produce arrow shafts. Coppicing (cutting back the tree to the ground to stimulate long, straight branches) was the source of branches suitable for producing arrow shafts.

351. Sparkleberry; Farkleberry

Vaccinium arboreum Marshall
Vac-cí-ni-um ar-bò-re-um
Ericaceae (Heath Family)

DESCRIPTION: Erect shrub or small tree to 30′ tall with thin, smooth, cinnamon-colored peeling bark; branches crooked, forming an irregular crown; leaves simple, broadly elliptic or broadly ovate to obovate, alternate, shiny above, nearly evergreen to tardily deciduous; flowers April–June; berries black, maturing September–October.

RANGE-HABITAT: Widely distributed in southeastern US; common throughout SC; sandy or rocky woodlands, bluffs, and cliffs, usually xeric.

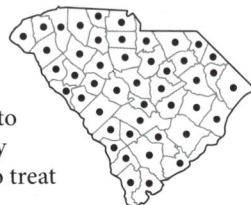

COMMENTS: The berries, often lasting through winter, are eaten by a wide variety of wildlife. The pulp is scanty but pleasant tasting and can be made into jelly or jam. The wood has little commercial value but is hard and used locally for tool handles and craft items. The roots, bark, and leaves were once used to treat diarrhea and dysentery.

352. Deerberry

Vaccinium stamineum L.
Vac-cí-ni-um sta-mí-ne-um
Ericaceae (Heath Family)

DESCRIPTION: Highly variable shrub to 5′ tall; leaves green or bluish-green above and whitish (glaucous) or green below; flowers white in the shape of an open bell; mature April–June; berries ripen greenish or purplish-green; fruits mature August–October.

RANGE-HABITAT: Widespread in the eastern US; in SC, common throughout in a variety of dry woodlands, margins of outcrops, and forest margins.

TAXONOMY: The deerberries (*Vaccinium* section *Polycodium*) are still under critical revision, and there are several unnamed varieties. The majority of the varieties in our area have glaucous undersides to the leaves. For a key and discussion of the taxonomy, see Weakley (2020).

353. Nestronia

Nestronia umbellula Rafinesque
Nes-trò-ni-a um-bél-lu-la
Santalaceae (Sandalwood Family)

DESCRIPTION: A colony-forming hemiparasitic shrub, 1.5–4.5′ tall; stems dark brown, dichotomous branching; leaves opposite, entire, and 0.75–2″ long; male and female flowers produced on different plants (dioecious); the small greenish flowers consist of 4–5 petal-like sepals; the female flowers are borne singly in the axils of leaves, the male flowers are in umbels; the fruit is a greenish drupe, to 0.5″ in diameter; flowers April–May; fruits mature July.

RANGE-HABITAT: VA and KY south to GA and MS; in SC, rare in the Piedmont, Sandhills, and inner Coastal Plain; dry oak-hickory forests with a somewhat open canopy and some component of pine species.

COMMENTS: Not a showy species, it is easily overlooked. Because of this, it has recently been found to be less rare than once thought. Like all members of the Sandalwood family, this species is partially parasitic, primarily on pines. Many populations consist of all male or all female plants, making reproduction from seeds impossible and a prime cause of the species' rarity. In recent years, a fungus that causes the early dropping of leaves has attacked plants in many populations.

CONSERVATION STATUS: SC-Vulnerable

354. May White

Rhododendron eastmanii Kron & Creel
Rho-do-dén-dron east-màn-i-i
Ericaceae (Heath Family)

DESCRIPTION: Shrub or small tree to 16′ tall, nonrhizomatous; corolla white with a yellow blotch on the upper lobe and pink-tinged lobes on newly opened flowers; outer surface of corolla covered with nonglandular and glandular hairs; corolla tube narrow; flower fragrance strong, spicy, and sweet; stamens 5, bending downwards, strongly exerted; flowers in May, after the leaves have expanded.

RANGE-HABITAT: Endemic to the Piedmont and inner Coastal Plain of SC; rare; deciduous forests on north-facing slopes with well-drained, slightly acidic soils, but often near calcium-rich soils.

TAXONOMY: May White can be separated from the other white-flowered species in SC by the following: from *R. alabamense,* by the flowers opening after the leaves have expanded; and from *R. arborescens, R. atlanticum, R. serrulatum,* and *R. viscosum,* by the yellow blotch on the upper corolla lobe, which the other four lack. The report of *R. alabamense* from SC (Radford et al., 1968) probably refers to *R. eastmanii.*

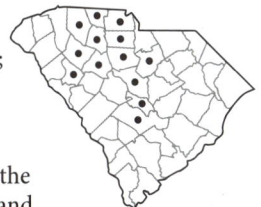

COMMENTS: Kathleen A. Kron and SC native Michael A. Creel described May White in 1999 (Kron and Creel, 1999). The specific epithet honors Charles Eastman of Columbia, SC, who is given credit for discovering the species while birdwatching in Santee State Park. It makes an outstanding specimen for the woodland garden and has recently become available from specialty native plant nurseries. Dr. Charlie Horn and his students at Newberry College have now documented this species from at least 70 populations in 13 counties.

CONSERVATION STATUS: SC-Vulnerable

355. Yellow Jessamine; Carolina Jessamine

Gelsemium sempervirens (L.) Aiton f.

Gel-sè-mi-um sem-pér-vi-rens

Gelsemiaceae (Jessamine Family)

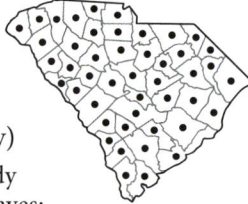

DESCRIPTION: Twining (left to right), woody vine with opposite, pointed, evergreen leaves; trailing or high-climbing; blades smooth, entire; flowers fragrant; sepals with tips obtuse to acute; flowers February–early May; fruit a capsule, mature September–November, persistent into spring.

RANGE-HABITAT: VA south to FL, west to TX and AR, and south into Central America; in SC, common throughout the Piedmont, Sandhills, Coastal Plain, and maritime strand; Longleaf Pine flatwoods and savannas, oak-hickory, beech, pine-mixed hardwood and maritime forests, lowland woods where water stands only for short periods, fencerows, thickets, and along roadsides.

SIMILAR SPECIES: Swamp Jessamine (*G. rankinii* Small) is a very similar species, best distinguished by the acuminate tips to the sepals, flowers with a lack of odor, and the peculiar habit of flowering in both the spring and again in the autumn. It is occasional in blackwater swamp forests and swamp forest margins at scattered locations throughout the Coastal Plain.

COMMENTS: This is the state flower of SC and a harbinger of spring. All parts are poisonous. It is one of the most common source of plant poisonings in SC. According to Foster and Duke (1990), it can cause contact dermatitis. Children have been poisoned by sucking nectar from the flowers, probably mistaking them for Japanese Honeysuckle. F. P. Porcher (1869) and Foster and Duke (1990) list many historical medicinal uses of this plant; however, no current uses of the plant appear in the literature.

Yellow Jessamine is often cultivated and when grown in full sun becomes denser and more attractive than when growing naturally. It can be used on trellises and as a groundcover.

356. Coral Honeysuckle; Woodbine

Lonicera sempervirens L.

Lo-níc-er-a sem-pér-vi-rens

Caprifoliaceae (Honeysuckle Family)

DESCRIPTION: Native, trailing or high-climbing, woody vine with simple, opposite leaves dark green above and glaucous (bluish-green) below; partially evergreen; twining is from left to right; last 1–2 pairs of leaves below the inflorescence joined at their bases; flowers March–July; berries red, maturing July–September.

RANGE-HABITAT: CT to OK, south to FL and TX; in SC, common throughout; thin oak-hickory and pine-hardwood forests, thickets, and fencerows.

COMMENTS: Coral Honeysuckle does well under cultivation. It is quite vigorous and flowers abundantly in full sun. A classic combination, termed a Carolina Fence Garden, combines Coral Honeysuckle with Carolina Jessamine producing a startling display. In cultivation it may flower throughout the growing season. The bluish-green (glaucous) undersides of the leaves are a reliable way to tell this native species from Japanese Honeysuckle (*L. japonica* Thunberg), which is an invasive exotic.

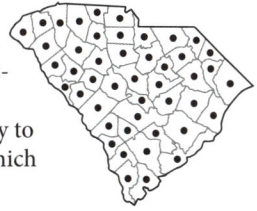

357. Muscadine

Muscadinia rotundifolia (Michaux)
Small var. *rotundifolia*
Mus-ca-dí-ni-a ro-tun-di-fò-li-a
var. ro-tun-di-fò-li-a
Vitaceae (Grape Family)

SYNONYM: *Vitis rotundifolia* Michaux—RAB, PR

DESCRIPTION: High-climbing (100′ in trees) or trailing woody vine, climbing by simple tendrils; bark adherent (except on large stems), with prominent lenticels; drooping aerial roots originate from the stem; leaves round or widely ovate with coarsely dentate margins; male and female flowers on different plants (dioecious); fruit a berry (grape), dark reddish-brown to black when ripe, August–October.

RANGE-HABITAT: DE south to FL and west to MO and TX; common throughout SC, except in the mountains where it is occasional; wide variety of habitats, such as coastal dunes, forests, swamps, low woods, in thickets, and along roadsides.

COMMENTS: Muscadine can be distinguished from all other grapes by the simple tendrils; all other grapes have branching tendrils. The tendrils are derived from modified inflorescences in the grape family and the inflorescence of Muscadine also has a single axis, without branches. Two types of domestic grapes originated from this species, both have male and female flowers on the same plant, and many cultivars are self-fertile. Plants with amber-green fruits are called Scuppernongs, and those with purple fruits are called Muscadines (as are the wild, dark-fruited plants). Few fruits have been used for so long and in so many ways. The fruits can be eaten plain or made into wine, jelly, juice, preserves, used in pies, or sun dried for future use. Muscadine leaves can be stuffed or rolled with a wide assortment of foods, and then boiled. Muscadines are rich in vitamins B and C and iron. Many species of wildlife eat muscadines, and they are excellent plants to cultivate to attract wildlife.

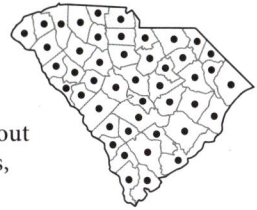

358. American Cancer-root; Bearcorn

Conopholis americana (L.) Wallroth
Co-nó-pho-lis a-me-ri-cà-na
Orobanchaceae (Broomrape Family)

DESCRIPTION: Herbaceous holoparasite on various oak species; 3–10″ tall; leaves reduced to brown scales; entire plant pale brown or yellow brown; flowers March–June.

RANGE-HABITAT: Nova Scotia west to WI, south to peninsular FL, AL, and TN; common throughout SC; under species of red oak in rich, moist forests, and oak-hickory forests.

COMMENTS: This plant starts its life cycle as an underground gall-like mass (combination of host and parasite tissue) on small oak roots. It takes about 4 years for the underground structure to mature, after

which it produces aboveground flowering stems for many years. The lifecycle and host-specificity is detailed in Baird and Riopel (1986). The stems die back in June, but the dried remains can be seen for months. Bear and deer eat this plant in large quantities in the spring, as evidenced by the large quantity of stems in their droppings.

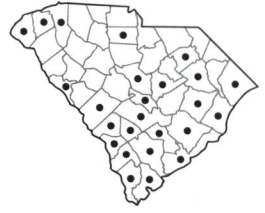

359. Coastal Plain Dwarf Iris; Sandhill Dwarf Iris

Iris verna L. var. *verna*
  Ì-ris vér-na var. vér-na
  Iridaceae (Iris Family)

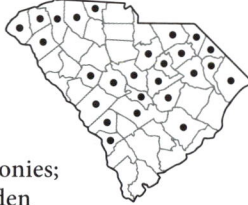

DESCRIPTION: Erect, perennial herb from densely scaly rhizomes forming loose colonies; stems (in SC) to 6″ tall, unbranched, hidden by sheathing bracts; leaves essentially straight, narrow; flowers with sepals not crested and with a sharply contrasting spot of orangeish-yellow toward the base; flowers very fragrant with a sweet, grapelike odor; flowers late March–early April.

RANGE-HABITAT: Eastern MD south to GA; in SC, common in the Piedmont and Coastal Plain and in the lower elevations in the mountains; sandy or rocky woods, dry pinelands, Longleaf Pine sandhills, swamp edges, and edges of pocosins.

TAXONOMY: Two varieties are recognized: *I. verna* var. *verna* with long and slender rhizomes, 2–6″ long between offshoots, and Upland Dwarf Iris (*I. verna* var. *smalliana* Fernald), with short and stocky rhizomes, 0.5–1.3″ long between offshoots and forming very dense clumps. Variety *smalliana* is restricted to the mountains and upper Piedmont.

COMMENTS: This handsome wildflower is found exclusively on dry, acidic to highly acidic soils, while the similar Dwarf Crested Iris (*Iris cristata* Aiton, plate 116) is found in rich, moist forests with less acidic soils. Dwarf Crested Iris has the contrasting spot on the sepal that is white and crested sepals. According to Kingsbury (1964), "Various iris have been found to contain an irritant principle in the leaves or particularly in the rootstocks which produce gastroenteritis if ingested in sufficiently, relatively large amounts."

360. Eastern Lousewort; Wood-betony

Pedicularis canadensis L.
  Pe-di-cu-là-ris ca-na-dén-sis
  Orobanchaceae (Broomrape Family)

DESCRIPTION: Perennial, hairy herb from thickened, fibrous roots; plant 6–12″ tall; lower leaves in a basal cluster; stem leaves alternate, reduced; all leaves deeply divided into toothed segments; corolla in various color combinations of yellow, red, and purplish; flowers April–May.

RANGE-HABITAT: Widespread and common in eastern North America; in SC, chiefly in the mountains and Piedmont; scattered and rare in the Coastal Plain; moist to dry forests, woodlands, and meadows; pine savannas and flatwoods in the Coastal Plain.

COMMENTS: The generic name comes from the Latin *pediculus,* "a louse," from an early European belief that cattle feeding where these plants abounded became

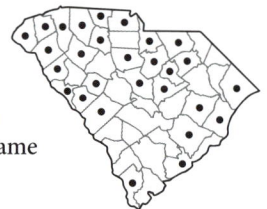

infected with lice. This is one of the species of figworts believed to be a hemiparasite; however, no studies have documented the host plant(s). Due to their lacy appearance and heavy dissection, the leaves are often mistaken for a fern by novice wildflower enthusiasts.

361. Small Solomon's-seal

Polygonatum biflorum (Walter) Elliott
var. *biflorum*
Po-ly-gó-na-tum bi-flò-rum var.
bi-flò-rum
Ruscaceae (Ruscus Family)

DESCRIPTION: Perennial herb from elongate, white rhizomes; stem unbranched, arching, to 2′ long; leaves whitish below; flowers in all but the uppermost leaf axils, usually in pairs from a common stalk, but sometimes up to 9 may be present; flowers April–June.

RANGE-HABITAT: Quebec south to FL and west to Manitoba and NM; in SC, common in the mountains and Piedmont; uncommon and scattered in the Coastal Plain and Sandhills; in a variety of dry-moist forests.

COMMENTS: This species extends into drier habitats than its vegetative look-alike, False Solomon's-seal. Robust plants that reach heights of 4′ tall, typically have more than 2 flowers per node and occur in the rich cove forests of the mountains are referred to as Large Solomon's-seal (*P. biflorum* var. *commutatum* (J. A & J. H. Schultes) Morong). In the mountains, both varieties may grow in close proximity. The rhizome is jointed, and where a leaf breaks off, it leaves a distinct scar said to resemble the seal of King Solomon. Native Americans and European colonists used the starchy rhizomes as food.

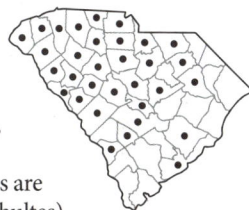

362. Carolina Spinypod

Matelea carolinensis (Jacquin) Woodson
Ma-tè-le-a ca-ro-li-nén-sis
Apocynaceae (Dogbane Family)

DESCRIPTION: Perennial, twining, herbaceous vine with milky sap; leaves opposite, ovate; petals dark maroon, 2–2.5x as long as wide, with lobes ovate-rounded and held in a horizontal plane; flowers April–June.

RANGE-HABITAT: DE south to FL and west to MO and TX; in SC, chiefly in the Piedmont and common but scattered in the outer Coastal Plain; moist nutrient-rich forests, especially along streams.

SIMILAR SPECIES: Two, possibly three, species of *Matelea* occur in SC, and it can be difficult to distinguish between the three. The best way to distinguish *M. caroliniensis* from other species in the genus is by the petals; the lobes are ovate-rounded, 2–2.5x as long as wide, maroon, and held in a horizontal plane. A plant that has been referred to as Deceptive Spinypod (*M. decipiens* (Alexander) Woodson) has longer, narrower petals which are ascending, rather than held horizontally and generally produces many more flowers per inflorescence. Recent evidence seems to suggest that what we have called Deceptive Spinypod in the Carolinas is not the same plant as is represented in the midwestern US and may either be an undescribed species or part of the natural variation in Carolina Spinypod. Yellow Spinypod (*M. flavidula* (Chapman) Woodson, plate 754) has greenish-yellow petals and is restricted to high-calcium soils in the Coastal Plain. Carolina Anglepod (*Gonolobus suberosus* (L.) R. Brown, plate 335) has petals that are generally outlined in a lighter color or green, pods that are angled rather than spiny, and has leaves that when crushed produce a strong, rancid, peanut butter like odor. The leaves of *Matelea* species do not produce this odor when crushed.

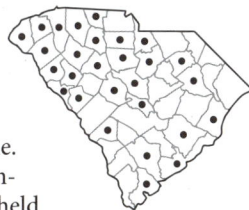

363. White Milkweed

Asclepias variegata L.
As-clè-pi-as va-ri-e-gà-ta
Apocynaceae (Dogbane Family)

DESCRIPTION: Perennial herb with milky sap; stem simple, solitary, 1–3′ tall; leaves broad, in 2–5 pairs; corolla bright white with a rose-colored ring at the base of the petals and corona; flowers May–June; follicles mature July–September.

RANGE-HABITAT: CT to IL and OK and south to the FL Panhandle and TX; throughout SC, but primarily in the mountains and Piedmont, where it is common; occasional elsewhere; oak-hickory forests; sandy, dry, open woods; and upland woodland margins and road banks.

COMMENTS: The following account applies to White Milkweed and most other species. When almost mature but still solid, the cooked seedpods make a palatable vegetable that is comparable to okra. The shoots and young leaves are used like spinach, after washing with water has removed the bitter taste from the milky juice. The raw shoots of any milkweed may be poisonous. All milkweed species are important as a host for the caterpillars of Monarch Butterfly and as an important nectar source for butterflies.

364. Whorled Loosestrife

Lysimachia quadrifolia L.
Ly-si-má-chi-a qua-dri-fò-li-a
Primulaceae (Primrose Family)

DESCRIPTION: Perennial, rhizomatous herb 1–3′ tall; stems erect, hairy, rarely branched; leaves in whorls of 3–6; flowers solitary from the axil of each leaf in the upper 2–6 whorls; corolla streaked with black, marked with red around the center; flowers May–July.

RANGE-HABITAT: ME west to MN and south to SC, AL, and TN; in SC, common in the mountains and Piedmont, occasional in the Sandhills and Coastal Plain; open, oak-hickory forests, forest margins, roadsides, and other moist to dry forests in the Piedmont and mountains; pocosins and wet pine flatwoods and savannas in the Coastal Plain.

COMMENTS: This species is the most distinctive of our common species of *Lysimachia*. Only Fraser's Loosestrife, (*L. fraseri* Duby, plate 213) and Pocosin Loosestrife (*L. asperulifolia* Poiret, plate 507) display whorled leaves but both have terminal racemes of flowers rather than axillary solitary flowers. Whorled Loosestrife often forms large colonies along road banks in our mountains.

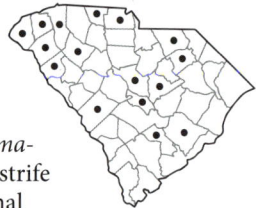

365. Appalachian Red Pinesap

Hypopitys lanuginosa
(Michaux) Small
Hy-pó-pi-tys la-nu-gi-nò-sa
Ericaceae (Heath Family)

SYNONYM: *Monotropa hypopithys*
L.—RAB, PR, in part

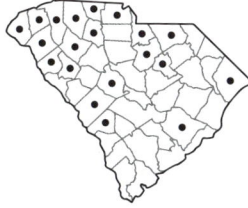

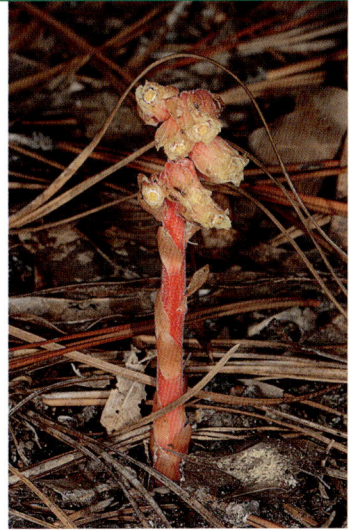

DESCRIPTION: Mycoparasitic herbaceous perennial herb without chlorophyll; 4–16″ tall; leaves reduced to scales; early season flowering plants are usually yellow or tawny, while the fall flowering plants are mostly pink to red, often marked with yellow; multiple flowers per stem; flowers May–October.

RANGE-HABITAT: Wide ranging across most of the eastern US; in SC, occasional in forests and upland woods in the mountains and Piedmont; rare in the Coastal Plain in upland woods and xeric ridges.

TAXONOMY: What was once thought of as one species appears to be several. None of the plants in eastern North America are the same species as the Eurasian Pinesap (*H. monotropa* Crantz). Based upon molecular data, there appear to be several cryptic species that need to be described. In SC, Pinesaps should be divided into two separate species, the Appalahcian Red Pinesap (*Hypotitys lanuginosa* (Michaux) Small), which flowers in the autumn and is more common throughout SC, and Eastern Pinesap, which has not yet been described, with pale stems and summer flowering. Eastern Pinesap is known from scattered locations throughout the state.

COMMENTS: Appalachian Red Pinesap obtains its nourishment from a fungus associated with its roots that in turn feeds on decaying organic matter and transfers nourishment to the plant. This species may be abundant in an area one year and not appear the next.

366. Indian Pipes

Monotropa uniflora L.
Mo-nó-tro-pa u-ni-flò-ra
Ericaceae (Heath Family)

DESCRIPTION: Mycoparasitic perennial herb without chlorophyll; stem 2–8″ tall; leaves reduced to scales; specimens vary from white, pale pink, or pale-yellow to lavender or a combination of two of these; flowers solitary at end of stem, drooping; petals inconspicuously and sparsely ciliate margined and sparsely pubescent within; filaments sparsely pubescent; flowers June–October; capsules become erect as they mature in August–November.

RANGE-HABITAT: Widespread in North America; also found in South America and east Asia; common throughout SC; upland forests in the mountains and Piedmont and in sandy forests and pocosin borders in the Sandhills and Coastal Plain.

COMMENTS: Plants are most commonly completely white. Occasional pink or salmon-colored forms may be confused with the rare Scrub Ghost Flower (*M. brittonii* Small, plate 532). Scrub Ghost Flower is confined to the Coastal Plain in SC where it is found in dry, sandy, Longleaf Pine scrub and sandhills.

367. Carolina Moonseed; Coralbeads

Cocculus carolinus (L.)
    Augustin de Candolle
      Cóc-cu-lus ca-ro-lì-nus
Menispermaceae (Moonseed
    Family)

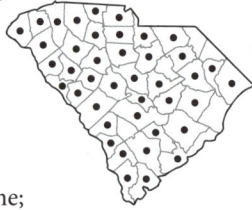

DESCRIPTION: Perennial, twining woody vine; sometimes only woody basally or woody stems long, reaching high into trees; leaves deciduous, entire to 3-lobed; male and female flowers on separate plants; fruit a brilliant red drupe, mature June–August.

RANGE-HABITAT: VA south to FL and west to MO and TX; common throughout SC (but uncommon in the mountains); thickets, sandy woods, oak-hickory forests, fields, along roadsides, and among shrubbery in home landscapes; most common in calcium-rich soils.

COMMENTS: The common name "moonseed" alludes to the shape of the stone within the fruit. Moonseed is ideal for cultivation on fences, arbors, and trellises. In full sun it produces numerous red drupes and grows in a variety of soils.

368. Pale Indian-plantain

Arnoglossum atriplicifolium (L.)
    H. E. Robinson
      Ar-no-glós-sum a-trip-li-ci-
      fò-li-um
Asteraceae (Aster Family)

SYNONYM: *Cacalia atriplicifolia* L.—RAB

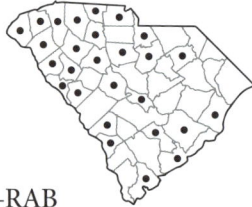

DESCRIPTION: Large perennial herb, 4–9′ tall; stems smooth, with a whitish bloom; leaves large, basal, triangular kidney-shaped, thick, with glaucous undersides; flowers June–October.

RANGE-HABITAT: Widespread in eastern North America; in SC, common in the mountains and Piedmont, scattered in the Coastal Plain and Sandhills; woodland margins, mesic deciduous forests, and clearings.

SIMILAR SPECIES: Great Indian-plantain (*Arnoglossum reniforme* (Hooker) H. E. Robinson) is similar but has very broad, bright-green, thin leaves and is not glaucous in any portion of the plant. It is found in rich cove forests in the mountains.

COMMENTS: Native Americans used the leaves as a poultice for cancers, cuts, and bruises and to draw out blood or poisonous materials.

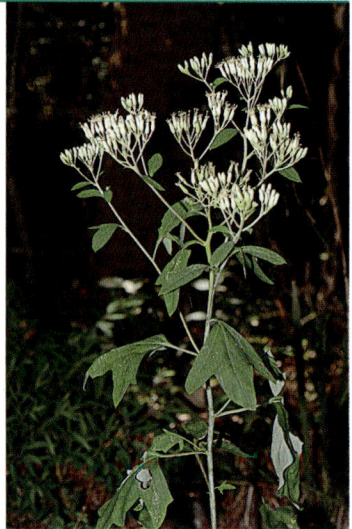

369. Hog Peanut; Groundnut

Amphicarpa bracteata (L.)
Fernald
Am-phi-cár-pa brac-te-à-ta
Fabaceae (Pea or Bean Family)

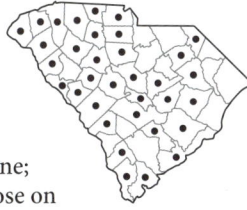

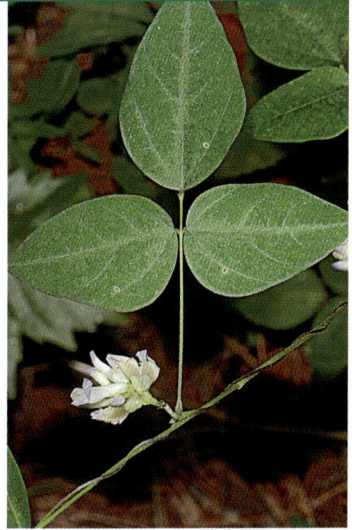

DESCRIPTION: Twining, climbing, annual vine; leaves trifoliolate; flowers of two types: those on upper branches open for pollination (chasmogamous) with petals pale purple or lilac to white; those on creeping, lower branches never open (cleistogamous), inconspicuous, without petals; fruit (legume) from cleistogamous flowers fleshy, often subterranean, 1-seeded; flowers July–October; legumes ripe August–October.

RANGE-HABITAT: Throughout most of eastern North America; common throughout SC; dry to moist forests and thickets.

COMMENTS: Native Americans boiled the subterranean fruits. The seeds from the fruit, when seasoned with salt and pepper, are not unlike garden beans. Boiling easily removes the shell. The subterranean fruits retain their vitality throughout the winter, so they may be dug into the spring. The seeds of the aerial fruits are inedible. Birds feed on the seeds of both fruit types, and hogs relish the subterranean fruits (hence the common name).

370. Cranefly Orchid

Tipularia discolor (Pursh) Nuttall
Ti-pu-là-ri-a dís-co-lor
Orchidaceae (Orchid Family)

DESCRIPTION: Perennial from a swollen underground stem bearing only a few spongy roots; single leaf produced in autumn, with a purplish lower surface, withering and disappearing at flowering time; flowers July–September.

RANGE-HABITAT: Widespread in eastern North America and Central America; common throughout SC in a variety of moist to dry habitats, usually on acidic soils.

COMMENTS: The flower resembles a cranefly—hence, the common name. *Tipula* is a genus of cranefly. Both Cranefly Orchid and Puttyroot (*Aplectrum hyemale* (Muhlenberg ex Willdenow) Torrey, plate 125) have single basal leaves that are purplish on the lower surface. Leaves of puttyroot are plicate, as in an incompletely unfolded fan, and the veins are distinctly raised and whitened. Cranefly Orchid is one of several species of dense deciduous forests that produce their leaves in the cold season (wintergreen). The dark purple backs to the leaves, which is similar to wearing a dark coat in winter; it allows them to absorb heat from solar radiation and raise the surface temperature of the leaf to allow for more efficient photosynthesis. Plants are rarely seen flowering because of the absence of the leaves and the coloration of the flowers. The authors suggest marking the location where you see the leaves in the winter and returning during July to look for the flowers.

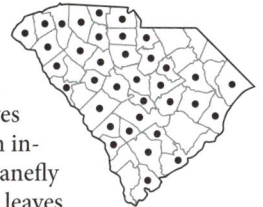

371. Appalachian Oak-leech (A)

Aureolaria laevigata (Rafinesque)
   Rafinesque
   Au-re-o-lá-ri-a lae-vi-gà-ta

Downy Oak-leech (B)

Aureolaria virginica (L.) Pennell
   Au-re-o-lá-ri-a vir-gí-ni-ca
   Orobanchaceae (Broomrape Family)

DESCRIPTION: Hemiparasites on members of the White Oak group; perennial herbs to 3′ tall or more; in *A. laevigata,* plant essentially smooth; in *A. virginica,* pubescent throughout; stem leaves opposite; in *A. laevigata,* leaves narrowly lanceolate, lower leaves entire to serrate; in *A. virginica,* leaves lanceolate to ovate lanceolate, lower leaves weakly pinnately lobed, particularly towards the leaf base; flowers yellow with very short pedicels in *A. virginica* and pedicels to 1″ long in *A. laevigata;* both species flower August–September.

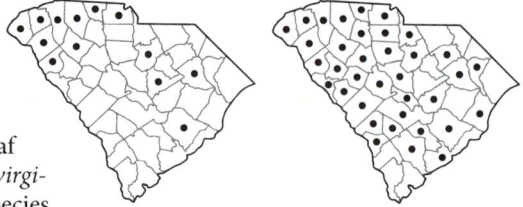

RANGE-HABITAT: Both species are wide-ranging in the eastern US; in SC, *A. laevigata* is common in the mountains, occasional in the Piedmont, rare in the Coastal Plain; *A. virginica* is common throughout; both species occur in moist-to-dry deciduous upland forests and woodlands.

SIMILAR SPECIES: Smooth Oak-leech (*A. flava* (L.) Farwell) is smooth and resembles a robust Appalachian Oak-leech. It has lower leaves pinnately lobed and is scattered throughout SC.

COMMENTS: The common name Oak-leach refers to its role as a parasite on members of the White Oak group.

Early Successional Habitats

372. Piedmont prairie

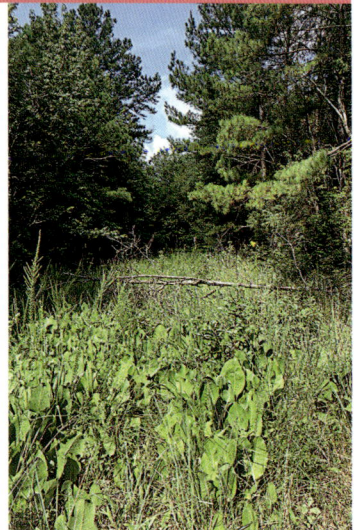

373. Xeric hardpan forest

374. Oak savanna

375. Blackjack Oak

Quercus marilandica Muenchhausen
  var. *marilandica*
    Quér-cus ma-ri-lán-di-ca var.
    ma-ri-lán-di-ca
Fagaceae (Beech Family)

DESCRIPTION: Small- to medium-sized deciduous tree with dark, deeply ridged blocky bark; twigs are sparsely pubescent, becoming smooth in the 2nd year; leaves obovate, frequently with three shallow lobes, bristle-tipped toward the apex, smooth above and pubescent to smooth below; female flowers 1–2 on a short stalk, separate from the male flowers, which are in catkins; acorns small, pubescent when young with a cup that covers one-half or more of the acorn; flowers March–April; fruits produced September–November of the following year.

RANGE-HABITAT: NY west to NE and south to the FL Panhandle and TX; common throughout SC in dry, rocky oak-hickory forest, shallow soils near outcrops, xeric hardpan forests (where often the dominant canopy), fine-textured soils within Longleaf Pine sandhills, flatwoods, and savannas (particularly where clay is present).

COMMENTS: The common name may have been derived from the skull and crossbones flag, also called a blackjack. It is abundant at the appropriately named Rock Hill Blackjacks Heritage Preserve in York County. Early accounts of the extensive savanna and prairie lands surrounding the Mecklenburg County and York County areas referred to them as "blackjack lands," in reference to the abundance of this tree.

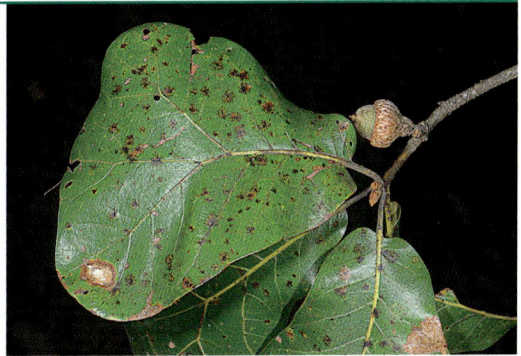

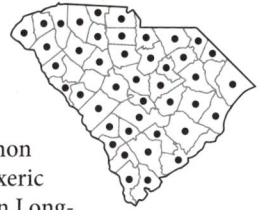

376. Post Oak

Quercus stellata Wangenheim
Quér-cus stel-là-ta
Fagaceae (Beech Family)

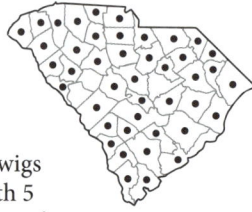

DESCRIPTION: Medium to large tree with bark that is light, ridged, and checkered; twigs pubescent when young; leaves obovate with 5 primary lobes and the two most distal lobes often directly opposing and appearing to form a crude cross; upper leaf surface covered with stellate hairs when expanding, but becoming glabrous with age, undersides persistently covered with stellate trichomes; female flowers 1–2 on a short stalk, separate from the male flowers, which are in catkins; acorn with some stellate trichomes toward the apex and with a cup that covers approximately one-fourth of the acorn; flowers March–April; fruit matures September–November of the first year.

RANGE-HABITAT: MA west to KS and south to FL and TX; in SC, common throughout the state in dry, rocky, oak-hickory forests, shallow soils near outcrops and glades, exposed ridges with heavy clay soils, Longleaf Pine flatwoods and sandhills.

COMMENTS: The common name alludes to the fact that the wood was often used for fence posts, because it is resistant to decay. The specific epithet *stellata*-refers to the tiny star-shaped hairs (trichomes) on the leaves. The species was apparently much more abundant and dominant or codominant with Blackjack Oak in many of the Piedmont prairie and savanna habitats prior to conversion to agriculture and fire suppression. An excellent natural area still exhibits a classic Post Oak savanna in the Long Cane District of the Sumter National Forest. Directions to this site can be obtained from the Forest Service website.

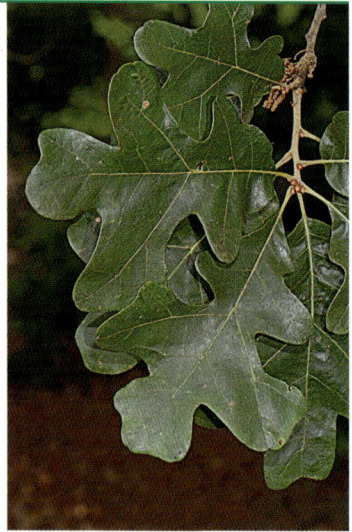

377. Carolina Shagbark Hickory

Carya carolinae-septentrionalis
(W. W. Ashe) Engler & Graebner
Cà-ry-a ca-ro-lì-nae-sep-ten-
tri-o-nà-lis
Juglandaceae (Walnut Family)

DESCRIPTION: Medium to large tree with pale shaggy bark, shedding in long strips; twigs slender, reddish when young and turning dark by end of the first year; leaves alternate, undersides with a few scattered tiny scales and sometimes with tufts of pubescence in the axils of veins; pinnately compound, typically with 5 leaflets; the terminal leaflet less than 2.3″ wide; upper leaf surface smooth, undersides with a few scattered tiny scales and sometimes with tufts of pubescence in the axils of veins; fruit is round, smooth, and up to 2″ wide; flowers March–May; fruits mature in October.

RANGE-HABITAT: VA south to GA and west to KY and MS; in SC, uncommon and mostly restricted to xeric hardpan forests and upland depression swamp forests in the Piedmont, rarely in other bottomland forest communities or basic-mesic forests, extending (barely) into the Coastal Plain along the Great Pee Dee River floodplain.

SIMILAR SPECIES: Carolina Shagbark Hickory is ecologically distinct and typically easy to distinguish from Common Shagbark Hickory (*Carya ovata* (P. Miller) K. Koch, plate 269). The terminal leaflets are much narrower, with those of Common Shagbark Hickory being 2.5 to 6″ wide, versus less than 2.3″ wide in Carolina Shagbark Hickory. Common Shagbark Hickory has much stouter twigs and buds and is pubescence and scaly on the undersides of the leaves. Additionally, Common Shagbark Hickory is found in rich upland habitats such as basic-mesic forests and marl forests and bluffs in the Piedmont and Coastal Plain.

COMMENTS: The specific epithet *carolinae-septentrionalis* is Latin for "North Carolina."

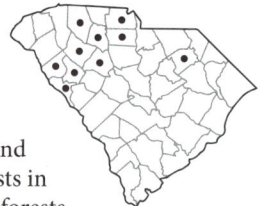

378. Creamy Wild Indigo

Baptisia bracteata Elliott
Bap-tí-si-a brac-te-à-ta
Fabaceae (Bean Family)

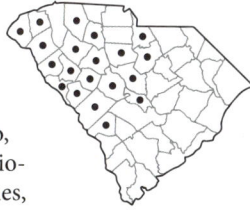

DESCRIPTION: Rhizomatous, perennial herb, 12–24″ tall; plant softly hairy; leaves trifoliolate; flowers in drooping, one-sided racemes, reflexed downward with large bracts; corolla pale yellow or cream; flowers March–April.

RANGE-HABITAT: NC south to GA and west to AL; in SC, common in the Piedmont, Sandhills, and Inner Coastal Plain where found on dry road banks; open oak-hickory woodlands; rocky woods; Longleaf Pine sandhills; and sandy, dry, open woodlands.

COMMENTS: Creamy Wild Indigo is distinctive due to the inflorescence being held below the leaves. It is easy to observe on dry road banks with clay soils in the Piedmont in the spring when the bushy clumps are highly visible.

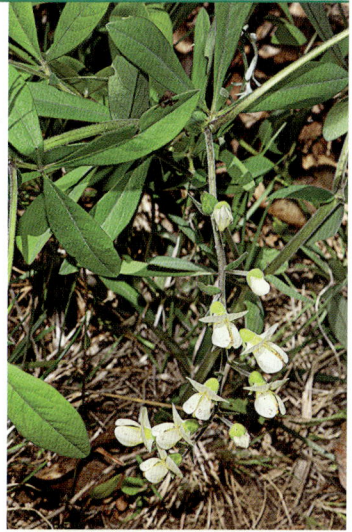

379. Curlyheads

Clematis ochroleuca Aiton
Clé-ma-tis o-chro-leù-ca
Ranunculaceae (Buttercup Family)

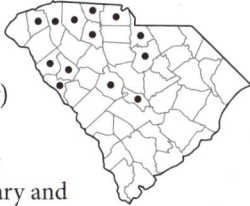

DESCRIPTION: Perennial herb with erect, unbranched stems to 2′ tall; leaves densely covered with soft hair below; flowers solitary and terminal; petals absent; sepals petal-like, bluish; flowers April–June; fruits an aggregate of achenes with persistent styles that are plumose (featherlike); fruiting May–July.

RANGE-HABITAT: MD to GA and disjunct in NY; in SC, uncommon in the Piedmont, rare in the Coastal Plain and Sandhills; restricted to dry upland forests in calcium-rich soils associated with gabbro or diabase rocks, xeric hardpan forests, and Piedmont prairie remnants.

COMMENTS: The common name derives from the fruit with its headlike clusters of achenes, each with a long, curling, and feathery style. Rock Hill Blackjacks Heritage Preserve harbors a large population of Curlyheads.

CONSERVATION STATUS: SC-Vulnerable

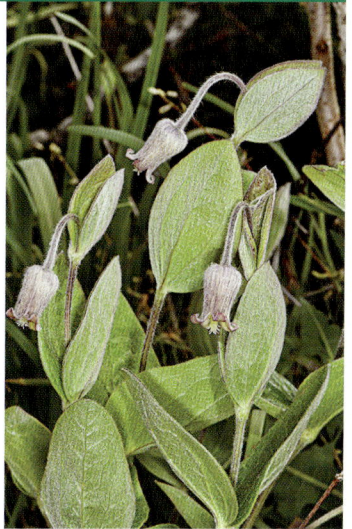

380. Clasping Milkweed

Asclepias amplexicaulis
J. E. Smith
As-clè-pi-as am-plex-i-caù-lis
Apocynaceae (Dogbane Family)

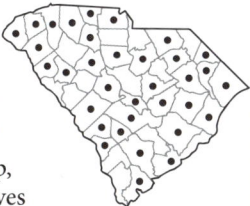

DESCRIPTION: Perennial herb with milky sap, with erect, unbranched stems 1–3′ tall; leaves opposite, smooth, auriculate-clasping with a wavy margin and whitish (glaucous) below; flowers in a single terminal umbel; flowers May–July.

RANGE-HABITAT: NH west to MN and south to FL and TX; in SC, occasional throughout; dry, open oak-hickory forests, rocky slopes, dry road banks, forest margins, Piedmont prairie remnants, Longleaf Pine flatwoods, sandhill habitats, and other dry and open habitats.

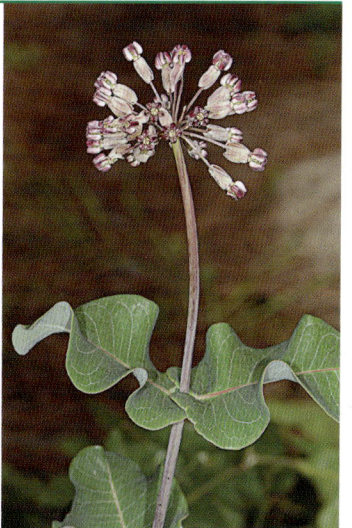

COMMENTS: This attractive species, like most milkweeds, is widespread but often exists as a few scattered individuals. It is only likely to be confused with Sandhills Milkweed (*A. humistrata* Walter, plate 521), which has stems that are held close to the ground at a low ascending angle and has bluish leaves with pink veins. Sandhills Milkweed is confined to dry, sandy habitats in the Sandhills and Coastal Plain. The genus *Asclepias* is named for the Greek god of medicine, Asclepius.

381. Low Wild-petunia

Ruellia humilis Nuttall
Ru-él-li-a hù-mi-lis
Acanthaceae (Acanthus Family)

DESCRIPTION: Short, highly branched, perennial herb, forming small colonies; leaves opposite, elliptic to narrowly ovate to 2.5″ long, short pubescent, sessile or nearly so; flowers held upright and sessile in the axils of the leaves; petals fused at the base, white or pale blue; flowers May–September.

RANGE-HABITAT: PA south to SC and west to MN and TX; in SC, very rare and restricted to the Piedmont in prairie remnants, powerline rights-of-way, xeric hardpan forests, and roadsides.

SIMILAR SPECIES: Low Wild-petunia is a short, very attractive plant that makes an ideal addition to the native garden in sunny locations. It is available from many nurseries specializing in native plants. It should never be disturbed in the wild in SC. It is restricted to York County, as a native plant, and is known from Rock Hill Blackjacks Heritage Preserve.

CONSERVATION STATUS: SC-Critically Imperiled

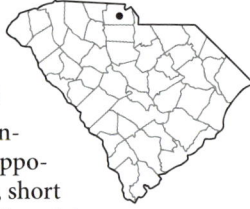

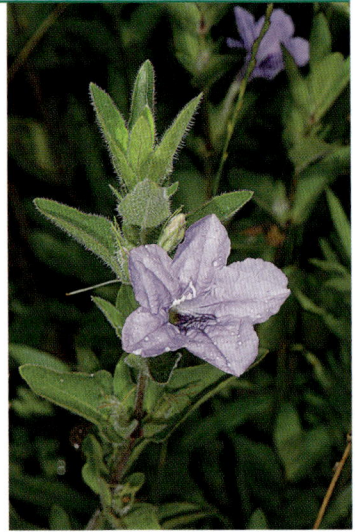

382. Carolina Rose

Rosa carolina L.
Rò-sa ca-ro-lì-na
Rosaceae (Rose Family)

DESCRIPTION: Low, semi-upright shrub to 3′ tall; stems armed with straight prickles; leaves compound, with 5, rarely 7, coarsely serrate leaflets; flowers May–June and sporadically later; hips (fruits) mature August–October.

RANGE-HABITAT: New Brunswick and Ontario south to FL and TX; throughout SC (except the maritime strand); common; open oak-hickory forests, margins of granitic domes and flatrocks, xeric hardpan forests, upland pastures, forest margins, and pinelands.

SIMILAR SPECIES: Carolina Rose is easily distinguished from Swamp Rose (*Rosa palustris,* plate 833); the former has straight prickles, the latter has prickles curved downward.

COMMENTS: Carolina Rose is suitable for cultivation and requires full sun and well-drained soil. It does especially well on rocky places unsuitable for most plants.

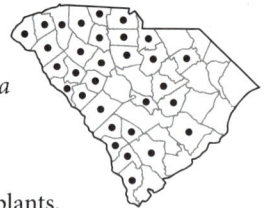

383. Piedmont Buckroot

Pediomelum piedmontanum J. R. Allison, M. W. Morris, & A. N. Egan
Pe-di-o-mè-lum pied-mon-tà-tum
Fabaceae (Bean Family)

DESCRIPTION: Perennial herb from a tuberous rootstock; stems with ascending branches, pubescent and with glandular spots; leaves palmately trifoliolate; leaflets ovate up to 2″ long and 1″ wide; flowers produced in dense racemes, with the axis of the stem not readily visible; flowers pale lavender or cream with a slight purple tinge; fruits are legumes to 0.5″ long with a pronounced beak and with their surfaces also covered with glandular spots; flowers May–July; fruits mature July–September.

RANGE-HABITAT: Known only from the lower Piedmont of GA (1 county) and SC (2 counties) where it occurs in rocky, open oak-hickory forests, road banks, and prairielike habitats.

COMMENTS: This distinctive species was described in 2006 and is known from only 3 populations. All populations are on private property, mostly along roadsides, and none has received protection. The plant should never be disturbed in the wild, and any additional populations should be reported to the SC Department of Natural Resources, Heritage Trust Program.

CONSERVATION STATUS: SC-Critically Imperiled

384. Northern Leatherflower; Vase-vine

Clematis viorna L.
Clé-ma-tis vi-ór-na
Ranunculaceae (Buttercup Family)

DESCRIPTION: Ascending, climbing to sprawling perennial, herbaceous vine; leaves opposite, bipinnately compound with leaflets entire and smooth; flowers purplish, very thick in texture and vaselike, produced in groups of 1–6 in cymes arising from the axils of leaves; flowers May–September; fruits are aggregates of achenes with persistent styles that are plumose (featherlike) when dry and carry the seed in the wind; fruits mature August–October.

RANGE-HABITAT: PA south to GA and west to MO and AR; infrequent in the mountains and Piedmont in open, oak-hickory forests, rich woods, margins of rock outcrops and forest margins; rare in the Coastal Plain in similar situations.

SIMILAR SPECIES: Netleaf Leatherflower (*Clematis reticulata* Walter, plate 547) is very similar but is a slightly more gracile plant with leaves that are thick and leathery with evident reticulate venation. It occurs in dry sandy woods, Longleaf Pine sandhills, and other open and sandy habitats in the Sandhills and Coastal Plain.

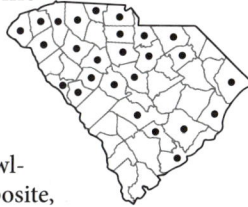

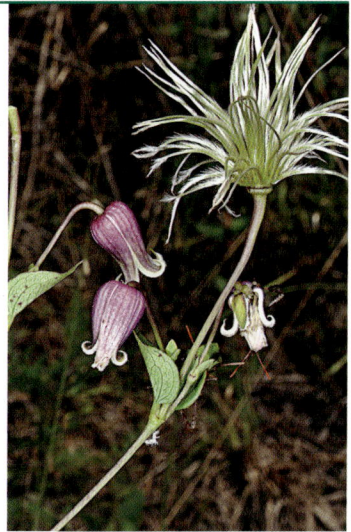

385. Gray-headed Coneflower

Ratibida pinnata (Ventenat)
Barnhart
Ra-tí-bi-da pin-nà-ta
Asteraceae (Aster Family)

DESCRIPTION: Clump-forming perennial with short tough rhizomes; stems to 4′ tall; basal and stem leaves produced; leaves pinnatifid with 3–9 lobes, narrowly lanceolate to ovate; flowers produced in involucrate heads held on long peduncles with long, drooping yellow ray flowers and disk flowers that are greenish to purplish; flowers May–October, but most prolifically in early summer.

RANGE-HABITAT: Primarily a midwestern species found from PA south to the FL Panhandle and west to SD and TX; in SC, highly disjunct and rare, restricted to Piedmont prairie remnants and xeric hardpan forests in York County.

COMMENTS: The native occurrence of this plant in York County is remarkable. It is a standard component of tall-grass prairies in the American Midwest. It makes an outstanding garden plant that supports numerous pollinators, is very easy to cultivate, and is drought tolerant. It should never be disturbed in the wild; it is widely available from native plant nurseries.

CONSERVATION STATUS: SC-Critically Imperiled

386. Smooth Coneflower

Echinacea laevigata (C. L. Boynton & Beadle) Blake
E-chi-nà-ce-a lae-vi-gà-ta
Asteraceae (Aster Family)

DESCRIPTION: Perennial herb from fibrous roots; stem 24–48″ tall, unbranched, often in clumps; basal leaves with few, spaced teeth, smooth, or rough above and smooth below, to 6″ long and 3″ wide; stem leaves similar but smaller; flowers produced in solitary involucrate heads of pale purple or violet, drooping ray flowers and purple disk flowers; flowers late May–early July and sporadically later.

RANGE-HABITAT: From se. PA south to sw. SC and ne. GA; in SC, rare in the mountains, upper Piedmont, Sandhills, and inner Coastal Plain; found in dry, open woodlands with soils rich in calcium or magnesium.

COMMENTS: This species has declined due to the loss of traditional burning practices that maintained open oak-pine savanna habitats and Piedmont prairie. All of the endemic species of Piedmont prairie and savanna are rare today. Early colonial period explorers describe the habitat as extensive and the species was no doubt more widespread prior to colonization, cultivation, and fire suppression. The South Carolina Department of Natural Resources and United States Forest Service have worked to reintroduce fire and keep the habitats of the remaining populations open. Today, it is still threatened by loss of habitat and invasive species that modify its habitat. *Echinacea* species are widely used as an herbal remedy for preventing and treating coughs and colds. Duke (1997) recommends it as an herbal remedy for viral infections, flu, and yeast infections. Smooth Coneflower should never be picked or disturbed in any way.

CONSERVATION STATUS: Federal Endangered

387. Starry Rosin-weed

Silphium asteriscus L. var. *dentatum* (Elliott) Chapman
Síl-phi-um as-te-rís-cus var. den-tà-tum
Asteraceae (Aster Family)

SYNONYM: *Silphium dentatum* Elliott— RAB, PR

DESCRIPTION: Coarse, erect, perennial herb, 2–6′ tall; stems smooth, rough, or densely spreading and hairy; lower stem leaves well developed, opposite; upper stem leaves alternate; basal leaves usually absent; bracts on receptacle with small, stalked glands on the back near the tip; flowers May–August.

RANGE-HABITAT: NC south to FL and west to KY and MS; common throughout SC; dry oak-hickory forests, old fields, and dry thin woodlands.

COMMENTS: Starry Rosin-weed has three varieties known from SC; this is the most common. It makes a very attractive and bold landscape plant when grown in full sun.

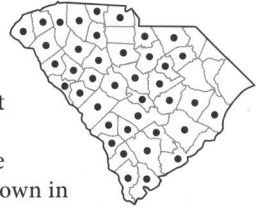

388. Green Comet Milkweed

Asclepias viridiflora Rafinesque
As-clè-pi-as vi-ri-di-fò-li-a
Apocynaceae (Dogbane Family)

DESCRIPTION: Erect, perennial herb to 20–32″ tall; stems and leaves minutely pubescent or becoming smooth with age; leaves opposite, linear-lanceolate to suborbicular, 1–4″ long, margins entire; flowers produced in 1–6 umbels from the upper nodes; corolla pale green, lobes reflexed, horns absent; fruit is a follicle; flowers June–August; fruits mature August–September.

RANGE-HABITAT: Throughout much of the central and eastern US and southern Canada, though predominantly midwestern; in SC, uncommon in the Piedmont but found in Piedmont prairie remnants, forest margins, margins of outcrops, and other open and grassy habitats, almost always over mafic or ultramafic rock.

COMMENTS: Green Comet Milkweed is one of the strangest looking milkweeds in SC. The round green umbels of slender flowers are unlike any other species of *Asclepias* in the piedmont. Pineland Milkweed (*A. obovata* Elliott) is a similar species. The genus is named for the Greek god of medicine, Asclepius.

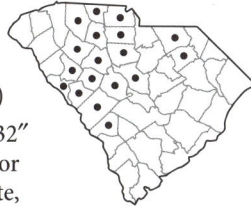

389. Wild Quinine

Parthenium integrifolium
L. var. *integrifolium*
Par-thè-ni-um in-teg-ri-fò-
li-um var. in-teg-ri-fò-li-um
Asteraceae (Aster Family)

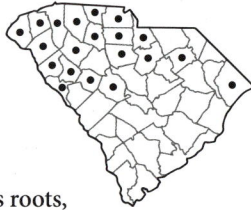

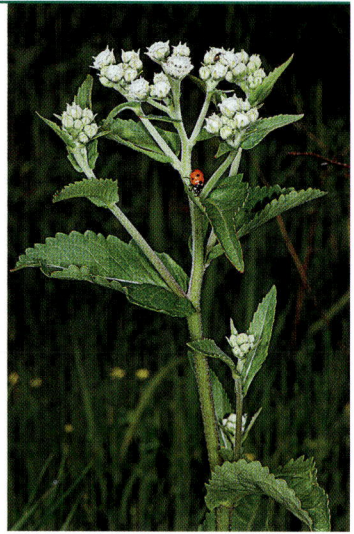

DESCRIPTION: Perennial herb with tuberous roots, 2–3′ tall; leaves large, oval, lanceolate, to 1′ long, rough, blunt-toothed; basal leaves with long leaf stalks; stem leaves reduced upward, the upper often sessile and clasping; flowers in involucrate heads composed of whitish disk flowers and a few sterile, whitish ray flowers; flowers late May–August.

RANGE-HABITAT: VA south to GA and west to MN and MS; in SC, occasional in the mountains, common throughout the Piedmont and inner Coastal Plain; dry open woods and forests, old fields, forest margins, Piedmont prairie remnants, and along roadsides.

TAXONOMY: Two varieties occur in SC. Sandhills Wild Quinine (*P. integrifolium* var. *mabryanum* Mears) has fewer flowers per head and narrower and less coarsely toothed leaves. It is found in Longleaf Pine sandhills and dry oak-hickory woodlands in the lower Piedmont and Sandhills regions.

COMMENTS: The flowering tops were once used for "intermittent fevers" like malaria—hence, the common name. Wild Quinine may cause dermatitis or allergies.

390. Spurred Butterfly-pea

Centrosema virginianum (L.)
Bentham
Cen-tro-sè-ma vir-gi-ni-à-
num
Fabaceae (Bean Family)

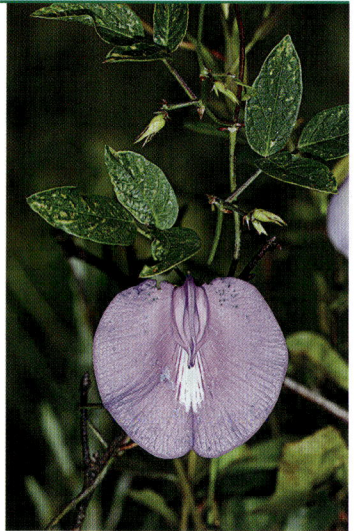

DESCRIPTION: Twining or trailing perennial, herbaceous vine to 5′ long from a tough, elongate root; leaves with 3 narrow leaflets; leaflets dark green, widest toward the base with acute to short acuminate tips, somewhat rough in texture above, coarsely reticulately veined below; standard petal very wide, held splayed open, with a nectar spur at the base; flowers June–October.

RANGE-HABITAT: From s. NJ south to FL and west to KY, AR, and TX; common throughout SC; oak-hickory forests, Longleaf Pine flatwoods, sandy and dry open woods, woodland openings, maritime forests, and coastal dunes.

SIMILAR SPECIES: Butterfly-pea (*Clitoria mariana* L., plate 391) is similar and often confused. These two species both produce resupinate flowers. The flowers of Butterfly-pea have a narrower standard without a nectar spur, are often lighter in coloration, and are not as widely open as Spurred Butterfly-pea. Additionally, Butterfly-pea is often a much shorter plant and has lighter colored leaves that are smooth, with less evident venation below, and thicker in texture than Spurred Butterfly-pea.

391. Butterfly-pea; She-pea

Clitoria mariana L.
  Cli-tò-ri-a ma-ri-à-na
 Fabaceae (Bean Family)

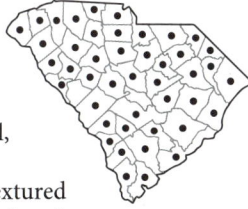

DESCRIPTION: Twining or trailing perennial, herbaceous vine to 3′ long, from a tough, elongate root; leaves with 3 rather thick-textured leaflets widest toward the middle or just below the middle; flowers resupinate (twisted upside down before flowering so that the broadest petal, the standard, is lowermost) without a nectar spur; the petals arranged so that the standard is not splayed wide-open; flowers June–October.

RANGE-HABITAT: From s. NJ south to FL and west to KY, AR, and TX; common throughout SC; oak-hickory forests, Longleaf Pine flatwoods, sandy and dry open woods, woodland openings, maritime forests, and coastal dunes.

SIMILAR SPECIES: Spurred Butterfly-pea (*Centrosema virginianum* (L.) Bentham, plate 390) is very similar; see discussion of that species (above).

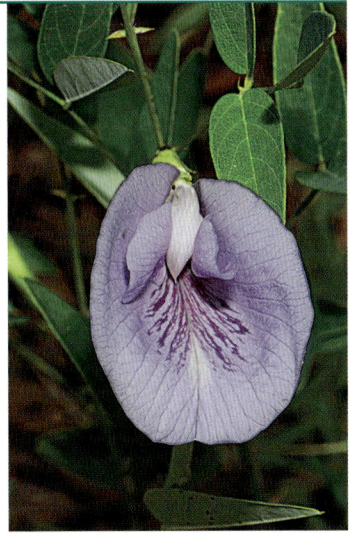

392. Pale Spiked Lobelia

Lobelia spicata Lamarck
  Lo-bèl-i-a spi-cà-ta
 Campanulaceae (Bellflower
  Family)

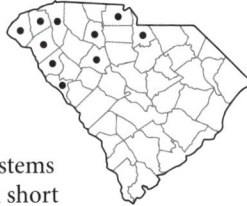

DESCRIPTION: Perennial herb with solitary stems and basal offshoots; stems smooth or with short pubescence, to 3′ tall; leaves oblong to elliptic, sessile or nearly so, with crenate to denticulate margins; flowers white or very pale blue, produced in elongate, dense, spikelike racemes; the flowers 0.3 to 0.55″ long; flowers May–August.

RANGE-HABITAT: Widespread in eastern North America; in SC, occasional in the Piedmont where found in open, nutrient-rich soils of forest margins, margins of rock outcrops., stream sides, and powerline clearings.

COMMENTS: The elongate flowering stems of pale flowers are distinctive. The genus honors Matthias de l'Obel (1538–1616), a Flemish herbalist.

393. Northern Rattlesnake-master

Eryngium yuccifolium Michaux var.
 yuccifolium
  E-rýn-gi-um yuc-ci-fò-li-um var.
  yuc-ci-fò-li-um
 Apiaceae (Carrot Family)

DESCRIPTION: Clump-forming perennial to 4′ tall; leaves basal and on the stem, linear, "yuccalike," thick and leathery with spinelike bristles on the margins, typically bluish-green; marginal bristles solitary; stems with ascending branches terminating in a dense headlike inflorescences; heads nearly round to ovoid; petals greenish-white; flowers June–August.

RANGE-HABITAT: NJ south to FL and west to MN and TX; in SC, common throughout in Longleaf Pine savannas and flatwoods, forest margins, margins of outcrops, and Piedmont prairie remnants.

COMMENTS: The common name comes from the false claim that it is able to cure snakebite or perhaps for a mystical power to keep rattlesnakes away. Native Americans traditionally used the fibers in the leaves to make rope and string (Ajilvsgi, 1984).

Two varieties of this species are recognized in SC. Southern Rattlesnake-master (*E. yuccifolium* var. *synchaetum* Gray ex Coulter & Rose) has more narrow leaves and has marginal bristles grouped, with 2–3 together. It is common in moist to wet savannas in the Coastal Plain.

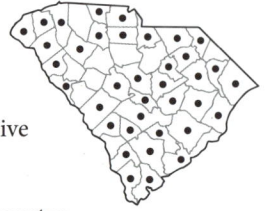

394. Northern Wild Senna

Senna hebecarpa (Fernald)
  H. S. Irwin & Barneby
  Sén-na he-be-cár-pa
  Fabaceae (Bean Family)

SYNONYM: *Cassia hebecarpa*
  Fernald—RAB

DESCRIPTION: Robust herbaceous perennial, to 5′ tall, appearing shrublike but dying to the ground in winter; leaves evenly pinnate with 6–10 pairs of elliptic to oblong, smooth leaflets; flowers produced in racemes of 5–10 flowers from the axils of leaves; fruit a legume with nearly square segments and pubescent when young; flowers June–August; fruits mature July–November.

RANGE-HABITAT: MA west to WI and south to SC and TN; in SC, uncommon, scattered throughout the Piedmont and Inner Coastal Plain, often in gladelike woodlands; Piedmont prairie remnants; fields; roadsides; and other open, grassy, moist habitats.

SIMILAR SPECIES: Maryland Wild Senna (*Senna marilandica* (L.) Link) is very similar but produces a legume that has narrow segments that are much shorter than broad and smooth, even when young. It grows in similar habitats scattered throughout the Piedmont.

COMMENTS: Both Northern Wild Senna and Maryland Wild Senna serve as hosts to the caterpillars of sulphur butterflies. Both species may be used as an alternative to non-native *Senna* species to attract sulphur butterflies to the garden.

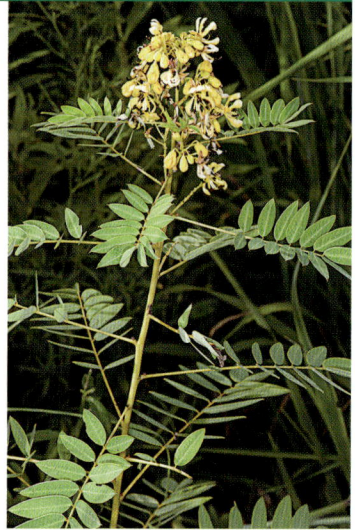

395. Pencil-flower

Stylosanthes biflora (L.) Britton, Sterns
  & Poggenberg
  Sty-lo-sán-thes bi-flò-ra
  Fabaceae (Bean Family)

DESCRIPTION: Low, prostrate, perennial herb with stems 4–20″ long; leaves trifoliolate; leaflets dark green, elliptic to oblanceolate, up to 1.5″ long; petal orangish or yellow; flowers June–August; fruit a short, pubescent legume, produced July–October.

RANGE-HABITAT: NY west to KS south to FL and TX; common throughout SC in open, sunny habitats such as Longleaf Pine flatwoods, glades, barrens, margins of rock outcrops, Piedmont prairie remnants, roadsides, and forest margins.

SIMILAR SPECIES: When this common species is encountered along road banks or powerline rights-of-way, it serves as an excellent indicator of previously fire-maintained, open habitats. It is now surviving in remnants of a formerly more open habitat.

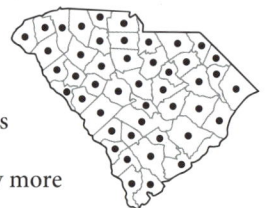

396. Prairie Dock

Silphium terebinthinaceum
Jacquin
Síl-phi-um te-re-bin-thi-nà-
ce-um
Asteraceae (Aster Family)

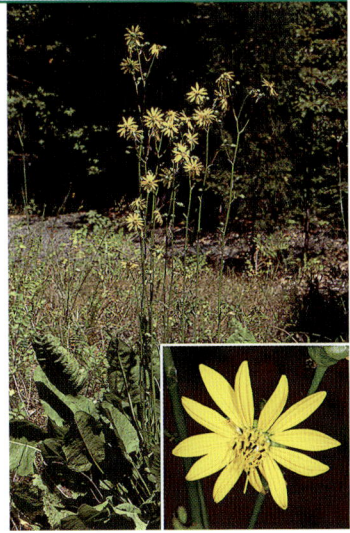

DESCRIPTION: Perennial herb from a taproot; leaves mostly basal and with long stalks and blades that are heart-shaped at the base and very large (mostly 12–16″ long); flowering stems often 7–10′ tall; flowers produced in involucrate heads of yellow ray and yellow disk flowers; heads numerous, large; flowers July–September.

RANGE-HABITAT: Ontario and MN, south to MS and AL; restricted in the Carolinas to the Piedmont; rare; Piedmont prairie remnants and glades in xeric hardpan forests, roadsides, and powerline rights-of-way.

COMMENTS: One of many SC plants with prairie affinities that is becoming more uncommon because fire is no longer part of the Piedmont landscape. This species makes a dramatic and long-lived addition to the sunny garden and is very attractive to pollinators, particularly butterflies. It is commonly available from mail order nurseries specializing in native plants and should never be disturbed in the wild in SC.

CONSERVATION STATUS: SC-Critically Imperiled

397. Southern Obedient-plant

Physostegia virginiana ssp.
praemorsa (Shinners)
Cantino
Phy-sos-té-gi-a vir-gi-ni-à-na
ssp. prae-mór-sa
Lamiaceae (Mint Family)

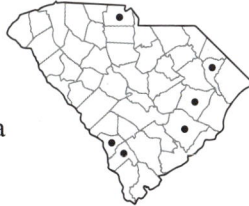

DESCRIPTION: Perennial herb to about 4′ tall, forming small clumps with short rhizomes; stems 4-angled; leaves opposite, with stalks or sessile, the largest sharply serrate; flowers produced in terminal racemes; petals tubular, bilabiate, purplish-pink; flowers June–October.

RANGE-HABITAT: From OH south to FL and west to IL and NM; in SC, restricted to a few locations in the Piedmont and Coastal Plain where found in Piedmont prairie remnants, open, sunny habitats on rich soils derived from mafic or calcareous bedrock, and open pinelands.

COMMENTS: The common name, Obedient-plant, comes from the fact that the flowers tend to stay in a new position for a while after they are twisted to one side. The weedy Northern Obedient-plant has been introduced into SC from horticulture and has escaped in many places. Northern Obedient-plant produces copious amounts of lateral rhizomes from the main rhizome and forms large masses. The more refined and rare Southern Obedient Plant is not well-known but is a beautiful and well-behaved landscape plant.

CONSERVATION STATUS: SC-Imperiled

398. Whorled-leaf Tickseed;
Woodland Coreopsis

Coreopsis major Walter
Co-re-óp-sis mà-jor
Asteraceae (Aster Family)

DESCRIPTION: Perennial herb, 20–40″ tall; rhizomatous, but with stems commonly tufted; upper and middle nodes with opposite leaves; leaves deeply divided into three segments, giving the appearance of 6 leaves in a whorl; middle leaflet of median leaves 0.2–1.2″ wide; flowers June–August.

RANGE-HABITAT: From PA and OH, south to FL, and west to MS; common throughout SC; dry woodland margins, dry oak-hickory forests, Longleaf Pine flatwoods, sandy and dry open woods, and along roadsides.

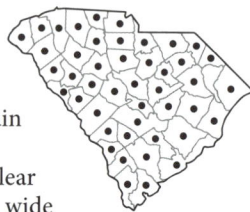

SIMILAR SPECIES: Threadleaf Tickseed (*C. verticillata* L., plate 467) is similar, but its leaf divisions are under 0.2″ wide, and it is restricted to the Coastal Plain and Piedmont of SC.

TAXONOMY: Various authors have described several varieties of *C. major*. It is clear that there are two main forms. One has downy pubescent, thin leaves that are wide and another has narrower, leathery, thick leaves that are smooth. There are diploid and polyploid complexes within this group, and it is likely that several segregate taxa will one day be recognized as distinct. Whorled-leaf Tickseed is great for naturalizing along clay banks and in difficult soils.

399. Hairy Sunflower

Helianthus hirsutus Rafinesque
He-li-án-thus hir-sù-tus
Asteraceae (Aster Family)

DESCRIPTION: Rhizomatous, perennial herb with rough pubescent stems to 6′ tall, forming large colonies; basal leaves absent; stem leaves opposite, lanceolate, roughly pubescent and scabrous (like sandpaper) above; the leaf base widely rounded to cuneate; leaves with short petioles; flowers produced in involucrate heads more than 3″ wide from ray-tip to ray-tip, with yellow ray flowers and yellow disk flowers; flowering June–August.

RANGE-HABITAT: Widespread in eastern North America from PA and MN, south to FL and TX; in SC, common in the mountains and Piedmont and uncommon in the Sandhills and Coastal Plain; open, rocky woodlands, margins of rock outcrops, oak-hickory forests, forest margins, Piedmont prairie remnants and road banks.

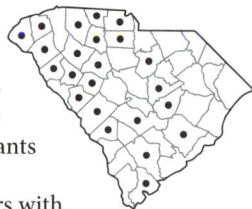

COMMENTS: This is one of the two earliest flowering of our many tall sunflowers with large flowers and without basal leaves present at flowering. It is often misidentified by native plant enthusiasts. The most similar species is Spreading Sunflower (*H. divaricatus* L.), which is similar in stature and has sessile leaves and glabrous to sparsely pubescent stems. Spreading Sunflower is widely distributed in similar habitats throughout the state.

400. Georgia Savory; Georgia Calamint

Clinopodium georgianum
R. M. Harper
Cli-no-pò-di-um geor-gi-à-num
Lamiaceae (Mint Family)

SYNONYM: *Satureja georgiana* (Harper) Ahles—RAB, PR

DESCRIPTION: Loosely sprawling, highly aromatic, freely branched, semi-evergreen shrub to 18″ tall; leaves opposite; corolla pink to lavender; flowers July–September.

RANGE-HABITAT: From s. NC south to FL and west to LA; in SC, primarily in the Piedmont (occasional) and rare in the Sandhills and Coastal Plain; dry, sandy, or rocky woods and forests.

COMMENTS: Georgia Savory is a good landscape plant for problem areas with dry, thin soil in full sunlight. The plants have one of the most pungent and pleasant minty odors of any species.

401. Sweet Goldenrod; Licorice Goldenrod

Solidago odora Aiton
So-li-dà-go o-dò-ra
Asteraceae (Aster Family)

DESCRIPTION: Erect perennial herb, 2–3′ tall; plant essentially smooth; leaves alternate, smooth, narrow, sessile, with small translucent dots; flowers yellow, in tiny involucrate heads of a few yellow ray flowers and yellow disk flowers; heads produced along one side of slightly arching branches; flowers July–October.

RANGE-HABITAT: New England south to FL, west to TX, and north to OH; common throughout SC; dry forests and woodlands, forest margins and roadsides, xeric Longleaf Pine sandhills, xeric hardpan forests, and Longleaf Pine flatwoods and savannas.

COMMENTS: The crushed leaves give off a licorice (anise) odor that readily identifies this goldenrod from similar species. A pleasant tea can be made from the leaves. The leaves were formally used as a digestive stimulant and for a variety of other uses, including treating toothache. A cluster of leaves can be chewed and then packed near an aching tooth to relieve pain, as a substitute for clove oil.

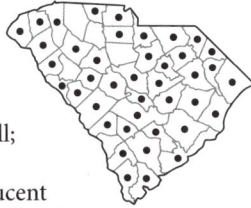

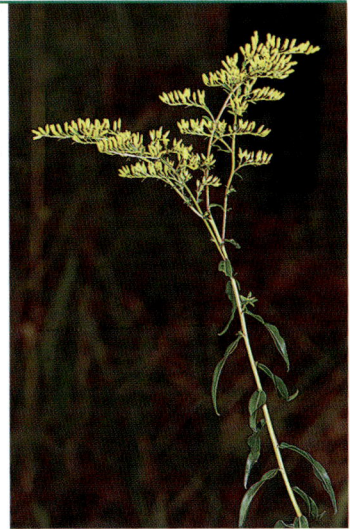

402. Bitter-bloom

Sabatia angularis (L.) Pursh
Sa-bà-ti-a an-gu-là-ris
Gentianaceae (Gentian Family)

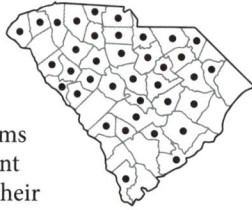

DESCRIPTION: Annual herb, 12–20″ tall; stems quadrangular, usually winged; a bushy plant with leaves that "clasp" the stem between their basal lobes; leaves opposite, entire; inflorescence branches paired; flowers rose-pink with a greenish center, occasionally white; flowers July–August.

RANGE-HABITAT: NY south to the FL Panhandle and west to MI, IL, and TX; in SC, common throughout in deciduous forests and woodlands, forest margins, freshwater marshes, along roadsides, granite outcrops, and fields.

COMMENTS: The genus is dedicated to Liberato Sabbati (1714–78), an Italian botanist.

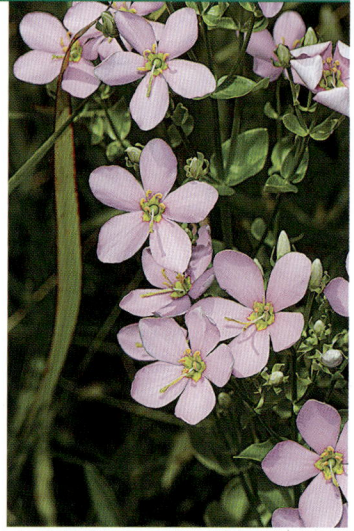

403. Bearsfoot

Smallanthus uvedalia (L.) Mackenzie
Smal-lán-thus uve-dà-li-a
Asteraceae (Aster Family)

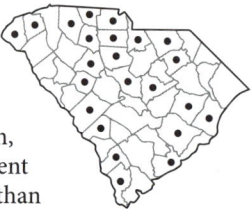

SYNONYM: *Polymnia uvedalia* L.—RAB, PR

DESCRIPTION: Perennial herb with thick, fleshy roots; stems erect, 3–10′ tall, hollow; glandular or with spreading hairs beneath the stems; leaves opposite, palmately lobed or cut, 4–12″ long, their shape suggesting the common name, Bearsfoot; flowers July–October.

RANGE-HABITAT: NY west to IL and KS, south to FL and TX; in SC, common throughout in moist forests and disturbed sites such as woodland borders and pastures; most abundant on calcium-rich or magnesium-rich soils; mostly on shell deposits or other calcareous sites in the maritime strand.

COMMENTS: The specific epithet honors the English horticulturalist Robert Uvedale (1642–1722), who had the plant in his garden.

404. Scaly Blazing Star

Liatris squarrosa (L.) Michaux
Li-à-tris squar-rò-sa
Asteraceae (Aster Family)

DESCRIPTION: Perennial herb from a swollen, bulbous, rootstock; stems 1–3′ tall, pubescent upwards; lowest leaves broadly linear, less than 0.6″ wide, smooth to pubescent; leaf size reduced upward; flowers produced in involucrate heads usually held on distinct and fairly long peduncles, occasionally nearly sessile; individual heads quite large, purplish, showy; the involucral bracts subtending the head green or purplish, without a light margin and with strongly divergent and recurved tips; flowers June–August.

RANGE-HABITAT: MD south to FL and west to MO and TX; in SC, common, found throughout the state in open, sunny, dry habitats

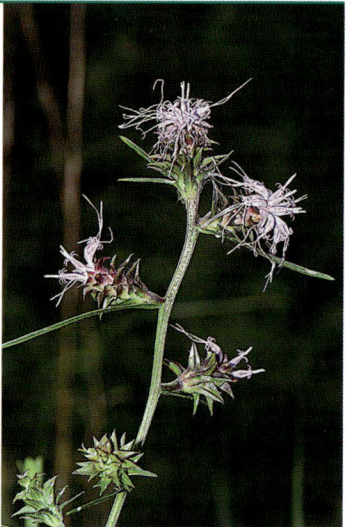

such as Piedmont prairie remnants, roadsides, forest margins, dry, rocky oak-hickory woodlands, margins of rock outcrops, Longleaf Pine sandhills, and dry and sandy oak woodlands.

COMMENTS: This species is unmistakable because of its strongly reflexed involucral bracts ("squarrose") and the source of the specific epithet and "scaly" in the common name. The name is similar to *Liatris squarrulosa* Michaux, plate 408, but it does not resemble that plant at all.

405. Southern Crownbeard

Verbesina occidentalis (L.) Walter
Ver-be-sì-na oc-ci-den-tà-lis
Asteraceae (Aster Family)

DESCRIPTION: Single-stemmed perennial herb 6–9′ tall; stems smooth or minutely hairy; leaves opposite, with leaf stalks decurrent on stem as a wide wing; flowers in involucrate heads of yellow disk and yellow ray flower; ray flowers unevenly spaced around the head; flowers late August–October.

RANGE-HABITAT: MD south to FL and west to OH, MO, and MS; common throughout SC; forests, woodlands, pastures, and along moist roadsides; especially abundant in alluvial areas or calcium-rich soils.

COMMENTS: This is the only crownbeard found in SC with a yellow flower and opposite leaves. The common name crownbeard comes from the bracts that subtend each flower, appearing like a crown to some observers and like a beard to others.

406. Small-headed Sunflower

Helianthus microcephalus Torrey & Gray
He-li-án-thus mic-ro-cé-pha-lus
Asteraceae (Aster Family)

DESCRIPTION: Rhizomatous, perennial herb with glabrous or sparsely pubescent stems to 8′ tall, forming colonies; basal leaves absent; stem leaves lanceolate with acuminate tips, glabrous to pubescent below, scabrous (like sandpaper) above; leaf arrangement opposite below, occasionally becoming subopposite to alternate toward the inflorescence; petioles 0.20–1″ long; flowers produced in involucrate heads on highly branching stems, small, less than 3″ from ray tip to ray tip; ray and disk flowers yellow; flowers August–October.

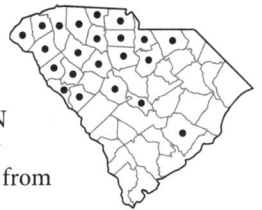

RANGE-HABITAT: Widespread in the eastern US; NJ south to FL and west to MN and LA; in SC common in the mountains and Piedmont in open oak-hickory forests, margins of rock outcrops, roadsides, and other open habitats ranging from mesic to dry.

COMMENTS: This is the most common small-flowered sunflower in SC. Like all sunflowers, it is beneficial to a wide variety of wildlife, both as a nectar source and for the nutritious achenes.

407. Appalachian Sunflower

Helianthus atrorubens L.
    He-li-án-thus a-tró-ru-bens
Asteraceae (Aster Family)

DESCRIPTION: Robust, perennial herb to 5′ tall with rough, pubescent stems and leaves; basal leaves rhombic to triangular ovate, often less than twice as long as broad; lower stem leaves opposite, upper leaves few and subopposite to alternate; flowers in involucrate heads of yellow ray flowers and dark purple disk flowers; late August–October.

RANGE-HABITAT: VA south to FL and west to TN and LA; in SC, common throughout the state in dry, sandy, or rocky woodlands with an open canopy; Longleaf Pine sandhills and flatwoods; and roadsides.

COMMENTS: Appalachian Sunflower is distinctive because of the nearly leaf-less stems with showy inflorescences with contrasting yellow ray flowers and purplish disk flowers. Only Savanna Sunflower (*Helianthus heterophyllus* Nuttall, plate 661) is similar; it has much more narrow basal leaves (2–5 times as long as broad) and only occurs in moist flatwoods and Longleaf Pine savannas in the Coastal Plain.

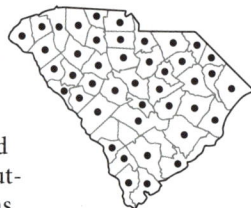

408. Southern Blazing Star; Appalachian Blazing Star

Liatris squarrulosa Michaux
    Li-à-tris squar-ru-lò-sa
Asteraceae (Aster Family)

SYNONYM: *Liatris earlei* (Greene)
    K. Schumann—RAB

DESCRIPTION: Perennial herb from a swollen, bulbous, rootstock; stems short pubescent, 2–6′ tall; lowest leaves elliptic, often nearly 2″ wide, smooth to short pubescent; leaf size and width reduced upward; flowers produced in involucrate heads produced in a long spikelike arrangement; individual heads quite large; disk flowers only, pink to purplish, showy; at least some of the involucral bracts subtending the inflorescence recurved; flowers August–October.

RANGE-HABITAT: WV west to OK and south to TX and FL; in SC, uncommon in the Piedmont in open, sunny habitats such as Piedmont prairie remnants, forest margins, often on high magnesium or high calcium soils; rare in the Sandhills and Coastal Plain in flatwoods, dry sandy oak forests, shell middens and forest margins in sands with a high content of clay or silt.

SIMILAR SPECIES: There are two species of similar blazing stars in the Piedmont of SC that are frequently confused with Southern Blazing Star, even by professional botanists. Northern Blazing Star (*Liatris scariosa* (L.) Willdenow) is similar but has heads produced on evident peduncles and with all of the involucral bracts ascending rather than reflexed. It occurs scattered throughout the mountains and Piedmont. Rough Blazing Star (*Liatris aspera* Michaux) is very similar, but it has light-colored margins to the swollen (bullate) involucral bracts. It is also found scattered in similar habitats across the Piedmont.

COMMENTS: Though one of the common names is Appalachian Blazing Star, this is a species most common in the Piedmont, Sandhills, and Coastal Plain. It is a large and showy species that makes a good garden specimen, if staked to support the tall stems.

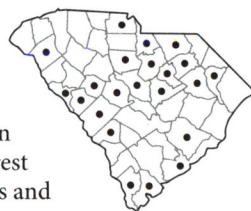

409. Grass-leaved Blazing Star

Liatris pilosa (Aiton) Willdenow
Li-à-tris pi-lò-sa
Asteraceae (Aster Family)

SYNONYM: *Liatris graminifolia*
Willdenow—RAB, PR

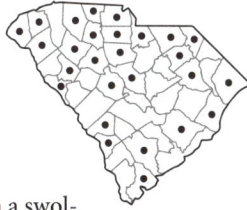

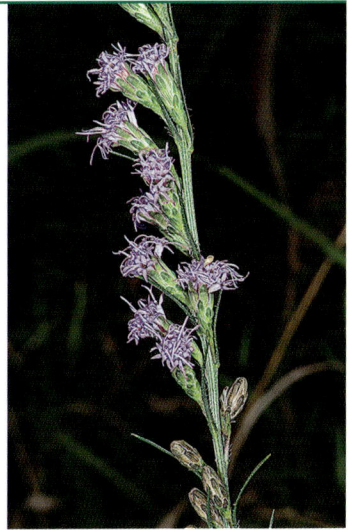

DESCRIPTION: Perennial herb to 5′ tall from a swollen, bulbous rootstock; leaves alternate, numerous, reduced in size upward, narrow, linear and 2–8″ long with ciliate margins; flowers produced in narrow involucrate heads of disk flowers only; heads in a narrow spikelike arrangement; involucral bracts rounded, without a pointed tip; heads longer than broad, sessile or on short stalks with 7–13 purple disk flowers per head; pappus bristles nearly as long as the corolla tube; flowers August–November.

RANGE-HABITAT: NJ and PA south to SC; in SC, common throughout; Longleaf Pine sandhills, oak savannas, Piedmont prairie remnants, roadsides, old fields, thinly wooded habitats, and glades.

SIMILAR SPECIES: The numerous species of *Liatris* with thin leaves and heads of purple flowers in a spikelike arrangement are difficult to distinguish. Grass-leaved Blazing Star is by far the most abundant narrow-leaved, small-flowered species in the state. Wand Blazing Star (*L. virgata* Nuttall) is very similar but has involucral bracts with a sharply acute tip. It is found in dry, open habitats throughout SC. Small-head Blazing Star (*L. microcephala* (Small) K. Schumann) is rare in SC, being known from acidic rock outcrops and rocky, open woodlands on thin soil at scattered locations in the Piedmont. It may be distinguished by the smaller heads with only 4–5 florets per head and more congested spikelike head arrangement of heads. Sandhills Blazing Star (*L. cokeri* Pyne & Stucky, plate 488) has involucral bracts that have a sharp acute tip and produces more congested arrangements of heads that are frequently all directed upwards. It is found in the Sandhills and Coastal Plain in Longleaf Pine sandhills and Longleaf Pine-Turkey oak xeric ridges.

COMMENTS: All blazing star species that grow in SC make excellent choices for the sunny garden and support numerous pollinators.

410. Schweinitz's Sunflower

Helianthus schweinitzii Torrey & Gray
He-li-án-thus schwei-nítz-i-i
Asteraceae (Aster Family)

DESCRIPTION: Perennial herb from a tuberous rootstalk; stems to 4.5′ tall; leaves sessile, 6–10x as long as wide, covered with dense, short, soft hairs below; heads small (the disc 0.2–0.6″ across), ray and disc flowers yellow; flowers late August–October.

RANGE-HABITAT: Restricted to the Piedmont of NC and SC; in SC, very rare in Piedmont prairie remnants, glades in xeric hardpan forests with diabase or gabbro rocks, mowed power lines, roadsides, and field margins.

COMMENTS: A fire-adapted species now found typically in mowed areas under power lines, roadsides, and field margins, as fires are no longer a part of the Piedmont landscape. Its rarity is probably due to its inability to persist as its habitat becomes shaded during intervals between disturbances. The specific epithet honors Lewis David von Schweinitz (1780–1834), a Moravian botanist, born in Pennsylvania and settling in Salem, NC; his work on fungi established him as the "patron saint of North American mycology."

CONSERVATION STATUS: Federally Endangered

411. Stiff-leaf Goldenrod; Bold Goldenrod

Oligoneuron rigidum (L.) Small
O-li-go-neù-ron rí-gi-dum
Asteraceae (Aster Family)

DESCRIPTION: Robust perennial herb, 3–6′ tall from short rhizomes, forming tight clumps; basal leaves lanceolate to elliptic-lanceolate to 2.5″ wide, scabrous above and pubescent to glabrous below; stem leaves reduced upward; flowers produced in involucrate heads consisting of yellow ray and yellow disk flowers; heads in a flat-topped, corymbiform arrangement; flowers August–October.

RANGE-HABITAT: Generally midwestern in distribution, extending into the Southeast in prairie remnants; in SC, rare and restricted to Piedmont prairie remnants, roadsides and powerline right-of-ways in the Piedmont.

COMMENTS: This is a typical species of the prairie remnants at Rock Hill Blackjacks Heritage Preserve in York County. It is rare in SC and should not be disturbed in the wild; it is readily available from native plant nurseries and makes an outstanding addition to the meadow garden. Two varieties are recognized in SC: *O. rigidum* var. *glabratum* (E. L. Braun) G. L. Nesom, with smooth backs to the leaves and smooth involucral bracts; and *O. rigidum* var. *rigidum,* which has pubescent backs to the leaves and pubescent involucral bracts. The validity, range, and abundance of each variety in SC requires clarification.

CONSERVATION STATUS: SC-Critically Imperiled

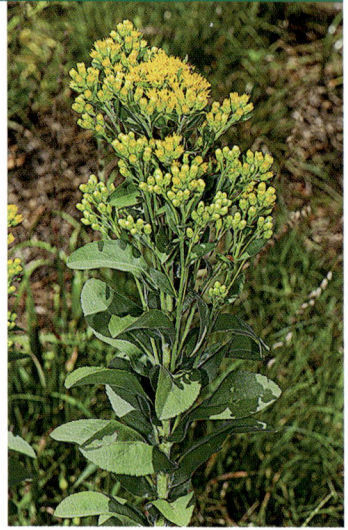

412. White Prairie-goldenrod

Oligoneuron album (Nuttall)
G. L. Nesom
O-li-go-neù-ron ál-bum
Asteraceae (Aster Family)

DESCRIPTION: Herbaceous perennial from short rhizomes, forming dense clumps; stems 4–15″ tall and highly branched; basal and low cauline leaves dark green, essentially smooth, narrow, linear to linear-elliptic, reduced upwards; flowers produced in involucrate heads composed of white ray flowers and yellow disk flowers in a flat-topped corymbiform arrangement; flowers July–October.

RANGE-HABITAT: VT south to GA and west to Saskatchewan and CO; in SC, rare and restricted to Piedmont prairie remnants and openings in xeric hardpan forest in the Piedmont.

COMMENTS: White Prairie-goldenrod is more likely to be mistaken for an "aster" than a goldenrod. The plants resemble the white-topped asters in the genus *Sericocarpus* but have much longer, narrower lower and basal leaves. The distribution east of the Appalachians is amazingly disjunct, as this is a plant of midwestern prairies. It should not be disturbed in the wild in SC. It is readily available from native plant nurseries and makes a wonderful addition to the meadow garden.

CONSERVATION STATUS: SC-Imperiled

413. Piedmont Wand Goldenrod; Southern Bog Goldenrod

Solidago austrina Small
So-li-dà-go aus-trì-na
Asteraceae (Aster Family)

DESCRIPTION: Herbaceous perennial from slender rhizomes; stems to 3′ tall, with a single main axis and wandlike or with short branches, if the stem is branched, the lower branches are typically arching upward; leaves narrow, elliptic to linear-elliptic with a serrate and scabrous margin; flowers arranged in involucrate heads of 2–7 flowers composed of yellow ray flowers and yellow disk flowers; the heads are arranged in a spikelike or narrow racemose manner; if the stem is branched, the lower branches are typically arching upward; flowers July–October.

RANGE-HABITAT: Ranges in the Piedmont from central NC south through AL and north into sc. TN; also found in the Coastal Plain of southwestern GA and the FL Panhandle; in SC, common in the Piedmont and extending into the Inner Coastal Plain in Piedmont prairie remnants, oak savannas, oak flatwoods, openings in xeric hardpan forests, upland depressions, forest margins, roadsides, and powerline rights-of-way.

SIMILAR SPECIES: This species is very similar to and has been confused in the past with Graceful Bog Goldenrod (*Solidago gracillima* Torrey & Gray) and Wand Goldenrod (*Solidago virgata* Michaux, plate 664). Neither of these species occur in upland clay soils or in the Piedmont.

COMMENTS: Piedmont Wand Goldenrod is a graceful species that is especially common in areas that were formerly prairielike habitat. It is especially abundant in the Sumter National Forest in the Long Cane district and can be observed at the Post Oak Savanna Natural Area on the Greenwood/Saluda County line as well as at Rock Hill Blackjacks Heritage Preserve in York County.

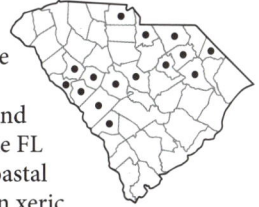

414. Downy Goldenrod

Solidago petiolaris Aiton var. *petiolaris*
So-li-dà-go pe-ti-o-là-ris var. pe-ti-o-là-ris
Asteraceae (Aster Family)

DESCRIPTION: Herbaceous perennial from a short rhizome with densely pubescent stems to 4′ tall; basal leaves absent at flowering; stem leaves sessile or nearly so, ovate to lanceolate-elliptic, with smooth or with a few teeth toward the tip; upper leaf surface scabrous or almost smooth, lower surface pubescent; flowers produced in involucrate heads with ray and disk flowers present; heads in paniclelike arrangement; individual heads are larger than most similar goldenrods, typically more than 0.5″ across; flowers August–November.

RANGE-HABITAT: NC south to FL and west to MO and NM; in SC, common throughout; dry road banks, open oak-hickory forests, forest margins.

COMMENTS: The relatively large involucral heads and relatively broad leaves help to make this species recognizable from other goldenrods with narrow inflorescences.

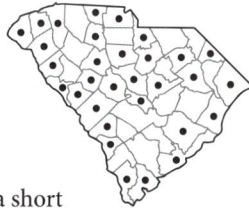

415.   Midwestern Tickseed-sunflower

Bidens aristosa (Michaux) Britton
Bì-dens a-ris-tò-sa
Asteraceae (Aster Family)

DESCRIPTION: Robust, often highly branched, annual herb from a taproot; stems glabrous to sparsely pubescent to 4′ tall; leaves opposite, deeply pinnately compound with 5–7 lanceolate leaflets; flowers produced in involucrate heads of yellow ray and yellow disk flowers; ray flowers 0.7–1.2″ long, showy; fruits are flattened nutlets or achenes with triangular, barbed awns that stick in clothing or on hair of passing animals; flowers September–October.

RANGE-HABITAT: DE south to FL and west to MO and TX; common in SC, primarily in the Piedmont, but found throughout; forest margins, roadsides, utility rights-of-way, and moist ditches.

COMMENTS: The large showy flowers and pinnately compound leaves help to distinguish this species from other species of *Bidens* in our area. During years of good rainfall, they often form huge swathes of yellow along roadsides.

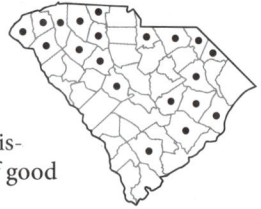

416.   Common Clasping Aster

Symphyotrichum patens (Aiton) G. L. Nesom var. *patens*
Sym-phy-o-trì-chum pà-tens var. pà-tens
Asteraceae (Aster Family)

SYNONYM: *Aster patens* Aiton—RAB

DESCRIPTION: Erect, perennial herb, to 4′ tall, from slender rhizomes forming colonies; softly pubescent; leaves lanceolate to oblanceolate, clasping the stem, often with auricles extending around the stem, scabrous (like sandpaper) and stiff; flowers produced in involucrate heads of blue ray flowers and yellow disk flowers in an open, paniclelike arrangement; flowers September–November.

RANGE-HABITAT: Widespread in the eastern US from VT south to FL and west to KS and TX; in SC, common in the mountains and Piedmont and rare in the Coastal Plain; dry, open woodlands; road banks; Piedmont prairie remnants; oak-hickory forests; margins of outcrops; and other open, sunny, dry habitats.

SIMILAR SPECIES: Appalachian Clasping Aster (*Symphyotrichum phlogifolium* (Muhlenberg ex Willdenow) G. L. Nesom), is similar but has softly pubescent, thin, and pliable leaves and involucrate heads with blue ray flowers and white disk flowers that are tinged with purple. It is found in similar habitats in the mountains. Georgia Aster (*Symphyotrichum georgianum* (Alexander) G. L. Nesom, plate 417) is vegetatively similar but has very large and showy involucrate heads with dark purple ray flowers and white disk flowers that turn purple with age. Georgia Aster is found in similar habitats in the Piedmont.

COMMENTS: Common Clasping Aster makes a good landscape plant, particularly in dry, clay soils. It attracts many pollinators and is an important nectar source for Monarch Butterflies in the autumn.

417. Georgia Aster

Symphyotrichum georgianum
(Alexander) G. L. Nesom
Sym-phy-o-trì-chum geor-gi-à-num
Asteraceae (Aster Family)

DESCRIPTION: Colony-forming perennial herb from slender rhizomes; stems to 4′ tall, rough pubescent; leaves oblanceolate, clasping the stem, very scabrous (like sandpaper) and stiff; flowers produced in large and showy involucrate heads of deep purple ray flowers and disk flowers that are white tinged with purple and turn purple with age; flowers September–January.

RANGE-HABITAT: NC south to GA and west to AL, primarily in the Piedmont; also in the Coastal Plain of GA and FL; in SC, restricted to the Piedmont where local and rare; open oak-hickory forests, Piedmont prairie remnants, oak savannas, roadsides, and powerline rights-of-way.

SIMILAR SPECIES: See comments under Common Clasping Aster above.

COMMENTS: Georgia Aster is extremely rare and localized. The largest populations are located on the Clemson Experimental Forest. It is an excellent indicator of where fire-maintained savanna and Piedmont prairie habitat previously existed. It makes the finest landscape plant of any native species of "aster" in SC, flowers very late in the season, and is readily available for purchase from native plant nurseries. Large displays may be seen at the South Carolina Botanical Garden.

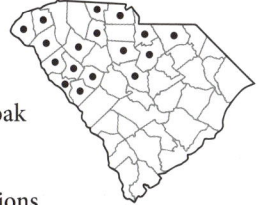

418. Piedmont springhead seepage forests

419. Poison Sumac

Toxicodendron vernix (L.) Kuntze
Tox-i-co-dén-dron vér-nix
Anacardiaceae (Cashew Family)

SYNONYM: *Rhus vernix* L.—RAB, PR

DESCRIPTION: Smooth, tall shrub or small tree, up to 20–30′ tall; leaves deciduous, produced simultaneously with flowers, alternate, pinnately compound, smooth, with 7–13 leaflets, the leaf stalks, rachis and young twigs nearly always reddish; flowers May–early June; fruit are drupes, white when mature August–September.

RANGE-HABITAT: Common and widespread in eastern North America; throughout SC, rare in the Piedmont and mountains in montane bogs and seepage forests; common in the Coastal Plain and Sandhills in streamhead pocosins, sandhill seepage bogs, pocosins, and adjacent, acidic swamps.

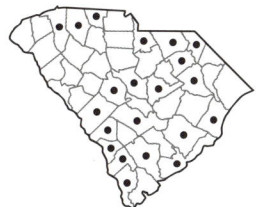

COMMENTS: The oils in all parts of the plant can cause severe dermatitis, much more severe than poison ivy, in susceptible individuals. Smoke from burning leaves or twigs can carry the volatile oil. Habitat and reddish leaf rachis and stems are helpful aids to identification and avoidance. The whitish, waxy fruits ripen in early fall but may persist throughout the winter, when they are fed upon by numerous birds.

420. Silky Dogwood

Swida amomum (P. Miller) Small
 Swí-da a-mò-mum
Cornaceae (Dogwood Family)

SYNONYM: *Cornus amomum*
 P. Miller—RAB, PR

DESCRIPTION: Shrub, 4–10′ tall; older plants usually with several erect, arching, or leaning stems from the base; young branches green to dark red, pith brown; leaves opposite, blades mostly ovate, with arcuate venation; flowers with 4 petals, white to cream, in open cymes; flowers May–June; drupes blue with areas of cream, mature August–September.

RANGE-HABITAT: Southern MA south to FL and west to IL and MS; in SC, chiefly in the mountains and Piedmont; common; scattered in the Coastal Plain; seepage forests, river and streamsides, wet thickets and clearings, borders of alluvial swamps, and marshes.

SIMILAR SPECIES: Swamp Dogwood (*S. foemina* (P. Miller) Rydberg) is very similar, but has pith that is white and there are 3–4 veins on each side of the leaf as compared to 5 or more in Silky Dogwood. It is common in swamp forests and marshes in the Coastal Plain and lower Piedmont.

COMMENTS: Silky Dogwood can be distinguished from *Viburnum* spp. that also have opposite leaves by the arcuate venation; all veins arch toward the tip. If you break a leaf carefully, the silky material in the veins will hold both broken portions together.

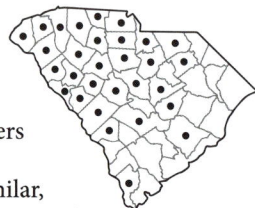

421. Dwarf-flowered Heartleaf

Hexastylis naniflora Blomquist
 Hex-ás-ty-lis na-ni-flò-ra
Aristolochiaceae (Birthwort Family)

DESCRIPTION: Perennial herb from a short, underground stem; leaves rounded, heart-shaped at the base, 1.5–2.5″ long and wide, dark green with lighter mottling; flowers solitary on short stalks, usually under leaf litter; calyx tube cylindrical, less than 0.45″ long and with an opening less than 0.2″ across; flowers February–May.

RANGE-HABITAT: Endemic to the upper Piedmont of SC and NC; found in acidic soils (usually Pacolet sandy loam) on bluffs and in ravines or on hummocks in Piedmont seepage forests (where Pacolet soils occupy the adjacent upland). It often grows beneath Mountain Laurel.

COMMENTS: *Hexastylis* is a difficult genus taxonomically. Proper identification requires fresh flowering material. However, the habitats of this and many other species are usually distinctive and aid in identification. This species may be told from the similar Variable-leaf Heartleaf (*H. heterophylla* (Ashe) Small, plate 169) by the very short calyx lobes that are less than 0.43″ long.

CONSERVATION STATUS: Federal Threatened

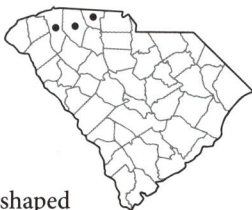

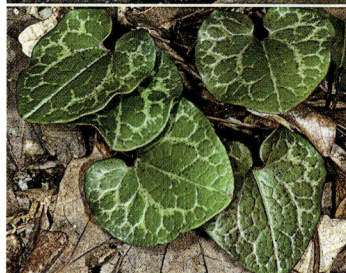

422. Small Jack-in-the-pulpit

Arisaema pusillum (Peck) Nash
A-ri-saè-ma pu-síl-lum
Araceae (Arum Family)

SYNONYM: *Arisaema triphyllum* (L.)
Schott—RAB, in part

DESCRIPTION: Erect, perennial herb to 6–13″ tall from a corm; leaves 1 or 2, trifoliolate, green above and below; flowers on a fleshy spadix, male above, female below; the spadix appendix, the sterile portion extending beyond the flowers, cylindrical; spathe (the pulpit) with a tube and a hood that arches over the spadix (jack); spathe hood solid purple, solid green, green-striped or purple-striped; flowers March–May; fruits are red berries, maturing June–August.

RANGE-HABITAT: CT south to FL and west to IN and TX; in SC, common throughout in moist forests, particularly bottomlands, seepage forests, and swamps.

SIMILAR SPECIES: Jack-in-the-pulpits have been traditionally treated as a single species. Recent re-evaluation confirms that at least five species were lumped under a single species; three of these occur in SC. Common Jack-in-the-pulpit (*Arisaema triphyllum* (L.) Schott, plate 803) is also widespread and produces leaves that are glaucous (whitish) on the back and has a spadix appendix that is often club-shaped; it is a larger plant and produces spathe hoods that are green striped with white or green striped with purple. Southern Jack-in-the-pulpit (*Arisaema quinatum* (Nuttall) Schott, plate 123) also has glaucous undersides of the leaves, typically produces 5 leaflets on flowering plants, and has a spathe hood that is green and blunt tipped.

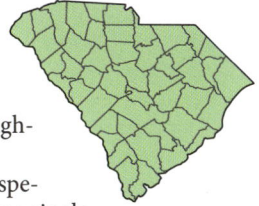

423. Bunched Arrowhead

Sagittaria fasciculata E. O. Beal
Sa-git-tà-ri-a fas-ci-cu-là-ta
Alismataceae (Water-plantain Family)

DESCRIPTION: Perennial herb with leafless flowering stems and basal leaves; emergent rosette of leaves spoon-shaped, usually 4–8″ long; submerged, winter rosette of leaves dark green and oblong; flowers white, in whorls of 3 at 2–4 nodes; lower flowers are female, the upper male; flowers mid-May–July.

RANGE-HABITAT: Restricted to the upper Piedmont of Greenville County, SC, and the mountains of Henderson and Buncombe Counties in NC; springhead seepage forests; rare.

COMMENTS: The rarity of this plant is due to the rarity of seepage headwaters where it is found. Habitat destruction, alteration of the water table, runoff from development, extended droughts, and invasive species, particularly Mud-Annie (*Murdannia keisak* (Hasskarl) Handel-Mazzetti, plate 424), are significant threats. Ducks foraging on the tubers are a threat to those few populations that have expanded into seepages at pond margins. The species is easily seen at Bunched Arrowhead Heritage Preserve in Greenville County. It should never be disturbed in the wild.

CONSERVATION STATUS: Federal Endangered

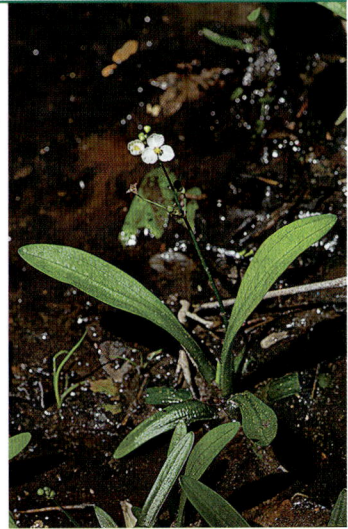

424. Mud-Annie; Murdannia

Murdannia keisak (Hasskarl)
Handel-Mazzetti
Mur-dánn-i-a keì-sak
Commelinaceae (Spiderwort Family)

SYNONYM: *Aneilema keisak*
Hasskarl—RAB

DESCRIPTION: Prostrate annual herb, usually rooting at the nodes; leaves alternate, linear to lanceolate; flowers solitary or in 2–4-flowered racemes borne in the upper leaf axils; flowers May–October.

RANGE-HABITAT: Native to Asia and a serious invasive-exotic in our area; widespread in the southeastern US; common throughout SC; seepage areas, ditches, streamsides, marshes, swamp forests, and wet disturbed places.

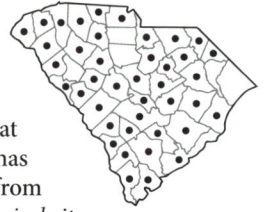

SIMILAR SPECIES: *Murdannia nudiflora* (L.) Brenan differs from *M. keisak* in that it has fruits (capsules) on stalks about as long as the fruit, whereas *M. keisak* has fruits on stalks much longer than its fruits. This species was also introduced from Asia, and although it is a Coastal Plain plant found in similar habitats to *M. keisak,* it is less common.

COMMENTS: Mud-Annie is an invasive exotic becoming increasingly common in SC wetland habitats, displacing native species.

425. Small Green Wood-orchid; Club-spur Orchid

Platanthera clavellata
(Michaux) Luer
Pla-tán-the-ra cla-vel-là-ta
Orchidaceae (Orchid Family)

SYNONYM: *Habenaria clavellata*
(Michaux)
Sprengel—RAB

DESCRIPTION: Perennial herb to 16″ tall, from a cluster of thickened fibrous roots; 1, sometimes 2 leaves on the lower half of the stem; flowers dull white or tinged with green or yellow, in an open terminal raceme; lip petal neither fringed nor lobed, twisted so that the 0.4–0.6″ long spur is lateral; flowers June–September.

RANGE-HABITAT: Widespread in eastern North America; common but scattered throughout SC; montane fens, swamps, seepages, and other wet places; especially common in Southern Appalachian fens and Piedmont seepage swamps.

COMMENTS: In the northern part of its range, this species is pollinated by mosquitoes.

Bottomland forests

426. Wild Allspice; Northern Spicebush

Lindera benzoin (L.) Bloom
Lín-der-a bén-zoin
Lauraceae (Laurel Family)

DESCRIPTION: Deciduous, aromatic shrub, much branched and usually 5–9′ tall; leaves alternate, thin; flowers yellow, in dense axillary umbels, before the leaves appear; flowers March–April; drupes mature August–September.

RANGE-HABITAT: Widespread in eastern North America; common throughout SC (except absent in the maritime strand); rich cove forests, floodplain forests, beech forests, basic-mesic forests, and other forest types on rich soils.

COMMENTS: The presence of this species on slopes away from the floodplain is a good indicator of calcium-rich soils. All parts are aromatic and are often described as lemonlike, but to the authors it smells like lemon furniture polish. Pioneers used Wild Allspice as a substitute for allspice. The specific epithet is from an old English word meaning of "balsamic resin." The genus name honors Johann Linder (1676–1723), a Swedish botanist.

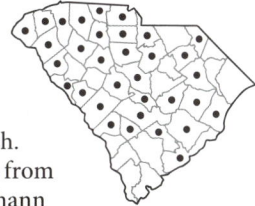

427. Common Pawpaw

Asimina triloba (L.) Dunal
A-sí-mi-na trí-lo-ba
Annonaceae (Pawpaw Family)

DESCRIPTION: Large shrub or small tree, to 16–33′ tall forming large colonies by rhizomes; twigs covered with fine, rust-colored hairs; leaves deciduous, oblanceolate, alternate, simple, malodorous when crushed; flowers borne on the wood produced the previous year; winter buds flattened, covered with rust-colored hairs; flowers do not open from a closed bud but gradually enlarge until fertile; flowers March–April; fruits mature in fall.

RANGE-HABITAT: NJ west to NY and Ontario, west to MN and NE, and south to FL Panhandle, LA, and ne. TX; throughout SC (except the maritime strand); common in the Piedmont and mountains; occasional in the Coastal Plain; rich hardwood forests, alluvial forests, and other moist, nutrient-rich forests.

COMMENTS: Fruits are rare, as they are only found on mature trees and require cross-pollination with a different genetic individual. Many SC stands consist of a single genetic individual that has spread via rhizomes. The ripe fruit of pawpaw is sweet; it can be eaten raw, baked as pie filling, used as a delicious flavoring for ice-cream, or made into a variety of other foods. The fruits are collected when green (often from the ground) and kept until ripe. Early settlers made a yellow dye from the ripe pulp. A wide variety of wildlife readily eat the fruits. All pawpaw species serve as the host for caterpillars of Zebra Swallowtails.

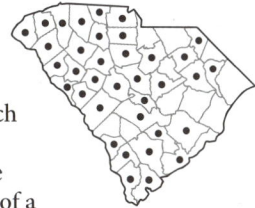

428. Bladdernut

Staphylea trifolia L.
    Sta-phy-lè-a tri-fò-li-a
    Staphyleaceae (Bladdernut Family)

DESCRIPTION: Shrub or small tree to 25′ tall; leaves trifoliolate; flowers white or greenish white, in drooping clusters from the leaf axils; fruit inflated and podlike, persisting well into the winter; flowers April.

RANGE-HABITAT: Widespread in the eastern US; in SC, rare in the mountains and Coastal Plain, occasional in the Piedmont; rich deciduous woods, including hardwood bottoms, basic-mesic forests, and cove forests.

COMMENTS: The opposite, compound leaves with 3, fine-toothed leaflets are distinctive. Bladdernut is an excellent indicator of high-calcium soils.

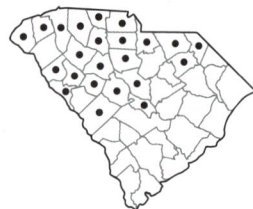

429. Spring Beauty

Claytonia virginica L.
    Clay-tòn-i-a vir-gí-ni-ca
    Portulacaceae (Purslane Family)

DESCRIPTION: Perennial herb usually less than 8″ tall, from a swollen underground stem (corm); a few leaves in a basal tuft and one pair, usually opposite, on the stem; leaves with linear blades, not distinct from the leaf stalk, together 3–8″ long and to 0.4″ wide; flowers white to pink; flowers February–April.

RANGE-HABITAT: Nova Scotia south to GA and west to MN and TX; in SC, rare in the Coastal Plain, common in the Piedmont and mountains; rich forests, often on floodplains in the Piedmont and Coastal Plain and in basic-mesic forests or rich cove forests in the Piedmont and mountains.

COMMENTS: Some authors now separate this species into two varieties, but as Weakley (2020) suggests, this seems unwarranted without more information on distribution and habitat. Sometimes called "fairy spuds" because of its edible (when boiled) corms. The genus honors John Clayton (1694/5–1773), one of the earliest North American botanists, who contributed to Gronovius the materials for his *Flora Virginica*.

430. Bulbous Bittercress

Cardamine bulbosa (Schreiber ex
    Muhlenberg) BSP
    Car-dá-mi-ne bul-bò-sa
    Brassicaceae (Mustard Family)

DESCRIPTION: Showy, perennial herb from 1–several tubers; stem smooth, strictly erect, to 20″ tall; leaves basal and on the stem, unlobed, lower often rounded; flowers white to rarely pink; flowers March–May.

RANGE-HABITAT: Widespread in eastern North America; in SC, uncommon and scattered in the Piedmont and Coastal Plain; hardwood bottoms, wet flats with prairie affinities, swamp forests, and bogs, primarily periodically wet sites with calcium-rich soils.

COMMENTS: The roots reportedly can be ground and used as a substitute for horseradish. This species makes a beautiful addition to the woodland garden.

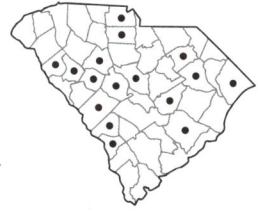

431. Early Buttercup; Thick-root Buttercup

Ranunculus fascicularis Muhlenberg
ex Bigelow
Ra-nùn-cu-lus fas-ci-cu-là-ris
Brassicaceae (Mustard Family)

DESCRIPTION: Erect or ascending perennial herb, 1–2′ tall, from fibrous and tuberous roots; basal leaves 3–5-foliolate, the terminal leaflet larger than the lateral leaflets; flowers with sepals spreading or reflexed backward; petals yellow; flowers February–June.

RANGE-HABITAT: MA west to MN and south to GA and TX; in SC, rare in the Piedmont where found in rich bottomland hardwood forests, wet flats with prairie affinities, glades and margins of outcrops with high magnesium or high calcium soils.

SIMILAR SPECIES: Two other native buttercups closely resemble Early Buttercup. Hairy Buttercup (*Ranunculus hispidus* Michaux) is similar, but with basal leaves trifoliolate and the terminal leaflet is about the same size as the lateral leaflets. It is found scattered throughout the state. Carolina Buttercup (*R. septentrionalis* Poiret in Lamarck, plate 750) is similar, but forms large colonies via abundant stolons produced during the flowering and fruiting periods. It also has trifoliolate leaves with the terminal leaflet about the same size as the lateral.

COMMENTS: This showy buttercup is often found in association with other species with ranges primarily centered west of the Appalachians.

CONSERVATION STATUS: SC-Critically Imperiled

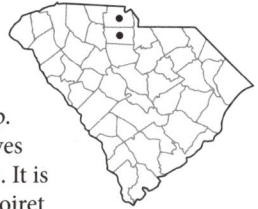

432. Wild Hyacinth

Camassia scilloides Rafinesque
Ca-más-si-a scil-loì-des
Agavaceae (Agave Family)

DESCRIPTION: Perennial herb from a bulb, up to 2′ tall; leaves basal only, to 20″ long and 0.8″ wide; flowers pale blue to whitish, in terminal, bracteat racemes to 12″ long; flowers April–May.

RANGE-HABITAT: WI and PA south to GA and west to KS and TX; a typical prairie species of the Midwest; in SC, a rare Piedmont species in upland depressions (wet flats) with circumneutral soils.

COMMENTS: This common species of midwestern prairies is one of many SC species with prairie affinities. It grows on calcium-rich soils, usually Elbert loam; known from only York and Chester Counties. This species makes a spectacular addition to the wildflower garden and is readily available from native plant nurseries. It should never be disturbed in the wild.

CONSERVATION STATUS: SC-Imperiled

433. Violet Wood-sorrel

Oxalis violacea L.

Óx-a-lis vi-o-là-ce-a

Oxalidaceae (Wood-sorrell Family)

DESCRIPTION: Herbaceous perennial from a bulbous base, forming large colonies from stolons; leaves basal, shamrocklike, trifoliolate; flowers with long peduncles exceeding the leaves and topped with pink, purple flowers with 5 petals; flowers March–May.

RANGE-HABITAT: Throughout the eastern half of the US; in SC, common throughout in bottomland forests, oak-hickory forests, marl forests, and other rich, mesic to dry habitats.

The Coastal Plain:
The Fall-Line Sandhills

434. Longleaf Pine-Scrub Oak sandhill

435. Longleaf Pine-Turkey Oak sandhill

436. Longleaf Pine

Pinus palustris P. Miller
Pì-nus pa-lús-tris
Pinaceae (Pine Family)

DESCRIPTION: Large, evergreen tree; needles 10–16″ long, in fascicles of 3; female cones at maturity 6–10″ long; flowers March–April; cones mature September–October.

RANGE-HABITAT: From se. VA south to FL, and west to se. TX; in SC, common in the Coastal Plain and Sandhills and rare in the outer Piedmont; sandy soils of Longleaf Pine flatwoods and savannas, Longleaf Pine-Turkey Oak sandhills, Longleaf Pine-scrub oak sandhills, bay rims, sandy ridges, fluvial ridges, pine-mixed hardwood forests, and disturbed sites.

COMMENTS: Longleaf Pine was the dominant pine in the woodlands of the Coastal Plain at the time of European colonization. Its thick bark makes it resistant to fires, and, unlike other southern pines, it is naturally resistant to fusiform rust.

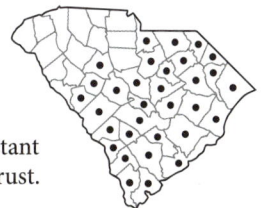

Growing tall and slender to a height of over 105′ and living 200–300 years, the original Longleaf Pine forests quickly fell to the ax of lumbermen. Fire suppression, feral hogs that consumed the young seedlings, competition from Loblolly Pine (a more prolific seeder), and the lack of replanting all contributed to Longleaf Pine's current occupation of less than 10% of its original acreage.

Longleaf once was the basis of the naval stores industry; its trunks were used for sailing ship masts, and its wood was and is still used for beams, floors, and general construction. Many of the Carolina Lowcountry mansions are made of "heart pine." The high resinous content makes the wood indestructible by insects and fungi. It is still an important lumber tree.

437. Turkey Oak

Quercus laevis Walter
Quér-cus laè-vis
Fagaceae (Beech Family)

DESCRIPTION: Small tree; bark thick, with deep, irregular furrows and scaly, rough ridges; inner bark reddish; leaves alternate, deciduous, generally with 3 wide-spreading, main lobes (like a turkey's foot) that are pointed and bristle-tipped; female flowers 1–2 on a short stalk, separate from the male flowers, which are in catkins; fruits are acorns that mature September–October of the following year.

RANGE-HABITAT: A Coastal Plain endemic ranging from VA south to FL, west to LA; in SC, common in the Coastal Plain and Sandhills; Longleaf Pine-Turkey Oak sandhills, Longleaf Pine-scrub oak sandhills, Carolina bay rims, sand ridges, sandy, Longleaf Pine flatwoods, and other sites with dry and sandy soil.

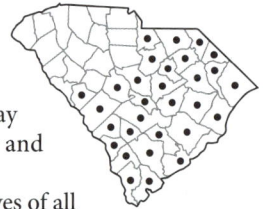

COMMENTS: This is one of the few oaks that will grow in sandy, sterile soil. Leaves of all ages, including seedlings, are orientated at right angles to the ground; this reduces heat absorption that is reflected from bare sand and heat absorption from direct sunlight. The tree is not large enough for lumber, but the seasoned wood is excellent as firewood. It is often used as the fuel in hog barbecue pits. The acorns are an important food source for deer, turkey, and small rodents.

438. Bluejack Oak

Quercus incana Bartram
Quér-cus in-cà-na
Fagaceae (Beech Family)

DESCRIPTION: Small tree; bark dark gray to blackish, thick, and broken into squares; twigs pubescent; leaves alternate, deciduous, elliptic to obovate, pale bluish-green and smooth above, pubescent and whitish below; female flowers 1–2 on a short stalk, separate from the male flowers, which are in catkins; flowers March–April; acorns with a cup covering about one-third of the acorn; fruit matures September–October of the following year.

RANGE-HABITAT: From VA south to FL and west to OK and TX, primarily on the Coastal Plain; in SC, common in the Coastal Plain and Sandhills; Longleaf Pine-scrub oak sandhills, Carolina bay rims, sand ridges and sandy, Longleaf Pine flatwoods; typically found on sites that have more loam in the soil than Turkey Oak, though the two may be found in mixed stands where conditions overlap.

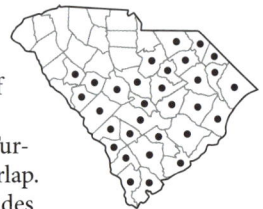

COMMENTS: The bluish-green hue to the upper leaf surface and lighter undersides make this oak easy to recognize.

439. Ravenel's Hawthorn

Crataegus ravenelii Sargent
Cra-taè-gus ra-ve-nél-i-i
Rosaceae (Rose Family)

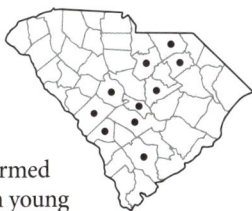

DESCRIPTION: Shrub or small tree to 30′ tall with slightly drooping branches; branches armed with spines; leaves obovate, pubescent when young and becoming smooth with age, with a narrow, cuneate to attenuate base and a rounded leaf tip with a short point; leaf margin toothed over most of the surface with small, sharp-pointed to wavy (crenate) teeth; flowers produced in 2–3-flowered clusters; inflorescence pubescent; flowers late March–May; fruit a yellowish, orange, or red pome, maturing August–September.

RANGE-HABITAT: NC south to FL and west to central AL; in SC, occasional, growing in deep sands or well-drained clay soils of the Sandhills region and extending into the Piedmont and Coastal Plain.

TAXONOMY: *Crataegus* is a large genus and the subject of considerable controversy among taxonomists as to species definition and number. This species is included so the reader will be aware of this diverse and important genus. This species belongs to the large group of often weeping species, known as Series *Lacrimatae,* and there are many other similar species in this series growing in similar habitats throughout the state. For the reader who is interested in a thorough treatment of the group, we recommend *Haws, a Guide to Hawthorns of the Southeastern United States,* by Ron Lance (2014).

COMMENTS: The flowers of hawthorns attract a huge diversity and number of pollinators ranging from flies to beetles and bees. The fruits of *Crataegus* are consumed by wildlife, but not as readily as many other plants with similar fruit and typically not until very late in the year (winter). All species have an edible fruit that can be used to make jam or jelly. Duke (1997) recommends all species of *Crataegus* as an herbal remedy for angina. The species is named in honor of SC botanist, Henry William Ravenel (1814–87), native of Berkeley County and long-time resident of Aiken, SC.

440. Sand-myrtle

Kalmia buxifolia (P. J. Bergius)
Gift & Kron
Kálm-i-a bux-i-fò-li-a
Ericaceae (Heath Family)

SYNONYM: *Leiophyllum buxifolium*
(P. J. Bergius) Elliott—RAB, PR

DESCRIPTION: Low, upright, widely branching evergreen shrub with crowded, tiny, dark green, leathery leaves; to around 3′ tall; flowers white or pinkish; flowers late March–April.

RANGE-HABITAT: NJ, NC, SC, ne. GA, e. TN, and se. KY; in SC, locally common in the Fall-Line Sandhills in Longleaf Pine scrub-oak sandhills and sandy flats, often along the margins of barren areas and edges of pocosins or in areas with perched water tables due to underlying sandstone; uncommon in the mountains on the margins of expansive outcrops.

COMMENTS: The range of Sand-myrtle is very interesting: throughout the range it is found in the Coastal Plain and then the mountains without populations in the majority of the Piedmont. Similar range patterns are exhibited by Swamp Pink (*Helonias bullata* L.), Turkey Beard (*Xerophyllum asphodeloides* (L.) Nuttall) and Pinebarrens Death Camas (*Stenanthium leimanthoides* (Gray) Zomlefer & Judd). Good populations of Sand-myrtle occur in Peachtree Rock Heritage Preserve in Lexington County and in Table Rock State Park in Pickens County. Sand-myrtle is difficult to transplant into cultivation and, because of its restricted range, should not be collected from the wild. The genus honors Peter Kalm (1761–79), student of Linnaeus who traveled and collected in North America.

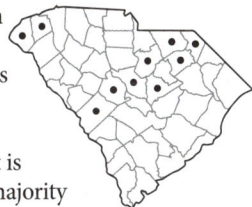

441. Northern Golden-heather

Hudsonia ericoides L.
    Hud-sòn-i-a er-i-coì-des
Cistaceae (Rockrose Family)

DESCRIPTION: Low, spreading, freely branched shrub, rarely more than 12–16″ tall; leaves evergreen, alternate, crowded, needlelike, about 0.25″ long; flowers yellow, numerous, on short stalks, solitary, at the ends of short spur shoots or terminating normal branches; flowers in May.

RANGE-HABITAT: Newfoundland south to MA, NH, and DE; disjunct in SC in Chesterfield County; very rare in Longleaf Pine-Turkey Oak and Longleaf Pine-scrub oak sandhills and sandy flats; rare.

COMMENTS: Northern Golden-heather was originally known in SC only from Cheraw State Park. Being disjunct so far from the closest site (in Delaware) prompted some field botanists to consider whether it might have been introduced. However, Richard D. Porcher Jr. and Patrick D. McMillan found a second site with a large population in Chesterfield County, adding support that it is a rare native disjunct. The genus honors William Hudson (1730–93), an English botanist.

CONSERVATION STATUS: SC-Critically Imperiled

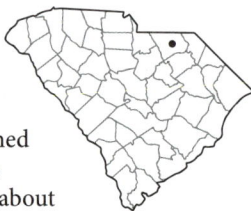

442. Southern Dwarf Huckleberry

Gaylussacia dumosa
    (Andrzejowski) Torrey & Gray
    Gay-lus-sàc-i-a du-mò-sa
Ericaceae (Heath Family)

DESCRIPTION: Deciduous or semi-evergreen shrub, 4–16″ tall, forming large colonies by rhizomes; leaves alternate, oblanceolate with a thick cuticle on the upper leaf surface, dark green; distinct, small resinous glands on lower leaf surface; flowers March–June; ripe berry black, mature June–October.

RANGE-HABITAT: NJ south to FL and west to LA; in SC, one of the most common shrubs of the Coastal Plain and Sandhills; sandy dry to moist habitats; also in the Piedmont and mountains, where rare in highly acidic soils such as pine-oak-heath.

COMMENTS: All species of *Gaylussacia* have small, golden, resinous glands on the lower leaf surface. This species is readily distinguished by its small size. The genus is named for Joseph Luis Gay-Lussac (1778–1850), a French chemist.

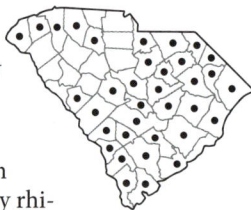

443. Ground Juniper; Common Juniper

Juniperus communis L. var. *depressa*
    Pursh
    Ju-ní-pe-rus com-mù-nis var.
    de-prés-sa
Cupressaceae (Cypress Family)

DESCRIPTION: Creeping shrub with ascending branches, forming dense colonies, to 2′ tall in SC, taller elsewhere in the range; leaves in whorls of 3, flat with sharp tips and a broad white stripe above; leaves 0.2–0.6″ long; male cones shedding

pollen March–April; female cones, fleshy, berrylike, bluish or blackish with a whitish bloom; female cones mature during the following year.

RANGE-HABITAT: Widespread, with a circumpolar distribution; in the eastern US, New England south to SC and GA, primarily in the Appalachians; in SC, very rare, found in the Piedmont and a single location in the Sandhills.

COMMENTS: Henry William Ravenel (1814–87) collected a specimen from "dry sandy hills" in Aiken, South Carolina in 1870. This population is almost certainly the same as exists today in Hitchcock Woods in Aiken, where it grows in a colony on a sandy ridge not far from the trail to the Vaucluse Udorthent outcrop. This population is amazingly disjunct from the heart of the species range. The species is sometimes cultivated but is not well-adapted to the hot climate of SC.

CONSERVATION STATUS: SC-Critically Imperiled

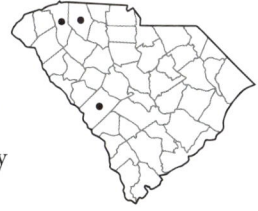

444. Whiteleaf Greenbriar; Wild Sarsaparilla

Smilax glauca Walter
Smì-lax glaù-ca
Smilacaceae (Greenbriar Family)

DESCRIPTION: Woody, slender-stemmed vine, with long underground runners with knotty, jointed, tuberous thickenings; climbing fairly high, but mostly forming thickets closer to the ground; stems round, green or brown, often with a whitish bloom, and with green spinelike projections; leaves partially evergreen, lighter green above, glaucous (whitish) beneath, turning reddish to purple in the fall; flowers in umbels April–June; berries bluish-black, with a whitish bloom, usually persistent, maturing September–November.

RANGE-HABITAT: Widespread and common in the southeastern US; common throughout SC in swamp forests, pocosins, Longleaf Pine sandhills, many types of upland and alluvial woods, old fields, and fencerows.

COMMENTS: F. P. Porcher (1869) and Fernald and Kinsey (1958) expound on the many uses of *Smilax* species as food, including the making of jelly from the cordlike rootstocks. Unlike most *Smilax* species, the young, spring shoots of Whiteleaf Greenbrier are bitter and are seldom cooked as a vegetable and never eaten raw.

445. Southern Wiregrass (A)

Aristida beyrichiana Trinius & Ruprecht
A-rís-ti-da bey-rich-i-à-na

Carolina Wiregrass (B)

Aristida stricta Michaux
A-rís-ti-da stríc-ta
Poaceae (Grass Family)

DESCRIPTION: Densely clumping perennial grasses forming dense tussocks; leaves extremely narrow, long, wiry, and appearing rounded (involute); leaves predominantly basal; the base of the leaf blade and collar has a tuft of long shaggy pubescence and the remainder of the leaf blade is smooth in *A. beyrichiana;* the base of the leaf blade and collar lacks a tuft of long shaggy pubescence and displays two loose lines of pubescence

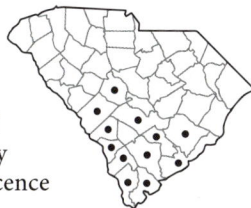

along nearly the entire length of the leaf blade in *A. stricta;* flowering stems to 4′ tall, typically only produced after burning; flowers produced in a panicle; fruit is a caryopsis (grain), tightly enclosed by a scale (lemma), with three prominent awns (long bristlelike projections); flowers September–November.

RANGE-HABITAT: *A. beyrichiana* is found from central SC south to FL and west to MS; in SC, locally abundant in the central and southern Coastal Plain and Sandhills region in Longleaf Pine flatwoods and savannas, Longleaf Pine-Turkey Oak sandhills and scrub; *A. stricta* is found from the northern Coastal Plain of NC south to northern and north-central SC; in SC, locally abundant in Longleaf Pine flatwoods and savannas and Longleaf Pine-Turkey Oak sandhills and scrub.

COMMENTS: The two species of wiregrass are very similar but consistently differ in their morphology and their ranges do not overlap. SC is the only state with both species and there is a narrow "gap" in the ranges of each in the central Coastal Plain. Peet (1993) gives a discussion of the two species and their ecology. Wiregrass is highly flammable and is one of the structural components that drives the incredible diversity of our Longleaf Pine-wiregrass savannas and flatwoods. Both species flower only after fire or occasionally after mowing.

446. Sandhills Pyxie-moss

Pyxidanthera brevifolia B. W. Wells
    Pyx-i-dan-thè-ra bre-vi-fò-li-a
    Diapensiaceae (Diapensia Family)

SYNONYM: *Pyxidanthera barbulata*
    Michaux var. *brevifolia* (B. W. Wells)
    Ahles—RAB, PR

DESCRIPTION: Creeping, perennial subshrub; leaves evergreen, lanceolate, about 0.06–0.19″ long, ovate, pubescent over the entire surface; flowers sessile, about 0.25″ across; stamens conspicuous, arising between the petals; flowers December–March.

RANGE-HABITAT: Endemic to the Carolina Sandhills; xeric Longleaf Pine sandhills, particularly on shallow sandy soils near ridgelines over sandstone or dense clays; globally rare, but can be locally abundant.

TAXONOMY: The taxonomic status of Sandhills Pyxie-moss has been controversial; some authors consider it a species, some a variety of Common Pyxie-moss (*P. barbulata* Michaux), and others consider it an ecotype not worthy of taxonomic status. Plants appearing to be intermediate between the two species are apparently a result of shading by leaf litter or physical disturbance of the soil. The two species are consistently separated by the following: Sandhills Pyxie-moss has leaves that are thin in texture, ovate, and 0.06–0.19″ long and hairy on the entire surface; Common Pyxie-moss has leaves that are succulent (thick), lanceolate, and 0.19–0.31″ long; it is hairy only toward the base, at least on sterile shoots.

COMMENTS: The name of the genus refers not to some relationship with pyxies (fairies) but to the way the anthers open to release pollen (pyxislike, the top coming off like a lid). The Greek *pyxie* means "small box," and *anthera* means "anther." A good site to view this species is at Sugarloaf Mountain in the Carolina Sandhills State Forest in Chesterfield County.

CONSERVATION STATUS: SC-Imperiled

447. Pineland Phlox

Phlox nivalis Loddiges ex Sweet
Phlóx ni-và-lis
Polemoniaceae (Phlox Family)

DESCRIPTION: Prostrate, evergreen, semi-woody perennial forming small dense mats; leaves opposite, subulate to linear-lanceolate; flowering shoots erect, 1–4″ tall, deciduous; flowers March–April.

RANGE-HABITAT: From NC south to the FL Panhandle; in SC, common in the Piedmont, Sandhills, and Inner Coastal Plain; Longleaf Pine sandhills, pinelands, margins of rock outcrops, dry deciduous woods and forests, and along road banks.

TAXONOMY: Two varieties are recognized. The other variety, Trailing Phlox (*P. nivalis* var. *hentzii* (Nuttall) Wherry) has flowering shoots 6″ or more tall, oblong-lanceolate leaves, and occurs in dry deciduous woods, primarily in the northeastern Piedmont. Pineland Phlox resembles Moss Pink, also called Thrift (*Phlox subulata* L.), which is not native to SC but is widely planted as a groundcover. Pineland Phlox is much more difficult in cultivation than Moss Pink; it generally does not persist long in cultivation.

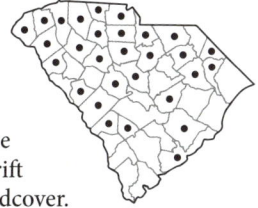

448. Carolina Ipecac

Euphorbia ipecacuanhae L.
Eu-phòr-bi-a i-pe-ca-
cu-àn-hae
Euphorbiaceae (Spurge Family)

DESCRIPTION: Perennial herb with white milky sap; taproot deep, many feet long, branched near the top into many stems with only the tips above ground; aboveground stems smooth, dichotomously branched, forming low, dark green tufts or small mats; leaves variable, from linear to round; flowers March–May.

RANGE-HABITAT: From Long Island, NY, and NJ, south to GA; in SC, common in the Sandhills and Inner Coastal Plain; rare in the Outer Coastal Plain; deep sands of sand ridges, Carolina bay rims, Longleaf Pine-scrub oak sandhills, and Longleaf Pine-Turkey Oak sandhills.

COMMENTS: F. P. Porcher (1869) reported that this plant was "tolerably certain emetic; but liable sometimes to produce excessive nausea by accumulation." It is an extremely strong laxative, and the juice from the fresh plant may cause blistering. The variability in the shape of leaves is remarkable, often leading to misidentification. The "flower" of members of the genus *Euphorbia* is a collection of tiny male and female flowers, an inflorescence termed a "cyathium." It is a "false flower," the colored parts are lobes of nectar glands, with petals and sepals lacking. The colored nectar glands serve to attract pollinators.

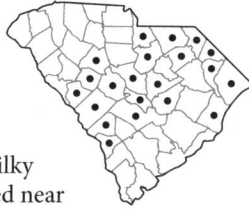

449. Tread-softly; Spurge-nettle

Cnidoscolus stimulosus (Michaux)
  Engelmann & Gray
    Cni-dós-co-lus sti-mu-lò-sus
Euphorbiaceae (Spurge Family)

DESCRIPTION: Erect or reclining, perennial herb, 6–36″ tall; entire plant covered with stinging hairs; leaves alternate, palmately lobed or dissected; petals absent; sepals white; flowers late March–August.

RANGE-HABITAT: From VA south to FL and west to TX; common throughout the Coastal Plain, Sandhills, and Piedmont of SC; Longleaf Pine-Turkey Oak and Longleaf Pine-scrub oak sandhills; Carolina bay rims; fluvial ridges; sandy, dry, open woods; sandy, fallow fields; and stable coastal dunes.

COMMENTS: Morton (1974) reports that the root was used as an aphrodisiac in rural SC; locally it is called the "courage" plant. The hairs can inflict a painful sting on contact and cause a severe reaction in some people.

450. Carolina Sandwort

Mononeuria caroliniana (Walter)
  Dillenberger & Rabeler
    Mo-no-neù-ri-a ca-ro-li-ni-à-na
Caryophyllaceae (Pink Family)

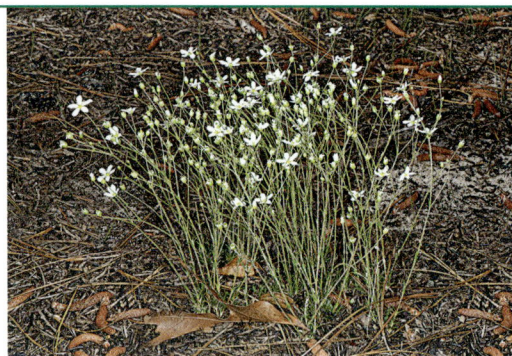

SYNONYM: *Arenaria caroliniana* Walter—RAB, PR

DESCRIPTION: Perennial herb with a deep taproot, from which extends a basal cushion of decumbent to prostrate stems; leaves tiny, opposite, linear-subulate, rigid, 0.1–0.5″ long, densely arranged and overlapping the ones above; flowering stems erect, to 12″ tall (usually shorter), with many small glands on clear stalks; flowers April–June.

RANGE HABITAT: Primarily a Coastal Plain species from RI to FL; in SC, common in the Sandhills and Inner Coastal Plain; dry, sandy communities such as Longleaf Pine-Turkey Oak and Longleaf Pine scrub-oak sandhills.

COMMENTS: The small cushion of stems is supported by a very deep and large taproot, which allows this dainty species to survive in deep, droughty, sandy soils.

451. Coastal Plain Puccoon

Lithospermum caroliniense (Walter ex
  J. F. Gmelin) MacMillan
    Li-tho-spér-mum ca-ro-li-ni-én-se
Boraginaceae (Borage Family)

DESCRIPTION: Herb 12–40″ tall, from a taproot; taproot with strongly-staining red tint; stems very leafy, rough; leaves alternate; flowers in dense cymes, leafy-bracted; flowers April–June.

RANGE-HABITAT: From SC south to FL and west to TX (absent in NC and disjunct in se. VA); in SC, common in the Sandhills and southeastern Coastal Plain; xeric sandhills, sandy roadsides, and fields.

COMMENTS: Puccoon is a Native American name for a number of plants that yield dyes.

---

452. Sandhills Milk-vetch;
Michaux's Milk-vetch

Astragalus michauxii (Kuntze)
F. J. Hermann
As-trá-ga-lus mi-chaùx-i-i
Fabaceae (Bean Family)

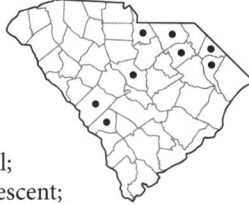

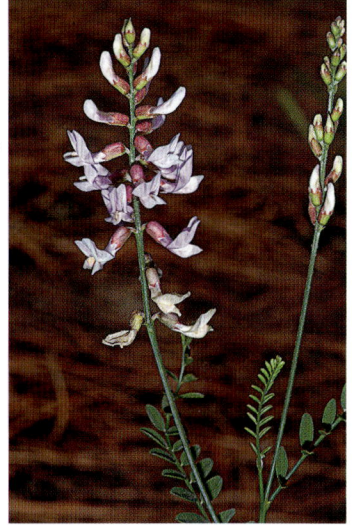

DESCRIPTION: Erect, perennial herb 1–3′ tall; stems greenish, glabrous to appressed pubescent; leaves alternate, pinnately compound, up to 6″ long; flowers in racemes up to 6″ long; petals white to lavender; fruit a glabrous, curved legume; flowers late April–June; fruits mature June–August.

RANGE-HABITAT: Restricted to the Sandhills and Coastal Plain of NC, SC, and GA; in SC, rare, in Longleaf Pine-Turkey Oak and Longleaf Pine-scrub oak sandhills and dry, sandy flatwoods in the sandhills and Coastal Plain.

COMMENTS: Sandhills Milk-vetch is vegetatively similar to several other species such as Goat's-rue but is a more upright, taller, and slender plant. It has declined dramatically due to loss of habitat and fire suppression. Sizable populations still exist in Aiken Gopher Tortoise Heritage Preserve in Aiken County.

CONSERVATION STATUS: SC-Critically Imperiled

---

453. Carolina Wild Indigo

Baptisia cinerea (Rafinesque) Fernald
& Schubert
Bap-tí-si-a ci-né-re-a
Fabaceae (Bean Family)

DESCRIPTION: Rhizomatous, perennial herb, 1–2′ tall, with appressed hairs; leaves trifoliolate; flowers showy, yellow, produced in racemes without large subtending bracts; flowers May–June; legume matures June–August.

RANGE-HABITAT: VA to SC; in SC, common in the Sandhills and northeastern Coastal Plain; Longleaf Pine-Turkey Oak sandhills, Longleaf Pine flatwoods, and sandy, dry, open woodlands.

SIMILAR SPECIES: Creamy Wild Indigo (*B. bracteata* Elliott) is similar but has creamy yellowish flowers produced in a raceme with persistent, large bracts subtending the individual flowers. It is found in the Piedmont and Sandhills region, mostly in the Savannah River drainage.

COMMENTS: The genus name comes from the Greek *baptizien,* which means "to dip in or under water," in reference to a former use of some species as sources of dyes. The stems dry in the autumn with all of the leaves attached. The dark stems often break and roll around the sandhills like tumbleweed.

454. Catbells; Gopherweed

Baptisia perfoliata (L.) R. Brown
   Bap-tí-si-a per-fo-li-à-ta
Fabaceae (Bean Family)

DESCRIPTION: Rhizomatous, erect or ascending, smooth, branched perennial herb, 20–36″ tall; leaves simple, rounded, bluish-green, perfoliate and smooth; flowers yellow, produced in the axils of the leaves, April–May; legume, short, rounded, less than 0.6″ long, matures May–July.

RANGE-HABITAT: A southeastern Coastal Plain endemic ranging from SC south to FL; in SC, locally common in the Coastal Plain and Sandhills mostly in the Savannah River drainage; Longleaf Pine-Turkey Oak and Longleaf Pine-scrub oak sandhills, and sandy, dry, open woods.

COMMENTS: Gopherweed looks more like a small *Eucalyptus* than a *Baptisia*. Gopherweed can be found in the Tillman Sand Ridge Heritage Preserve in Jasper County and Aiken Gopher Tortoise Heritage Preserve in Aiken County. The gopher referred to in the common name is the Gopher Tortoise, which shares the same habitat. The whole, dried plants have been traditionally used in dried arrangements. It makes an attractive ornamental in dry, sandy soils and is easily grown from seed.

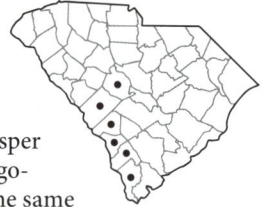

455. Lance-leaf Gopherweed

Baptisia lanceolata (Walter) Elliott var.
   lanceolata
      Bap-tí-si-a lan-ce-o-là-ta var. lan-
      ce-o-là-ta
Fabaceae (Bean Family)

DESCRIPTION: Rhizomatous, minutely pubescent, perennial herb, 20–32″ tall; leaves trifoliolate with thick-textured leaflets; flowers yellow, solitary, or in short racemes at the tips of branches; flowers April–May; fruit a legume to 1.4″ long, matures May–July.

RANGE-HABITAT: Endemic to the se. Coastal Plain from SC south to FL; in SC, rare, restricted to the Coastal Plain and Sandhills; Longleaf Pine-Turkey Oak and Longleaf Pine-scrub oak sandhills, and sandy, dry, open woods.

COMMENTS: Lance-leaf Gopherweed makes an attractive ornamental in dry, sandy soils. It is easily grown from seed but is rare in SC, and wild plants should not be disturbed.

CONSERVATION STATUS: SC-Vulnerable

456. Sandhills Barbara's-buttons

Marshallia obovata (Walter)
  Beadle & F. W. Boynton var.
  scaposa Channell
  Mar-shǎll-i-a ob-o-và-ta var.
  sca-pò-sa
Asteraceae (Aster Family)

DESCRIPTION: Cespitose, perennial herb, 4–24″ tall; basal leaves numerous; stems with 0–3 leaves, restricted to lower region; leaves 3-nerved, entire; flowers arranged in involucrate heads with only disk flowers; flowers light pink, produced late April–May.

RANGE-HABITAT: NC south to se. AL in the Coastal Plain; in SC primarily a plant of the Sandhills and Inner Coastal Plain; Longleaf Pine-Turkey Oak sandhills, Longleaf Pine-scrub oak sandhills and dry, sandy pine flatwoods; common.

SIMILAR SPECIES: This variety is similar to *M. obovata* (Walter) Beadle & Boynton var. *obovata,* which occurs in the Piedmont in old fields, meadows, and forest margins and extends rarely into the Coastal Plain. Variety *obovata* has 4–7 stem leaves extending one-fourth or more up the stem. The distinctiveness of the two varieties may indicate they would be better treated as distinct species.

COMMENTS: The genus honors Pennsylvania plantsman Humphry Marshall (1722–1801) and his nephew, Dr. Moses Marshall (1758–1813).

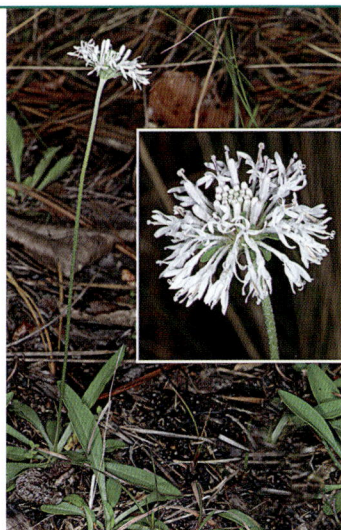

457. Hairy Phlox; Chalice Phlox

Phlox amoena Sims
  Phlóx a-moè-na
Polemoniaceae (Phlox Family)

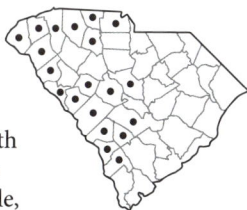

DESCRIPTION: Pubescent perennial herb with erect sterile and fertile shoots, the latter to 12″ tall; leaves opposite in 5–9 pairs, simple, linear to oblong-elliptic to lanceolate, approximately 5–10 times as long as wide; flowers in compact cymes; corolla pink to lavender; stamens shorter than the corolla tube; flowers April–June.

RANGE-HABITAT: FL to MS, north to NC; in SC chiefly in the mountains, Piedmont, and Sandhills in the mid-central and western counties; Longleaf Pine sandhills, dry woodlands, and open banks; common.

COMMENTS: Though this phlox is very attractive, it is often short-lived in cultivation and is best appreciated in the wild. Some authors recognize two species in what we are considering a single species. Lighthipe's Phlox (*P. lighthipei* Small) would be distinguished by having linear leaves approximately 10 times as long as wide and lacking bracts that hide the calyx. Hairy Phlox (*P. amoena*) would be distinguished by having leaves approximately 5 times as long as wide. Even Small (1933), who described *P. lighthipei,* doubted whether the two were separate species.

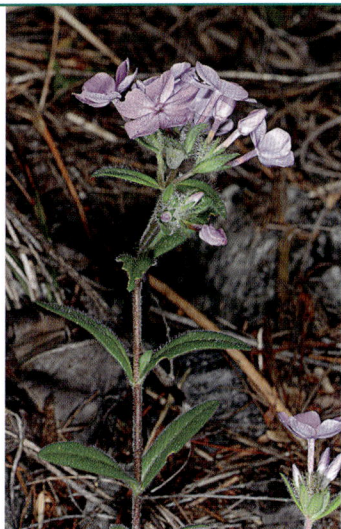

458. Cuthbert's Onion

Allium cuthbertii Small
       Ál-li-um cuth-bért-i-i
       Alliaceae (Onion Family)

DESCRIPTION: Perennial, bulbous, scapose, smooth herb; basal leaves 6–18″ long, linear, flat; flowering stems 8–24″ tall, terminated by an erect umbel composed of white to pinkish flowers with acuminate tepals to 0.4″ long; ovaries bright green, crested with tiny projections; flowers May.
RANGE-HABITAT: From SC south to FL and west to AL; in SC, occasional, but often locally abundant in the Sandhills and extending into the lower Piedmont and Coastal Plain, mostly in the Savannah River drainage; Longleaf Pine-Turkey Oak Longleaf Pine sandhills, dry, sandy woodlands, and dry road banks.

COMMENTS: Cuthbert's Onion is an attractive species that makes a good garden plant in sandy, well-drained soils. Like most onion species, the bulb is edible. It can be distinguished from other wild onions with upright flowers by the acuminate tepals as well as the tiny projections on the ovary. The species honors Alfred Cuthbert (1857–1932), who collected plants primarily in GA and FL.

459. Grassleaf Roseling

Cuthbertia graminea Small
       Cuth-bért-i-a gra-mí-ne-a
       Commelinaceae (Spiderwort Family)

SYNONYM: *Tradescantia rosea* var. *graminea* (Small) Anderson and Woodson—RAB, PR

DESCRIPTION: Smooth perennial herb, 5–10″ tall; narrow basal leaves tufted, less than 0.13″ wide; flowers subtended by minute bracts; petals 3, brilliant pink; fertile stamens 6, bearded; flowers May–July.
RANGE-HABITAT: From VA south to FL; in SC, common in the Inner Coastal Plain and Sandhills; Longleaf Pine-Turkey Oak sandhills, Longleaf Pine-scrub oak sandhills, sandy, dry, open woods, Carolina bay rims, and sand ridges.
SIMILAR SPECIES: Common Roseling (*C. rosea* (Ventenat) Small) is similar. It occurs more frequently in the Piedmont and Outer Coastal Plain and in moister habitats, but the ranges of the two species overlap. The leaves of Common Roseling are more than 0.13″ wide versus less than 0.13″ wide.
COMMENTS: The genus honors Alfred Cuthbert (1857–1932), who collected plants primarily in GA and FL. The narrow leaf blades and somewhat succulent stems and leaves are good adaptations to the droughty habitat. Grassleaf Roseling makes a beautiful and compact specimen when planted in well-drained soils.

460. Coastal Plain Wireplant

Stipulicida setacea Michaux
Sti-pu-lí-ci-da se-tà-ce-a
Caryophyllaceae (Pink Family)

DESCRIPTION: Dichotomously branched, smooth, wiry annual or short-lived perennial, with a tap-root and a tiny overwintering basal rosette; stem 2–8″ tall; stem leaves opposite, reduced to scales; flowers white, very small, in clusters of 1–6; flowers May–August.

RANGE-HABITAT: From VA south to FL and west to LA; in SC in the Sandhills, Coastal Plain, and maritime strand; common in all xeric communities in its range, including sandhill habitats, sand ridges and flats, dry pine flatwoods, and sandy openings in maritime forests.

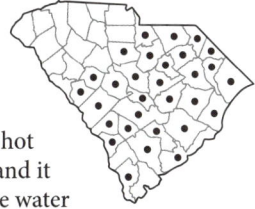

COMMENTS: The lifecycle of the wireplant is timed to miss the hot, dry summer days. During spring and early summer, it grows and flowers. During the hot summer months, it survives in the seed stage. Its seeds germinate in the fall, and it overwinters as a rosette. The minute size of its stem leaves also helps to reduce water loss from transpiration.

461. Narrowleaf Dawnflower

Stylisma angustifolia (Nash) House
Sty-lís-ma an-gus-ti-fò-li-a
Convolvulaceae (Morning Glory Family)

SYNONYMS: *Stylisma patens* ssp. *angustifolia* (Nash) Myint—PR; *Bonamia patens* (Desrousseaux) Shinners var. *angustifolia* (Nash) Shinners—RAB

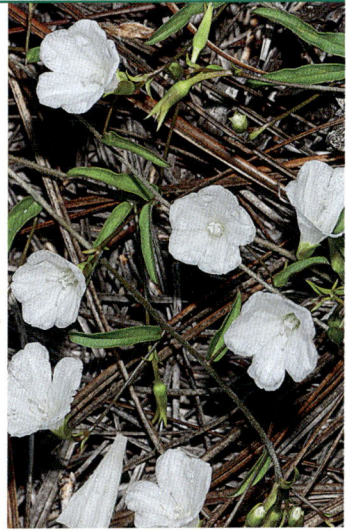

DESCRIPTION: Herbaceous, vinelike perennial, prostrate or spreading with no tendency to twine; larger leaves 0.04–0.1″ wide, 7–15x as long as wide; flowers solitary or in few-flowered cymes; sepals smooth; flowers May–August.

RANGE-HABITAT: Endemic to the Coastal Plain from NC south to FL; in SC, common in the Sandhills and Coastal Plain; dry Longleaf Pine sandhills and other relatively dry, sandy areas, including roadsides.

SIMILAR SPECIES: Common Dawnflower (*S. patens* (Desrousseaux) Myint) is similar, is found throughout SC, and has similar habitat preferences. It can be distinguished by its villous pubescent sepals and leaves (0.1–0.4″ wide) that are 4–6x as long as wide.

462. Squarehead

Tetragonotheca helianthoides L.
Tet-ra-go-no-thè-ca he-li-an-thoì-des
Asteraceae (Aster Family)

DESCRIPTION: Perennial, erect herb with 1–several stems, to 3′ tall; leaves opposite, elliptic to ovate, coarsely dentate to serrate; flowers in involucrate heads composed of yellow ray flowers and

brownish-purple disk flowers; involucre consists of 4 large ovate bracts; flowers April–July.

RANGE-HABITAT: From se. VA south to FL and west to TN and MS; in SC, occasional, scattered throughout; dry sandhill habitats, sandy woods, roadsides, and thickets.

COMMENTS: The common name and generic name *Tetragonotheca* (meaning "4-angled case") refers to the 4 large bracts that subtend the flower head. The size of the flower as well as the peculiar bracts make this species unmistakable.

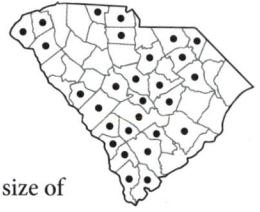

463. Woolly-white; Old Plainsman

Hymenopappus scabiosaeus L'Heritier
de Brutelle var. *scabiosaeus*
Hy-me-no-páp-pus sca-bri-o-saè-us
var. sca-bri-o-saè-us
Asteraceae (Aster Family)

DESCRIPTION: Perennial herb from a thick taproot, 1–2′ tall; leaves more or less clothed with whitish hairs on the underside; basal rosette of leaves present; stem leaves alternate, finely twice-dissected; flowers arranged in involucrate heads of white disk flowers only; subtending bracts with petal-like, whitish tips; flowers late April–June.

RANGE-HABITAT: From sc. SC to FL, west to AR and OK, and north to IL and MO; in SC, locally abundant in the Sandhills region in the vicinity of Aiken; Longleaf Pine-Turkey Oak and Longleaf Pine-scrub oak sandhills, sandy fields, and along sandy roadsides.

464. Narrowleaf Rose-pink

Sabatia brachiata Elliott
Sa-bà-ti-a bra-chi-à-ta
Gentianaceae (Gentian Family)

DESCRIPTION: Annual herb, with a single stem (rarely 2 or 3), 6–20″ tall, with a basal rosette of leaves; branches opposite; stem round below, sometimes lined or finely ridged; stem leaves nearly oblong, 3x or more longer than wide, not clasping; corolla lobes 5, pale pink to dark rose, with a greenish yellow "eye" bordered by a reddish line at the base; flowers late May–July.

RANGE-HABITAT: VA south to GA, west to MO and LA; in SC, infrequent in the Sandhills and rare in the Coastal Plain; dry sandhill habitats, Longleaf Pine flatwoods, and Longleaf Pine savannas.

SIMILAR SPECIES: Narrowleaf Rose-pink is similar to Bitter-bloom (*S. angularis* (L.) Pursh, plate 402), which occurs throughout SC in old fields, pastures, pine flatwoods, ditches, and meadows. Bitter-bloom has ovate stem leaves, their bases clasp the stem, and the stem is angled, with membranous wings on the angles.

COMMENTS: The genus honors Liberato Sabbati (1714–78), an Italian botanist.

465. Buckroot; Eastern Prairie-turnip

Pediomelum canescens (Michaux)
     Rydberg
      Pe-di-o-mè-lum ca-nés-cens
Fabaceae (Bean Family)

SYNONYM: *Psoralea canescens*
     Michaux—RAB

DESCRIPTION: Erect, perennial herb from a tuberous root; stems branched, 1–2′ tall, densely pubescent with gray hairs; leaves trifoliolate; leaflets elliptic to obovate, to 2.5″ long, less than 2x as long as wide; flowers produced in loosely flowered racemes to 2.5″ long; petals blue to light violet; flowers May–July.

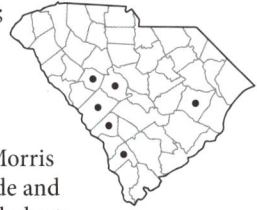

RANGE-HABITAT: Endemic to the Coastal Plain from VA south to FL and west to AL; in SC, rare in the Sandhills and Coastal Plain; Longleaf Pine-Turkey Oak sandhills, dry, sandy flatwoods, sandy roadsides.

SIMILAR SPECIES: Piedmont Buckroot (*P. piedmontanum* J. R. Allison, M. W. Morris & A. N. Egan, plate 383) is similar but has leaves more than 2x as long as wide and dense racemes. Piedmont Buckroot is one of the rarest and most endangered plants in SC.

COMMENTS: The name "prairie-turnip" is given to members of the genus because of the large, edible tuberous roots. All species are rare in SC and should never be disturbed in the wild.

CONSERVATION STATUS: SC-Critically Imperiled

466. Lupine Scurfpea

Orbexilum lupinellus (Michaux) Isley
      Or-béx-i-lum lu-pi-nél-lus
Fabaceae (Bean Family)

SYNONYM: *Psoralea lupinellus*
     Michaux—RAB

DESCRIPTION: Perennial herb 8–24″ tall from slender rhizomes; leaves palmately compound with 5–7 leaflets; leaflets threadlike to linear and to 3″ long; flowers produced in short, loosely-flowered racemes to 2″ long; petals blue to violet; flowers May–July.

RANGE-HABITAT: Endemic to the Coastal Plain from NC south to FL and west to AL; in SC, very rare in the Sandhills and Coastal Plain in Longleaf Pine-Turkey Oak sandhills, especially in loamier soils of "bean dips," which is a local term used to describe these swalelike areas with a high abundance of plants in the bean family.

COMMENTS: This delicate, and beautiful species derives the common and specific epithet from the resemblance of the leaves to a lupine. It was first reported for SC by McMillan et al. (2002).

CONSERVATION STATUS: SC-Critically Imperiled

467. Threadleaf Coreopsis

Coreopsis verticillata L.
Co-re-óp-sis ver-ti-cil-là-ta
Asteraceae (Aster Family)

DESCRIPTION: Perennial herb with 1–several, erect stems forming colonies by slender rhizomes; stems 1–3′ tall; leaves ternately divided to the base; segments pinnatifid (highly divided) with 11–25 divisions per leaf, divisions extremely narrow; flowers in involucrate heads with yellow ray and disk flowers; involucre in two distinct series; achenes with a narrow winged margins, pappus lacking; flowers May–July.

RANGE-HABITAT: MD and WV south to SC; in SC, occasional in the Piedmont and Sandhills, rare in the Coastal Plain; dry, clay soils of open oak-hickory forests (often rocky), forest margins, Longleaf Pine-Turkey Oak sandhills, and other dry, sandy habitats.

SIMILAR SPECIES: Larkspur Coreopsis (*C. delphinifolia* Lamarck) is similar but has far fewer divisions (5–11) per leaf and the divisions are wider and generally very thick in texture. Larkspur Coreopsis appears to be intermediate between Threadleaf Coreopsis and Woodland Coreopsis (*C. major* Walter, plate 398) and many specimens of Larkspur Coreopsis are misidentified as these species.

COMMENTS: The specific epithet describes the leaf arrangement, which appears to be whorled (verticillate). Threadleaf Coreopsis is one of the best native plants for use in the home landscape where it forms tight colonies that are very attractive to humans as well as pollinators. Horticultural varieties are weak and do not perform as well as the wild species in our area.

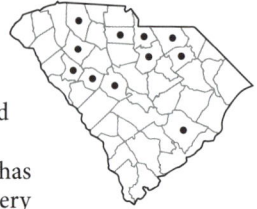

468. Erect Milkpea

Galactia erecta (Walter) Vail
Ga-lác-ti-a e-réc-ta
Fabaceae (Bean Family)

DESCRIPTION: Erect to ascending perennial herb 8–16″ tall; leaves trifoliolate with linear-oblong to oblong or elliptic, smooth leaflets to 1.5″ long; flowers produced in a short, 1–6 flowered raceme; petals white to pale purple; fruit a short-pubescent legume to 1.5″ long; flowers May–July.

RANGE-HABITAT: NC south to FL and west to TX; scattered but common in the Sandhills and Coastal Plain; Longleaf Pine-Turkey Oak sandhills; dry, sandy flatwoods; and other dry, sandy habitats.

COMMENTS: This is SC's only erect milkpea.

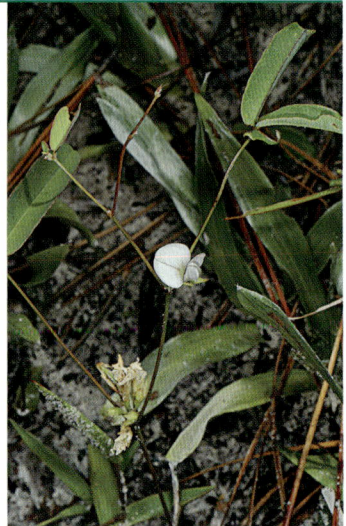

469. Soft Milkpea

Galactia mollis Michaux
Ga-lác-ti-a mól-lis
Fabaceae (Bean Family)

DESCRIPTION: Twining or trailing perennial herb; stems 1–5′ long; leaves trifoliolate, with leaflets to 1.5″ long, broadly oblong to elliptic, with soft, downy (pilose) pubescence; flowers produced in a short raceme of 1–3 flowers; petals reddish to rose-purple; fruit a legume to 1.5″ long with long shaggy pubescence; flowers May–July.

RANGE-HABITAT: Endemic to the Coastal Plain; NC south to FL and west to AL; in SC, occasional in the Sandhills and Coastal Plain in Longleaf Pine-Turkey Oak sandhills and dry, sandy flatwoods.

COMMENTS: The specific epithet, *mollis,* is Latin for soft. The copious, soft, downy pubescence of this species distinguishes it from all other trailing or climbing milkpeas.

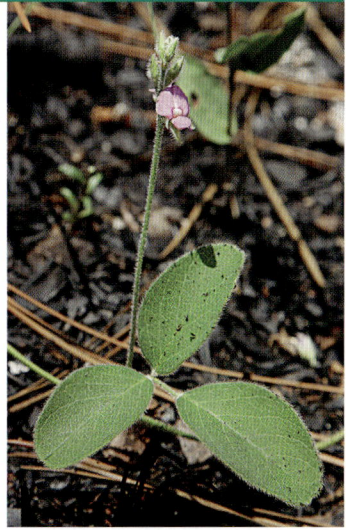

470. Dollarweed

Rhynchosia reniformis A. P. de Candolle
Rhyn-chò-si-a re-ni-fór-mis
Fabaceae (Bean Family)

DESCRIPTION: Erect, perennial herb; stems 2–10″ tall; leaves 1-foliolate, with a round leaflet up to 2″ long and broad, pubescent; flowers produced in short axillary racemes with densely arranged flowers; petals yellow; flowers June–September.

RANGE-HABITAT: NC south to FL and west to TX; in SC, common in the Sandhills and Coastal Plain in Longleaf Pine-Turkey Oak sandhills; sandy, dry Longleaf Pine flatwoods; and other dry, sandy habitats.

COMMENTS: The round leaves that give this species its common name also make it unmistakable.

471. Eastern Green-eyes

Berlandiera pumila (Michaux) Nuttall var. *pumila*
Ber-lan-di-èr-a pù-mi-la var. pù-mi-la
Asteraceae (Aster Family)

DESCRIPTION: Perennial herb, 1–3.5′ tall, with thick, fleshy roots; leaves alternate, ovate, pubescent on both surfaces with crenate (scalloped) margins; flowers in involucrate heads composed of yellow ray flowers and disk flowers that are green in bud—hence, the common name; bracts below flower heads wide; flowers late May–frost.

RANGE-HABITAT: Primarily a Coastal Plain plant from SC to FL and west to TX; in SC, common in the lower Piedmont, Sandhills, and Inner Coastal Plain; Longleaf Pine-Turkey Oak sandhills; Longleaf Pine-scrub oak sandhills; sandy fields; and sand ridges.

COMMENTS: The genus honors Jean Louis Berlandier (1805–55), a Swiss botanist who collected in Texas and Mexico. Eastern Green-eyes makes a good addition to the garden in well-drained, sunny locations.

472. Whorled Milkweed

Asclepias verticillata L.
As-clè-pi-as ver-ti-cil-là-ta
Apocynaceae (Dogbane Family)

DESCRIPTION: Perennial herb with milky sap; stem erect, simple, or branching in upper third; 12–32″ tall; leaves numerous, whorled, or subwhorled, linear; flowers in 2–8 umbels, from upper nodes; flowers June–September.

RANGE-HABITAT: From MA south to FL and west to ND and AZ; in SC, occasional throughout; Longleaf Pine sandhills communities; xeric hardpan forests; sandy, dry, open woods; margins of rock outcrops; rocky slopes; and dry clay roadsides.

COMMENTS: The common name and the specific epithet describe the leaf arrangement, which is often whorled (verticillate). Like all members of its genus, this species can serve as a host for Monarch Butterflies. The genus *Asclepias* is named for the Greek God of Medicine, Asclepius.

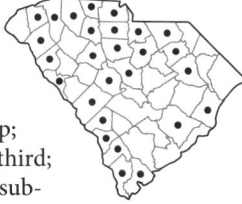

473. Sandhills Milkweed

Asclepias tomentosa Elliott
As-clè-pi-as to-men-tò-sa
Apocynaceae (Dogbane Family)

DESCRIPTION: Perennial herb with milky sap; stem pubescent, erect, 8–24″ tall; leaves lightly pubescent, opposite, typically broadly elliptic to obovate, sometimes linear-lanceolate, 2–3″ long; flowers in 1–3 umbels, from upper nodes; corolla yellowish-green; flowers June–September.

RANGE-HABITAT: NC south to FL, west to AL, and disjunct in TX; in SC, rare, in the Sandhills and Coastal Plain where found in Longleaf Pine-Turkey Oak sandhills, and other dry and sandy habitats.

COMMENTS: This often-overlooked milkweed is never abundant. It is found most often as a single plant with no others for many feet in any direction.

CONSERVATION STATUS: SC-Imperiled

474. Spiked Hoarypea

Tephrosia spicata (Walter) Torrey &
Gray
Te-phrò-si-a spi-cà-ta
Fabaceae (Bean Family)

DESCRIPTION: Perennial herb with short-trailing to
ascending, pubescent stems, 1–2′ long; pubescence long and rusty-colored; leaves pinnately
compound with 9–17 finely pubescent leaflets;
flowers produced in racemes from the upper
nodes or appearing terminal; petals white, turning pink and then red with age; fruit a pubescent
legume 1–2″ long; flowers June–August; fruits mature July–October.

RANGE-HABITAT: DE south to FL and west to KY and LA; in SC, common
throughout in a variety of Longleaf Pine sandhill and flatwoods habitats,
open oak-hickory forests, forest margins, margins of rock outcrops, Piedmont
prairie remnants, and road banks.

SIMILAR SPECIES: Spiked Hoarypea is one of three similar, small species of *Tephrosia*
found in SC. The smaller and more delicate Sprawling Hoarypea (*T. hispidula* (Michaux) Persoon, plate 573) lacks the long, rusty-colored pubescence typical of the stems of Spiked Hoarypea. Sprawling Hoarypea is found in moister habitats mostly in the Coastal Plain. Florida Hoarypea
(*T. florida* (F. G. Dietrich) C. E. Wood) has longer petioles (2–4 times as long as the lowest leaflet vs.
less than 1x as long as the lowest leaflet) and very short pubescence on the legume. Florida Hoarypea
is found in the Sandhills and Coastal Plain in dry Longleaf Pine-Turkey Oak sandhills, Longleaf Pine-
Turkey Oak scrub ridges, and dry flatwoods.

475. Hairy False Foxglove;
Southern Oak-leech

Aureolaria pectinata (Nuttall) Pennell
Au-re-o-lá-ri-a pec-ti-nà-ta
Orobanchaceae (Broomrape Family)

DESCRIPTION: Annual herb, densely covered with
glandular (sticky) hairs; stems to 3′ tall, often
highly branched; leaves deeply dissected and
also densely pubescent and glandular with acute
tips to the lobes; flowers produced in the axils of
leaves; corolla tubular, yellow, drying black; flowers May–September.

RANGE-HABITAT: From VA south to FL and west to MO and TX; locally common throughout SC but primarily in the Piedmont, Sandhills, and Inner
Coastal Plain; Longleaf Pine-Turkey Oak and Longleaf Pine-scrub oak sandhills; sandy, dry, open woods; and Carolina bay ridges, dry roadsides, and
other dry, sandy, open habitats.

COMMENTS: Hairy False Foxglove is a hemiparasite on the roots of Turkey Oak or
other members of the red oak group. The similar Annual Oak-leech (*A. pedicularia*
(L.) Rafinesque ex Pennell) is generally glandular pubescent only on the lower portions
of the stem and plant and has leaf lobes that are rounded rather than acute. It is found in open oak-
hickory forests throughout the state where it parasitizes oaks and some members of the heath family.

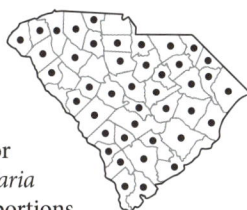

476. Sandhills St. John's-wort

Hypericum lloydii (Svenson) P. Adams
Hy-pé-ri-cum llóyd-i-i
Hypericaceae (St. John's-wort Family)

DESCRIPTION: Decumbent, usually matted shrub, 4–20″ tall; stems angled; leaves needlelike, 0.5–1.2″ long; flowers approximately 0.5″ wide, with 5 showy yellow petals; flowers June–September.

RANGE-HABITAT: NC south to AL; in SC, common in the Sandhills and lower Piedmont; Longleaf Pine-Turkey Oak and Longleaf Pine-scrub oak sandhills, and margins of rock outcrops.

COMMENTS: Good populations of this plant occur in Lexington County at Shealy's Pond Heritage Preserve, in the Longleaf Pine-Turkey Oak sandhills west of the pond, and at Peachtree Rock Heritage Preserve. There are many confusingly similar species of St John's-wort in SC with needlelike leaves, but this is the only common short species within its range that occupies dry habitats. The specific epithet honors Francis E. Lloyd (1868–1947), an American botanist, born in England.

477. Pickering's Dawnflower

Stylisma pickeringii (Torrey ex M.A. Curtis) Gray var. *pickeringii*
Sty-lís-ma pick-e-ríng-i-i var. pick-e-ríng-i-i
Convolvulaceae (Morning Glory Family)

SYNONYM: *Bonamia pickeringii* (Torrey) Gray—RAB

DESCRIPTION: Herbaceous perennial vine; stems numerous, arching from a central point, then trailing radially away, sometimes forming a mound 3–6′ in diameter; leaves linear, held nearly vertically; flowers 1–3, on axillary peduncles; flowers June–August (September).

RANGE-HABITAT: From NC through SC, GA, and AL and disjunct into the NJ pine barrens; in SC, rare in the Sandhills and Coastal Plain where growing in the driest, most barren, deepest sands.

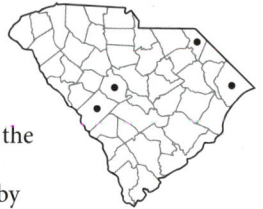

COMMENTS: This species is easily distinguished from other species of *Stylisma* by its narrow, linear leaves borne vertically and by its numerous stems arching from a central point, then trailing radially. Fire is necessary for seed germination. Bob McCartney of Woodlander's, Inc., in Aiken, SC, has induced germination by placing the seeds in sub-boiling water for a few minutes. The species may be seen in the environs of Shealy's Pond Heritage Preserve in Lexington County and Henderson Heritage Preserve in Aiken County. The specific epithet honors Charles Pickering (1805–78), a Pennsylvanian botanist.

CONSERVATION STATUS: SC-Imperiled

478. Georgia Beargrass

Nolina georgiana Michaux
No-lì-na geor-gi-à-na
Ruscaceae (Ruscus Family)

DESCRIPTION: Perennial herb to 5′ tall; leaves in a basal rosette, linear, 12–18″ long, reduced upward, gracefully arching away from the stem; inflorescence a large panicle; flowers white; flowers late May–June; fruits a 3-lobed, bladderlike capsule, with thin "wings" on each angle; maturing June–August.

RANGE-HABITAT: From nc. SC south to sc. GA; in SC, rare in the Sandhills and Inner Coastal Plain; Longleaf Pine-Turkey Oak and Longleaf Pine-scrub oak sandhills.

COMMENTS: *Nolina* is a genus of plants that is common in the desert regions of the southwestern US and Mexico. Though it does not appear to be related to Solomon's-seal, they are now known to be closely related. A good population of Georgia Beargrass occurs in the Aiken Gopher Tortoise Heritage Preserve in Aiken County.

CONSERVATION STATUS: SC-Vulnerable

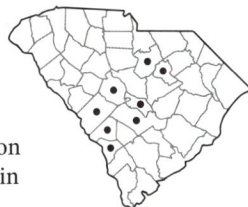

479. Southern Jointweed

Polygonella americana (Fischer & C. A. Meyer) Small
Po-ly-go-nél-la a-me-ri-cà-na
Polygonaceae (Buckwheat Family)

DESCRIPTION: Perennial subshrub with numerous, short, leafy branches, 24–32″ tall; appearing as a depressed, matted shrub early in the growing season; leaves linear to linear-spatulate to 0.5″ long with pale tips; petals absent; sepals white; flowers June–September.

RANGE-HABITAT: From sc. NC to s. GA, west to TX and NM, and north to the interior to MO and AR; in SC, common in the Sandhills and Inner Coastal Plain where it is found in Longleaf Pine-scrub oak sandhills, Longleaf Pine-Turkey Oak sandhills, and on dry, sandy roadsides.

COMMENTS: Large populations of this species are easily observed on the roadsides through Manchester State Forest in Sumter County.

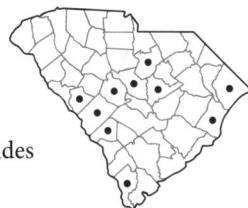

480. Eastern Sensitive-briar

Mimosa microphylla Dryander
Mi-mò-sa mi-cro-phýl-la
Fabaceae (Bean Family)

SYNONYM: *Schrankia microphylla* (Solander ex Smith) Macbride—RAB

DESCRIPTION: Trailing perennial herb; stems prickly, 3–6′ long; leaves, sensitive to touch and quickly closing, bipinnately compound with 6–16 pinnae, each divided again with 20–32 leaflets; flowers produced in globular heads from the axils

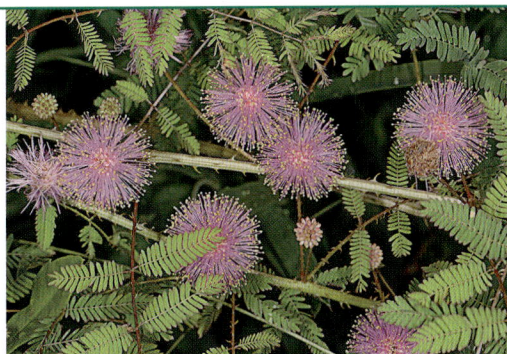

of leaves; flowers with insignificant sepals and petals and long, the flowering head appearing like a pink pom-pom; flowers June–September.

RANGE-HABITAT: DE south to FL and west to KY and TX; in SC, common throughout in a variety of Longleaf Pine sandhill habitats, Longleaf Pine flatwoods; open, rocky, oak-hickory forests; forest margins; road banks; and other open, dry habitats with clay or sandy soil.

COMMENTS: Eastern Sensitive-briar is a popular plant familiar to many children because of the remarkably fast closing of the leaflets and leaves in response to touch (seismonastic movement). The leaves also close slowly at night (nyctinasty) and in response to fire. The rapid movement is controlled by changes to the pulvini, which are the swollen bases of the leaves and leaflets. The movement of ions such as potassium into and out of the cell forces/allows water to move and results in a loss or gain of turgor in the pulvini, causing movement. The movement is thought to deter herbivorous insects from eating the leaf. Closing the leaf at night reduces the surface area available to predators while photosynthesis is not occurring.

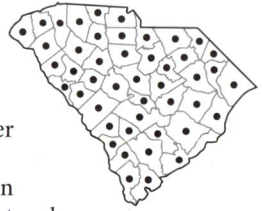

481. Carolina Sandhill Ironweed

Vernonia angustifolia Michaux var.
  angustifolia
    Ver-nòn-i-a an-gus-ti-fò-li-a var.
    an-gus-ti-fò-li-a
  Asteraceae (Aster Family)

DESCRIPTION: Perennial herb with 1–several, erect stems; stems 2–4′ tall; basal leaves absent; stem leaves narrow, linear to linear-elliptic, 2–5″ long; flowers produced in involucrate heads in a flat-topped (corymbose) arrangement; heads with disk flowers only; disk flowers purple; pappus ranging from white to purple; flowers June–September.

RANGE-HABITAT: Endemic to the Coastal Plain from NC south to GA; in SC, common in the Sandhills and Coastal Plain in a variety of sandhill and dry, Longleaf Pine flatwoods habitats.

COMMENTS: Carolina Sandhill Ironweed makes a stunning landscape plant in dry soils, achieving a much larger stature than it does in the wild. It is visited by a large number of native pollinators including many butterflies. A second variety, Georgia Sandhill Ironweed (*V. angustifolia* var. *scaberrima* (Nuttall) Gray), is known from the southern Coastal Plain of SC; it has wider leaves and involucral bracts with a long acuminate tip. The genus *Vernonia* is named in honor of William Vernon (1666–1711), an English botanist who traveled and collected plants in Maryland.

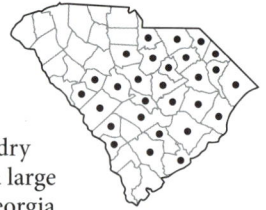

482. Rose Purslane; Kiss-me-quick

Portulaca pilosa L.
    Por-tu-là-ca pi-lò-sa
  Portulacaceae (Purslane Family)

DESCRIPTION: Prostrate or erect, many-branched annual, 2–8″ tall; leaves fleshy, with tufts of whitish hairs in the axils; flowers dark pink to purple; fruit a capsule with a "lid" that comes off near the middle to expose the smooth, red seeds for dispersal; flowers June–October.

RANGE-HABITAT: NC south to FL and west to OK and NM; also found in Central America; in SC, occasional in the Sandhills, the lower Piedmont, and Inner Coastal Plain; disturbed, sandy soils, Longleaf Pine-scrub oak sandhills, yards, and other disturbed sites.

COMMENTS: This attractive, weedy species is common throughout the South, but its native range is obscure, and it has certainly benefited from soil disturbance due to human activity. The similar, very rare, native Small's Portulaca (*P. smallii* P. Wilson) is confined to granite flatrocks in the Piedmont. It has much paler pink flowers. It has been confused with the introduced *P. amilis* Spegazzini that has leaves flattened in cross section.

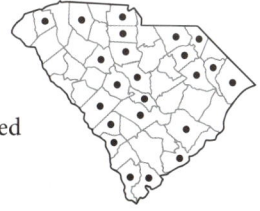

483. Sandhills Bean

> *Phaseolus sinuatus* (Nuttall) Torrey
> Pha-sè-o-lus si-nu-à-tus
> Fabaceae (Bean Family)

DESCRIPTION: Trailing perennial vine with stems 3–12′ long; leaves trifoliolate; leaflets with 3 prominent lobes and evident, heavy reticulate venation; flowers produced in few-flowered racemes, 4–12″ long; each flower subtended by a bract and two smaller bractlets; flowers pinkish to light purple; fruit a curved, smooth legume 1–2″ long; flowers July–September; fruits mature August–October.

RANGE-HABITAT: Endemic to the Coastal Plain from NC south to FL and west to MS; in SC, rare in the Sandhills and Coastal Plain where found in Longleaf Pine-scrub oak sandhills, dry Longleaf Pine flatwoods, and other dry, sandy habitats.

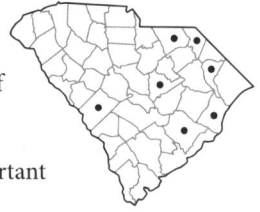

COMMENTS: *Phaseolus* is in the same genus as many of the commercially important beans. The specific epithet refers to the sinuous lobes of the leaflets.

CONSERVATION STATUS: SC-Imperiled

484. Sandhill Chaffhead

> *Carphephorus bellidifolius* (Michaux)
> Torrey & Gray
> Car-phé-pho-rus bel-li-di-fò-li-us
> Asteraceae (Aster Family)

Perennial herb, with ascending stems 6–20″ tall; basal leaves numerous, 2–8″ long, somewhat spoon-shaped; stem leaves reduced upward; flowers in heads, pink to lavender; flowers July–October.

RANGE-HABITAT: Endemic to the se. Coastal Plain from VA south to GA; in SC, common in the Sandhills and Coastal Plain in a variety of sandhill habitats and sandy, dry, open woods and other dry, sandy communities.

COMMENTS: The only other *Carphephorus* commonly found in SC is Carolina Chaffhead (*C. tomentosus* (Michaux) Torrey & Gray, plate 657). The lower stems of Sandhill Chaffhead are smooth or with short appressed hairs; Carolina Chaffhead has lower stems with obvious, long spreading hairs and also frequently occurs in moister habitats.

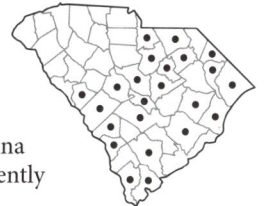

485. Sandhills Gaillardia

Gaillardia aestivalis (Walter) H. Rock.
Gail-lárd-i-a aes-ti-và-lis
Asteraceae (Aster Family)

DESCRIPTION: Erect, annual herb from a taproot; leaves oblanceolate, entire or with a few teeth; flowers arranged in an involucrate head composed of ray and disk flowers or disk flowers only; ray flowers, when present, 3-lobed, yellow, sometimes infused with red, to 0.5″ long; disk flowers dark reddish-purple; receptacle without long bristles; flowers July–October.

RANGE-HABITAT: NC west to KS and south to FL and TX; in SC, occasional in the Sandhills and Coastal Plain; Longleaf Pine-scrub oak sandhills; dry, sandy flatwoods; and other dry, sandy habitats.

COMMENTS: In SC this species tends to lack ray flowers and displays a simple purple head of reddish-purple disk flowers. The genus is named in honor of eighteenth-century French patron of botany, M. Gaillard de Charentonneau.

486. One-sided Blazing Star

Liatris secunda Elliott
Li-à-tris se-cún-da
Asteraceae (Aster Family)

DESCRIPTION: Perennial herb from a globose rootstock; stems pubescent, arching to reclining, 1–3′ long; lower leaves linear to narrowly elliptic; flowers produced in involucrate heads composed of disk flowers only, usually 5-flowered; arrangement of heads spikelike, with all the heads directed to the same side (secund); corolla lobes pink to whitish, widely spreading; flowers late July–October.

RANGE-HABITAT: Endemic to the Coastal Plain from NC to FL and AL; in SC, occasional in the Sandhills and Coastal Plain; a variety of sandhill habitats, sand ridges, Carolina bay ridges, and other dry, sandy sites.

COMMENTS: This species is distinctive because of the light-colored disk flowers and the secund arrangement of the heads—all arranged in one direction. The only other species that commonly displays a secund arrangement of heads is Sandhills Blazing Star (*L. cokeri* Pyne & Stucky, plate 488), which has smooth stems and smaller heads with purple disk flowers produced in a very congested spikelike arrangement. Occasionally other species of blazing star will produce secund heads if the stem falls over or is bent.

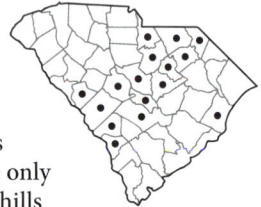

487. Shortleaf Blazing Star

Liatris tenuifolia Nuttall
Li-à-tris te-nu-i-fò-li-a
Asteraceae (Aster Family)

DESCRIPTION: Perennial herb from a globose underground stem (corm); stems smooth, erect, to 6′ tall; fibers of previous years basal leaves often conspicuous at the base of the plant; basal and lower leaves filiform to linear, to 6″ long; upper leaves greatly reduced; flowers produced in involucrate heads composed of disk flowers only, usually 5-flowered, the involucral bracts with blunt tips and typically pink-margined; heads arranged in a spike-like structure; corolla lobes lavender; flowers August–November.
RANGE-HABITAT: Endemic to the Coastal Plain from SC south to FL and west to AL; in SC, occasional in the Sandhills and Coastal Plain in Longleaf Pine-scrub oak sandhills, Longleaf Pine-Turkey Oak sandhills, sand ridges, Carolina bay ridges, and other dry, sandy habitats.
COMMENTS: This species is distinctive in its blunt involucral bracts with a pinkish margin and very narrow tuft of basal and lower leaves.

488. Sandhills Blazing Star

Liatris cokeri Pyne & Stucky
Li-à-tris cò-ker-i
Asteraceae (Aster Family)

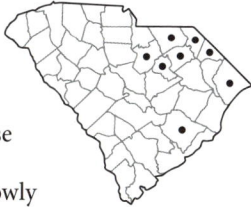

DESCRIPTION: Perennial herb from a globose underground stem (corm); stems smooth, erect, 1–4′ tall; lower leaves linear to narrowly elliptic, to 8″ long; flowers produced in involucrate heads composed only of disk flowers, usually 4–7 flowered, the involucral bract tips sharply acuminate to acute; arrangement of heads spikelike, with heads very densely arranged and often all held on the same side of the stem (secund); corolla lobes lavender to purple; flowers late August–November.
RANGE-HABITAT: Endemic to the Coastal Plain of the Carolinas; in SC, infrequent in the Sandhills and Coastal Plain where found in Longleaf Pine-Turkey Oak sandhills and Longleaf Pine-scrub oak sandhills.
SIMILAR SPECIES: This is one of two species with a secund arrangement of heads in SC. The smooth stems and purplish flowers, which are produced in a very congested spikelike arrangement, distinguish it from One-sided Blazing Star (*L. secunda* Elliott, plate 486). It is more likely to be confused with species that typically do not display secund arrangements of heads such as Grass-leaved Blazing Star (*Liatris pilosa* (Aiton) Willdenow, plate 409) or Wand Blazing Star (*L. virgata* Nuttall). From Grass-leaved Blazing Star it can be distinguished by the sharp-pointed involucral bracts and more congested inflorescence. From Wand Blazing Star it can be distinguished by the more congested inflorescence and 4–7 flowers per head (vs. 7–10 flowers per head in Wand Blazing Star).
COMMENTS: The specific epithet honors William Chambers Coker (1872–1953), from Hartsville, SC, and long-time professor at the University of North Carolina, Chapel Hill.

489. Sandhill Goldenaster

Pityopsis pinifolia (Elliott) Nuttall
   Pi-ty-óp-sis pi-ni-fó-li-a
Asteraceae (Aster Family)

SYNONYM: *Heterotheca pinifolia* (Elliott)
   Ahles—RAB

DESCRIPTION: Smooth perennial herb with short stolons, to 20″ tall; basal leaves shorter than the stem leaves; leaves grasslike, entire, ascending; flowers produced in involucrate heads of narrow yellow ray flowers and yellow disk flowers; heads few to many; flowers late August–September.

RANGE-HABITAT: Endemic to the Coastal Plain from NC south to GA and west to AL; in SC, locally common at scattered localities in the Sandhills and Coastal Plain; Longleaf Pine-Turkey Oak sandhills, sand ridges, and other dry, sandy habitats.

COMMENTS: This plant is locally common but has a very restricted range. It is common at Peachtree Rock Heritage Preserve in Lexington County. There is a disjunct occurrence on Sandy Island in the Waccamaw River National Wildlife Refuge in Georgetown County, the only site outside the Sandhills.

CONSERVATION STATUS: SC-Imperiled

490. Senna Seymeria

Seymeria cassioides (J. F. Gmelin)
   Blake
      Sey-mèr-i-a cas-si-oì-des
Orobanchaceae (Broomrape Family)

DESCRIPTION: Erect, profusely branched annual herb, 20–40″ tall, hemi-parasitic; glandular-hairy; fresh plants green, drying dark; leaves opposite, deeply dissected, segments filiform, less than 0.5″ long; corolla smooth, lemon yellow, sometimes marked with purple within; flowers August–October.

RANGE-HABITAT: VA south to FL, west to LA; in SC, common in the Sandhills and Coastal Plain and rare in the Piedmont; Longleaf Pine savannas, flatwoods and sandhill habitats, pocosin margins, and along roadsides.

SIMILAR SPECIES: Comb Seymeria (*Seymeria pectinata* Pursh ssp. *pectinata*) is very similar but has leaf segments that are lanceolate and pubescent corollas. It is infrequent in dry Longleaf Pine habitats throughout the Coastal Plain of SC.

COMMENTS: This species is parasitic on pines. The genus honors Henry Seymer (1745–1800), an English naturalist.

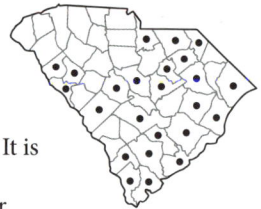

491. Woody Goldenrod

Chrysoma pauciflosculosa
(Michaux) Greene
Chry-sò-ma pau-ci-flos-cu-
lò-sa
Asteraceae (Aster Family)

SYNONYM: *Solidago pauciflosculosa*
Michaux—RAB

DESCRIPTION: Evergreen shrub to 5′ tall; trunk short and stocky with ascending branches; branches forming a flat-topped arrangement; overwintering leaves grayish-green, densely clustered on the terminal portions of the older, woody branches; during summer new flowering branches are produced which quickly elongate and have wide-spaced leaves and flowers and die back in the winter to the crown of vegetative branches; flowers produced in involucrate of 5 yellow disk flowers; heads produced in a paniclelike arrangement, with each head appearing to be attached on one side (secund); flowers late July–October.

RANGE-HABITAT: NC south to n. FL and west to s. MS; in SC, very rare, found only in the Sandhills in Lexington and Chesterfield Counties and Longleaf Pine-Turkey Oak and Longleaf Pine-scrub oak sandhills.

COMMENTS: Woody Goldenrod can be seen in Peachtree Rock Heritage Preserve in Lexington County and the Hudsonia Flat in Cheraw State Park in Chesterfield County.

CONSERVATION STATUS: SC-Critically Imperiled

492. Sandhill Wild-buckwheat

Eriogonum tomentosum Michaux
E-ri-ó-go-num to-men-tò-sum
Polygonaceae (Buckwheat Family)

DESCRIPTION: Perennial herb to 3′ tall; stem erect or often leaning, hairy, freely branched, usually from a basal rosette; basal leaves often numerous, evergreen, sometimes dying with age or because of drought; stem leaves with dense white or tan hairs beneath, in whorls of 3 or 4; flowers July–September.

RANGE-HABITAT: From NC south to FL and west to AL; in SC, fairly common in the Sandhills in Longleaf Pine-Turkey Oak and Longleaf Pine-scrub oak sandhills.

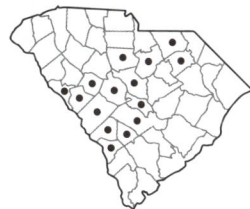

493. Stiff-leaved Aster

Ionactis linariifolia (L.) Greene
I-o-nác-tis li-na-ri-i-fò-li-a
Asteraceae (Aster Family)

SYNONYM: *Aster linariifolius*
L.—RAB

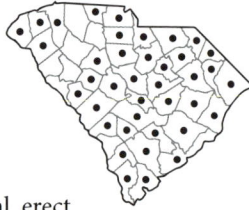

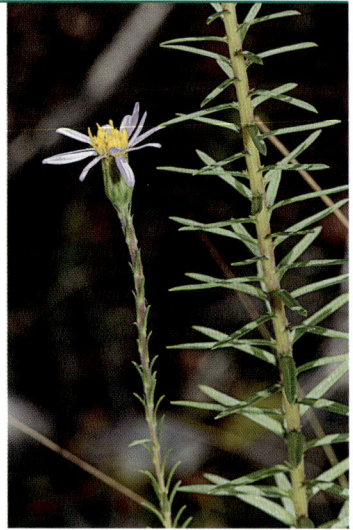

DESCRIPTION: Perennial herb with 1–several, erect stems forming a dense clump; stems 1–2.5′ tall; leaves alternate, linear, very stiff, and crowded along the stem; flowers in involucrate heads composed of ray and disc flowers; ray flowers bluish to violet, disc flowers yellow to red; flowers August–November.
RANGE-HABITAT: Widespread in eastern North America from Quebec west to WI and south to FL and TX; in SC, common throughout; Longleaf Pine sandhills, Longleaf Pine flatwoods, forest margins, margins of rock outcrops, old fields, openings in dry oak-hickory forests, roadsides, and other dry, open habitats.
COMMENTS: Stiff-leaved Aster is unique among SC asters in having very stiff, needlelike leaves. It is an attractive and well-behaved addition to the sunny, dry, meadow garden.

494. Eastern Silvery Aster

Symphyotrichum concolor (L.)
G. L. Nesom var. *concolor*
Sym-phy-o-trì-chum cón-co-
lor var. cón-co-lor
Asteraceae (Aster Family)

SYNONYM: *Aster concolor* L.—RAB, PR

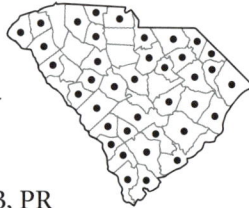

DESCRIPTION: Perennial herb covered with a fine, silvery down; stems 1–several, erect, 1–3′ tall; leaves alternate, about 2″ long, elliptic; bracts subtending flower head whitish with green apices; plant covered with a fine, silvery down; flowers September–October.
RANGE-HABITAT: From VA to FL, west to LA and north to TN and KY; common and essentially throughout SC (rare in the mountains); Longleaf Pine sandhills and flatwoods, woodland margins, old fields, thickets, and openings in dry oak-hickory forests.
COMMENTS: The flowers of Eastern Silvery Aster are produced in long "wands," which look more like a blazing star than an aster from a distance. All native American species that were formerly in the genus *Aster* have now been moved to segregate genera because they are not closely related to the Old World genus *Aster*.

495. Streamhead pocosin

496. Coastal Plain Serviceberry

Amelanchier obovalis
(Michaux) Ashe
A-me-lán-chi-er ob-o-và-lis
Rosaceae (Rose Family)

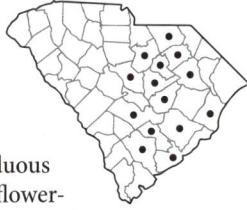

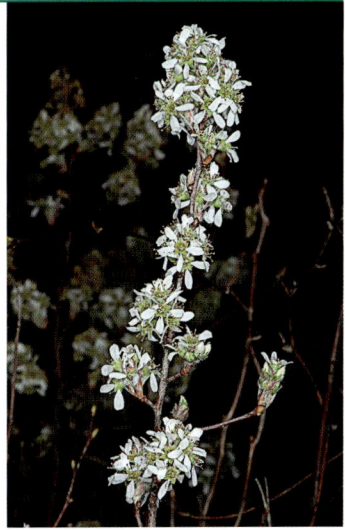

DESCRIPTION: Rhizomatous, colonial, deciduous
shrub, to 5′ tall; leaves scarcely evident at flowering; leaves simple, alternate, elliptic to oblong with
serrate margins; flowers March–April; fruit a purple, soft, and
sweet pome, mature May–June.

RANGE-HABITAT: NJ and PA south to GA; in SC, common throughout the Sandhills and Coastal Plain in pocosins, bays, low woods,
and Longleaf Pine flatwoods and savannas.

COMMENTS: This is the only shrub-sized, strongly rhizomatous
species in our area. This species and other species of *Amelanchier*
make excellent landscape plants when used in natural situations.
They bloom early, announcing the arrival of spring. The orange fall
foliage is outstanding, and the fruits are excellent food for birds.
Native Americans used the fruits in making bread and pemmican; the fruits can be eaten raw,
made into preserves, cooked as sauces or in pies, or dried for winter use.

497. Bog Spicebush

Lindera subcoriacea B. E.
Wofford
Lín-der-a sub-co-ri-à-ce-a
Lauraceae (Laurel Family)

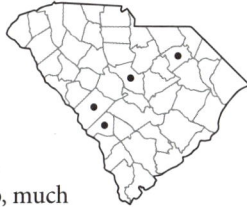

DESCRIPTION: Deciduous, male and female
flowers on separate plants, aromatic shrub, much
branched, and usually 5–12′ tall; young twigs
pubescent; leaves alternate, thick-textured, elliptic to oblanceolate,
usually with some rounder leaves produced early in the season;
sun-grown plants with distinctive whitish undersides of the leaves;
young leaves with the odor of lemon when crushed, older leaves
less aromatic; flowers yellow, in dense axillary umbels, before
the leaves appear; flowers March–April; drupes mature August–
September.

RANGE-HABITAT: Endemic to the Coastal Plain and edge of the
Piedmont from VA south to FL and west to LA; in SC, uncommon
in the Sandhills and Coastal Plain in streamhead pocosins, bay
forests, hillside seeps, Atlantic White Cedar forests and floodplains of blackwater streams.

COMMENTS: This species occupies more acidic habitats than the related Northern Spicebush
(*L. benzoin* (L.) Blume). In some parts of SC, it grows in shaded floodplains and displays
wider than normal leaves. The species is dioecious (male and female flowers on separate
plants) and is sexually dimorphic with male plants displaying larger leaves. The genus name
honors Johann Linder (1676–1723), a Swedish botanist. Robert McCartney, of Woodlanders,
Inc., in Aiken, first documented the species for SC. It is easily overlooked when not in flower
and is likely to be found in additional counties and locations.

CONSERVATION STATUS: SC-Vulnerable

498. Red Chokeberry

Aronia arbutifolia (L.) Persoon
A-rò-ni-a ar-bu-ti-fò-li-a
Rosaceae (Rose Family)

DESCRIPTION: Deciduous, rhizomatous shrub to 8′ tall, often forming large colonies; leaves alternate, elliptic to elliptic-lanceolate, with crenate margins and pubescent beneath; flowers in corymbs; white; fruit are red pomes; flowers February–April; fruits mature September–November.

RANGE-HABITAT: Widespread in eastern North America from Newfoundland south to FL and west to TX, predominantly in the Coastal Plain; in SC, common throughout but most abundant in the Sandhills and Coastal Plain in pocosins, bay forests, wet savannas, margins of swamps, bogs and fens, and other moist habitats.

COMMENTS: The dramatic red fruit, as well as fine red coloration to the leaves in the autumn and a brief but showy display of flowers in the spring, make this a great landscape plant for SC. The fruit are eaten by birds and mammals but often not until very late in the winter or early spring.

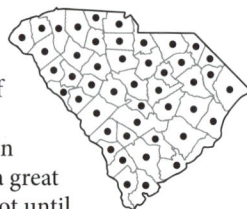

499. Coastal Witch-alder

Fothergilla gardenii Murray
Fo-ther-gíll-a gar-dèn-i-i
Hamamelidaceae (Witch-hazel Family)

DESCRIPTION: Deciduous colonial shrub 1–3′ tall; leaves spreading, alternate, narrowly ovate to ovate, pubescent with microscopically starlike hairs; leaves with oblique or rounded bases, rarely shallowly cordate; leaf margins crenate to serrate from near the middle to the tip; inflorescence terminal, in erect spikes appearing before the leaves; flowers April–May.

RANGE-HABITAT: Primarily a Coastal Plain species from se. NC south to FL Panhandle and west to AL; in SC in the Sandhills and Coastal Plain; occasional; pocosins and pocosin margins, wet Longleaf Pine savannas, and sandhill seepages.

SIMILAR SPECIES: Small-leaf Witch-alder (*F. parvifolia* Kearney in Small) was formerly considered part of a broader concept of *F. gardenii.* It was recognized in a recent revision of the genus (Haynes et al., 2020). It is a taller plant with drooping leaves that are distinctly cordate at the base. It is known from similar habitats in Aiken County, SC, and is more common in GA.

COMMENTS: Coastal Witch-alder is planted as an ornamental because of its small size, showy white spikes in the spring, and brilliant yellowish-orange to scarlet foliage in the fall. The genus honors John Fothergill (1712–80), a London physician and botanist. The specific epithet honors Alexander Garden, M.D. (1730–92), a Charleston physician and botanist.

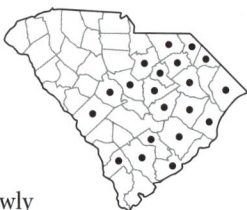

500. Coastal Sweet Pepperbush

Clethra alnifolia L.
Clèth-ra al-ni- fò-li-a
Clethraceae (Clethra Family)

DESCRIPTION: Deciduous, colonial shrub 3–8′ tall; leaves alternate, obovate to oblong, serrate, lower leaf surface slightly pubescent to smooth, petioles 1–2.3″ long; flowers produced in racemes, fragrant, petals white; flowers June–July.

RANGE-HABITAT: Primarily on the Coastal Plain from Nova Scotia south to FL and west to TX; disjunct in TN; in SC, common in the Coastal Plain and Sandhills; pocosins, wet flatwoods, savannas, and swamp forests.

SIMILAR SPECIES: Downy Sweet Pepperbush (*C. tomentosa* Lamarck, plate 564) is similar but has highly pubescent undersides to the leaves and shorter petioles that are also pubescent. It is common in the Sandhills and Coastal Plain in similar situations. Downy Sweet Pepperbush is the more abundant of the two species in the central and southern Coastal Plain of SC, largely replacing Coastal Sweet Pepperbush.

COMMENTS: Coastal Sweet Pepperbush has become a very popular ornamental because of the size, showy and sweet-scented flowers, and the fact that it is extremely attractive to pollinators.

501. White Wicky

Kalmia cuneata Michaux
Kálm-i-a cu-ne-à-ta
Ericaceae (Heath Family)

DESCRIPTION: Deciduous, colonial shrub to 6′ tall; leaves alternate, oblanceolate to narrowly elliptic; flowers in short racemes or axillary fascicles near the tip of branches of the previous season; corolla greenish white with a red band within, the pedicels drooping; flowers May–June.

RANGE-HABITAT: Endemic to the Coastal Plain of se. NC and adjacent SC; in SC, rare in the Sandhills region in streamhead pocosins, particularly along the margins, in pocosin to savanna or pocosin to sandhill ecotones.

COMMENTS: The flowers of White Wicky are similar to other species of *Kalmia*, but it can easily be separated by the following characteristics: deciduous leaves; a solid, red band on the inside of the petals; and a curved, fruiting stalk. This species can be extremely difficult to find among all the other similar pocosin shrubs. It is most easily recognized throughout the year by the clusters of drooping capsules on last season's growth.

CONSERVATION STATUS: SC-Critically Imperiled

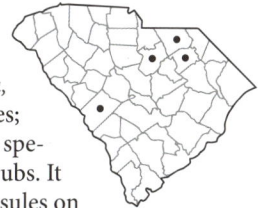

502. Swamp Azalea

Rhododendron serrulatum (Small)
Millais
Rho-do-dén-dron ser-ru-là-tum
Ericaceae (Heath Family)

SYNONYM: *Rhododendron viscosum*
(L.) Torrey var. *serrulatum* (Small)
Ahles—RAB

DESCRIPTION: Deciduous shrub to 20′ tall; leaves alternate, elliptic to obovate, smooth with some pubescent along the midrib; bud scales 15–20, the inner aristate; flowers produced well after the leaves have expanded; petals white with the corolla tube glabrous within; flowers late June–October.
RANGE-HABITAT: VA south to FL and west to Louisiana; in SC, common in bay forests, pocosins, and blackwater swamp forests.
COMMENTS: Many authors have treated Swamp Azalea as a variety or form of Clammy Azalea (*R. viscosum* (L.) Torrey, plate 727). It appears to be quite distinct. It typically forms a tall shrub, rather than a dwarf shrub as in Clammy Azalea. The bud scales are more numerous, 15–20 (vs. 8–12 in Clammy Azalea) and the corolla tube is smooth within (vs. pubescent within in Clammy Azalea). Additionally, Swamp Azalea often grows in densely shaded habitats and flowers much later in the season. Swamp Azalea makes a great addition to the landscape where the later flowering period greatly extends the color provided by our native azaleas.

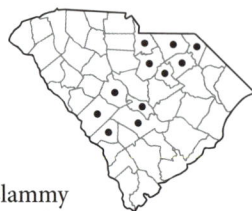

503. Sandhill Heartleaf

Hexastylis sorriei L. L. Gaddy
Hex-ás-ty-lis sòr-rie-i
Aristolochiaceae (Birthwort Family)

DESCRIPTION: Prennial herb from a short underground stem, forming small clumps; leaves rounded, heart-shaped at the base, 1.5–2.5″ long and wide; dark green often with some lighter mottling along the veins; flowers solitary on short stalks, usually under the leaf litter; calyx tube broad cylindric and flared just below the opening calyx tube only slightly constricted just before the orifice; 0.3–0.8″ long; flowers February–April.
RANGE-HABITAT: Restricted to the Fall-line Sandhills of NC and SC; in SC, locally common along the margins of streamhead pocosins and in herbaceous-dominated seepage bogs.
COMMENTS: The habitat serves to distinguish this species from all related species. The most morphologically similar species is Little Heartleaf (*Hexastylis minor* (Ashe) Blomquist), which is found along hardwood bluffs and oak-hickory forests with acidic soils in the Piedmont, mostly adjacent to the NC line. The species was recently described by SC botanist L. L. Gaddy. The specific epithet honors botanist Bruce Sorrie (b. 1944), whose work has focused on Longleaf Pine ecosystems.
CONSERVATION STATUS: SC-Imperiled

504. Herbaceous seepage slope

505. Southern Purple Pitcherplant;
Frog's Breeches; Hunter's Cup

Sarracenia purpurea L.
var. *venosa* (Rafinesque)
Fernald
Sar-ra-cèn-i-a pur-pú-re-a var.
ve-nò-sa
Family Sarraceniaceae (Pitcherplant
Family)

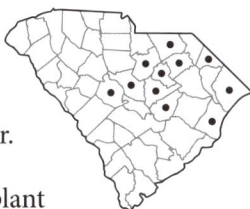

DESCRIPTION: Rhizomatous perennial, evergreen, carnivorous
herb with leaves held horizontally to ascending, near the ground
(decumbent), and modified into hollow tubes that are effective as
passive insect traps; flowering stalks 8–16″ tall; the hood of the
leaves held away from the opening of the trap; flowers April–May.
RANGE-HABITAT: Endemic to the Atlantic Coastal Plain Province of
the se. US from VA south to GA; in SC, rare in the Sandhills and
Coastal Plain; favors sphagnum openings in pocosins where it
grows more robust; also found in moist, Longleaf Pine savannas,
and herbaceous seepage slopes.
COMMENTS: Frog's Breeches differ from other Sandhills and Coastal Plain pitcherplants by
having leaves that lie horizontally but curve upward, an erect hood that does not cover the
mouth, and the inner surface of the hood bearing many stiff hairs that point downward
toward the mouth. The open mouth permits the pitcher to fill with rainwater where insects
that fall in are drowned. It is believed that glands secrete a wetting agent into the water
that denies the insect buoyancy, so it cannot fly off the water's surface. For distinctions
from Southern Appalachian Pitcherplant (*S. purpurea* var. *montana* Schnell & Determann,
plate 59), see discussion under that variety.

506. Bantam-buttons; Yellow Hatpins

Syngonanthus flavidulus (Michaux)
Ruhland
Syn-go-nán-thus fla-ví-du-lus
Eriocaulaceae (Pipewort Family)

DESCRIPTION: Perennial herb, with separate male
and female flowers on the same plant, forming very congested basal rosettes of numerous
short linear to hairlike leaves generally less than
3″ long; flowers produced in dense heads of
minute flowers on elongate extremely narrow
peduncles 6–18″ tall; heads yellowish and often
with a dry texture; flowers May–October.
RANGE-HABITAT: Endemic to the Coastal Plain from NC south to FL and west
to MS; in SC, rare in the Sandhills and Coastal Plain in herbaceous seepage
slope; ecotones of pocosins; and wet, sandy, Longleaf Pine and Pond Cypress
savannas.
SIMILAR SPECIES: The pale yellowish heads, the extremely floriferous nature, and
extremely narrow leaves arranged into tight clumps help to identify this species from
similar species of *Lachnocaulon* and *Eriocaulon*.
CONSERVATION STATUS: SC-Imperiled

507. Pocosin Loosestrife

Lysimachia asperulifolia Poiret
Ly-si-má-chi-a as-pe-ru-li-
fò-li-a
Primulaceae (Primrose Family)

DESCRIPTION: Slender, erect, herbaceous perennial, 1–3′ tall; leaves, dark bluish-green, in whorls
of 3–4 (sometimes 2 and opposite), sessile, ovate to
lanceolate with acute tips and smooth; flowers produced in terminal, leafy racemes; sepals 5, green with dark streaks (resin canals);
petals 5, yellow also with dark streaks; flowers May–June.

RANGE-HABITAT: Endemic to the Coastal Plain of NC and SC; in
SC, very rare and known from Darlington (where extirpated)
and Richland counties, in herbaceous seepage slopes and pocosin
ecotones.

COMMENTS: This is one of the rarest species in SC. It is very distinctive and not easily confused with any other species. It had not
been seen in SC for more than 100 years when it was discovered
on a firing range at Fort Jackson by John Nelson, botanist at the A. C. Moore Herbarium. It is
more frequent in southeastern NC and should be sought in the northeastern Coastal Plain of
SC. It should never be disturbed in the wild. New populations should be photographed and a
report made to the South Carolina Heritage Trust Program.

CONSERVATION STATUS: Federally Endangered

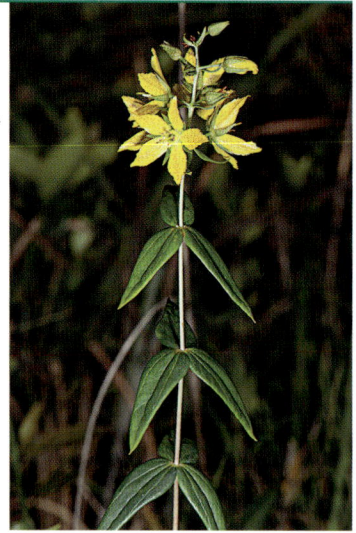

508. Sandhills Bog Lily

Lilium pyrophilum M. W.
Skinner & Sorrie
Lí-li-um py-ró-phi-lum
Liliaceae (Lily Family)

DESCRIPTION: Erect, branching, perennial herb
from a scaly bulb with rhizomes, 2–5′ tall; leaves
in 1–12 whorls, narrowly elliptic, 0.75–6.25″ long,
tips acute; leaves held ascending (toward the stem) or horizontal
but not arching downward at the tip; flowers produced in a raceme
of pendant nonfragrant flowers with tepals that are deep red or
orange at the base fading to yellow or light orange toward the tips;
flowers July–August.

RANGE-HABITAT: Endemic to the Sandhills and Inner Coastal Plain
region of NC, SC, and GA; in SC, rare in the Sandhills and at least
historically at one location in the Inner Coastal Plain, in herbaceous seepage slopes, and the ecotones of pocosins.

COMMENTS: One of the most stunning wildflowers in the Carolinas.
This species was only recently described (Skinner and Sorrie, 2002). Its habitat is uncommon and threatened by fire suppression; it should never be disturbed in the wild. The reddish
flowers are visited by hummingbirds and swallowtail butterflies.

CONSERVATION STATUS: SC-Critically Imperiled

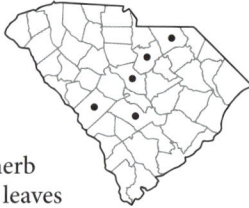

509. Water Sundew

Drosera intermedia Hayne
    Dró-se-ra in-ter-mè-di-a
    Droseraceae (Sundew Family)

DESCRIPTION: Perennial herb; leaves in basal rosettes; flowering stalks with or without erect leaves; leaves narrowly spoon-shaped; leaf stalks and blades with tentaclelike, glandular hairs, the secretion of each gland contributing to the insect-catching function of the leaf; flower stalk smooth, 2–4″ tall, strongly curved at base and standing away from the rosette; flowers white or tinged with pink; flowers July–September.

RANGE-HABITAT: Newfoundland south to FL and west to MN and TX; also into tropical America; in SC common in the Sandhills and Coastal Plain, rare in the mountains; infrequent; bogs, savannas, edges of Pond Cypress ponds, seepage areas, pocosins, cataract fens and seepages over granitic flatrocks, and margins of pools or streams, often in standing water.

COMMENTS: Water Sundew is the most robust of the common species in SC. The only larger species is Threadleaf Sundew (*Drosera filiformis* Rafinesque), which is known from a single historical collection in Orangeburg County.

510. Roundleaf Sundew

Drosera rotundifolia L.
    Dró-se-ra ro-tun-di-fò-li-a
    Droseraceae (Sundew Family)

DESCRIPTION: Perennial, rosette-forming, herb; leaves with glandular hairs, the secretion of each gland contributing to the insect-catching function of the leaf; leaf blades round to weakly kidney-shaped, wider than long, about 0.25″ across, tapering abruptly to a distinct leaf stalk; flowering stem 2–6″ tall; flowers with a white corolla often tinged with pink; flowers July–September.

RANGE-HABITAT: Roundleaf Sundew is the most widely distributed sundew in the cooler temperate regions; throughout North America, Europe, and Asia; in SC, rare and primarily found in the mountains; disjunct in the Sandhills in habitats where cool seepage water provides suitable habitat and one location in the Coastal Plain; sphagnum and cataract bogs, herbaceous seepage slopes, Atlantic White-cedar forests, and vertical seepages on rock or clay.

COMMENTS: Roundleaf Sundew usually grows among sphagnum moss. It is primarily a self-pollinating species. Throughout most of its range it does not produce flowers that open. These tiny, white buds produce seeds through cleistogamy (self-pollinating only).

CONSERVATION STATUS: SC-Imperiled

511. Tawny Cottonsedge

Eriophorum virginicum L.
E-ri-ó-pho-rum vir-gí-ni-cum
Cyperaceae (Sedge Family)

DESCRIPTION: Perennial herb, forming small clumps with flowering stems 2–3.5′ tall; leaves grasslike; basal and low stem leaves to 12″ long and 0.15″ wide, with smooth blades with a sharp, scaberulous margin; upper stem leaves reduced; flowers produced in a dense tawny head of spikelets with elongate, cottonlike bristles; the bristles become light brown to whitish when fruits are mature; fruit is an achene; flowers July–September; fruits mature August–October and often persisting through the winter.

RANGE-HABITAT: Widespread in eastern North America from Labrador south to SC and west to Ontario and KY; disjunct in the Okefenokee Swamp in GA; in SC, very rare in the mountains, Sandhills, and Outer Coastal Plain; herbaceous seepages, montane fens, and sphagnous openings in pocosins.

COMMENTS: Cottonsedges (genus *Eriophorum*) are often called "cottongrasses," but they are true sedges. They are among the most beautiful members of the family and are most abundant in the far north in the boreal forest and tundra regions. Tawny Cottonsedge is the only species that ranges as far south as SC, where it is near its southern limit. Although it is very rare in SC, it is startling when encountered and is unmistakable. The tuft of "cotton" is formed by the bristles of the fruits and fades from orangish-brown (tawny) when young to nearly white as winter approaches.

CONSERVATION STATUS: SC-Critically Imperiled

512. Southern White Beaksedge

Rhynchospora macra (C. B. Clarke) Small
Rhyn-chós-po-ra mác-ra
Cyperaceae (Sedge Family)

DESCRIPTION: Perennial herb, with short pale rhizomes, forming tight clumps; stems 15–30″ tall; leaves grasslike, less than 1/15″ wide, three-angled toward the tip; inflorescence terminal or with 2–3 well-spaced clusters in the upper one-third of the flowering stem; flowers occur in spikelets; perianth reduced to 16–25 bristles that are microscopically retrorsely barbed; spikelet clusters are pale white when young and fade to a tawny brown; flowers August–October.

RANGE-HABITAT: From NC to FL, west to se. TX; also in Nicaragua and Puerto Rico; in SC, rare, in the Sandhills in herbaceous seepage slopes.

SIMILAR SPECIES: Southern White Beaksedge is one of many species in the genus found in the herbaceous seepage slope community. It is easily distinguished by the attractive white clusters of spikelets. Pale Beaksedge (*R. pallida* M. A. Curtis) is similar and can be even more robust but does not display the perianth bristles of Southern White Beaksedge. In SC, Pale Beaksedge is restricted to the edges (ecotones) of pocosins and herbaceous seepage slopes in the Sandhills.

CONSERVATION STATUS: SC-Critically Imperiled

513. Carolina Bog Asphodel

Tofieldia glabra Nuttall
   To-field-i-a glà-bra
   Tofieldiaceae (False-asphodel Family)

DESCRIPTION: Perennial herb from short rhizomes forming small clumps; flowering stems smooth, 6–27″ tall; leaves conduplicate (irislike), glossy deep green, smooth; flowers produced in a raceme; tepals white; flowers September–November.
RANGE-HABITAT: Endemic to NC and SC; in SC, rare, edges (ecotones) of pocosins and herbaceous seepage slopes in the Sandhills and northeastern Coastal Plain.
SIMILAR SPECIES: Southern Bog Asphodel (*Triantha racemosa* (Walter) Small, plate 648) is a much more abundant species. It is easily differentiated from Carolina Bog Asphodel by the pubescent and scabrous stems and flowers produced in a thyrse in June–August.
COMMENTS: The genus honors English botanist Thomas Tofield (1730–79).
CONSERVATION STATUS: SC-Critically Imperiled

514. Atlantic White-cedar forests

515. Atlantic White-cedar

Chamaecyparis thyoides (L.) Britton,
   Sterns, & Poggenberg var. *thyoides*
   Cha-mae-cý-pa-ris thy-oì-des var.
   thy-oì-des
Cupressaceae (Cypress Family)

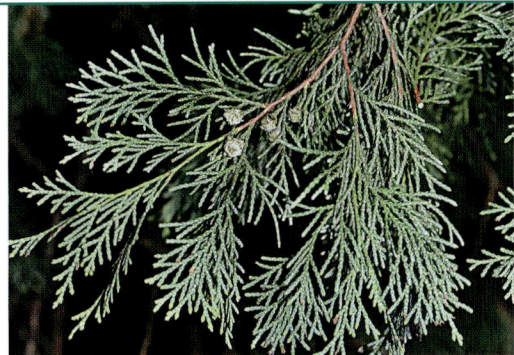

DESCRIPTION: Moderate to slow-growing tree that may live for more than 1,000 years; to 92′ tall; leaves small, evergreen, opposite, scalelike; male and female cones on same tree, produced in the spring; seed cones green with a bluish white bloom, turning bluish purple at maturity; seeds winged, shed in the fall.
RANGE-HABITAT: ME south to FL and west to MS; in SC, uncommon in the Sandhills and Inner Coastal Plain; when Atlantic White-cedar dominates a community, the community is named for it; also found in acid swamps, hillside seepages, wet sands, and streamhead pocosins.

COMMENTS: The light brown wood is unmatched in its resistance to decay. In the 1700s, the wood was used for log cabins, roof shingles, barrels, and boats. Today, it is much reduced in abundance and is not a major commercial tree. It is a hardy tree and is planted as an ornamental. Atlantic White-cedar can be seen in Shealy's Pond Heritage Preserve in Lexington County. In local usage and on topographic maps, juniper refers to *Chamaecyparis thyoides* (i.e., Juniper Creek), while cedar refers to Eastern Red Cedar (*Juniperus virginiana* L.).

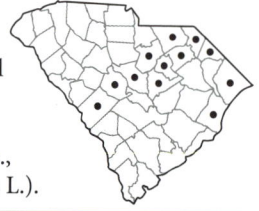

516. Rayner's Blueberry

Vaccinium sempervirens
     Rayner & Henderson
          Vac-cí-ni-um sem-pér-vi-rens
     Ericaceae (Heath Family)

DESCRIPTION: Evergreen perennial; stems to 16″ tall, erect to ascending in shade and tending to arch and creep in full sun; rooting at nodes where in contact with the ground; leaves elliptic to obovate, with fine, glandular, rounded to pointed teeth most obvious toward the tip; flowers white, urn-shaped, late April–early May; fruit mature in the fall.

RANGE-HABITAT: Endemic to the Sandhills in Lexington County, SC, and known from only a few sites; boggy openings in Atlantic White-cedar forests, especially along the headwaters of Scouter Creek.

TAXONOMY: This species is clearly related to Creeping Blueberry (*Vaccinium crassifolium* Andrews, plate 558) and it is sometimes reduced to a subspecies of *crassifolium* (*V. crassifolium* Andrews ssp. *sempervirens* (Rayner & Henderson) Kirkman & Ballington). The authors agree with Weakley (2020) and retain it as a species because it is allopatric (the two do not inhabit the same area) and distinctive morphologically. A population with arching to creeping stems occurs at Peachtree Rock Heritage Preserve in Lexington County.

COMMENTS: Douglas A. Rayner (one of the authors this book) and J. Henderson were the first to describe this species of blueberry (Rayner and Henderson, 1980). A population of Rayner's Blueberry occurs at Shealy's Pond Heritage Preserve in Lexington County.

CONSERVATION STATUS: SC-Critically Imperiled

517. Collins's Sedge

Carex collinsii Nuttall
          Cà-rex col-líns-i-i
     Cyperaceae (Sedge Family)

DESCRIPTION: Tightly clumping evergreen perennial sedge; flowering stems 10–30″ tall; basal and lower stem leaves narrow, dark green, grasslike, to 0.10″ wide; pistillate spikes 1–4, toward the top of the flowering stem, few-flowered and very narrowly lanceolate; flowers June–July.

RANGE-HABITAT: Found on the Coastal Plain from RI south to GA with disjunct occurrences in the mountains of NC, NJ, and PA; in SC, uncommon, restricted to Atlantic White-cedar forests and bay forests in saturated, highly acidic soils.

COMMENTS: Collins's Sedge is one of many species of *Carex* that occupy our bay forests and Atlantic White-cedar forests. It is included here because of the distinct character of the spike allowing for easy identification and as a representation of the genus *Carex,* which is abundant and diverse in SC acidic wetlands. It may be seen at Shealy's Pond Heritage Preserve in Lexington County.

CONSERVATION STATUS: SC-Imperiled

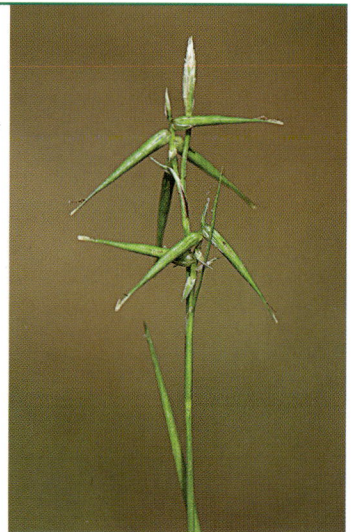

The Coastal Plain:
The Inner and Outer Coastal Plain

518. Longleaf Pine-Turkey Oak xeric ridge

519. Sand Live Oak

Quercus geminata Small
Quér-cus ge-mi-nà-ta
Fagaceae (Beech Family)

DESCRIPTION: Small- to medium-sized evergreen tree, occasionally shrublike in frequently burned areas; leaves resembling an overturned boat— the blade curved downward and the margins rolled under; the upper surface of the leaves with strongly impressed veins, the lower surface with veins raised; leaf undersides stellate pubescent (with starlike trichomes); female flowers 1–2 on a short stalk, separate from the male flowers, which are in catkins; flowers April; fruits mature September–October of the same year.

RANGE-HABITAT: Endemic to the Coastal Plain from NC south to FL and west to MS; in SC, common in the Coastal Plain where growing in Longleaf Pine-Turkey Oak xeric ridges, sandhills, and dry flatwoods.

COMMENTS: This species is sometimes lumped with Live Oak (*Quercus virginiana* P. Miller, plate 903), which is a much larger tree, with larger leaves that are not curved and boat-shaped, nor do they display the deeply impressed veins typical of Sand Live Oak.

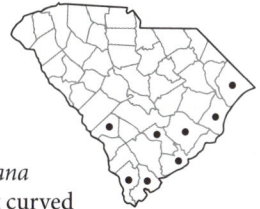

520. Florida Rosemary

Ceratiola ericoides Michaux
Ce-ra-ti-ò-la e-ri-coì-des
Ericaceae (Heath Family)

DESCRIPTION: Many-branched, dense shrub, 2–8′ tall with male and female flowers on separate plants (dioecious); branches with 4 rows of short, slender, needlelike leaves; flowers, sessile in leaf axils; sepals 2 and petals 2, but reduced and minute; stamens reddish-brown; flowers October– November.

RANGE-HABITAT: SC south to FL and west to MS; in SC, occasional in the Sandhills and the Coastal Plain where found in Longleaf Pine-Turkey Oak sandhills, Longleaf Pine-Turkey Oak xeric ridges of Carolina bay rims, and sand ridges along blackwater rivers.

COMMENTS: This is not the herb, rosemary, and has none of the pungent odor of that species. It derives the common name from the overall resemblance of the plant to rosemary. It is highly flammable because of the aromatic compounds it contains. Adult plants are killed by fire, but the bare and exposed sand following a fire opens up space for seedling recruitment. Plants do not produce seeds until they are at least 10–15 years old, and too-frequent a fire return time may eliminate populations. Much research has been done of the potential allelopathic (toxic) chemicals exuded by these plants. Many species cannot grow directly beneath the adult plants, and so a barren zone is created around them where competition from other plants is reduced.

CONSERVATION STATUS: SC-Imperiled

521. Sandhill Milkweed

Asclepias humistrata Walter
As-clè-pi-as hu-mis-trà-ta
Apocynaceae (Dogbane Family)

DESCRIPTION: Herbaceous perennial with milky sap from a deep, narrow, tapering root; stems 1–several, stiff and nearly prostrate to spreading upward at a low angle, up to 2' long; leaves usually 5–8 pairs, opposite, broad, sessile, clasping, gray-green to bluish-green with pink, reddish, or cream-colored venation; flowers May–June; fruits erect follicles, maturing June–July.

RANGE-HABITAT: From NC south to FL and west to LA; in SC, common in the Sandhills and Coastal Plain; Longleaf Pine sandhills, Longleaf Pine-Turkey Oak xeric ridges, and other sandy, dry, open woodlands.

COMMENTS: This milkweed and Sandhills Butterflyweed (*A. tuberosa* L. ssp. *rolfsii* (Vail) Woodson) have the habit of lying nearly flat on the ground, which is unique among SC milkweeds. Sandhills Butterflyweed has green leaves, clear sap, and orange flowers, and it is uncommon in the Sandhills and Coastal Plain.

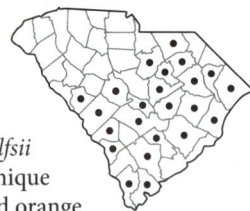

522. Sandhill Thistle

Cirsium repandum Michaux
Cír-si-um re-pán-dum
Asteraceae (Aster Family)

SYNONYM: *Carduus repandus* (Michaux) Persoon—RAB

DESCRIPTION: Deep-rooted, perennial herb, 8–24" tall; stems with cobweblike hairs, leafy to the summit; leaves alternate, margins spiny, with small spines in addition to the scattered, larger spines; flowers in involucrate heads of purple disk flowers only, held on a short peduncle; heads solitary or more often terminating short branches from near the summit; bracts below the flower heads spine-tipped; flowers May–July.

RANGE-HABITAT: From se. VA south to e. GA; in SC, common in the Coastal Plain and Sandhills, rarely into the adjacent edge of the Piedmont; nearly endemic to the Carolinas; xeric sandhills and other dry, sandy habitats.

523. Southern Bogbuttons

Lachnocaulon beyrichianum
Sporleder ex Körnicke
Lach-no-caù-lon bey-rich-i-
à-num
Eriocaulaceae (Pipewort Family)

DESCRIPTION: Clump-forming herb, with tufts of
leaves aggregated into dense mats of rosettes which often have substantial stems, many years old, held on or just below the ground;
leaves narrowly linear and tapering, dull green; flowers produced
in heads at the end of a long, very narrow, weak, leafless stem
that often lays flat on the ground due to the weight of the head;
separate male and female flowers in the same head; heads seldom
broader than 0.2″; seeds dark brown, very lustrous, the longitudinal ribs obscure; flowers May–September.
RANGE-HABITAT: Endemic to the Coastal Plain from NC to FL; in
SC, uncommon in the Coastal Plain on white, wind-sorted, barren
sands of ecotones between pocosins and Longleaf Pine-Turkey
Oak xeric ridges such as are found on Carolina bay rims and sand ridges along blackwater
rivers; also in white sands of dry flatwoods in the ecotone to pocosins.
SIMILAR SPECIES: Identification of the three species of *Lachnocaulon* that occur in coastal SC
takes practice but is not difficult. This species is distinct because of the long-lived, expansive
colonies of dense basal rosettes with dull, narrow green leaves and long slender flowering
stems. Common Bogbuttons (*L. anceps* (Walter) Morong, plate 628) is similar but occupies
Longleaf Pine savannas and has brighter green leaves, forms less extensive clumps, and has
shorter flowering stems. Brown Bogbuttons (*L. minus* (Chapman) Small) is a smaller plant
than the other two species, is short-lived, does not form the large clumps typical of the other
two species, and has short flowering stems with a head that is elongated like the tip of a
drumstick.
COMMENTS: Recent fieldwork by Patrick D. McMillan has documented this species as more
common than reported, especially on Carolina bay rims. It can be seen at Lewis Ocean Bay
Heritage Preserve in Horry County.
CONSERVATION STATUS: SC-Imperiled

524. Large-fruited Beaksedge

Rhynchospora megalocarpa Gray
Rhyn-chós-po-ra me-ga-lo-cár-pa
Cyperaceae (Sedge Family)

DESCRIPTION: Coarse, rhizomatous perennial from
scaly rhizomes 12–50″ tall; leaf blades mostly
basal forming a dense tuft; flowers produced
in spikelets, arranged in 2–6 clusters per stem;
spikelets 0.15–0.35″ long, 1–2 fruited; achenes
very large for a beaksedge (0.15″ long); perianth
reduced to microscopic bristles that are an-
trorsely barbed; flowers June–August.
RANGE-HABITAT: Endemic to the Coastal Plain from NC south to FL and west
to MS; in SC, locally common but restricted to barren sands of Longleaf
Pine-Turkey Oak xeric ridges on Carolina bay rims and sand ridges along
blackwater rivers.
COMMENTS: This beaksedge is distinct both in its habitat and its spikelets—which
are enormous for a beaksedge. It grows in the driest habitats of any SC species.
Although some manuals list it as rare, it appears to be more abundant than once
thought and is locally abundant in its preferred habitat.

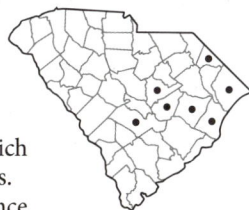

525. Summer-farewell

Dalea pinnata (J. F. Gmelin)
  Barneby var. *pinnata*
  Dàle-a pin-nà-ta var. pin-nà-ta
Fabaceae (Bean Family)

SYNONYM: *Petalostemum pinnatum*
(Walter ex J. F. Gmelin) Blake—RAB

DESCRIPTION: Perennial, aromatic herb, 12–24″ tall, from a taproot; stems smooth, branched above; leaves odd-pinnately compound, glandular-dotted, with 3–11 linear-filiform leaflets; flowers arranged in densely-flowered, short, involucrate spikes, petals and petaloid staminodes (sterile stamens) white; fruit a 1- or 2-seeded legume; flowers August–frost.

RANGE-HABITAT: Endemic to the Coastal Plain from NC south to FL; in SC, occasional in the Coastal Plain and Sandhills where found in Longleaf Pine-Turkey Oak xeric ridges; Longleaf Pine sandhills; and sandy, dry, open woodlands.

COMMENTS: The short spikes with numerous tiny bracts forming an involucre appear to be more similar to a member of the aster family than the bean family. Good populations in SC exist in the Tillman Sand Ridge Heritage Preserve and environs in Jasper County. The genus honors Samuel Dale (1659–1739), an English botanist.

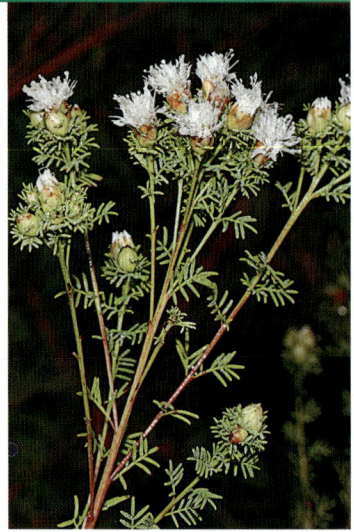

526. Soft-haired Coneflower

Rudbeckia mollis Elliott
  Rud-béck-i-a mól-lis
Asteraceae (Aster Family)

DESCRIPTION: Annual, biennial, or occasionally short-lived perennial; 1–3′ tall, with a deep taproot; stems branched and with dense pubescence; stem leaves alternate, sessile, oblong, with dense, soft, woolly pubescence; flowers produced in an involucrate head of yellow ray flowers and blackish-purple disk flowers; flowers late August–October.

RANGE-HABITAT: Endemic to the Coastal Plain from se. SC south to FL; in SC, very rare in the Coastal Plain where it is restricted to Longleaf Pine-Turkey Oak xeric ridges with unsorted, "brown" sands and adjacent sandy woodlands and road banks.

COMMENTS: Soft-haired Coneflower is found in SC only in Jasper County in the Tillman Sand Ridge Heritage Preserve and environs. It is conspicuous along the roadsides through this habitat when in flower.

CONSERVATION STATUS: SC-Critically Imperiled

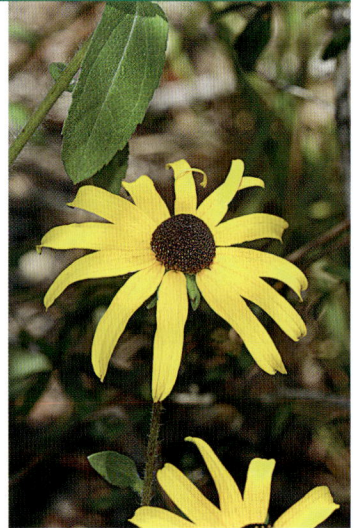

527. Carolina Warea

Warea cuneifolia (Muhlenberg ex
  Nuttall) Nuttall
  Wà-re-a cu-nei-fò-li-a
Brassicaceae (Mustard Family)

DESCRIPTION: Annual herb with erect stems to 2′ tall, branched above; leaves alternate, oblanceolate, 0.75–1.5″ long and 0.2–0.4″ wide; flowers

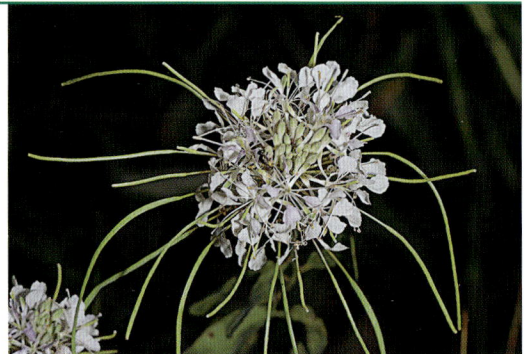

with pink corollas; flowers July–September; fruit a silique, sickle-shaped and curved downward, long-stalked, matures in August–September.
RANGE-HABITAT: Endemic to the Coastal Plain from NC south to the FL Panhandle, west to se. AL; in SC, rare in the Coastal Plain in Longleaf Pine-Turkey Oak xeric ridges and Longleaf Pine sandhills.
COMMENTS: In SC, this beautiful and rare species can be seen at the Tillman Sand Ridge Heritage Preserve and environs in Jasper County.
CONSERVATION STATUS: SC-Vulnerable

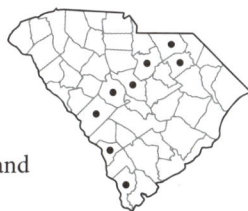

528. Sandhills Gerardia

Agalinis setacea (J. F. Gmelin) Rafinesque
A-ga-lì-nis se-tà-ce-a
Orobanchaceae (Broomrape Family)

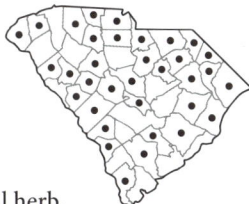

DESCRIPTION: Dark green or reddish annual herb, 8–24″ tall; leaves long and very slender; flowers with pedicels slender, 0.2–0.75″ long; the corolla pink, the throat with yellow lines; upper corolla lobes erect or reflexed, not bent over the stamens; flowers September–October.
RANGE-HABITAT: NY south to FL and west to AL; in SC, common in the Sandhills and Coastal Plain and rare in the Piedmont; Longleaf Pine-Turkey Oak xeric ridges; Longleaf Pine-Turkey Oak sandhills; sandy, dry, open woodlands; and dry clay banks and margins of rock outcrops.
COMMENTS: All species of *Agalinis* are thought to be parasitic (to various degrees) on roots of grasses or other herbs. Many other species of *Agalinis* occur in the Coastal Plain. Sandhills Gerardia is the only species growing in the extreme xeric sandhill habitat.

529. Harper's Scrub-balm; Rose Dicerandra

Dicerandra odoratissima Harper
Di-ce-rán-dra o-do-ra-tís-si-ma
Lamiaceae (Mint Family)

DESCRIPTION: Freely branched, aromatic annual herb; stems erect, 4-angled, 8–20″ tall; leaves opposite, linear to narrowly elliptic; flowers arranged in a bracteate thyrse; corolla bilabiate with three lobes above and two below, pink to lavender; flowers September–October.
RANGE-HABITAT: Endemic to the Coastal Plain from extreme se. SC to se. GA; in SC, rare in the Coastal Plain where restricted to Longleaf Pine-Turkey Oak xeric ridges with unsorted "brown" sands and adjacent sandy, dry, open woodlands and road banks.
COMMENTS: In SC, Harper's Scrub-balm is known only from the Tillman Sand Ridge Heritage Preserve and environs in Jasper County. This is the most northern species of a genus that is far more diverse in the scrub and sandhills of GA and FL. The odor of Harper's Scrub-balm is extremely pungent and smells like a mixture of spicy mint and cinnamon. It is among the most beautiful to the eye, and nose, of any native species in SC.
CONSERVATION STATUS: SC-Critically Imperiled

530. Woolly Golden-aster

Chrysopsis gossypina (Michaux) Elliott
Chry-sóp-sis gos-sý-pi-na
Asteraceae (Aster Family)

SYNONYM: *Heterotheca gossypina*
(Michaux) Shinners—RAB

DESCRIPTION: Perennial herb, 12–28″ tall; reclining, ascending, or erect; dense, woolly hairs throughout, except sometimes not on the bracts below the flower heads; flowers produced in involucrate heads in a corymbose structure; heads composed of yellow ray and yellow disk flowers; flowers September–October.

RANGE-HABITAT: From VA south to FL and west through GA into AL; in SC, common in the Coastal Plain and Sandhills and adjacent areas of the Piedmont, where found in Longleaf Pine-Turkey Oak xeric ridges; Longleaf Pine sandhills; and other sandy, dry, open woodlands.

COMMENTS: The woolly, soft pubescence of the leaves and stem help to distinguish this species of dry, sandy soils from other golden aster species in SC.

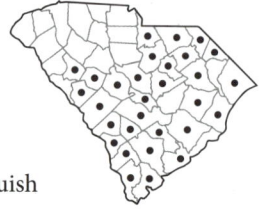

531. Gopher-apple

Geobalanus oblongifolius (Michaux)
Small
Ge-o-ba-là-nus ob-lon-gi-fò-li-us
Chrysobalanaceae (Coco-plum Family)

SYNONYMS: *Chrysobalanus oblongifolius*
Michaux—RAB; *Licania michauxii*
Prance—PR

DESCRIPTION: Low colonial shrub to 16″ tall; fire adapted, with an extensive underground stem system; leaves simple, evergreen, alternate, and oblanceolate; flowers produced in a panicle of cymes at the ends of branches; petals white; fruit a drupe; flowers May–June; fruit matures September–October.

RANGE-HABITAT: From se. SC south to FL and west to LA; abundant in dry sandy habitats in the southern part of its range; in SC, rare in the Coastal Plain where restricted to Longleaf Pine-Turkey Oak xeric ridges with unsorted, "brown" sands and adjacent sandy woodlands and road banks.

COMMENTS: Gopher-apple is easily observed in the Tillman Sand Ridge Heritage Preserve and environs in Jasper County. The "gopher" referred to in the name is probably the gopher tortoise, which shares the habitat.

CONSERVATION STATUS: SC-Critically Imperiled

532. Scrub Ghost Flower

Monotropa brittonii Small
Mo-nó-tro-pa brit-tón-i-i
Ericaceae (Heath Family)

DESCRIPTION: Mycoparasitic perennial herb without chlorophyll; stem 2–8″ tall; leaves reduced to scales; stems and flowers range from pink or salmon to whitish with a light pink hue; flowers solitary at end of stem, drooping; petals conspicuously and densely ciliate margined and pubescent

within; filaments densely pubescent; flowers September–November; capsules become erect as they mature in October–November.

RANGE-HABITAT: Endemic to the Coastal Plain, most abundant in FL, ranging sporadically north to NC and west to LA; in SC, rare in the Coastal Plain where found in Longleaf Pine-Turkey Oak scrub ridges.

COMMENTS: This exquisitely colored relative of the Indian Pipe (*M. uniflora* L., plate 366) is most easily distinguished by the intense coloration. Indian Pipes may also vary to pink, but the flowers do not have the dense ciliate margins and are not densely pubescent within.

CONSERVATION STATUS: SC-Critically Imperiled

533. Sandy, dry, open woodlands

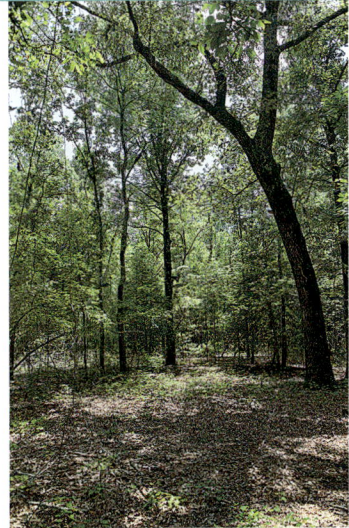

534. Hog Plum; Flatwoods Plum

Prunus umbellata Elliott
Prù-nus um-bel-là-ta
Rosaceae (Rose Family)

DESCRIPTION: Small deciduous tree or shrub to 20′ tall, forming small colonies; leaves alternate, glabrous or slightly pubescent, elliptic, 1–3″ long, not abruptly tapered to an acuminate tip, with simple crenate to crenate-serrate margins; flowers white, produced before the leaves appear; fruit a drupe that is very dark purple, sometimes varying to red or yellow, with a whitish sheen; flowers February–April; fruits mature August–September.

RANGE-HABITAT: NC south to FL and west to AR and TX; in SC, common throughout but most abundant in the Coastal Plain, usually in sandy or rocky upland woods, thickets, roadsides, and riverbanks.

SIMILAR SPECIES: Chickasaw Plum (*P. angustifolia* Marshall, plate 936) is similar to Hog Plum. The teeth on the leaves of Chickasaw Plum have a small red gland on the tip; the leaves of hog plum lack glands. American Plum (*P. americana* Marshall) is also similar but has leaves that abruptly taper to an acuminate tip and have doubly serrate leaves.

COMMENTS: The fruits are very tart but are excellent in pies, jams, or jellies. The wood has little commercial value. This is a fantastic landscape shrub or small tree that is equally as impressive in the spring as the invasive Bradford Pear and is an excellent native replacement. The SC native *Prunus* species provide a valuable food source for a myriad of pollinators and are some of the best plants one can include in their landscape for supporting pollinators, wildlife, and birds, as well as providing edible and useful fruit.

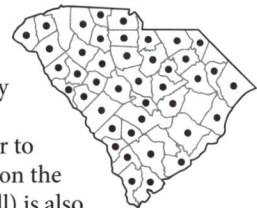

535. Poison Oak

Toxicodendron pubescens P. Miller
Tox-i-co-dén-dron pu-bés-cens
Anacardiaceae (Cashew Family)

SYNONYM: *Rhus toxicodendron*
L.—RAB, PR

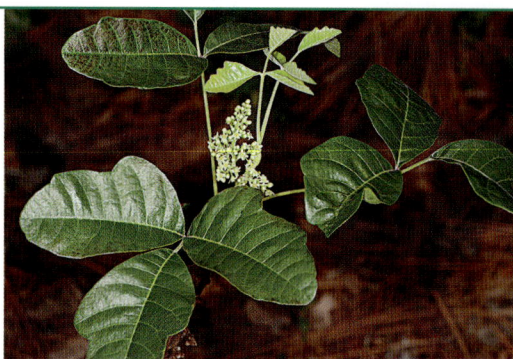

DESCRIPTION: Erect shrub to 6′ tall; leaves alternate, thick, compound with 3 leaflets; leaflets pubescent on both surfaces, often velvety when young, shallowly lobed, coarsely serrate, or rarely entire; flowers April–May; drupes densely hairy, mature August–October.

RANGE-HABITAT: NY south to FL and west to KS and TX; in SC, common throughout in xeric habitats such as Longleaf Pine-Turkey Oak sandhills, dry rock outcrops, and other dry habitats.

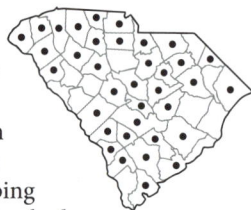

SIMILAR SPECIES: Although the name Poison Oak is sometimes used for Poison Ivy (*T. radicans* (L.) Kuntze, plate 878) and they are sometimes confused with each other, the two species are quite distinct. Poison Ivy is either a high-climbing vine with numerous aerial roots, a trailing plant, or is reduced to a subshrub in shady, forested habitats. The leaves of Poison Ivy are thin and its fruits hairless; the fruits of Poison Oak are velvety pubescent and its leaves are thick. Poison Ivy is as allergenic as Poison Oak.

COMMENTS: All parts of Poison Oak contain an oleoresin (urushiol) that can cause severe skin inflammation, itching, and blistering on contact. Poison Oak is toxic all year, and the oleoresin can also be contracted when it is borne on smoke or dust particles, clothes, and animal hair. Washing thoroughly with soap and hot water or swabbing with alcohol within a few hours after contact prevents the allergic reaction.

Poison Oak and its close relatives have some significant beneficial traits. They are important wildlife foods: the fruits are eaten and dispersed by birds, and deer, bear, raccoons and other animals eat all above ground parts. On exposure to air, urushiol is transformed into a lacquer that is very heat resistant and is used by Asian artisans.

536. Common Chinquapin

Castanea pumila (L.) P. Miller
Cas-tà-ne-a pù-mi-la
Fagaceae (Beech Family)

DESCRIPTION: Small, deciduous tree or shrub, often with several main stems; bark smooth; twigs smooth or pubescent; leaves alternate, elliptic to obovate, 2.5–12″ long, smooth above, pubescent below; flowers produced in long, upward arching creamy white catkins; flowers June–July; fruits mature in September.

RANGE-HABITAT: NJ and PA south to FL and west to MO and TX; in SC, occasional throughout in dry, rocky woodlands; margins of rock outcrops; sandy, dry woodlands; Longleaf Pine sandhills; and flatwoods.

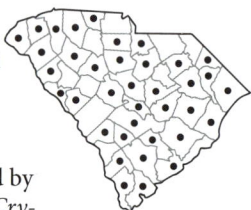

COMMENTS: Though Chinquapin flowers are produced in catkins, they are not wind-pollinated and are attractive to many pollinators, particularly beetles. The nuts are extremely tasty and can be eaten fresh or roasted and are relished by wildlife. Chinquapin are susceptible to chestnut blight, caused by the fungus *Cryphonectria parasitica*. Their ability to flower and fruit at a small size and young age has allowed them to persist in spite of the fungus. Despite persisting, they are not as abundant as they were prior to the blight, particularly in the Upstate.

537. Southern Woolly Violet

Viola villosa Walter
Vì-o-la vil-lò-sa
Violaceae (Violet Family)

DESCRIPTION: Small, perennial herb without a leafy stem; rhizome elongate and stocky; leaves approximately as long as wide, often held close to the ground, evergreen, densely hairy on both surfaces, heart-shaped; both chasmogamous (openly pollinated) and cleistogamous (self-pollinating) flowers present; chasmogamous flower solitary with short pedicels, often hidden by leaves, appearing late February–early April, followed throughout the season by stalked cleistogamous flowers, lacking petals, that produce abundant seeds.

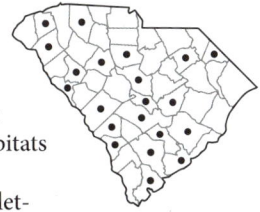

RANGE-HABITAT: MD south to FL and west to OK and TX; in SC, common but scattered in the Sandhills, Coastal Plain, and Piedmont where growing in habitats with dry, open, sandy or rocky soils, including roadsides.

COMMENTS: Southern Woolly Violet is the earliest flowering of the blue or violet-flowered native violets. It is often overlooked because of its small size.

538. Blue Sandhill Lupine; Sky-blue Lupine

Lupinus diffusus Nuttall
Lu-pì-nus dif-fù-sus
Fabaceae (Bean Family)

DESCRIPTION: Clump-forming perennial herb with decumbent stems, 8–16″ long; with dense, short pubescence throughout; leaves 1-foliolate, evergreen; flowers produced in racemes; petals light to deep blue, with a conspicuous white to cream spot on the standard; fruit a legume with appressed pubescence; flowers March–May; fruits mature June–July.

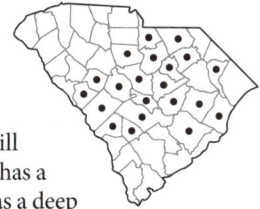

RANGE-HABITAT: Endemic to the Coastal Plain from NC south to FL and west to MS; in SC, common in the Coastal Plain and Sandhills; Longleaf Pine-Turkey Oak sandhills; dry Longleaf Pine flatwoods; and sandy, dry, open woodlands.

SIMILAR SPECIES: The range of Blue Sandhill Lupine overlaps with Pink Sandhill Lupine (*L. villosus* Willdenow, plate 539). The standard of Blue Sandhill Lupine has a conspicuous white to cream spot, while the standard of Pink Sandhill Lupine has a deep reddish-purple spot.

COMMENTS: American lupines are complex taxonomically and poorly differentiated into acceptable taxa. In the Southeast, however, the species are few and not hard to identify. Some of the western species are known to be poisonous to livestock. The cultivated lupines are mostly hybrids.

539. Pink Sandhill Lupine; Lady Lupine

Lupinus villosus Willdenow
Lu-pì-nus vil-lò-sus
Fabaceae (Bean Family)

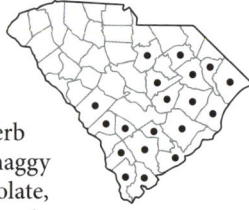

DESCRIPTION: Clump-forming perennial herb with decumbent stems, 8–16″ long, and shaggy long pubescence throughout; leaves 1-foliolate, evergreen; flowers produced in racemes; petals light to dark pink, with a conspicuous dark reddish-purple spot on the standard; fruit a shaggy pubescent legume; flowers March–May; fruits mature June–August.

RANGE-HABITAT: Endemic to the Coastal Plain from NC south to FL and west to LA; in SC, common in the Coastal Plain and Sandhills in xeric Longleaf Pine sandhills; dry flatwoods; sandy, dry, open woodlands; and dry, sandy roadsides.

SIMILAR SPECIES: See discussion under Blue Sandhill Lupine (*L. villosus* Willdenow, plate 538).

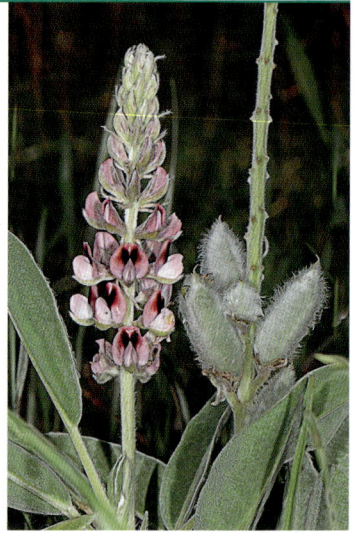

540. Northern Sundial Lupine

Lupinus perennis L. ssp. *perennis*
Lu-pì-nus pe-rén-nis ssp. pe-rén-nis
Fabaceae (Bean Family)

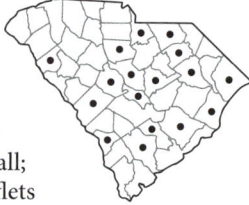

DESCRIPTION: Perennial, erect herb 8–24″ tall; leaves palmately compound with 7–11 leaflets radiating in all directions (like a sundial); flowers produced in a terminal raceme; petals blue, rarely pink or white; fruit a short-pubescent legume; flowers April–May; fruits mature June–July.

RANGE-HABITAT: ME south to SC and west to MN and IL; in SC, uncommon in the Sandhills and Coastal Plain where found in sandy, dry, open woodlands, sandhills, and clearings.

COMMENTS: The genus name *Lupinus* comes from the Latin *lupus* (wolf) because they were once thought to deplete or "wolf" the soil of minerals. This is unfounded; members of the pea family actually enrich the soil by harboring bacteria in their roots that fix atmospheric nitrogen. The palmately compound leaves in the formation of a sundial is diagnostic for this species in SC. A more southern subspecies *L. perennis* ssp. *gracilis* (Nuttall) Dunn, is a smaller plant with short stems and leaves that are tightly clustered. It is found in GA adjacent to the Savannah River and may yet be located in SC.

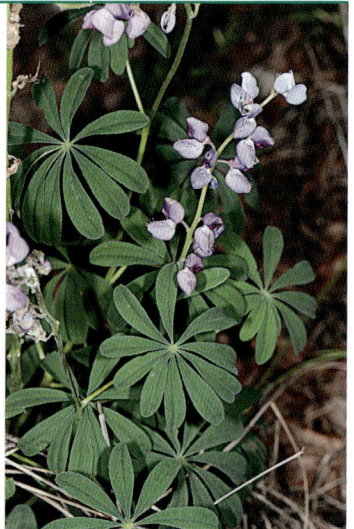

541. Sandhill Bluestar

Amsonia ciliata Walter
Am-sòn-i-a ci-li-à-ta
Apocynaceae (Dogbane Family)

DESCRIPTION: Perennial herb with pubescent stems; 12–28″ tall; leaves alternate, lower leaves lanceolate, the upper leaves filiform to linear; flowers produced in clusters at the terminus of stems; corolla star-shaped, light blue, smooth on the outer surface; fruit a slender follicle; flowers late April; fruits mature September–October.

RANGE-HABITAT: NC south to FL, west to TX, and north to MO; in SC, common in the Sandhills and Inner Coastal Plain; sandy, dry, open woodlands; dry pinelands;;and xeric sandhills.

COMMENTS: The genus is named for John Amson (1698–1765?), a physician of Williamsburg, VA. A very slender-leaved and smooth form with decumbent stems, considered at this time to be conspecific with the plant described above, is known from southern GA and may represent the plants on which the original description was derived. This form has been given the name of "Georgia Pancake" in the horticultural trade. It appears that this plant and the ones known from the Carolinas represent two distinct species, and their taxonomy is likely to change.

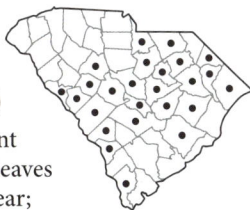

542. South Carolina Wild Pink

Silene caroliniana Walter var. *caroliniana*
Si-lè-ne ca-ro-li-ni-à-na var. ca-ro-li-ni-à-na
Caryophyllaceae (Pink Family)

DESCRIPTION: Tufted perennial up to 8″ tall from a thin, deep taproot; leaves oblong, well separated, up to 2″ long; leaves pubescent over both surfaces with appressed, white hairs; petals vary from pink to white, about 0.5″ long; flowers April–July.

RANGE-HABITAT: NC south to GA; in SC, locally common in the Piedmont, Sandhills, and Coastal Plain where found in sandy, dry, open woods; dry forests over mafic bedrock; and around granitic flatrocks.

COMMENTS: This is one of the most beautiful native plants in SC and has become available in the nursery trade. It is a short-lived and temperamental perennial. It is best to leave it to its wild haunts and visit it there.

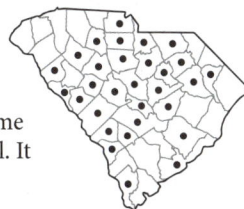

543. Rosemary Sunrose

Crocanthemum rosmarinifolium (Pursh) Janchen
Cro-cán-the-mum ros-ma-ri-ni-fò-li-um
Cistaceae (Rockrose Family)

SYNONYM: *Helianthemum rosmarinifolium* Pursh—RAB, PR

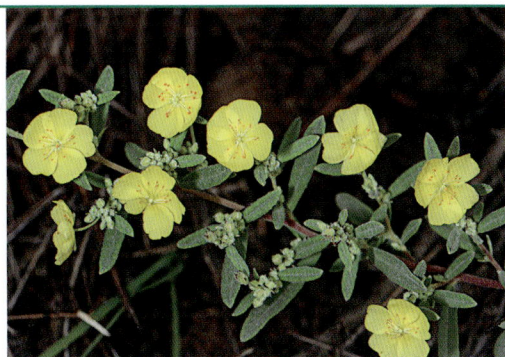

DESCRIPTION: Small pubescent, pale gray-green perennial herb up to 16″ tall; leaves very narrow, 0.13″ wide or less, pubescent; chasmogamous (openly pollinated) and cleistogamous (self-pollinating) flowers present, the former with showy, yellow petals, the latter without petals; the two types of flowers mature at the same time; both types of flowers produce fertile fruits, but fruits from the chasmogamous flowers generally have more and larger seeds; flowers May–June.

RANGE-HABITAT: From NC south to the FL Panhandle and west to OK and TX; in SC, primarily the Sandhills and Coastal Plain; common; sandy, dry, open woodlands; along sandy roadsides; and in fields.

COMMENTS: Rosemary Sunrose had chasmogamous flowers (with petals) open for a day and are followed by (or mingled with) the cleistogamous flowers (without petals), which do not open but still produce fruit. This is the most common species of sunrose in SC, often forming showy masses along disturbed, sandy roadsides through the Sandhills.

544. Carolina Piriqueta; Pitted Stripeseed

Piriqueta caroliniana (Walter) Urban
  var. *caroliniana*
    Pi-ri-què-ta ca-ro-li-ni-à-na var.
    ca-ro-li-ni-à-na
  Turneraceae (Turnera Family)

DESCRIPTION: Perennial pubescent herb to 20″ tall; spreading and forming colonies from rhizomes; leaves alternate, oblong to lanceolate, 1–2″ long, sessile or nearly so; flowers produced in terminal racemes; flowers 1.2–1.6″ wide, remaining open only on sunny days; petals fall off easily when touched; fruits are small, nearly round capsules; flowers May–September.

RANGE-HABITAT: SC south to FL and through the Neotropics; in SC, occasional in the Coastal Plain in Longleaf Pine sandhills and sandy, dry, open woodlands.

COMMENTS: This is the only member of the Turneraceae family in SC. It is sometimes confused with smaller flowered, spring-flowering Sunroses (*Crocanthemum* spp.). The distinctive round capsule of Carolina Piriqueta, rather than angled capsules, serves to distinguish this species. It is easily observed at Tillman Sand Ridge Heritage Preserve and environs in Jasper County.

545. Queen's-delight

Stillingia sylvatica Garden ex L.
    Stil-líng-i-a syl-vá-ti-ca
  Euphorbiaceae (Spurge Family)

DESCRIPTION: Perennial herb, usually with several stems from a rootstock; to 32″ tall; leaves alternate, simple, smooth, with minutely crenate margins; male and female flowers on the same plant, in a terminal spike, male uppermost, female at base; spike rachis with numerous, large glands; capsule 3-locular, 1 seed per locule; flowers May–July.

RANGE-HABITAT: VA south to FL and west to KS and NM, south into Mexico; in SC, common in the Coastal Plain and Sandhills in Longleaf Pine sandhills; sandy, dry, Longleaf Pine flatwoods; open pine-mixed hardwood forests; and sandy, dry, open woodlands.

COMMENTS: Queen's-delight is a component in S.S.S. Tonic, still sold today. It has been used in a variety of home remedies. In the Old South, it was used to treat

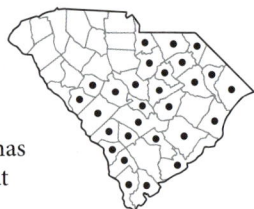

constipation and to induce vomiting. Like all members of this family, it contains toxins and can lead to accidental poisoning. The genus honors Dr. Benjamin Stillingfleet (1702–71), an English naturalist. The species description is attributed, by Linnaeus, to Charleston naturalist Alexander Garden (1730–91).

546. Netleaf Leatherflower

Clematis reticulata Walter
  Clé-ma-tis re-ti-cu-là-ta
  Ranunculaceae (Buttercup Family)

DESCRIPTION: Perennial herbaceous vine, ascending or sprawling; leaves opposite, thick and leathery, pinnately compound, with 3–9 leaflets; venation on the leaflets forming evident reticulations (net patterns) on the upper leaf surface; flowers terminal and from axils, solitary, held on pedicels up to 6″ long; petals absent, calyx urceolate with showy bluish or purplish sepals; flowers May–August.

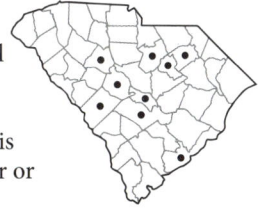

RANGE-HABITAT: SC and TN south to FL, west to TX and AR; in SC, occasional in the Sandhills and Inner Coastal Plain growing in sandy, dry, open woodlands; Longleaf Pine sandhills; and open, hammock forests.

COMMENTS: All species of *Clematis* are reported to be poisonous. This species is an attractive, drought-tolerant addition to the native garden as a groundcover or low vine.

547. Southeastern Gaura

Oenothera simulans (Small)
  W. L. Wagner & Hoch
  Oe-no-thè-ra sí-mu-lans
  Onagraceae (Evening-primrose Family)

SYNONYM: *Gaura angustifolia* Michaux—RAB

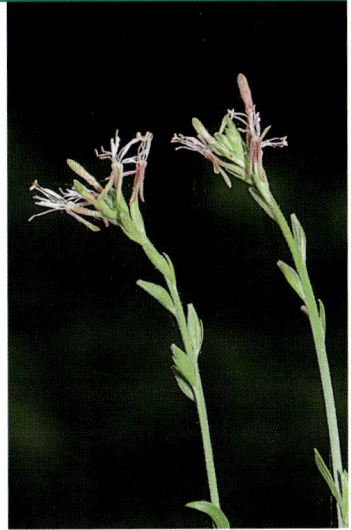

DESCRIPTION: Short-lived perennial herb to 6′ tall with pubescent stems; leaves narrowly elliptic to lanceolate, pubescent or becoming smooth with age; flowers produced in branched or simple spikes; flowers very small, whitish or pinkish with reflexed petals; flowers May–September.

RANGE-HABITAT: Endemic to the Coastal Plain from NC south to FL and west to TX; in SC, common in the Coastal Plain growing in sandy, dry, open woodlands; Longleaf Pine sandhills; and dry flatwoods, roadsides, and other open, sunny habitats.

COMMENTS: Southeastern Gaura is very attractive to pollinators, both bees and butterflies. It is also a host for White-lined Sphinx Moths and Clouded Crimson Moths. This species, along with other similar and related species formerly in the genus *Gaura,* are now treated as the same genus as the evening primroses, *Oenothera.* All species of *Oenothera* contain gamma-linolenic acid, one of the essential fatty acids, and one of the most popular herbal remedies for a wide variety of uses.

548. Snowy Black-anthers

Melanthera nivea (L.) Small
Me-lán-the-ra ní-ve-a
Asteraceae (Aster Family)

SYNONYM: *Melanthera hastata*
Michaux—RAB, PR

DESCRIPTION: Coarse, erect, rough-pubescent perennial, 3–6′ tall; stems 4-angled, heavily mottled with purple; leaves opposite, lobed and irregularly serrate; flowers arranged in involucrate heads composed only of white disk flowers with blackish anthers; flowers June–October.

RANGE-HABITAT: SC and KY south to FL and west to LA; in SC, common in the Coastal Plain in sandy, dry, open woodlands; moist to dry live oak woods; shell middens; roadsides with marl hash; and maritime dunes. It is almost always on circumneutral soils.

COMMENTS: This attractive species is frequently associated with high-calcium soils. It makes a robust, stately addition to the native garden and is attractive to a wide range of pollinators.

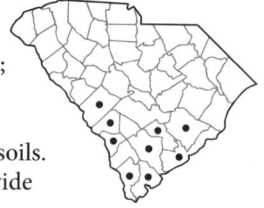

549. Doubleform Snoutbean

Rhynchosia difformis (Elliott) A.P. de Candolle
Rhyn-chò-si-a dif-fór-mis
Fabaceae (Bean Family)

DESCRIPTION: Prostrate or twining herbaceous vine; stem angled, tawny, and pubescent; earliest leaves with a single kidney-shaped leaflet; later leaves with 3 rounded-to-ovate or elliptic leaflets; flowers produced in axillary racemes; petals yellow; fruit a pubescent legume; flowers June–August; fruits mature August–October.

RANGE-HABITAT: From VA south to FL and west to TX; in SC, common in the Coastal Plain and Sandhills growing in sandy, dry, open woodlands; Longleaf Pine sandhills; and clearings.

SIMILAR SPECIES: Dollarweed (*Rhynchosia reniformis* de Candolle, plate 470) has 1 foliolate leaves and grows in similar habitats; *R. tomentosa* (L.) Hooker & Arnott is similar and grows throughout SC but is an erect plant and doesn't trail or twine.

550. Southern Dawnflower

Stylisma humistrata (Walter) Chapman
Sty-lís-ma hu-mis-trà-ta
Convolvulaceae (Morning Glory Family)

SYNONYM: *Bonamia humistrata* (Walter) Gray—RAB

DESCRIPTION: Prostrate, herbaceous, perennial vine; stems with a tendency to twine, at least near the growing tip; larger leaves 1–2″ long and

0.5–1″ wide, reduced upward; flowers usually in 3- to 7-flowered cymes; flowers white; anthers slightly villous toward the base; flowers June–August.

RANGE-HABITAT: VA south to FL and west to AR and e. TX; in SC, common in the Coastal Plain, Sandhills, and Piedmont where found in Longleaf Pine sandhills; dry flatwoods; sandy, dry, open woodlands; waste places; and along roadsides.

SIMILAR SPECIES: Narrowleaf Dawnflower (*S. angustifolia* (Nash) House, plate 461) is similar but with much more narrow leaves (0.04–0.1″ wide) and with solitary or few-flowered cymes. Common Dawnflower (*S. patens* (Desrousseaux) Myint) is very similar but also has more narrow leaves (0.1–0.4″ wide) and solitary or few-flowered cymes.

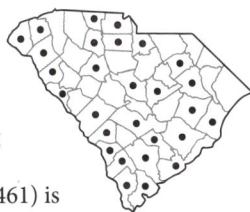

---

551. Coastal Plain Sida

Sida elliottii Torrey & Gray var. *elliottii*
Sì-da el-li-ótt-i-i var. el-li-ótt-i-i
Malvaceae (Mallow Family)

DESCRIPTION: Erect, perennial herb to 1–3′ tall; spreading and forming colonies from root sprouts; leaves minutely pubescent, linear to narrowly elliptic, 0.75–3″ long with dentate margins; flowers produced in the axils of leaves on slender pedicels to 1.5′ long, 1–2″ broad, from tip of petal to tip of petal; petals bright yellow-orange; flowers July–October.

RANGE-HABITAT: VA south to FL and west to MO and LA; in SC, rare in the Piedmont and uncommon in the Coastal Plain where found in dry, sandy, woodlands, often with circumneutral soils; clay road banks; open, rocky woodlands; openings in maritime forest; and pond margins.

COMMENTS: This is our only native species of *Sida*. The large flowers and upright growth form make it an attractive ornamental for the native garden, but it is seldom seen in cultivation. The narrowly elliptic to linear leaves help distinguish it from the common weedy Arrowhead Sida (*Sida rhombifolia* L.) that was introduced from the tropics.

---

552. Pineland Agrimony

Agrimonia incisa Torrey & Gray
Ag-ri-mò-ni-a in-cì-sa
Rosaceae (Rose Family)

DESCRIPTION: Tuberous-rooted, pubescent, perennial herb, 2–3′ tall; stems pubescent and with glistening glands; leaves pinnately compound with larger leaflets obovate to elliptic with an obtuse to acute tip and coarsely serrate margins; flowers produced in a terminal panicle, the individual flowers very small with tiny yellowish petals; flowers July–September.

RANGE-HABITAT: SC south to FL and west to TX; in SC, uncommon and sporadic in the Coastal Plain where growing in sandy, dry, woodlands; Longleaf Pine flatwoods; and areas with soil disturbance within pinelands.

SIMILAR SPECIES: Six species of agrimony grow in SC. All have very similar inflorescences. This species is distinctive as it grows in dry, sandy habitats and has very coarsely serrate leaves.

COMMENTS: Agrimony species are well-known as traditional herbal remedies for sore throat, upset stomach, and diarrhea. The high level of tannins in the plant are apparently responsible for the astringent properties.

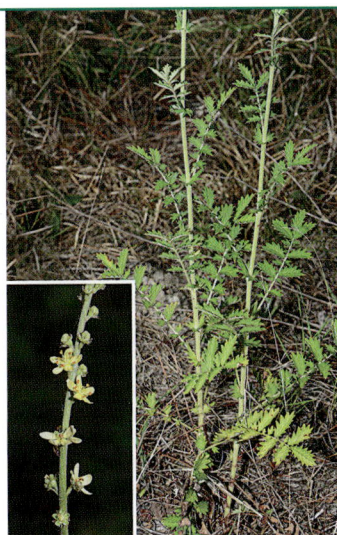

553. Eastern Horsemint

Monarda punctata L. var.
   punctata
      Mo-nár-da punc-tà-ta var.
      punc-tà-ta
   Lamiaceae (Mint Family)

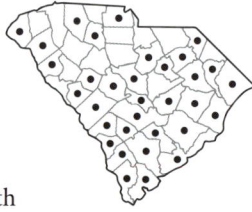

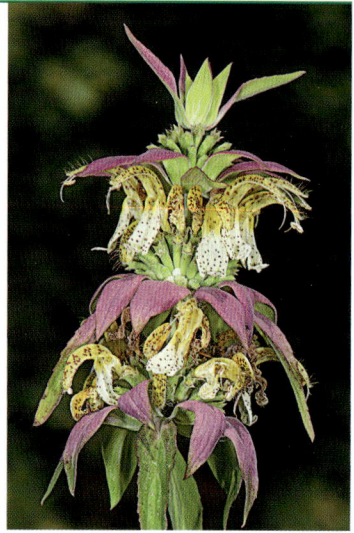

DESCRIPTION: Aromatic, perennial herb with 4-angled stems 16–40″ tall; leaves opposite; flowers in tight clusters subtended by several wholly or partially pink to lavender bracts; corolla yellow, spotted with purple; flowers July–September.

RANGE-HABITAT: VT to MN and south to FL, TX, and AZ; in SC, common in the Sandhills and Coastal Plain, occasional in the Piedmont; sandy, dry, open woodlands; rocky and open woodlands; roadsides; and fields.

COMMENTS: F. P. Porcher (1869) gives numerous home remedies and medicinal uses of horse mint. The plant contains thymol, which is toxic and potentially fatal in large doses. The term "horse" when used in a common name signifies "coarse." The genus honors Nicolás Monardes (1493–1588), a Spanish physician and botanist who authored many tracts on the medicinal application and other uses of plants, especially of the New World.

554. Common Elegant Blazing Star

Liatris elegans (Walter)
   Michaux var. *elegans*
      Li-à-tris é-le-gans var. é-le-gans
   Asteraceae (Aster Family)

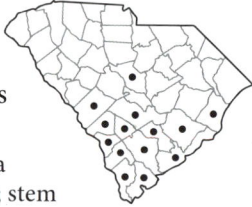

DESCRIPTION: Erect, perennial herb, from a globose rootstock, generally under 3′ tall; stem leaves alternate, linear to narrowly elliptic, up to 0.25″ wide; flowers produced in involucrate heads of pink disk flowers only; heads arranged in a spikelike formation; bracts below the heads dilated, pink; flowers September–October.

RANGE-HABITAT: SC to FL and west to TX and OK; in SC, common in the Coastal Plain where found in Longleaf Pine sandhills; dry flatwoods; dry, sandy, open woods; margins of maritime forests; and clearings such as sandy roadsides.

TAXONOMY: Kral's Elegant Blazing Star (*L. elegans* var. *kralii* Mayfield) differs in having bracts subtending the flowers with yellow or cream-colored tips. It is rare in sandhill habitats in the southeastern Coastal Plain of SC.

COMMENTS: The attractive pink bracts subtending the flower heads make this the showiest of SC's Blazing Stars. The bracts retain their color for weeks and extend the show produced by this species. It makes a striking addition to the native garden in well-drained soils.

555. Longleaf Pine flatwoods

556. Running Oak

Quercus elliottii Wilbur
Quér-cus el-li-ótt-i-i
Fagaceae (Beech Family)

SYNONYM: *Quercus pumila* Walter—
RAB, PR

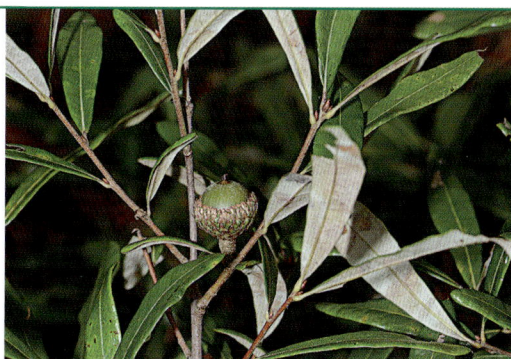

DESCRIPTION: Suckering shrub from roots or underground stems, often forming large colonies; rarely exceeding 3′ tall; leaves densely white and hairy beneath; female flowers 1–2 on a short stalk, separate from the male flowers, which are in catkins; flowers March–April; acorns mature in September.

RANGE-HABITAT: A southeastern Coastal Plain endemic; NC to FL and west to MS; in SC, common in the Coastal Plain and Sandhills in moist to dry loamy Longleaf Pine flatwoods, moister sandy Longleaf Pine flatwoods, and loamy swales in sandhill habitats.

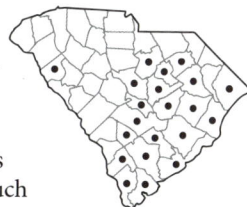

COMMENTS: Running Oak is readily identifiable in the Longleaf Pine flatwoods since it is the only narrow-leafed, low-growing oak that produces acorns of such small size. It can burn to the ground and then quickly produce new growth. It produces a crop of acorns the year following a fire. Wildlife readily eat the acorns.

557. Southern Blueberry

Vaccinium tenellum Aiton
Vac-cí-ni-um te-nél-lum
Ericaceae (Heath Family)

DESCRIPTION: Low-growing, rhizomatous, colonial shrub, 4–12″ tall; leaves deciduous, usually green with reddish glandular hairs below; flowers in 5–15-flowered clusters, before or during the expansion of leaves; corolla slenderly subcylindrical, pink to milk-white; flowers late March–early May; berries small, black, rather dry, maturing June–July.

RANGE-HABITAT: From VA south to FL and west to MS; common from se. VA to GA; scattered and rare in the rest of its range; in SC, common in the lower Piedmont, Sandhills, and Coastal Plain; Longleaf Pine flatwoods and sandhills, dry oak-hickory forests, and dry forest margins.

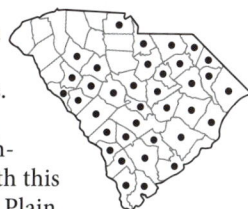

COMMENTS: Southern Blueberry is distinctive because of its low habit, being the only rhizomatous knee-high or shorter blueberry in the flatwoods. Highland Blueberry (*Vaccinium pallidum* Aiton, plate 228) sometimes occurs with this species in the Sandhills, such as at Peachtree Rock and rarely in the Coastal Plain. It can be distinguished by the smooth, often bluish-green undersides of the leaves, and dark blue or purplish (rather than black) fruit.

558. Creeping Blueberry

Vaccinium crassifolium Andrews
Vac-cí-ni-um cras-si-fò-li-um
Ericaceae (Heath Family)

DESCRIPTION: Long, trailing shrub, with short, upright branches, usually rooting at the nodes; leaves evergreen less than 0.75″ long with thickened and slightly involute margins; flowers 3–7, in fascicles or racemes on older growth; corolla pink, greenish, or whitish, urceolate; flowers April–May; berry black, maturing June–July.

RANGE-HABITAT: Nearly endemic to the Coastal Plain of the Carolinas, barely extending into adjacent GA and VA; in SC, common in the Coastal Plain and Sandhills but often overlooked because it grows close to the ground, hidden in vegetation; pocosin-sandhill ecotones and Longleaf Pine flatwoods and savannas.

COMMENTS: The berry is edible but not palatable. This species is typical of sandy flatwoods in the Carolinas. It can also be found on larger hills growing from the peak to the base where sandstone perches water, such as at Sugarloaf Mountain in the Sandhills State Forest in Chesterfield County, where it is especially abundant.

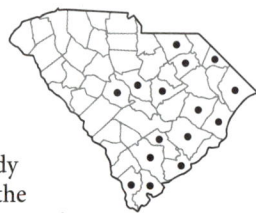

559. Stagger-bush

Lyonia mariana (L.) D. Don
Ly-òn-i-a ma-ri-à-na
Ericaceae (Heath Family)

DESCRIPTION: Deciduous to semi-evergreen, erect, rhizomatous shrub 1–5′ tall; leaves 1.5–3.5″ long, broadly elliptic, with entire margins; axillary buds are pink to red; flowers produced in axillary racemes; flowers nodding; corolla white, urceolate (urn-shaped), to 0.5″ long; flowers April–May.

RANGE-HABITAT: RI south to FL and west to TX, MO, and OK; in SC, common in the Coastal Plain and Sandhills; Longleaf Pine flatwoods and savannas; pocosin-sandhill ecotones; and sandy, dry, open woods.

COMMENTS: The genus honors John Lyon (1765–1814), a Scottish botanist and explorer of the southern Allegheny Mountains. There are many similar rhizomatous shrubs found in the flatwoods. This species is most easily recognized when not in flower by the reddish, rounded buds in the axils of the leaves.

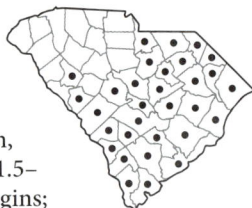

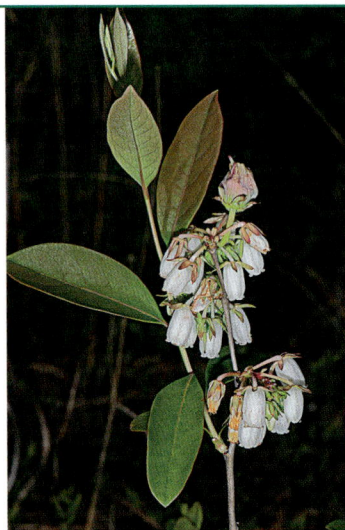

560. Dwarf Azalea

Rhododendron atlanticum (Ashe) Rehder
Rho-do-dén-dron at-lán-ti-cum
Ericaceae (Heath Family)

DESCRIPTION: Deciduous, low, erect shrub, 3–5′ tall; often forming large colonies from rhizomes; flowers appearing before the leaves; corolla white, often with a yellow spot on the largest corolla lobe, rarely lavender to deep pink, sticky from glandular hairs; flowers April–May and often later following fires.

RANGE-HABITAT: Endemic to the southeastern Coastal Plain; NJ and PA to s. GA; in SC, common in the Sandhills and Coastal Plain; pocosin ecotones and Longleaf Pine flatwoods and savannas.

SIMILAR SPECIES: Dwarf Azalea is similar to Piedmont Azalea (*R. canescens* (Michaux) Sweet, plate 795), which also occurs in the Coastal Plain. Piedmont Azalea does not form large colonies by rhizomes and is a larger plant, with pale pink to pink flowers; it only rarely shares the same habitat.

COMMENTS: Dwarf Azalea makes an attractive landscape plant and achieves a greater height in cultivation than in the wild.

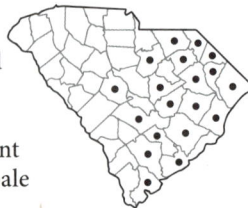

561. Common Dangleberry

Gaylussacia frondosa (L.) Torrey &
Gray ex Torrey
Gay-lus-sàc-i-a fron-dò-sa
Ericaceae (Heath Family)

DESCRIPTION: Rhizomatous, deciduous shrub,
to 6′ tall; leaves and twigs smooth or essentially
so; leaves with sessile, resinous dots on lower
surface, often also bluish-green (glaucous); corolla
greenish-white, drooping, urn-shaped; flowers
late March–April; fruit a berry, with 10 seeds,
glaucous-blue, matures June–August.

RANGE-HABITAT: Primarily southeastern Coastal Plain; from NH south to SC,
less commonly inward to w. NY, w. PA, and w. VA; in SC, common in the
Coastal Plain and Sandhills, Longleaf Pine flatwoods, sandy or rocky woods,
sandhill-pocosin, and savanna-pocosin ecotones.

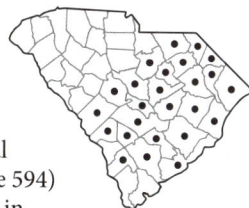

SIMILAR SPECIES: Two other similar species of dangleberry occur in the Coastal
Plain. Hairy Dangleberry (*Gaylussacia tomentosa* (Gray) Pursh ex Small, plate 594)
has twigs and lower surfaces of the leaves densely hairy and green. It is found in
similar habitats from the central Coastal Plain of SC south to FL and west to AL. Dwarf Dangleberry
(*Gaylussacia nana* (Gray) Small) also has short-pubescent twigs and undersides of the leaves, but
the underside of the leaf is glaucous, and it has smaller leaves (0.75–1.5″ long in *G. nana* vs. 1.2–2.4″
long in *G. tomentosa*). Dwarf Dangleberry is also found in similar habitats and can be found from the
Coastal Plain of NC south to FL and west to AL.

COMMENTS: Dangleberries have delicious fruits and makes excellent desserts, being juicy with a rich,
spicy, sweet flavor. The fruits are consumed by a wide variety of wildlife.

562. Dwarf Wax-myrtle; Dwarf Bayberry

Morella pumila (Michaux) Small
Mo-rél-la pù-mi-la
Myricaceae (Bayberry Family)

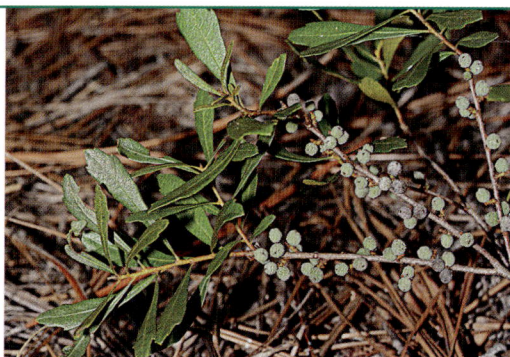

DESCRIPTION: Evergreen, rhizomatous shrub to
3′ tall; leaves elliptic to oblanceolate, to 2.5″ long,
smelling strongly of *Eucalyptus* when crushed;
flowers insignificant, produced in catkins March–
April; fruit bluish with a whitish bloom (glaucous) produced August–October.

RANGE-HABITAT: Endemic to the Coastal Plain
from VA south to FL, west to LA; in SC, common
in the Coastal Plain in Longleaf Pine flatwoods, savannas, and ecotones of
pocosins.

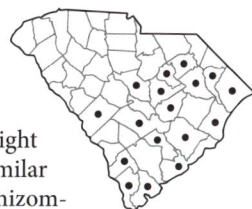

COMMENTS: Though it has only recently been widely recognized as distinct,
Dwarf Wax-myrtle is very different from Common Wax-myrtle (*M. cerifera*
(L.) Small, plate 909). It is a dwarf, rhizomatous shrub, never attaining the height
of Southern Wax-myrtle, even in cultivation. It has a distinctive odor, very similar
to *Eucalyptus,* when bruised. Like many Longleaf Pine flatwood shrubs, the rhizomatous habit and low growth form are an adaptation to frequent fires. Dwarf Wax-
myrtle makes an exceptional low shrub in home landscapes. The fruits of all wax-myrtles are
eaten by birds, including the Yellow-rumped Warbler, also known as the Myrtle Warbler.

563. Inkberry; Bitter Gallberry

Ilex glabra (L.) Gray
Ì-lex glà-bra
Aquifoliaceae (Holly Family)

DESCRIPTION: Evergreen, rhizomatous shrub to 9′ tall; leaves with widely spaced teeth toward the tip, smooth, shiny above; flowers May–June; drupe black, bitter, mature September–November, persistent through the winter.

RANGE-HABITAT: Nova Scotia south to FL, west to TX; in SC, common in the Coastal Plain and Sandhills; Longleaf Pine flatwoods and savannas, pocosins, and seepage areas in woodlands.

SIMILAR SPECIES: Sweet Gallberry (*Ilex coriacea* (Pursh) Chapman) is similar but often taller (to 15′) and mostly found in pocosins. The leaves of Sweet Gallberry are 1.5–3x as long as wide, with a few, tiny, irregularly spaced, marginal teeth. The leaves of inkberry are 3–4x as long as wide, with teeth in the apical one-half to one-third of the blade.

COMMENTS: Inkberry is an important honey plant in the Southeast. The short, tidy form and evergreen leaves have made this a widely used landscape plant in SC.

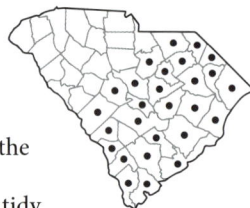

564. Downy Sweet-pepperbush

Clethra tomentosa Lamarck
Clèth-ra to-men-tò-sa
Clethraceae (Clethra Family)

SYNONYM: *Clethra alnifolia* L. var. *tomentosa* (Lamarck) Michaux—RAB

DESCRIPTION: Deciduous, colonial shrub 3–8′ tall, with densely pubescent twigs; leaves alternate, obovate to oblong, serrate, lower leaf surface densely pubescent, petioles 0.2–0.4″ long; flowers produced in racemes, fragrant, petals white; flowers June–July.

RANGE-HABITAT: Endemic to the Coastal Plain from SC south to FL and west to LA; in SC, common in the Coastal Plain and Sandhills in wet flatwoods, savannas, pocosin margins, and swamp forests.

SIMILAR SPECIES: See Coastal Sweet-pepperbush (*C. alnifolia* L., plate 500).

COMMENTS: Downy Sweet-pepperbush has become a very popular ornamental. It is showy, remains a short shrub, produces showy and fragrant flowers, and is highly attractive to pollinators. The cultivar "Cottondale," selected by SC plantsman Robert McCartney, is common in the landscape trade and has extremely long and showy racemes.

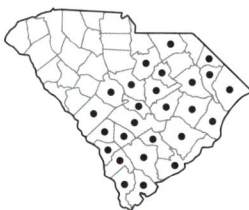

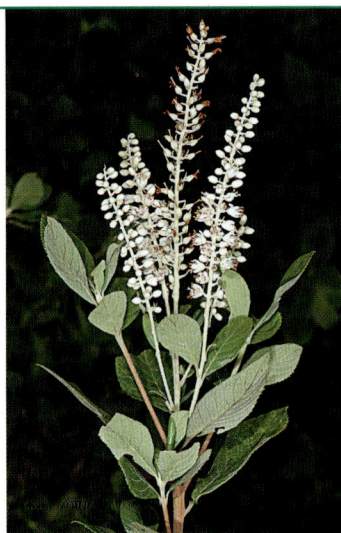

565. Dryland White Bluestem

Andropogon capillipes Nash
    An-dro-pò-gon ca-pil-li-pes
Poaceae (Grass Family)

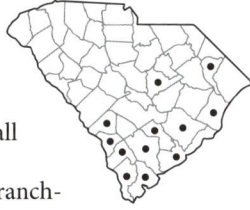

DESCRIPTION: Perennial grass, forming small
clumps, with bright, chalky, whitish-blue
leaves; leaves 4–15″ long; summit of the branchlet below the raceme sheath glabrous; inflorescence
slender, not bushy; flowering September–October.
RANGE-HABITAT: Endemic to the Coastal Plain from NC south to FL
and west to AL; in SC, common in the Coastal Plain in dry to moist,
sandy, Longleaf Pine flatwoods and sandy, exposed roadsides.
SIMILAR SPECIES: Three bluestems in SC have chalky, whitish leaves.
Dryland White Bluestem is by far the most brilliant white. Wetland White Bluestem (*Andropogon dealbatus* (C. Mohr) Weakley
& LeBlond) is very similar but has pubescence at the summit of
the branchlet below the raceme sheath and occurs in wet Longleaf
Pine and Pond Cypress savannas and depression meadows in the
Coastal Plain. Chalky Bluestem (*Andropogon cretaceus* Weakley & Schori) has the inflorescence densely clustered near the top of the stem, has much longer basal leaves (1–2′ long),
and occurs in savanna and pocosin habitats in wet soils in the Coastal Plain.
COMMENTS: Dryland White Bluestem makes a very attractive ornamental grass in sandy, well-
drained soils. The chalky white wax of the leaves easily wipes off onto the fingers when run
over the plant.

566. Dwarf Butterwort

Pinguicula pumila Michaux
    Pin-guí-cu-la pù-mi-la
Lentibulariaceae (Bladderwort
    Family)

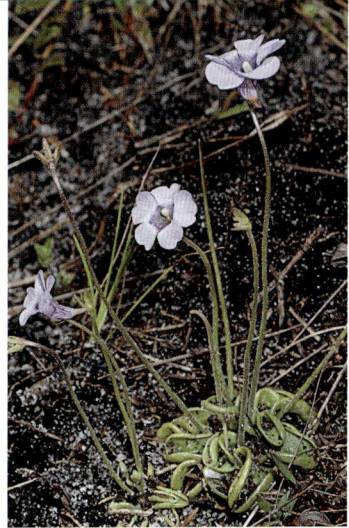

DESCRIPTION: Tiny, perennial herbaceous
carnivorous plant, with a dense basal rosette;
flowering stems 2–6″ tall; basal leaves less than
1.2″ long, sticky with glands and with rolled margins; flowers
white upon opening and fading to bluish or lavender; flowers
February–May.
RANGE-HABITAT: Endemic to the Coastal Plain from NC south to
FL and west to TX; also in the Bahamas; in SC, rare in the Coastal
Plain in sandy, seasonally moist Longleaf Pine flatwoods with
some exposed sand, often along roadsides through flatwoods.
COMMENTS: This is the rarest, smallest, and earliest-flowering species of butterwort in SC. It is often overlooked because of its early
flowering and tiny size. The genus name derives from the Latin
pinguis, meaning "somewhat fat," and refers to the greasy texture of the leaf.
CONSERVATION STATUS: SC-Imperiled

567. Cleft-leaved Violet; Southern Coastal Violet

Viola septemloba Le Conte
Vì-o-la sep-tem-lò-ba
Violaceae (Violet Family)

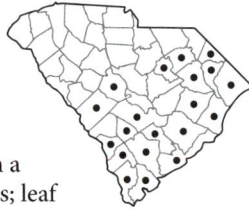

DESCRIPTION: Perennial, smooth herb, with a short, erect rhizome and thick, fleshy roots; leaf blades variable, some unlobed, others 3-, 5-, 7-, or 9-lobed nearly as wide as long; stalks of early spring flowers very variable in length, up to 8″ tall; spurred petal densely bearded within; flowers late March–early May.
RANGE-HABITAT: Coastal Plain from se. VA to peninsular FL and west to TX; in SC, common in seasonally wet to well-drained Longleaf Pine flatwoods.
SIMILAR SPECIES: Cleft-leaved Violet is similar to Salad Violet (*Viola edulis* Elliott). The spurred petal of the former is bearded, while the spurred petal of the latter is hairless. Unlike Cleft-leaved Violet, Salad Violet grows in the floodplains of blackwater streams in the Coastal Plain.

568. Dwarf Milkwort

Polygala nana (Michaux) A.P. de Candolle
Po-lý-ga-la nà-na
Polygalaceae (Milkwort Family)

DESCRIPTION: Biennial herb to 6″ tall; leaves mostly confined to a basal rosette; leaves fleshy; flowers lemon yellow, turning a dark bluish green on drying; flowers March–October and sporadically through the winter.
RANGE-HABITAT: Endemic to the Coastal Plain from SC south to FL and west to TX; in SC, rare in the Sandhills and Coastal Plain; edges (ecotones) of streamhead pocosins, wet Longleaf Pine flatwoods, and open, wet sands.
CONSERVATION STATUS: SC-Critically Imperiled

569. Carolina Sunrose

Crocanthemum carolinianum (Walter) Spach
Cro-cán-the-mum ca-ro-li-ni-à-num
Cistaceae (Rockrose Family)

SYNONYM: *Helianthemum carolinianum* (Walter) Michaux—RAB, PR

DESCRIPTION: Erect, herbaceous perennial, 4–12″ tall, from roots with tuberous thickenings; stems

hairy, arising from a basal rosette of broad leaves; leaves with starlike hairs on both surfaces; petals lasting a day or less; flowers April–May.
RANGE-HABITAT: From NC south to FL and west to AR and TX; in SC, occasional in the Sandhills and Coastal Plain growing in Longleaf Pine savannas, old fields, and dry Longleaf Pine flatwoods.
COMMENTS: This species is easily separated from the other sunroses found in SC. Distinctive features for this species are a stem with 2–4 leaves below the flower, a basal rosette of broad leaves, and large flowers (almost 2″ wide).

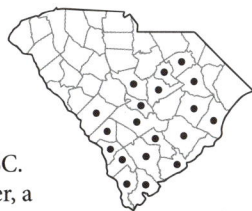

570. Horsefly Weed; Yellow False-indigo

Baptisia tinctoria (L.) R. Brown
    Bap-tí-si-a tinc-tò-ri-a
Fabaceae (Bean Family)

DESCRIPTION: Bushy, branched, rhizomatous perennial up to 3′ tall; usually blackening upon drying; leaves alternate, trifoliolate; leaflets less than 1″ long; flowers produced in few-flowered racemes terminating most of the branches; flowers April–August.
RANGE-HABITAT: Common and widespread in eastern US; common throughout SC; Longleaf Pine sandhills, Longleaf Pine flatwoods, open woods and clearings, ridges, dry oak-hickory forests, and along road banks.
COMMENTS: Horsefly Weed was cultivated in colonial times as a source of a blue dye. It was not a quality dye and could not replace true indigo obtained from Asian and tropical members of the genus *Indigofera*. The specific epithet *tinctoria* refers to "tincture," meaning "a dyeing substance." F. P. Porcher (1869) gives a long discourse on the use of this plant as a medicine. He also relates how the fresh plant was attached to the harness of horses to keep flies off—hence, one of its common names. He further states: "I have noticed that they [flies] will not remain upon the plant."

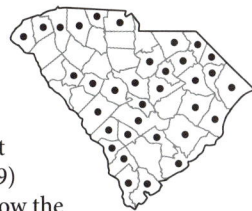

571. Florida Bluehearts

Buchnera floridana Gandoger
    Bùch-ner-a flo-ri-dà-na
Orobanchaceae (Broomrape Family)

DESCRIPTION: Erect, pubescent, hemi-parasitic perennial 15–30″ tall; leaves lanceolate-elliptic, pubescent; flowers produced in a spike terminating the stem; flowers tubular, to 0.2″ long, with 5 lobes, two of which are smaller than the remaining 3, purple; flowers April–October.
RANGE-HABITAT: NC south to FL and west to TX; also found in tropical America; in SC, common in the Coastal Plain in Longleaf Pine flatwoods, savannas, and moist roadsides.
COMMENTS: Florida Bluehearts is a common, small, yet attractive wildflower that often is abundant on roadsides in Longleaf Pine habitats. It is a hemiparasite on a wide variety of hosts. The genus is named for Johann Gottfried Buchner (1695–1749), a German botanist.

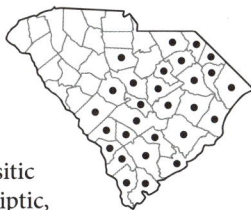

572. Northern White Colicroot

Aletris farinosa L.
Á-le-tris fa-ri-nò-sa
Nartheciaceae (Bog-asphodel Family)

DESCRIPTION: Perennial herb with thick, short rhizomes; leaves arranged in a low basal rosette, narrowly lanceolate, smooth; stem leaves greatly reduced; flower stalk 1–3′ tall; perianth white, covered with small, wartlike projections, giving a mealy appearance; flowers April–June.

RANGE-HABITAT: Widespread in the eastern US, from ME to MN, and south to FL and TX; in SC, common throughout; Longleaf Pine flatwoods and savannas, roadsides, mountain bogs, seepage fens, and margins of rock outcrops.

SIMILAR SPECIES: The only other white-flowered species in SC is Southern White Colicroot (*A. obovata* Small). It has smaller, rounded flowers and is restricted to the southeastern corner of the Coastal Plain in SC. Occurring in the same habitats, and area, is Golden Colicroot (*A. aurea* Walter, plate 623) with yellow flowers produced several weeks later than Northern White Colic-root.

COMMENTS: This plant was used medicinally to treat colic in colonial times. Historically an infusion of the roots in vinegar was used as a purgative. Morton's (1974) research indicates rural people still chew the rootstock of both species to stop toothache. The genus name *Aletris* is Greek for "a female slave who grinds corn," in reference to the mealy texture of the flowers.

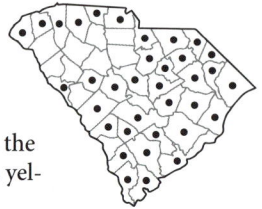

573. Sprawling Hoarypea

Tephrosia hispidula (Michaux) Persoon
Te-phrò-si-a his-pí-du-la
Fabaceae (Bean Family)

DESCRIPTION: Perennial herb from a taproot; stems, with inconspicuous, appressed pubescence, trailing to erect, to 20″ long; leaves pinnately compound with 9–23 acute-tipped leaflets, smooth to short-pubescent; flowers produced in a raceme opposite the leaves; flowers white when young, turning pink then red with age; fruit a short-pubescent legume; flowers May–July; fruits mature July–October.

RANGE-HABITAT: Endemic to the Coastal Plain from NC south to FL and west to LA; in SC, common in the Coastal Plain in Longleaf Pine flatwoods and savannas.

SIMILAR SPECIES: See discussion under Spiked Hoarypea (*T. spicata* (Walter) Torrey & Gray, plate 474).

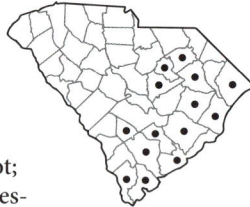

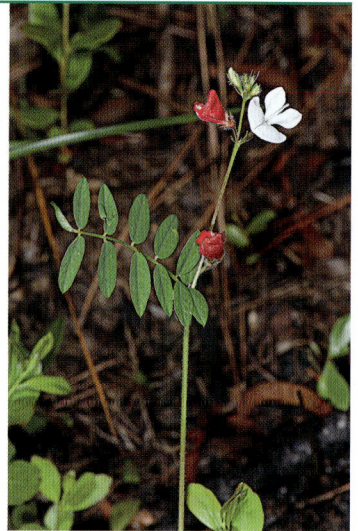

574. Eastern Sampson's-snakeroot

Orbexilum psoralioides (Walter) Vincent
Or-béx-i-lum pso-ra-li-oì-des
Fabaceae (Bean Family)

SYNONYM: *Psoralea psoralioides* (Walter) Cory var. *psoralioides*—RAB, PR

DESCRIPTION: Perennial herb, 1–3′ tall, arising from a cigar-shaped taproot; leaves trifoliolate; leaflet

surfaces, bracts, calyces, and pods with conspicuous punctate glands; purple to lavender flowers produced in spikelike racemes; flowers May–July; legume matures July–September.

RANGE-HABITAT: VA south to FL; in SC, common in the Coastal Plain and Sand-hills, rare in the Piedmont; moist Longleaf Pine flatwoods and savannas, loamy swales in Longleaf Pine sandhills, and open woodlands.

SIMILAR SPECIES: See discussion under Western Sampson's-snakeroot (*Orbexilum pedunculatum* (P. Miller) Rydberg, plate 86).

COMMENTS: This is one of many species historically believed to act against snake venom.

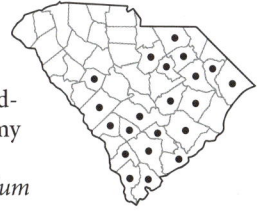

---

575. Dwarf Indigo-bush; Lead Plant

Amorpha herbacea Walter var. *herbacea*
A-mór-pha her-bá-ce-a var. her-bá-ce-a
Fabaceae (Bean Family)

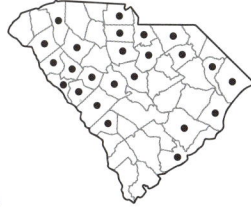

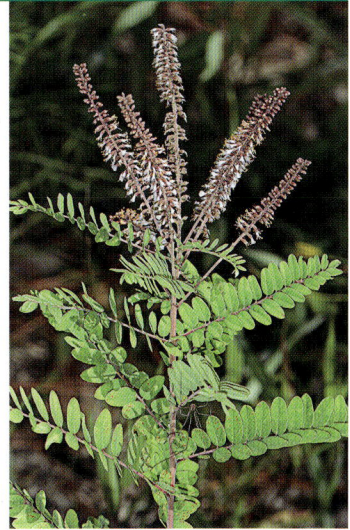

DESCRIPTION: Short shrub, 1–4.5′ tall, with densely pubescent stems; leaves pinnately compound, densely short pubescent, and punctate-glandular; flowers produced in racemes; flowers light blue to pink with sharply contrasting orange anthers; flowers May–July.

RANGE-HABITAT: Endemic to the Coastal Plain and Piedmont from NC south to FL; in SC, uncommon in the Piedmont, common in the Coastal Plain in Longleaf Pine flatwoods and savannas, open fields, Longleaf Pine sandhills, and dry oak-hickory forests in the Piedmont.

COMMENTS: The genus name is from the Greek *amorphous,* meaning "without form," in reference to the flowers having only one petal. Though the specific epithet refers to an herb, this plant is a short shrub, often dying back close to the ground in winter and thus mimicking an herbaceous species.

---

576. Pineland Yellow-eyed-grass

Xyris caroliniana Walter
Xỳ-ris ca-ro-li-ni-à-na
Xyridaceae (Yellow-eyed-grass Family)

DESCRIPTION: Perennial herb forming small, tight clumps of basal leaves, the persistent bases of the leaves creating a bulblike thickening; leaves grasslike, elongate, twisted, shiny, very thick and coriaceous, dark green, 12–30″ tall; flowering stem 8–20″ tall with the flowers produced in a dense, terminal, headlike cylindrical to elliptic spike with a sharp tip; lustrous, brown, overlapping bracts subtending the flowers; flowers yellow, opening in the late afternoon; flowers June–July.

RANGE-HABITAT: Primarily in the Coastal Plain from VA south to FL and west to TX with disjunct occurrences in NJ; in SC, common in the Coastal Plain and Sandhills in Longleaf Pine flatwoods, moist but not saturated areas of Longleaf Pine savannas, ecotones of pocosins, and sandhills or flatwoods habitats.

COMMENTS: At least 19 species of *Xyris* occur in SC; all have similar morphology but range from 4″ to over 4′ tall. Pineland Yellow-eyed-grass grows in the driest situation of any of the tall species and is ubiquitous in the Longleaf Pine flatwoods community. The bulblike base and narrow, dark green, lustrous, thick, and twisted leaves are diagnostic. The name *Xyris* comes from the Greek, meaning "gladen," or sword-shaped, in reference to the leaves.

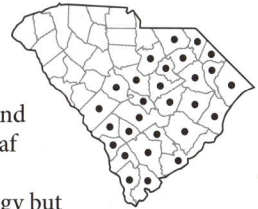

577. Coastal Plain Rattlebox; Pursh's Rattlebox

Crotalaria purshii A.P. de Candolle
Cro-ta-là-ri-a púrsh-i-i
Fabaceae (Bean Family)

DESCRIPTION: Perennial herb 8–20″ tall; leaves simple, linear to lanceolate or spatulate on the lower portion of the stem; stipules of the leaves decurrent (fused to the stem and extending downward along the stem) and forming the shape of an upside down arrow; petals yellow; fruit a smooth, oblong-cylindric legume up to 0.5″ long, inflated, and with seeds rattling when shaken when dry; flowers May–July; fruits mature July–September.

RANGE-HABITAT: Primarily in the Coastal Plain from VA south to FL and west to LA; in SC, common in the Coastal Plain and Sandhills in Longleaf Pine flatwoods, Longleaf Pine sandhills, and other dry, sandy habitats; rare in the Piedmont and mountains on the margins of rock outcrops and other dry, open habitats.

COMMENTS: Coastal Plain Rattlebox is distinctive because of the strange stipules that form an arrowheadlike configuration. The common name is derived from the fact that the swollen fruits, when dry, produce a distinctive rattle when shaken. This rattle can often be heard walking through the pinelands and gives away the presence of this plant when not in flower. The species is named for its discoverer, Frederick Taugott Pursh (1774–1820).

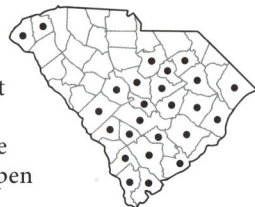

578. Southern Sneezeweed

Helenium flexuosum Rafinesque
He-lé-ni-um flex-u-ò-sum
Asteraceae (Aster Family)

DESCRIPTION: Perennial herb, 16–36″ tall; leaves alternate, with the base of the blades extending down the stems (decurrent), making them winged; flowers produced in involucrate heads with 3-lobed yellow ray flowers, and reddish disk flowers; flowers May–August.

RANGE-HABITAT: Common throughout much of eastern US; in SC, common in the Coastal Plain, uncommon in the Piedmont and mountains; Longleaf Pine flatwoods and savannas, roadsides, ditches and alluvial pastures, wet meadows, and riverbanks.

COMMENTS: The common name comes from the use of its dried leaves in making snuff, which was inhaled to cause sneezing in the belief it would rid the body of evil spirits. Though this species is common in many habitats, it is especially abundant along roadsides and disturbed areas in Longleaf Pine flatwoods.

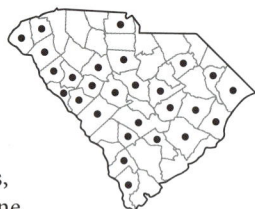

579. Black-root; Pineland Wingstem

Pterocaulon pycnostachyum
(Michaux) Elliott
Pte-ro-caù-lon pyc-nos-tà-chy-um
Asteraceae (Aster Family)

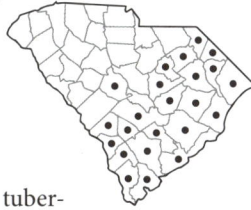

DESCRIPTION: Perennial herb from a black, tuberous, and thickened root; stems erect, 16–32″ tall, with dense, matted, white pubescence; stem winged by the base of the leaf stalks extending down the stem (decurrent); vegetative parts often remain green through the winter; flowers are produced in involucrate heads arranged in a whitish, spikelike configuration at the terminus of the stem; heads composed of white disk flowers only, the individual flowers tiny; flowers May–June.

RANGE-HABITAT: Endemic to the Coastal Plain from NC south to s. FL and west to AL; in SC, common in the Coastal Plain in Longleaf Pine flatwoods, Longleaf Pine sandhills, pocosin ecotones, and sandy fields.

COMMENTS: The root was used in a variety of folk remedies according to F. P. Porcher (1869) and Morton (1974). The latter states that even today a decoction of the root is taken for colds and menstrual cramps, or the whole root is boiled, and the tea taken for backache.

580. American Chaff-seed

Schwalbea americana L.
Schwál-be-a a-me-ri-cà-na
Orobanchaceae (Broomrape Family)

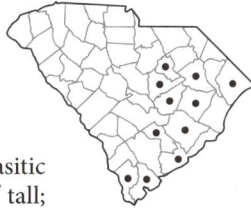

DESCRIPTION: Erect, unbranched, hemiparasitic perennial herb, with pubescent stems 1–2′ tall; leaves alternate, lanceolate to elliptic-lanceolate, pubescent, the largest at the base, and gradually diminishing in size upward; flowers produced in axils of reduced leaves in a raceme; calyx and corolla pubescent; corolla yellowish, reddish, or purplish; flowers April–June.

RANGE-HABITAT: *Schwalbea* is primarily a Coastal Plain species of the Atlantic and Gulf coasts, formerly ranging from MA south to FL and west to LA, and historically in TN and KY; currently there are populations in NJ, NC, SC, GA, and FL; in SC, rare in Longleaf Pine flatwoods and savannas; ecotones between peaty wetlands and xeric sandy soils; and other open, grass-sedge systems.

COMMENTS: Formerly widespread, this species has suffered greatly because of the loss of habitat due to fire suppression. Field studies in SC during the 1990s documented more than 50 populations in nine coastal counties. Because of its fire-dependent nature, its populations must be carefully monitored. Any significant increase in fire suppression throughout its range could jeopardize its continued survival. Chaff-seed is a root-parasite, the host species representing components of its habitat. The genus name is dedicated to Christian G. Schwalbe, who wrote on botany in the early eighteenth century

Research on the colonial-era collections of SC has revealed that the earliest specimen collected was made by the Moravian pastor and naturalist Joseph Lord (b. 1672), one of the first residents of Dorchester, SC, where the specimen was collected (Blackwell and McMillan, 2013).

CONSERVATION STATUS: Federally Endangered

581. Virginia Goat's-rue;
Devil's Shoestrings; Hoary-pea

Tephrosia virginiana (L.) Persoon
Te-phrò-si-a vir-gi-ni-à-na
Fabaceae (Bean Family)

DESCRIPTION: Perennial herb, 8–28″ tall, with dense grayish hairs; leaves pinnately compound, 2–5.5″ long, covered with grayish hairs; inflorescence terminal, in a raceme; flowers bicolored, with a cream-colored standard and pink wings and keel; flowers May–June, occasionally later following fire.

RANGE-HABITAT: NH south to FL and west to MN and TX; common throughout SC; pine-oak heaths, dry oak-hickory forests, Longleaf Pine flatwoods, Longleaf Pine sandhills, dry or sandy open woods, clearings, outcrops, woodlands, and along roadsides in the Coastal Plain.

COMMENTS: This species is an excellent indicator of formerly fire-maintained habitats. It persists for decades as small plants in woodlands that have become shaded by fire suppression and occurs along road banks and open areas bordering sites that were formerly more open. F. P. Porcher (1869) states that Native Americans used Goat's Rue and that it was later used as a vermifuge. The common name, Devil's Shoestrings, comes from its long, stringy roots; the common name, Hoary-pea, refers to its grayish hairs. It was once fed to goats to increase milk production. When it was discovered that the roots contain rotenone, the practice was discontinued. Rotenone is used as an insecticide and fish poison.

582. St. Peter's-wort

Hypericum crux-andreae (L.) Crantz
Hy-pé-ri-cum crúx-an-drè-ae
Hypericaceae (St. John's-wort Family)

SYNONYM: *Hypericum stans* (Michaux) P. Adams & Robson—RAB, PR

DESCRIPTION: Erect shrub, 1–3′ tall; leaves leathery, broadly elliptic to ovate or obovate 0.5–1.5″ long and 0.2–0.75″ wide, rounded or cordate-clasping at the base; sepals 4 and unequal, the two outer, larger ones enclosing the two much narrower, inner ones; corolla with 4 petals; flowers June–October.

RANGE-HABITAT: Widespread in the eastern US, from NY south to FL and west to KS and TX; in SC, common in the Coastal Plain, Sandhills, and Piedmont; Longleaf Pine flatwoods and savannas; ditches; and sandy, open woods.

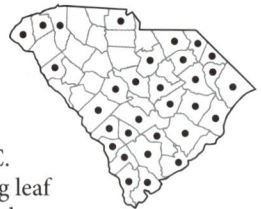

COMMENTS: This is one of four *Hypericum* species with four petals found in SC. St. Peter's-wort is the only one of these species with rounded, cordate-clasping leaf bases. The chemical hypericin is found in many species of *Hypericum* and has been shown to modify levels of neurotransmitters in ways that are similar to some antidepressant drugs. Clinical studies in humans suggest hypericins might be useful for some types of depression. There are reports of interactions with several other prescribed drugs, including increasing or decreasing the action of these drugs and leading to serious consequences (Dauncey and Howes, 2020).

583. Flatwoods Ironweed

Vernonia acaulis (Walter)
Gleason
Ver-nòn-i-a a-caù-lis
Asteraceae (Aster Family)

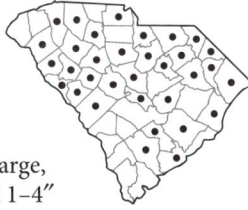

DESCRIPTION: Perennial herb; basal leaves large, lanceolate to oblanceolate, 4–12″ long and 1–4″ wide, rough-textured (scabrous) on the upper surface; stem leaves much smaller and reduced upward; flowers produced in an involucrate head with purple disk flowers and no ray flowers, the heads in a flat-topped, corymblike arrangement; flowers June-August.

RANGE-HABITAT: Found in the Coastal Plain and Piedmont provinces from NC south to GA; in SC, common in the Coastal Plain, Sandhills, and Piedmont where found in Longleaf Pine flatwoods, swales in Longleaf Pine sandhills, ecotones of pocosins, and dry and rocky woodlands and roadsides.

COMMENTS: A hybrid, Georgia Ironweed (*Vernonia* x *georgiana* Bartlett), is found sporadically in SC. It is a hybrid of this species and Narrowleaf Ironweed. Vegetatively, Flatwoods Ironweed is more likely to be confused with a species of *Elephantopus* than with another ironweed. It can be distinguished from *Elephantopus* by the fact that the basal leaves are rough above and smooth to slightly pubescent with long hairs below, while those of *Elephantopus* are densely pubescent below and not rough like sandpaper (scabrous) above. Flatwoods Ironweed makes an excellent addition to the meadow garden in home landscapes.

584. Spiked Medusa

Orthochilus ecristatus (Fernald)
Bytebier
Or-thò-chi-lus e-cris-tà-tus
Orchidaceae (Orchid Family)

SYNONYM: *Eulophia ecristata* (Fernald) Ames—RAB; *Pteroglossaspis ecristata* (Fernald) Rolfe—PR

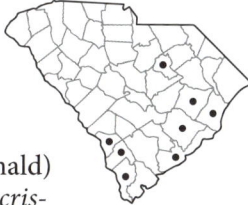

DESCRIPTION: Perennial herb from a thickened cormlike pseudobulb; leaves several, arising from the pseudobulb, linear-lanceolate, 6–28″ long, distinctively plicate (pleated into longitudinal folds, like a fan); single flowering stem grows from the pseudobulb, near the leaves, 20 to 40″ tall; flowers June–September; fruit a capsule maturing by November.

RANGE-HABITAT: Endemic to the Coastal Plain, occurring from NC south to FL and west to LA; in SC, rare in Longleaf Pine flatwoods and savannas, especially where Blackjack Oak occurs, and other open, moist, grassy areas.

COMMENTS: Spiked Medusa is easily overlooked because it grows among tall grasses such as species of *Andropogon* and *Paspalum*. The fieldwork of Patrick D. McMillan, John F. Townsend, Richard D. Porcher Jr., and Eric J. Kiellmark has greatly expanded its known range. It is best located by looking for the pleated leaf among the grasses. It is extremely difficult to spot and may be more abundant than previously thought. The plant grows in extremely low densities at any one site and requires large areas of fire-maintained habitat for its continued survival.

CONSERVATION STATUS: SC-Imperiled

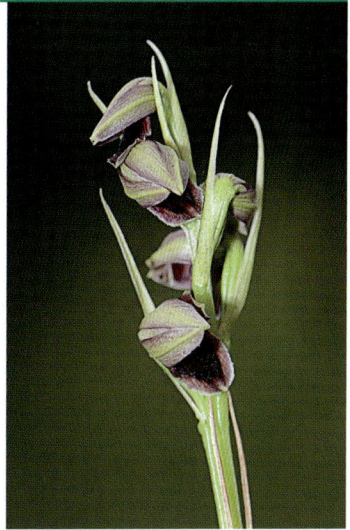

585. Stalked Milkweed;
Savanna Milkweed

Asclepias pedicellata Walter
As-clè-pi-as pe-di-cel-là-ta
Apocynaceae (Dogbane Family)

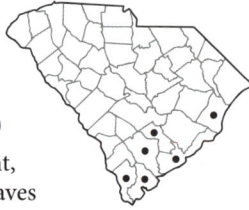

DESCRIPTION: Erect, unbranched, pubescent, herbaceous perennial; stems 4–12″ tall; leaves opposite, linear to narrowly lanceolate; flowers produced in 1–3 umbels at the top of the stem; corolla yellow to greenish yellow, all lobes erect; fruit a follicle; flowers June-August; fruits mature August–October.

RANGE-HABITAT: Endemic to the Coastal Plain from NC south to FL; in SC, rare in the Coastal Plain where found in sandy, open flatwoods and sandy ecotones of bay rims and sand ridges with pocosin.

COMMENTS: This is perhaps the most unusual and distinctive of our milkweed species. It is easily distinguished by the upright, yellow-ish corolla lobes when in flower. It is very small for a milkweed and generally occurs as solitary plants or widely spaced individuals. All known populations consist of a very few individuals.

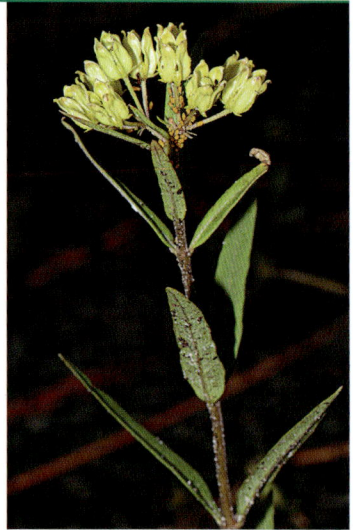

586. Zornia

Zornia bracteata Walter
ex J. F. Gmelin
Zórn-i-a brac-te-à-ta
Fabaceae (Bean Family)

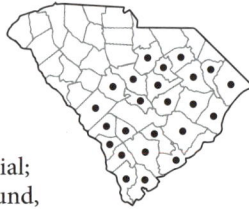

DESCRIPTION: Prostrate herbaceous perennial; stems numerous; leaves palmately compound, mostly with 4 leaflets; golden to orangish-yellow flowers are enclosed by 2 conspicuous bracts; fruit a legume that disarticulates into bristly segments; flowers June–August; fruits mature July–October.

RANGE-HABITAT: VA south to FL and west to TX and the Gulf coast of Mexico; in SC, common in the Coastal Plain and Sandhills, very rare in the Piedmont; Longleaf Pine flatwoods; sandy, dry, open woods; Longleaf Pine sandhills; margins of outcrops; and sandy roadsides.

COMMENTS: The palmately 4-foliolate leaves are unique. The specific epithet *bracteata* refers to the two bracts that enclose the flower. The genus honors Johannes Zorn (1739–99), a German apothecary.

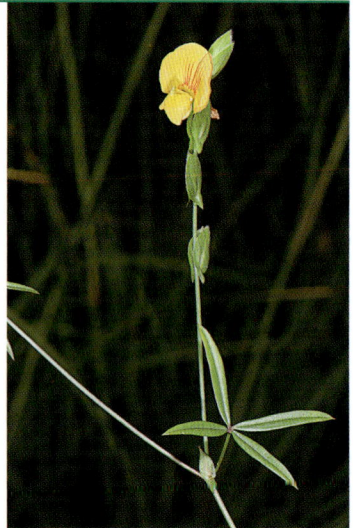

587. Vanilla Plant; Deer's-tongue

Trilisa odoratissima (J. F. Gmelin) Cassini
Tri-lìs-a o-do-ra-tís-si-ma
Asteraceae (Aster Family)

SYNONYM: *Carphephorus odoratissimus* (J. F. Gmelin) Herbert—PR

DESCRIPTION: Perennial herb to 3′ tall, smooth throughout; leaves alternate, the basal ones to 4″ long, and 3″ wide; stem leaves reduced upward; leaves usually purple toward the base and white toward the tip; flowers produced in involucrate heads of purplish disk flowers; heads arranged into corymblike structures, the lateral branches usually overtopping the terminal; flowers late July–October.

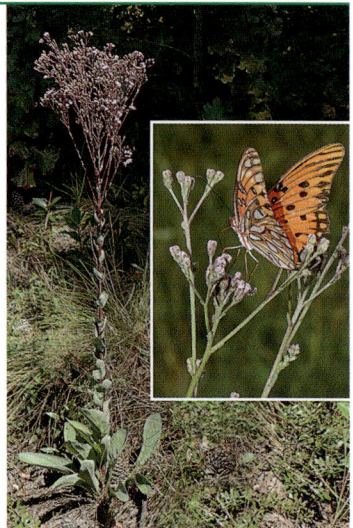

RANGE-HABITAT: Endemic to the southeastern Coastal Plain; from NC south to FL and west to LA; in SC, common in the Coastal Plain in moist Longleaf Pine flatwoods and savannas.

COMMENTS: The leaves contain coumarin, giving them a distinct and pleasant odor of vanilla. They are collected from the wild and sold for flavoring smoking tobacco.

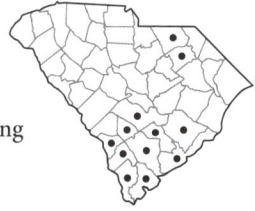

---

588. Coastal Plain Elephant's-foot (A)

Elephantopus nudatus Gray
E-le-phán-to-pus nu-dà-tus

Common Elephant's-foot (B)

Elephantopus tomentosus L.
E-le-phán-to-pus to-men-tò-sus
Asteraceae (Aster Family)

DESCRIPTION: Perennial herbs with a crown of large basal leaves, held flat on the ground; basal leaves 4–12″ long, densely pubescent below, oblanceolate, and less than 3″ wide in *E. nudatus,* and elliptic to oblanceolate and greater than 3″ wide in *E. tomentosus;* stem leaves much smaller than basal leaves and inconspicuous; flowering stems 1–2′ tall, branched with flowers produced in involucrate heads subtended by three large, foliaceous bracts; involucre less than 0.35″ long in *E. nudatus* and 0.4–0.5″ long in *E. tomentosus;* flowers July–September

RANGE-HABITAT: *E. nudatus* ranges from DE south to FL and west to AR and TX; in SC, common in the Coastal Plain and Sandhills in Longleaf Pine flatwoods, Longleaf Pine sandhills, and dry and open woodlands and forest margins; *E. tomentosus* ranges from MD south to FL and west to KY and TX; in SC, common throughout in Longleaf Pine flatwoods, Longleaf Pine sandhills, open woodlands, and forest margins.

COMMENTS: The basal leaves held nearly flat on the ground give the impression they have been stepped on by an elephant–hence, the common name. The two species included here are very similar, but *E. tomentosus* is a larger plant and common in the Piedmont.

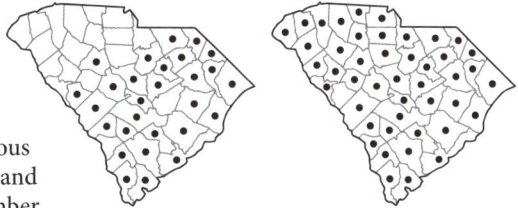

---

589. Mohr's Eupatorium (A)

Eupatorium mohrii Greene
Eu-pa-tò-ri-um mòhr-i-i

Recurved Eupatorium (B)

Eupatorium recurvans Small
Eu-pa-tò-ri-um re-cúr-vans
Asteraceae (Aster Family)

DESCRIPTION: Erect, rhizomatous, perennial herbs; *E. mohrii* is 2–5′ tall, not branching at the base, flowering stems typically solitary; *E. recurvans* is 1–2′ tall, branching from the base to give rise to a clump of similar flowering stems; leaves in both species sessile, opposite toward the base of the stem becoming alternate toward the top, narrow, elliptic to elliptic-oblanceolate, 1–2″ long and 0.1–0.8″ wide, serrate with tapered bases and acute to obtuse tips; flowers in both species produced in involucrate heads composed of cream-colored disk flowers only,

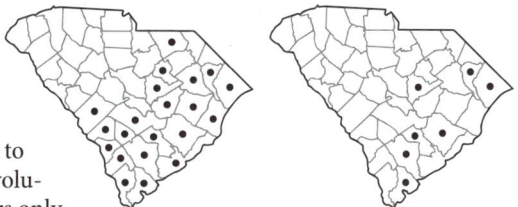

ray flowers absent; outer involucral bracts rounded at the tip and inner bracts acute at the tip in *E. mohrii;* all involucral bracts rounded at the tip in *E. recurvans;* flowering August–October.

RANGE-HABITAT: *E. mohrii* ranges from VA south to FL and west to AR and TX; in SC, common in the Coastal Plain and Sandhills in Longleaf Pine flatwoods, savannas, Pond Cypress savannas, depression meadows, and pocosin ecotones; *E. recurvans* ranges from NC south to FL; in SC, rare, in the Coastal Plain in similar habitats.

SIMILAR SPECIES: Savanna Eupatorium (*Eupatorium leucolepis* (A. P. de Candolle) Torrey & Gray, plate 662), is very similar to Mohr's Eupatorium, but it has leaves that are very narrow and opposite throughout the stem and long, tapered (acuminate) tips to the involucral bracts.

COMMENTS: These two species look very similar. The branching at the base of the plant and smaller stature in *E. recurvans,* as well as the significantly smaller involucres that are always rounded at the tip, are the best distinguishing features. *Eupatorium recurvans* is a diploid species (2 sets of chromosomes); *E. mohrii* is a tetraploid (4 sets of chromosomes) that was originally derived through hybridization of *E. recurvans* and *E. rotundifolium* and now exists as a distinct species.

590. Pinebarren Gentian

Gentiana autumnalis L.
   Gen-ti-à-na au-tum-nà-lis
Gentianaceae (Gentian Family)

DESCRIPTION: Perennial herb 8–24″ tall; stems single (usually) or few, erect or arching; leaves sessile, opposite, dark green, glossy, linear to narrowly oblanceolate, twisted and curved; petals usually intensely indigo blue, but on individual plants may be greenish white, white and blue, or purple; corolla tube brown-spotted within, plaited between the lobes (clearly seen in the photograph); flowers late September–early December.

RANGE-HABITAT: Two widely separated areas: the pine barrens of NJ and DE, and se. VA south through to SC; in SC, rare in the Coastal Plain and Sandhills; found in sandy Longleaf Pine savannas, flatwoods, and sandhill habitats.

COMMENTS: Pinebarren Gentian may be underreported in part because of its late fall bloom, when field studies are usually at a minimum and because it is inconspicuous except when in flower. Duke (1997) recommends all species of *Gentiana* as an herbal remedy for ulcers. This beautiful, rare gentian should never be disturbed in the wild.

CONSERVATION STATUS: SC-Imperiled

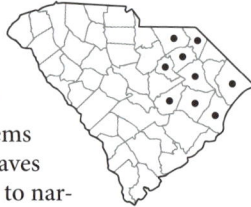

591. Southern Bracken

Pteridium pseudocaudatum (Clute)
   Christenhusz
      Pte-rí-di-um pseu-do-cau-dà-tum
Dennstaedtiaceae (Bracken Family)

SYNONYM: *Pteridium aquilinum* (L.)
   Kuhn—RAB, in part

DESCRIPTION: Deciduous fern from long, slender rhizomes; fronds 1–4′ long, triangular in outline, thick and leathery in texture, 3-pinnatifid to 3-pinnate; the pinnae thick, coriaceous and with the tips of the blades with long, tail-like pinnae 6–15 times as long as broad; spores produced in sori, seen as lines along the margins of the undersides of the pinnae; spores produced July–September.

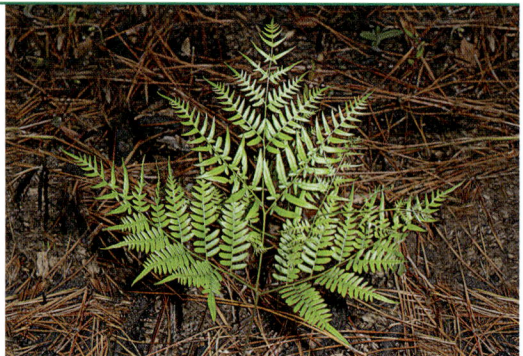

RANGE-HABITAT: MA west to MI and MO and south to FL and TX; in SC, abundant in the Sandhills and Coastal Plain, uncommon in the Piedmont and mountains; found in Longleaf Pine flatwoods; savannas and sandhill habitats; rocky, dry, open woodlands; and road banks.

SIMILAR SPECIES: Two species of bracken are known from SC. This species is distinguished by the long tail-like tips of the pinnae and smooth fronds. The related Eastern Bracken (*Pteridium latiusculum* (Desvaux) Hieronymus) has shorter tipped pinnae and pinnae with pubescent margins. It is more abundant in the mountains and Piedmont.

COMMENTS: Southern Bracken is often extremely abundant in the herbaceous layer of Longleaf Pine flatwoods. The abundance may be increased by prolonged periods of dormant season burns and by the reluctance of most grazing and browsing animals to eat it. The fronds are carcinogenic and poisonous, and yet the plant, or close relatives, have been eaten for thousands of years across the globe.

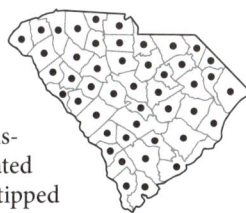

592. Pine/Saw Palmetto flatwoods

593. Slash Pine

Pinus elliottii Engelmann
Pì-nus el-li-ótt-i-i
Pinaceae (Pine Family)

DESCRIPTION: Tall, straight-trunked tree with ridged bark that often is sheared into smooth plates on the ridges that peel off in paperlike sheets; leaves in bundles of 2–3, to 10″ long; mature cones typically 6″ or less long, often lustrous and reddish brown and with slender prickles.

RANGE-HABITAT: Coastal Plain from SC south to FL and west to LA; in SC, common, restricted as a native species to the maritime fringe and Coastal Plain, close to the coastline and inland along the Savannah River drainage where found in maritime forests and flatwoods habitats; this species is extensively planted and naturalized throughout the Coastal Plain and Piedmont.

COMMENTS: Slash Pine is an extremely important source of wood for construction. It is frequently confused with Loblolly Pine (*P. taeda* L.), which is one of the most common and widespread trees in SC. It is most easily distinguished by the lustrous, reddish-brown cones, which have a short stalk, rather than being dull gray and sessile, as in Loblolly Pine. It also frequently shows freshly peeled areas of the bark that are reddish in color and appear to have been "slashed," while the bark of Loblolly Pine is simply gray.

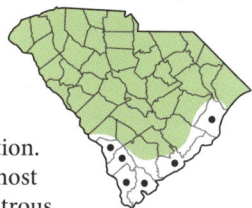

594. Hairy Dangleberry

Gaylussacia tomentosa (Gray)
    Pursh ex Small
        Gay-lus-sàc-i-a to-men-tò-sa
    Ericaceae (Heath Family)

DESCRIPTION: Rhizomatous, deciduous shrub, to 6′ tall; leaves and twigs densely short pubescent; leaves with sessile, resinous dots on lower surface; corolla greenish-white, drooping, urn-shaped; flowers late March–April; fruit a berry, with 10 seeds, glaucous-blue, matures June–August.

RANGE-HABITAT: Endemic to the Coastal Plain from SC south to FL and west to AL; in SC, uncommon but locally abundant in the southern portions of the Coastal Plain and up the Savannah River drainage to Aiken County; Longleaf Pine flatwoods, sandhill-pocosin, and flatwoods/savanna-pocosin ecotones.

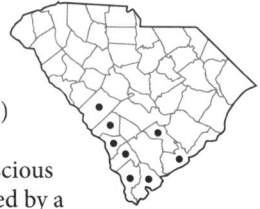

SIMILAR SPECIES: See discussion under Common Dangleberry (*G. frondosa* (L.) Torrey & Gray ex Torrey), plate 561).

COMMENTS: Dangleberries have delicious fruits and makes one of the most luscious of desserts, being juicy, with a rich, spicy, sweet flavor. The fruits are consumed by a wide variety of wildlife.

595. Rusty Lyonia; Tree Lyonia

Lyonia ferruginea (Walter)
    Nuttall
        Ly-òn-i-a fer-ru-gí-ne-a
    Ericaceae (Heath Family)

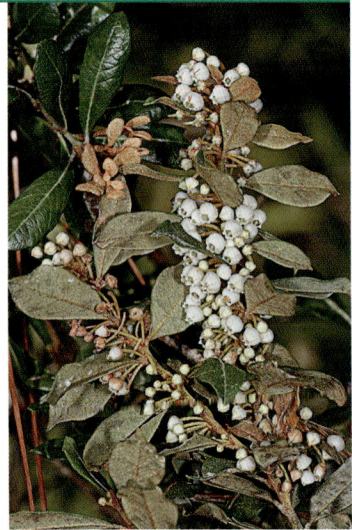

DESCRIPTION: Evergreen shrub, 3–6′ tall, sometimes much taller; often a low shrub in frequently burned sites; in long unburned locations, stems crooked and contorted, forming a small, irregular, open crown; occasionally reaches tree size under favorable conditions in the southern part of its range becoming 20–25′ tall; leaves evergreen, simple, alternate, with lower surface rusty and scaly; flowers white, small, urn-shaped; flowers March–May.

RANGE-HABITAT: Coastal Plain from SC to FL; in SC, rare and known only from the southeastern counties; Pine/Saw Palmetto flatwoods, pocosins, and bay forests.

COMMENTS: The specific epithet *ferruginea* refers to the rusty color of the lower leaf surface due to the presence of rusty-colored scales. The genus honors John Lyon (1765–1814), a Scottish botanist and explorer of the southern Allegheny Mountains.

CONSERVATION STATUS: SC-Critically Imperiled

596. Buckwheat Tree

Cliftonia monophylla (Lamarck)
  Britton ex Sargent
  Clif-tón-i-a mo-no-phýl-la
Cyrillaceae (Ti-ti Family)

DESCRIPTION: Evergreen shrub or small tree to 30′ tall, often much shorter and shrublike; leaves elliptic to oblanceolate, smooth, green above, light green, or with a bluish-white tinge below; flowers in upright racemes produced toward the tips of the previous season's growth; flowers white to pink; fruit is samaralike, with 2–5 wings, ellipsoid; flowers March–May; fruits mature June–August.

RANGE-HABITAT: Endemic to the Coastal Plain from SC south to FL and west to LA; in SC, very rare in the Coastal Plain, where it is found in pocosin or depression pond ecotones with Longleaf Pine flatwoods and savannas and blackwater swamp forests.

COMMENTS: Though superficially similar to Titi (*Cyrilla racemiflora* L., plate 728), it is easily distinguished by the shorter, broader, elliptic leaves, and distinctive flowers. It makes a beautiful shrub in the home landscape and is especially attractive to pollinators. The genus is named in honor of William Clifton, attorney general of GA (1754–64).

CONSERVATION STATUS: SC-Critically Imperiled

597. Saw Palmetto

Serenoa repens (Bartram) Small
  Se-re-nò-a rè-pens
Arecaceae (Palm Family)

DESCRIPTION: Evergreen palm, with branched, trailing stems obtaining the stature of a shrub; occasionally as large as a small, branched tree; leaves to 3′ across, palmately divided into numerous segments without filaments along the margins; leaf stalks armed with prickles; flowers produced on elongated branches to 3′ long; flowers May–July; fruit a bluish black drupe, maturing October–November.

RANGE-HABITAT: Endemic to the Coastal Plain from SC to FL and west to e. LA; in SC, restricted to the southeastern Coastal Plain counties where uncommon; Pine/Saw Palmetto flatwoods, Longleaf Pine-scrub oak sand ridges, maritime forests, and coastal dunes.

COMMENTS: Saw Palmetto often forms dense, almost impenetrable stands in flatwoods. It is highly resistant to fire. Prescribed fire in winter promotes its growth. The drupes were an important food for Native Americans and are an important food for whitetail deer. It can be seen in Victoria Bluff Heritage Preserve in Beaufort County.

Saw Palmetto is used to treat enlarged prostate glands. Although it appears to significantly improve urine flow in men with enlarged prostates, reviewers of the drug cautioned that studies showed that the effects of Saw Palmetto lasted only a few weeks, too short a time to determine long-term effects. The genus honors botanist Sereno Watson (1826–92), long-time curator of the Gray Herbarium.

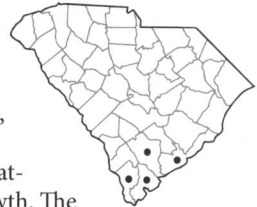

598. Southern Evergreen Blueberry

Vaccinium myrsinites Lamarck
Vac-cí-ni-um myr-si-nì-tes
Ericaceae (Heath Family)

DESCRIPTION: Low, evergreen, branched, colonial shrub, usually less than 2′ tall; leaves evergreen, lustrous, smooth, elliptic, 0.2–0.5″ long; flowers 2–8 in short, axillary racemes produced before the season's growth; corolla white to deep pink, urn-shaped; flowers March–April; berry black, matures May–June.

RANGE-HABITAT: Endemic to the Coastal Plain from SC south to peninsular FL and west to AL; in SC, occasional, but locally abundant in the Coastal Plain, in Pine/Saw Palmetto flatwoods, and Longleaf Pine flatwoods and sandhill to pocosin ecotones.

COMMENTS: The berries are edible but not very palatable. Duke (1997) recommends all *Vaccinium* species as an herbal remedy for bladder infections, macular degeneration, and multiple sclerosis. Though it is not widespread in SC, this species is one of the most abundant members of its habitat. The Horry County location is definitely native and quite disjunct from the remainder of the range. Southern Evergreen Blueberry makes a fine landscape plant.

599. Hairy Wicky

Kalmia hirsuta Walter
Kálm-i-a hir-sú-ta
Ericaceae (Heath Family)

DESCRIPTION: Low, evergreen shrub, 6–20″ tall; twigs and leaves sessile or nearly so, pubescent with stiff hairs; leaves alternate, elliptic to elliptic-oblanceolate or ovate, 0.25–0.6″ long; flowers solitary or in small clusters from the axils of the leaves of new growth of the season; petals fused, pink with indented pockets in which the stamens are held; fruit a capsule; flowers June–July; fruits mature September–October.

RANGE-HABITAT: Endemic to the Coastal Plain from SC to FL and west to MS; in SC, occasional in the southeastern Coastal Plain counties in Pine/Saw Palmetto flatwoods and Longleaf Pine flatwoods and savannas.

COMMENTS: In newly opened flowers, the 10 stamens are seated in small pockets of the corolla. An insect such as a bumblebee visiting the mature flowers may cause the stamens to pop out of the pockets, spraying the pollen onto the insect's back, which in turn is deposited on the next flower the insect visits, thus effecting cross-pollination. Hairy Wicky may appear to be herbaceous in frequently burned areas, rapidly springing back from burned off stems and flowering on the current season's growth.

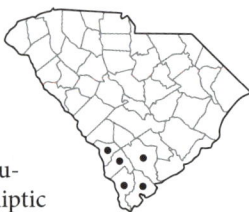

600. Dwarf Sundew

Drosera brevifolia Pursh
    Dró-se-ra bre-vi-fò-li-a
Droseraceae (Sundew Family)

SYNONYM: *Drosera leucantha* Shinners—RAB

DESCRIPTION: Annual, herbaceous carnivorous plant with leaves in a basal rosette; rosette 0.3–1.5″ wide; leaves covered with stalked glands that exude a clear, sticky material that aids in catching insects; blades obovate to spatulate with cuneate bases; flowering stalk glandular-hairy, 0.8–2.5″ tall; flowers white sometimes pinkish; flowers April–May.

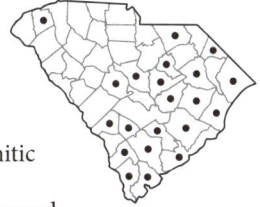

RANGE-HABITAT: MD south to FL and west to TX, AR, and OK, mostly in the Coastal Plain; in SC, common in the Coastal Plain and Sandhills and rare in the upper Piedmont; Pine/Saw Palmetto flatwoods, moist Longleaf Pine flatwoods, Longleaf Pine savannas, sandy roadside ditches, and seepages on granitic domes and granitic flatrocks.

COMMENTS: The common name of the sundews comes from the sticky, dew-covered tentacles on the leaves that in the early morning shine and glitter with the color of the sun's spectrum. Though most abundant in seasonally moist sandy openings in flatwoods, it is locally very abundant on seasonally moist margins of rock outcrops in Pickens County, just below the Blue Ridge Escarpment, where it shares habitat with numerous other Coastal Plain disjuncts.

601. Pineland Dyschoriste

Dyschoriste oblongifolia
    (Michaux) Kuntze
    Dys-cho-rís-te ob-lon-gi-fò-li-a
Acanthaceae (Acanthus Family)

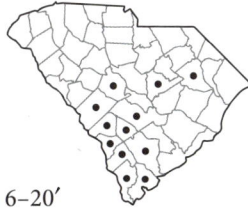

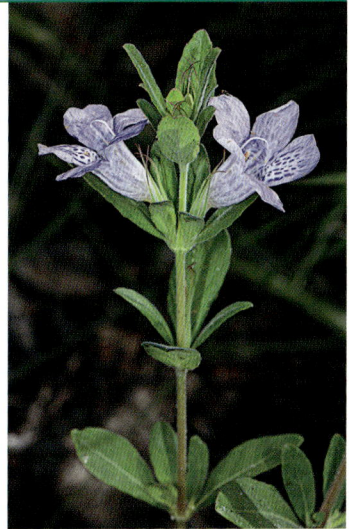

DESCRIPTION: A dull green, perennial herb 6–20′ tall; stems soft and hairy; leaves opposite, oblanceolate to elliptic; flowers solitary or in small clusters in the axils of the leaves; corolla tubular with 5 short lobes, bluish purple with dark spots; flowers April–May, sometimes later following burning.

RANGE-HABITAT: Endemic to the Coastal Plain from SC to FL; in SC, locally abundant in the Coastal Plain and Sandhills in Pine/Saw Palmetto flatwoods and Longleaf Pine flatwoods.

SIMILAR SPECIES: Swamp Dyschoriste (*Dyschoriste humistrata* (Michaux) Kuntze) is similar but has corollas less than 0.6″ long and is found in swamp forests and low ditches with calcium-rich soils. It is known from Berkeley and Charleston Counties, where it is very rare.

602. Pine-barren Aster

Oclemena reticulata
(Pursh) G. L. Nesom
Oc-le-mè-na re-ti-cu-là-ta
Asteraceae (Aster Family)

SYNONYM: *Aster reticulatus* Pursh—
RAB

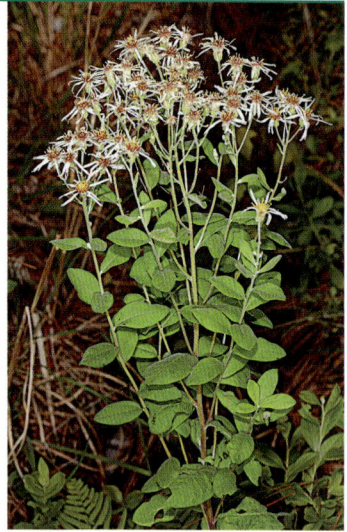

DESCRIPTION: Perennial herb forming small clumps, nonrhizomatous, with glandular pubescent stems, to 15–30″ tall; stem leaves 12–30, alternate, elliptic to elliptic-ovate, glandular pubescent, 1–3″ long and 0.5–1″ wide, sessile or nearly sessile, bases rounded to cuneate; flowers produced in involucrate heads of white ray and yellow disk flowers in corymbiform (flat-topped) arrangements; flowers April–June and sporadically later in response to fire.

RANGE-HABITAT: Endemic to the Coastal Plain from SC south to FL; in SC, rare but locally abundant in the southeastern Coastal Plain counties where found in Pine/Saw Palmetto flatwoods.

COMMENTS: This species is abundant in its habitat but has a very restricted range in SC. It can flower later in the season in response to fire, often flowering within 3–4 weeks of a burn.

CONSERVATION STATUS: SC-Critically Imperiled

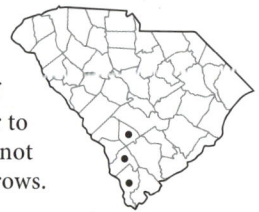

603. Walter's Milkweed

Asclepias cinerea Walter
As-clè-pi-as ci-né-re-a
Apocynaceae (Dogbane Family)

DESCRIPTION: Perennial herb 1–2′ tall; stem unbranched and typically solitary; leaves opposite, linear, 2.0–3.5″ long; corolla lavender with the lateral hood margins displaying a tooth that extends beyond the hood apex; flowers June–July; follicles mature August–September.

RANGE-HABITAT: Endemic to the Coastal Plain from SC south to FL; in SC it occurs only in Jasper and Hampton Counties; Longleaf Pine savannas and Pine/Saw Palmetto flatwoods, often where wet or boggy.

COMMENTS: This milkweed is distinctive among the Coastal Plain species of SC due to the extremely narrow leaves that are opposite and do not appear to be whorled. The plant can be easily overlooked as the stem, and leaves are not noticeably larger than the associated grasses and sedges among which it grows.

CONSERVATION STATUS: SC-Critically Imperiled

604. White Meadow-beauty

Rhexia mariana L. var. *exalbida*
Michaux
Rhéx-i-a ma-ri-à-na var. ex-ál-
bi-da
Melastomataceae (Melastome Family)

DESCRIPTION: A short, rhizomatous, perennial herb forming colonies; stems to 1.5′ tall, angled with unequal faces; leaves opposite, linear to narrowly linear-elliptic; flowers with an urnlike

hypanthium 0.2–0.4″ long, smooth or sparsely hairy with a narrow neck; corolla white, rarely light pink; flowers May–October.

RANGE-HABITAT: Endemic to the Coastal Plain from NC south to FL and west to MS; in SC, locally abundant in Longleaf Pine flatwoods, Pine/Saw Palmetto flatwoods, ecotones of pocosins, upper areas of herbaceous seepages, and Longleaf Pine savannas, mostly on sandy (spodosol) soils.

COMMENTS: This variety is often abundant in moist sands of the Coastal Plain and may be found in association with several other species, including the typical variety of *Rhexia mariana* L. It is easily distinguished by the extremely narrow, untwisted leaves and the typically bright white petals. Like all meadow-beauties, the flowers are open only for a few hours from morning to mid-day and begin to drop their petals in the afternoon.

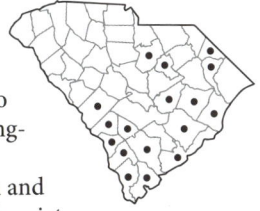

---

605. Savanna Honeycomb-head

Balduina uniflora Nuttall
Báld-uin-a u-ni-flò-ra
Asteraceae (Aster Family)

DESCRIPTION: Perennial herb; stem ribbed, stiffly erect, or with 2–20 stiffly erect branches, each bearing a single terminal head; plant 16–32″ tall; basal leaves in a rosette; stem leaves reduced upward; flowers produced in involucrate heads with yellow ray flowers and yellow disk flowers; receptacle honeycomblike; flowers late July–September.

RANGE-HABITAT: Endemic to the Coastal Plain from NC to FL and west to se. LA; in SC, rare in the southeastern portion of the Coastal Plain and in Horry County; wet Longleaf Pine flatwoods, Pine/Saw Palmetto flatwoods, and Longleaf Pine savannas.

COMMENTS: The common name refers to the honeycomblike appearance of the receptacle. The genus name honors William Baldwin, M.D. (1779–1819), an American surgeon and botanist, who collected extensively in GA. The only similar species in SC is Purple Honeycomb-head (*Balduina atropurpurea* R. M. Harper), which is extremely rare in herbaceous seepages in the Sandhills in SC. It can be easily distinguished by the purplish disk flowers.

CONSERVATION STATUS: SC-Imperiled

---

606. Elliott's Milk Pea

Galactia elliottii Nuttall
Ga-lác-ti-a el-li-ótt-i-i
Fabaceae (Bean Family)

DESCRIPTION: Twinging, climbing, herbaceous vine, sometimes somewhat woody at the base; leaves pinnately compound with 7 to 9 leaflets; leaves evergreen or nearly so; petals white or tinged with red; flowers July–September.

RANGE-HABITAT: Endemic to the Coastal Plain from SC south to FL; in SC, locally common in the southern Coastal Plain where found in Longleaf Pine flatwoods, Pine/Saw Palmetto flatwoods, thickets, and low woods.

COMMENTS: This is the only species of *Galactia* in SC that has leaves with more than 3 leaflets. The specific epithet honors SC botanist Stephen Elliott (1771–1830).

CONSERVATION STATUS: SC-Critically Imperiled

607. Smooth Mountain-mint

Pycnanthemum nudum Nuttall
Pyc-nán-the-mum nù-dum
Lamiaceae (Mint Family)

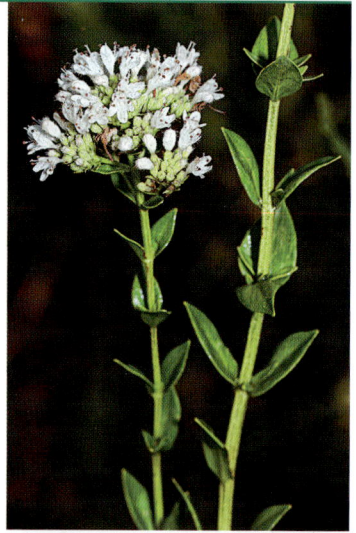

DESCRIPTION: Erect, smooth, perennial herb 12–28″ tall; stems square; leaves sessile, opposite, ovate or elliptic, thick textured, less than 0.75″ long; corolla lips whitish, purple-punctate on both surfaces; flowers July–September.

RANGE-HABITAT: Endemic to the Coastal Plain from se. SC south to n. FL and se. AL; in SC, rare in the Coastal Plain in wet Longleaf Pine flatwoods, Pine/Saw Palmetto flatwoods, and hillside bogs in pinelands.

COMMENTS: This species is distinctive among SC mountain mints in its completely smooth stems and leaves and very thick-textured leaves.

CONSERVATION STATUS: SC-Critically Imperiled

608. Rayless Sunflower; Roundleaf Sunflower

Helianthus radula (Pursh)
Torrey & Gray
He-li-án-thus rá-du-la
Asteraceae (Aster Family)

DESCRIPTION: Herbaceous perennial from a basal rosette; basal leaves broadly elliptic to orbicular, pubescent, held flat against the ground; stem leaves restricted to the lower portion of flowering stem; flowering stems 1–5, 2–3′ tall; flowers produced in involucrate heads consisting of purplish or reddish disk flowers, ray flowers absent; flowers August–October.

RANGE-HABITAT: Endemic to the Coastal Plain from se. SC to FL and west to LA; in SC, rare but locally abundant in the Coastal Plain in Pine/Saw Palmetto flatwoods, sand ridges, and Longleaf Pine flatwoods.

COMMENTS: This is SC's only sunflower without ray flowers. The nearly round basal leaves allow for easy identification when not in flower.

CONSERVATION STATUS: SC-Critically Imperiled

609. Georgia Blazing Star

Liatris patens G. L. Nesom
& Kral
Li-à-tris pà-tens
Astreraceae (Aster Family)

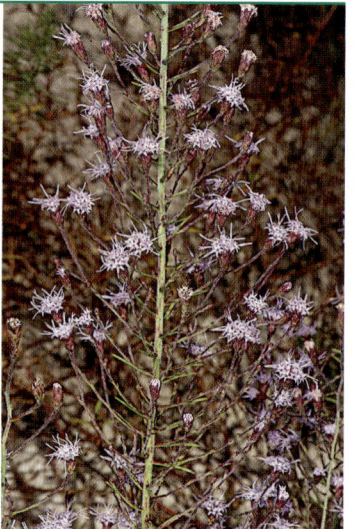

DESCRIPTION: Perennial herb from a swollen, bulbous rootstock; stems pubescent, 1–3′ tall; lowest leaves narrowly oblanceolate to linear-oblanceolate, smooth, 3.5–7″ long and less than 0.3″ wide; leaf size reduced upward; flowers produced in involucrate heads usually held on distinct and fairly long pubescent peduncles that are 0.4–1″ long and curve upward; heads containing disk flowers only, purplish, showy; the involucral bracts subtending the head with a blunt tip; flowers August–October.

RANGE-HABITAT: Endemic to the Coastal Plain from SC south to the FL Panhandle; in SC, uncommon in the southeastern Coastal Plain counties in Pine/Saw Palmetto flatwoods, Longleaf Pine flatwoods, and mesic, open maritime forests.

COMMENTS: This recently described species is very similar to Slender Blazing-star (*L. gracilis* Pursh. Slender Blazing Star has shorter peduncles that are less than 0.4″ long and ascending and is known only from a single location in Jasper County (Kral and Nesom, 2003).

610. Longleaf Pine savanna

611. Toothache Grass; Orange-grass

Ctenium aromaticum
    (Walter) Wood
      Cté-ni-um a-ro-má-ti-cum
Poaceae (Grass Family)

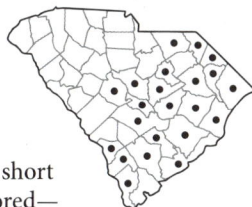

DESCRIPTION: Tufted, erect perennial from short rhizomes; stems 2–4′ tall; leaves are bicolored—blue-green on the upper surface and bright green below; the flowering spike is distinctive and comblike, twisting as it dries and releases the grains; flowers June–August.

RANGE-HABITAT: Endemic to the Coastal Plain from VA south to FL and west to LA; in SC, common in the Sandhills and Coastal Plain in Longleaf Pine savannas, wet areas in Longleaf Pine flatwoods, ecotones of pocosins, and herbaceous seepages.

COMMENTS: The fresh herbage, inflorescence, and rhizome, when bruised or crushed, produces a strong orange-citrus aroma. The common name, Toothache Grass, refers to the numbing sensation felt on the mouth, tongue, and lips when the plant is chewed. Like many plants in this habitat, it flowers primarily following fire. Do not mistake Large Death-camas (*Zigadenus glaberrimus* Michaux, plate 646) for sterile plants of Toothache Grass. Large Death-camas has somewhat grasslike leaves, keeled on the undersides, that are green on top and bluish-green below. It also does not smell like oranges when crushed.

612. Spring Bartonia; White Bartonia

Bartonia verna (Michaux) Rafinesque
    ex Barton
      Bar-tòn-i-a vér-na
Gentianaceae (Gentian Family)

DESCRIPTION: Erect, inconspicuous annual or biennial herb, 2–8″ tall; usually in colonies; stem wiry, purplish (rarely yellowish); leaves essentially opposite, minute and scalelike, erect or

appressed; flowers mostly solitary, in racemes, or panicles in very robust plants; flowers February–April.

RANGE-HABITAT: From NC south to FL and west to TX; in SC, uncommon in the Coastal Plain; wet Longleaf Pine savannas, shores of depression ponds, and pocosin ecotones.

COMMENTS: Weakley (2020) states that this genus has coral-like mycorrhizae and lacks root hairs and is thus presumably partially mycotrophic. Most references list Spring Bartonia as either rare or uncommon, but it is probably more common than has been reported; most likely it is overlooked because of its diminutive nature and early flowering period. The genus is named for Benjamin S. Barton (1766–1815), a Philadelphia botanist.

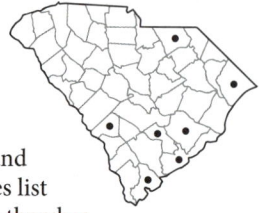

613. Sun-bonnets

Chaptalia tomentosa Ventenat
Chap-tàl-i-a to-men-tò-sa
Asteraceae (Aster Family)

DESCRIPTION: Fibrous-rooted, perennial herb with a basal cluster of leaves; leaves with dense cover of white hairs beneath, glossy green above; flowering stems 1–several, 3–16″ tall; flowers arranged into solitary involucrate heads of white ray and white disk flowers, nodding at first, then erect, but again nodding after flowering; flowers February–April.

RANGE-HABITAT: Endemic to the Coastal Plain from NC south to FL and west to TX; in SC, common in the Coastal Plain in Longleaf Pine savannas, moist flatwoods, and pocosin ecotones and along moist roadsides.

COMMENTS: The genus name honors Jean-Antoine Chaptal (1756–1832), a French chemist.

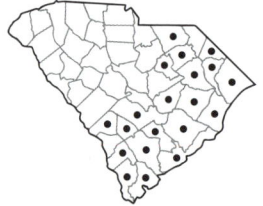

614. Leopard's-bane

Arnica acaulis (Walter) Britton, Sterns & Poggenburg
Ár-ni-ca a-caù-lis
Asteraceae (Aster Family)

DESCRIPTION: Perennial herb; stems erect, 6–32″ tall, usually one from a crown; principal leaves basal, 4 to 8, pubescent, ovate to broadly elliptic, forming a rosette; stem leaves few and reduced; both types of leaves glandular-hairy; flowers produced in involucrate heads consisting of yellow ray and yellow disk flowers; heads in a corymbose arrangement; flowers late March–early June.

RANGE-HABITAT: DE and PA, south to FL; in SC, common throughout the Coastal Plain, Sandhills, and lower Piedmont and at least one location in the upper Piedmont; Longleaf Pine savannas and flatwoods, moist areas within Longleaf Pine sandhills, open woodlands, margins of granitic flatrocks, and along moist roadsides.

COMMENTS: According to Duke (1997), the European species, *Arnica montana,* has pain-relieving, antiseptic, and anti-inflammatory properties and has been approved by the German Commission E for external treatment of bruises, sprains, and muscle and joint complaints and for disinfecting cuts. *Arnica acaulis* may have similar properties and uses.

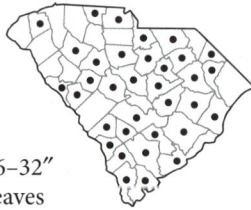

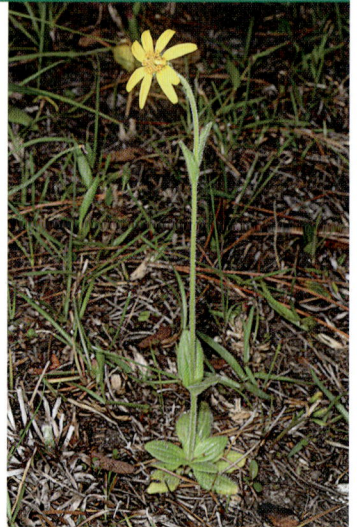

615. Bearded Grass-pink (A)

Calopogon barbatus (Walter) Ames
    Ca-lo-pò-gon bar-bà-tus

Many-flowered Grass-pink (B)

Calopogon multiflorus Lindley
    Ca-lo-pò-gon mul-ti-flò-rus
Orchidaceae (Orchid Family)

DESCRIPTION: Erect perennials, arising from a corm; flowering stalk 6–15″ tall; leaves 1 or 2, narrowly linear and grasslike, sheathing the flowering stalk and appressed to the stem at anthesis in *C. barbatus;* flowers rose-pink (*C. barbatus*), pink, or rarely white, mostly opening simultaneously; flowers of *C. barbatus* 1 to 12 per stem (mostly 3–6) with petals equal or narrower toward the tip than toward the base, not fragrant, and the lip longer than wide; flowers of *C. multiflorus* 2 to 15 per stem, with petals wider toward the tip than the base, strongly fragrant, and the lip as wide or wider than long; flowers March–May.

RANGE-HABITAT: Endemic to the Coastal Plain; NC south to FL and west to LA; *C. barbatus* in SC, widely scattered and occasional in the Coastal Plain in Longleaf Pine savannas and flatwoods, pocosin ecotones, and herbaceous seepages; *C. multiflorus* in SC, very rare in the Coastal Plain in moist Longleaf Pine savannas and flatwoods and pocosin ecotones.

COMMENTS: Although the flowering period is given as March–May, the actual flowering period for either species, in any one given year, is probably only 2–4 weeks. In addition, it is sensitive to changes in environmental parameters; one year it may be scarce, the next year more common. Many-flowered Grass-pink appears to flower only following a winter or spring burn.

   Many-flowered Grass-pink was first reported for SC by McMillan et al. (2002). It has since been located in only a handful of locations, mostly within the Francis Marion National Forest. Both species are considerably rare and should never be disturbed in the wild.

CONSERVATION STATUS: *C. barbatus:* SC-Critically Imperiled; *C. multiflorus:* SC-Critically Imperiled

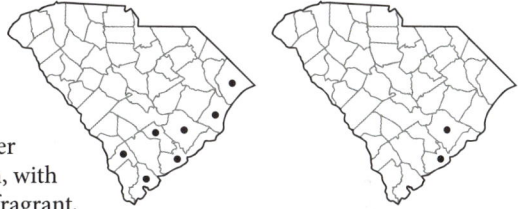

616. Violet Butterwort

Pinguicula caerulea Walter
    Pin-guí-cu-la cae-rù-le-a
Lentibulariaceae (Bladderwort Family)

DESCRIPTION: Perennial, carnivorous plant from a small basal rosette; flowering stems to 15″ tall; upper leaf surface covered with stalked, glandular hairs in which insects become mired and sessile glands that complete the digestive process; basal leaves retained over the winter; flowers more than 0.7″ across, blue, lavender, light purple or whitish; flowers April–May.

RANGE-HABITAT: Endemic to the Coastal Plain from NC south to FL; in SC, common in the Coastal Plain, uncommon in the Sandhills; Longleaf Pine savannas and moist, Longleaf Pine flatwoods, herbaceous seepages, and sandhill-pocosin ecotones.

SIMILAR SPECIES: Yellow Butterwort (*P. lutea* Walter, plate 617), with yellow flowers, also occurs in the Coastal Plain of SC. Dwarf Butterwort (*P. pumila* Michaux, plate 566) is much smaller, less than 6″ tall at flowering, with paler flowers that are less than 0.6″ across; it is rare in the Coastal Plain.

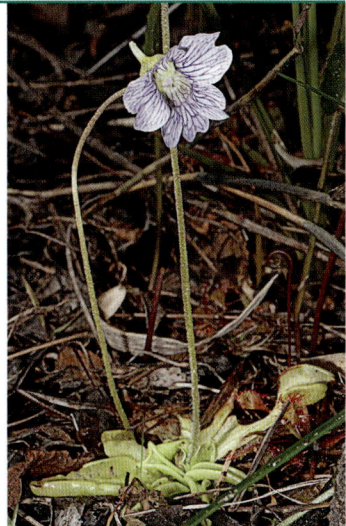

COMMENTS: The genus name derives from the Latin *pinguis,* meaning "somewhat fat," and refers to the greasy texture of the leaf. The butterworts were used as crude bandages during the Civil War; the upper leaf surface was placed on the wound.

617. Yellow Butterwort

Pinguicula lutea Walter
Pin-guí-cu-la lù-te-a
Lentibulariaceae (Bladderwort Family)

DESCRIPTION: Perennial, carnivorous plant from a small basal rosette with flowering stems to 18″ tall; upper leaf surface covered with stalked, glandular hairs in which insects become mired and sessile glands that complete the digestive process; leaves appear in a basal cluster and are retained over the winter; flowers more than 0.7″ across, yellow; flowers March–May

RANGE-HABITAT: Endemic to the Coastal Plain from NC south to FL and west to MS; in SC, common in the Coastal Plain and uncommon in the Sandhills; Longleaf Pine savannas, moist Longleaf Pine flatwoods, and herbaceous seepages.

COMMENTS: See Comments under *P. caerulea* Walter, above.

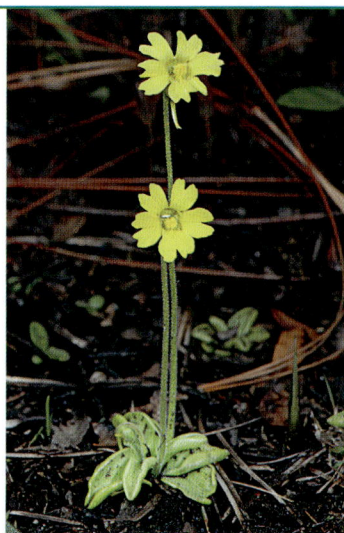

618. Orange Milkwort

Polygala lutea L.
Po-lý-ga-la lù-te-a
Polygalaceae (Milkwort Family)

DESCRIPTION: Biennial or short-lived perennial to 16″ tall; roots with a strong wintergreen odor; stems usually unbranched, smooth, 1 to several from base; leaves oblanceolate to spatulate, smooth; flowers in dense headlike racemes, bright orange, drying yellow; flowers April–October, with peak flowering in the spring.

RANGE-HABITAT: NY south to FL and west to LA; in SC, common in the Coastal Plain and Sandhills, very rare in the Piedmont; Longleaf Pine savannas and flatwoods and adjacent ditches, along roadsides, pocosin ecotones, herbaceous seepages, and boggy borders of Pond Cypress savannas and gum ponds.

COMMENTS: The specific epithet *lutea* is Latin for "yellow" and refers to the flower's color when dried; when fresh, the flowers are bright orange. The plant is also locally called "candyroot," due to the wintergreen scent of the roots. The wintergreen odor is due to methyl salicylate, which is poisonous when ingested, so this "candy" should not be consumed. The plant makes an excellent companion to carnivorous plants in the bog garden.

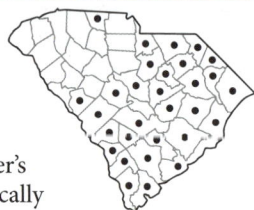

619. Hooded Pitcherplant

Sarracenia minor Walter
Sar-ra-cèn-i-a mì-nor
Sarraceniaceae (Pitcherplant Family)

DESCRIPTION: Evergreen, rhizomatous, carnivorous, herbaceous perennial; leaves 6–24″ tall, modified into hollow, tubular structures to catch insects; leaves winged; hood arching closely over the opening of the pitcher; upper portion of hood

spotted with white or translucent blotches (the windows); flowering stalk usually shorter than the leaves; flowers April–May.

RANGE-HABITAT: Endemic to the Coastal Plain from NC south to FL; in SC, formerly common, now uncommon due to habitat loss; Longleaf Pine savannas, Pond Cypress savannas, and wet ditches; less frequent in moist Longleaf Pine flatwoods.

COMMENTS: Hooded Pitcherplant gets its common name from the hood arching over the opening into the pitcher. The arching hood acts to keep flying insects from exiting the pitcher. After an insect has entered the pitcher and drunk its fill of nectar, it attempts to exit. The hood blocks most of the light entering the pitcher. The insect, seeking escape, flies toward the brightest light, which comes from the back of the pitcher. Here, pigment-free areas (the windows) allow light to enter. As the insect tries to fly through the "windows," it repeatedly bumps into the back of the hood, becomes stunned, and ultimately falls down the pitcher and drowns in the water and enzyme mixture at the base.

Morton (1974) reports an unusual folk remedy using Hooded Pitcherplant that is still practiced in the Coastal Plain: "Rootstock is boiled, and the decoction kept in a jar, applied warm on skin rashes or eruptions. People say the spots on the leaves are a sign the plant is a remedy for skin troubles, a belief which harks back to the old 'Doctrine of Signatures.'"

CONSERVATION STATUS: SC-Vulnerable

620. Crow-poison; Black Snakeroot

Stenanthium densum (Desrousseaux) Zomlefer & Judd
Ste-nán-thi-um dén-sum
Melanthiaceae (Bunchflower Family)

SYNONYM: *Zigadenus densus* (Desrousseaux) Fernald—RAB

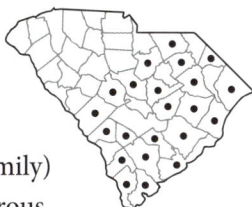

DESCRIPTION: Perennial herb from a bulblike base; stems 1–3′ (rarely 5′) tall; leaves mostly basal, 1–3, less than 0.25″ wide, enclosed by a purplish bladeless sheath; flowers produced in racemes; tepals 6, white, turning pink with age; fruit a conical capsule; flowers April–June; fruits mature June–August.

RANGE-HABITAT: Endemic to the Coastal Plain from VA south to FL and west to MS; in SC, common in the Coastal Plain and Sandhills in Longleaf Pine savannas, wet flatwoods, pocosin ecotones, and burned areas within pocosins.

SIMILAR SPECIES: Fly-poison (*Amianthium muscaetoxicum,* plate 240), which is more abundant in the mountains, ranges into the Coastal Plain. However, it has more numerous basal leaves, which are not enclosed by a purplish bladeless sheath; wider leaves (0.25–0.9″ wide); and more globular capsule.

COMMENTS: Many members of the family are poisonous, and this species is particularly toxic when ingested.

621. Pineland Plantain

Plantago sparsiflora Michaux
Plan-tà-go spar-si-flò-ra
Plantaginaceae (Plantain Family)

DESCRIPTION: Perennial herb from a basal rosette; flowering stem to 18″ long; leaves lanceolate to lance-elliptic, pubescent; flowers produced in narrow spikes, loosely flowered with the rachis visible over most of the length of the spike between flowers; individual flowers with reduced and inconspicuous perianth; flowers April–October.

RANGE-HABITAT: Endemic to the Coastal Plain from NC south to FL; in SC, uncommon in the Coastal Plain and restricted to wet, Longleaf Pine savannas over limestone substrate; more frequently in ditches and roadsides with marl hash.

COMMENTS: This species is a member of one of the rarest natural communities in SC, the "marl savanna." These savannas were formerly present in St. Johns Parish in Berkeley County but are now destroyed due to the construction of Lake Moultrie or reduced to fragments that are present as degraded habitats in other parts of the state. Today, this species has colonized many ditches in the Francis Marion National Forest along roadways lined with marl gravel. This is one of the most obvious species indicating the presence of limestone-influenced substrates, and it is a good indicator of many other extremely rare species such as *Eryngium ravenelii* Gray, *Scleria verticillata* Muhlenberg ex Willldenow, *Steironema hybridum* (Michaux) Rafinesque ex B.D. Jackson, and *Rhynchospora pinetorum* Britton & Small.

CONSERVATION STATUS: SC-Vulnerable

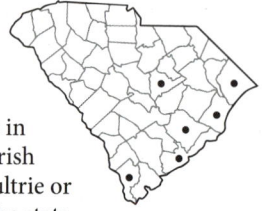

---

622. Venus' Fly Trap

> *Dionaea muscipula* Ellis
> Di-o-naè-a mus-cí-pu-la
> Dionaeaceae (Venus' Fly Trap Family)

DESCRIPTION: Perennial, carnivorous herb from a short rhizome; leaves in basal clusters, modified into trapping structures consisting of 2 hingelike, touch-sensitive lobes, each with stiff, marginal bristles; each lobe contains 3 to 6 trigger hairs by which the trap can be sprung; almost the entire upper surface of lobes covered with glands of two kinds: nectar producing and enzyme producing; flowering stalk 4–12″ tall; flowers white; flowers May–June.

RANGE-HABITAT: Endemic to the Coastal Plain and Sandhills of the Carolinas; in SC, rare, in Horry, Georgetown, and formerly Charleston Counties; wet, sandy ditches; Longleaf Pine savannas; pocosin ecotones; and openings in pocosins.

COMMENTS: Populations of Venus' Fly Trap occur in Lewis Ocean Bay Heritage Preserve and Cartwheel Bay Heritage Preserve in Horry County. Habitat destruction, fire suppression, and collection from the wild have considerably reduced the populations of this carnivorous plant. State law in SC protects it from harvest on public lands.

William Bartrum reported that the Native American name for this plant was "Tippitiwitchet" (Reveal, 1996).

CONSERVATION STATUS: SC-Imperiled

---

623. Golden Colicroot

> *Aletris aurea* Walter
> Á-le-tris àu-re-a
> Nartheciaceae (Bog-asphodel Family)

DESCRIPTION: Perennial herb with leaves arranged in a basal rosette, from thick, short rhizomes; basal leaves to 3.2″ long and 0.4–0.8″ wide, lanceolate; stem leaves greatly reduced; flower stalk 1–3′ tall; perianth yellow, rounded and blunt tipped, covered with small, sticky projections, giving a mealy appearance; flowers mid-May–July.

RANGE-HABITAT: From MD south to peninsular FL and west to TX and OK; in SC, common in the Coastal Plain and Sandhills; Longleaf Pine flatwoods, savannas, and pocosin borders.

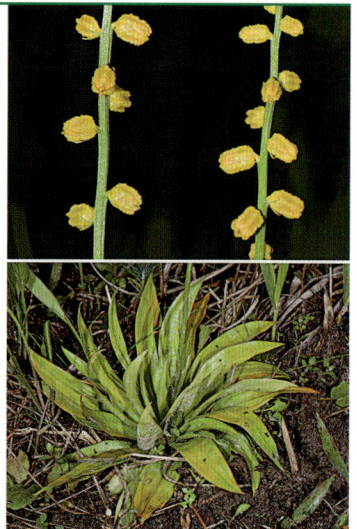

SIMILAR SPECIES: Northern White Colicroot (*A. farinosa* L., plate 572) occurs in the same areas but has longer leaves and white flowers that appear 2–4 weeks before Golden Colicroot.
COMMENTS: See Comments under Northern White Colic-root (plate 572).

624. Large Spreading Pogonia

Cleistesiopsis divaricata
(L.) Ames
Cleis-te-si-óp-sis di-va-ri-cà-ta
Orchidaceae (Orchid Family)

SYNONYM: *Cleistes divaricata* (L.)
Ames, in part—RAB, PR

DESCRIPTION: Perennial herb, 1–2.5' tall; entire plant has a bluish green color with a fine, frosty-white coating; leaf solitary, inserted above the middle of the stem; sepals widely spreading, dark maroon to brownish; petals 1–2" long, projected forward, pink; the basal ¾ of the central keel of the lip with 1–3 parallel ridges; flowers have the distinct odor of daffodils when fresh; May–July.
RANGE-HABITAT: Endemic to the Coastal Plain from NJ south to FL; in SC, occasional in the Sandhills and Coastal Plain where found in Longleaf Pine savannas, wet flatwoods, herbaceous seepages, and ecotones of pocosins.
SIMILAR SPECIES: Three species are now recognized from what was formerly considered a single species. Small Coastal Plain Spreading Pogonia (*C. oricamporum* P.M. Brown, plate 734), is smaller in all respects and has flowers that have a strong vanillalike odor when fresh, with paler pink to whitish petals and a lip with the basal ¾ of the central keel with 5–7 discontinuous ridges. Appalachian Small Spreading Pogonia (*C. bifaria* (Fernald) Pansarin & F. Barros, plate 241) is restricted to the mountains and Piedmont and is very similar to *C. oricamporum* but has odorless flowers and bracts that exceed the length of the pedicellate flower.
COMMENTS: Large Spreading Pogonia is among the largest flowered and most attractive orchids in North America. It should never be picked or disturbed in the wild.

625. Common Grass-pink (A)

Calopogon tuberosus (L.) Britton, Sterns
& Poggenburg var. *tuberosus*
Ca-lo-pò-gon tu-be-rò-sus
SYNONYM: *Calopogon pulchellus* (Salisbury) R. Brown—RAB

Pale Grass-pink (B)

Calopogon pallidus Chapman
Ca-lo-pò-gon pál-li-dus
Orchidaceae (Orchid Family)

DESCRIPTION: Erect perennial, arising from a corm; flowering stalk 6–24" tall; leaves (when present) 1 or 2, narrowly linear and grasslike, sheathing the flowering stalk at the base; flowers open successively up the raceme, each flower lasting for about 5 days; in *C. tuberosus,* often 3–4 flowers open at once, and in *C. pallidus,* rarely more than 2, and with racemes containing fewer flowers overall; lateral sepals 0.4–0.6" long, strongly sickle-shaped and widely spreading in *C. pallidus,* and 0.6–1.1" long and widely sickle-shaped and held straight in *C. tuberosus;* lip strongly dilated and bearded with numerous yellow hairs mimicking anthers in the

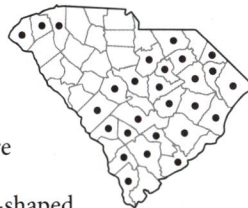

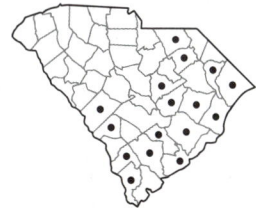

center; in *C. tuberosus,* flowers pink to rose-purple or magenta-crimson, rarely white, and in *C. pallidus* flowers white to pale pink; *C. tuberosus* flowers April–July; *C. pallidus* flowers May–June.

RANGE-HABITAT: *C. tuberosus* is widespread in North America from Newfoundland west to MT and south to FL and TX; in SC, common in the Coastal Plain and Sandhills and rare in the mountains and Piedmont; Longleaf Pine savannas, pocosin ecotones, herbaceous seepages, cataract fens, mountain bogs, and seepages on granitic flatrocks; *C. pallidus* is endemic to the Coastal Plain from VA south to FL and west to LA; in SC, common in the Coastal Plain and Sandhills in Longleaf Pine savannas, herbaceous seepages, and pocosin ecotones.

COMMENTS: These two common species of grass-pink are often found in the same location and are easily separated by the smaller and paler flowers of *C. pallidus,* which display slender and spreading sickle-shaped lateral sepals. A most unusual feature of the grass-pinks is the inverted perianth where the lip is uppermost. In most other orchid genera, the lip, which is actually the top petal, is twisted to be lowermost = resupinate. The upper lip acts like an elevator: The weight of an insect that lands on the lip causes it to bend downward, bringing the insect in contact with the stigma. Cross-pollination is thus affected.

The Grass Pinks have no nectar to entice insects. Instead, they mimic flowers that offer a re-ward for pollinators. Coastal grass-pinks grow in association with Smooth Meadow-beauty (*Rhexia alifanus,* plate 630), which has pink petals and long, yellow stamens that provide pollen as food for visiting insects. The pink petals and yellow hairs (the beard) of the lip, mimicking the pink petals and yellow stamens of the meadow-beauty, possibly fool insects into mistaking the orchid flower for the meadow-beauty.

626. Snowy Orchid

Platanthera nivea (Nuttall) Luer
Pla-tán-the-ra ní-ve-a
Orchidaceae (Orchid Family)

SYNONYM: *Habenaria nivea* (Nuttall) Sprengel—RAB

DESCRIPTION: Erect, herbaceous perennial, to 1' tall; leaves 2–3, near the base, rigidly suberect, linear-lancolate, reduced upward to as many as 10 slender, erect bracts; flowers produced in a raceme, snowy white; all petals and sepals entire, not fringed; the long nectar spur extending sideways but curved upward; flowers May–September.

RANGE-HABITAT: NJ south to FL and west to TX; disjunct in TN; in SC, rare in the Coastal Plain where found in Longleaf Pine savan-nas, Longleaf Pine flatwoods, and pocosin ecotones and very rarely in the upper Piedmont, with a single known location on seepages along the edge of a granitic dome at the edge of the Blue Ridge Escarpment.

COMMENTS: Like many orchids, Snowy Orchid may cover a site one year and then be absent for several years before again making an appearance. Many populations have been eliminated because of alteration of drainage patterns or fire suppression of SC savannas. The common name is derived from the Latin *niveus,* meaning "white as snow," in reference to the intense whiteness of the flowers. The occurrence of a large population of this species along the edge of a granitic dome in seepage at the edge of the Blue Ridge Escarpment is remarkable. This granitic dome supports numerous species otherwise restricted to the Coastal Plain.

CONSERVATION STATUS: SC-Vulnerable

627. Carolina Loosestrife

Lysimachia loomisii Torrey
    Ly-si-má-chi-a loo-mís-i-i
Primulaceae (Primrose Family)

DESCRIPTION: Perennial herb from a rhizome with stiffly, erect stems 12–48″ tall; leaves opposite or in whorls of 3–4, sessile, linear, and tapered to both ends; flowers in racemes, corolla yellow, with several maroon streaks and a ring of purplish color at the throat; flowers May–June.

RANGE-HABITAT: Endemic to the Coastal Plain of NC, SC, and GA; occasional; moist to wet Longleaf Pine savannas and pocosin ecotones.

COMMENTS: Two groups of plants are called loosestrife: *Lysimachia* of the primrose family (Primulaceae) and *Lythrum* of the loose-strife family (Lythraceae).

628. Common Bogbuttons

Lachnocaulon anceps (Walter)
    Morong
     Lach-no-caù-lon án-ceps
Eriocaulaceae (Pipewort Family)

DESCRIPTION: Clump-forming, perennial herb, with tufts of leaves aggregated into small, dense mats of rosettes; leaves linear and tapering toward the tips, bright green; flowers produced in heads at the end of a long, leafless stem that is typically erect to ascending while in flower; separate male and female flowers in the same head; heads 0.15–0.27″ across; seeds with longitudinal ribs apparent under magnification; flowers May–October.

RANGE-HABITAT: NJ south to FL and west to TX; also in the Ca-ribbean and Central America; disjunct in TN; in SC, common in the Coastal Plain in Longleaf Pine savannas, wet flatwoods, and pocosin ecotones.

SIMILAR SPECIES: See discussion under Southern Bogbuttons (*L. beyrichianum* Sporleder ex Körnicke, plate 523).

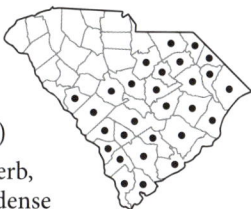

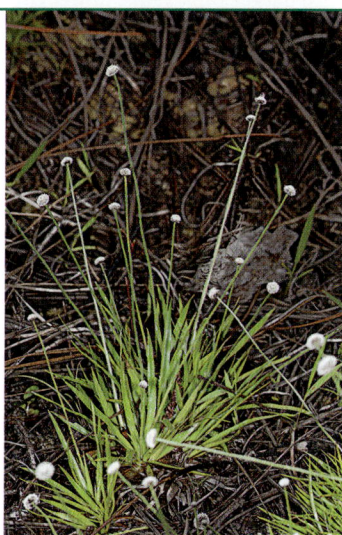

629. Yellow Meadow-beauty

Rhexia lutea Walter
    Rhéx-i-a lù-te-a
Melastomataceae (Melastome Family)

DESCRIPTION: Highly branched, pubescent, perennial herb, 4–24″ tall, with angled, ciliate stems; leaves opposite, elliptic, oblanceolate, or obovate, mostly less than 0.25″ wide, with ciliate margins; petals 4, yellow; hypanthium urn-shaped, smooth, with a constricted neck and flared upward; sepals ciliate; flowers April–July.

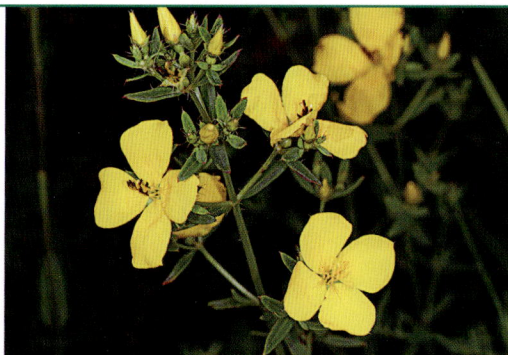

RANGE-HABITAT: Endemic to the Coastal Plain from NC south to FL and west to TX; in SC, common in the Coastal Plain and uncommon in the Sandhills in Longleaf Pine savannas and herbaceous seepages.

COMMENTS: This is the only species of meadow-beauty with yellow flowers. It generally occurs in some of the wettest areas of savannas.

---

630. Smooth Meadow-beauty

Rhexia alifanus Walter
   Rhéx-i-a a-li-fà-nus
   Melastomataceae (Meadow-
      beauty Family)

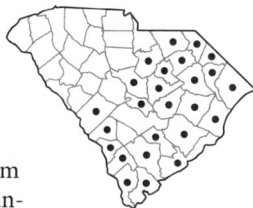

DESCRIPTION: Perennial herb to 40″ tall from an enlarged root, not rhizomatous; stem un-branched or sparingly branched below, smooth; leaves opposite, bluish-green, prominently 3-nerved; petals pink with darker pink veins; flowers open in the morning and shatter by late afternoon; the 8 bright yellow anthers prominently curved; hypanthium urn-shaped with gland-tipped hairs; flowers May–September.

RANGE-HABITAT: Endemic to the Coastal Plain from NC south to FL and west to TX; in SC, common in the Coastal Plain and Sandhills; Longleaf Pine savannas and Longleaf Pine flatwoods, pocosin ecotones, and moist roadbanks and ditches.

COMMENTS: *Alifanus* refers to an earthenware cup, which the hypanthium resembles. This is SC's most spectacular species of meadow-beauty. The tall, smooth, unbranched stems with ascending, bluish green leaves set this species apart from all other meadow-beauties. It is extremely floriferous following fire.

---

631. Nash's Meadow-beauty

Rhexia nashii Small
   Rhéx-i-a násh-i-i
   Melastomataceae (Melastome
      Family)

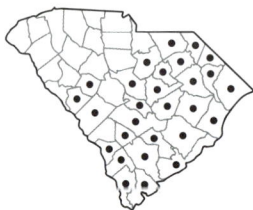

SYNONYM: *Rhexia mariana* L. var.
   purpurea Michaux—RAB

DESCRIPTION: Perennial herb to 30″ tall from rhizomes, forming extensive colonies; stem branched, pubescent, angled with 2 faces narrow and two broad; leaves opposite, green, elliptic to ovate; petals 0.8–1″ long, bright pink with glandular hairs on the back surface; flowers open in the morning and shatter by late afternoon; the 8 bright yellow anthers prominently curved; hypanthium 0.4–0.6″ long, urn-shaped, smooth; flowers May–October.

RANGE-HABITAT: VA south to FL and west to LA; in SC, common in the Coastal Plain and Sandhills in Longleaf Pine savannas and Longleaf Pine flatwoods, pocosin ecotones, and moist roadbanks and ditches.

COMMENTS: Nash's Meadow-beauty is similar but larger in all respects to Maryland Meadow-beauty (*R. mariana* L., plate 87). Maryland Meadow-beauty has paler pink flowers and pubescent and smaller hypanthia.

632. Ciliate Meadow-beauty

Rhexia petiolata Walter
    Rhéx-i-a pe-ti-o-là-ta
Melastomataceae (Melastome
    Family)

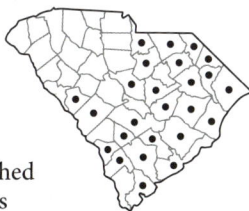

DESCRIPTION: Perennial herb with unbranched or basally branched stems 4–20″ tall; stems 4-angled, narrowly winged, pubescent; leaves opposite, broadly elliptic to ovate, with acute to acuminate tips, sessile or subsessile, 3-nerved with ciliate margins; flowers held facing upward, clustered toward top of the stem; petals pink; anthers straight; flowers open in the morning and shutter by late afternoon; hypanthium smooth or pubescent, with a short, constricted neck; flowers May–October.

RANGE-HABITAT: Endemic to the Coastal Plain from VA south to FL and west to TX; in SC, common in the Sandhills and Coastal Plain; wet pine flatwoods and savannas, pocosin borders, and ditches.

COMMENTS: This is the only SC species with flowers held upright. Nuttall's Meadow-beauty (*R. nuttallii* C.W. James) is very similar but has acute or blunt-tipped leaves with subentire margins that are not ciliate. It is known from GA very close to the SC line and should be sought in SC's southeastern Coastal Plain.

---

633. Coastal Plain Yellow-eyed Grass

Xyris ambigua Beyrich ex Kunth
    Xỳ-ris am-bíg-u-a
Xyridaceae (Yellow-eyed Grass
    Family)

DESCRIPTION: Perennial herb with only basal leaves; leaves linear, straight (not twisted) 6–18″ long, conduplicate, like an *Iris,* distichous, lustrous green, and striped with purple at the base (if peeled back) and with copious amounts of mucouslike clear liquid at the base when leaves are removed; flowering stalk 12–39″ tall, straight to slightly twisted; flowers yellow, produced in a compact, terminal, round to drum-head-shaped spike, with each flower subtended by a woody-textured, lustrous brown scale that hides the flower bud and fruit; flowers ephemeral, opening in the early morning and closed by noon; flowers May–August.

RANGE-HABITAT: VA south to FL and west to TX; disjunct in s. NJ, DE, and c. TN; in SC, common in the Sandhills and Coastal Plain in Longleaf Pine savannas, Pond Cypress savannas, upper edges of depression ponds, pocosin ecotones, herbaceous seepages, ditches, and wet roadsides.

SIMILAR SPECIES: No fewer than 19 species of *Xyris* are known in SC. Strict Yellow-eyed Grass (*Xyris stricta* Chapman) is very similar but has narrower leaves that are generally brown or reddish-brown at the base and longer, more narrow spikes with duller brown scales. It is rare in SC, growing in wet Longleaf Pine savannas and depression pond margins.

COMMENTS: This is the most abundant of the larger, perennial straight-leaved species of this genus in SC. It is typical of most savanna habitats. The name *Xyris* comes from the Greek, meaning "gladen" (sword-shaped), in reference to the leaves.

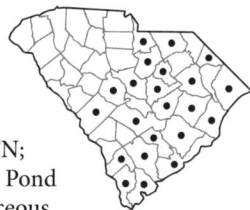

634. Bulbous Yellow-eyed Grass

Xyris platylepis Chapman
Xỳ-ris pla-tý-le-pis
Xyridaceae (Yellow-eyed Grass
Family)

DESCRIPTION: Perennial herb with only basal leaves; leaves linear, slightly twisted, 8–24″ long, conduplicate, like an *Iris,* distichous, lustrous green and markedly swollen at the base to appear nearly bulblike; if leaf bases are peeled back they contain copious amounts of mucouslike clear liquid; flowering stalk 12–39″ tall, straight to slightly twisted; flowers in a compact, terminal, round to drum-head-shaped spike, with each flower subtended by a woody-textured, lustrous brown scale that hides the flower bud and fruit; flowers ephemeral, opening in the late afternoon; flowers May–August.

RANGE-HABITAT: VA south to FL and west to TX; in SC, common in the Sandhills and Coastal Plain, rare in the Piedmont and mountains; Longleaf Pine savannas, Pond Cypress savannas, upper edges of depression ponds, pocosin ecotones, herbaceous seepages, cataract fens, seepage over granitic domes and flatrocks, sandy-bottomed small escarpment streams, ditches, and wet roadsides.

COMMENTS: The bulblike swollen base helps to identify this species from other large, perennial *Xyris.* Its occurrence in the upper Piedmont and escarpment of SC is remarkable, and here it occurs with many other species that have distributions primarily in the Coastal Plain. The name *Xyris* comes from the Greek, meaning "gladen" (sword-shaped), in reference to the leaves.

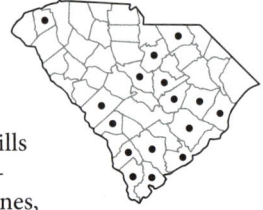

635. Longleaf Milkweed; Savanna Milkweed (A)

Asclepias longifolia Michaux
As-clè-pi-as lon-gi-fò-li-a

Michaux's Milkweed (B)

Asclepias michauxii Decaisne
As-clè-pi-as mi-chaùx-i-i
Apocynaceae (Dogbane Family)

DESCRIPTION: *A. longifolia:* perennial herb with milky sap; stems simple, 8–27″ long; leaves linear, subopposite (almost opposite but slightly offset), 3–5.5″ long; flowers produced in 1–4 umbels at the terminus of and in the upper axils of the stem; flowers greenish white; the corona without horns (projections); fruit a smooth follicle; flowers May–June; fruits mature June–July; *A. michauxii* is similar but is a smaller plant with more narrow leaves, a solitary terminal umbel, and horns on the corona; flowers May–June; fruits mature June–July.

RANGE-HABITAT: *A. longifolia:* DE south to FL and west to TX; in SC, common in the Coastal Plain in Longleaf Pine savannas and pocosin ecotones; *A. michauxii:* SC south to FL and west to LA; in SC, uncommon in Longleaf Pine savannas and pocosin ecotones.

COMMENTS: Both of these species are often overlooked due to the linear leaves and the arching stems, often hidden in the dense grasses of savannas. Like all milkweeds, they are hosts for monarch butterflies. The genus name honors Asclepius, an ancient Greek god of medicine.

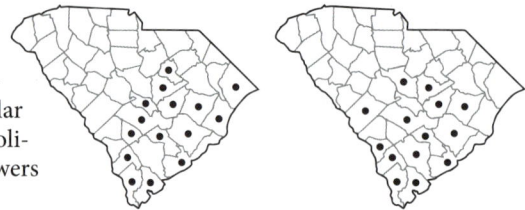

636. Red Milkweed

Asclepias lanceolata Walter
As-clè-pi-as lan-ce-o-là-ta
Apocynaceae (Dogbane Family)

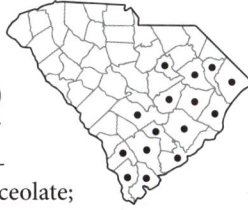

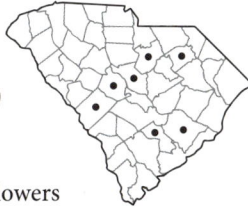

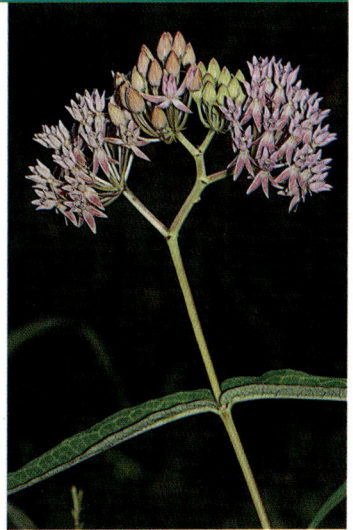

DESCRIPTION: Perennial, smooth herb, 3–5′ tall; stem erect, rarely branched; leaves opposite, in 3–6 pairs, linear to narrowly lanceolate; flowers in 2–4 umbels produced toward the top of the stem; corolla red, crown usually orange; flowers June–August; fruit is a follicle maturing August–September.

RANGE-HABITAT: Endemic to the Coastal Plain from NJ south to FL and west to e. TX; in SC, common in wet Longleaf Pine savannas, fresh to brackish marshes, cypress depressions, swamps, ditches, and wet roadsides.

COMMENTS: This tall orange-flowered milkweed with very narrow leaves is unmistakable. It is frequent in ditches along roadways through Longleaf Pine habitats. Several species of milkweed are known to be poisonous when eaten raw; it is likely that most, if not all, raw milkweeds are toxic. Duke (1997) recommends *Asclepias* species as an herbal remedy for warts. The genus name honors Asclepius, an ancient Greek god of medicine.

637. Purple Savanna Milkweed

Asclepias rubra L.
As-clè-pi-as rù-bra
Apocynaceae (Dogbane Family)

DESCRIPTION: Perennial herb with simple stem, 16–40″ tall; leaves opposite, in 3–5 pairs, simple, sessile, ovate to lanceolate; flowers produced in 1–4 umbels toward the top of the stem; corolla pink to lavender; flowers June–July; follicles mature July–September.

RANGE-HABITAT: Endemic to the Coastal Plain from NJ south to FL and west to TX; in SC, rare in the Coastal Plain and Sandhills where found in pocosin ecotones, herbaceous seepages, wet Longleaf Pine savannas, flatwoods, and swamps.

COMMENTS: Though the specific epithet *rubra* means red and one of the widely used common names for this species is Red Milkweed, the flowers are never red; they are more frequently pinkish purple. This is a remarkably beautiful species that is seldom seen in the wild and could make a beautiful landscape plant if brought into commercial production. Like all milkweeds, this species is a host for Monarch Butterflies. The genus name honors Asclepius, an ancient Greek god of medicine.

CONSERVATION STATUS: SC-Vulnerable

638. Savanna Seedbox

Ludwigia virgata Michaux
Lud-wíg-i-a vir-gà-ta
Onagraceae (Evening-primrose
Family)

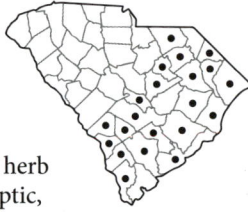

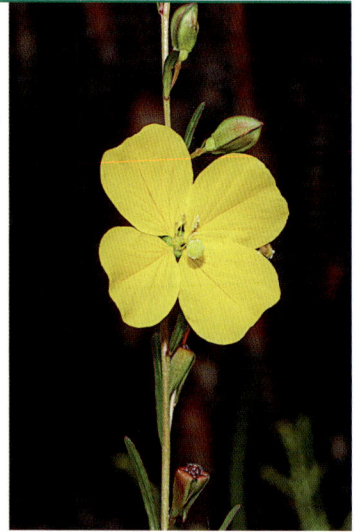

DESCRIPTION: Erect to ascending perennial herb to 3′ tall; leaves alternate, lanceolate to elliptic, truncate to rounded at the base, smooth; flowers solitary on short pedicels in the upper leaf axils; petals 4, yellow, flowering in the morning and shattering by late afternoon; flowers June–September.

RANGE-HABITAT: Endemic to the Coastal Plain from VA south to FL and west to AL; in SC, common in Longleaf Pine savannas, Pond Cypress savannas, and upper margins of depression ponds.

COMMENTS: Savanna Seedbox is one of the most common species of *Ludwigia* in the pinelands of the Coastal Plain. It may be differentiated from the closely related *L. hirtella* Rafinesque and *L. maritima* Harper by the smooth stem and leaves as both of the similar species are pubescent.

639. White Sabatia; Lanceleaf Rose-gentian

Sabatia difformis (L.) Druce
Sa-bà-ti-a dif-fór-mis
Gentianaceae (Gentian Family)

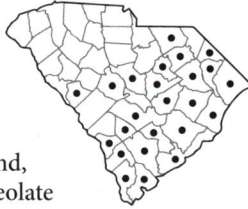

DESCRIPTION: Erect, herbaceous perennial 1–3′ tall, from a short rhizome; stems round, smooth, not winged; leaves opposite, lanceolate to ovate, 0.8–2″ long; branches opposite, forming a convex to flat-topped inflorescence; flowers with petals white; flowering May–September.

RANGE-HABITAT: NJ south to FL and west to AL; in SC, common in the Coastal Plain and Sandhills in Longleaf Pine savannas, Pond Cypress savannas, upper edges of depression ponds, herbaceous seepages, and pocosin ecotones.

SIMILAR SPECIES: Two species of white-flowered *Sabatia* are found in SC. White Sabatia is a perennial with round or slightly angled stems, and the annual Four-angled Sabatia (*S. quadrangulata* Wilbur) has sharp, winged angles to the stem. The genus is named for Italian botanist Liberato Sabbati (1714–78).

640. Slender Marsh-pink

Sabatia campanulata
(L.) Torrey
Sa-bà-ti-a cam-pa-nu-là-ta
Gentianaceae (Gentian Family)

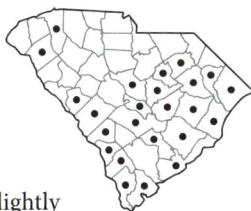

DESCRIPTION: Erect, herbaceous perennial 12–27″ tall, from a short rhizome; stems slightly angled, smooth, not winged; leaves opposite, linear to elliptic, 0.4–1.4″ long, with acute tips; branches alternate; flowers arranged in a diffuse panicle, petals 5, pink, rarely white with a small yellow base separated from the pink portion by a dark red or pink ring; flowering June–August.

RANGE-HABITAT: MA south to FL and west to KY, AR, and LA; in SC, common in the Coastal Plain and Sandhills in Longleaf Pine savannas, Pond Cypress savannas, upper margins of depression ponds, herbaceous seepages, and ecotones of pocosins.

COMMENTS: This is the most common pink-flowered *Sabatia* in the savannas of the Coastal Plain pinelands. Other pink-flowered species in similar habitats have more than 5 petals. The genus is named for Italian botanist Liberato Sabbati (1714–78).

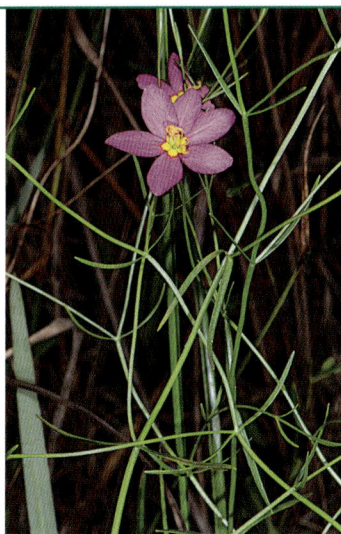

641. Crested Fringed Orchid

Platanthera cristata (Michaux) Lindley
Pla-tán-the-ra cris-tà-ta
Orchidaceae (Orchid Family)

SYNONYM: *Habenaria cristata* (Michaux) R. Brown—RAB

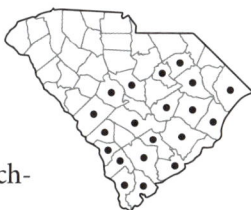

DESCRIPTION: Perennial, stout herb, 7–35″ tall; leafy below, bracteate above; leaves oblong-lanceolate to linear-lanceolate; flowers in a cylindrical raceme, orange; lip fringed; nectar spur 0.15–0.67″ long, often requiring close examination to see; flowers June–September.

RANGE-HABITAT: Primarily limited to the Coastal Plain from MA south to FL and west to TX; also inland in KY, TN, AR, and NC; in SC, common in the Coastal Plain, infrequent in the Sandhills; Longleaf Pine savannas, moist flatwoods, pocosins, and moist roadsides.

SIMILAR SPECIES: Yellow Fringed Orchid (*Platanthera ciliaris* (L.) Lindley, plate 244) is similar in color and form but is a much larger species with a much longer and obvious nectar spur 0.8–1.3″ long. It is common in similar habitats but extends into the Piedmont and mountains, where it is occasional.

COMMENTS: The "crested" in the common name comes from the Latin term *cristata* in reference to the fringed lip.

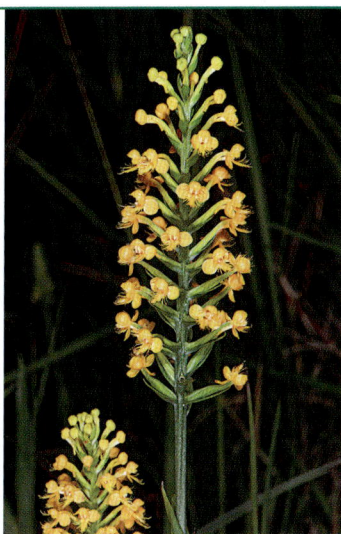

642. Redroot

Lachnanthes caroliniana
    (Lamarck) Dandy
        Lach-nán-thes ca-ro-li-ni-à-na
Haemodoraceae (Bloodwort
    Family)

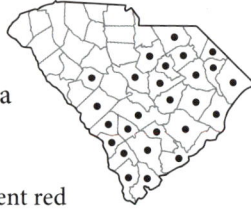

DESCRIPTION: Perennial herb with prominent red rhizomes and fibrous roots, both with red sap; flowering plants 1–3′ tall; leaves mostly basal, linear, conduplicate, *Iris*-like, rapidly reduced upward; perianth pubescent with white trichomes on the exterior, yellowish and smooth within; flowers June–early September.

RANGE-HABITAT: Nova Scotia south to FL and west to LA; disjunct inland to w. VA and c. TN; in SC, common in the Coastal Plain and Sandhills in Longleaf Pine savannas, ecotones of pocosins, ditches, and managed freshwater impoundments along rivers.

COMMENTS: Some sources list Redroot as poisonous. Kingsbury (1964), however, disputes this with good evidence. Redroot is a prized food for waterfowl along the coastal rivers; both the seeds and rhizomes are used as food. Freshwater impoundments, such as those at the Savannah National Wildlife Refuge, are often managed to encourage the growth of Redroot to attract ducks.

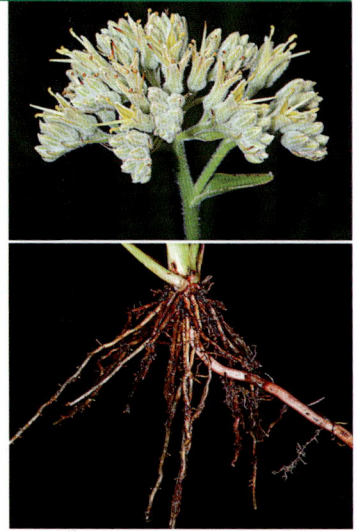

643. Pineland Hibiscus; Comfort-root

Hibiscus aculeatus Walter
        Hi-bís-cus a-cu-le-à-tus
Apocynaceae (Dogbane Family)

DESCRIPTION: Perennial herb to 3′ tall; stems spreading to ascending or erect; stems and leaves roughly pubescent; leaves deeply dissected, with 3–5 primary lobes; corolla with a purple center, the lobes cream-colored fading to yellow and then pink as the petals wither; flowers June–August.

RANGE-HABITAT: NC south to FL and west to LA; in SC, common in Longleaf Pine savannas, wet flatwoods, margins of depression ponds, ditches, and wet roadsides.

COMMENTS: Of the several species of *Hibiscus* found in SC, this is the only species with deeply dissected leaves and commonly occurring in fire-maintained pineland habitats.

644. Rose Fleabane

Pluchea baccharis (P. Miller) Pruski
        Plù-che-a bác-cha-ris
Asteraceae (Aster Family)

SYNONYM: *Pluchea rosea* R. K.
    Godfrey—RAB

DESCRIPTION: Perennial herb emitting a strong, camphorlike odor; stem 12–30″ tall, densely short-pubescent; leaves alternate, oblong to elliptic-oblong, sessile and auriculate-clasping;

inflorescence congested with very short branches; flowers arranged in involucrate heads of light pink disk flowers only; flowers June–July.

RANGE-HABITAT: NC to FL and west to TX; also found in the Neotropics; in SC, common in the Coastal Plain in wet, Longleaf Pine savannas and Pond Cypress savannas, ditches, and wet roadsides.

SIMILAR SPECIES: Camphorweed (*P. camphorata* (L.) A. P. de Candolle) also has pink flowers but has petiolate leaves and is a course, weedier species of wetlands and marshes. Saltmarsh Fleabane (*P. odorata* (L.) Cassini) has intense pink flowers and petiolate leaves and inhabits the borders of salt marshes. Stinking Fleabane (*P. foetida* (L.) A. P. de Candolle) has a more open inflorescence with longer branches, cream-colored to pale pinkish flowers, foul odor, and sessile or clasping leaves; it inhabits savannas, ditches, and marshes.

COMMENTS: Plants in this genus were once used as a flea repellant. The genus is dedicated to Noël-Antoine Pluche (1688–1761), a French naturalist and priest.

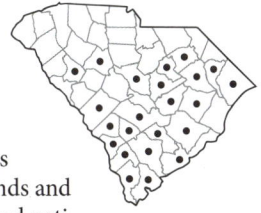

645. Pine lily; Catesby's lily

Lilium catesbyi Walter
Lí-li-um càtes-by-i
Liliaceae (Lily Family)

DESCRIPTION: Perennial herb with unbranched, erect stems from scaly bulbs; stems 20–28″ tall; flower solitary, erect; leaves alternate, ascending or appressed to the stem, the lower about 3.5″ long, reduced upward; flowers solitary, held upright with the tepals recurved, brilliant orangish-red, fading to yellow toward the base of the tepals, which also display purplish spots; flowers mid-June–mid-September.

RANGE-HABITAT: Endemic to the Coastal Plain from VA to FL and west to LA; in SC, occasional in the Sandhills and Coastal Plain where growing in Longleaf Pine savannas, wet flatwoods, ecotones of pocosins, and herbaceous seepages.

COMMENTS: This lily grows in a delicate balance with its environment and has been declining in SC due to habitat loss, alteration of drainage patterns, and fire suppression. It should never be disturbed in the wild. The specific epithet honors Mark Catesby (1683–1749), English naturalist who visited SC (1722–24) and published *The Natural History of Carolina, Florida and the Bahama Islands.*

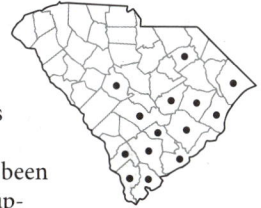

646. Large Death-camas

Zigadenus glaberrimus Michaux
Zi-ga-dè-nus gla-bér-ri-mus
Melanthiaceae (Bunchflower Family)

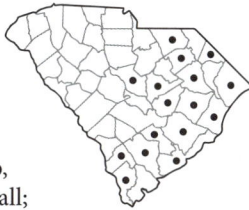

DESCRIPTION: Rhizomatous, perennial herb, forming clumps; flowering stems 20–30″ tall; leaves linear, 6–16″ long, 0.2–0.8″ wide, keeled on the undersides, green above, bluish-green beneath, smooth; flowers produced in a panicle of racemes; tepals white, with two distinct green glands; flowers late June–September.

RANGE-HABITAT: Endemic to the Coastal Plain from VA south to FL and west to MS; in SC, common in the Sandhills and Coastal Plain in Longleaf Pine savannas, ecotones of pocosins, and herbaceous seepages.

COMMENTS: The common name implies the toxicity of this plant. Many members of the bunchflower family are potently toxic if

ingested. When not flowering, the plant is sometimes mistaken for a nonflowering Toothache Grass (*Ctenium aromaticum* (Walter) Wood, plate 611), which also has bicolored leaves. The leaves of Toothache Grass are bluish-green above and green below and lack the distinct keel on the undersides.

647. Savanna Mountain-mint

Pycnanthemum flexuosum (Walter)
  Britton, Sterns & Poggenburg
  Pyc-nán-the-mum flex-u-ò-sum
Lamiaceae (Mint Family)

DESCRIPTION: Erect, perennial herb 16–44″ tall; stems 4-angled, the angles sharp to rounded; leaves opposite, 0.2–0.6″ wide, elliptic to elliptic-lanceolate; flowers in flat-topped clusters; calyx lobes attenuate-aristate, stiff and whitish; corolla bilabiate, white to lavender, lower lip with purple dots; flowers June–August.

RANGE-HABITAT: VA south to FL; disjunct inland, in bogs in sw. NC and TN; in SC, common in the Sandhills and Coastal Plain in moist to wet Longleaf Pine savannas, pocosins, and herbaceous seepages.

COMMENTS: The English common name is not appropriate as this species is much more common in the lowlands. Mountain mints are some of the best plants to attract and support pollinators. This species forms small clumps and does not spread as much as other species and makes a very well-behaved ornamental. Though it is aromatic, it does not have as intense an odor as many other members of the genus.

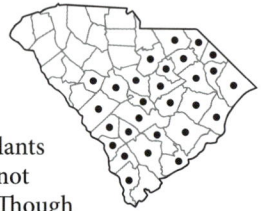

648. Southern Bog-asphodel; Coastal False-asphodel

Triantha racemosa
  (Walter) Small
  Tri-án-tha ra-ce-mò-sa
Tofieldiaceae (False-asphodel
  Family)

SYNONYM: *Tofieldia racemosa* (Walter) Britton, Sterns & Poggenburg—RAB, PR

DESCRIPTION: Perennial herb from a short rhizome; flowering stalk 12–28″ tall; basal leaves conduplicate, *Iris*-like, erect, linear, to 16″ long; usually 1 bractlike leaf inserted somewhere below the middle of the flowering stalk; flowering stalk pubescent; flowers produced in a thyrse; tepals white; flowers June–early August.

RANGE-HABITAT: Primarily on the Coastal Plain from NJ south to FL, west to TX, and disjunct inland in TN, VA, and AL; in SC, common in the Sandhills and Coastal Plain in Longleaf Pine savannas and flatwoods, herbaceous seepages, and ecotones of pocosins.

SIMILAR SPECIES: Carolina Bog-asphodel (*Tofieldia glabra* Nuttall, plate 513), a much less common plant, is similar but has smooth flowering stalks and flowers much later in autumn (September–November).

649. Yellow Fringeless Orchid;
Frog-arrow

Platanthera integra (Nuttall)
Gray ex Beck
Pla-tán-the-ra ín-te-gra
Orchidaceae (Orchid Family)

SYNONYM: *Habenaria integra* (Nuttall)
Sprengel—RAB

DESCRIPTION: Perennial herb to 2′ tall, with smooth stems; leaves several below, oblong-lanceolate to narrowly lanceolate, reduced to bracts above; flowers produced in a raceme, initially conical but soon becoming cylindrical; flowers saffron to orange with the lip irregularly lobed or serrate (not fringed); flowers July–September.
RANGE-HABITAT: NJ south to FL and west to TX; disjunct inland in NC, TN, and n. AL; in SC, very rare; Longleaf Pine savannas and wet Longleaf Pine flatwoods.
COMMENTS: Yellow Fringeless Orchid is often mistaken for Crested Fringed Orchid (*P. cristata* (Michaux) Lindley, plate 641). Yellow Fringeless Orchid can be identified by its entire lip; the lip of *P. cristata* is fringed. This orchid appears one year at a site and then is absent the next year (or possibly longer); the reason for this periodicity is unknown. Though formerly abundant, recent fieldwork in the Coastal Plain has documented a precipitous decline with the plant now nearly extirpated in SC. The primary cause for decline is likely fire suppression and habitat loss. The name *integra* is Latin for "entire," in reference to the nearly entire, fringeless lip.
CONSERVATION STATUS: SC-Critically Imperiled

650. Grassleaf Barbara's-buttons

Marshallia graminifolia (Walter) Small
Mar-sháll-i-a gra-mi-ni-fò-li-a
Asteraceae (Aster Family)

DESCRIPTION: Fibrous-rooted, perennial herb, forming small clumps; stems 16–32″ tall; basal leaves firm and ascending, linear to narrowly elliptic, 2–8″ long; stem leaves not strongly different from basal leaves except in size, usually less than 1.2″ long; fibrous base of old basal leaves persistent; flowers arranged in involucrate heads consisting solely of pink disk flowers at the tips of the stems and long nearly erect branches; flowers late July–September.
RANGE-HABITAT: Endemic to the Coastal Plain from NC south to GA; in SC, common in the Coastal Plain and Sandhills in Longleaf Pine savannas and wet flatwoods, ecotones of pocosins, herbaceous seepages, and ditches.
COMMENTS: Grassleaf Barbara's-buttons makes a good companion for carnivorous plants in the bog garden. It has low, unobtrusive leaves and does not rapidly spread. The genus was named in honor of Dr. Moses Marshall (1758–1813), a Pennsylvania botanist.

651. Savanna Eryngo

Eryngium integrifolium Walter
E-rýn-gi-um in-te-gri-fò-li-um
Apiaceae (Carrot Family)

DESCRIPTION: Slender, erect perennial herb; stems 8–20″ tall with ascending branches terminating in a headlike, nearly round inflorescence; basal leaves petiolate, lanceolate to deltoid, 0.8–4″ long, with crenate or serrate margins; flowering heads 0.15–0.35″ long; petals blue; flowers August–October.

RANGE-HABITAT: VA south to FL and west to OK and TX; in SC, common in the Coastal Plain and Sandhills in Longleaf Pine savannas, wet flatwoods, ecotones of pocosins, herbaceous seepages, blackwater swamp forests, and ditches; very rare in the Piedmont in moist pastures and boggy areas.

COMMENTS: The flowers are quite small, but their bright blue coloration attracts notice among the savanna grasses and sedges. This species is considerably smaller than other *Eryngium* species found in the savanna habitat.

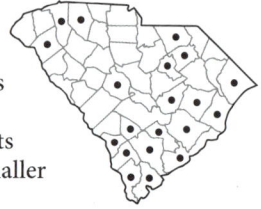

652. Ravenel's Eryngo

Eryngium ravenelii Gray
E-rýn-gi-um ra-ve-nél-i-i
Apiaceae (Carrot Family)

DESCRIPTION: Robust perennial herb, forming clumps, 1–3′ tall; basal leaves narrowly lance-elliptic to nearly linear, with few short, irregular crenate or short-serrate teeth along the margin; stems with ascending branches terminating in flowers that are produced in subglobose to hemispherical heads; flowers light to deep blue at maturity; bractlets subtending the flowers with middle cusp essentially equal in length to the lateral cusps; flowers July–September.

RANGE-HABITAT: A Coastal Plain endemic found from SC south to FL; in SC, very rare and restricted to the Coastal Plain where found in wet savannas and ditches through savannas where coquina limestone (marl) is near the surface and the soil has an elevated calcium level.

SIMILAR SPECIES: Marsh Eryngo (*E. aquaticum* L., plate 841) is similar but has bractlets that subtend the flowers with the middle cusp approximately the same length as the lateral cusps. The styles are short (3–4 mm long) and only about as long as the bractlets, while in Ravenel's Eryngo the style is 4–6 mm long and exceeds the bractlets. Marsh Eryngo also is a biennial or short-lived perennial that rarely forms sizeable clumps and has basal leaves that are typically oblanceolate but vary to nearly linear. Marsh Eryngo is common in the Coastal Plain where it is found in a variety of tidal and nontidal marsh and swamp systems, wet savannas, and ditches.

COMMENTS: Ravenel's Eryngo is named in honor of Henry William Ravenel (1814–87) a prolific plant collector, botanist, and mycologist from Berkeley County, SC. He originally collected this species in Berkeley County and sent it to Asa Gray for identification. The species was known only from Ravenel's nineteenth century collections in SC until rediscovered for the state by Patrick McMillan in 2001 (McMillan, 2003). It is known today in SC from a single location in Charleston County but is much more abundant in FL where marl savannas and marl prairies are more common.

CONSERVATION STATUS: SC-Critically Imperiled

653. Rose-gentian

Sabatia gentianoides Elliott
Sa-bà-ti-a gen-ti-a-noì-des
Gentianaceae (Gentian Family)

DESCRIPTION: Perennial herb without rhizomes; stems strict or branched above, 6–36″ tall; basal leaves, when present, much broader and shorter than the stem leaves; leaves opposite, 20–60x as long as wide; flowers sessile or subsessile, in groups of 1–5; calyx subtended by linear bracts that usually exceed the corolla lobes; petals more than 6 (typically 10), bright pink; flowers July–August.

RANGE-HABITAT: NC south to FL and west to TX; in SC, rare in the Coastal Plain in wet Longleaf Pine savannas, wet flatwoods, and herbaceous seepages.

COMMENTS: This species has a linear bract (reduced leaf) that projects beyond the corolla, a feature that along with the sessile flowers distinguishes this *Sabatia* from other Coastal Plain species.

CONSERVATION STATUS: SC-Imperiled

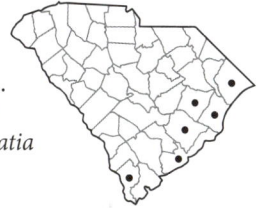

654. Coastal Plain Tickseed

Coreopsis gladiata Walter
Co-re-óp-sis gla-di-à-ta
Asteraceae (Aster Family)

DESCRIPTION: Perennial herb, stem erect to reclining, 20″–36″ long; leaves all alternate, thick and leathery, the lower stem and basal leaves with well-differentiated petioles and blades, smooth to short-pubescent; basal leaves present when flowering; flowers in involucrate heads of dark reddish disk flowers and yellow ray flowers that are conspicuously toothed; flowers August–October.

RANGE-HABITAT: NC south to FL and west to MS; disjunct inland occurrences in the Blue Ridge and Ridge and Valley; in SC, scattered and uncommon in the Coastal Plain, Sandhills, upper Piedmont, and mountains where found in wet, Longleaf Pine savannas—particularly over limestone substrate, ditches through savannas, margins of swamp forests, herbaceous seepages, cataract fens, and seepages on granitic domes.

SIMILAR SPECIES: Pool Coreopsis (*C. falcata* F.E. Boynton, plate 681) is very similar but easily distinguished by flowering May–June, long before Coastal Plain Tickseed. Beadle's Coreopsis (*C. palustris* Sorrie) flowers at approximately the same time as Coastal Plain Tickseed but lacks basal leaves at the time of flowering, has acute-tipped and wider (elliptic) mid-cauline leaves, and grows in the drawdown zones of blackwater rivers and margins of blackwater swamp forests. Savanna Coreopsis (*C. linifolia* Nuttall, plate 655) is a smaller plant, found in slightly drier sites in savannas, with a similar flowering time, but has all but the lowest-most stem leaves all opposite.

COMMENTS: Recent field work has revealed that this species is more widespread and abundant than once thought. It is most abundant in the ditches along marl roads in the Francis Marion National Forest. It is also easily observed at Ashmore and Eva Russell Chandler Heritage Preserves in Greenville County. The specific epithet, *gladiata,* refers to a sword, apparently in reference to the very slender upper stem leaves.

CONSERVATION STATUS: SC-Vulnerable

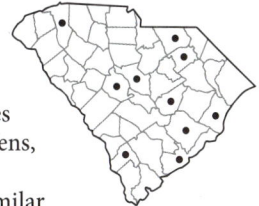

655. Savanna Coreopsis; Savanna Tickseed

Coreopsis linifolia Nuttall
    Co-re-óp-sis li-ni-fò-li-a
Asteraceae (Aster Family)

SYNONYM: *Coreopsis angustifolia* Aiton—RAB

DESCRIPTION: Perennial herb, stem erect to reclining, 12″–36″ long; stem leaves opposite above, lowermost alternate, thick and leathery, the lowest stem and basal leaves with well-differentiated petioles and blades, smooth; upper stem leaves linear and reduced; all leaves thick and leathery with dark glands internally that are highly visible when viewed with light from behind the leaf; flowers in involucrate heads of dark reddish disk flowers and yellow ray flowers that are conspicuously toothed; flowers July–October.

RANGE-HABITAT: Primarily on the Coastal Plain from VA south to FL and west to TX; in SC, common in the Coastal Plain and Sandhills where found in Longleaf Pine savannas and wet flatwoods, ecotones of pocosins, and herbaceous seepages.

SIMILAR SPECIES: See Comments under *C. gladiata* Walter, plate 654.

COMMENTS: This is the most abundant of the *Coreopsis* species in our Longleaf Pine savannas. Before flowering it can easily be distinguished by holding the basal leaves up to the light and looking through the leaf to see the dark glands. Plants in SC are apparently tetraploid (4 sets of chromosomes); those of the Gulf Coastal Plain are diploids.

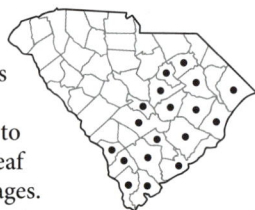

656. Rayless-goldenrod

Bigelowia nudata (Michaux) A.P. de Candolle var. *nudata*
    Bi-ge-lòw-i-a nu-dà-ta
Asteraceae (Aster Family)

SYNONYM: *Chondrophora nudata* (Michaux) Britton—RAB

DESCRIPTION: Smooth, perennial herb with fibrous roots, forming a small clump; stems 12–30″ tall; basal leaves elliptic to oblanceolate; stem leaves, 5–15, reduced upward; flower arranged in involucrate heads of yellow disk flowers only; heads in a corymbose (flat-topped) arrangement; flowers August–October.

RANGE-HABITAT: Endemic to the Coastal Plain from NC south to FL and west to LA; in SC, common in the Coastal Plain and Sandhills in Longleaf Pine savannas and wet flatwoods, ecotones of pocosins, herbaceous seepages, and seasonally wet roadside ditches.

COMMENTS: The common name is quite descriptive. This plant looks like a goldenrod that has a flat-topped arrangement of heads and lacks disk flowers. It is very common in SC Longleaf Pine savannas.

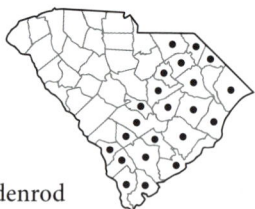

657. Carolina Chaffseed

Carphephorus tomentosus (Michaux)
Torrey & Gray
Car-phé-pho-rus to-men-tò-sus
Asteraceae (Aster Family)

DESCRIPTION: Erect perennial herb, 15–30″ tall; stems purplish, densely pubescent; leaves basal and cauline, elliptic to oblanceolate, pubescent, occasionally sparsely so or becoming smooth with age; basal leaves petiolate, upper stem leaves sessile; flowers arranged in involucrate heads of purple disk flowers only; heads in a corymbose (flat-topped) arrangement; flowers August–October.

RANGE-HABITAT: Endemic to the Coastal Plain from VA south to GA; in SC, common in the Coastal Plain, uncommon in the Sandhills; Longleaf Pine savannas and wet flatwoods, herbaceous seepages, ecotones of pocosins, moist swales in Longleaf Pine sandhills, and moist roadsides.

COMMENTS: This highly attractive species is recognizable, even at a distance, at least in the typical, highly pubescent individuals, by the dull or gray-green color of the entire vegetative portions of the plant. It is only likely to be confused with Sandhills Chaffseed (*C. bellidifolius* (Michaux) Torrey & Gray, plate 484), which is found in much drier sandhill and xeric Longleaf Pine ridges in the Sandhills and Coastal Plain. Sandhills Chaffseed is smooth throughout.

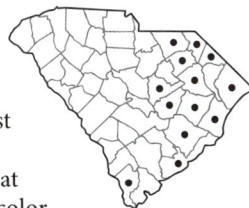

658. Trilisa

Trilisa paniculata (J. F. Gmelin)
Cassini
Trí-lis-a pa-ni-cu-là-ta
Asteraceae (Aster Family)

SYNONYM: *Carphephorus paniculatus*
(J. F. Gmelin) Herbert—PR

DESCRIPTION: Perennial, herb, 15–30″ tall, with a dense basal rosette; stems pubescent; basal leaves narrowly elliptic to oblanceolate; stem leaves reduced upward and lanceolate; flowers arranged in involucrate heads of purple disk flowers only; heads in a paniculate arrangement; flowers September–November.

RANGE-HABITAT: Endemic to the Coastal Plain from NC south to FL and west to AL; in SC, common in the Coastal Plain and Sandhills in Longleaf Pine savannas and wet flatwoods, ecotones of pocosins, herbaceous seepages, and moist road banks.

COMMENTS: One of the most recognizable and attractive late autumn flowers in the Longleaf Pine savannas. The pubescent stem distinguishes it from the much larger Vanilla-leaf (*C. odoratissima* (J. F. Gmelin) Cassini, plate 587). The common name and the genus name may have been an attempt at humor, being an anagram of the similar genus *Liatris*.

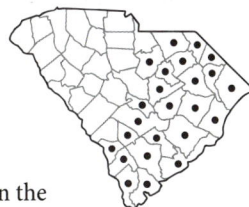

659. Savanna Blazing Star

Liatris spicata (L.) Willdenow
var. *resinosa* (Nuttall) Gaiser
Li-à-tris spi-cà-ta var. re-si-
nò-sa
Asteraceae (Aster Family)

DESCRIPTION: Perennial herb, 3–5′ tall; stem erect, pubescent to glandular pubescent; basal leaves and lower leaves linear to narrowly elliptic, stem leaves abruptly reduced upward and linear; flowers arranged in involucrate heads of pinkish to purple disk flowers only; heads arranged in a spikelike structure; invoucral bracts pink or purplish; flowers August–October.
RANGE-HABITAT: Primarily on the Coastal Plain from NJ south to FL and west to LA; in SC, common in the Coastal Plain and Sandhills in Longleaf Pine savannas and herbaceous seepages.
COMMENTS: This variety is readily distinguishable from its upper Piedmont and mountain cousin, *L. spicata* var. *spicata*. Savanna Blazing Star is generally less robust and with a glandular-pubescent stem and a later flowering season. This distinctive Blazing Star is likely worthy of recognition as a species rather than a variety.

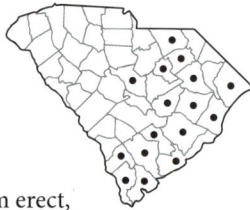

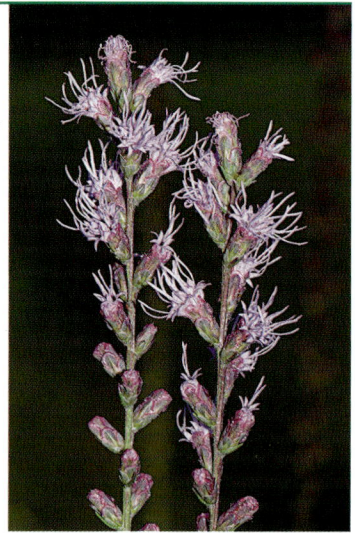

660. Narrowleaf Sunflower; Swamp Sunflower

Helianthus angustifolius L.
He-li-án-thus an-gus-ti-fò-li-us
Asteraceae (Aster Family)

DESCRIPTION: Perennial, herb to 3–7′ tall, without extensive rhizomes; stems rough and pubescent, often many branched in upper half; leaves occurring all along the stem, 10–30x as long as wide, up to 6″ long, dark green and rough above, margins rolled downward; flowers arranged in involucrate heads of yellow ray flowers and dark reddish disk flowers; outer involucral bracts acuminate to acute; flowers July–frost.
RANGE-HABITAT: Primarily a Coastal Plain species from NY south to FL and west to MO and TX; in SC, common in the Coastal Plain, Sandhills, and lower Piedmont, scattered in the upper Piedmont and mountains; Longleaf Pine savannas and flatwoods, ditches, cataract fens, bogs, seepage on granitic flatrocks, swales, and along roadsides.
SIMILAR SPECIES: Muck Sunflower (*Helianthus simulans* E. Watson) is similar but often much larger, greater than 7′ tall, and is aggressively rhizomatous. It is native to the Coastal Plain of SC. Muck Sunflower is often sold as *H. angustifolius* in the horticultural trade. It rapidly spreads in the garden to become quite invasive.
COMMENTS: Narrowleaf Sunflower is highly attractive late in the autumn and makes an excellent pollinator-friendly landscape plant. The problem with including it in the landscape is that most of what is sold as this species in the horticultural trade is the aggressively colonial Muck Sunflower. Buyer beware.

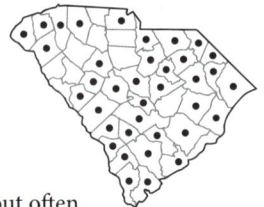

661. Savanna Sunflower

Helianthus heterophyllus Nuttall
He-li-án-thus he-te-ro-phýl-lus
Asteraceae (Aster Family)

DESCRIPTION: Rough pubescent perennial herb, with short rhizomes; 12–36″ tall; leaves primarily basal and on the lowest portions of the stem; flowering stem subscapose, with only a few reduced leaves; basal and low cauline leaves 2–5 times as long as wide, opposite, elliptic; upper leaves lanceolate and alternate; flowers arranged in involucrate heads of yellow ray flowers and dark reddish disk flowers; flowers August–November.

RANGE-HABITAT: Endemic to the Coastal Plain from NC south to FL and west to LA; in SC, common in the Coastal Plain and Sandhills in Longleaf Pine savannas, pocosin ecotones, and herbaceous seepages.

SIMILAR SPECIES: Appalachian Sunflower (*H. atrorubens* L., plate 407) has lower stem and basal leaves that are often less than 2 times as long as wide and rhombic to triangular-ovate. It occurs in much drier habitats throughout SC, where it is common.

COMMENTS: This attractive, late-flowering sunflower makes a good addition to moist soil in the sunny, native garden. It also makes a nice autumn companion planting in carnivorous plant bog gardens.

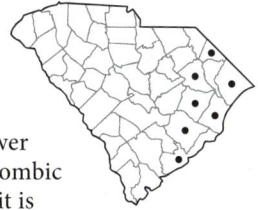

662. Savanna Eupatorium; Justiceweed

Eupatorium leucolepis (A.P. de Candolle) Torrey & Gray
Eu-pa-tò-ri-um leu-có-le-pis
Asteraceae (Aster Family)

DESCRIPTION: Erect, perennial herb from a crown or very short rhizome; stems 2–4′ tall, not branching at the base, flowering stems typically solitary; leaves sessile, opposite, narrow, elliptic to elliptic-oblanceolate, 1–3″ long, 0.23–0.5″ wide, reflexed-spreading, partially longitudinally folded, with inconspicuously serrate and strigose margins, tapered bases, and acute to obtuse tips; flowers produced in involucrate heads composed of cream-colored disk flowers only; outer involucral bracts acuminate to attenuate, pubescent; flowering August–October.

RANGE-HABITAT: NY south to FL and west to LA; in SC, common in the Coastal Plain and Sandhills in Longleaf Pine savannas and wet flatwoods, upper margins of depression ponds, Pond Cypress savannas, ecotones of pocosins, herbaceous seepages, and moist roadside ditches.

SIMILAR SPECIES: Mohr's Eupatorium (*E. mohrii* Greene, plate 589) and Recurved Eupatorium (*E. recurvans* Small, plate 589) are both quite similar to Savanna Eupatorium. The involucral bracts on these two species are acute to obtuse rather than acuminate to attenuate as in *E. leucolepis*. The leaves of the Mohr's and Recurved Eupatorium are not partially longitudinally folded, and the leaf arrangement toward the tip of the stem is subopposite or alternate rather than opposite as in *E. leucolepis*. The three species have similar ranges and habitats.

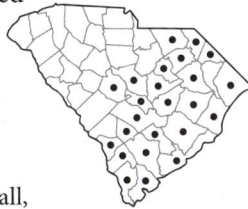

663. Common Roundleaf Eupatorium

Eupatorium rotundifolium L.
Eu-pa-tò-ri-um ro-tun-di-fò-
li-um
Asteraceae (Aster Family)

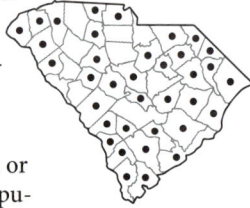

DESCRIPTION: Perennial herb from a crown or
short rhizomes; stem 15–48″ tall, densely pubescent throughout; leaves opposite, nearly sessile
to sessile, rhombic-ovate with crenate to short-serrate margins;
flowers arranged in involucrate heads of disk flowers only; corolla
white; heads in a corymbose (flat-topped) arrangement; flowers
August–October.

RANGE-HABITAT: MA south to FL and west to OK and TX; in SC,
common in the Coastal Plain and Sandhills, scattered in the Piedmont and mountains; Longleaf Pine savannas, flatwoods, ecotones
of pocosins, herbaceous seepages, cataract fens, seepage over
granitic flatrocks, ditches, and moist roadsides.

SIMILAR SPECIES: Several similar species are found in SC, but Common Roundleaf Eupatorium has the widest leaves. Ragged Eupatorium (*E. pilosum* Walter)
has narrower, often broad-lanceolate leaves and has branching of the upper stem alternate,
rather than opposite, as in Common Roundleaf Eupatorium.

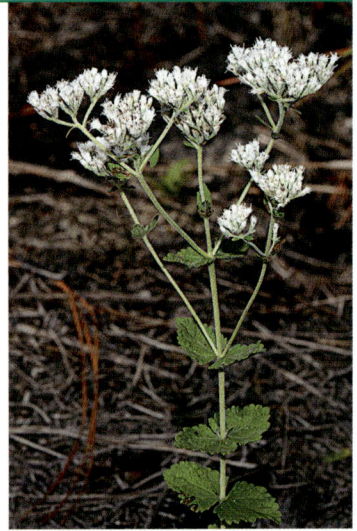

664. Wand Goldenrod

Solidago virgata Michaux
So-li-dà-go vir-gà-ta
Asteraceae (Aster Family)

SYNONYM: *Solidago stricta* Aiton—
RAB

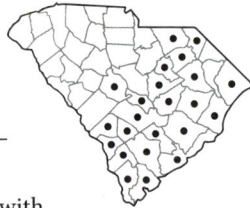

DESCRIPTION: Perennial rhizomatous herb with
erect stems to 6′ tall; leaves basal and cauline; lower leaves and
basal leaves elliptic, smooth, margins crenate; upper leaves
reduced and appressed to the stem with margins entire; flowers
arranged into involucrate heads composed of yellow ray flowers
and yellow disk flowers; heads arranged in a narrow, spikelike, or
racemelike panicle with branches stiffly erect, the entire structure
wandlike; flowers September–October.

RANGE-HABITAT: NC south to FL and west to TX; also in the
Neotropics; in SC, common in the Coastal Plain and Sandhills in
Longleaf Pine savannas, upper edges of depression ponds, pocosin
ecotones, herbaceous seepages, and moist roadsides.

SIMILAR SPECIES: Southern Bog Goldenrod (*S. gracillima* Torrey & Gray) is similar but can
be distinguished by having scabrous leaf margins on the upper leaves and branches that are
more spreading, though the overall effect is still a narrow structure. It is found in Longleaf
Pine savannas and herbaceous seepages in the Sandhills and Coastal Plain. Beautiful Goldenrod (*S. pulchra* Small) is also similar. It is smaller, seldom reaching 3′ tall, and has margins of
all leaves, including the basal leaves, without serrations and not scabrous. It is very rare in SC,
being found in Longleaf Pine savannas and herbaceous seepages in Horry and Chesterfield
Counties.

COMMENTS: The stiffly erect branches make this species appear like golden wands in the
savanna in late autumn, hence the common name. It makes a good ornamental in a sunny
native garden, if the soil is not too rich. Rich soil results in overly lanky and large plants that
tend to flop over and become unsightly.

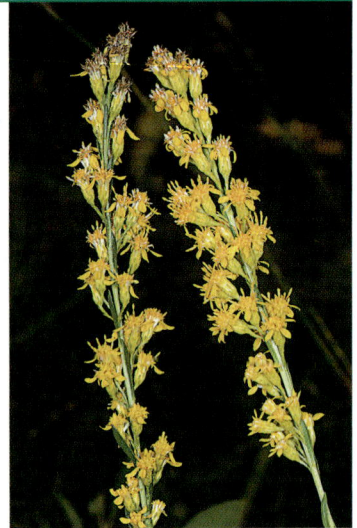

665. Savanna Grass-leaved Aster

Eurybia paludosa (Aiton) G. L. Nesom
Eu-rý-bi-a pa-lu-dò-sa
Asteraceae (Aster Family)

SYNONYM: *Aster paludosus* Aiton—RAB

DESCRIPTION: Rhizomatous, perennial, smooth or sparsely pubescent herb, forming loose colonies; basal and cauline leaves linear to narrowly elliptic, 0.2–0.6″ wide; leaves gradually reduced upward, with only the midvein evident; flowers arranged in involucrate heads composed of 15–35 bluish ray flowers and yellow disk flowers; flowers July–October.

RANGE-HABITAT: Endemic to the Coastal Plain from NC south to ne. FL; in SC, common in the Coastal Plain and Sandhills in Longleaf Pine savannas, pocosin ecotones, and herbaceous seepages.

COMMENTS: This species is distinctive among "asters" in the savanna habitat because of the long, narrow basal and stem leaves, large inflorescences, and rhizomatous habit. Only Slender Aster (*E. compacta* G. L. Nesom) is similar. Slender Aster has only 9–14 ray flowers that are 0.2–0.6″ long, whereas Savanna Grass-leaved Aster has 15–35 ray flowers that are 0.4–1.0″ long, resulting in much larger and showier involucrate heads. Many species formerly in the genus *Aster* have now been moved to new genera because the genus *Aster* is based on an Old World species and the New World "asters" are not as closely related to the Old World species as they are to other genera.

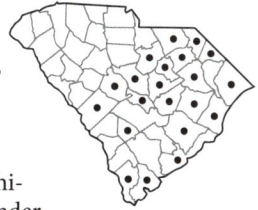

666. Scale-leaf Gerardia

Agalinis aphylla (Nuttall)
Rafinesque
A-ga-lì-nis a-phýl-la
Orobanchaceae (Broomrape
Family)

DESCRIPTION: Annual herb 12–30″ tall; stems smooth, with few branches; leaves absent or inconspicuous and scalelike; flowers in terminal racemes, held on short pedicels, corolla 0.5–0.8″ long, brilliant purplish-pink with yellow lines and purplish spots inside the corolla tube; flowers September–November.

RANGE-HABITAT: Endemic to the Coastal Plain from NC to FL and west to LA; in SC, very rare in the Coastal Plain where it is found in wet Longleaf Pine savannas and flatwoods and edges of Pond Cypress-gum ponds and depressions.

COMMENTS: The scalelike leaves distinguish this species from all other species of *Agalinis* in SC. The specific epithet *aphylla* comes from the Greek *a* meaning "without" and *phylla* meaning "leaves." This plant has declined throughout its range and is now very rare; it is known from a single extant population. The last recorded observation was in the early 2000s in Berkeley County. Any suspected populations should be photographed and the record sent to the South Carolina Heritage Trust Program.

CONSERVATION STATUS: SC-Critically Imperiled

667. Carolina Grass-of-Parnassus;
Eyebright

Parnassia caroliniana Michaux
Par-nás-si-a ca-ro-li-ni-à-na
Parnassiaceae (Grass-of-Parnassus
Family)

DESCRIPTION: Smooth, perennial herb with rhizomes; leaves primarily basal, usually ovate, with long leaf stalks; stem leaves similar to basal but smaller and heart-shaped; flowers solitary on elongated stalks, 8–20″ tall; petals 5, white with conspicuous green veins; stamens 5, shorter than the 5 staminodia, which bear glands; ovary whitish; flowers September–November.

RANGE-HABITAT: Endemic to the Coastal Plain of NC, SC, and the FL Panhandle; in SC, rare in wet Longleaf Pine or Pond Cypress savannas, often over calcareous substrates.

COMMENTS: Grass-of-Parnassus is one of the most striking flowers of the wet savannas of the coastal area and is rare in SC. Throughout its range, its distribution is fragmented and disjunct. Much of its habitat has been altered by agriculture, conversion to tree farms, fire suppression, and ditching. It is very similar to Big-leaf Grass-of-Parnassus (*P. grandifolia* A. P. de Candolle, plate 67), which is confined to the mountains in SC.

CONSERVATION STATUS: SC-Imperiled

668. Walter's Aster

Symphyotrichum walteri (Alexander)
G. L. Nesom
Sym-phy-o-trì-chum wál-ter-i
Asteraceae (Aster Family)

SYNONYM: *Aster squarrosus* Walter—
RAB

DESCRIPTION: Perennial, herb, 8–20″ tall, without basal leaves at the time of flowering; stem erect to sprawling, branched; leaves ovate to lanceolate, tiny, 0.08–0.5″ long, with the tips reflexed; flowers arranged in involucrate heads of blue ray flowers and yellow disk flowers; heads in paniclelike arrangements at the ends of branches; flowers October–December.

RANGE-HABITAT: Endemic to the Coastal Plain from NC south to FL; in SC, common in the Coastal Plain and Sandhills where found in Longleaf Pine savannas, flatwoods, and mesic areas within Longleaf Pine sandhills.

COMMENTS: This is one of the most distinct "asters" found in SC. The bizarre, reflexed, and tiny leaves give the impression of a spiny stick. This strange sticklike, branched plant suddenly bursts into flower with large, showy blue and yellow inflorescences very late in the autumn—a very welcome surprise. The specific epithet is in honor of Thomas Walter (ca. 1740–89), a British born American planter and naturalist who lived just south of the Santee River in Berkeley County, SC. In 1788 he published *Flora Caroliniana*.

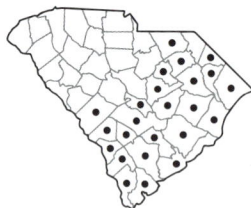

669. Slender Rattlesnake-root

Nabalus autumnalis (Walter)
　Weakley
　　Ná-ba-lus au-tum-nà-lis
Asteraceae (Aster Family)

SYNONYM: *Prenanthes autumnalis*
　Walter—RAB, PR

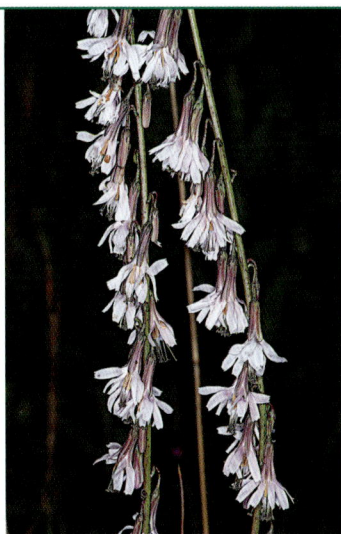

DESCRIPTION: Slightly whitish, smooth, perennial herb with milky sap; plants 12–48″ tall; leaves pinnately lobed, alternate, principal ones on the lower part of the stem and abruptly reduced upward; heads borne on a narrow, terminal spikelike panicle; flowers arranged in nodding involucrate heads composed of only pink 5-lobed ray flowers; flowers late September–December.

RANGE-HABITAT: Endemic to the Coastal Plain from NJ south to GA; in SC, common in the Coastal Plain in Longleaf Pine savannas, wet flatwoods, and pocosin ecotones.

COMMENTS: All species of *Nabalus* are recognizable by the nodding flower heads. The arrangement of flower heads in a narrow, spikelike panicle is distinctive to this species. This is our showiest species of *Nabalus*.

670. Catesby's Gentian;
　Coastal Plain Gentian

Gentiana catesbyi Walter
　Gen-ti-à-na càtes-by-i
Gentianaceae (Gentian Family)

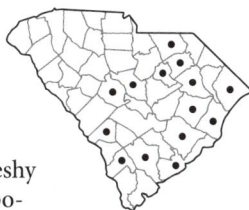

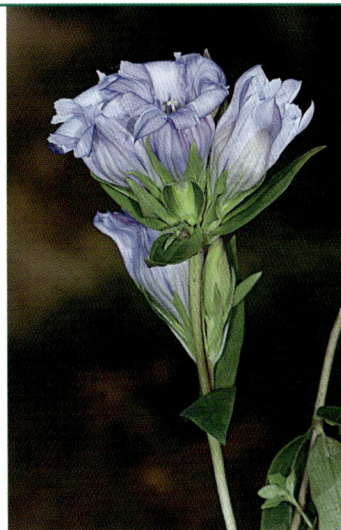

DESCRIPTION: Perennial herb with thick, fleshy roots; stems rough, 12–32″ tall; leaves opposite, entire, ovate; flowers sessile, 1–9, in compact cymes; corolla blue to violet, or whitish, with plaits in the sinuses, narrower and shorter than the lobes; corolla lobes spreading; flowers late September–November.

RANGE-HABITAT: From NJ south to FL; in SC, occasional in the Coastal Plain and Sandhills; Longleaf Pine savannas, pocosins, and edges of moist hardwood forests.

SIMILAR SPECIES: Catesby's Gentian is similar to Soapwort Gentian (*Gentiana saponaria* L., plate 194), which is found scattered throughout SC. Catesby's Gentian has bright green, ovate leaves widest near the base, and calyx lobes longer than the calyx tube and spreading. Soapwort Gentian has dark green, linear to elliptic leaves widest near the middle and calyx lobes shorter than or about equal to the calyx tube. The specific epithet honors Mark Catesby (1683–1749), English naturalist who visited SC (1722–24) and published *The Natural History of Carolina, Florida and the Bahama Islands*.

671. Foxtail Clubmoss

Lycopodiella alopecuroides
  (L.) Cranfill
    Ly-co-po-di-él-la a-lo-pe-cu-
    roì-des
Lycopodiaceae (Clubmoss Family)

SYNONYM: *Lycopodium alopecuroides*
  L.—RAB, PR

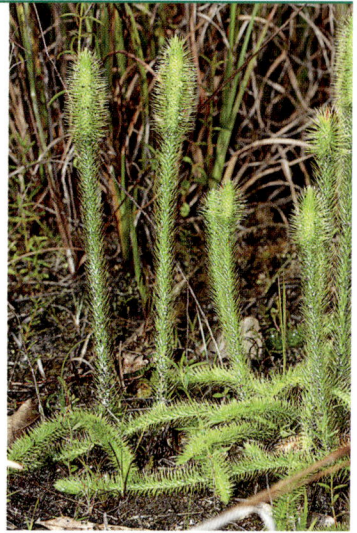

DESCRIPTION: Stems arching and rooting at the tips; leaves many-ranked, narrow, awl-shaped, with sharp, distinct, divergent teeth; spores produced in bushy strobili (cones), on upright stalks with spreading leaves, arising from the upper surface of the stems; spores mature July–September.

RANGE-HABITAT: MA south to FL and west to TX; in SC, common in the Coastal Plain and Sandhills, rare in the Piedmont; Longleaf Pine savannas, ecotones of pocosins, herbaceous seepages, ditches, and wet roadsides.

SIMILAR SPECIES: Southern Bog Clubmoss (*Lycopodiella prostrata* (Chapman) Cranfill) is a very similar species with narrower stems; it grows flat on the ground, rooting all along its length. It is found in similar habitats in the Coastal Plain and Sandhills.

COMMENTS: Clubmosses are primitive vascular plants that reproduce by spores. The spores of clubmosses were involved in a number of fields of early commerce. Their small and uniform size made them ideal for microscopic measurements. Because they are water-repellent and dustlike, the spores were used as soothing powders for chafes and wounds. They were also used for fireworks and for photographic flashes, because their rapid ignition gives off a flash. They are still sometimes used as a lubricant.

Clubmoss stems are all "annual" in that the current year's growth dies back (or really dies forward) toward a perennial stem tip that roots into the ground and resumes growth with the next growing season.

672. Pond Cypress savanna

673. Depression meadow

674. Limesink pond

675. Pond Cypress

Taxodium ascendens Brongniart
Tax-ò-di-um as-cén-dens
Cupressaceae (Cypress Family)

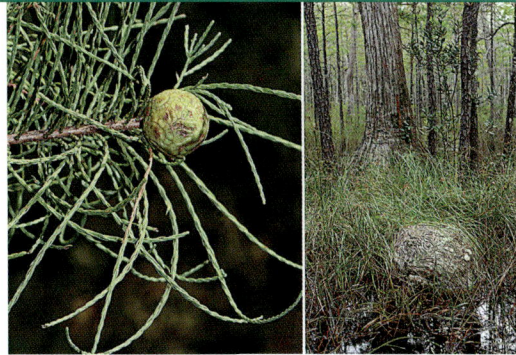

DESCRIPTION: Medium-sized, deciduous tree with rounded "knees" that are mostly less than 15″ tall, moundlike, with rounded, blunt tops; bark thick and spongy, dark-brown, not exfoliating; leafy branchlets ascending; leaves needlelike, pressed against the ascending twig, except on seedlings and fast growing shoots where they are 2-ranked; male and female cones on the same tree.

RANGE-HABITAT: Endemic to the Coastal Plain from NC south to FL and west to LA; in SC, common in the Sandhills and Coastal Plain where found in isolated depression ponds, non-alluvial swamp forests, blackwater rivers, and shores of natural blackwater lakes.

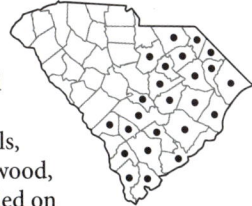

COMMENTS: When Pond Cypress grows in conditions of fluctuating water levels, the base tends to grow larger, forming a buttress of lighter and more porous wood, and the roots send up shoots ("knees"). The knees of Pond Cypress are rounded on top; the knees of Bald Cypress (*Taxodium distichum,* plate 769) are pointed on top. Pond Cypress is adapted to frequent, low-intensity fires and grows primarily in isolated wetlands where it forms the dominant canopy in Pond Cypress savanna habitats when burned regularly. In these habitats, trees are often very stunted and grow very slowly. A 20-foot-tall tree may be hundreds of years old.

Pond Cypress wood contains an essential oil that gives it natural durability and makes it valuable for fence posts, shingles, and paneling. Few merchantable stands exist today because it is very slow-growing, and reproduction has not kept up with lumbering. It is often used as an ornamental since it grows well in upland habitats under cultivation.

676. Corkwood; Water Toothleaf

Stillingia aquatica Chapman
Stil-líng-i-a a-quá-ti-ca
Euphorbiaceae (Spurge Family)

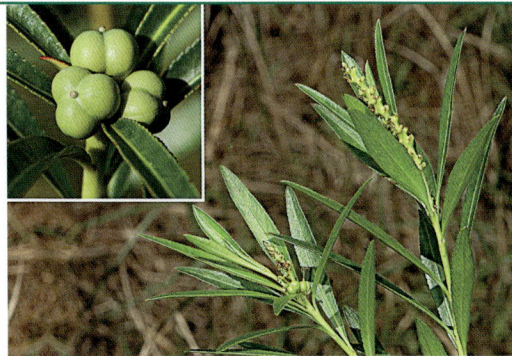

DESCRIPTION: Short, semi-woody to woody shrub, with a solitary, often corky, stem to 3′ tall from a tap root; stems with few, alternate branches; leaves linear-lanceolate 1–2.5″ long and less than 0.4″ wide, minutely serrate; flowers produced in terminal yellowish spikes; fruit a 3-lobed capsule; flowers March–October; fruits mature June–October.

RANGE-HABITAT: Endemic to the Coastal Plain from SC south to FL and west to AL; in SC, rare in the Coastal Plain where found in open, depression pond habitats such as Pond Cypress savannas and roadside ditches through such habitats.

COMMENTS: Where present, this distinctive species often forms large colonies. It is more common farther south and restricted to the southeastern Coastal Plain of SC, mostly along the Savannah River corridor. The genus honors Benjamin Stillingfleet (1702–71), a British botanist.

CONSERVATION STATUS: SC-Imperiled

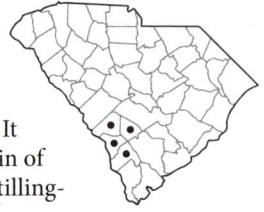

677. Peelbark St. John's-wort

Hypericum fasciculatum Lamarck
Hy-pé-ri-cum fas-ci-cu-là-tum
Hypericaceae (St. John's-wort Family)

DESCRIPTION: Erect shrub, 2.5–5′ tall, many-branched above and spongy and thickened below; bark peeling in thin sheets; leaves linear and needlelike, 0.2–.6″ long; inflorescence produced toward ends of branches, short, 1–3 internodes; petals 5, yellow; fruit a 3-locular capsule; flowers May–September; fruits mature August–October.

RANGE-HABITAT: Endemic to the Coastal Plain from NC south to FL and west to LA; in SC, occasional in the Coastal Plain and Sandhills in Pond Cypress savannas, upper margins of depression ponds, wet Longleaf Pine flatwoods, herbaceous seepages, wet powerline clearings, and ditches.

SIMILAR SPECIES: Bedstraw St. John's-wort (*H. gallioides* Lamarck) is distinguished by having larger oblanceolate to linear-oblanceolate leaves that are up to 0.2″ wide; its leaves are not needlelike and it is a less robust shrub. Carolina St. John's-wort (*H. nitidum* Lamarck) does have needlelike leaves but has tight bark that does not readily exfoliate, has more elongate inflorescences produced in 3–7 internodes, and is found mostly along the margins of flowing blackwater streams.

COMMENTS: This is one of the largest and most charismatic of the St. John's-worts typical of fire-maintained wetland habitats of the Coastal Plain. A beautiful display can be observed in several of the intact Pond Cypress savannas located along Halfway Creek Road in the Francis Marion National Forest.

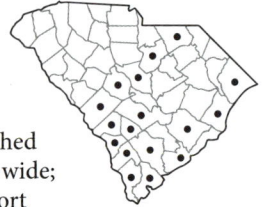

678. Yellow Pitcherplant; Trumpets

Sarracenia flava L.
Sar-ra-cèn-i-a flà-va
Sarraceniaceae (Pitcherplant Family)

DESCRIPTION: Perennial, carnivorous herb from rhizomes; 1.5–3.5′ tall; leaves modified into hollow tubes (pitchers) as passive traps to catch insects; flowers develop in March–April before or with the new leaves; flowers with yellow dangling petals and a strong, unusual scent that is unpleasant to some; leaves die back during the fall and winter.

RANGE-HABITAT: In the Coastal Plain and isolated Piedmont locations from VA south to FL and west to MS; in SC, occasional, but still locally common in the Coastal Plain and Sandhills where found in Longleaf Pine savannas, Pond Cypress savannas, ecotones of pocosins, openings within pocosins, boggy margins of Sandhills ponds, and herbaceous seepages.

COMMENTS: This pitcherplant is becoming less common due to habitat destruction; good populations still exist in the Francis Marion National Forest and other protected lands such as the Carolina Sandhills National Wildlife Refuge, Santee Coastal Reserve, and Shealy's Pond Heritage Preserve.

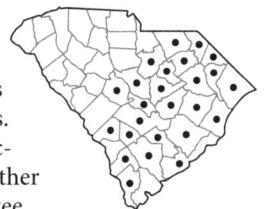

The leaves (pitchers) occur in four color forms: (1) pale green to bright yellow in full sun, with a large maroon splotch on the inside of the column from which red veins radiate; (2) bright to deep red color on the external surface of the lid and column, with a weak maroon spot; (3) uniformly golden-yellow in full sunlight, with coarse and prominent veins all over, with the interior column spot weak; and (4) no red pigment at all, the mature pitchers being pale green to yellow. The different forms often grow mixed in the same site. The nectar that is produced at the top of the inside of the trap has recently been shown to produce the toxic alkaloid coniine, which is also found in Poison Hemlock, which may aid in capturing the unlucky insects attracted to the pitcher.

679. Southeastern Sneezeweed

Helenium pinnatifidum (Nuttall)
Rydberg
He-lén-i-um pin-na-tí-fi-dum
Asteraceae (Aster Family)

DESCRIPTION: Fibrous-rooted, erect, perennial herb, 8–40″ tall; commonly with only 1 flowering head; basal leaves tufted and persistent; stem leaves few and reduced upward; mid-stem leaves barely decurrent (extending down the stem from the point of blade attachment); peduncles (stalks of the inflorescence) pubescent; flowers produced in solitary involucral heads composed of yellow disk and yellow ray flowers; flowers April–May.

RANGE-HABITAT: Endemic to the Coastal Plain from NC south to FL and west to AL; in SC, local and uncommon in the Coastal Plain where found in Pond Cypress savannas, depression meadows, wet Longleaf Pine savannas, and ditches through these habitats.

SIMILAR SPECIES: Savanna Sneezeweed (*Helenium vernale* Walter) is very similar and often confused with Southeastern Sneezeweed. It may be distinguished by the smooth peduncle and leaves that are 4–6 times as long as wide and decurrent for more than 0.8″ down the stem. It is found in similar habitats throughout the Coastal Plain.

CONSERVATION STATUS: SC-Imperiled

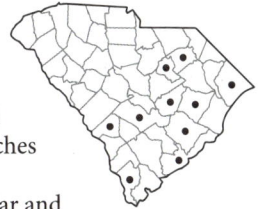

680. Soft-headed Pipewort

Eriocaulon compressum Lamarck
E-ri-o-caù-lon com-prés-sum
Eriocaulaceae (Pipewort Family)

DESCRIPTION: Perennial herb, forming dense basal rosettes; flowering stems 8–28″ tall, scapose (leafless) with a sheath at the base that is longer than the basal leaves; basal leaves pale green, linear-attenuate (gradually narrowed to a long, tapered tip), 2–12″ long; flowers produced in dorsiventrally flattened chalky-white heads, 0.4–0.8″ wide; flowers minute, densely arranged in the heads; flowers April–June, rarely later.

RANGE-HABITAT: NJ south to FL and west to TX, primarily on the Coastal Plain; in SC, common in the Coastal Plain in Pond Cypress savannas, depression meadows, Pond Cypress-Swamp Gum wetlands, wet ditches, and borrow pits.

COMMENTS: Soft-headed Pipewort is the common species in portions of depression ponds that are flooded in the spring. The leaves are generally submerged or still have their bases in water during flowering. The flowering heads are soft and easily compressed between the fingers, distinguishing it from Ten-angled Pipewort (*Eriocaulon decangulare,* plate 695), which occupies areas that are less wet.

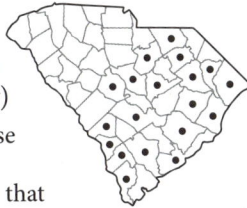

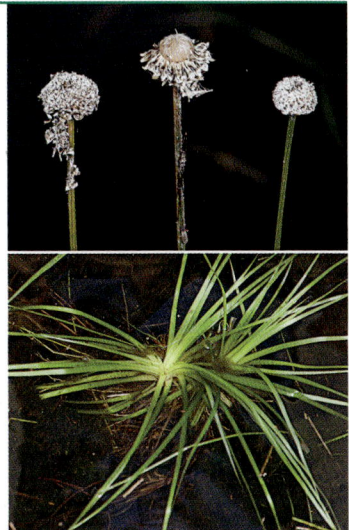

681. Pool Coreopsis

Coreopsis falcata Boynton
Co-re-óp-sis fal-cà-ta
Asteraceae (Aster Family)

DESCRIPTION: Smooth, perennial herb, 25–80″ tall; leaves all alternate, entire, up to 8″ long, the lower sometimes with a pair of narrow lobes at the base of the blade; basal leaves present at flowering; flowers arranged in involucrate heads of dark purplish-red disk flowers and yellow ray flowers that are conspicuously lobed; flowers May–July.

RANGE-HABITAT: Endemic to the Coastal Plain from VA south to GA; in SC, common in the Coastal Plain where found in Pond Cypress savannas, upper margins of depression ponds, wet Longleaf Pine savannas, ditches, and borrow pits.

SIMILAR SPECIES: See discussion under Coastal Plain Coreopsis (*C. gladiata* Walter, plate 654).

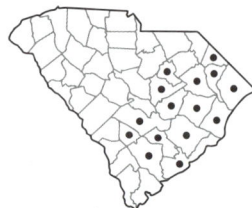

682. Broadleaf Whitetop Sedge

Rhynchospora latifolia (Baldwin ex Elliott) Thomas
Rhyn-chós-po-ra la-ti-fò-li-a
Cyperaceae (Sedge Family)

SYNONYM: *Dichromena latifolia* Baldwin ex Elliott—RAB, PR

DESCRIPTION: Perennial herb from elongate rhizomes; flowering stems solitary, usually 10–27″ tall; leaves appearing mostly basal; inflorescence of crowded clusters of spikes subtended by 6–10 widely linear to lanceolate bracts; bracts conspicuously white at the base, tapering gradually to green at the apex—the white portion 0.9–2.2″ long; fruit an achene with a short, broad tubercle that is decurrent on the achene; flowers May–September.

RANGE-HABITAT: Endemic to the Coastal Plain from VA south to FL and west to TX; in SC, common in the Coastal Plain in Pond Cypress savannas, depression meadows, wet Longleaf Pine savannas, ditches, and borrow pits.

SIMILAR SPECIES: Narrowleaf Whitetop Sedge (*R. colorata* (L.) H. Pfeiffer) is a much smaller plant that has 3–6 shorter bracts with much less extensive white at the base, quickly and abruptly changing to green toward the tip, and the white portion is 0.35–0.9″ long. Narrowleaf Whitetop Sedge is almost never found in the same habitats and is much more weedy. It is found in ditches, particularly along marl gravel roads, interdune wetland swales along the coast, and disturbed areas such as pond margins, deep ditches, and borrow pits.

COMMENTS: Broadleaf Whitetop Sedge forms spectacular masses in our wetland savannas. It is seldom cultivated, but its relative, Narrowleaf Whitetop Sedge, is widely cultivated as an ornamental pond plant. It is frequently mislabeled as *R. latifolia* in the horticulture trade.

683. Savanna Obedient-plant

Physostegia purpurea (Walter) Blake
　　Phy-sos-té-gi-a pur-pú-re-a
　　Lamiaceae (Mint Family)

SYNONYM: *Dracocephalum purpureum*
(Walter) McClintock—RAB

DESCRIPTION: Perennial herb with erect stem,
16–40″ tall; leaves opposite, produced in 6–12
pairs below the inflorescence, nearly sessile or
sessile, oblanceolate, base cuneate to rounded,
obtuse to acute at the tip, reduced upward with
those immediately below the inflorescence hardly
larger than the bracts; inflorescence a bracteate raceme; corolla pink to light
purple; flowers May–July.

RANGE-HABITAT: Endemic to the Coastal Plain from NC south to FL; in SC,
occasional in the Coastal Plain in wet Longleaf Pine savannas, Pond Cypress
savannas, savanna-swamp ecotones, and ditches through these habitats.

SIMILAR SPECIES: Two other species may be encountered in the Coastal Plain. Tidal
Marsh Obedient-plant (*P. leptophylla* Small, plate 836), as the name suggests, is found
in freshwater tidal marshes and margins of tidal swamp forests. It has leaves that are widest
below the middle of the leaf, frequently has leaves with petioles more than 0.8″ long, and
is a coarser plant. Northern Obedient-plant (*P. virginiana* (L.) Bentham, plate 1008) forms
large colonies by many slender, pale secondary rhizomes and also has leaves widest below the
middle. It has escaped in ruderal habitats throughout SC and is native farther north.

COMMENTS: The common name, Obedient-plant, comes from the fact that the flowers tend to
briefly stay in a new position when manipulated.

CONSERVATION STATUS: SC-Critically Imperiled

684. Tall Milkwort

Polygala cymosa Walter
　　Po-lý-ga-la cy-mò-sa
　　Polygalaceae (Milkwort Family)

DESCRIPTION: Biennial herb with a solitary,
smooth, flowering stem, 16–48″ tall; leaves present at flowering, mostly basal, linear to lanceolate, 15–20 times as long as wide; stem leaves
rapidly reduced upward, linear-subulate and becoming bractlike toward the inflorescence; flowers in spikelike racemes arranged in a flat-topped
corymb; petals brilliant yellow; flowers May–July.

RANGE-HABITAT: Endemic to the Coastal Plain from DE south to FL and west
to LA; in SC, common in the Coastal Plain in wet Longleaf Pine savannas,
Pond Cypress savannas, shallow waters of Pond Cypress and Swamp Gum
upland swamps, borrow pits, and wet ditches.

COMMENTS: This is the only yellow-flowered milkwort that grows more than
2′ tall. Short Pinebarren Milkwort (*P. ramosa* Elliott) looks like a stunted form of
this species, has several stems from the base and less reduced stem leaves, and occurs
in drier portions of savannas and flatwoods.

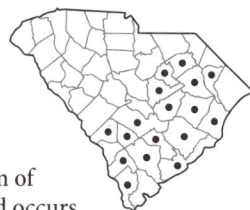

685. Lace-lip Ladies'-tresses

Spiranthes laciniata
(Small) Ames
Spi-rán-thes la-ci-ni-à-ta
Orchidaceae (Orchid Family)

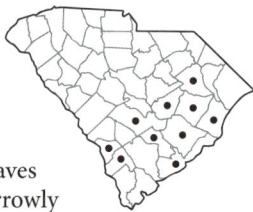

DESCRIPTION: Perennial herb, 8–40″ tall; leaves basal and on stem, linear-lanceolate to narrowly oblanceolate; flowers in a single spike with a pubescent rachis (some of the trichomes glandular), strongly spiraled, or with all flowers arranged on one side (secund); flowers cream to white, tubular at the base and gaping open at the apex; the lip petal with a distinct yellowish blotch; flowers May–July.

RANGE-HABITAT: Endemic to the Coastal Plain from NJ south to FL and west to TX; in SC, rare in the Coastal Plain where found in Pond Cypress savannas, margins of depression ponds, and depression meadows.

COMMENTS: This rare ladies'-tresses is often overlooked due to the short bloom period, as plants frequently are in flower in any one population for only about 2 weeks. It is frequently associated with other rare members of this habitat such as Awned Meadow-beauty and Boykin's Lobelia. It is similar to Spring Ladies'-tresses (*S. vernalis* Engelmann & Gray), which does not have any glandular hairs on the rachis of the spike. Spring Ladies'-tresses also flowers a couple weeks earlier and grows in savannas, flatwoods, and road banks.

CONSERVATION STATUS: SC-Imperiled

686. Bay Blue-flag Iris

Iris tridentata Pursh
Ì-ris tri-den-tà-ta
Iridaceae (Iris Family)

DESCRIPTION: Perennial herb from slender rhizomes with flowering stems 10–28″ tall, usually unbranched; leaves conduplicate (folded lengthwise to have only one surface exposed), 6–16″ long, 0.1–0.4″ wide, smooth; flowers produced at the ends of stems, subtended by bracts; sepals 3, showy, violet with a yellow or white blotch at the base, petals 3, much shorter than the sepals and often nearly hidden, much smaller than the style, which has lobes that are often mistaken for petals; fruit an angled capsule 1–1.5″ long; flowers late May–June; fruits mature August–October.

RANGE-HABITAT: Endemic to the Coastal Plain from NC south to FL and west to AL; in SC, common in Pond Cypress savannas, depression meadows, upper edges of depression ponds, ditches, and borrow pits.

COMMENTS: This species is readily distinguished from other tall, blue-flowered *Iris* by the flower that looks flat, as if missing its petals, which are barely as long as the claws (stalks) of the sepals and often hidden. This is generally the only *Iris* found in its natural Pond Cypress savanna habitat. Southern Blue Flag (*I. virginica* L., plate 779) is a larger plant with upright, showy petals and grows in swamps and marshes throughout the state.

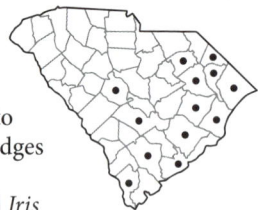

687. Boykin's Lobelia

Lobelia boykinii Torrey & Gray
Lo-bèl-i-a boy-kín-i-i
Campanulaceae (Bellwort
Family)

DESCRIPTION: Smooth, perennial herb from rhizomes; stem 20–32″ tall, the base generally spongy and thickened; stem leaves very narrow, filiform; corolla blue to white; flowers produced in a raceme, with individual flowers generally directed toward one side; bracts present but smaller bracteoles subtending these absent; corolla tubular at the base, with 2 upright narrow lobes, three wider lower lobes, blue to light violet with white toward the throat; flowers May–June.
RANGE-HABITAT: Endemic to the Coastal Plain from NJ south to FL and west to MS; in SC, rare in the Coastal Plain in Pond Cypress savannas, depression meadows, and upper margins of depression ponds.
SIMILAR SPECIES: The three species of *Lobelia* that occur in the Coastal Plain in similar and overlapping habitats can be difficult to separate. Only Boykin's Lobelia is rhizomatous, but this identification tool requires uprooting a very rare plant and should not be done. Both of the similar species have flowers subtended by bracts and bracteoles, and Boykin's Lobelia has bracts only. Nuttall's Lobelia (*L. nuttallii* Roemer & J. A. Schultes) is abundant in moist, sunny habitats in the Coastal Plain and has similar flowers with white toward the throat of the corolla. It lacks rhizomes and has very narrow stems and leaves that are linear; it never has spongy-thickened bases to the stem. It is, overall, a less showy plant with smaller flowers than Boykin's Lobelia. Canby's Lobelia (*L. canbyi* Gray) has a corolla without white toward the throat, often has minute pubescence on the stems, and flowers during the late summer and autumn.
COMMENTS: Boykin's Lobelia has always been considered rare throughout its range. It is frequently overlooked. Field work during the 1990s documented numerous additional locations. Unfortunately, this species appears to be declining at an alarming rate, and many of the known locations have disappeared. Habitat loss due to drainage, conversion and fire suppression are the primary threats to this species. The genus is named in honor of Flemish botanist Matthias de Lobel (1538–1616). The specific epithet honors its discoverer Samuel Boykin (1786–1848), a GA botanist.
CONSERVATION STATUS: SC-Vulnerable

688. Sclerolepis

Sclerolepis uniflora (Walter)
Britton, Sterns & Poggenburg
Scle-ró-le-pis u-ni-flò-ra
Asteraceae (Aster Family)

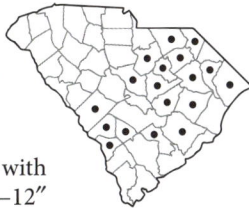

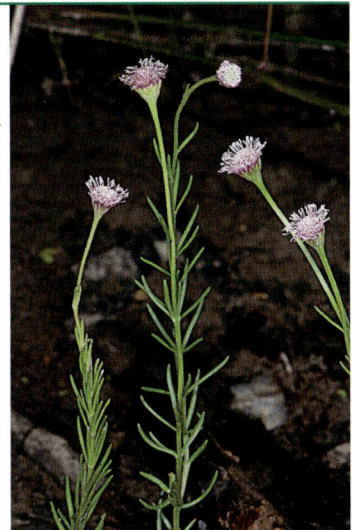

DESCRIPTION: Mat-forming perennial herb with slender rhizomes; flowering stems erect, 4–12″ tall (in water to 24″ tall); leaves linear, produced in whorls of 3–5; flowers arranged in involucrate heads consisting of disk flowers only, which are pink to lavender or rarely white; heads are solitary at the apex of the stem; flowers May–August.
RANGE-HABITAT: NH south to FL and west to MS; in SC, scattered and rare in the Coastal Plain and Sandhills where found in depression meadows, Pond Cypress savannas, other depression ponds, ditches, and borrow pits.

COMMENTS: There is only one species in the genus *Sclerolepis*. This odd plant is rare throughout most of its range and habitat alteration including drainage, conversion, and fire suppression has dramatically reduced its populations in SC. Many of the counties listed in the range here are probably only historical today, as fire-maintained wetland habitats in SC continue to be degraded.

689. Shrubby Seedbox

Ludwigia suffruticosa Walter
    Lud-wíg-i-a suf-fru-ti-cò-sa
Onagraceae (Evening-primrose
    Family)

DESCRIPTION: Smooth, erect, rhizomatous perennial forming long, leafy stolons in the fall; flowering stems smooth, erect, 12–32″ tall; leaves alternate, lanceolate, smooth, 1–4″ long and up to 0.4″ wide; inflorescence a terminal, headlike cluster of sessile or nearly sessile flowers; sepals 4, widely ovate, creamy white; petals absent; fruit is a 4-angled capsule; flowers June–October

RANGE-HABITAT: Endemic to the Coastal Plain from NC to south FL and west to AL; in SC, common in the Coastal Plain in Pond Cypress savannas, depression meadows, upper edges of other depression pond habitats, wet ditches, and borrow pits.

COMMENTS: Though it lacks petals, the showy sepals attract pollinators such as bumblebees and wasps. The genus was named in honor of Christian Gottlieb Ludwig (1709–73), a German botanist.

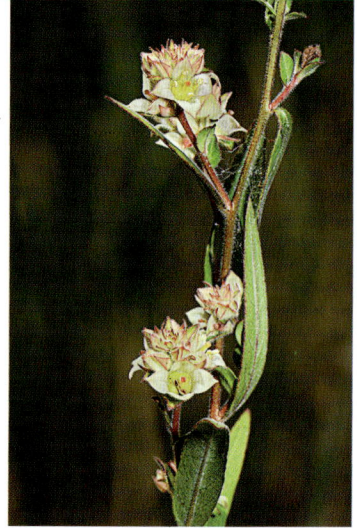

690. Awned Meadow-beauty

Rhexia aristosa Britton
    Rhéx-i-a a-ris-tò-sa
Melastomataceae (Meadow-beauty
    Family)

DESCRIPTION: Erect perennial herb, to 28″ tall, usually branched only in the top portions of the stem; stems 4-angled, pubescent; nodes with stiff hairs; leaves opposite, mostly broadly lanceolate, 3-nerved, with scattered hairs along the toothed margin; flowers with an urn-shaped hypanthium, from which the sepals, petals, and stamens arise; the sepal lobes are awned and with bristlelike, yellowish, stiff hairs along the margin; corolla with 4 petals, brilliant pink; flowers June–September.

RANGE-HABITAT: Endemic to the Coastal Plain from NJ to GA and west to AL; in SC, rare but locally abundant in Pond Cypress savannas, depression meadows, and upper margins of other open, sunny depression ponds.

COMMENTS: Awned Meadow-beauty was once considered very rare throughout its range. Fieldwork in the 1990s–early 2000s uncovered numerous populations in suitable habitat. The species is especially abundant in the intact Carolina bays and limesink ponds in the Francis Marion National Forest, where management with fire is still common. Though locally common, this species is quite rare outside SC. It is almost never found in human modified habitats and is much less aggressively weedy than many species of meadow-beauty.

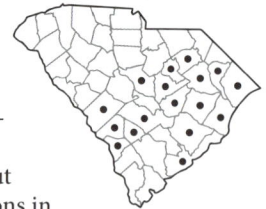

691. Cuban Meadow-beauty

Rhexia cubensis Grisebach
Rhéx-i-a cu-bén-sis
Melastomataceae (Melastome
Family)

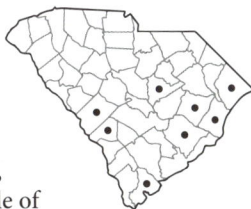

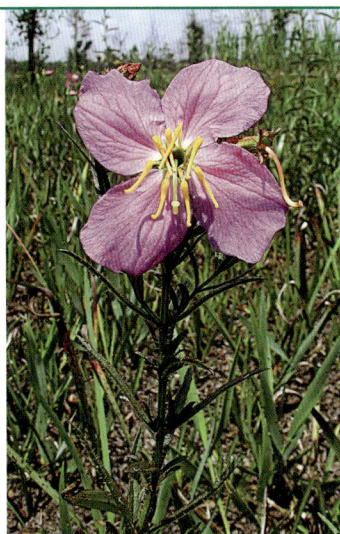

DESCRIPTION: Perennial, rhizomatous herb,
often freely branched from near the middle of
the stem; flowering stems hirsute, to 27″ tall, four-
angled, with unequal sides; leaves generally 1-nerved, very narrow,
less than 0.31″ wide at the midstem, linear to linear-elliptic or narrowly oblong, hirsute; flower with a distinct glandular-pubescent
hypanthium 0.4–0.6″ long; petals bright lavender-rose, 0.6–1.0″
long; anthers bright yellow, distinctly curved; flowers June–
September.

RANGE-HABITAT: NC south to FL and west to MS; also in the West
Indies; in SC, scattered and locally common in the Coastal Plain
and Sandhills in Pond Cypress savannas; depression meadows; upper margins of other open, sunny, depression pond habitats; deep
swales; and ditches through savannas and in borrow pits.

SIMILAR SPECIES: Nash's Meadow-beauty (*R. nashii* Small, plate 631) has similarly colored
showy flowers but has much wider mid-stem leaves that are evidently 3-veined. Maryland
Meadow-beauty (*R. mariana* L., plate 87) is also similar and may have leaves varying in width
to that typical of West Indian Meadow-beauty. The flowers are faded, pale pink to white, with
smaller petals that are less than 0.6″ long. White Meadow-beauty (*R. mariana* L. var. *exalbida*
Michaux, plate 604) is vegetatively similar but is a much smaller plant with much smaller
petals that are white (rarely pink), a mostly glabrous hypanthium, and typically inhabits drier
sites.

COMMENTS: West Indian Meadow-beauty is one of the showiest *Rhexia* species found in SC.
It is much more common and widespread farther south, where it may be found in a wider
variety of moist and wet habitats.

CONSERVATION STATUS: SC-Critically Imperiled

692. Water Dawnflower

Stylisma aquatica (Walter) Rafinesque
Sty-lís-ma a-quá-ti-ca
Convolvulaceae (Morning Glory
Family)

DESCRIPTION: Prostrate, perennial herb with stems
up to 4′ long, often forming mats; leaves oblong
to oblong-lanceolate, 0.8–1.4″ long, pubescent;
flowers produced in clusters of 1–3 on peduncles
that exceed the leaves in length; petals pink; flowers June-July.

RANGE-HABITAT: Endemic to the Coastal Plain
from NC south to FL and west to TX and AR; in SC, rare in the Coastal Plain
where found in Pond Cypress savannas, depression meadows, upper edges
of other depression ponds, wet swales within Longleaf Pine savannas, and
ditches through such habitat.

COMMENTS: This is SC's only species of dawnflower with pink petals. The populations of this species, like many of those that share its habitat, are in decline because
of drainage, conversion, and fire suppression.

CONSERVATION STATUS: SC-Vulnerable

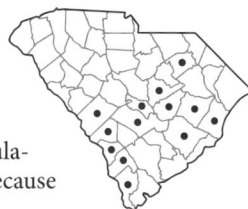

693. Tracy's Beaksedge

Rhynchospora tracyi Britton
Rhyn-chós-po-ra trà-cy-i
Cyperaceae (Sedge Family)

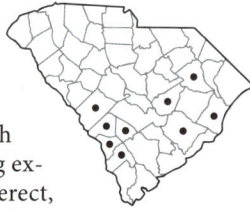

DESCRIPTION: Slender, erect, perennial with slender, scaly stolons, commonly forming extensive colonies; stems 1–3.5′ tall; leaves erect, nearly equaling the flowering stem in height, very narrow, to 0.1″ wide, linear, channeled or revolute; spikelets in 1–2 globose clusters that are 0.4–0.6″ in diameter; flowers June–September.
RANGE-HABITAT: Endemic to the Coastal Plain from NC south to FL, west to LA, and in the West Indies; in SC, occasional, but locally common in the Coastal Plain where found in Pond Cypress savannas, depression meadows, and margins of other open and sunny habitats in depression ponds.
COMMENTS: Tracy's Beaksedge can be found in many of the fire-maintained Carolina bays and limesink depressions in the Francis Marion National Forest. It is often one of the dominant species in its restricted habitat.
CONSERVATION STATUS: SC-Imperiled

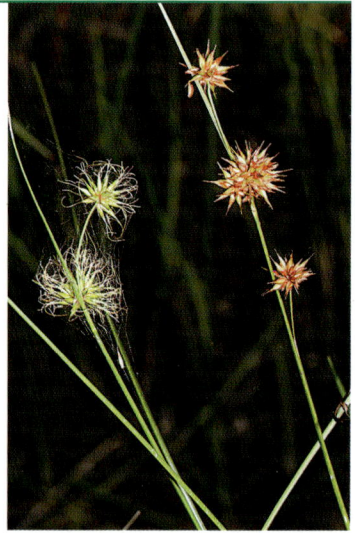

694. Depression Meadow Arrowhead

Sagittaria isoetiformis J. G. Smith
Sa-git-tà-ri-a i-so-ë-ti-fór-mis
Alismataceae (Water-plantain Family)

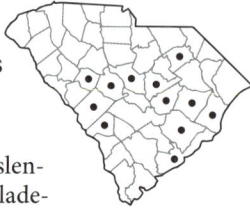

DESCRIPTION: Aquatic perennial with very slender rhizomes, stolons, and tubers; leaves bladeless or with very narrow blades, narrowly straplike, occasionally somewhat dilated at the tips; petiole round to slightly triangular; flowers arranged in terminal panicles, racemes, or rarely umbels; male and female flowers on the same plant; petals white, quickly falling after opening; flowers June–September.
RANGE-HABITAT: From NC south to FL and west to MS; in SC, rare in the Coastal Plain where found in depression meadows, Pond Cypress savannas and Pond Cypress-Swamp Gum depression swamps.
COMMENTS: Several species of narrow-leaved arrowheads may be found in depression pond habitats; this is the only one with narrow leaves appearing to lack an expanded blade and that lacks thick rhizomes. Other similar species also lack the stolons and tubers typical of Depression Meadow Arrowhead. The specific epithet refers to the resemblance of the narrow, round leaves to those of a quillwort (*Isoëtes*).
CONSERVATION STATUS: SC-Vulnerable

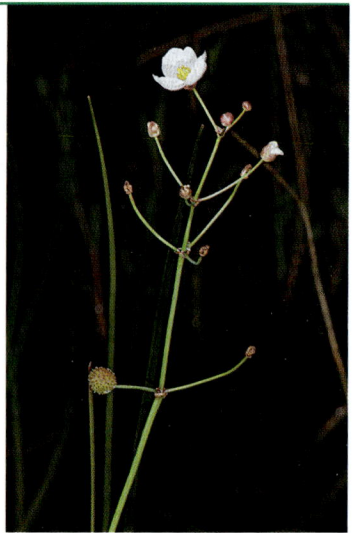

695. Ten-angled Pipewort

Eriocaulon decangulare L. var. *decangulare*
E-ri-o-caù-lon de-can-gu-là-re var. de-can-gu-là-re
Eriocaulaceae (Pipewort Family)

DESCRIPTION: Perennial herb with flowering stalks 12–32″ tall, finely 8–12 ridged; leaves mostly basal, dark green, lanceolate-linear with acute to

obtuse tips; flowers white, in dense, hard, rounded heads (buttons) at the tip of leafless stalks; flowers June–October.

RANGE-HABITAT: NJ south to FL and west to AR and TX; also in Mexico; in SC, common in the Coastal Plain and Sandhills, rare in the mountains and upper Piedmont; wet Longleaf Pine savannas, ecotones of pocosins, Pond Cypress savannas, depression meadows, upper margins of other depression pond habitats, herbaceous seepages, seepages over granitic flatrocks, Piedmont springhead seepages, cataract fens, and montane fens.

SIMILAR SPECIES: See discussion in Soft-headed Pipewort (*Eriocaulon compressum* Lamarck, plate 680).

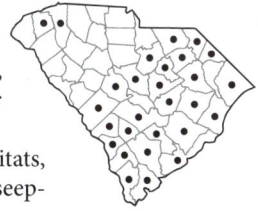

696. Feathery Mermaid-weed

Proserpinaca pectinata Lamarck
Pro-ser-pi-nà-ca pec-ti-nà-ta
Haloragaceae (Water-milfoil
Family)

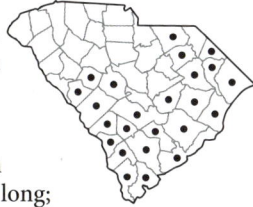

DESCRIPTION: Perennial, wetland herb with weak, trailing, or submerged stems to 20" long; submerged leaves pectinate (featherlike), emergent leaves also pectinate; flowers produced in the leaf axils of emergent leaves, tiny and inconspicuous, with 3 stamens and 3 carpels; flowers June–October.

RANGE-HABITAT: Nova Scotia south to FL and west to LA; in SC, common in the Coastal Plain and Sandhills where found in Pond Cypress savannas, depression meadows, Pond Cypress-Swamp Gum depression swamps, margins of impoundments, ditches, and other wet and nutrient-poor habitats.

SIMILAR SPECIES: Common Mermaid-weed (*Proserpinaca palustris* L.) is similar but has pectinate submerged leaves and simple, serrate-margined emergent leaves. It is found in swamp forests and many other aquatic habitats in the Coastal Plain and Piedmont. It is seldom found in the nutrient-poor isolated depressions that are typical of Feathery Mermaid-weed. The name is derived from the Greek goddess Proserpina who was abducted by Hades, ruler of the underworld.

697. Blue Sedge; Southern Waxy Sedge

Carex glaucescens Elliott
Cà-rex glau-cés-cens
Cyperaceae (Sedge Family)

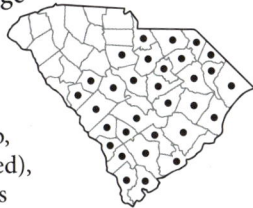

DESCRIPTION: Rhizomatous, perennial herb, 28–44" tall; leaf sheaths glaucous (whitened), blades bluish-green, 0.1–0.3" wide; flowers produced in spikes; the female spikes with bluish fruits with a reddish tinge, the lowest drooping; flowers and fruits July–September.

RANGE-HABITAT: Coastal Plain from MD to FL and west to TX; disjunct in TN; in SC, common throughout the Coastal Plain and Sandhills; backwater swamps, pocosins, wet pine savannas, seepage bogs, and Pond Cypress savannas.

SIMILAR SPECIES: Warty Sedge (*C. verrucosa* Muhlenberg) is very similar and also occurs in depression ponds. It may be distinguished by the spikes of female flowers having the lowest spike upright, not drooping, and having even more intense bluish-green foliage.

COMMENTS: *Carex glaucescens* is one of about 144 species of *Carex* in SC and is one of the very few with bluish-green foliage throughout and glaucous leaf sheaths.

698. Mud-babies; Dwarf Burhead

Helanthium tenellum (Martius) Britton
He-lán-thi-um te-nél-lum
Alismataceae (Water-plantain Family)

SYNONYM: *Echinodorus tenellus* (Martius) Buchenau—RAB

DESCRIPTION: Small, annual, wetland herbs without rhizomes; leaves submerged and emergent; submerged leaves linear, to 3″ long and less than 0.2″ wide; emergent leaves with slightly dilated blades that are narrowly elliptical in outline, less than 0.8″ wide; flowering stems to 4″ long; flowers produced in umbels or racemes in 1–2 whorls, each whorl subtended by 3 bracts and smaller bracteoles, with 4–6 flowers in a whorl; petals white; flowers August–September.
RANGE: MA south to FL and west to MN and TX; scattered and uncommon throughout its range; in SC, rare in the Coastal Plain where found in depression meadows, margins of other depression pond habitats, and along the banks of blackwater rivers.
SIMILAR SPECIES: Mud-babies are in the same family and similar to some arrowhead species (*Sagittaria* spp.). When in flower, they can be distinguished from arrowheads by thebracts and bracteoles that subtend the whorls of flowers and the rounded (rather than flattened) achenes (fruits).
COMMENTS: Mud-babies are indicative of a high-quality habitat that has an intact hydrology.
CONSERVATION STATUS: SC-Vulnerable

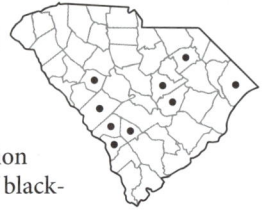

699. Creeping St. John's-wort

Hypericum adpressum Barton
Hy-pé-ri-cum ad-prés-sum
Hypericaceae (St. John's-wort Family)

DESCRIPTION: Erect, rhizomatous, colonial, perennial herb, 16–32″ tall; stem only sparingly branched in the upper portions; plants growing in water with lower stems spongy, not spongy on plants not growing in the water; leaves with depressed midrib on upper surface, strongly elevated beneath, margins revolute; lower leaf surface pale green, upper surface glossy green; flowers produced in a terminal flat-topped arrangement (cymes); petals 5, yellow, styles united into a beak; flowers July–August.
RANGE-HABITAT: Primarily in the Coastal Plain from MA south to GA; also disjunct inland in WV, IL, and c. TN; in SC, rare in the Coastal Plain; found in depression meadows, Pond Cypress savannas, margins of high ponds, and other wet and open areas.
COMMENTS: This species has a curious distribution, being rare throughout its large range and absent from many sites that would otherwise seem optimal for its growth. Where it does occur, it is often abundant. With less than 6 locations in SC, it is among the most imperiled plants in SC. Habitat alteration by disturbing or destroying upland depression ponds, such as Carolina bays, are the greatest threat. It may be seen at Dingle Pond in the Santee National Wildlife Refuge in Clarendon County.
CONSERVATION STATUS: SC-Imperiled

700. Skyflower

Hydrolea corymbosa Macbride ex
   Elliott
      Hy-drò-le-a co-rym-bò-sa
Hydrophyllaceae (Waterleaf Family)

DESCRIPTION: Erect, slender perennial herb 24–32″
tall; leaves elliptic to elliptic-lanceolate, alternate,
entire; flowers in a narrow, few-flowered, flat-
topped arrangement (corymb); sepals with long
trichomes, petals showy, brilliant violet to bluish-
violet to purplish pink; flowers July–September.
RANGE-HABITAT: Endemic to the Coastal Plain
from SC south to FL; in SC, rare in the Coastal Plain in Pond Cypress savannas and depression meadows.
COMMENTS: Skyflower is among the most vibrant, beautiful flowers in SC.
It is also, unfortunately, one of the rarest. Several populations have been lost
because of habitat alteration through draining and conversion of its depression
pond habitat into impoundments for hunting ducks.
CONSERVATION STATUS: SC-Critically Imperiled

701. Southern Bladderwort

Utricularia juncea M. Vahl
      U-tri-cu-là-ri-a jún-ce-a
Lentibulariaceae (Bladderwort
   Family)

DESCRIPTION: Terrestrial carnivorous herb;
subterranean leaves dissected, with small, trapping bladders; above ground leaves, when present,
linear and flat; flowering stalk 4–18″ tall, leafless, but with small
scales and bracts; flowers 2–15 per stem, widely spaced on the
stem, yellow, at least 0.6″ long, with a spur 0.3″ long or less; flowers
May–September.
RANGE-HABITAT: NY south to FL and west to AR and TX; in SC,
common in the Coastal Plain and Sandhills in Pond Cypress savannas, depression meadows, and borrow pits.
SIMILAR SPECIES: See discussion under Horned Bladderwort
(*Utricularia cornuta* Michaux, plate 61).
COMMENTS: *Utricularia* get their name from their distinctive trapping bladders (utriculus = small bladder).

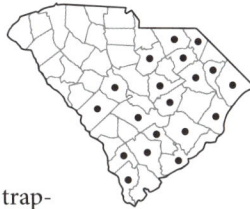

702. Bay Boneset

Eupatorium paludicola E.E. Schilling &
   LeBlond
      Eu-pa-tò-ri-um pa-lu-dí-co-la
Asteraceae (Aster Family)

DESCRIPTION: Erect, perennial herb from a
rhizome 0.8–8.0″ long; often forming large, dense
colonies; stems 2–5′ tall, not branching at the
base; flowering stems typically solitary; leaves
sessile, opposite, narrow, linear-elliptic to
linear-oblanceolate, 1–5″ long, less than 0.18″
wide, ascending to erect-recurved; flowers

produced in involucrate heads composed of cream-colored disk flowers only; outer involucral bracts acuminate to attenuate, pubescent; flowers August–October.

RANGE-HABITAT: Endemic to the Coastal Plain of NC and SC; in SC, rare in the Coastal Plain where restricted to depression meadows and Pond Cypress savannas.

SIMILAR SPECIES: Savanna Eupatorium (*E. leucolepis* (A.P. de Candolle) Torrey & Gray, plate 662) is similar, but either lacks a rhizome or has a very short one, and it has leaves 0.23–0.5″ wide and reflexed-spreading to spreading-ascending. It is common in Longleaf Pine savannas, ecotones of pocosins, and other moist and open areas in the Coastal Plain and Sandhills.

COMMENTS: This species was first discovered for SC by Richard D. Porcher Jr., Patrick McMillan, and John F. Townsend in Florence County in the 1990s. The species was first described in 2007 (LeBlond et al., 2007). The largest population, dominating acres of a depression meadow in Florence County, has been possibly lost due to alteration of the Carolina bay. The South Carolina Native Plant Society recently took over management of the bay and is attempting to restore the habitat. In the late 1990s this species was abundant in several locations; today, it is nearly extinct.

CONSERVATION STATUS: SC-Critically Imperiled

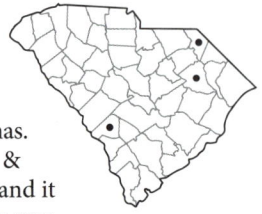

703. Flax-leaf Agalinis;
Flax-leaf False-foxglove

Agalinis linifolia (Nuttall) Britton
A-ga-lì-nis li-ni-fò-li-a
Orobanchaceae (Broomrape
Family)

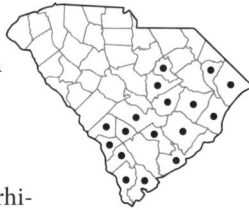

DESCRIPTION: Perennial herb with slender rhizomes; stems smooth, round, 28–48″ tall; leaves opposite, linear; flowers relatively widely spaced in racemes with 8–20 flowers, 1 or 2 flowers per node; corolla 1–1.6″ long, light violet-purple to pink, with the tube lacking yellow lines; flowers August–September.

RANGE-HABITAT: Endemic to the Coastal Plain from NC south to FL and west to LA; disjunct in DE; in SC, occasional in the Coastal Plain where found in Pond Cypress savannas, depression meadows, upper portions of other open, depression pond habitats, wet Longleaf Pine savannas, and ditches through such habitats.

COMMENTS: This is SC's only *Agalinis* species that is a perennial; it has the largest flowers of any of the SC *Agalinis*.

704. Fringed Yellow-eyed-grass

Xyris fimbriata Elliott
Xỳ-ris fim-bri-à-ta
Xyridaceae (Yellow-eyed Grass
Family)

DESCRIPTION: Perennial herb with only basal leaves; leaves linear, straight, conduplicate (like an *Iris*), distichous, dull green 12–24″ long, not swollen or bulbous at the base; flowering stalk 20–48″ tall, rough (scabrous) toward the inflorescence; flowers in a compact, terminal, round spike, with each flower subtended by a woody scale that hides the flower bud and fruit; lateral sepals fringed and extending from the bract; flowers ephemeral, projecting from behind the woody scale; flowers August–October.

RANGE-HABITAT: From NJ south to FL and west to TX, mostly in the Coastal Plain but with disjunct inland populations to TN; in SC,

common in the Coastal Plain and Sandhills in mucky or sandy soils, often in shallow water of Pond Cypress savannas, depression meadows, other depression pond habitats, herbaceous seepages, and ditches through such areas.

COMMENTS: This species is easily distinguished from all other *Xyris* in wet habitats by the long and exerted lateral sepal, which has a long fringe on the keel. It flowers on sunny days from mid-morning to early afternoon, after which the flowers quickly fade. It may also be distinguished by the scabrous upper stem, rough when one runs a hand up the stem. The name *Xyris* comes from the Greek, meaning "gladen" (sword-shaped) in reference to the leaves.

705. Canby's Dropwort; Canby's Cowbane

Tiedemannia canbyi (J. M. Coulter & Rose) Feist & S. R. Downie
  Tiede-mánn-i-a cán-by-i
Apiaceae (Carrot Family)

SYNONYM: *Oxypolis canbyi* (Coulter and Rose) Fernald—RAB, PR

DESCRIPTION: Perennial, rhizomatous herb; stems hollow, to 4′ tall, often forming large colonies; rhizomes stoloniferous, very slender, and 4–12″ long; leaves reduced to round phyllodes; flowers in compound umbels, white; fruits with corky-thickened ribs; flowers August–October.

RANGE-HABITAT: Coastal Plain from DE south to GA; in SC, rare in the Coastal Plain in depression meadows, Pond Cypress savannas, and upper margins of other depression pond habitats.

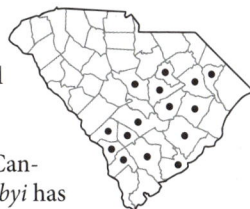

SIMILAR SPECIES: Water Dropwort (*T. filiformis* (Walter) Britton) is similar to Canby's Dropwort. The following features are useful in separating the two: *O. canbyi* has mature fruits with corky, thickened ribs or wings, long narrow rhizomes, and lower nodes that lose their leaves by flowering time; *T. filiformis* has fruits with thin peripheral ribs or wings, stout, very short rhizomes or no rhizomes, and lower nodes retaining leaves until flowering. The corky, thickened, winged fruit of *O. canbyi* is clearly evident in the photograph.

COMMENTS: Canby's Dropwort is threatened by habitat loss and alteration throughout its range. The largest population in the world was mostly destroyed when a site in Clarendon County was recently converted to a duck pond. SC still harbors the largest concentrations of Canby's Dropwort in one state. One site for viewing of this species is the Crosby Oxypolis Heritage Preserve in Colleton County.

The genus honors Friederich Tiedemann (1781–1861), a French zoologist. The specific epithet honors William M. Canby (1831–1904), its discoverer.

CONSERVATION STATUS: Federally Endangered

706. Pond Cypress-Swamp Gum upland swamps

707. Swamp Gum; Swamp Tupelo

Nyssa biflora Walter
 Nýs-sa bi-flò-ra
 Nyssaceae (Tupelo Family)

SYNONYM: *Nyssa sylvatica* Marshall var. *biflora* (Walter) Sargent—RAB

DESCRIPTION: Small to large wetland tree, commonly found where water stands much of the time; bases of trunks swollen; leaves elliptic to lance-elliptic, widest at the middle, with long-tapering tips; male and female flowers on separate trees; mature fruits are blue-black drupes, produce mostly 2 per peduncle; flowers April–May; fruits mature August–October.

RANGE-HABITAT: NJ south to FL and west to TX; in SC, common in the Coastal Plain and Sandhills; blackwater and brownwater river swamps, depressions in pinelands (where it grows alone or in association with Pond Cypress), pocosins, pond and lake margins, and wet Longleaf Pine savannas and flatwoods.

SIMILAR SPECIES: Swamp Gum is similar to Black Gum (*N. sylvatica* Marshall). Swamp Gum has leaves widest beyond the middle with obtuse apices and fruits usually 1–2 per stalk; Black Gum has most leaves widest at the middle with long-tapering apices and fruits 1–5 per stalk. Black Gum is mostly an upland tree in well-drained soils, but it does occur in wet sites.

COMMENTS: The SC state champion tree was measured in 1995 at 113′ tall and 4.6′ in diameter; it is in the Congaree National Park in Richland County.

708. Ogeechee Lime; Ogeechee Plum

Nyssa ogeche Marshall
 Nýs-sa o-gè-che
 Nyssaceae (Sourgum Family)

DESCRIPTION: Small to medium-sized tree, usually less than 40′ tall; typically with swollen buttresses and multiple, crooked trunks; leaves to 6″ long, alternate, simple, deciduous, variable in shape but usually broadest at or above the middle; male and female flowers on same tree; flowers April; drupes red, mature August–October.

RANGE-HABITAT: Endemic to the Coastal Plain from SC south to FL and west to AL; in SC, rare in the southern Coastal Plain where it grows in Pond Cypress-Swamp Gum upland swamps; it occurs in blackwater river swamps farther south in FL and GA.

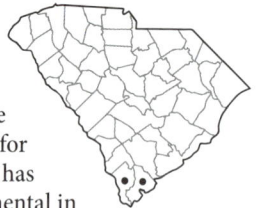

COMMENTS: The common name, Ogeechee Lime, is derived from the Ogeechee River in GA and from the use of the acid juice of its ripe fruits as a substitute for lime. The pulp of the juice makes a good preserve. Because of its small size, it has never been economically important for timber but makes an attractive ornamental in the landscape. Birds and mammals eat the fruit.

CONSERVATION STATUS: SC-Critically Imperiled

709. Climbing Fetterbush

Pieris phillyreifolia (Hooker)
A. P. de Candolle
Pì-e-ris phil-ly-rei-fò-li-a
Ericaceae (Heath Family)

DESCRIPTION: Evergreen shrub or woody vine; as a vine, unique in climbing in bark crevices or beneath the outer bark of Pond Cypress; as a shrub, it grows on cypress knees or stumps; leaves alternate, leathery; lateral branches project from under the bark; flowers urn-shaped; flowers April–May.
RANGE-HABITAT: Endemic to the Coastal Plain from SC south to FL and west to AL; in SC, rare; in the Coastal Plain, found in Pond Cypress-Swamp Gum upland swamps.
COMMENTS: Climbing Fetterbush grows as a creeping woody vine under the bark of Pond Cypress with its branches exerted through the cypress bark. Fire will kill the stem under the bark; however, it will regrow upward under the bark from the main stem at the base of the tree. The stem under the bark assumes a flattened condition. In SC, Climbing Fetterbush has been observed only growing on Pond Cypress. In FL, it also climbs under the bark of Atlantic White-cedar (*Chamaecyparis thyoides*). The SC locations are remarkably disjunct from the nearest locations in southern GA. The majority of known SC populations are protected within the Francis Marion National Forest.
CONSERVATION STATUS: SC-Critically Imperiled

710. Pondspice

Litsea aestivalis (L.) Fernald
Lìt-se-a aes-ti-và-lis
Lauraceae (Laurel Family)

DESCRIPTION: Many-branched aromatic shrub to 15′ tall; the twigs zigzag; leaves deciduous, elliptic with entire margins, 0.5–1.2″ long; male and female flowers on separate plants; flowers appearing before the leaves from overwintering buds; flowers February–April; fruit is a red drupe; fruits mature May–June.
RANGE-HABITAT: Endemic to the Coastal Plain from MD south to FL and west to LA; rare throughout much of its range; in SC, uncommon in the Coastal Plain where growing along the margins of limestone sinks and Carolina bays, pineland depressions, and Pond Cypress-Swamp Gum upland swamps.
COMMENTS: Pondspice is locally abundant in the Francis Marion National Forest. The strange, narrow, dark brown zigzag stems help to distinguish it from other shrubs in the habitat.

711. Pondberry; Jove's Fruit

Lindera melissifolia (Walter) Blume
Lìn-der-a me-lis-si-fò-li-a
Lauraceae (Laurel Family)

DESCRIPTION: Deciduous, aromatic shrub with the odor of Sassafras, 3–4′ tall, colonial from rhizomes; leaves entire, alternate, narrowly ovate, thin-textured with bases cuneate to rounded; surfaces of blade with reticulate-rugose venation;

male and female flowers on separate plants; flowers March–April, before the leaves expand; the red drupes mature August–September.

RANGE-HABITAT: Endemic to the Coastal Plain from NC south to the FL Panhandle and west to LA; disjunct in MO and AR; in SC, rare in the Coastal Plain in wet depressions in pine flatwoods, along the margins of Pond Cypress-Swamp Gum upland swamps, open boggy areas, and sandy sinks.

COMMENTS: The largest concentration of Pondberry in SC occurs in the Honey Hill region of the Francis Marion National Forest in Berkeley County. The numerous lime sinks in this area provide critical habitat for this rare shrub. It is very short and seldom vigorous. It is federally protected and should never be disturbed in the wild. The genus name honors Johan Linder (1676–1724), a Swedish botanist.

CONSERVATION STATUS: Federally Endangered

712. Cassena; Dahoon Holly

Ilex cassine L.
Ì-lex cas-sì-ne
Aquifoliaceae (Holly Family)

DESCRIPTION: Small to medium-sized evergreen tree, rarely exceeding 25′ tall; leaves thick and leathery, simple, obovate, oblanceolate, or narrowly lanceolate, with remotely serrate margins, 1.5–4″ long and over 0.25″ wide, with petioles 0.2–0.6″ long; male and female flowers on separate trees; flowers May–June; fruit a drupe, red or sometimes yellow, mature October–November, persisting until spring.

RANGE-HABITAT: Endemic to the southeastern Coastal Plain from NC to FL and west to TX; occasional in upland Pond Cypress-Swamp Gum forests (particularly along blackwater streams), pocosins, depressions in flatwoods, wet hummocks, and edges of spring-fed rivers and streams; almost always on acidic peaty soils.

SIMILAR SPECIES: Myrtle-leaved Holly (*I. myrtifolia* (Walter) Sargent, plate 713) is similar and has, in the past, been considered a variety of Cassena. It can be distinguished by the much narrower lanceolate to narrowly oblong leaves with very short petioles. It is found in depression ponds throughout the Coastal Plain of SC.

COMMENTS: Cassena is often grown as an ornamental for its attractive red berries and leaves. Birds consume the fruits in the late winter through early spring. It is particularly attractive to Cedar Waxwings and American Robins.

713. Myrtle-leaved Holly

Ilex myrtifolia (Walter) Sargent
Ì-lex myr-ti-fò-li-a
Aquifoliaceae (Holly Family)

DESCRIPTION: Small evergreen tree, or more often a shrub; leaves narrow, stiff, lanceolate to narrowly oblong, less than 1.5″ long, and less than 0.25″ wide; petioles less than 0.15″ long; male and female flowers on separate plants; fruit a red drupe, mature October–November, persisting until spring.

RANGE-HABITAT: Endemic to the Coastal Plain from NC south to FL and west to LA; in SC, common in the Coastal Plain in upland Pond Cypress-Swamp Gum upland swamps, margins of sandy ponds, and Pond Cypress savannas.
SIMILAR SPECIES: See notes under Cassena above.
COMMENTS: Myrtle-leaved Holly grows exclusively in isolated depressions. It is readily distinguished from Cassena, but the two sometimes hybridize. It makes a good landscape plant producing fruits that are attractive to birds in the late winter–spring. It is particularly attractive to Cedar Waxwings and American Robins.

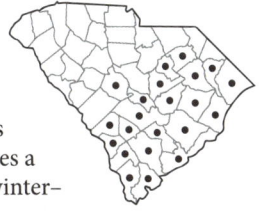

---

714. Small Floating Bladderwort

Utricularia radiata Small
  U-tri-cu-là-ri-a ra-di-à-ta
Lentibulariaceae (Bladderwort Family)

SYNONYM: *Utricularia inflata* Walter var. *minor* (L.) Chapman—RAB

DESCRIPTION: Aquatic, carnivorous herb: stems with leaves of two types; submerged portions with long, forked leaves with filiform segments and bladder traps; floating leaves in a whorl of 6–8, less than 2″ long, fused basally, swollen, inflated, and modified into a "float;" inflorescence of 3–4 flowers, corolla yellow with a spur that is entire; flowers May–November.
RANGE-HABITAT: Nova Scotia south to FL and west to OK and TX; in SC, occasional in the Coastal Plain and Sandhills in standing water of Pond Cypress savannas, upland Pond Cypress-Swamp Gum wetlands, depression meadows, ditches, impoundments, and borrow pits.
SIMILAR SPECIES: Swollen Bladderwort (*Utricularia inflata* Walter, plate 864) is similar but is a much more robust plant, with floating leaves that are not fused basally, each leaf more than 2″ long, and with flowers having a spur that is bifid at the apex. It is found in similar sites but in a wider variety of aquatic habitats, throughout the Coastal Plain.
COMMENTS: The floats that develop in the spring on Small Floating Bladderwort and Inflated Bladderwort are unique to the genus in SC. The empty traps are held under water where, when triggered, they quickly open and inflate with water, snaring small aquatic insects such as mosquito larvae and crustaceans such as *Daphnia.*
CONSERVATION STATUS: SC-Vulnerable

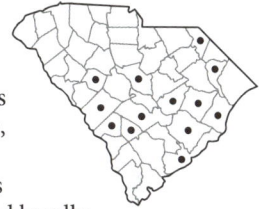

---

715. Loose Water-milfoil

Myriophyllum laxum
  Shuttleworth ex Chapman
  My-ri-o-phýl-lum láx-um
Haloragaceae (Water-milfoil Family)

DESCRIPTION: Perennial, aquatic herb; leaves submerged and emergent; submerged leaves, alternate to appearing whorled (actually with short internodes), to 1.2″ long, with 3–7 pairs of segments; flowers produced in erect spikes, extending above the water, with bracts shorter than the internodes; flowers August–October.
RANGE-HABITAT: Endemic to the Coastal Plain from VA south to FL and west to MS; in SC, rare, in the Coastal Plain and Sandhills where growing in standing water of depression ponds, often in upland Pond Cypress-Swamp Gum swamps in Carolina bays and limesink depressions; also found in Sandhills streams and in impoundments.

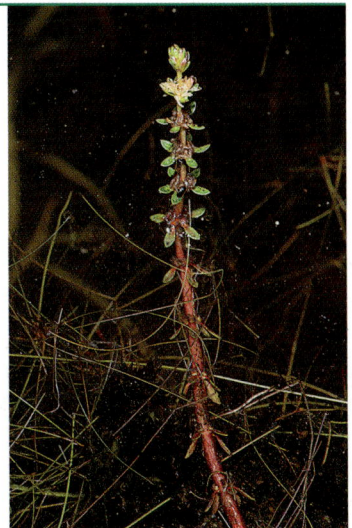

SIMILAR SPECIES: There are several species of water-milfoil in SC; this is the only one with emergent leaves that are spatulate and less than 0.2″ wide and emergent spikes of flowers with bracts shorter than the internodes.

COMMENTS: Loose Water-milfoil is rare in SC but not as uncommon as once thought. It may be present in large numbers during wet years and completely absent in especially dry years.

CONSERVATION STATUS: SC-Vulnerable

---

716. Chapman's Arrowhead

Sagittaria chapmanii
(J. G. Smith) C. Mohr
Sa-git-tà-ri-a chap-mán-i-i
Alismataceae (Water-plantain
Family)

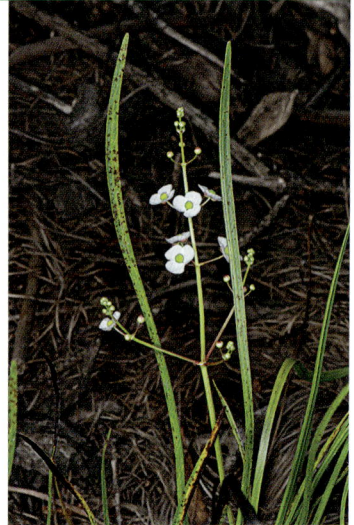

DESCRIPTION: Perennial, wetland herb; rhizomes present, but without slender stolons or corms; leaves broadly linear to narrowly ovate, or phyllodial, to 0.4″ wide, sometimes wider; inflorescence paniculate, multibranched at the lower nodes of the inflorescence; bracts of the inflorescence not fused (connate); flowers May–September.

RANGE-HABITAT: Endemic to the Coastal Plain from NC south to FL and west to LA; in SC, occasional in the Coastal Plain where found in depression ponds in habitats such as upland Pond Cypress-Swamp Gum swamps, Pond Cypress savannas, depression meadows, and ditches through such habitats.

SIMILAR SPECIES: This species has been confused with (and sometimes been treated as a variety of) Grassleaf Arrowhead (*S. graminea* Michaux). Recent molecular evidence indicates it is distinct. It is easily distinguished in flower from Grassleaf Arrowhead by the branching at the lower nodes of the inflorescence and the bracts of the inflorescence, which are not connate (fused).

---

717. Rose Coreopsis

Coreopsis rosea Nuttall
Co-re-óp-sis rò-se-a
Asteraceae (Aster Family)

DESCRIPTION: Smooth perennial with stolons; stem ascending, to 32″ tall; leaves linear, opposite; flowers produced in involucrate heads of purplish-pink to light rose or white ray flowers and yellow disk flowers; flowers July–September.

RANGE-HABITAT: From Nova Scotia south to GA and TN; in SC, rare and known from two distinct habitats: upper margins of upland depression ponds ("high ponds") in the Inner Coastal Plain, and drawdown zones of blackwater rivers in the Outer Coastal Plain.

COMMENTS: Rose Coreopsis can be found in the Janet Harrison High Pond Heritage Preserve in Aiken County. No other species of coreopsis in SC has purplish-pink to light-rose or white flowers.

CONSERVATION STATUS: SC-Imperiled

718. Violet Burmannia (A)

Burmannia biflora L.
Bur-mánn-i-a bi-flò-ra
Burmanniaceae (Burmannia Family)

White Burmannia (B)

Burmannia capitata (Walter ex J. F.
Gmelin) Martius
Bur-mánn-i-a ca-pi-tà-ta
Burmanniaceae (Burmannia Family)

DESCRIPTION: Diminutive herbs with erect, filiform stems and threadlike roots; to 6″ tall; leaves
scalelike, widely spaced; *B. biflora* with 1 flower
terminal, the remainder in 2 lateral racemes or
spikes; floral tube 3-winged, lavender to purple;
B. capitata with flowers in a capitate (headlike)
cluster; floral tube 3-angled, white to greenish; flowers
July–November.

RANGE-HABITAT: VA (*B. biflora*) and NC (*B. capitata*) south
to FL and west to TX and AR, mostly in the Coastal Plain; in SC, occasional in the Coastal Plain and
Sandhills, very rare in the upper Piedmont in Pickens county; Pond Cypress-Swamp Gum swamps,
Pond Cypress savannas, herbaceous seepages, and wet Longleaf Pine savannas; also in seepage over a
granitic flatrocks in the Piedmont; easily overlooked because of its small size and because it blooms in
late fall after many floristic studies are completed.

COMMENTS: Easily overlooked because of their small size, these species may be abundant in a site one
year and then absent for the next few years. The genus honors Johannes Burmann (1706–79), a Dutch
botanist. The Pickens County location, where both species are present, is remarkably disjunct and
contains many other primarily Coastal Plain species.

CONSERVATION STATUS: SC-Imperiled (*B. biflora*)

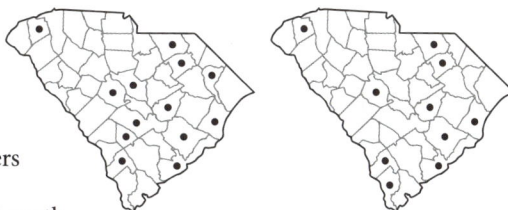

719. Pocosin

720. Pond Pine; Pocosin Pine

Pinus serotina Michaux
Pì-nus se-rò-ti-na
Pinaceae (Pine Family)

DESCRIPTION: Medium-sized, evergreen tree; cones
top-shaped or almost globe-shaped; needles
mostly in 3s, 6–8″ long, persistent for 3–4 years;
young cones develop March–April.

RANGE-HABITAT: NJ south to FL and AL; in SC,
common in the Coastal Plain and Sandhills;

pocosins, bay forests, Pond Cypress savannas, swamps of small blackwater streams, and Longleaf Pine savannas.

COMMENTS: Pond Pine is a "serotinous" species that depends on fire for regeneration. Serotinous means "late to open." Most of its cones remain closed with viable seeds for years until fire softens the resinous seal, which allows the cone to open. Fire also removes the underbrush, presenting open, sunny conditions, ideal for germination and establishment of the seedlings. Pond Pine also has another adaptation to fire: it sprouts new branches along the trunk from latent axillary buds, termed "epicormic sprouting." Although fire may destroy the crown, the new epicormic shoots are tied to an old, healthy root system and quickly develop into new branches. Pond Pine also sprouts from roots or stumps after serious fire injury. Unlike most pines, Pond Pine thrives in saturated soils and can tolerate wide fluctuations in water level. For this reason, it is common in peat-based Carolina bays.

Pond Pine is not an important lumber species because it has poor growth form and grows in wet areas not easily accessible for timbering.

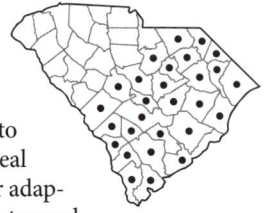

---

721. Swamp Bay

Persea palustris (Rafinesque) Sargent
Pér-se-a pa-lús-tris
Lauraceae (Laurel Family)

DESCRIPTION: Medium to large, evergreen, aromatic tree to 65′ tall; twigs pubescent; leaves alternate, elliptic, with entire margins and acute tips; undersurface of leaf with ascending to spreading rusty hairs; leaves often with galls; flowers produced in compact cymes from the axils of leaves; peduncles 1.5–2.8″ long; petals yellowish; fruit a dark purplish-black drupe; flowers May–June; fruits mature September–October.

RANGE-HABITAT: DE south to FL and west to TX; also in the Bahamas; in SC, common in the Coastal Plain and Sandhills in pocosins, swamp forests, and bay forests, generally on saturated, highly acidic soils.

SIMILAR SPECIES: Red Bay (*P. borbonia* (L.) Sprengel, plate 906) is very similar but easily distinguished by its appressed, rather than spreading silvery, golden, or dark hairs on the undersides of the leaves, smooth or much less pubescent twigs, and peduncles that are short (0.4–1.0″ long). Red Bay does not occur in the highly acidic pocosin and blackwater swamp forest habitats and is restricted to maritime forests.

COMMENTS: The term "bay" is in reference to the aromatic leaves that can be used as a cooking substitute for bay leaf, which is produced by Bay Laurel (*Laurus nobilis* L.). Though several species in the pocosin resemble Swamp Bay with elliptic, entire leaves, the presence of copious galls, caused by the aphid relative Red Bay psyllid (*Trioza magnoliae*), are often diagnostic. The psyllids apparently have little impact on the vigor of the plants. Swamp Bay, along with Red Bay (and to a lesser extent, other members of the Lauraceae), are the primary host for the caterpillars of Palamedes Swallowtail, one of the most common swallowtail butterflies in the Coastal Plain. Swamp Bay and Red Bay have recently been devastated by the accidental introduction in the 1990s of an Asian Ambrosia Beetle (*Xyleborus glabratus*), which bore into the trees and infected them with a fungus in the genus *Ophiostoma* that causes the death of the stem. Swamp Bay continues to be common in SC pocosins where its smaller size, copious sprouting, and ability to flower on young stems help it to deal with this new threat.

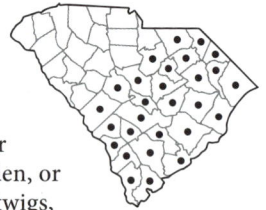

722. Sweet Bay

Magnolia virginiana L.
Mag-nòl-i-a vir-gi-ni-à-na
Magnoliaceae (Magnolia Family)

DESCRIPTION: Small shrub or tree, sometimes reaching 30–80′ tall; often a bushy stump-sprout in burned or cutover areas; leaves entire, evergreen or semi-evergreen, persisting into winter, with a few remaining until spring; leaves white beneath, easily noticeable from a distance; flowers with a strong sweet odor; flowers April–July; fruits (an aggregate of follicles) mature July–October, with stalks keeping the bright red seeds attached to the open follicles.

RANGE-HABITAT: From MA south to FL and west to TX and AR; in SC, common in the Coastal Plain and Sandhills in pocosins, Longleaf Pine savannas, Pond Cypress savannas, swamp and bay forests, and low wet woodlands.

TAXONOMY: Two varieties are frequently distinguished. *M. virginiana* var. *virginiana* is deciduous, has twigs that are pubescent on the previous season's growth, and has flowers that open in the afternoon. It is a shorter, often multitrunked tree. It ranges south to SC and GA. *M. virginiana* var. *australis* Sargent has evergreen leaves, twigs that are smooth on the previous season's growth, and flowers that open at sunset. It forms a large tree when not burned. The distribution of the two varieties in SC needs clarification.

COMMENTS: Sweet Bay is often cultivated because of its showy, fragrant flowers. It makes a great landscape plant that is adaptable, stays relatively small, and has a good form. The plant is the primary host for Sweet Bay Silkmoth.

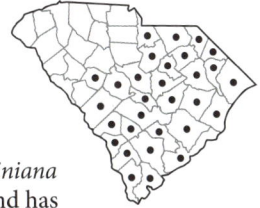

723. Leatherleaf

Chamaedaphne calyculata (L.) D. Don
Cha-mae-dáph-ne ca-ly-cu-là-ta
Ericaceae (Heath Family)

SYNONYM: *Cassandra calyculata* (L.)
D. Don—RAB, PR

DESCRIPTION: Low, rhizomatous shrub, 1–3′ tall; leaves evergreen, 1–2″ long, elliptic to oblanceolate, both surfaces tawny and scurfy, becoming smooth above with age; urn-shaped, white flowers on an arching stem (raceme), produced in axils of previous year's leaves; flowers March–April.

RANGE-HABITAT: Circumpolar in distribution; in SC, occasional and widely scattered in the Coastal Plain and Sandhills in pocosins.

COMMENTS: This species is a common component of cold bogs in the boreal forest region but extends southward on extremely nutrient-poor, acidic soils to the Coastal Plain of SC—its southern limit. A good site to see Leatherleaf is Lewis Ocean Bay Heritage Preserve in Horry County.

724. Carolina Wicky;
Southern Sheep-kill

Kalmia carolina Small
    Kálm-i-a ca-ro-lì-na
Ericaceae (Heath Family)

SYNONYM: *Kalmia angustifolia* var.
    caroliniana (Small) Fernald—RAB

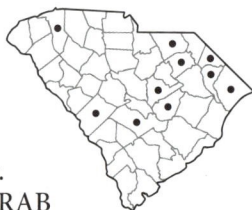

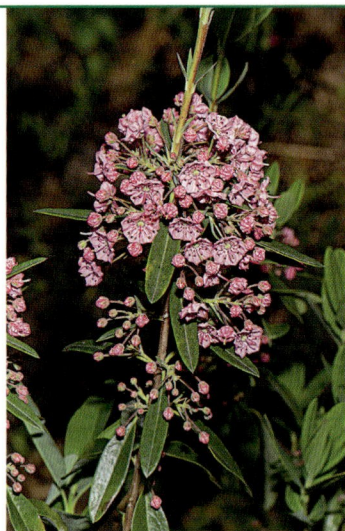

DESCRIPTION: Evergreen, rhizomatous shrub, 1–3′ tall, forming sizable colonies; leaves usually in whorls of 3, opposite or alternate on some branches, elliptic to elliptic-lanceolate, 1–2.75″ long; dark green or dark bluish-green above, light bluish-green below; flowers produced in axillary and/or terminal racemes; petals fused at least one-half of their length; corolla 0.4–0.6″ broad; anthers fitted into pockets in the corolla; flowers April–early July.

RANGE-HABITAT: Two disjunct areas: the Coastal Plain from se. VA, south to SC; and the southern Appalachians, from VA south to GA; in SC, occasional in the Coastal Plain in pocosins, particularly the transition from barren sand ridges to dense pocosin, and ecotones between sand ridges and blackwater swamp forests; very rare in the mountains where known from a single montane fen.

COMMENTS: The genus honors Pehr Kalm (1716–79), a pupil of Linnaeus, who traveled and collected in America. It is poisonous to livestock, hence the common name. The flowers are miniatures of the larger Mountain Laurel (*Kalmia latifolia* L., plate 226). The 10 anthers are tucked into pockets of the corolla and pop out when touched by insects, causing pollen to spray the insect, an adaptation that assists in cross-pollination.

Sizeable populations are found in Cartwheel Bay Heritage Preserve, Little Pee Dee River Heritage Preserve, and Bennett's Bay Heritage Preserve.

725. Shining Fetterbush

Lyonia lucida (Lamarck) K. Koch
    Ly-òn-i-a lù-ci-da
Ericaceae (Heath Family)

DESCRIPTION: Rhizomatous, evergreen shrub to 8′ tall; usually forming dense colonies; first-year twigs strongly angled; leaves alternate, elliptic to obovate, 1–3″ long, smooth, glossy, leathery, with a distinct marginal vein that is thick and yellowish; urn-shaped white or pinkish flowers produced in fascicles from the axils of the leaves; April–early June.

RANGE-HABITAT: Endemic to the Coastal Plain from VA south to FL and west to LA; in SC, common in the Coastal Plain and Sandhills in pocosins, Longleaf Pine and Pond Cypress savannas, blackwater swamps, and wet woodlands.

COMMENTS: Fetterbush is often the dominant shrub in pocosins, forming dense colonies that make walking difficult–hence, it "fetters" one's path. It sprouts vigorously from rhizomes after a fire, quickly becoming reestablished. The thick, yellowish veins along the margins of the leaf are an easy way to distinguish it from the many other evergreen shrubs in SC pocosins.

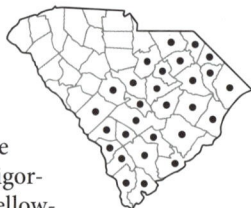

726. Honey-cups

Zenobia pulverulenta (Bartram) Pollard
Ze-nò-bi-a pul-ve-ru-lén-ta
Ericaceae (Heath Family)

DESCRIPTION: Rhizomatous, deciduous shrub to 6′ tall; leaves of two forms: either green on both surfaces or green above and bluish-white (glaucous) beneath; both forms occur together; leaves 1.5–2.5″ long, elliptic to elliptic-ovate, thick and coriaceous with a scalloped (crenate) margin; the broad urn-shaped flowers are highly fragrant, like honey, produced in racemelike clusters grouped in an elongate arrangement on the upper part of twigs of the preceding season; flowers April–June.

RANGE-HABITAT: Endemic to the Coastal Plain from VA south to GA; in SC, common in the Coastal Plain and Sandhills in pocosins and margins of pineland ponds.

COMMENTS: This is one of SC's most spectacular ericaceous shrubs when in flower. The genus name comes from Zenobia, the queen of Palmyra, Syria, in the third century of the Common Era.

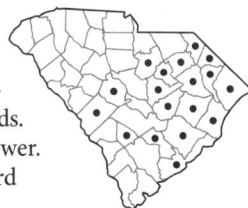

727. Clammy Azalea; Swamp Honeysuckle

Rhododendron viscosum (L.) Torrey
Rho-do-dén-dron vis-cò-sum
Ericaceae (Heath Family)

DESCRIPTION: Deciduous shrub with hairy twigs, to 6′ tall; leaves ovate to obovate, 0.75–2.75″ long, margins serrulate and ciliate, green above, green or bluish-white (glaucous) below; corolla covered with reddish, sticky hairs; petals white or rarely pink; calyx and leaf stalks densely covered with stalked glands; flowers fragrant, appearing after the leaves; flowers May–July.

RANGE-HABITAT: Widespread in the eastern US; common in the Sandhills and Coastal Plain and uncommon in the Piedmont and mountains; pocosins, moist streamsides, rocky streamsides, cataract fens, and montane bogs.

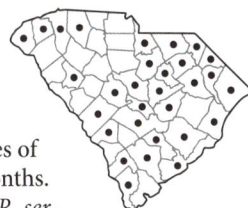

COMMENTS: Clammy Azalea flowers after most other species, and it makes a nice addition to the garden. When planted with other earlier flowering species of native deciduous azaleas, it extends the bloom time well into the summer months. The common name refers to the sticky nature of the flowers. Swamp Azalea (*R. serrulatum* (Small) Millais, plate 502) is similar but the bud scales are more numerous, 15–20 (vs. 8–12 in Clammy Azalea), and the corolla tube is smooth within (vs. pubescent in Clammy Azalea). Additionally, Swamp Azalea often grows in densely shaded habitats and flowers much later, in June–October.

728. Titi; Leatherwood

Cyrilla racemiflora L.
Cy-ríl-la ra-ce-mi-flò-ra
Cyrillaceae (Titi Family)

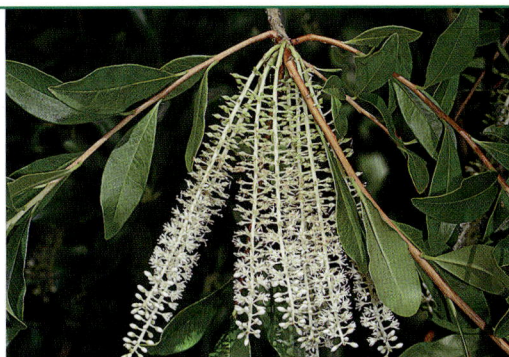

DESCRIPTION: Small shrub or medium-sized tree to 25′ tall; commonly reproducing vegetatively by sprouts from shallow roots and forming dense thickets; leaves to 4″ long, smooth, thick and leathery, alternate, obovate, oblanceolate or spatulate, with entire margins, semi-evergreen, some falling throughout the winter, but a few remaining until new ones appear in the spring; white flowers, in racemes, clustered near the end of the previous year's twig; flowers May–July. **RANGE-HABITAT**: Common Coastal Plain species from VA south to FL and west to TX; in SC, common in the Sandhills and Coastal Plain in pocosins, swamp edges, Longleaf Pine flatwoods, savannas, bogs, and streamsides. **COMMENTS**: Titi is a good ornamental because of its attractive flowers and leaves, which turn orange and scarlet in the fall. The trees are good honey plants because the flowers produce large quantities of nectar. Under various ecological conditions, Titi can range from a small shrub to a medium-sized tree. The genus honors Domenico Cirillo (1734–99), a professor of medicine in Naples, Italy.

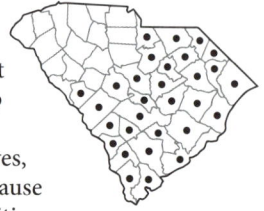

729. Loblolly Bay

Gordonia lasianthus (L.) Ellis
Gor-dòn-i-a la-si-án-thus
Theaceae (Tea Family)

DESCRIPTION: Evergreen shrub or small- to medium-sized tree to 75′ tall; leaves leathery, smooth, simple, alternate, elliptic to oblanceolate, 3–6″ long, shallowly serrate; generally with a few orange-red leaves visible at any season; flowers resemble a white camelia, produced in the axils of leaves; flowers July–September; fruit a capsule; matures September–October.
RANGE-HABITAT: Endemic to the Coastal Plain from NC south to FL and west to MS; in SC, common in the Coastal Plain and Sandhills in pocosins, bay forests, blackwater swamp forests, wet Longleaf Pine savannas, and Atlantic White-cedar forests.
COMMENTS: Loblolly Bay is among SC's most beautiful flowering trees. The beauty of the flowers would make it an exceptional landscape plant, but it is notoriously difficult in cultivation. The genus honors George Gordon (1806–79), a British botanist.

Loblolly Bay often grows in association with two other bay trees, Sweet Bay (*Magnolia virginiana* L., plate 722) and Swamp Bay (*Persea palustris* (Rafinesque) Sargent, plate 721). The leaves of Loblolly Bay are odorless when crushed, while the leaves of Swamp Bay and Sweet Bay are aromatic. The leaves of Sweet Bay are strongly whitish beneath, while the leaves of Loblolly Bay are olive-green beneath. The shallow serrate teeth on the margin of Loblolly Bay also help to distinguish it from both of these other "bay" species.

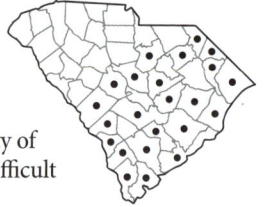

730. Blaspheme-vine; Bamboo-vine

Smilax laurifolia L.
Smì-lax lau-ri-fò-li-a
Smilacaceae (Greenbrier Family)

DESCRIPTION: Stout, evergreen, high-climbing vine; dead stems often intermixed with the living, forming impenetrable thickets; stems armed with large, very hard, dark thornlike projections; leaves thick and leathery, smooth, oblong to oblong-linear, with entire margins, and an abruptly acute tip; flowers in umbels in the leaf axils; perianth is greenish; flowers July–August; fruit a bluish-black berry with a whiteish waxy tinge, maturing September–October of the second year.
RANGE-HABITAT: NJ south to FL and west to AR and TX; in SC, common in the Coastal Plain and Sandhills, rare in the Piedmont and mountains; pocosins, blackwater swamp forests, seepage over granitic flatrocks, cataract fens, stream margins, and bottomlands.

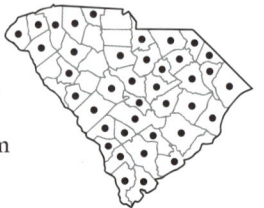

COMMENTS: The term Blaspheme-vine seems very appropriate to anyone who has had the "pleasure" of walking through a section of pocosin. The tough stems with abundant thornlike projections quickly shred the skin and may elicit a few choice words–hence, the blaspheme! This species is predominantly found in the Coastal Plain but appears in the mountains and Piedmont, often with other typical Coastal Plain species.

731. Pocosin Sedge; Walter's Sedge

Carex striata Michaux var.
 brevis L. H. Bailey
  Cà-rex stri-à-ta var. bré-vis
Cyperaceae (Sedge Family)

SYNONYM: *Carex walteriana*
 Bailey—RAB

DESCRIPTION: Rhizomatous, perennial herb, 20–28″ tall; blades green or yellowish-green, less than 0.1″ wide; flowers produced in spikes; the female spikes with yellowish-green perigynia, all held erect; flowers and fruits May–June.

RANGE-HABITAT: MA south to FL; in SC, common throughout the Coastal Plain where found in pocosin (particularly in openings and flooded areas), Pond Cypress-Swamp Gum upland swamps, and other open, boggy areas.

COMMENTS: Pocosin Sedge is the only common *Carex* with light green leaves that forms extensive colonies by the elongate rhizomes in the pocosin and depression pond habitat. It is ecologically important in this habitat. The species is often separated into two different varieties in SC. *C. striata* var. *brevis* L. H. Bailey has smooth perigynia and reaches its southern range limit in SC. *C. striata* var. *striata* has pubescent perigynia and reaches its northern range limit in SC. It is found in similar habitats but is especially abundant in Pond Cypress-Swamp Gum upland swamps.

732. Carolina Sweet Pitcherplant

Sarracenia rubra Walter ssp. *rubra*
 Sar-ra-cèn-i-a rù-bra ssp. rù-bra
Sarraceniaceae (Pitcherplant Family)

DESCRIPTION: Carnivorous, rhizomatous, perennial herb; leaves modified into slender hollow tubes (pitchers), 4–20″ tall, as passive traps that catch insects; flowering stalk usually exceeding the leaves; petals maroon on outer surface, greenish on inner; flowers April–May.

RANGE-HABITAT: Endemic to the Coastal Plain from NC south to GA; in SC, occasional in the Coastal Plain and Sandhills where found in openings and along the edges of pocosins, herbaceous seepages, and in wet Longleaf Pine savannas.

SIMILAR SPECIES: Mountain Sweet Pitcherplant (*S. jonesii* Wherry, plate 57) is restricted to cataract fens and seepage over granitic flatrocks in the mountains and upper Piedmont. It has larger pitchers with the opening exposed and with a distinct notch. Georgia Sweet Pitcherplant (*S. rubra* ssp. *viatorum* B. Rice) is confined to the Sandhills region where it is found in streamhead pocosins, White Cedar swamp forests and herbaceous seepages in the Fall-line Sandhills of Aiken and Lexington Counties. It has taller leaves with a wider opening and the pitcher "lid" is cordate (weakly heart shaped); it is straplike in *S. rubra* ssp. *rubra*.

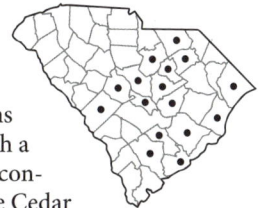

COMMENTS: Carolina Sweet Pitcherplant gets the name from the pleasant odor of the flowers. It appears to grow more robust in seepage areas along the edge of pocosins and in sphagnum openings in pocosins. In the pine savannas, where it is also found, it is often diminutive. The genus honors Dr. Michel Sarrasin de l'Étang (1659–1734), physician at the Court of Quebec, who sent a sample of the genus to Europe.

733. Rose Pogonia

Pogonia ophioglossoides
   (L.) Ker-Gawler
      Po-gò-ni-a o-phi-o-glos-
      soì-des
Orchidaceae (Orchid Family)

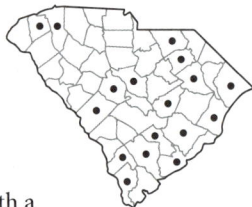

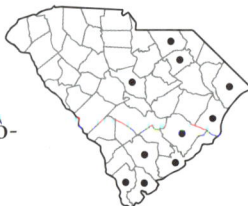

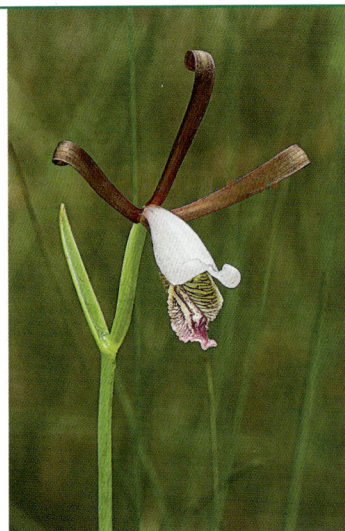

DESCRIPTION: Slender orchid, 3–24″ tall with a single, green leaf about halfway up the stem; stem supports a single flower subtended by a leaflike bract; flowers rose pink to whitish, fading to pink, the tepals up to 2.5″ long; lip bears 3 rows of fleshy hairs tipped with yellow or brown; flowers May–June.

RANGE-HABITAT: Widespread in eastern North America; in SC, occasional in the Coastal Plain and Sandhills growing along the margins of pocosins, in openings within pocosins, herbaceous seepages, Pond Cypress savannas, wet Longleaf Pine savannas, and ditches through these habitats; rare in the Piedmont and mountains in cataract fens, seepage over granitic flatrocks, and montane fens.

COMMENTS: *Pogon* is Greek for "beard," referring to the hairs on the lip. This is a very beautiful orchid, often forming large, diffuse colonies in good habitat via shoots arising from the roots. The flowers are very small, and many people are surprised at their size when comparing them to photographs in wildflower books that don't include scale. Due to their rapid rate of reproduction and ease of cultivation in bog gardens, this makes an excellent companion for plantings of carnivorous plants. It should not be disturbed in the wild. Plants can be obtained from nursery-propagated sources including at the South Carolina Botanical Garden in Clemson.

734. Small Coastal Plain Spreading Pogonia

Cleistesiopsis oricamporum
   P. M. Brown
      Cleis-te-si-óp-sis o-ri-cam-pò-
      rum
Orchidaceae (Orchid Family)

SYNONYM: *Cleistes divaricata* (L.) Ames, in part—RAB, PR

DESCRIPTION: Perennial herb, 8–24″ tall; entire plant has a bluish green color with a fine, frosty-white coating; typically one leaf, inserted above the middle of the stem and a bract produced just below the flower; bract is equal to or shorter than the flower; sepals widely spreading and often curled backward toward the tips, dark maroon to brownish-purple; petals 1–1.5″ long, projected forward, white to pale pink; the lip with the basal three-fourths of the central keel with 5–7 discontinuous ridges; flowers have the distinct odor of vanilla when fresh; April–June.

RANGE-HABITAT: Endemic to the Coastal Plain from VA south to FL and west to LA; in SC, occasional in the Coastal Plain and Sandhills where found in Longleaf Pine savannas and wet flatwoods, herbaceous seepages, and ecotones of pocosins.

SIMILAR SPECIES: See discussion under Large Spreading Pogonia (*C. divaricata* (L.) Pansarin & F. Barros, plate 624).

COMMENTS: Small Coastal Plain Spreading Pogonia is like an earlier flowering, smaller version of the showier Large Spreading Pogonia. Where they co-occur, it flowers approximately 10 days earlier. Populations containing both species occur at several sites in the Francis Marion National Forest, and they appear distinct in size, flowering time, and odor.
CONSERVATION STATUS: SC-Vulnerable

735. Spoonflower; White Arrow-arum

Peltandra sagittifolia (Michaux)
   Morong
      Pel-tán-dra sa-git-ti-fò-li-a
Araceae (Arum Family)

DESCRIPTION: Perennial herb from a short, stout vertical rhizome; leaves sagittate (arrowhead-shaped), green or bluish-green above and bluish-green beneath; inflorescence a spadix subtended by a flared, open, white spathe; flowers of two sexes, with the male flowers on the upper part of the spadix and the female flowers on the lower part; fruits are brilliant reddish-orange berries; flowers July–August.
RANGE-HABITAT: Endemic to the Coastal Plain from NC south to FL and west to MS; in SC, rare, found in the Coastal Plain in sphagnum-dominated openings in pocosins.
COMMENTS: All parts of the plant contain crystals of calcium oxalate, which can cause irritation of the mucus membranes of the mouth and throat. The genus comes from the Greek *pelte*, meaning shield, and *andros*, meaning male and refers to the shape of the spathe. The specific epithet refers to the arrowhead shape of the leaves (sagittate). It can be locally quite common following large fires that create openings in extensive pocosins.
CONSERVATION STATUS: SC-Imperiled

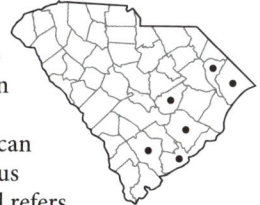

736. Large White Fringed Orchid

Platanthera conspicua
   (Nash) P. M. Brown
      Pla-tán-the-ra con-spí-cu-a
Orchidaceae (Orchid Family)

SYNONYM: *Habenaria blephariglottis*
   (Willdenow) Hooker, in part—RAB

DESCRIPTION: Robust perennial orchid, 13–40″ tall; leaves ovate-lanceolate to linear-lanceolate; flowers produced in a loosely flowered bracteate raceme; flowers white with 1–2″ long nectar spurs and with a lip that is ornately long-fringed; flowers May–June.
RANGE-HABITAT: Endemic to the Coastal Plain from NC south to FL and west to TX; in SC, uncommon in the Coastal Plain and Sandhills where found in edges of pocosins, openings in pocosins, herbaceous seepages, and wet savannas.
SIMILAR SPECIES: Small White Fringed Orchid (*P. blephariglottis* (Willdenow) Lindley) is a smaller plant with a congested raceme that has flowers with a shorter nectar spur (0.6–1.0″ long) and a lip that has a short, irregular fringe. It is also uncommon in similar habitats in the Coastal Plain and Sandhills. Until recently, the two species were thought to be varieties of the same species. Small White Fringed Orchid is a more northern plant ranging from Newfoundland south to GA.
COMMENTS: Large White Fringed Orchid is one of the most startling and beautiful orchids in SC. It should never be disturbed in the wild.

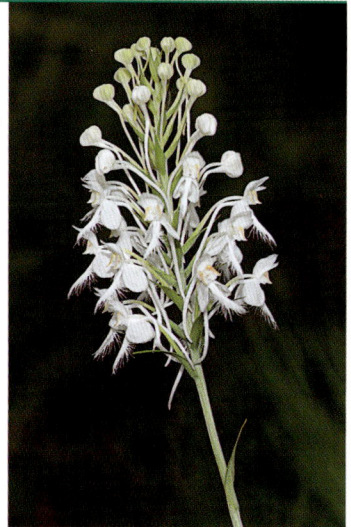

737. Virginia Chain Fern

Anchistea virginica (L.) C. Presl
An-chís-te-a vir-gí-ni-ca
Blechnaceae (Deer Fern Family)

SYNONYM: *Woodwardia virginica* (L.)
Smith—RAB

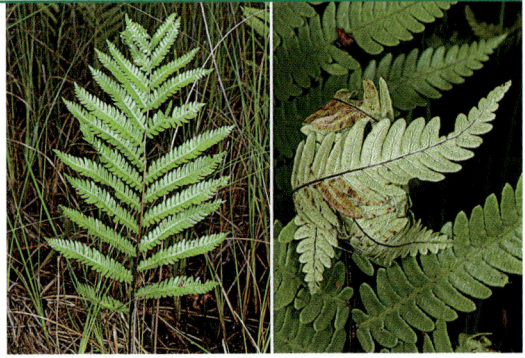

DESCRIPTION: Colonial, rhizomatous, deciduous fern from stout rhizomes covered with lanceolate reddish-brown scales; leaves 2–3.5′ long, held erect, petioles similar in length to the blade, very dark; blade 1-pinnatifid with pinnae divided to near the midrib; spores produced in linear sori along the midrib and lobes; spores produced June–September.

RANGE-HABITAT: Nova Scotia south to FL and west to MI and TX; in SC, common in the Coastal Plain and Sandhills, often forming large colonies in pocosins, particularly in openings within pocosins, wet savannas, pocosin ecotones, and depression pond habitats.

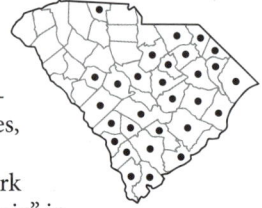

COMMENTS: Virginia Chain Fern is distinctive because of its upright leaves, dark blackish petiole, strongly rhizomatous habit, and habitat preferences. The "chain" in the common name comes from the arrangement of the veins along the midrib of the leaflets, which appear to form chains. Virginia Chain Fern is often a host to leaf-roller moths. The caterpillars roll the leaflets into balls and house themselves inside as they feed.

738. Calcareous bluff forest

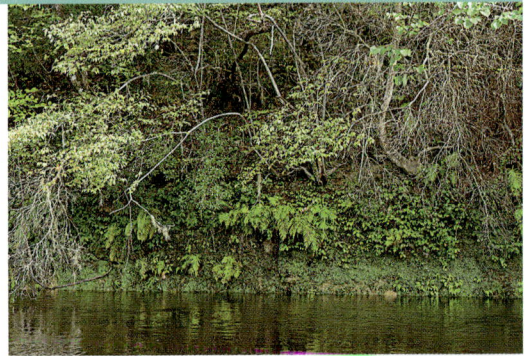

739. Wet, flat, calcareous forest
(marl forest)

740. Bluff Oak

Quercus austrina Small
Quér-cus aus-trì-na
Fagaceae (Beech Family)

DESCRIPTION: Large deciduous tree with gray-ish, ridged, and scaly bark; leaves obovate 3–6″ long and 1–3″ wide, 3–7 lobed with blunt apices that are not bristle-tipped; the sinuses of the lobes indented from one-fourth to one-half of the way to the midrib; female flowers 1–2 on a short stalk, separate from the male flowers, which are in catkins; fruit is an acorn, the cup with scales thin and appressed, maturing the first year; flowers March–April; fruits mature in October.

RANGE-HABITAT: NC south to FL and west to MS; disjunct in AR; in SC, rare in the Coastal Plain where restricted to high-pH soils of calcareous bluff forests, marl forests on well-drained sites, and shell hammocks.

SIMILAR SPECIES: White Oak (*Q. alba,* plate 340) is similar and may be found growing with Bluff Oak but has leaves that have 7–11 lobes with deep sinuses (indented two-thirds to five-sixths of the way to the midrib). It is widespread and common in SC. Bastard Oak (*Q. durandii* Buckley var. *durandii*) is restricted to the Piedmont where it occurs on slopes, bluffs, glades, and hardpan woodlands. It has significantly smaller leaves (1.5–4.0″ long) with 1–5 shallow undulations rather than distinct lobes.

COMMENTS: Bluff Oak is an often overlooked but typical species of well-drained sites on high pH substrate in the Coastal Plain. It is often associated with other rare calcium-loving plants (calcicoles).

CONSERVATION STATUS: SC-Vulnerable

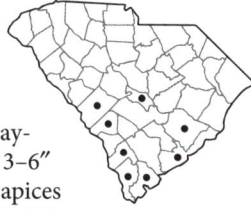

741. Swamp Post Oak

Quercus similis Ashe
Quér-cus sí-mi-lis
Fagaceae (Beech Family)

DESCRIPTION: Large deciduous tree with grayish-brown, scaly bark; twigs grayish and stellate pubescent; leaves obovate, 3–5″ long and 2–3″ wide, with 2–3 lobes on each side, not forming a distinct "cross," green above with sparse stellate pubescence, undersides of leaves stellate pubescent; female flowers 1–2 on a short stalk, separate from the male flowers, which are in catkins; fruit is an acorn, maturing the first year; flowers March–April; fruits mature in October.

RANGE-HABITAT: SC south to GA and west to TX; disjunct in central TN; in SC, rare in the Coastal Plain and confined to wet, flat, calcareous forests (marl flats).

SIMILAR SPECIES: Post Oak (*Q. stellata* Wangenheim, plate 376) is similar. It has at least some leaves where the lobes form a distinct cross and has no stellate pubescence on the upper surface of the mature leaves. It occupies much drier habitats in the uplands throughout SC.

COMMENTS: Swamp Post Oak is a distinctive member of the White Oak group that grows in saturated or moist soils with high pH. It is often a good indicator for locating other rare calcium-loving species (calciphiles).

CONSERVATION STATUS: SC-Imperiled

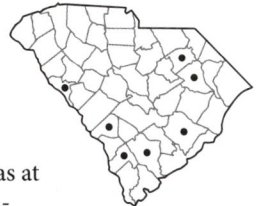

742. Red Buckeye

Aesculus pavia L. var. *pavia*
  Aès-cu-lus pà-vi-a var. pà-vi-a
Hippocastanaceae (Buckeye Family)

DESCRIPTION: Deciduous shrub or small tree, seldom over 20–25′ tall; leaves palmately compound with 5 leaflets; flowers scarlet red; stamens longer than the lateral petals; flowers March–May; capsule leathery, with 1–3 large, dark brown, shiny seeds; matures July–August.

RANGE-HABITAT: NC south to FL and west to TX; in SC, common in the Coastal Plain and Sandhills and extending into the edge of the Piedmont; wet, flat, calcareous forests; calcareous bluff forests; shell hammocks; shell middens; calcium-rich maritime forests; bottomland hardwood forests; and basic-mesic forests.

SIMILAR SPECIES: Painted Buckeye (*Aesculus sylvatica* Bartram, plate 313) frequently has salmon-colored flowers in the upper Piedmont of SC. These plants apparently have a few genes that have been transported via pollen from Red Buckeye to create hybrid plants that are genetically mostly Painted Buckeye, while maintaining a tinge of color through genes obtained from Red Buckeye (Thomas et al., 2008). Painted Buckeye is a shorter plant with a more congested inflorescence and flowers that have stamens as long as to shorter than the lateral petals. Painted Buckeye is restricted to the Piedmont and lower elevations of the mountains in SC.

COMMENTS: Red Buckeye is an excellent ornamental in shady situations. The flowers are pollinated primarily by hummingbirds but are also visited by butterflies and other pollinators. The crushed fruits were once used to stun fish for easy harvest. Like all buckeyes, the fruits are toxic to humans.

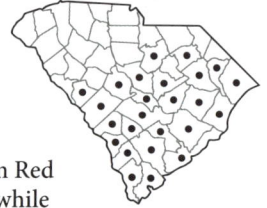

743. Southern Sugar Maple; Florida Maple

Acer floridanum (Chapman) Pax
  À-cer flo-ri-dà-num
Aceraceae (Maple Family)

SYNONYMS: *Acer saccharum* ssp. *floridanum* (Chapman) Desmarais—RAB; *Acer barbatum*—PR

DESCRIPTION: Small to large, deciduous tree, usually 40–60′ tall, typically with a single trunk; leaves simple, opposite, almost circular in outline, 1.5–4″ across, lower surface whitish and pubescent; lobes 3–5, the terminal lobes of some leaves broader toward the apex than toward the base; sinuses rounded; flowers April–May; fruits June–October.

RANGE-HABITAT: VA south to FL and west to TX and OK; in SC, common in the Coastal Plain and Piedmont; basic-mesic forests, calcareous bluff forests, oak-hickory forests with circumneutral or magnesium-rich soils, bottomland hardwood forests, and shell hammock forests.

SIMILAR SPECIES: Chalk Maple (*A. leucoderme* Small, plate 268) is similar and occurs in the mountains, Piedmont, and Inner Coastal Plain. Chalk Maple usually has multiple trunks and leaves greenish yellow beneath, and the terminal lobe is narrower toward the apex than the base, with lobe tips pointed and often drooping.

COMMENTS: Southern Sugar Maple is not tapped commercially for sap. The wood is of good quality but is not much used because of the limited supply. It is often planted as a small shade or lawn tree because of its bright yellow-to-red autumn leaves and makes an excellent replacement in the South for the less heat tolerant Sugar Maple or Norway Maple. It is also resistant to wind and ice damage. The state champion Southern Sugar Maple was 135′ tall and 31″ in diameter in 1991.

744. Nutmeg Hickory

Carya myristiciformis Michaux f.
Cà-ry-a my-ris-ti-ci-fór-mis
Juglandaceae (Walnut Family)

DESCRIPTION: Large, deciduous tree with grayish to brownish bark, fissured and sometimes exfoliating into long curly strips or broad plates; buds with imbricate bronze, ovoid, densely punctate scales; leaves odd-pinnately compound, alternate, with 7–9 leaflets; leaflets with serrate margins and undersides with peltate scales that are silver when young and turn bronze with age; fruit a nut; fruit husks with the sutures winged from the base to the tip.

RANGE-HABITAT: NC south to GA and west to OK and TX; in SC, rare in the Coastal Plain, but locally abundant in wet, flat, calcareous forests, almost always on Meggett series soils.

SIMILAR SPECIES: Bitternut Hickory (*C. cordiformis* (Wangenheim) K. Koch), with which Nutmeg Hickory often grows, can be easily confused with this species. The buds of Bitternut Hickory are generally sulfur yellow, without obvious scales (naked), and the husk has sutures with wings confined to the upper portions. It also typically has more slender leaflets. The dense, persistent, peltate scales on the undersides of the leaflets also serve to distinguish Nutmeg Hickory.

COMMENTS: As the name suggests, the nut is edible and as tasty as pecan nuts. The scales on the buds, twigs, and undersurfaces of leaves often have a metallic-bronze sheen that is associated with the numerous peltate scales. The distribution of this species is curious as it is disjunct hundreds of miles between centers of populations in the South. The range is restricted by high-pH soils, and the large, heavy fruits are not conducive to long-distance dispersal. How it found the isolated high-pH soil locations in the Carolinas is not clear. The disjunct occurrence of Shagbark Hickory (*C. ovata* (Miller) K. Koch, plate 269) is similarly curious, with the populations in Berkeley County disjunct many miles from the closest Piedmont populations. Nutmeg Hickory can be easily seen along the floodplain of Huger Creek near the bridge on SC 402 in the Francis Marion National Forest.

CONSERVATION STATUS: SC-Imperiled

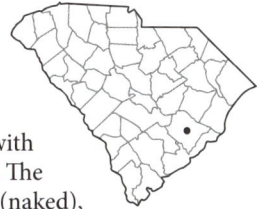

745. Black Walnut

Juglans nigra L.
Jùg-lans nì-gra
Juglandaceae (Walnut Family)

DESCRIPTION: Large tree, to 100–130′ tall; distinguished in winter by twigs with a light-brown, chambered pith; leaves deciduous, alternate, odd-pinnately compound, with 15–23 leaflets; fruit is actually a drupe though often referred to as a nut; drupe with thick husks, green to yellow-green, dark brown at maturity; shells deeply furrowed; kernel oily, sweet; fruits mature in October.

RANGE-HABITAT: Virtually throughout eastern North America and in states west and adjacent to the Mississippi River; common throughout SC; primarily a tree of moist, nutrient-rich forests, including bottomlands and floodplains.

COMMENTS: The beautifully grained, brownish-colored wood makes Black Walnut one of the finest lumber trees in North America. It is used for paneling, furniture, and gun stocks. Today, because the supply of mature trees is limited, it is mostly used for veneer. The wood's ability to absorb the recoil of a gun discharging without damage, and the fact that the wood does not shrink or warp with age, makes it ideal for custom-crafted gunstocks.

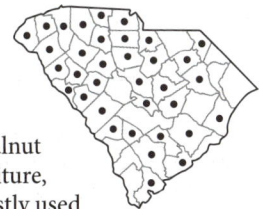

Though many references refer to the fruit as a nut, it is a single-seeded indehiscent fleshy fruit and thus a drupe. Black Walnut fruits are used in candies, confections, and ice creams. The fruits are an important wildlife food. Historically, the inner bark was used as a mild laxative, especially during the American Revolution. The peel and juice of the fruit were used as a vermifuge for intestinal worms. Duke (1997) recommends walnut as an herbal remedy for athlete's foot. Extracts are available as tinctures and capsules as herbal remedies for a variety of ailments.

Black Walnut seldom occurs in pure stands in the wild because they produce juglone in their leaves and roots, which inhibits germination and growth of new walnut trees (and many other plants), and thus limits competition for soil nutrients. Black Walnut was often planted around homes, where it often persists after abandonment.

746. Prickly-ash

Zanthoxylum americanum
  P. Miller
  Zan-thóx-y-lum a-me-ri-
  cà-num
  Rutaceae (Rue Family)

DESCRIPTION: Deciduous, aromatic shrub or small tree with short, paired spines at the base of the leaves (stipular spines); plants in SC rarely exceed 3′ in height; leaves thin-textured, odd-pinnately compound, with a spiny rachis; leaves bitter-aromatic; flowers in small axillary clusters; flowers March–April, before the leaves.
RANGE-HABITAT: From Quebec south to SC and GA and west to ND and OK; in SC, very rare in the Coastal Plain in wet, flat, calcareous forests.
COMMENTS: The wood has no commercial value. The oil that occurs throughout the plant, especially the bark, has medicinal qualities. F. P. Porcher (1869) gives a long discourse on its medicinal properties. Foster and Duke (1990) restate much of Porcher's information: "Bark tea tincture [was] historically used by American Indians and herbalists for chronic rheumatism, dyspepsia, dysentery, kidney trouble, heart trouble, colds, coughs, lung ailments, and nervous debility. . . . Bark [was] chewed for toothache." This species and Hercules'-club (*Zanthoxylum clava-herculis* L., plate 904) were used medicinally to numb the mouth, either with an extract from the fruits and leaves or by chewing the inner bark. Four populations of this rare shrub have been located in wet, flat, calcareous forests in Berkeley County west of the Western Branch of the Cooper River.

If the genus was made into syllables according to classical Latin, as used in this book, it would be Zan-thox-y-lum, since x is always placed with the vowel preceding it. The authors choose the common usage since the genus is derived from the Greek *Xanthos*, yellow, and *xylon*, wood.
CONSERVATION STATUS: SC-Critically Imperiled

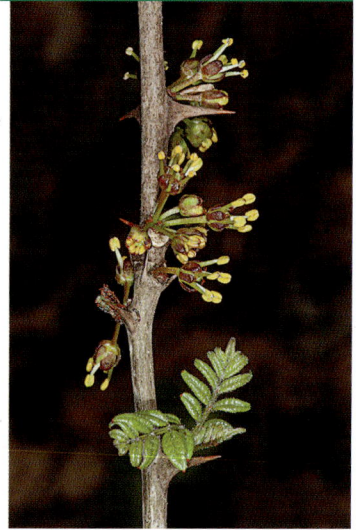

747. Eastern Roughleaf Dogwood

Swida asperifolia (Michaux) Small
  Swí-da as-pe-ri-fò-li-a
  Cornaceae (Dogwood Family)

SYNONYM: *Cornus asperifolia*
  Michaux—RAB, PR

DESCRIPTION: Deciduous shrub to 15′ tall; twigs rough to the touch; leaves simple, opposite, lance-elliptic, rough-pubescent on both surfaces; flowers produced in terminal cymes; petals 4, white; fruits are bluish drupes; flowers May–June; fruits mature August–September.

RANGE-HABITAT: Endemic to the Coastal Plain from NC south to FL and west to AL; in SC, occasional in wet, flat, calcareous forests; margins of hardwood bottoms; shell hammocks; and shell middens; always in areas with elevated calcium in the soil.

COMMENTS: Native dogwood species have recently been shown to be distinct from the type species of the genus *Cornus* and thus have been transferred to different genera, no doubt causing considerable consternation among horticulturalists and botanists alike. Eastern Roughleaf Dogwood is among the easiest of the species to distinguish, because the rough, pubescent leaves are unique among SC's species.

748. Needle Palm

Rhapidophyllum hystrix (Pursh) H. Wendland & Drude
Rha-pi-do-phýl-lum hýs-trix
Arecaceae (Palm Family)

DESCRIPTION: Small- to medium-sized palm, with the stature of a shrub, never producing an elongated above-ground trunk; stem with conspicuous, dark, elongated spines to 18″ long, with the spine base obscured by fibers; leaves palmately lobed and pleated, less than 30″ wide; flowers yellowish, produced at the base of the leaves; fruits are ellipsoid brownish drupes; flowers April–May; fruits produced July–September.

RANGE-HABITAT: SC south to FL and west to MS; in SC, restricted to the southern Coastal Plain where found in wet, flat, calcareous forests; calcareous bluff forests; shell hammocks; and middens.

COMMENTS: Though this is the hardiest species of palm in North America, it is not found as far north as other palm species. It makes a spectacular ornamental in the shade where the leaves are more spread out, highlighting the distinctive, attractive leaves. It will also tolerate full sun but becomes dense and crowded, losing the elegant appearance of the individual fronds.

CONSERVATION STATUS: SC-Imperiled

749. Mottled Trillium

Trillium maculatum Rafinesque
Tríl-li-um ma-cu-là-tum
Trilliaceae (Trillium Family)

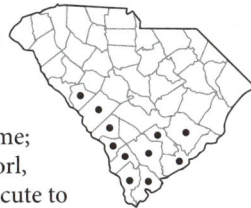

DESCRIPTION: Perennial herb from a rhizome; stems 4–12″ tall; leaves simple, 3, in a whorl, mottled with 2 or 3 shades of green, tips acute to acuminate; flower sessile, with the scent of spicy bananas; petals 3, held erect, dark red, brownish or yellow, narrowly spoon-shaped to linear, length 4–5 times the width; flowers January–March.

RANGE-HABITAT: SC south to FL and west to AL; in SC, scattered but locally common in the Coastal Plain and maritime strand, rare in the lower Piedmont; calcareous bluff forests; wet, flat, calcareous forests; shell hammocks and middens; beech forests; and basic mesic forests.

COMMENTS: This species is immediately recognizable in the SC Coastal Plain as the only sessile-flowered *Trillium* species. Yellow-flowered forms and those with brownish-colored flowers are simply part of the genetic variation in flower color. In the lower Piedmont it may be distinguished from the similar Little Sweet Betsy (*Trillium cuneatum* Rafinesque, plate 108) by the longer petals that are brighter in color and the odor. Little Sweet Betsy either has a foul odor or smells similar to sweet concord grapes, like Sweetshrub (*Calycanthus floridus* L., plate 80).

750. Carolina Buttercup

Ranunculus septentrionalis Poiret in
    Lamarck
        Ra-nùn-cu-lus sep-ten-tri-o-nà-lis
    Ranunculaceae (Buttercup Family)

SYNONYM: *Ranunculus carolinianus* A. P.
    de Candolle—RAB, PR

DESCRIPTION: Weakly spreading, smooth perennial herb with rhizomes and cordlike roots, forming colonies; stems to 30″ long; leaves typically
3-foliolate; terminal leaflet approximately the
same size as the lateral leaflets; sepals reflexed; petals bright to pale yellow;
fruit an aggregate of achenes; flowers April–July.

RANGE-HABITAT: NY south to FL and west to MN and TX; in SC, scattered
but common in the Coastal Plain and Piedmont; wet, flat, calcareous
forests; swamps; low woods and banks of wooded streams; and floodplain
forests.

SIMILAR SPECIES: Early Buttercup (*R. fascicularis* Muhlenberg ex Bigelow) is similar
but does not form the large, open colonies of Carolina Buttercup. It also has leaves
that are typically 5–7 foliolate with the terminal leaflet considerably larger than the lateral
leaflets. It is found in cove forests and oak-hickory forests in the mountains and Piedmont.

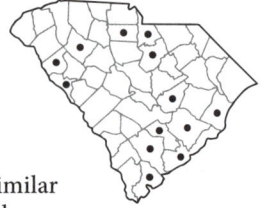

751. Green Dragon

Arisaema dracontium (L.) Schott
        A-ri-saè-ma dra-cón-ti-um
    Araceae (Arum Family)

DESCRIPTION: Smooth, perennial herb, 1–3′
tall; leaf solitary, divided with 7–15 segments
arranged along a semi-circular axis; the segments lanceolate with long-tapering tips; flowers
produced in a spadix with male flowers on the
upper portions and female flowers below; spadix
subtended by a narrow green spathe; the sterile
extension of the spadix axis greatly exceeding the
spathe; fruits are red berries; flowers May; fruits mature in July.

RANGE-HABITAT: Widespread in eastern North America, from Quebec south to
FL and west to WI and TX; in SC, occasional throughout in floodplain forests; basic mesic forests; rich cove forests; wet, flat, calcareous forests; and other
moist forests with circumneutral soils.

COMMENTS: Green Dragon is easily distinguished from all other species of *Arisaema*
by the much more divided leaves and the curious extension of the axis of the spadix
into a "tail-like" structure. There are many species of *Arisaema* in Asia that exhibit the structure of Green Dragon and our more familiar Jack-in-the-pulpits.

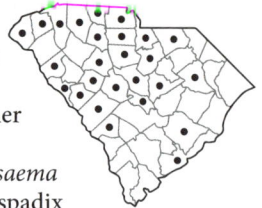

752. Spring Coralroot

Corallorhiza wisteriana Conrad
Co-ral-lo-rhì-za wis-ter-i-à-na
Orchidaceae (Orchid Family)

DESCRIPTION: Small, slender, leafless orchid, without chlorophyll, mycoparasitic; stems yellowish or purplish, to 17″ tall; flowers produced in a raceme that is loosely flowered; individual flowers small with brownish or reddish sepals and lateral petals; the lip petal is whitish with purple spots; flowers April.

RANGE-HABITAT: NJ south to FL and west to OK and TX; in SC, scattered and uncommon throughout; basic-mesic forests, calcareous bluff forests, rich cove forests, and other forests with circumneutral soils.

SIMILAR SPECIES: Autumn Coralroot (*Corallorhiza odontorhiza* (Willdenow) Poiret, plate 174) is a smaller species without a showy lip and flowers in the autumn. It is occasional throughout SC in hardwood forests.

COMMENTS: Several species of orchids in SC lack chlorophyll (have no green parts) and are not photosynthetic. These orchids are mycoparasites that use fungi to gain their carbohydrates. The fungus obtains the carbohydrates through the decomposition of organic material. Virtually all orchids, whether they are photosynthetic or not, have fungi associated with their roots. Since all orchids produce tiny seeds without stored food, they depend on root-associated fungi for germination and early growth. The species is named for its discoverer, Charles Wister (1782–1865), a Pennsylvanian botanist.

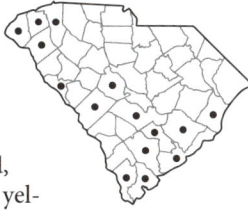

753. Coastal Plain Oxeye

Heliopsis gracilis Nuttall
He-li-óp-sis grá-ci-lis
Asteraceae (Aster Family)

DESCRIPTION: Erect, perennial herb with stems 1–2.5′ tall; leaf blades ovate to ovate-lanceolate, 1.5–3.5″ long, with serrate margins and acuminate tips; flowers arranged in involucrate heads of yellow ray and yellow disk flowers; receptacle conic; 1–5 heads per plant; flowers April–July.

RANGE-HABITAT: SC south to FL and west to LA; in SC, occasional in the Coastal Plain in wet, flat, calcareous forests; calcareous bluff forests; and roadsides through these habitats.

SIMILAR SPECIES: Common Oxeye (*H. helianthoides* (L.) Sweet) is a much larger plant flowering in the summer through early fall. It is found in the Piedmont and mountains in rich, moist forests and forest margins.

COMMENTS: Coastal Plain Oxeye is one of several species formerly thought to be endemic to the Gulf Coastal Plain but that has been found in the central Coastal Plain of SC. The species was first reported by McMillan and Porcher (2002) for SC. Members of the genus *Heliopsis* are similar to sunflowers (*Helianthus*) but have a distinctly conic receptacle.

CONSERVATION STATUS: SC-Vulnerable

754. Yellow Spinypod

Matelea flavidula (Chapman) Woodson
Ma-tè-le-a fla-ví-du-la
Apocynaceae (Dogbane Family)

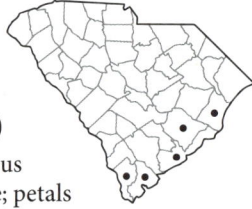

DESCRIPTION: Perennial, twining, herbaceous vine with milky sap; leaves opposite, ovate; petals yellow or greenish-yellow, less than 1.5x as long as wide; corolla lobes ovate with rounded tips and held in a horizontal plane; fruit a spiny "pod" that is a follicle; flowers April–June.

RANGE-HABITAT: Endemic to the Coastal Plain from SC south to the FL Panhandle; in SC, rare and local in calcareous bluff forests; wet, flat, calcareous forests; shell middens; shell hammock forests; and roadsides through such areas.

SIMILAR SPECIES: Eastern Anglepod (*Gonolobus suberosus* (L.) R. Brown, plate 335) is vegetatively very similar but has dark maroon petals with greenish tips that are lanceolate with sharp tips and often has the petals reflexed. It also has a smooth, angled pod. It is common throughout SC in mesic forests and forest margins. Carolina Spinypod (*M. carolinensis* (Jacquin) Woodson, plate 362) is similar but has reddish or maroon petals and is common in mesic forests and forest margins throughout SC.

COMMENTS: Yellow Spinypod is always restricted to high-pH, high-calcium soils and is considered rare throughout its range.

CONSERVATION STATUS: SC-Imperiled

755. Eastern Yellow Passionflower

Passiflora lutea L.
Pas-si-flò-ra lù-te-a
Passifloraceae (Passionflower Family)

DESCRIPTION: Slender, herbaceous, deciduous vine; stems to 15′ long, with tendrils; leaves alternate, palmately 3-lobed, green, often with silver mottling, 0.8–4″ wide; flowers produced in the axils of the leaves; corolla yellowish-green, 0.7–1.2″ wide; fruit a small blackish berry; flowers June–September; fruits mature August–October.

RANGE-HABITAT: PA south to FL and west to KS and TX; in SC, common throughout in forests with rich soils and woodland borders, generally in soils with high-magnesium or high-calcium content.

COMMENTS: This species is often overlooked. It is a much smaller plant than the showier Maypops (*Passiflora incarnata* L., plate 990). It is never abundant at any one site and occurs as scattered plants. Like Maypops, this species is a host for many of SC's showy butterfly caterpillars such as the zebra Llongwing, gulf fritillary, and variegated fritillary.

756. Ocmulgee Skullcap

Scutellaria ocmulgee Small
Scu-tel-là-ri-a oc-múl-gee
Lamiaceae (Mint Family)

DESCRIPTION: Perennial herbs, stems square, 1–3′
tall with the second node below the inflorescence
stipitate glandular; leaves ovate with a cordate
base and bluntly serrate to crenate margins;
flowers produced in racemes; flowers with a light
purple tubular corolla, the corolla with a ring of
hairs at the bend in the tube; flowers June–July.
RANGE-HABITAT: Endemic to SC and GA; in SC,
very rare along the fall-line, in calcareous bluff forests, and other forest systems with high-calcium soils.
SIMILAR SPECIES: Mellichamp's Skullcap (*Scutellaria mellichampii* Small) is
similar but has stems that are soft pubescent below the inflorescence and leaves
with truncate to cuneate bases. It is found on shell hammocks, calcareous bluff
forests, and other high-calcium areas in the Coastal Plain.
COMMENTS: Ocmulgee Skullcap is a large and showy member of the genus that is
ideal for cultivation in the home landscape. It is very rare in the wild and should not be
disturbed if located. It is available from native plant mail-order nurseries and the South
Carolina Botanical Garden.
CONSERVATION STATUS: SC-Critically Imperiled

757. Carolina Elytraria; Carolina Scalystem

Elytraria carolinensis (J. F. Gmelin)
Persoon var. *carolinensis*
E-ly-trá-ri-a ca-ro-li-nén-sis var.
ca-ro-li-nén-sis
Acanthaceae (Acanthus Family)

DESCRIPTION: Perennial herb from long and slender rhizomes; basal leaves obovate to spatulate;
flowering stem leafless, 4–20″ tall with scalelike
bracts and a bracteate, 0.8–2.0″ long, terminal
spike of flowers; corolla white, funnel-shaped
with lobes arranged in a weakly 2-lipped manner;
flowers June–August.
RANGE-HABITAT: Endemic to the Coastal Plain from SC south to FL; in SC,
infrequent in the Coastal Plain in wet, flat, calcareous forests.
COMMENTS: Carolina Elytraria is one of the most beautiful of the plants that are
endemic to the wet, flat, calcareous forests. Though it is still present in good numbers in SC, the species is threatened by development and habitat loss in the state.
The name "Elytraria" comes from the Greek *elytron*, meaning "a covering," which
refers to the bracts of the inflorescence nearly covering the flowers.

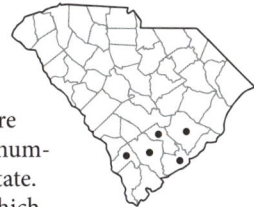

758. Florida Adder's-mouth

Malaxis spicata Swartz
Ma-láx-is spi-cà-ta
Orchidaceae (Orchid Family)

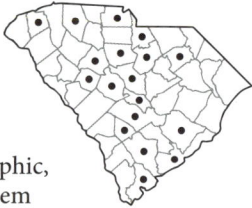

DESCRIPTION: Terrestrial herbaceous orchid with stems 3–18″ tall; leaves 2 (rarely more), subopposite, glossy, ovate, produced below the middle of the stem; flowers produced in a raceme 0.5–8″ long; flowers with the lip petal uppermost (nonresupinate), pale yellow to orange; flowers July–August.

RANGE-HABITAT: VA south to FL; also in the West Indies; in SC, rare in the Coastal Plain where found in wet, flat, calcareous forests; small stream floodplains; and hummocks in swamp forests, nearly always where the soil has elevated calcium.

SIMILAR SPECIES: Green Adder's-Mouth (*Malaxis unifolia* Michaux) has only a single leaf and much smaller, green flowers. It is found scattered throughout SC in a variety of forest habitats.

COMMENTS: Though this species is considered rare in SC, it may be present in large numbers at some locations during good years and absent or nearly so in others. The common name refers to the resemblance of the flower to the open mouth of an adder, a common, venomous European snake.

CONSERVATION STATUS: SC-Imperiled

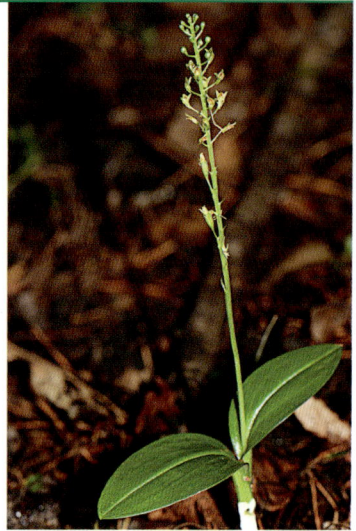

759. Crested Coral-root

Hexalectris spicata (Walter)
Barnhart
Hex-a-léc-tris spi-cà-ta
Orchidaceae (Orchid Family)

DESCRIPTION: Nonphotosynthetic, mycotrophic, perennial herbaceous orchid, 6–30″ tall; stem flesh-colored; leaves reduced to sheathing, scalelike bracts; flowers with the lip lowermost (resupinate), yellowish with purple striations; July–August.

RANGE-HABITAT: MD south to FL and west to MO and TX; in SC, infrequent in widely scattered locations throughout in calcareous or circumneutral soils such as shell hummocks; shell middens; calcareous bluff forests; wet, flat, calcareous forests; rich cove forests; basic-mesic forests; and calcareous oak-hickory forests.

COMMENTS: These plants are mycoparasites that use fungi to gain their carbohydrates. The fungus obtains the carbohydrates through the decomposition of organic material. Like most species of orchid that are completely dependent on fungal associations, it does not easily survive transplantation and should not be disturbed in the wild. The preference for high-calcium or high-magnesium substrates is demonstrated by its widely scattered populations.

760. Florida Horsebalm

Collinsonia punctata Elliott
Col-lin-sòn-i-a punc-tà-ta
Lamiaceae (Mint Family)

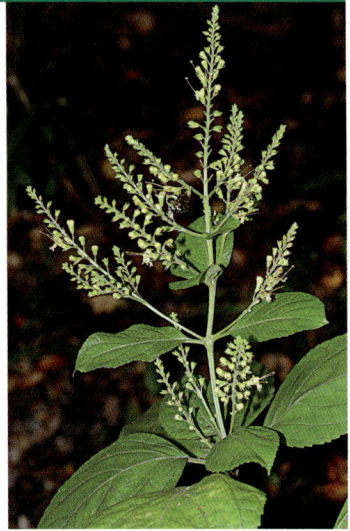

DESCRIPTION: Robust, aromatic perennial herb, smelling licoricelike or like *Perilla frutescens* (L.) Britton; stems 1–several, square, to 4′ tall, from thick stout rhizomes; leaves opposite, broadly elliptic to obovate, soft pubescent below, scabrous and slightly pubescent above; flowers produced in a panicle of racemes; individual flowers 0.5–0.7″ long, light yellow; sepals 0.2–0.3″ long, broadly lanceolate; flowers August–October.

RANGE-HABITAT: SC south to FL and west to LA; in SC, very rare in the Coastal Plain growing in calcareous woodlands and river bluff forests.

SIMILAR SPECIES: Richweed or Northern Horsebalm (*C. canadensis* L., plate 141) is similar and is found in the mountains and Piedmont of SC in a variety of rich woodlands. It is distinguished by the smooth leaves and stem as well as the shorter sepals, which are lance-subulate to narrowly lanceolate. Stoneroot (*C. tuberosa* Michaux, plate 761) is also similar but is a smaller plant, typically with a single stem and a white corolla with pinkish or purplish lines. It is found in high-calcium forests throughout the state.

CONSERVATION STATUS: SC-Imperiled

761. Stoneroot

Collinsonia tuberosa Michaux
Col-lin-sòn-i-a tu-be-rò-sa
Lamiaceae (Mint Family)

DESCRIPTION: Perennial herb with square stems, 1–3′ tall, from a tuberlike crown; leaves 1.5–4″ long; opposite, ovate with an acuminate tip and with 5–15 serrate teeth along the margin; inflorescence a panicle of racemes; flowers cream colored, often with pink or purplish streaks or spots; flowers July–September.

RANGE-HABITAT: NC south to GA and west to TN and MS; in SC, common in the Coastal Plain in wet, flat, calcareous forests; shell middens; calcareous bluffs; and other forests with calcareous substrate. Although uncommon in the Piedmont and mountains, it is found, in rich cove forests, basic-mesic forests, and oak-hickory forests with circumneutral soils.

SIMILAR SPECIES: Richweed or Northern Horsebalm (*C. canadensis* L., plate 141) is more robust, with thicker stems, and is found in the mountains and Piedmont of SC in a variety of rich woodlands. It is distinguished by the larger leaves with more teeth on the margin, flowers without the striping or spotting with purple, and the enlarged rhizomes rather than a tuberlike crown.

762. Pink Thoroughwort

Fleischmannia incarnata (Walter)
  King & H. E. Robinson
    Fleisch-mán-i-a in-car-nà-ta
  Asteraceae (Aster Family)

SYNONYM: *Eupatorium incarnatum*
  Walter—RAB

DESCRIPTION: Herbaceous perennial with erect to spreading stems 1.5–6′ long; leaves opposite, ovate with crenate margins, pubescent and often also glandular on the underside; flowers arranged in involucrate heads of pink to light pink disk flowers only; heads in a corymblike (flat-topped) arrangement; receptacle flat; florets 18–25 per head; flowers August–October.

RANGE-HABITAT: VA south to FL and west to OK and TX; also found in Mexico; in SC, occasional in the Coastal Plain and Piedmont in wet, flat, calcareous forests; calcareous bluff forests; basic-mesic forests; and other nutrient-rich forests and forest margins on circumneutral soils.

SIMILAR SPECIES: Mistflower (*Conoclinium coelestinum* (L.) A. P. de Candolle) has blue flowers arranged into heads with conical receptacles and 25–70 florets per head. It is common in ditches, disturbed areas, and moist roadsides throughout the state.

COMMENTS: Pink Thoroughwort is an extremely attractive addition to the garden. It produces copious numbers of flowers that attract a wide variety of pollinators. It can be planted in combination with yellow- or white-flowered species such as Camphorweed (*Heterotheca subaxillaris* (Lamarck) Britton & Rusby) or various Aster species for a stunning effect. The distribution of Pink Thoroughwort is very spotty in the southeastern United States where it is always found on high-pH soils. The genus honors the German botanist, Gottfried Fleischmann (1777–1850).

763. Shadow-witch; Ponthieu's Orchid

Ponthieva racemosa (Walter) Mohr
    Pon-thiè-va ra-ce-mò-sa
  Orchidaceae (Orchid Family)

DESCRIPTION: Perennial, terrestrial orchid with leafless flowering stems, usually about 1′ tall; leaves in a basal cluster, oblong-elliptic, oblanceolate or obovate, 0.8–6.7″ long; flowers whitish green, in a terminal raceme; flowering September–October.

RANGE-HABITAT: Found in the Coastal Plain from se. VA to FL and west to TX; also in the West Indies and Central and South America; in SC, occasional in the Coastal Plain where found in wet, flat, calcareous forests; bases of slopes; floodplains; and moist ravines, nearly always on calcium-rich soils.

COMMENTS: Most sources (e.g., Radford et al., 1968) list this orchid as rare; however, recent fieldwork in the Coastal Plain indicates that it is more abundant and a listing of occasional is justified. The genus honors Henry de Ponthieu (1731–1808), a Huguenot living in London who collected Caribbean plants. Farther south in FL it grows on logs and hummocks in swamp forests.

CONSERVATION STATUS: SC-Vulnerable

764. Mountain Catchfly

Silene ovata Pursh
Si-lè-ne o-và-ta
Caryophyllaceae (Pink Family)

DESCRIPTION: Erect perennial herb to 4′ tall, short pubescent throughout; leaves 2–5″ long, opposite, ovate with a rounded, slightly clasping base; flowers bracteate, white, numerous, and produced in a complex terminal arrangement; petals so deeply divided as to appear ragged; flowers August–October.

RANGE-HABITAT: VA and KY west to IL and south to AR, MS, and GA; in SC, very rare in the Coastal Plain where growing in calcareous bluff forests and ravines.

COMMENTS: Mountain Catchfly is known in SC from a single population in Florence County. The majority of the range is in the Appalachian region–hence, the common name. It is distinct from all other native *Silene* in its autumn flowering. It is often still flowering when the first frost arrives at the South Carolina Botanical Garden in Clemson. It makes a great addition to the native garden. It tolerates dry shaded conditions while providing ample color at a time of year not much else is flowering. The species ranges mostly in the Appalachian area and west of the mountains and its occurrence in SC is surprising. Due to its rarity in SC, it should not be disturbed in the wild but is available from the South Carolina Botanical Garden for use in the home landscape.

CONSERVATION STATUS: SC-Critically Imperiled

765. Venus'-hair Fern;
Southern Maidenhair

Adiantum capillus-veneris L.
A-di-án-tum ca-píl-lus-vé-ne-ris
Pteridaceae (Maidenhair Fern Family)

DESCRIPTION: Small fern with short and creeping rhizome; fronds up to 1.5′ long, often hanging vertically from ledges and rocks; pinnae shining, resistant to wetting, semi-evergreen in SC, deciduous in more northern areas; frond stalks lustrous dark purple, appearing almost black; spores produced in sori arranged in lines along the outer margin of the pinnules; spores produced January–December.

RANGE-HABITAT: Widespread on several continents; in eastern North America largely southern in distribution, from VA south to FL, then to TX and AZ, and north to MO, SD, and British Columbia; south to the tropics and in warmer parts of the Old World; in SC, uncommon in natural habitats because of a lack of suitable environments and occurring mostly in the Coastal Plain but also in the Piedmont in Greenwood County; calcareous bluff forests and coquina limestone (marl) outcrops and on limestone masonry of old cemeteries and buildings; especially prominent in Charleston.

COMMENTS: Montpellier, France, was the origin of Syrup of Capillaire, a cough medicine made by using fresh maidenhair fronds, orange-flower water, and honey. This and other interesting folklore about Venus'-hair Fern and about ferns in general appear in *Ferns of the Coastal Plain* (Dunbar, 1989). This species is far more common as an adventive on masonry than in natural settings. It is a fantastic garden plant for the shade.

CONSERVATION STATUS: SC-Imperiled (native, not adventive populations)

766. Southern Shield Fern; Kunth's Maiden Fern

Thelypteris kunthii (Desvaux) Christenhusz
The-lýp-te-ris kúnth-i-i
Thelypteridaceae (Marsh Fern Family)

DESCRIPTION: Medium-sized, colonial, rhizomatous, herbaceous fern; fronds 1-pinnate, 1–3′ long, with pubescent, pale green to yellow petioles; blade to 1.5′ long; pinnae deeply lobed, pubescent; veins from the adjacent lobes of the pinnae not united into a single vein extending to the sinus; sori produced in spots toward the center of the pinnae; the lowest pair of pinnae are the same size or only slightly shorter than the pinnae directly above it; spores produced January–December.

RANGE-HABITAT: Widespread in the Neotropics; in the US found from NC south to FL and west to TX and AR; in SC, common in the Coastal Plain and lower Piedmont and uncommon in the upper Piedmont and mountains; wet, flat, calcareous forests; calcareous bluff forests; basic-mesic forests; and around masonry and aging concrete structures.

SIMILAR SPECIES: Downy Maiden Fern (*Christella dentata* (Forskål) Brownsey & Jermy) and Hairy Maiden Fern (*C. hispidula* (Decaisne) Holttum) are adventive from the tropics and common in old masonry and limestone outcrops in the Coastal Plain. Both have the veins of the adjacent lobes of the pinnae united into a single vein extending to the sinus. Ovate Maiden Fern (*C. ovata* (R. P. St. John) Love & Love) is a rare native species that has the lowest pair of pinnae significantly smaller than the adjacent pair. It is found on limestone and masonry in the Coastal Plain.

COMMENTS: Recent studies indicate that Southern Shield Fern is better treated as a species of the genus *Christella*, but the appropriate combination has not been formally made. Southern Shield Fern is an adaptable and stately ornamental in the shade garden, if given the room to spread. It makes a good native groundcover. The specific epithet honors Carl Sigismund Kunth (1788–1850), a German botanist.

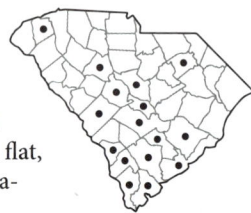

767. Wagner's Spleenwort (A)

Asplenium heteroresiliens Wagner
As-plè-ni-um he-te-ro-re-sí-li-ens

Blackstem Spleenwort (B)

Asplenium resiliens Kunze
As-plè-ni-um re-sí-li-ens
Aspleniaceae (Spleenwort Family)

DESCRIPTION: Small, evergreen ferns; rachis purplish-black; fronds erect to spreading, somewhat leathery, pinnately divided, 3–6″ long in *A. heteroresiliens*, and to 12″ long in *A. resiliens*; the pinnae opposite and entire in *A. resiliens*, and opposite, shallowly crenate with a shallow auricle at the base and obscure venation in *A. heteroresiliens*; spores produced in sori that are arranged in 2 lines on the lower surface of pinnae; spores produced April–October.

RANGE-HABITAT: *A. heteroresiliens* is endemic to the Coastal Plain from NC south to FL; rare throughout its range; moist, shady sites on marl or limestone outcrops and on masonry composed of tabby (a mixture of sand, lime, and oyster shells); *A. resiliens* is wide-ranging in

North America; also occurs in the Neotropics; in SC, rare in the mountains where restricted to outcrops of meta-sedimentary rock such as marble or amphibolite at lower elevations; rare in the Coastal Plain where found on limestone marl outcrops.

COMMENTS: Wagner's Spleenwort is a pentaploid (5 sets of chromosomes) and is a hybrid of *A. resiliens* and *A. heterochroum* Kunze. With an odd number of chromosome sets, it is unable to reproduce sexually. It reproduces instead via spores that are produced asexually (apogamous). Several species of spleenworts were derived in a similar manner.

CONSERVATION STATUS: *A. heteroresiliens:* SC-Critically Imperiled; *A. resiliens:* SC-Imperiled

768. Bald Cypress-Tupelo Gum swamp forests

769. Bald Cypress

Taxodium distichum (L.) Richard
Tax-ò-di-um dís-ti-chum
Cupressaceae (Cypress Family)

DESCRIPTION: Deciduous conifer with 2-ranked leaves (needles); large tree, 70–130′ tall; bark sloughing in thin, flaky scales; separate male and female cones on the same tree; female cones green, turning brown at maturity in the fall.

RANGE-HABITAT: Bald Cypress is a wide-ranging tree, from DE and e. MD south to FL and west to TX and OK; then north along the Mississippi River and its tributaries to IN and IL; in SC, common, primarily found on the Coastal Plain, but it does extend into the Piedmont along the Savannah and other rivers; common in alluvial swamp forests and hardwood bottoms.

SIMILAR SPECIES: Bald Cypress is distinguished from Pond Cypress (*T. ascendens* Brongniart, plate 675) by having flat, linear leaves mostly spreading in one plane; Pond Cypress has slender, needle-shaped to awl-shaped leaves spirally arranged and appressed to the branchlets. Bald Cypress also produces tall, sharp knees, while those of Pond Cypress are rounded and short.

COMMENTS: Historically, Bald Cypress was one of the most important lumber trees of the South. Its wood is very durable, due to essential oils, and was used for shingles, barrels, caskets, and beams. Native Americans and early settlers carved cypress logs into boats, troughs, and washtubs. Cypress was also used to make rice field trunks in the 1700s and 1800s, many of which persist today in the abandoned rice fields. Much fine, Charleston-made furniture has cypress as the secondary wood. The use of cypress as a major commercial timber source has decreased over the years because it does not reproduce well after clear-cutting. Large stands of Bald Cypress exist on protected state and federal lands.

The trunk of the majestic Bald Cypress, when growing in saturated or seasonally submerged soil, produces an enlarged base, the buttress, which has important survival value. The wide buttress gives the tree a base of support in the soft, swamp soil. Indeed, it is seldom that a Bald Cypress is blown down. Other swamp trees, such as Tupelo Gum, Swamp Gum, and Pond Cypress, also form buttresses.

When growing in sites with fluctuating water, the roots of Bald Cypress also produce knees projecting above water. The knees have recently been shown to possibly function as

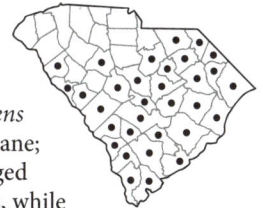

pneumatophores—providing air for the plant when the roots are submerged (Martin and Francke, 2015). But it is also clear that they function to provide stability in saturated soils via the interlocking roots and knees, which provide a strong base of support for the tree. One thing is certain: the knees do not grow into new trees.

Bald Cypress is widely planted as an ornamental because it grows well on a variety of upland soils, where it does not produce a buttress or knees. The SC state record Bald Cypress, in the Congaree Swamp National Park, is 131′ tall and 8.3′ in diameter. Bald Cypress is one of the longest-living trees and has been documented to reach an age of over 2,600 years in the Black River drainage of NC.

770. Coastal Fetterbush

Eubotrys racemosus (L.) Nuttall
　　Eù-bo-trys ra-ce-mò-sus
Ericaceae (Heath Family)

SYNONYM: *Leucothoë racemosa* (L.)
　　Gray—RAB, PR

DESCRIPTION: Deciduous shrub to 13′ tall; leaves alternate, simple, elliptic to elliptic-lanceolate, 1–3.5″ long, toothed, turning scarlet in fall; flowers produced in curved racemes, 2.0–4.8″ long; corolla urn-shaped, white, sometimes pink-tinged; fruit a capsule that is rounded on the sutures and breaking apart like sections from a citrus fruit; seeds are not winged; flowers March–early June.

RANGE-HABITAT: MA south to FL and west to LA; in SC, common in the Coastal Plain and lower Piedmont and rare in the mountains; swamp forests, streamsides, pocosins, Bald Cypress-Tupelo Gum swamps, montane bogs and fens, and Longleaf Pine savannas.

SIMILAR SPECIES: Mountain Fetterbush (*E. recurvus* (Buckley) Britton, plate 156) is similar but is restricted to the mountains and absent from the Coastal Plain. It may be distinguished by the larger and wider leaves, the shorter racemes (1–2″ long), angled sutures on the capsules, and winged seeds.

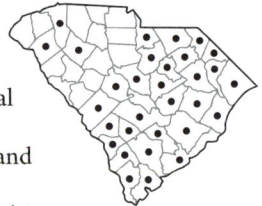

771. Florida Doghobble; Pipestem

Agarista populifolia (Lamarck) Judd
　　A-ga-rís-ta po-pu-li-fò-li-a
Ericaceae (Heath Family)

SYNONYM: *Leucothoë populifolia* (Lamarck) Dippel—RAB

DESCRIPTION: Large, evergreen shrub to 13′ tall with hollow stems; leaves smooth, lanceolate 1.5–4″ long with acute to acuminate tips, serrulate to entire; flowers white, urn-shaped in short axillary racemes approximately 1″ long; flowers March–May.

RANGE-HABITAT: Endemic to the Coastal Plain from SC south to FL; in SC, rare in the Coastal Plain in blackwater swamps.

COMMENTS: Florida Doghobble is a large and conspicuous shrub but is extremely limited in distribution in SC. It makes a good landscape shrub throughout our area and grows quickly. It is readily available in the landscape trade.

CONSERVATION STATUS: SC-Critically Imperiled

772. American Storax; Snowbell

Styrax americanus Lamarck var. *americanus*

Stý-rax a-me-ri-cà-nus var. a-me-ri-cà-nus

Styracaceae (Storax Family)

DESCRIPTION: Deciduous shrub or small tree to 16′ tall; leaves alternate, oblong-elliptic, the margins denticulate to serrate; buds naked; axillary bud adjacent to each terminal bud appearing terminal, but not actually terminal; flowers fragrant, semi-drooping; petals recurved; fruit a globose capsule; flowers April–June; fruits mature July–September.

RANGE-HABITAT: From WV and VA south to FL and west to TX, MO, and IL; in SC, found in the Coastal Plain, Piedmont, and mountains; common in the Coastal Plain, occasional in the Piedmont, rare in the mountains; montane bogs, Piedmont seepage forests, alluvial and nonalluvial swamp forests, floodplain forests, Bald Cypress-Tupelo Gum swamps, streamsides, cypress-gum depressions, and pocosins.

TAXONOMY: Downy Snowbell (*S. americanus* var. *pulverulentus* (Michaux) Perkins ex Rehder) is probably worthy of recognition as a distinct species. Downy Snowbell is a species of wet, Longleaf Pine savannas and isolated depressions where it seldom reaches 4′ in height. The obovate and densely pubescent leaves are often held ascending. Downy Snowbell flowers on new growth following burns. It is known from southeastern SC south to FL. Some plants of typical American Storax can be pubescent, and this has caused confusion with Downy Snowbell.

COMMENTS: Storax is an excellent landscape species, doing best in full sun or light shade.

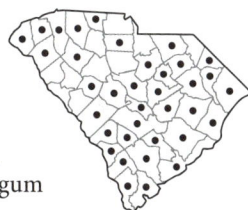

773. Sweetspire; Virginia Willow

Itea virginica L.

Ì-te-a vir-gí-ni-ca

Iteaceae (Sweetspire Family)

DESCRIPTION: Deciduous shrub, 3–9′ tall, forming colonies via stolons; leaves alternate, elliptic to oblong-lanceolate, serrulate; inflorescence a terminal, narrow raceme with white flowers; fruit a capsule; flowers May–June; fruits mature August–October.

RANGE-HABITAT: NJ south to FL and west to TX and OK; in SC, common throughout in Bald Cypress-Tupelo Gum swamps, small stream swamp forests, along streambanks and other swampy or saturated habitats.

COMMENTS: The sweet-smelling flowers of this species, attractiveness to pollinators, and small size have made it popular in the native plant industry. It is often short-lived or spreads to form thickets and is most useful naturalized along woodland borders. It is best appreciated in its native habitat and is not as desirable in the landscape as Titi or the various *Clethra* species and cultivars.

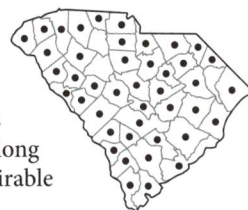

774. Fever-tree; Pinckneya

Pinckneya bracteata (Bartram) Rafin-
esque
Pínck-ney-a brac-te-à-ta
Rubiaceae (Madder Family)

SYNONYM: *Pinckneya pubens* Michaux—
RAB, PR

DESCRIPTION: Small deciduous tree or shrub;
leaves soft pubescent, opposite, and simple;
flowers borne in clusters; petals greenish yellow
with bright-red spots on the inside; flower made
conspicuous by enlargement of 1 (often two)
calyx lobe(s) that is/are bright rose pink to white; fruit a capsule; flowers
May–June; fruits mature August–September.

RANGE-HABITAT: Restricted to SC, GA, and FL; in SC, rare in the Coastal Plain
but can be found along the edges of hardwood bottoms or blackwater swamp
forests.

COMMENTS: This is the only species in the genus *Pinckneya*. Nineteenth-century
botanists, such as Joseph Hinson Mellichamp (1829–1903), reported fever tree from
sites in Beaufort County, especially near Bluffton. More recently, it was found in Jasper County. It
is certainly one of the more interesting plants in SC from a historical and biological perspective. It
is only conspicuous for a short time when it flowers. The common name, Fever-tree, comes from
the use of the bark as a substitute for quinine in treating malaria. André Michaux named this tree
in honor of SC statesman and revolutionary war veteran, General Charles Cotesworth Pinckney
(1746–1825). The stunning poinsettialike inflorescence makes this one of the most beautiful flowering trees in the world. It is short-lived and though it lives in saturated soils in the wild, in cultivation
it needs well-drained but moist soil. It achieves flowering size very quickly (2 years from seed) but
thrives for only 5–10 years and then declines. The species is very easily propagated from its copious
tiny seeds.

CONSERVATION STATUS: SC-Critically Imperiled

775. Sebastian-bush

Ditrysinia fruticosa (W. Bartram)
Govaerts & Frodin
Di-try-sín-i-a fru-ti-cò-sa
Euphorbiaceae (Spurge Family)

SYNONYM: *Sebastiana fruticosa* (W. Bartram) Fernald—RAB

DESCRIPTION: Small, semi-evergreen shrubs; leaves
smooth and glossy, alternate, simple, with entire
margins; inflorescence a raceme with separate
male and female flowers (monoecious); staminate
flowers with 3 tiny sepals and no petals; pistillate flowers with 3 sepals and no
petals; fruit a capsule; flowers April–June; fruits mature July–October.

RANGE-HABITAT: NC south to FL and west to TX and AR; in SC, common in
the Coastal Plain in shady, swamp forests, and stream and riverbanks.

COMMENTS: This shrub is often overlooked and frequently confuses botanists and
wildflower enthusiasts alike. It is in the same family as *Poinsettia* and spurges and
is one of the few woody members of the family in SC. The genus contains only one
species.

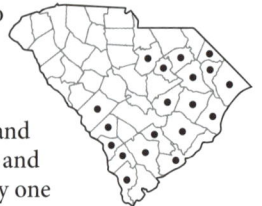

776. Coral Greenbrier

Smilax walteri Pursh
Smì-lax wál-ter-i
Smilacaceae (Greenbrier Family)

DESCRIPTION: High-climbing, deciduous to semi-evergreen, vine with perennial stems; leaves ovate-oblong to ovate-lanceolate with a rounded to somewhat heart-shaped base, green and smooth above and below; flowers in umbels, petals greenish-yellow; fruits are berries maturing to bright red; flowers April–May; fruits mature in September and persist through the winter.

RANGE-HABITAT: NJ south to FL and west to AR and TX; in SC, common throughout the Coastal Plain and rare in the lower Piedmont; alluvial and nonalluvial swamp forests and hardwood bottoms.

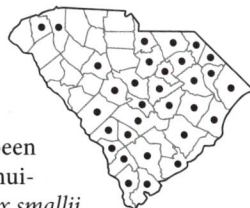

COMMENTS: The leafless vines with brilliant red berries serve to distinguish this species from all other greenbriers in SC. The berries have traditionally been used for Christmas decorations. Though greenbriers are often considered a nuisance due to the spinelike projections, this species, and Jackson-briar (*Smilax smallii* Morong, plate 798), have recently become popular as ornamental vines in the landscape due to their attractive foliage and fruit and lack of these "spines." The plants often grow from hummocks around the bases of trees or in areas that are flooded at least part of the year. The specific epithet honors Thomas Walter (1740–89), an English botanist who settled in Berkeley County, SC and the author of *Flora Caroliniana* (1788).

777. Cypress-knee Sedge

Carex decomposita Muhlenberg
Cà-rex de-com-pó-si-ta
Cyperaceae (Sedge Family)

DESCRIPTION: Clumping, perennial herb; culms 20–40″ long; inflorescence elongate, in a narrow, arched panicle, branched only at the base; flowers May–June.

RANGE-HABITAT: NY south to FL and west to MI and TX; in SC, rare in the Coastal Plain, Sandhills, and lower Piedmont where growing in backwaters of swamps and ox-bow lakes, frequently as an epiphyte on cypress knees and buttresses, and on fallen logs; often forms tussocks in shallow waters of swamps and pond margins; it is rarely found as an epiphyte in deep, frequently flooded swamps since moving water would dislodge it.

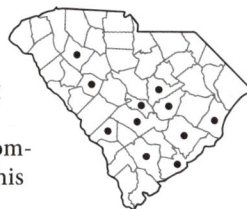

COMMENTS: The specific epithet, *decomposita,* refers to the branched or decompound (twice-compound) inflorescence, one of the key characteristics of this rare sedge.

CONSERVATION STATUS: SC-Imperiled

778. Southern Twayblade

Neottia bifolia (Rafinesque)
Baumbach
Ne-ót-ti-a bi-fò-li-a
Orchidaceae (Orchid Family)

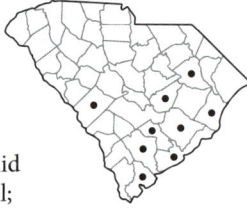

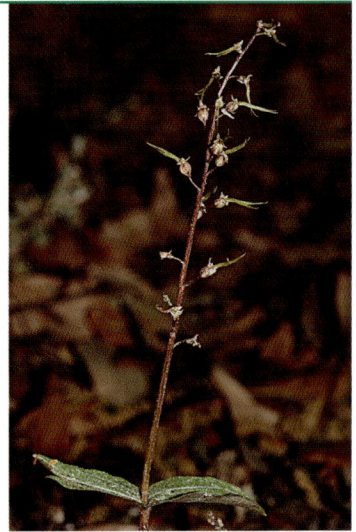

DESCRIPTION: Small, dainty, terrestrial orchid
with slender purplish-green stems to 1' tall;
leaves two, opposite, dark green, smooth, ovate to
elliptic, with a weakly heart-shaped base; flowers produced in a
raceme of 5–25 flowers; flowers January–May.

RANGE-HABITAT: Eastern Canada south to FL and west to TX and
OK; in SC, occasional in the Coastal Plain and Sandhills, where
found in blackwater swamp forests; beech forests; wet, flat, calcareous forests; and streamsides.

COMMENTS: This tiny orchid with bizarre flowers is overlooked and
rarely encountered due to its small size and early period of flowering. It is not as rare as many sources indicate and if you are in the
proper habitat at the right time of the year, it is not difficult to
locate in most regions of the Coastal Plain.

CONSERVATION STATUS: SC-Vulnerable

779. Southern Blue-flag Iris

Iris virginica L.
Ì-ris vir-gí-ni-ca
Iridaceae (Iris Family)

DESCRIPTION: Perennial herb from stout rhizomes,
0.4–0.8 cm wide; flowering stems to 2' tall,
unbranched; leaves conduplicate (folded lengthwise to have only one surface exposed), 15–30"
long, 0.4–1.2" wide, smooth; flowers produced at
the ends of stems subtended by bracts; sepals 3,
showy, 2.4–3.1" long, lavender to bluish-purple,
with a yellow or white blotch on the claws; pet-
als 3, lavender to bluish-purple, erect, 2–2.4" long; fruit an angled capsule
1.5–2.4" long; flowers late April–May; fruits mature July–September.

RANGE-HABITAT: VA south to FL and west to TN and MS; in SC, common
nearly throughout, most abundant on the Coastal Plain; Bald Cypress-Tupelo
Gum swamps, marshes, pond margins, ditches, and other wet and marshy
habitats.

COMMENTS: This is the most common wetland *Iris* species in SC. A second
species, Northern Blue-flag Iris (*I. shrevei* Small) is a larger plant, reaching 3' in height that
frequently has branches on the flowering stem. It is restricted to the upper Piedmont and
mountains in SC.

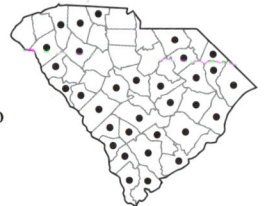

780. Pygmy Spiderlily;
Waccamaw Spiderlily

Hymenocallis pygmaea Traub
Hy-me-no-cál-lis pyg-maè-a
Amaryllidaceae (Amaryllis
Family)

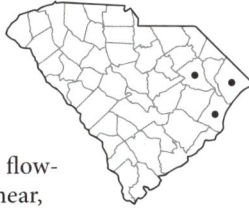

DESCRIPTION: Smooth, perennial herb with flowering stems to 1′ tall; leaves to 16″ long, linear, all basal; stamens with the lower portions of the filament united into a thin, membranous crown, with the upper filament extending beyond the crown; perianth segments narrow, linear, 2–2.5″ long; flowers May–June.

RANGE-HABITAT: Endemic to the Waccamaw and Pee Dee River drainages of NC and SC; in SC, rare but locally common; blackwater swamp forests, levee forests, and floodplains.

COMMENTS: Pygmy Spiderlily is the smallest of SC's four species of *Hymenocallis.* The plant, and its flowers, are significantly smaller than the similar Coastal Spiderlily (*H. crassifolia* Herbert, plate 835) but are still strikingly beautiful and large compared to other native flowering plants.

CONSERVATION STATUS: SC-Vulnerable

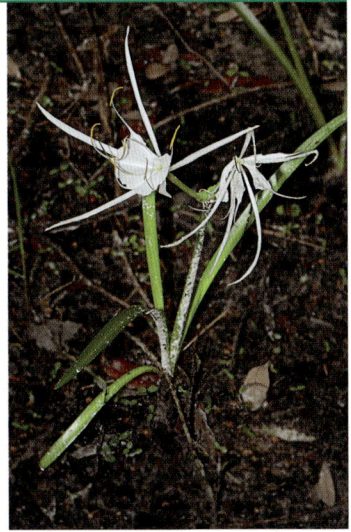

781. Lizard's-tail

Saururus cernuus L.
Sau-rù-rus cér-nu-us
Saururaceae (Lizard's-tail Family)

DESCRIPTION: Perennial emergent aquatic or terrestrial herb, often forming extensive colonies by rhizomes; leaves simple, cordate to 10″ long; flowers produced in a dense spikelike raceme of very small flowers that lack sepals or petals; racemes drooping in flower, becoming erect in fruit; flowers May–July.

RANGE-HABITAT: Quebec south to FL and west to MN and TX; in SC, common throughout the Coastal Plain and Piedmont in a variety of aquatic habitats including alluvial and nonalluvial swamp forests, streams, lake and pond margins, low woodlands, and ditches.

COMMENTS: The common name comes from the drooping raceme. There are no sepals and petals; the white color of the spike is due to the white stamen stalks. This is the only member of the family in SC; it is related to Black Pepper plants (order Piperales).

782. Yellow Hedge-hyssop

Gratiola lutea Rafinesque
Gra-tì-o-la lù-te-a
Plantaginaceae (Plantain
Family)

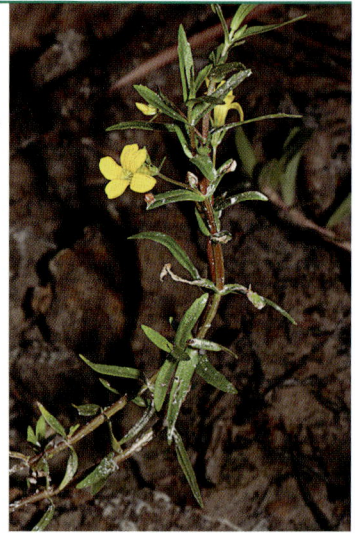

DESCRIPTION: Small, ascending to erect annual or short-lived perennial herb with stems 4–15″ long; leaves glandular-punctate, opposite, linear-lanceolate to lanceolate-ovate, 0.3–1″ long, the margins sparsely denticulate or serrulate, clasping at the base, glandular-punctate; flowers axillary on long stalks (pedicels); corolla brilliant golden-yellow with the two upper lobes fused for much of their length and three lower petals with longer, more widely spreading lobes; flowers May–September.

RANGE-HABITAT: Newfoundland south to FL, primarily in the Coastal Plain; also in the upper Midwest; in SC, uncommon in the Coastal Plain in drawdown zones of blackwater rivers, swamp forests, depression meadows, Pond Cypress savannas, and margins of ponds and ditches.

COMMENTS: Yellow Hedge-hyssop may be common in some years and absent in others. Its preferred habitat has greatly fluctuating water levels, and it is most common in drought years during drawdowns of riverbanks, such as along the Waccamaw River. The opposite, clasping leaves and brilliant yellow, weakly bilabiate flowers are distinctive.

783. Coastal Plain Water-willow

Justicia ovata (Walter) Lindau
Jus-tíc-i-a o-và-ta
Acanthaceae (Acanthus Family)

DESCRIPTION: Colonial, rhizomatous perennial herb; stems 4–20″ tall; leaves opposite, entire, elliptic to ovate, smooth and dark green; flowers in a bracteate spike produced in the uppermost leaf axils, often appearing terminal; petals violet or pinkish with a white throat outlined by darker purple lines and spots; flowers May–July.

RANGE-HABITAT: VA south to FL and west to IL and AL, mostly in the Coastal Plain; in SC, common in the Coastal Plain in a variety of swamp forests and marshes and uncommon in the Piedmont along streams, isolated upland depressions, and seeps in Post Oak savannas. The genus honors James Justice (1698–1763), a Scottish horticulturalist and botanist.

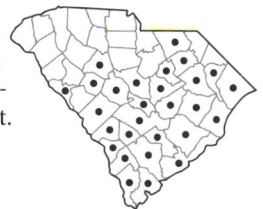

784. Plymouth Rose-gentian

Sabatia kennedyana Fernald
Sa-bà-ti-a ken-ne-dy-à-na
Gentianaceae (Gentian Family)

DESCRIPTION: Smooth, perennial, rhizomatous herb with erect stems, 1.5–4′ tall; primary branches are opposite; leaves opposite; flowers produced at the tips of branches, large and showy, with 10 or more pink petals; calyx lobes flat, not

cylindrical; the terminal flowers with short pedicels (shorter than the adjacent internode); flowers June–August.

RANGE-HABITAT: Nova Scotia, MA, and RI and disjunct to NC and SC; in SC, locally abundant but restricted to the Waccamaw River drainage in the Coastal Plain where found in the drawdown zone of rivers and streams, as well as adjacent swamp forests.

SIMILAR SPECIES: Three similar, large-flowered rose-gentian species are found in the Coastal Plain in similar habitats. Bartram's Rose Gentian (*S. decandra* (Walter) R. M. Harper), which is restricted to isolated depressions in the southeastern Coastal Plain in SC, may be distinguished by the terete (round) calyx lobes and upper stem leaves that are about as wide as the stem. *S. kennedyana* has flat calyx lobes and upper stem leaves wider than the stem. Large Marsh Rose-pink (*S. dodecandra* (L.) Britton, Sterns & Poggenburg, plate 840) has the primary branches alternate and terminal flower on long pedicels. It is common in tidal freshwater marshes but also occurs in openings in swamp forests and on drawdown zones of larger rivers. Blackwater Rose-pink (*S. foliosa* Fernald) also has alternate branches but displays plentiful surficial stolons and occurs in openings in blackwater swamp forests.

COMMENTS: The genus honors Liberato Sabbati (1714–78), an Italian botanist, and the specific epithet honors George Golding Kennedy (1841–1918), a physician and botanist from Massachusetts. This is one of the largest flowered and most beautiful species in the genus. It makes a fine addition to the bog garden and is available from specialty nurseries. It should not be disturbed in the wild. Plants from North and SC may be distinct from those in New England and are currently under study.

CONSERVATION STATUS: SC-Imperiled

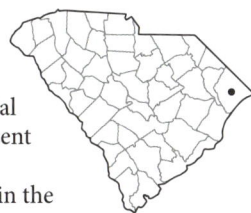

785. Ladies'-eardrops

Brunnichia ovata (Walter) Shinners
Brun-ních-i-a o-và-ta
Polygonaceae (Buckwheat Family)

SYNONYM: *Brunnichia cirrhosa* Banks ex Gaertner—RAB

DESCRIPTION: Large, partly woody vine, climbing by tendrils, reaching 40′ in length; leaves alternate, ovate; calyx modified into winglike structures, that enclose the fruits; flowers June–July; fruits mature August–September.

RANGE-HABITAT: SC south to FL, west to TX, and north to MO and IL; in SC, uncommon in the Coastal Plain where found in alluvial swamps and along riverbanks.

COMMENTS: Ladies'-eardrops is a little-known vine to botanists and laymen. The common name comes from the shape of the winged fruit. The genus honors M. T. Brünnich (1737–1827), a Danish naturalist.

786. Riverbank Sundrops

Oenothera riparia Nuttall
Oe-no-thè-ra ri-pà-ri-a
Onagraceae (Evening-primrose Family)

DESCRIPTION: Perennial herb to 3′ tall; lower stem usually winged, spongy, and thickened; leaves alternate, nearly smooth, linear-lanceolate, rather thick; flowers large with brilliant yellow petals, 0.6–1.2″ long with a notch at the tip; flowers June–late August; fruit is a club-shaped capsule, angular and long-stalked, widest above the middle, and mature in the fall.

RANGE-HABITAT: Endemic to the Coastal Plain from VA south to SC; in SC, locally common, growing on cypress knees, buttresses of hardwoods, and with driftwood just above the high tide line in large, tidal, blackwater rivers; in the Carolinas reported from the Northeast Cape Fear, Black, Lower Pee Dee, Waccamaw, and Edisto River systems.

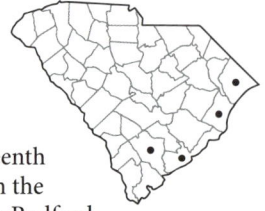

COMMENTS: Nuttall described *O. riparia* in 1818. Though recognized in nineteenth century treatments, it quickly faded into obscurity and was considered within the variation of the common Southern Sundrops (*O. fruticosa* L.) by G. B. Straley. Radford et al. (1968) and Weakley (2001) did not recognize it as distinct from *O. fruticosa* L. It has been shown to be an octoploid, and due to the complete lack of intermediates between *O. riparia* and *O. fruticosa,* the restrictive habitat, the ploidy, and distinct geographic range, it has since been recognized as distinct (McMillan et al., 2002). Richard Porcher and Patrick D. McMillan documented this plant in the Great Pee Dee, Waccamaw, and Edisto Rivers in 1997. The habitat, smooth and fleshy leaves, and swollen, spongy bases make this the most distinctive and easy to recognize sundrop in SC.

CONSERVATION STATUS: SC-Imperiled

787. Powdery-thalia; Powdery Alligator-flag

Thalia dealbata Fraser ex Roscoe
Thàl-i-a de-al-bà-ta
Marantaceae (Arrowroot Family)

DESCRIPTION: Robust, scapose, erect perennial to 7′ tall; frequently whitish throughout; leaves large, 12–20″ long, 6–8″ wide, similar to banana or canna leaves; leaf stalks to 32″ long; sepals 3, pale purple; petals 3, longer and darker purple than the sepals; flowers June–October.

RANGE-HABITAT: SC south to GA and west to TX and OK; in SC, rare and confined to the Coastal Plain; ditches, openings within and margins of swamps, pond edges, and freshwater marshes.

COMMENTS: The specific epithet *dealbata* means "whitewashed," in reference to the whitish cast to the plant. The genus honors Johann Thal (1542–83), a German physician and naturalist. Though uncommon in SC, the stature and striking appearance of this plant often attract attention, even when driving down a highway.

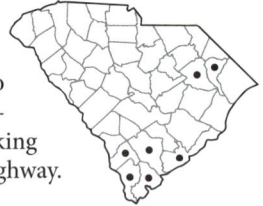

CONSERVATION STATUS: SC-Imperiled

788. Green-fly Orchid

Epidendrum conopseum R. Brown
E-pi-dén-drum co-nóp-se-um
Orchidaceae (Orchid Family)

DESCRIPTION: Epiphytic evergreen orchid; stems erect or ascending, smooth, to 16″ tall; leaves leathery; roots with well-developed, whitish velamen; flowers fragrant; flowers July–December.

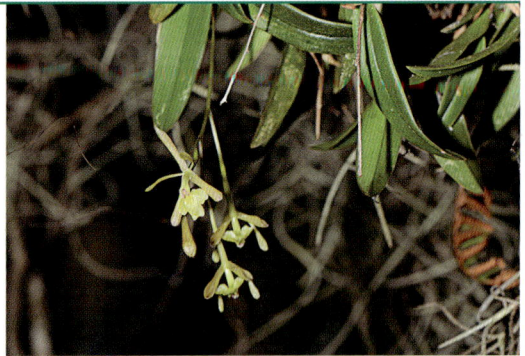

RANGE-HABITAT: From NC south to FL and west to LA; primarily an Outer Coastal Plain species; common; swamp forests and hardwood bottoms; commonly found growing on Live Oak trees in various habitats throughout the Coastal Plain.

COMMENTS: Green-fly Orchid is the only epiphytic orchid in the continental US that is found outside of FL. Based on the authors' field observations, it is common rather than rare (as most manuals report). It is often overlooked because it grows high up in trees where it is sometimes hidden in Resurrection Fern and/or Spanish Moss. Populations have been found in almost every river system in the Coastal Plain. It generally grows on horizontal limbs of Live Oak, Bald Cypress, Water Elm, and Tupelo Gum.

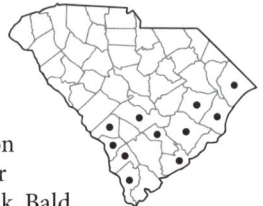

789. Carolina Birds-in-a-nest;
Carolina Macbridea

Macbridea caroliniana
(Walter) Blake
Mac-brìd-e-a ca-ro-li-ni-à-na
Lamiaceae (Mint Family)

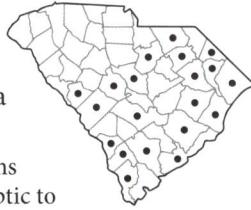

DESCRIPTION: Perennial herb to 3′ tall; stems square; leaves opposite, in 7–11 pairs, elliptic to lanceolate with margins shallowly serrate; flowers in 1 to 3 tight, bracteate, separated clusters; flowers July–August.
RANGE-HABITAT: Endemic to the Coastal Plain from NC south to FL and west to LA; rare throughout its range; in SC, openings in swamp forests, freshwater marshes, ditches, savanna edges, and bogs.
COMMENTS: The common name comes from the birds (flowers) that arise out of the nest (the subtending bracts). The genus honors Dr. James MacBride (1784–1817), a SC botanist of St. John's Parish, Berkeley County.
CONSERVATION STATUS: SC-Vulnerable

790. Southern Rein Orchid

Platanthera flava (L.) Lindley
Pla-tán-the-ra flà-va
Orchidaceae (Orchid Family)

SYNONYM: *Habenaria flava* (L.) R. Brown—RAB

DESCRIPTION: Smooth, terrestrial orchid, 5–25″ tall; leaves spreading to ascending, rapidly reduced upward; flowers in spikes, the lip petal lowermost (resupinate); tepals yellowish-green; flowers March-September.
RANGE-HABITAT: NJ south to FL and west to TX, OK, and IL; also found in Nova Scotia; in SC, common in the Coastal Plain and rare in the Piedmont and mountains; swamp forests, pond margins, and streambanks.
COMMENTS: Southern Rein Orchid is often locally common in extensive swamp forest systems, particularly tidal freshwater swamp systems. It is easily overlooked because its green flowers blend into the surroundings. Though it is similar to Small Green Woodland Orchid (*P. clavellata* (Michaux) Luer), the much longer spike of flowers and more than one stem leaf serve to easily distinguish it.

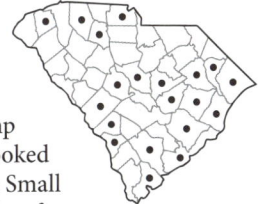

791. American Mistletoe

Phoradendron leucarpum (Rafinesque) Reveal & M. C. Johnson ssp. *leucarpum*
Pho-ra-dén-dron leu-cár-pum ssp. leu-cár-pum
Santalaceae (Sandalwood Family)

SYNONYM: *Phoradendron serotinum* (Rafinesque) M. C. Johnston—RAB, PR

DESCRIPTION: Evergreen shrub; obligate hemi-parasite on a variety of broadleaf, deciduous trees; flowers small; male and female flowers on separate plants (dioecious); flowering September–November and sporadically through the winter; berries white, maturing November–January, persisting into spring.

RANGE-HABITAT: NJ south to FL and west to TX and OK; common throughout SC and found on the branches of broadleaf, deciduous trees exposed to sun in swamp forests and other forested areas.

COMMENTS: There are numerous species of *Phoradendron* in North America; however, this is the only species in SC. This is the common mistletoe used in Christmas holiday decorations.

Mistletoes cause enormous economic loss in many parts of the world. In North America it is a pest in walnut, pear, and pecan plantations. In these plantations, injury results from broken branches that allow invasion of insects and fungi; tree growth is slowed due to loss of water and minerals. In SC, little economic loss results from mistletoe since it mostly affects trees that are of little commercial value.

The berries are covered with a sticky material poisonous to humans. Poisoning often occurs during Christmas when the plant is used for decorations and children eat the berries. Its one-seeded fruits are eaten by a wide variety of birds that use the pulp for food. The birds spread the seeds through their droppings and from wiping their beaks on branches. In both cases, germination occurs on the branch, the haustorium penetrates the host tissue, and a xylem bridge forms between the host and mistletoe.

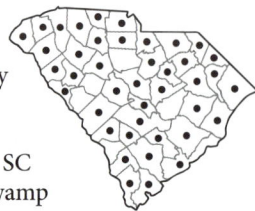

792. Hardwood bottom forests

793. Sweet Gum

Liquidambar styraciflua L.
Li-quid-ám-bar sty-ra-cí-flu-a
Altingiaceae (Sweet Gum Family)

DESCRIPTION: Large, deciduous tree; young branches usually with irregularly corky wings; leaves alternate, palmately 5-lobed and serrate; male and female flowers in separate clusters on the same plant; fruits are aggregates of capsules in hard, rounded ball-like structures; flowers March–April; fruits mature August–September and persist through winter.

RANGE-HABITAT: Widespread from CT west to OH and OK and south to FL and TX; also in Mexico and Guatemala; in SC, common throughout except at higher elevations; bottomland hardwoods and other moist forests, also invading almost all upland communities in the absence of fire.

COMMENTS: Sweet Gum is one of SC's most versatile trees. Although it is not strong enough for structural timber, its pink or ruddy heartwood shows handsome figures on the quarter-sawed cut and is used for veneer, furniture, and plywood panels. Today, Sweet Gum shows a higher commercial harvest than any other deciduous hardwood. A resin ("gum") is produced by the tree in response to a cut or stripping bark from the tree. During the colonial era in the South, this gum was used for treatment of sores and skin troubles, as chewing gum,

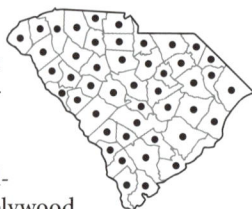

and by the Confederate armies to treat dysentery. During World Wars I and II, Sweet Gum was the base of soaps, drugs, and adhesives. This species was probably confined to floodplains and bottomland hardwood forests in colonial times and has since invaded many upland communities as widespread fire-suppression has become common. It is very invasive in fire-suppressed Longleaf Pine habitats. Amazingly, it is also a large canopy tree in the montane rainforests of Central America. The tallest known Sweet Gum is 157′ tall and is located in Congaree National Park in Richland County.

794. Ironwood; Blue-beech; American Hornbeam

Carpinus caroliniana Walter
    Car-pì-nus ca-ro-li-ni-à-na
Betulaceae (Birch Family)

DESCRIPTION: Usually a shade-tolerant, understory tree to 35′ tall and 8–10″ in diameter; trunk irregularly fluted and twisted; leaves deciduous, simple, symmetrical at base, alternate and 2-ranked, doubly serrate; male and female flowers on the same tree; male flowers in drooping catkins; female catkins appearing in early spring with emergence of leaves; fruits are nutlets subtended by 3-lobed, leaflike bracts, mature September–October.

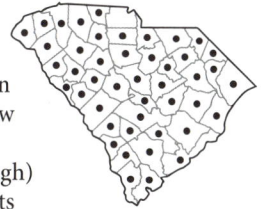

RANGE-HABITAT: Quebec west to MN and south to GA and OK; in SC, common throughout; rich cove forests, beech forests, bottomland hardwoods, other low and wet woodlands, floodplain forests, and extending upslope in rich soils.

COMMENTS: The common name, hornbeam, comes from "horn" (meaning tough) and "beam" (similar to the German "baum" for tree); it accurately describes its wood—close-grained, hard, and heavy. It was traditionally used for mallets, tool handles, wooden ware, dishes, and bowls. The tree is too small to be commercially used for lumber. Small mammals and songbirds eat the nutlets.

795. Piedmont Azalea

Rhododendron canescens (Michaux) Sweet
    Rho-do-dén-dron ca-nés-cens
Ericaceae (Heath Family)

DESCRIPTION: Deciduous shrub to 16′ tall; flowers fragrant, appearing before the leaves, or with the leafy shoots of the season; outer vegetative bud scales densely pubescent; leaves densely white pubescent (canescent) below; corolla pale to deep pink, without yellow markings; the corolla tube pubescent and with sticky glands; fruit a cylindrical capsule 4–5x as long as broad; flowers March–April; fruits mature September–October.

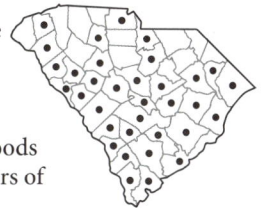

RANGE-HABITAT: DE and MD, south to FL, and west to TX and OK; in SC, common throughout the Coastal Plain and Piedmont; primarily a plant of moist, wooded slopes along edges of swamps and hardwood bottoms, low woods and thickets, riverbanks and streambanks, rocky open woodlands, and borders of pocosins and bogs.

SIMILAR SPECIES: Pinxterflower (*R. periclymenoides* (Michaux) Shinners, plate 348) is very similar but may be distinguished by its corolla tube that is pubescent, but not with sticky glands, and bud scales that are not densely pubescent. It essentially replaces Piedmont Azalea in the upper Piedmont and mountains.

COMMENTS: Piedmont Azalea makes a good landscape shrub that is easily cultivated throughout SC. The flowers are pollinated by butterflies and moths. All parts are poisonous.

796. Coastal Doghobble

Leucothoë axillaris (Lamarck)
D. Don
Leu-cóth-o-ë ax-il-là-ris
Ericaceae (Heath Family)

DESCRIPTION: Evergreen shrub to 5′ tall, with clustered, arching stems, usually forming dense colonies; leaves smooth, thick and glossy, lanceolate to oblanceolate or elliptic, with obtuse, acute, or short-acuminate tips; flowers produced in drooping racemes of urn-shaped (urceolate) flowers; flowers late March–May.

RANGE-HABITAT: VA south to FL and west to LA; in SC, common in the Coastal Plain and extending into the Piedmont; hardwood bottoms, blackwater swamps, pocosins, moist and acid slopes, and along streams on acidic soils.

SIMIALR SPECIES: Mountain Doghobble (*L. fontanesiana* (Steudel) Sleumer, plate 157) is similar. It is a larger plant with long acuminate leaf tips and is restricted to the mountains and upper Piedmont.

COMMENTS: The common name comes from the dense colonies that are almost impenetrable to move through, even for a hunting dog. The genus is from the mythical Leucothoë, daughter of Orchamus, King of Babylon, who was buried alive by her father but resurrected by Apollo as an incense shrub.

797. Possum-haw

Ilex decidua Walter
Ì-lex de-cí-du-a
Aquifoliaceae (Holly Family)

DESCRIPTION: Deciduous shrub or small understory tree, to 33′ tall; leaves gray-green, alternate, oblanceolate to obovate, 0.3–1″ wide, base cuneate; male and female flowers on separate plants (dioecious); fruits are red berries with pedicels that are very short (less than 0.25″ long); flowers March–May; fruits mature September–October and persist through winter.

RANGE-HABITAT: MD south to FL and west to MO and TX; in SC, common throughout the Coastal Plain and Piedmont; hardwood bottoms, floodplain forests, alluvial swamps, low woodlands along creeks, and wet thickets.

SIMILAR SPECIES: Cuthbert's Holly (*I. cuthbertii* Small) is endemic to SC and GA and is restricted to dry, oak-hickory forests on circumneutral soils. It is distinguished by the longer pedicels (0.4–1.0″ long) and smaller leaves. This rare species may be observed at Stevens Creek Heritage Preserve in McCormick County.

COMMENTS: Winter branches of Possum-haw, bearing the orange to red berries, are gathered for use in Christmas decorations. A variety of wildlife eat the fruits. The wood is hard and dense, but the trees are too small to be of commercial value. The tree is valued as a landscape plant because of its drought tolerance and conspicuous fruits that provide winter interest.

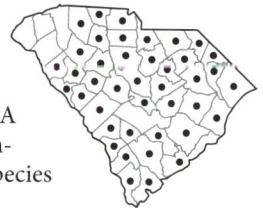

798. Jackson-briar

Smilax smallii Morong
Smì-lax smáll-i-i
Smilacaceae (Greenbrier Family)

DESCRIPTION: High-climbing, evergreen, vine; spinelike projections on stems few or absent; leaves thick, smooth, lance-elliptic to narrowly ovate, with cuneate bases; flowers in umbels, petals greenish-yellow; flowers April–May; fruits are blackish berries, mature in September and persisting through the winter.

RANGE-HABITAT: From the Coastal Plain of NC south to FL and west to AR and TX; in SC, common in the Coastal Plain and uncommon in the Piedmont; hardwood bottoms, swamp forests, and other moist forests.

COMMENTS: The narrow, evergreen leaves, stems without spinelike projections, and the dark fruits serve to distinguish this handsome *Smilax.* It has become a popular vine in landscape plantings, in part because the stems lack spinelike projections and because the greenery of this species is traditionally collected in the South for use in Christmas decorations. The specific epithet honors John Kunkel Small (1869–1938), botanist at the New York Botanical Garden and author of *Manual of the Southeastern Flora* (1933).

799. American Wisteria

Wisteria frutescens (L.)
Poiret var. *frutescens*
Wis-tèr-i-a fru-tés-cens var.
fru-tés-cens
Fabaceae (Bean Family)

DESCRIPTION: Deciduous woody vine, forming thickets or climbing on low trunks and branches; leaves alternate, 4–12″ long, pinnately compound; leaflet tips acute to short acuminate; flowers produced in drooping racemes; fruit a smooth legume; flowers April–May; fruit mature June–September.

RANGE-HABITAT: VA south to FL and west to AR and TX; in SC, common in the Coastal Plain and Piedmont in hardwood bottoms, swamp forests, and other wet areas, particularly along the margins or in openings of forests with abundant sunlight.

COMMENTS: American Wisteria makes a handsome and much less aggressive addition to the landscape than its cousin, the invasive Chinese Wisteria. American Wisteria is readily available from nurseries and does not reach great heights, so a lower trellis or fence is recommended to support the plant. The SC cultivar Amethyst Falls is particularly nice in the landscape. The genus honors Dr. Casper Wistar (1761–1818), physician and anatomist from Philadelphia.

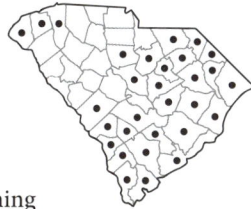

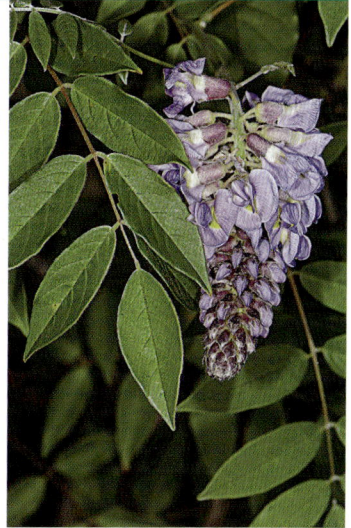

800. Supplejack; American Rattan

Berchemia scandens (Hill)
K. Koch
Ber-chèm-i-a scán-dens
Rhamnaceae (Buckthorn Family)

DESCRIPTION: High-climbing, deciduous, woody vine; young stems reddish and smooth, older stems with smooth, gray bark; leaves alternate, smooth, dark green with deeply impressed veins evident on the upper surface; leaf margins entire; inflorescence with 7–20 flowers; flowers small, greenish-white; fruit a dark blue to black drupe; flowers April–May; fruits mature August–October.

RANGE-HABITAT: VA south to FL and west to IL and TX; also found in Mexico; in SC, common in the Coastal Plain, uncommon in the Piedmont; swamp forests, hardwood bottoms, riverbanks, and other wet and forested habitats.

COMMENTS: This is among the largest of the SC woody vines, with stems reaching 8″ in diameter. It is often fed on by Yellowbellied Sapsuckers, and older vines have their characteristic smooth bark lined with tiny holes created by these birds to initiate sap flow and attract insects. The genus is presumably named for Jacob Pierre Berthoud von Berchem (1763–1832), a Dutch naturalist and mineralogist.

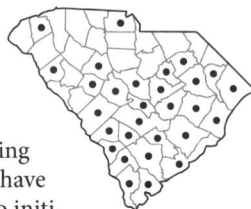

801. Peppervine

Nekemias arborea (L.) J. Wen &
Boggan
Ne-kè-mi-as ar-bò-re-a
Vitaceae (Grape Family)

SYNONYM: *Ampelopsis arborea* (L.)
Koehne—RAB, PR

DESCRIPTION: High-climbing, deciduous, woody vine with tendrils; leaves alternate, 2–3-pinnately compound; leaflets with serrate margins; flowers produced in corymblike cymes opposite the leaves; individual flowers tiny; fruits are grapelike, dark blackish-purple berries; flowers June–October; fruits mature September–October.

RANGE HABITAT: VA south to FL and west to IL and TX; also found in Mexico; in SC, common in the Coastal Plain and Piedmont in swamp forests, hardwood bottoms, maritime forests, and other wet and forested systems.

COMMENTS: Peppervine's relationship to grapes is not immediately apparent when looking at the bipinnately or tripinnately compound leaves. The tendrils and fruits are very similar to those of grapes.

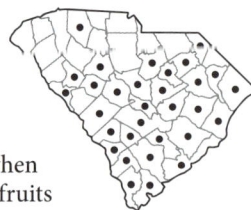

802. Dwarf Palmetto

Sabal minor (Jacquin) Persoon
Sà-bal mì-nor
Arecaceae (Palm Family)

DESCRIPTION: Shrublike palm with no above-ground trunk; petioles without a serrate margin; leaves fan-shaped, almost round in outline, without an obvious midrib, bluish green; flowers produced in panicles; flowers May–July; fruits black at maturity, mature in September–November.

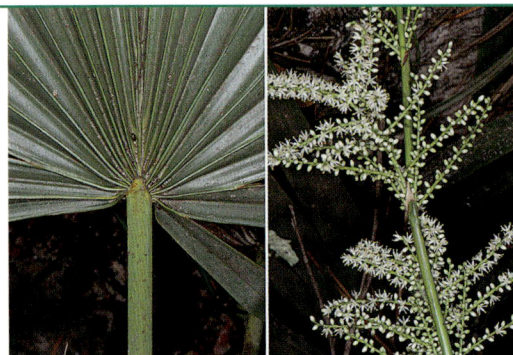

RANGE-HABITAT: NC south to FL and west to AR and TX; in SC, common in the Coastal Plain and extending into the Piedmont along rivers, especially in the Savannah River drainage system; hardwood bottoms, floodplain forests, swamps, commonly where flooded seasonally, and occasionally on sandy soils, including moist pinelands.

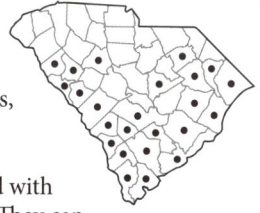

SIMILAR SPECIES: Where it grows with Cabbage Palmetto (*S. palmetto* (Walter) Loddiges ex J. A. & J. H. Schultes, plate 907), Dwarf Palmetto may be confused with an immature Cabbage Palmetto that has not produced an aboveground stem. They can be separated as follows: Dwarf Palmetto has no filaments on the leaf segments and is without a midrib on the blade except at the very base; Cabbage Palmetto has filaments on the leaf blades and a prominent midrib curving downward at the leaf tip.

COMMENTS: The flowers are a source of honey, and Native Americans used the fruits for food. Although the dried leaves are used occasionally for thatch roofs, Dwarf Palmetto has no significant economic value. It is a good landscaping plant throughout SC and is hardy far north of its native range. The flowers of our native palms are extremely attractive to pollinators, particularly bees, flies, and wasps.

---

803. Common Jack-in-the-pulpit; Indian Turnip

Arisaema triphyllum (L.) Schott
A-ri-saè-ma tri-phýl-lum
Araceae (Arum Family)

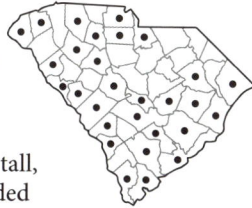

DESCRIPTION: Erect, perennial herb, 8–30″ tall, from a corm; leaves 1 or 2, palmately divided with 3 leaflets that are greenish-white below (glaucous); flowers on a fleshy spadix, male above, female below; spathe (the pulpit) with a tube and a hood that arches over the spadix (Jack); spathe hood striped throughout with dark purplish-black and white or green and white; flowers mature March–April; fruits are red berries, mature in July.

RANGE-HABITAT: Common and widespread in eastern North America; common throughout SC in rich woods, low woods, and bogs.

SIMILAR SPECIES: Only a single species of Jack-in-the-pulpit was recognized as growing in the eastern United States by most authors, until recently. The Common Jack-in-the-pulpit (*Arisaema triphyllum* (L.) Schott) is a tetraploid (4 sets of chromosomes) and is functionally sterile with the four other diploid taxa in the eastern US. The only other species in SC that produces leaves that are whitish-green (glaucous) beneath is Southern Jack-in-pulpit (*Arisaema quinatum* (Nuttall) Schott, plate 123), which is more abundant to the south and extends into SC only in the Savannah River watershed, where it reaches well into the mountains. The leaves in Southern Jack-in-the-pulpit are divided into five leaflets; it typically only produces green spathes.

COMMENTS: All parts of the plant, but especially the corm, contain crystals of calcium oxalate that can irritate the mucous membranes of the mouth and throat, causing a burning sensation. Death can result by asphyxiation if the air passages swell. However, Fernald and Kinsey (1958) report that the crystals can be broken up by heat and drying (but not by boiling) for a long period of time. Once done, the corm becomes mild and pleasant tasting. Native Americans used it as flour.

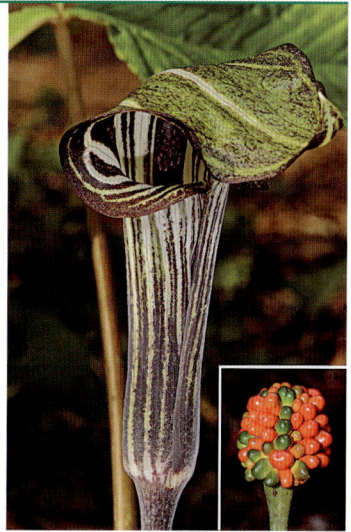

804. Carolina Least Trillium

Trillium pusillum Michaux
var. *pusillum*
Tríl-li-um pu-síl-lum var.
pu-síl-lum
Trilliaceae (Trillium Family)

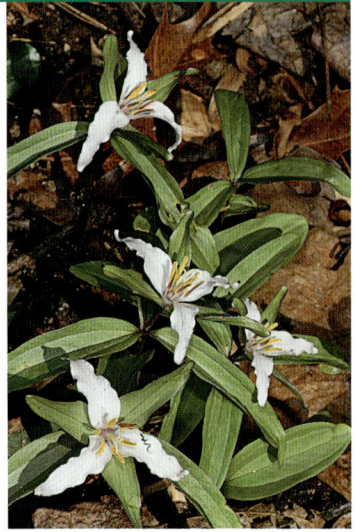

DESCRIPTION: Perennial, rhizomatous herb; stem 2–8″ tall; leaves 3, whorled, to 3.2″ long and 1.25″ wide, usually smaller, oblong to lanceolate, with obtuse tips; flower solitary, on short stalk; petals at first white, changing to pink; flowers March–May.

RANGE-HABITAT: Endemic to the Coastal Plain of NC and SC, with only a handful of locations; in SC, very rare and found in pocosin borders, bottomland forests among small streams in the Inner Coastal Plain and in hardwood bottoms and ecotones of calcareous savannas in the Outer Coastal Plain.

COMMENTS: The most significant population of Carolina Least Trillium in SC occurs in Francis Beidler Forest in Dorchester County. In 1989, Hurricane Hugo removed most of the canopy from the hardwood bottom where it grows. The population has been monitored since the canopy removal, but no significant decrease occurred in its numbers since the hurricane, despite growing in almost full sunlight. In fact, since the canopy has recovered, the population has reduced vigor and numbers. This species has apparently never been common; nineteenth-century collectors such as F. P. Porcher and J. H. Mellichamp have noted its rarity in their comments on herbarium specimens from the period.

TAXONOMY: The *Trillium pusillum* complex is under active review. Members of the complex have very spotty and highly disjunct ranges. The various disjunct populations are beginning to be described as segregate species, such as the recently described *Trillium georgianum* Farmer (Schilling et al., 2017).

CONSERVATION STATUS: SC-Critically Imperiled

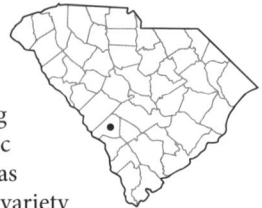

805. Aiken Least Trillium

Trillium sp. nov.
Tríl-li-um
Trilliaceae (Lily Family)

DESCRIPTION: Perennial, rhizomatous herb; stem 4–10″ tall; leaves ascending, short and stout; flower solitary, on short stalk; petals at first white, changing to pink with age; sepals broad ovate; flowers March–May.

RANGE-HABITAT: Known only from hummocks and seepages within a swamp forest in Barnwell County very near the Aiken County line.

COMMENTS: This species was first encountered and collected by Richard Porcher and Patrick McMillan in 1997. The population was subsequently studied by Susan Farmer, who has proposed that it be recognized as distinct from the very similar Carolina Least Trillium. It differs in the more ascending leaves, taller stem, and broader sepals. Farmer's research also indicates genetic divergence. Though the site is protected from development, the population has been badly damaged by hogs. Whether it is considered a distinct species or a variety of *Trillium pusillum,* it certainly is the rarest trillium in SC, perhaps in the world.

CONSERVATION STATUS: SC-Critically Imperiled

806. Common Atamasco-lily; Rain Lily; Easter Lily

Zephyranthes atamasco (L.) Greene
Ze-phy-rán-thes a-ta-más-co
Amaryllidaceae (Amaryllis Family)

DESCRIPTION: Perennial herb from a bulb; flowering stalk to 1′ tall, generally solitary, terminated by a single large white flower; leaves basal, linear; perianth usually white, rarely pink or fading to pink; tepals spreading and often curving downward; flowers March–April.

RANGE-HABITAT: VA south to FL and west to MS; common in the Coastal Plain and Piedmont of SC; bottomland hardwood forests, floodplains, road shoulders, and wet meadows.

SIMILAR SPECIES: Florida Atamasco-lily (*Z. simpsonii* Chapman, plate 941) is similar but has shorter tepals that are held erect to ascending. It is restricted to maritime forest margins and maritime grasslands in the northeastern Coastal Plain and to flatwoods margins, ditches, and roadsides in the southeastern Coastal Plain.

COMMENTS: Leaves, and especially the bulb, are highly poisonous to horses, cattle, fowl, and people. This is one of the showiest wildflowers in SC and is frequently cultivated by native plant enthusiasts. It often flowers a few days after rains, like many species in the genus, which gives rise to the colloquial name Rain Lily. The genus name refers to the Greek god of the west wind (Zephyrus) and flower (anthos).

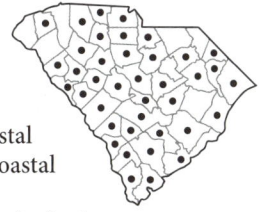

807. Swamp Leatherflower; Marsh Clematis

Clematis crispa L.
Clé-ma-tis crís-pa
Ranunculaceae (Buttercup Family)

DESCRIPTION: Short, perennial, deciduous vine, with stems to 9′ long, not dying back to the ground until winter; leaves thin, pinnately to bipinnately compound; leaflets 4–10, with a tendril-like terminal leaflet; flowers solitary, terminal, bell-shaped, with spreading to recurved blue to purplish sepals; petals absent; fruit an aggregate of achenes with plumose extensions; flowers April–August; fruits mature May–October.

RANGE-HABITAT: VA south to FL and west to IL and TX; in SC, common in the Coastal Plain, rare in the Piedmont; swamp forests, bottomland hardwoods, floodplain forests, marshes, stream margins, and other wet habitats.

COMMENTS: This beautiful and delicate flower is easily distinguished by the thin leaves, short, deciduous stems, and spreading to recurved sepals. It makes a beautiful specimen plant in the home landscape.

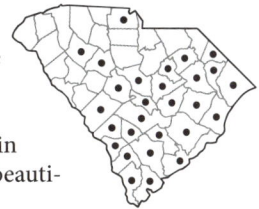

808. Lanceleaf Loosestrife (A)

Steironema lanceolatum (Walter) Gray
Stei-ro-nè-ma lan-ce-o-là-tum
SYNONYM: *Lysimachia lanceolata* Walter
var. *lanceolata*—RAB

Lowland Loosestrife (B)

Steironema hybridum (Michaux)
Rafinesque ex B. D. Jackson
Stei-ro-nè-ma hý-bri-dum
SYNONYM: *Lysimachia lanceolata* Walter
var. *hybrida* (Michaux) Gray—RAB

Primulaceae (Primrose Family)

DESCRIPTION: Perennial herbs with opposite leaves
and yellow, axillary flowers; *S. lanceolatum:* generally
6–12″ tall, with erect, unbranched stems; strongly rhizomatous with narrow, elongate rhizomes, often forming
thickets; leaves narrow, opposite, linear-lanceolate, lanceolate or narrowly elliptic; petioles ciliate along their entire
length; *S. hybridum:* 1–3′ tall, with erect to reclining stems, often branching; base of stem thickened
but not rhizomatous, generally occurring as single plants or in small patches; leaves elliptic-lanceolate
to lanceolate or linear-lanceolate; petioles generally ciliate only near the base; *S. lanceolatum* flowers
May–August; *S. hybridum* flowers May–September.

RANGE-HABITAT: *S. lanceolatum* ranges from NJ west to WI and south to GA and TX; in SC, infrequent
in wet to dry, semi-open forested habitats with rich soils, such as margins of bottomland hardwoods,
floodplains, margins of rock outcrops, and fens; *S. hybridum* ranges from Quebec west to WA and
south to FL and AZ; rare in SC; found on rich, circumneutral soils in margins of hardwood bottoms,
remnants of calcareous savannas, and ditches.

COMMENTS: Both of these similar loosestrife species are wide-ranging but neither is very common in SC. Populations of both tend to be transient and may not persist for many years. Both
make excellent garden plants and flower for a long period. Lanceleaf Loosestrife makes a
good "green mulch," forming a living carpet of groundcover into which other larger seasonal-
interest plants may be planted.

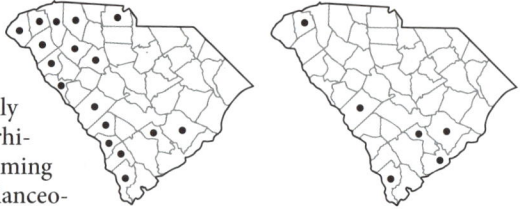

809. False Nettle

Boehmeria cylindrica (L.)
Swartz
Boeh-mèr-i-a cy-lín-dri-ca
Urticaceae (Nettle Family)

DESCRIPTION: Perennial herb without stinging
hairs; stems 1.5–3′ tall; leaves opposite, long-
stalked; flowers tiny, in small, headlike clusters,
arranged in continuous (female) or interrupted (male) spikes in
the axils of opposite leaves; spikes often terminated by small leaves;
flowers July–August.

RANGE-HABITAT: Quebec south to FL and west to MN and TX; in
SC, common throughout on fallen logs, stumps, cypress knees,
and buttresses in alluvial and nonalluvial swamp forests; also in
hardwood bottoms, freshwater marshes, wet thickets, canals, and
drainage or irrigation ditches.

COMMENTS: The genus honors Georg Boehmer (1723–1803), a German botanist and physician. Though this species resembles one of
the "stinging" nettles, it is harmless.

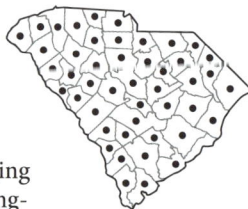

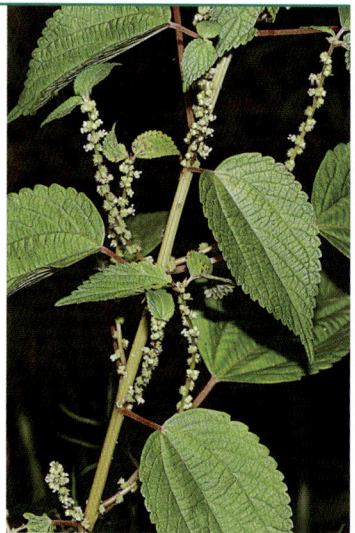

810. Virginia Dayflower

Commelina virginica L.
    Com-me-lìn-a vir-gí-ni-ca
Commelinaceae (Spiderwort Family)

DESCRIPTION: Rhizomatous, herbaceous perennial with erect stems 10–36″ tall, often forming extensive loose colonies; leaves lanceolate, with conspicuous basal sheaths; leaf blades 2.5–8″ long; flowers emerging from a green spathe; spathe margins fused at the base; all 3 petals blue, ephemeral, one slightly smaller; flowers July–October.

RANGE-HABITAT: NJ west to KS and OK and south to FL and TX; found throughout SC in bottomlands, swamp forests, and other moist to wet forests and forest edges; rare in the upper Piedmont and mountains.

COMMENTS: The dayflowers are known for the flowers that open for only one morning. This species is unique among SC native dayflowers in its significant rhizomes, robust size, and three blue petals rather than two blue petals and one white petal.

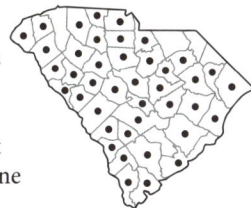

811. Netted Chain Fern

Lorinseria areolata (L.) C. Presl
    Lo-rin-sèr-i-a a-re-o-là-ta
Blechnaceae (Deer Fern Family)

SYNONYM: *Woodwardia areolata* (L.)
    T. Moore—RAB

DESCRIPTION: Rhizomatous, perennial fern, often forming large colonies; leaves strongly dimorphic; sterile fronds with blades 15–23″ long, pinnatifid with deep lobes and flat; fertile fronds with blades 20–28″ long, pinnately compound, with very narrow and stiff, thick pinnae, held erect; spores produced in sori on the undersides of the fertile pinnae, evident as distinct linear-oblong masses, which bulge upward on the upper leaf surface; spores produced May–September.

RANGE-HABITAT: Nova Scotia south to FL and west to MI and TX; in SC, common in swamp forests, bottomland hardwoods, floodplains, stream margins, Piedmont seepage forests, and other moist habitats.

COMMENTS: Molecular studies have shown that though this species, though traditionally known as a member of the genus *Woodwardia;* it is a more ancient lineage that is basal to *Woodwardia* and merits being placed in a separate genus. The "chain" in the common name refers to the unique pattern of venation on the leaves, which forms links, like links in a chain. This species makes a wonderful groundcover that can be used as a "green mulch" in the shade garden. The genus is named for Dr. Gustav Lorinser (1811–63), Austrian physician and botanist.

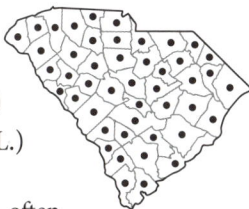

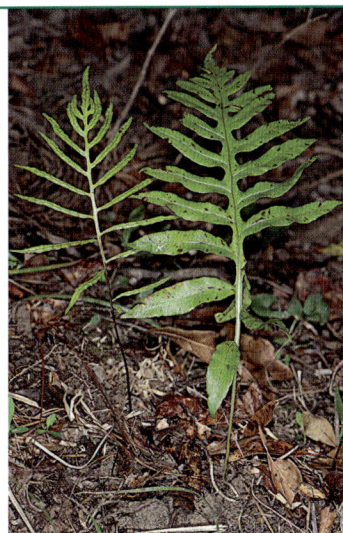

812. Southern Woodfern and Log Fern (A)

Dryopteris ludoviciana (Kunze) Small
Dry-óp-te-ris lu-do-vi-ci-à-na

Log Fern (B)

Dryopteris celsa (W. Palmer) Knowl-
ton, W. Palmer & Pollard
Dry-óp-te-ris cél-sa
Dryopteridaceae (Woodfern Family)

DESCRIPTION: Clump-forming, robust, evergreen (*D. ludoviciana*), and semi-evergreen (*D. celsa*) ferns with fronds 1–4′ tall; petioles with brown scales, particularly toward the base; *D. ludoviciana* with fronds fertile only toward the tip and the fertile pinnae distinctly more narrow than the fertile; *D. celsa* with fronds fertile nearly throughout; fertile pinnae not significantly more narrow than the sterile; the fertile fronds deciduous, the sterile fronds persisting through most of the winter; spores in both species produced in sori that appear as dots in two lines on the backs of the pinnae; spores produced June–September.

RANGE-HABITAT: *D. ludoviciana* ranges from NC south to FL and west to LA; in SC, common in the Coastal Plain in blackwater swamp forests and bottomland hardwoods; *D. celsa* ranges from NY south to GA and west to IL and TX; in SC, rare in the Coastal Plain and Piedmont; swamp forests; bottomland hardwoods; floodplains; wet, flat, calcareous forests; streamsides; and other moist habitats, often on rich, circumneutral soils.

COMMENTS: These two robust and attractive species make some of the best and most dramatic ferns for cultivation in the shady garden in SC. The dark green, bold fronds are attractive and very easy to cultivate. A third species, *D. australis* (Wherry) Small, was apparently originally derived from a hybridization event between Southern Woodfern and Log Fern and is intermediate in character. It is rare in the Coastal Plain.

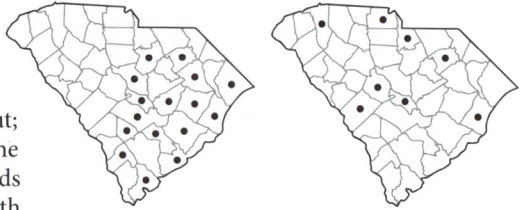

813. Levee forests

814. River Birch

Betula nigra L.
Bé-tu-la nì-gra
Betulaceae (Birch Family)

DESCRIPTION: Small- to medium-size deciduous tree, to 60–80′ tall; bark of trunk peels off in thin, light brown, paperlike layers; leaves doubly serrate, pubescent beneath when young; leaf bases wedge-shaped or truncate, not cordate;

male and female flowers in separate catkins on the same plant; female catkins mature in the fall, shedding winged seeds; flowers March–April; fruits mature May–June; seeds shed in the fall.

RANGE-HABITAT: Throughout the eastern US; throughout SC; common; generally associated with sunny, periodically wet areas such as streamsides, floodplains, levees, and sandbars.

COMMENTS: River birch is often cultivated because of its attractive peeling bark and high drought tolerance. The inner bark is edible and makes a good emergency trail food. Native Americans made use of the inner bark by drying it, then grinding it into flour. A refreshing sap can be drunk from the tree in the spring or boiled down into a sweet syrup. A tea can be made from the inner bark by simply boiling it in water. The dry, outer bark makes a good fire-starter. The wood is too knotty for lumber. The bark is the most identifiable feature.

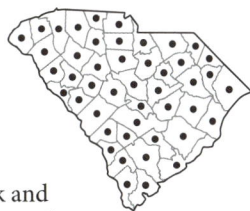

---

815. American Sycamore

Platanus occidentalis L.
 Plà-ta-nus oc-ci-den-tà-lis
 Platanaceae (Sycamore Family)

DESCRIPTION: Large, deciduous tree to 115′ tall; outer bark separating into large, thin scales that fall away and expose the lighter, inner bark; male and female flowers produced on the same plant in separate, dense heads in April–May; female cluster develops into a hard fruit ball of achenes that breaks apart in early spring of the following year; fruits mature October.

RANGE-HABITAT: Common throughout eastern North America; common throughout SC; rocky streamsides, along streams and rivers, and in bottomlands and floodplains; may also become established in abandoned fields and spoil banks.

COMMENTS: The seeds were a favorite food of the extinct Carolina Parakeet. Sycamore is often cultivated as an ornamental for its peeling outer bark that exposes the lighter, inner bark.

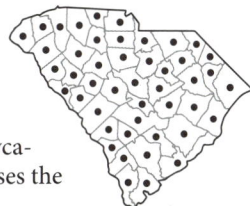

---

816. American Holly

Ilex opaca Aiton
 Ì-lex o-pà-ca
 Aquifoliaceae (Holly Family)

DESCRIPTION: Small- to medium-sized tree, 33–66′ tall; leaves evergreen, simple, alternate, with sharp marginal spines; male and female flowers on separate trees (dioecious); flowers April–June; fruits bright red or orange, containing 4 irregularly grooved pits (with enclosed seeds); fruits mature September–October, often persisting until the spring.

RANGE-HABITAT: Native to the eastern, southeastern, and southcentral sections of the US; common throughout SC in a wide variety of forests, ranging from xeric to wetland.

COMMENTS: This is the most frequently planted native holly because of its red berries and evergreen, spiny leaves. There are more than 300 varieties or cultivated forms. Songbirds eat the fruits and they are relished by Cedar Waxwings in the late winter. However, they are poisonous to humans. The trees are too small to yield commercial quantities of lumber. The tough, white wood (which turns brown with age) has been used to make small wooden ware.

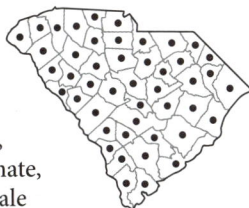

817. Wooly Dutchman's-pipe

Isotrema tomentosum (Sims) H. Huber

I-so-trè-ma to-men-tò-sum

Aristolochiaceae (Birthwort Family)

SYNONYM: *Aristolochia tomentosa* Sims—RAB

DESCRIPTION: Large woody vine that may climb high into trees; pubescent throughout; leaves large (to 12″ wide) and heart-shaped; flowers borne singly; corolla absent; calyx dull brown-purple and shaped like ornate tobacco pipes; flowers May–June.

RANGE-HABITAT: In south to FL and west to MO and TX; in SC, rare, restricted to margins and openings in levee forests and floodplains of large rivers in the Coastal Plain.

COMMENTS: Though this species is quite rare in SC, it is becoming common in cultivation. This species is superbly adapted to growing in our hot, humid climate and is grown both for the strange flowers and because it is an excellent host for the caterpillars of Pipevine Swallowtail butterflies. It thrives in full sun or afternoon shade and grows well in sandy or clay soils.

CONSERVATION STATUS: SC-Imperiled

818. Cross-vine

Bignonia capreolata L.

Big-nòn-i-a cap-re-o-là-ta

Bignoniaceae (Bignonia Family)

SYNONYM: *Anisostichus capreolata* (L.) Bureau—RAB

DESCRIPTION: Perennial, high-climbing, woody vine; leaves opposite, pinnately compound, with 2 lateral leaflets, and terminal leaflets that are modified into a pinnately compound tendrils; flowers produced in axillary clusters; corolla tubular, narrowly funnelform, red or orange, often with a lighter colored throat; flowers April–May.

RANGE-HABITAT: MD south to FL and west to OH, MO, and TX; in SC, common throughout; alluvial swamp forests, all upland wooded habitats, fencerows, thickets, and disturbed areas.

COMMENTS: The terminal leaflets that are modified into a tendril are highly branched. At the ends of the branches are small adhesive disks that are used for attachment, allowing Cross-vine to even climb the sides of buildings. The common name comes from the anatomy of the stem, an easy aid to identification, but one that should not be employed because it kills the vine. When the stem is viewed in cross-section, narrow bands of thick-walled cells separate the stem into four equal segments, producing a crosslike appearance. Cross-vine is a native vine that has spread widely into disturbed habitats. The genus honors Abbé Jean-Paul Bignon (1662–1743), a court librarian at Paris.

819. Switch Cane

Arundinaria tecta (Walter)
Muhlenberg
A-run-di-nà-ri-a téc-ta
Poaceae (Grass Family)

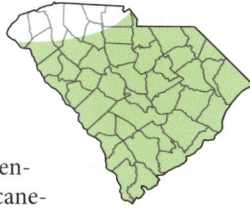

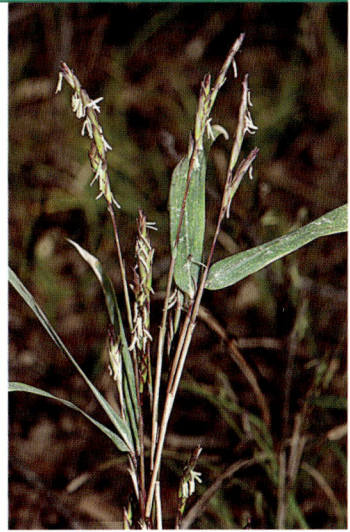

DESCRIPTION: Rhizomatous, evergreen, perennial bamboo forming extensive colonies (canebrakes); stems 2–12′ tall, slender, with internodes round (in cross-section, not grooved); plants flower every 3–4 years; flowers April–May

RANGE-HABITAT: MD south to FL and west to AL, primarily in the Coastal Plain; in SC, common in the Coastal Plain, Sandhills, and lower Piedmont; habitats include but are not confined to low lying, moist to wet places such as low woodlands, riverbanks and streambanks, hardwood bottoms, levees, shrub bogs, sloughs, bayous, and Longleaf Pine savannas.

SIMILAR SPECIES: See River Cane (*Arundinaria gigantea* (Walter) Muhlenberg, plate 77) for discussion

COMMENTS: *Arundinaria* is the only member of the bamboo tribe of the grass family (Poaceae) native to the United States. Numerous other species of the bamboo tribe are cultivated. Stock browse on the young leaves and spikelets of fruit. In coastal SC, during colonial times, lands were burned to produce an abundance of young plants for grazing. Where exposed to fire, such as in Longleaf Pine savannas (where this photograph was taken), it may only attain a height of 3′, but in levee forests and hardwood bottoms that are sheltered from fire, they attain heights of 10′ or more. Unlike *A. gigantea* it is commonly encountered in flower.

820. River Oats; Fish-on-a-pole

Chasmanthium latifolium (Michaux)
Yates
Chas-mán-thi-um la-ti-fò-li-um
Poaceae (Grass Family)

SYNONYM: *Uniola latifolia*
Michaux—RAB

DESCRIPTION: Rhizomatous, perennial grass; stems 2–5′ tall; stem leaves lanceolate, to 8″ long; flowers in elongate panicles of spikelets; the spikelets laterally flattened and held on long, dangling pedicels; flowers June–October.

RANGE-HABITAT: Widespread in eastern North America; throughout SC, except for the maritime strand; along riverbanks and streambanks, along ditches and levee forests, in hardwood bottoms, and in seepages and glades over mafic or calcareous rock.

COMMENTS: River Oats makes an exceptional native ornamental for partly shaded yards. It can be planted as an accent plant or as a dense groundcover. It spreads rapidly from seeds in the landscape. It is also dried and used for bouquets and floral arrangements.

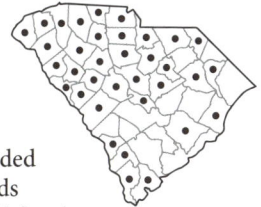

821. Butterweed

Packera glabella (Poiret)
C. Jeffrey
Páck-er-a gla-bél-la
Asteraceae (Aster Family)

SYNONYM: *Senecio glabellus* Poiret—
RAB, PR

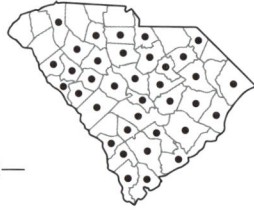

DESCRIPTION: Annual, erect herb with smooth, hollow stems to
3′ tall, often forming dense stands; leaves alternate, deeply lobed;
lateral lobes rounded and similar to the terminal lobe; flowers are
produced in involucrate heads of yellow ray and yellow disk flowers; flowers March–June.

RANGE-HABITAT: NC south to FL and west to SD and TX; in SC,
common in the Piedmont, Sandhills, and Coastal Plain; alluvial
and nonalluvial swamp forests, hardwood bottoms, levee forests,
and wet pastures.

COMMENTS: Butterweed has been suspected of poisoning cattle in
FL. In spring, our swamp and bottomland forests often turn golden with the masses of But-
terweed in flower. This is our only annual member of the genus *Packera*. All other species in
SC have distinctively perennial bases and basal leaves.

822. Creeping Spotflower; Spilanthes

Acmella repens (Walter) L. C. Richard
in Persoon
Ac-mél-la rè-pens
Asteraceae (Aster Family)

SYNONYM: *Spilanthes americana* (Mutis
ex Linnaeus f.) Hieronymus var.
repens (Walter) A. H. Moore—
RAB, PR

DESCRIPTION: Colonial, perennial herb with trailing or bent stems rooting at the nodes; stem usu-
ally purple, 12–40″ long; leaves simple, opposite; flowers in involucrate heads
of yellow ray and yellow disk flowers; receptacle conic, high-conic in fruit;
flowers late August–October.

RANGE-HABITAT: NC south to FL and west to MO and TX; in SC, common in
the Coastal Plain and rare in the Piedmont; wet pastures, swamp forests, levee
forests, seepage areas in woodlands, tidal freshwater marshes, and riverbanks.

COMMENTS: This species sometimes forms floating mats that extend into bodies of
water.

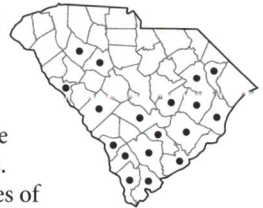

823. Tidal freshwater marshes

824. Southern Wild Rice; Giant Cutgrass

Zizaniopsis miliacea (Michaux)
Döll & Ascherson
Zi-za-ni-óp-sis mi-li-à-ce-a
Poaceae (Grass Family)

DESCRIPTION: Robust, rhizomatous perennial
forming large, dense colonies; stems to 10′ tall,
smooth; leaves with blades to 4′ long, smooth
with sharp (scaberulous) margins; spikelets in an open, drooping
panicle; flowers May–July.

RANGE-HABITAT: MD south to FL and west to MO and TX; also in
Mexico; in SC, common in the Coastal Plain, rare in the adjacent
Piedmont; freshwater tidal marshes and brackish marshes, slug-
gish rivers, and inland marshes.

COMMENTS: Southern Wild Rice is one of the most abundant
components of many of our tidal freshwater marshes. It is easily
distinguished from the robust, annual Wild Rice by its perennial
habit and the much earlier flowering period. The grains of South-
ern Wild Rice are much smaller and not significant for utilization as an edible grain, but the
rhizome has been noted by some sources to be edible.

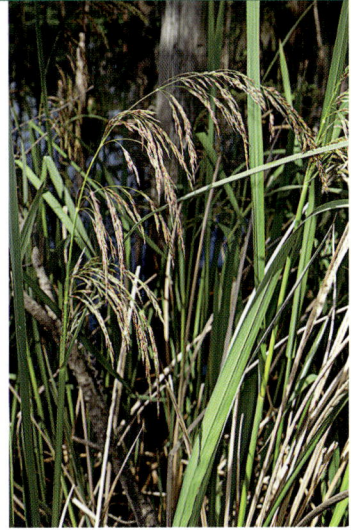

825. Wild Rice

Zizania aquatica L.
Zi-zà-ni-a a-quá-ti-ca
Poaceae (Grass Family)

DESCRIPTION: Coarse and robust annual with
smooth stems to 10′ tall; leaves with blades
to 4′ long, smooth with sharp (scaberulous)
margins; spikelets in panicles with ascending
branches; lower branches of panicle widely ascending, upper branches strictly ascending; male
spikelets hang from the lower branches, and the
female spikelets are erect on the upper branches;
flowers May–October.

RANGE-HABITAT: MA south to FL and west to MN and LA; in SC, common in
the Coastal Plain; freshwater marshes, brackish marshes, and inland marshes;
occurs in almost every river system along the coast, often locally abundant.

COMMENTS: Wild Rice is an important food source for animals and people. Native
Americans used the grain to thicken soup, to make flour for bread, and to cook
with game. Today it is marketed as Wild Rice. Its main distribution is the Great Lakes
and upper Mississippi region. The fruits are ready for harvest from late summer through

autumn. The yield is insufficient to make wild rice harvest feasible in SC. Wild Rice plants are not abandoned rice plants from the rice growing era but rather a distinct species. Commercial rice of Asian origin, Oryza *sativa* L., does not persist after cultivation.

826. Common Threesquare

Schoenoplectus pungens (Vahl)
   Palla var. *pungens*
      Schoe-no-pléc-tus pún-gens
      var. pún-gens
Cyperaceae (Sedge Family)

SYNONYM: *Scirpus americanus*
   Persoon—RAB, misapplied

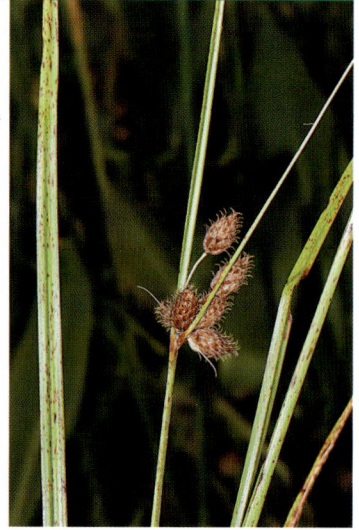

DESCRIPTION: Robust, rhizomatous sedge, forming colonies; stems to 4′ tall, triangular with shallowly concave to convex faces; inflorescence a cluster of ovoid spikelets appearing to rise directly from the triangular stem; involucral bract appearing like a continuation of the stem and extending up to 8″ beyond the inflorescence; apical notch of scales 0.5–1 mm deep; flowers and fruits June–September.

RANGE-HABITAT: Circumboreal; in the Americas from Newfoundland and AK south through South America; in SC, common in the Coastal Plain in tidal freshwater and brackish marshes, inland marshes, and other wetlands.

COMMENTS: Common Threesquare is a very distinctive and abundant component of our tidal freshwater marshes. The strange stems appear like triangular wands with the clusters of spikelets springing from the stem itself. A similar species, Olney Threesquare (*S. americanus* (Persoon) Volk ex Schinzius & R. Keller), is also common in similar habitats and can be told by the deeply concave, sharply triangular stems; the involucral bract that extends beyond the inflorescence only a short distance; the stem typically less than 1′ tall; and the shallow apical notch of the scales. Both Common Threesquare and Olney Threesquare were formerly in the genus *Scirpus.* There has been considerable confusion because the authors of the most popular treatment of the flora of SC (Radford et al., 1968) mistakenly applied the name *Scirpus americanus* to Common Threesquare. This name applies properly to Olney Threesquare; what was known as *Scirpus olneyi* is now *Schoenoplectus americanus* and what was *Scirpus americanus* is now *Schoenoplectus pungens.*

827. Sawgrass

Cladium jamaicense Crantz
      Clà-di-um ja-mai-cén-se
Cyperaceae (Sedge Family)

DESCRIPTION: Robust, perennial sedge, forming dense clumps; flowering stems to 10′ tall; leaf blades to 4′ long, sharply serrate along the margin and able to cut through skin and clothing; inflorescence an elongate compound cymose arrangement of spikelets; flowers July–October.

RANGE-HABITAT: VA south to FL and west to OK and TX; in SC, common in the Coastal Plain and maritime fringe in tidal freshwater and brackish marshes, edges of salt marsh, wet swales along the coastline, and other marshy areas.

COMMENTS: Sawgrass is unmistakable with its extremely dense, long, dangerously sharp-edged leaves that easily slice through skin. This is the same species that forms the famous Sawgrass prairies of the FL Everglades.

828. Common Reed

Phragmites australis
(Cavanilles)
Trinius ex Steudel
Phrag-mì-tes aus-trà-lis
Poaceae (Grass Family)

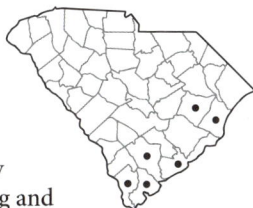

SYNONYM: *Phragmites communis*
Trinius—RAB

DESCRIPTION: Coarse and robust perennial grass to 10′ tall, forming vast monocultures; leaf blades to 20″ long and 1.2″ wide; flowers in a dense, ovoid "feathery" panicle; flowers August-October.

RANGE-HABITAT: Nearly worldwide; in SC, an invasive exotic, locally abundant in tidal freshwater marshes, spoil and dredge areas, and other wetlands, particularly in the maritime strand.

COMMENTS: Common Reed has apparently been repeatedly, though inadvertently, introduced into the eastern United States. It has become a serious biological threat to native vegetation in some areas, such as in the Santee Delta. There are native species of *Phragmites* found to our north and south; apparently all Common Reed in SC is introduced. Individual clones of this aggressively rhizomatous plant can be extremely long-lived. Some clones studied in other regions are known to be 1,000 years old (Haslam, 1971). In other parts of the world, such as Europe, the species has been utilized extensively as building materials for thatching.

829. Giant Foxtail-grass

Setaria magna Grisebach
Se-tà-ri-a mág-na
Poaceae (Grass Family)

DESCRIPTION: Robust, annual grass; stems to 9′ tall, with smooth nodes and minutely pubescent internodes; leaf blades to 2′ long and 1.6″ wide; flowering spikes to 20″ long, like a long "fox-tail" with a drooping tip; flowers August–October.

RANGE-HABITAT: NJ south to FL and west to TX, mostly along the Coastal Plain with disjunct inland populations west to NM; also found in the Caribbean and Central America; in SC, common in the Coastal Plain in tidal freshwater marshes, brackish marshes, inland marshes, freshwater swales along the coastline, and wet depressions on sea islands.

COMMENTS: This enormous foxtail-grass is unmistakable. The annual habit as well as the large, elongate, drooping inflorescence distinguishes it from all other foxtail-grass species in SC.

830. Giant Plume Grass

Erianthus giganteus (Walter) Muhlenberg

E-ri-án-thus gi-gan-tè-us

Poaceae (Grass Family)

DESCRIPTION: Robust perennial grass forming clumps; stems to 13′ tall, with bearded nodes, and internodes that are pubescent to nearly smooth; panicles of spikelets silvery to purplish-pink, to 16″ long; spikelets with a dense beard of callus hairs exceeding the spikelet in length; flowers September–October.

RANGE-HABITAT: NY south to FL and west to AR and TX; in SC, common in the Coastal Plain, uncommon in the Piedmont, and rare in the mountains; tidal freshwater marshes, inland marshes, ditches, wet savannas, cataract fens, and seepages over granitic flatrocks.

COMMENTS: This beautiful and impressively large grass resembles the cultivated Pampas Grass in appearance. It is particularly abundant along wet ditches in the Coastal Plain. Five additional species of plume grass are found in SC in habitats varying from dry road banks to marshes. This species, as the common and scientific names suggest, is the largest of the SC *Erianthus* species.

831. Water-spider Orchid

Habenaria repens Nuttall

Ha-be-nà-ri-a rè-pens

Orchidaceae (Orchid Family)

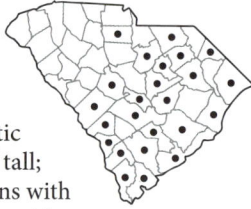

DESCRIPTION: Terrestrial or emergent aquatic orchid; stem slender or stout, leafy; 0.5–3′ tall; lower stem often producing elongate stolons with plantlets forming at the tips; leaves narrow, with three longitudinal ribs; flowers produced in a raceme, the individual flowers resembling a green spider; flowers April–frost.

RANGE-HABITAT: Endemic to the Coastal Plain from NC south to FL and west to TX; in SC, common in the Coastal Plain and Sandhills and rare in the Piedmont; tidal freshwater marshes, ditches and canals, muddy shores of lakes, ponds, and streams, often in floating mats of vegetation.

COMMENTS: Water-spider Orchid is often found unexpectedly while looking for some other plant. Since all parts are green, it is inconspicuous. In the tidal freshwater marshes, it is very difficult to spot because of the dense growth of emergent species. In the reservoirs created for the inland rice culture or similar impoundments, Water-spider Orchid often grows on large floating mats of vegetation in association with other aquatic plants. This is one of the few native orchids that can be described as "weedy," often forming large masses. Its range is expanding around man-made impoundments inland.

832. Heartleaf Pickerelweed

Pontederia cordata L. var. *cordata*
Pon-te-dèr-i-a cor-dà-ta var. cor-dà-ta
Pontederiaceae (Pickerelweed Family)

DESCRIPTION: Emergent, hollow-stemmed perennial from a thick, short rhizome; stems to 3' tall; one leaf not far below the inflorescence, the others basal; leaf bases deeply heart-shaped to truncate; flowers May–October; seeds mature late summer to early fall.

RANGE-HABITAT: Nova Scotia south to FL and west to MN and TX; common throughout SC in a variety of wetland habitats, including tidal freshwater marshes, lakes, ponds, swamp forests, and roadside ditches.

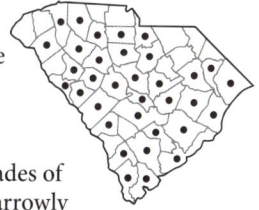

TAXONOMY: Two varieties of pickerelweed are recognized based on leaf shape. The leaf blades of Heartleaf Pickerelweed are deltoid (delta-shaped)-ovate to triangular-lanceolate, with bases deeply heart-shaped to truncate. The leaf blades of Lanceleaf Pickerelweed (*P. cordata* var. *lancifolia* (Muhlenberg) Torrey) are narrowly to broadly lanceolate, with typically unlobed bases. Lanceleaf Pickerelweed is found mainly in the Coastal Plain of SC.

COMMENTS: The seeds of Pickerelweed are a pleasant and hearty food. The young leaf stalks can be cooked as greens. The roots are inedible, producing a burning sensation if ingested. The flowers are attractive to a wide variety of pollinators; they are frequently swarming with bees, flies, wasps, and small butterflies.

The common name comes from a fish called the pickerel, which often occupies the same habitat as Pickerelweed. The genus honors Italian botanist Giulio Pontedera (1688–1757).

833. Swamp Rose

Rosa palustris Marshall
Rò-sa pa-lús-tris
Rosaceae (Rose Family)

DESCRIPTION: Upright, rhizomatous shrub to 7' tall; stems smooth with curved prickles; leaves pinnately compound with 5–9 leaflets; flowers with 5 pink petals, shallowly indented at the tip; flowers May–July; fruit (a hip) red, matures September–October.

RANGE-HABITAT: Nova Scotia south to FL and west to MN, AR, and MS; common throughout SC; tidal freshwater marshes and along streams, ponds, lakes, wet thickets, and swamp forests.

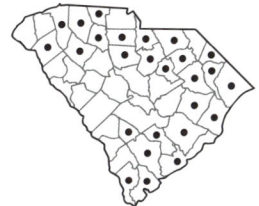

COMMENTS: The hip, like all rose hips, is rich in vitamin C and can be eaten raw, made into jams, or steeped to make rose hip tea. Swamp Rose is easily cultivated and does well in moderate shade.

834. Water Hemlock

Cicuta maculata L.
Ci-cù-ta ma-cu-là-ta
Apiaceae (Carrot Family)

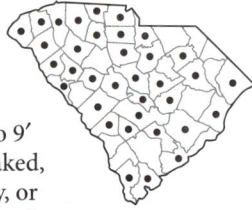

DESCRIPTION: Erect, branching perennial to 9′ tall; stem hollow above and magenta-streaked, hollow above; leaves pinnately, bipinnately, or tripinnately compound; leaflets serrate; the plant can be readily identified by cutting lengthwise through the stem base and root to reveal its diaphragmed nature (see photo); flowers white, produced in compound umbels; flowers May–August.

RANGE-HABITAT: Widespread throughout North America, ranging from Nova Scotia south to FL and west to TX, MO, and Ontario; common throughout SC in freshwater tidal marshes, swamps, streamsides, seepages, and low roadside ditches.

COMMENTS: All parts of the plant contain cicutoxin, a poisonous compound. The roots are particularly potent: a mouthful is sufficient to kill an adult. Children making peashooters from the hollow stems have been poisoned. The plant is not related to true hemlocks (*Tsuga* spp., tree species) but to Poison Hemlock (*Conium maculatum* L.), the plant that was used to kill Socrates.

835. Coastal Carolina Spiderlily

Hymenocallis crassifolia Herbert
Hy-me-no-cál-lis cras-si-fò-li-a
Amaryllidaceae (Amaryllis Family)

SYNONYM: *Hymenocallis floridana* (Rafinesque) Morton—PR

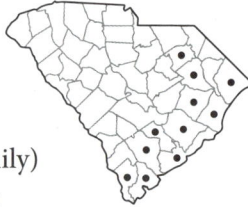

DESCRIPTION: Bulbous, smooth, perennial herb to 26″ tall; leaves linear, all basal; perianth segments white, narrow, linear 2.75–4.5″ long; stamens with the lower portion of the filament united into a thin, membranous crown, with the upper filament extending beyond the crown; flowers mid-May–June.

RANGE-HABITAT: NC south to FL and west into LA; in SC, common in the Coastal Plain where found in tidal freshwater marshes, wet riverbanks, brackish marshes, and swamp forests.

COMMENTS: Coastal Carolina Spiderlily is one of the most spectacular of the SC river marsh plants. It often hangs over the water from the bulbs embedded in the riverbank. It occurs in every freshwater river system in the coastal area of SC. The similar, but smaller, Pygmy Spiderlily (*H. pygmaea* Traub, plate 780) is confined to the Pee Dee and Waccamaw River drainages.

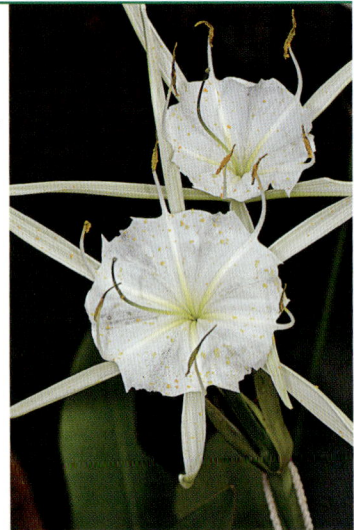

836. Tidal-marsh Obedient-plant

Physostegia leptophylla Small
Phy-sos-té-gi-a lep-to-phýl-la
Lamiaceae (Mint Family)

DESCRIPTION: Smooth, rhizomatous perennial
herb, forming colonies; stems square, 2–4′ tall;
leaves opposite, petiolate, bluntly toothed; flowers
tubular, bilabiate, vibrant purplish-pink, produced in bracteate racemes; flowers May–June
(August).

RANGE-HABITAT: Endemic to the Coastal Plain
from VA south to FL and west to AL; in SC,
locally abundant in tidal freshwater marshes and margins of tidal swamp
forests.

SIMILAR SPECIES: Northern Obedient-plant (*P. virginiana* (L.) Bentham var
virginiana, plate 1008) may be distinguished by the nearly sessile leaves and its
habitat—roadsides and disturbed habitats. Northern Obedient-plant is established from escapees from cultivation in SC and is only native farther to the north.

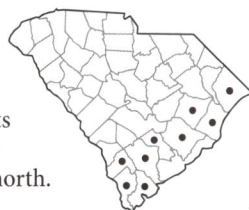

837. Bulltongue Arrowhead

Sagittaria lancifolia L. var.
media Micheli
Sa-git-tà-ri-a lan-ci-fò-li-a
var. mèd-i-a
Alismataceae (Water-plantain
Family)

SYNONYM: *Sagittaria falcata* Pursh—RAB

DESCRIPTION: Robust, emergent aquatic perennial herb; leaves
to 4′ long (including the petiole); blades elliptic, linear to ovate;
flowers in whorls on a leafless stem; male and female flowers on
the same plant (monoecious); sepals and bracts with small bumps
(papillose); fruits are aggregates of achenes; flowers June–October;
fruits mature July–November.

RANGE-HABITAT: DE south to FL and west to TX; also in Central
America; in SC, common in the Coastal Plain where found in tidal
freshwater marshes, margins of swamp forests, riverbanks, and
ditches.

COMMENTS: This large species is easily distinguished from all other *Sagittaria* without
arrowhead-shaped leaves by its size alone. A second variety, *S. lancifolia* L. var. *lancifolia,* is
far more abundant to the south and is widespread in the tropics. It is rare in the Coastal Plain
of SC, where it reaches its northern limit. The sepals and bracts of the flowers of var. *lancifolia*
are striate (lined) rather than covered with bumps (papillose).

838. Groundnut

Apios americana Medikus
Á-pi-os a-me-ri-cà-na
Fabaceae (Bean Family)

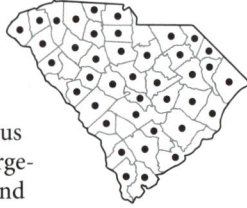

DESCRIPTION: Perennial, twining, herbaceous vine, 3–10′ long; roots with tuberous enlargements; leaves alternate, pinnately compound with 5–7 leaflets; flowers June–August; legumes mature July–September.
RANGE-HABITAT: Quebec south to FL and west to SD and TX; common throughout SC; tidal freshwater marshes, swamp forests, bottomland forests, wet thickets, wet meadows, wet pinelands, and stream sides.
COMMENTS: The root tubers are edible. Native Americans used them as a staple food source, and the Pilgrims relied on them during their first year. Eaten raw, they leave an unpleasant rubberlike coating in the mouth, which cooking removes. The tubers can be used in soups and stews or fried like potatoes, or they can be ground into flour and used for bread. The fruits are generally too scarce to supply much food, although they are edible.

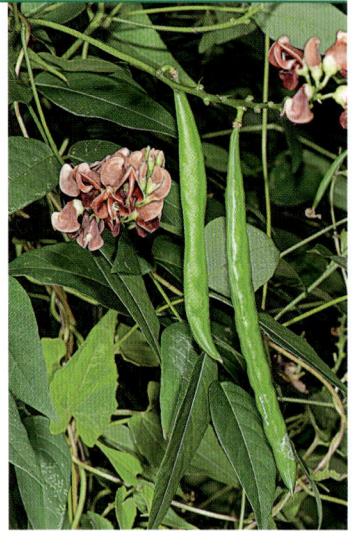

839. Swamp Rose-mallow

Hibiscus moscheutos L.
Hi-bís-cus mos-cheù-tos
Malvaceae (Mallow Family)

DESCRIPTION: Robust, herbaceous perennial from a crown; stems to 6′ tall, pubescent when young; leaves ovate to elliptic-lanceolate, serrate or crenate on the margin, densely pubescent beneath; petals white, less often pink, but always with a purple-reddish center; stamens monadelphous (with filaments fused into a tube surrounding the stigma); flowers June–September.
RANGE-HABITAT: MA south to FL and west to MI and TX; in SC, common throughout; tidal freshwater marshes, edges of swamp forests, brackish marshes, and roadside ditches.
SIMILAR SPECIES: Smooth Rose-mallow (*H. laevis* Allioni) is similar but has smooth stems and leaves that are halberd-lobed at the base. It is found in similar habitats throughout SC.
COMMENTS: The leaves and roots contain copious mucilage and historically were used as a soothing agent in dysentery and for lung and urinary ailments (Foster and Duke, 1990).

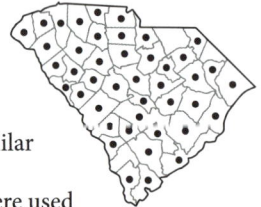

840. Large Marsh Rose-pink

Sabatia dodecandra (L.) Britton, Sterns, & Poggenburg
Sa-bà-ti-a do-de-cán-dra
Gentianaceae (Gentian Family)

DESCRIPTION: Perennial herb to 3′ tall, with slender, coarse rhizomes and lacking copious stolons; primary branches alternate and the internodes are longer than the length of the leaves; leaves opposite, entire, sessile; terminal

flower with a long stalk; petals 9–12, deep rose-purple, rose-pink, pink (or rarely white), with a mostly 3-lobed, yellow, red-margined patch at base; flowers June–August.
RANGE-HABITAT: CT south to FL; chiefly Outer Coastal Plain; in SC, common in tidal freshwater marshes, ponds, brackish marshes, riverbanks, streamsides, and ditches.
SIMILAR SPECIES: See discussion under Plymouth Rose-gentian (*Sabatia kennedyana* Fernald, plate 784).

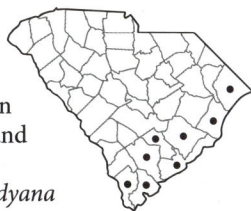

841. Marsh Eryngo

Eryngium aquaticum L.
　　E-rýn-gi-um a-quá-ti-cum
　　Apiaceae (Carrot Family)

DESCRIPTION: Robust biennial or short-lived perennial herb, 1–6′ tall; basal leaves petiolate, often oblanceolate; flowers in heads, subglobose to hemispherical; bractlets subtending each flower with middle cusp clearly longer than the lateral cusps; flowers July–September.
RANGE-HABITAT: NJ south to FL; primarily an Outer Coastal Plain species; in SC, common in the Coastal Plain in tidal freshwater marshes, riverbanks, ditches, ponds, brackish marshes, wet pine flatwoods, savannas, swamps, and depressions.
SIMILAR SPECIES: See discussion under Ravenel's Eryngo (*E. ravenelii* A. Gray, plate 652).

842. Seashore Mallow

Kosteletzkya pentacarpos (L.) Ledebour
　　Kos-te-létz-ky-a pen-ta-cár-pos
　　Malvaceae (Mallow Family)

SYNONYM: *Kosteletzkya virginica* (L.)
　　Presl ex Gray—RAB, PR

DESCRIPTION: Perennial herb to 5′ tall; pubescent with star-shaped hairs on all parts, varying from sparse to very dense and from harsh to soft-velvety; petals range from pink to lavender to white; stamens monadelphous (with filaments fused into a tube surrounding the stigma); each flower lasts only a day; flowers July–October.
RANGE-HABITAT: NY south to FL and west to TX, mostly on the Coastal Plain; in SC, common in the Outer Coastal Plain and maritime strand; freshwater tidal marshes, brackish marshes, sloughs, ditches, borders of swamps, and wet clearings.
COMMENTS: Plants of this species are highly variable in leaf shape, pubescence, and flower size, so much so that several varieties have been recognized. Intergrading forms are so common that it is difficult to make varietal determinations. The genus honors Vincenz Franz Kosteletzky (1801–87), a botanist from Bohemia, in what is now the Czech Republic.

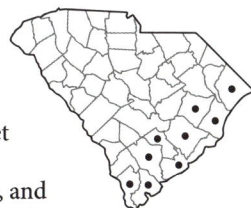

843. Cardinal Flower

Lobelia cardinalis L.
Lo-bèl-i-a car-di-nà-lis
Campanulaceae (Bellwort Family)

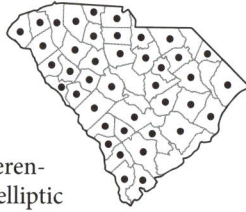

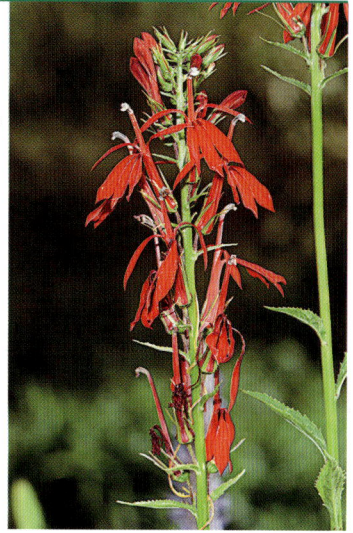

DESCRIPTION: Erect, usually unbranched perennial from basal offshoots, 2–6′ tall; leaves elliptic to lanceolate with serrate or crenate margins, the serrations alternating large and small; tubular red flowers produced in a raceme to 16″ long; flowers July–October.

RANGE-HABITAT: Widely distributed, from New Brunswick and Ontario south to FL and west to CA; common throughout SC; freshwater tidal marshes, swamp forests, riverbanks and streambanks, wet meadows, bogs, and low woods.

COMMENTS: Indigenous species of *Lobelia* were employed in medicines for various purposes; however, cases of death from overdoses of medicinal preparations were not infrequent. They are now best considered poisonous.

The genus is named in honor of Flemish botanist Mathias de l'Obel (1538–1616). The common name alludes to the bright red robes worn by Roman Catholic cardinals. Hummingbirds pollinate Cardinal Flower; most insects cannot reach the nectar at the bottom of the long, tubular flowers. Cardinal Flower is adaptable to cultivation and can be used in a variety of locations. Its cultivation can be extremely frustrating because it tends to grow where it wants to grow. A plant may look good one year and then disappear the next, but seeds may germinate in other areas and persist for some time.

844. Halberd-leaf Tearthumb

Persicaria arifolia (L.) Haraldson
Per-si-cà-ri-a a-ri-fò-li-a
Polygonaceae (Buckwheat Family)

SYNONYM: *Polygonum arifolium* L.— RAB, PR

DESCRIPTION: Freely branched perennial herb; stems slender, weak, to 6′ or longer, erect at first, then reclining; stems rib-angled, with backward-pointing prickles; leaves with hastate basal lobes; flowers few, pink, purplish, or white; flowers July–frost.

RANGE-HABITAT: New Brunswick to MN, generally southward to GA and MO; in SC, common in the Outer Coastal Plain, rare in the Inner Coastal Plain and Piedmont; tidal freshwater marshes and wet open places.

SIMILAR SPECIES: Growing in the same habitats (but throughout SC) is Arrowleaf Tearthumb (*P. sagittata* (Linaneus) H. Gross ex Nakai), which has sagittate leaf bases (with inward-pointing tips) rather than hastate leaf bases (with outward-pointing tips).

COMMENTS: The common name, Tearthumb, owes its origin to the stiff hooked prickles on the stem; they can easily tear one's thumb, or other body parts.

845. Climbing Aster

Ampelaster carolinianus (Walter) G.L. Nesom

Am-pe-lás-ter ca-ro-li-ni-à-nus

Asteraceae (Aster Family)

SYNONYM: *Aster carolinianus* Walter— RAB, PR

DESCRIPTION: Robust, branching, somewhat woody, sprawling perennial; often forming a tangle on low branches of woody vegetation; stems 3–12′ long, pubescent; leaves alternate, elliptic to lanceolate, 0.75–2.5″ long, pubescent; flowers produced in involucrate heads of blue ray flowers and yellow disk flowers, in a paniclelike arrangement; flowers late September–December.

RANGE-HABITAT: Endemic to the Coastal Plain from NC south to FL; in SC, common in the Coastal Plain where found in tidal freshwater marshes, river swamps, marshy shores, and streamsides, often in standing water.

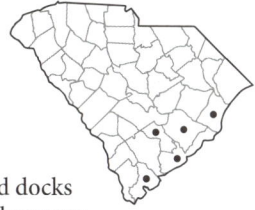

COMMENTS: Climbing Aster often forms robust growth on weathered posts and docks in abandoned tidal rice fields. It makes a stunning landscape plant and is easily grown away from wet soil. This species flowers extremely late, providing color in the landscape well into December and providing a good late-season nectar source for pollinators.

846. Showy Bur-marigold; Showy Beggar Ticks

Bidens laevis (L.) Britton, Sterns & Poggenburg

Bì-dens laè-vis

Asteraceae (Aster Family)

DESCRIPTION: Robust, smooth, annual herb; stems ascending, to 3′ tall, often creeping at the base, and rooting at the nodes, often forming dense colonies; leaves elliptic to lance-elliptic, unlobed, with serrate margins; flowers produced in involucrate heads of yellow ray and yellow disk flowers; rays are bright yellow with lighter yellow at the tips; fruits are achenes with awns that are retrorsely barbed and readily stick into clothing; flowers late September–November.

RANGE-HABITAT: Throughout the eastern and southwestern US; in SC, chiefly, but not exclusively, a Coastal Plain species; common; tidal freshwater marshes, shallow ponds, ditches, sluggish streams, and wet meadows.

847. Inland freshwater marshes

848. Golden-club; Never-wet

Orontium aquaticum L.
O-rón-ti-um a-quá-ti-cum
Araceae (Arum Family)

DESCRIPTION: Emergent, perennial herb from a thick rhizome; 8″ to 2′ tall; leaves in a basal cluster, either extending above or floating on water; flowers in a clublike spadix of reduced male and female flowers without a spathe; flowers March–April.

RANGE-HABITAT: Primarily on the Coastal Plain from MA south to FL and west to LA; also in central NY, WV, and KY; scattered throughout SC, but most abundant in the Sandhills and Coastal Plain; muddy sites in alluvial and nonalluvial swamp forests, edges of ponds, tidal freshwater marshes, and ditches.

COMMENTS: The common name, Never-wet, alludes to the fact that water drops quickly roll of the leaf surface like beads of mercury. When the submerged leaves come out of the water, they are dry. Golden-club is the only species in the genus. All parts of the plant contain crystals of calcium oxalate which, when ingested, may irritate the lining of the throat, causing it to swell, possibly producing asphyxiation. Native Americans used the roots and seeds for food by cooking them over dry heat for prolonged periods, which dissolves the calcium oxalate crystals.

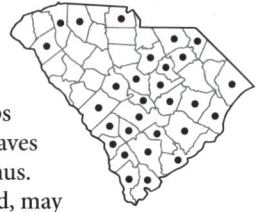

849. Green Arrow-arum; Tuckahoe

Peltandra virginica (L.) Schott
Pel-tán-dra vir-gí-ni-ca
Araceae (Arum Family)

DESCRIPTION: Rhizomatous perennial herb to 30″ tall; leaves all basal, sagittate (shaped like an arrowhead); flowers are produced on a spadix, surrounded by a spathe; spathe is green on the outer surface and light green on the inner surface; spadix consists of many tiny flowers reduced to separate clusters of anthers and pistils; flowers May–June.

RANGE-HABITAT: Quebec south to FL and west to MI and TX; in SC, common throughout; marshes, pond margins, ditches, and sluggish riverbanks.

SIMILAR SPECIES: This common wetland species is sometimes confused, when not in flower, with Common Arrowhead (*Sagittaria latifolia* Willdenow). It may be separated by the pinnate venation of the leaves. The veins on arrowheads are parallel. It also is similar to Spoonflower (*P. sagittifolia* (Michaux) Morong, plate 735), but the bright white spathes and restriction to acidic, nutrient poor habitats of Spoonflower serves to easily separate the two.

COMMENTS: Green Arrow-arum tubers are eaten by waterfowl. They are poisonous if not cooked properly. To remove the toxins, Native Americans cooked the roots over a dry heat for a day or more before eating them. Like many other members of the family, it contains raphides of calcium oxalate, which, when ingested, may irritate the lining of the throat, causing it to swell, and possibly producing asphyxiation.

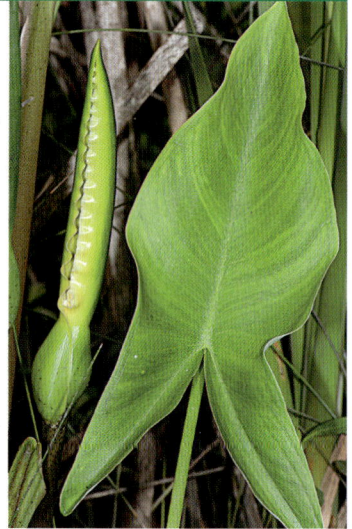

850. Carolina Hedge-nettle

> *Stachys caroliniana* J. B. Nelson
> & D. A. Rayner
> Stá-chys ca-ro-li-ni-à-na
> Lamiaceae (Mint Family)

DESCRIPTION: Perennial herb with pale rhizomes; stems erect to 1.5′ tall, 4-angled and pubescent throughout; leaves opposite, sessile or nearly so, lanceolate to lance-elliptic, pubescent; inflorescence a terminal series of verticels (flowers whorled around the stem); flowers May–July.

RANGE-HABITAT: Endemic to the Santee Delta of SC along the margins of ponds, marshes, and wet flatwoods.

COMMENTS: This is one of the few species found only in SC. It is currently known only from the Georgetown and Charleston County side of the Santee Delta. This distinctive species should be sought elsewhere in similar habitats. The nearly sessile leaves immediately distinguish it from the common and weedy Florida Hedge-nettle (*Stachys floridana* Shuttleworth ex Bentham, plate 982). This unique hedge-nettle was described by University of SC botanist John Nelson and one of the authors of this book, Douglas Rayner.

CONSERVATION STATUS: SC-Critically Imperiled

851. Carolina Water-hyssop

> *Bacopa caroliniana* (Walter)
> B. L. Robinson
> Ba-cò-pa ca-ro-li-ni-à-na
> Plantaginaceae (Plantain Family)

DESCRIPTION: Aromatic, rhizomatous, perennial herb, forming extensive mats when emersed in moist soil or in dense colonies with elongate stems when submerged; stems pubescent; erect portion of stem 4–12″ tall; leaves opposite, ovate to broadly elliptic, with 3–7 palmate veins; corolla pale to bright blue or violet-blue; flowers May–September.

RANGE-HABITAT: VA south to FL and west to TX; mostly in the Coastal Plain but with disjunct populations inland; in SC, common in the Coastal Plain and Sandhills; sandy, shallow ponds, marsh or stream margins, cypress-gum ponds, depressions, ditches, and canals.

COMMENTS: The strongly lemony-minty fragrance of its stems and leaves is distinctive among aquatic plants. It is often noticed by the smell of plants bruised while walking through a wetland. Its small stature and dense growth habit make it a great addition to ornamental ponds.

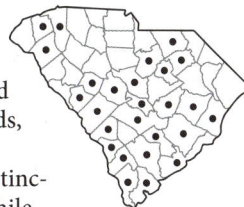

852. Golden Canna

Canna flaccida Salisbury
Cán-na flác-ci-da
Cannaceae (Canna Family)

DESCRIPTION: Perennial herb to 4′ tall; lower leaves large, to 2′ long, sheathing at the base; upper leaves smaller; sepals 3, greenish; petals 3, yellow, bases united into a tube; showy part of the flower consists of 3 yellow, petal-like staminodes, 2 larger, 1 smaller; a single stamen is attached to one of the largest staminodes; flowers May–early July; fruit covered with small, elongated warts, mature July–August.

RANGE-HABITAT: Coastal Plain from SC south to FL and west to MS; rare in the Coastal Plain of SC; pine savannas, inland freshwater marshes, swamp margins, and wet ditches.

COMMENTS: The showy flowers of Golden Canna make a stunning addition to the wetland garden. Although its non-native relatives are common in Southern gardens, this native is very rarely cultivated. Golden Canna can be seen on Bulls Island in the Cape Romain National Wildlife Refuge.

CONSERVATION STATUS: SC-Imperiled

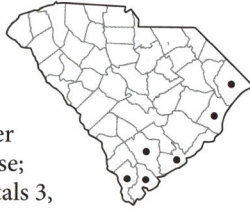

853. American Bur-reed

Sparganium americanum Nuttall
Spar-gà-ni-um a-me-ri-cà-num
Typhaceae (Cat-tail Family)

DESCRIPTION: Perennial herb; stems zigzag, erect, 1–3′ tall; leaves alternate, 2-ranked, sheathing at the base; male and female flowers in separate heads; smaller male heads above, female heads below; male flowers wither and die as soon as pollen is shed; flowers May–September.

RANGE-HABITAT: Newfoundland south to FL and west to MN and TX; in SC, common, chiefly in the Coastal Plain, Sandhills, and mountains; open, muddy areas of swamp forests, streams, roadside ditches, and shallow ponds.

COMMENTS: Bur-reed is an emergent plant that sometimes forms dense stands. Waterfowl and marsh birds eat the seeds.

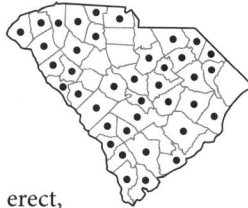

854. Creeping Burhead

Echinodorus cordifolius (L.)
Grisebach
E-chi-nó-do-rus cor-di-fò-
li-us
Alismataceae (Water-plantain
Family)

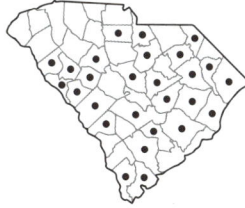

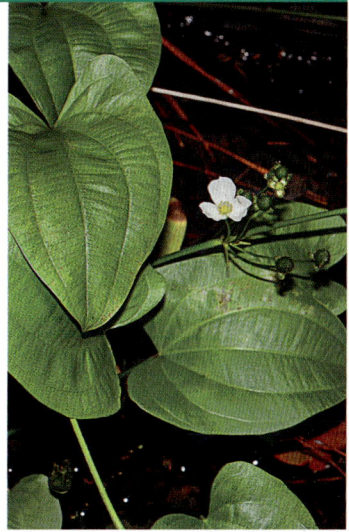

DESCRIPTION: Short-lived perennial herb; leaves with long-grooved stalks up to 2′ long; blades widely ovate with 5–9 prominent palmate veins; flowers in clusters of 5–15, on a long scape that is erect when young and decumbent with age; flowers June–November.

RANGE-HABITAT: MD south to FL, west to TX, and north in the interior to IL; in SC, common in the lower Piedmont, Sandhills, and Coastal Plain; swamps, streamsides, inland marshes, ditches, and wet thickets.

COMMENTS: The common name burhead refers to the headlike cluster of fruits. Decumbent stems often root where they contact the ground, thus "creeping" along the ground.

855. Long Beach Primrose-willow

Ludwigia brevipes (Long)
Eames
Lud-wíg-i-a bré-vi-pes
Onagraceae (Evening-primrose
Family)

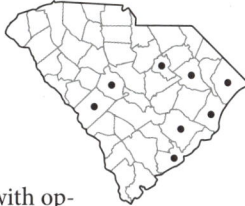

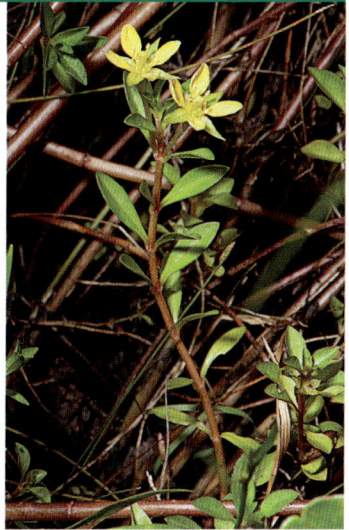

DESCRIPTION: Prostrate, matted perennial with opposite leaves; stems mostly smooth, to 20″ long or longer in deep water; pedicels of flowers more than 0.2″ long; petals 4, yellow, spoon-shaped, usually equaling the sepals; flowers June–October; fruit a curved capsule.

RANGE-HABITAT: Coastal Plain from NJ south to FL; in SC, rare and scattered in the Coastal Plain and Sandhills; depression meadows, "high ponds," pond shores, drawdown zones on blackwater rivers, low wet places, gravel pits, ditches, and inland marshes.

SIMILAR SPECIES: Five species of prostrate, opposite-leaved members of the genus *Ludwigia* occur in SC. They are separated as follows: *L. arcuata* Walter and *L. brevipes* have flowers or capsules distinctly stalked, with stalks over 0.2″ long; the petals of *arcuata* are longer than the sepals, while the petals of *brevipes* have petals equaling or shorter than the sepals. *L. repens* Forster, *L. spathulata* Torrey & Gray, and *L. palustris* (L.) Elliott have sessile flowers and fruits or stalks shorter than 0.13″. *L. repens* has petals; *L. spathulata* and *L. palustris* have no petals. *L. spathulata* is hairy, whereas *L. palustris* is smooth.

COMMENTS: The genus name honors Christian Ludwig (1709–73), a professor of botany at The University of Leipzig, Germany.

CONSERVATION STATUS: SC-Imperiled

856. Winged Monkey-flower

Mimulus alatus Aiton
Mí-mu-lus a-là-tus
Phrymaceae (Lopseed Family)

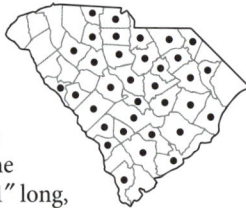

DESCRIPTION: Perennial, rhizomatous herb; stems 16–48″ tall, with narrow wings on the angles; leaves opposite, serrate; flowers to 1″ long, solitary in the axils of leaves or bracts; corolla 2-lipped, purplish with a yellow splotch, surrounded by a band of white in the throat; calyx with a relatively long, angled, prismatic tube somewhat oblique at the opening; flowers July–frost.
RANGE-HABITAT: New Brunswick south to FL and west to MI and TX; in SC, common in the Piedmont and Coastal Plain; floodplain forests, swamps, creek banks, marshy shores, stream margins, and wet ditches.
SIMILAR SPECIES: *M. alatus* is similar to Alleghany Monkeyflower (*M. ringens* L.), which occurs in the mountains and upper Piedmont. Their ranges overlap in the Piedmont and their habitats are similar. *M. alatus* has stalked leaves and winged stems; *M. ringens* has sessile leaves and wingless or obscurely winged stems.
COMMENTS: When viewed head on, the corolla resembles the face of a monkey.

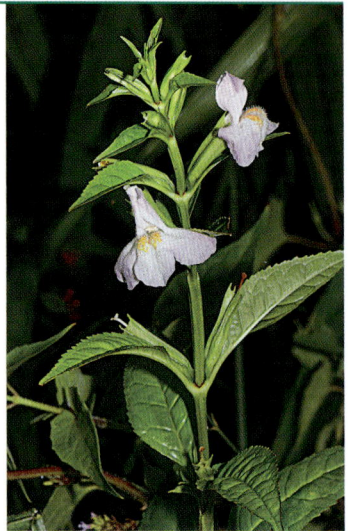

857. Southern Winged Loosestrife

Lythrum lanceolatum Elliott
Lỳth-rum lan-ce-o-là-tum
Lythraceae (Loosestrife Family)

DESCRIPTION: Smooth, perennial herb to 4′ tall; leaves opposite below and alternate above, lanceolate to elliptic; flowers produced in the axils of bracts at the ends of stems and branches in a wandlike arrangement; flowers brilliant dark pinkish-purple; fruit a capsule; flowers May–September.
RANGE-HABITAT: VA south to FL and west to OK and TX, mostly in the Coastal Plain; in SC, common in the Coastal Plain in marshes, margins of swamps, ditches, and other wetlands.
COMMENTS: Southern Winged Loosestrife is a beautiful and often overlooked native loosestrife that makes a good addition to the native garden. It does not require wet soil for growth in cultivation and is very attractive to smaller butterflies.

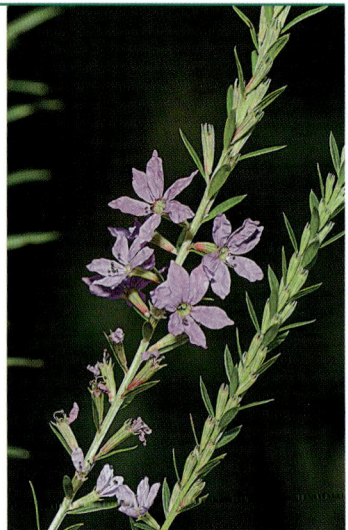

858. Water Oleander; Water Willow; Swamp Loosestrife

Decodon verticillatus (L.) Elliott
Dé-co-don ver-ti-cil-là-tus
Lythraceae (Loosestrife Family)

DESCRIPTION: Robust semi-woody perennial; stems pubescent, arching and often rooting when they contact soil or water; leaves whorled or opposite, lanceolate to elliptic; flowers produced in axillary clusters; petals pinkish-purple; flowers July–September.

RANGE-HABITAT: Nova Scotia south to FL and west to MN and TX; in SC, common in the Coastal Plain where found in isolated depression ponds, marshes, swamps, ditches, and other wetlands.

SIMILAR SPECIES: Water Oleander is nearly a shrub but has very weak stems and often forms a tangled mass in the shallow water in which it grows.

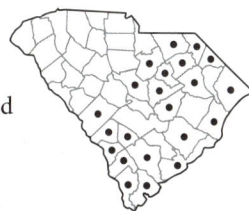

859. Elliott's Aster; Southern Swamp Aster

Symphyotrichum elliottii (Torrey & Gray) G. L. Nesom
    Sym-phy-o-trì-chum el-li-ótt-i-i
Asteraceae (Aster Family)

SYNONYM: *Aster elliottii* Torrey & Gray—RAB, PR

DESCRIPTION: Robust rhizomatous perennial to 5′ tall; stem glabrous to sparsely pubescent; leaves alternate, lanceolate to elliptic with acute tips; base attenuate and not clasping; flowers produced in involucrate heads, about 1″ across, composed of blue ray flowers and yellow disk flowers; heads in a corymblike arrangement; flowers September–December.

RANGE-HABITAT: Endemic to the Coastal Plain from VA south to FL and west to LA; in SC, common in the Coastal Plain; tidal freshwater marshes, inland marshes, margins of swamp forests, ditches, and other wetlands.

COMMENTS: Elliott's Aster essentially replaces the more northern and inland Swamp Aster (*S. puniceum* (L.) Love & Love) in the Coastal Plain. Swamp Aster is a much more strictly erect plant, without evidently pubescent stems, and a more corymblike arrangement of flowering heads. Elliott's Aster makes an exceptional ornamental in the garden because it rarely grows leggy and seldom flops over, as many other asters are prone to do. It also flowers very late in the season, bringing needed color to the garden and late-season nectar to insects.

860. Open water community

861. Mosquito Fern

Azolla caroliniana Willdenow
A-zól-la ca-ro-li-ni-à-na
Azollaceae (Mosquito Fern Family)

DESCRIPTION: Unmistakable free-floating aquatic fern, about 0.3″ wide.

RANGE-HABITAT: Widespread in the southeastern US, irregularly north into s. New England and MN and south into the tropics; in SC, common in the Coastal Plain and maritime strand, rare in the Piedmont; ponds and sluggish streams, swamps, lakes, and ditches.

COMMENTS: The common name comes from the belief that its dense growth on a body of water prevents mosquitos from laying eggs and the larvae from obtaining oxygen above the surface. Mosquito Fern harbors a symbiotic nitrogen-fixing cyanobacterium (*Anabaena azollae* Strasburger) in its fronds. This nitrogen-fixing ability has resulted in Mosquito Fern being used historically, and presently, as a green manure plant in rice paddies. Mosquito Fern fronds change from red in full sun to green in shade. The red color in full sun is due to the pigment anthocyanin and is an adaptation that prevents a reduction in photosynthesis due to excess light. It is spread by waterfowl and wading birds. Accordingly, it may appear one year in a pond where it has never been seen before. It is often most abundant in the cooler months.

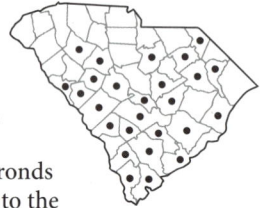

862. Broadleaf Pondlily (A)

Nuphar advena (Aiton)
R. Brown ex W. T. Aiton
Nù-phar ád-ve-na
SYNONYM: *Nuphar lutea* J. E.
Smith—RAB, PR

Narrowleaf Pondlily (B)

Nuphar sagittifolia (Walter)
Pursh
Nù-phar sa-git-ti-fò-li-a
Nymphaeaceae (Water-lily Family)

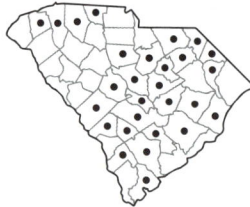

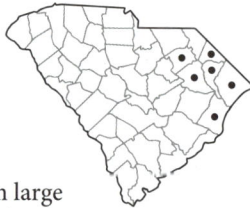

DESCRIPTION: Perennial, aquatic herbs from large rhizomes; in *N. advena* leaf blades 1–2x as long as wide and emergent or floating only; in *N. sagittifolia* leaf blades typically 3–6x as long as wide, up to 12″ long and 2–4″ wide, with numerous thin-textured submerged leaves and thick floating leaves; flowers April–October.

RANGE-HABITAT: *N. advena* is found from ME south to FL and west to WI and TX; in SC, common throughout where found in ponds, marshes, and sluggish backwaters; *N. sagittifolia* is endemic to the Coastal Plain from e. VA south to ne. SC; in SC, it occurs in the Pee Dee River and Waccamaw River basins; locally abundant on shallow bars in the Waccamaw River; rivers, bayous, sloughs, and blackwater streams, extending down rivers and reaching its greatest abundance in freshwater tidal areas.

COMMENTS: Native Americans used the seeds and rhizomes as food. The rhizomes were roasted or boiled, after which they could be easily peeled; the sweet interiors were then cut up for soups and stews.

CONSERVATION STATUS: SC-Imperiled

863. Big Floating Heart (A)

Nymphoides aquatica
(Walter ex J. F. Gmelin)
Kuntze
Nym-phoì-des a-quá-ti-ca

Little Floating Heart (B)

Nymphoides cordata
(Elliott) Fernald
Nym-phoì-des cor-dà-ta
Gentianaceae (Gentian Family)

DESCRIPTION: Perennial aquatic herbs; anchored from a thick rhizome; stems reddish purple-punctate in *N. aquatica,* not spotted in *N. cordata;* in *N. aquatica,* leaves 2–6″ wide, thick-textured, green above, usually purple below; in *N. cordata,* leaves 1–2.75″ wide, thin-textured, mottled dark green and purple above, purple below; stems terminate in an umbel of flowers and one floating leaf with a short stalk; late in the season the inflorescence is mixed with or subtended by tuberlike roots; flowers late April–September.

RANGE-HABITAT: *N. aquatica:* Coastal Plain from NJ south to FL and west to TX; in SC, frequent in the Coastal Plain in freshwater lakes and ponds, sluggish streams, swamps, and beaver ponds; *N. cordata:* Newfoundland south to FL and west to LA, primarily northern but disjunct in the South in various regions; in SC, common in the Sandhills and Inner Coastal Plain in Sandhills impoundments, sluggish streams, and other impoundments.

COMMENTS: The Asian Crested Floating Heart (*N. cristata* (Roxburgh) Kuntze) has been introduced into SC. It is an aggressive invasive and has become common in Lake Marion.

864. Swollen Bladderwort

Utricularia inflata Walter
U-tri-cu-là-ri-a in-flà-ta
Lentibulariaceae (Bladderwort Family)

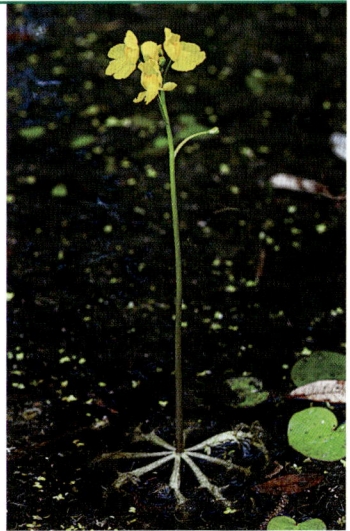

DESCRIPTION: Free-floating, carnivorous herb; upper leaves whorled, each consisting of an inflated petiole and rachis, forming a flotation device that supports the flowering stalk; 4–7 floats, not fused basally; submerged "leaves" bear the bladders that trap and digest aquatic animals, providing minerals; flowers May–November.

RANGE-HABITAT: Common Coastal Plain species that ranges from NJ south to FL and west to TX; in SC, common in the Coastal Plain in swamps, lakes, ponds, roadside ditches, and pools.

SIMILAR SPECIES: A smaller plant is Small Floating Bladderwort (*U. radiata* Small, plate 714), which also occurs in the Coastal Plain usually in isolated depression ponds. In *U. radiata* the inflated leaf stalk and rachis are less than 2″ long, it has 6–8 floats that are fused basally, and the racemes are mostly 3–4-flowered. In *U. inflata* the inflated leaf stalk and rachis are more than 2″ long, the floats are not fused basally, and the racemes are usually 9–14-flowered.

COMMENTS: An interesting feature of this bladderwort is the springtime development of the stalk and flotation device from the submerged part of the plant. The flotation structures develop at the end of the immature stalk while under water. As both grow, their buoyancy causes the whole plant to rise to the surface, by which time the flower stalk is well developed. The flowers are elevated above the water, allowing for cross-pollination to occur. Beholding a pond covered with a mass of floating bladderworts where none existed the previous day is one of the wonders of nature. This plant survives drought by producing drought-resistant tubers; when the water body fills with water, the tubers generate a new plant.

865. Water-shield

Brasenia schreberi J. F. Gmelin
  Bra-sèn-i-a schré-ber-i
  Cabombaceae (Water-shield Family)

DESCRIPTION: Perennial, aquatic herb from a rootstalk with abundant rhizomes; leaves floating, alternate, long-stalked, elliptic, and peltate; upper surface of leaf purplish, lower surface bright green; all submerged parts thickly coated with a gelatinous material; flowers emergent, dull purple, solitary on long stalks from the leaf axils; flowers June–October.

RANGE-HABITAT: Nova Scotia south to FL and west to CA; also found in tropical America and the Old World; in SC, common in the Coastal Plain and Sandhills in freshwater ponds, swamps, lake edges, and sluggish streams.

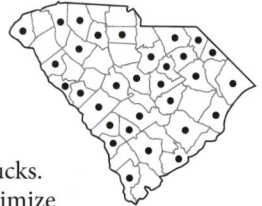

COMMENTS: Water-shield often grows in extensive stands that exclude other vegetation in small ponds. The rootstock is a favorite food of Ring-necked Ducks. The gelatinous covering of the submerged parts is likely an adaptation to minimize herbivory. The genus name honors Moravian missionary and surgeon, Christoph Brasen (1738–74), and the specific epithet honors Johann Christian Schreber (1739–1810).

866. Water Hyacinth

Eichhornia crassipes (Martius) Solms
  Ei-chhórn-i-a crás-si-pes
  Pontederiaceae (Pickerelweed Family)

DESCRIPTION: Free-floating, freshwater aquatic herb; basal cluster of leaves with inflated petioles that act as floats; often rooted in mud as the water recedes and may persist for several months; flowers June–September.

RANGE-HABITAT: Introduced from tropical America; naturalized from VA south to FL and west to TX and MO; throughout the Coastal Plain of SC; ponds, ditches, canals, and abandoned rice fields along coastal freshwater rivers.

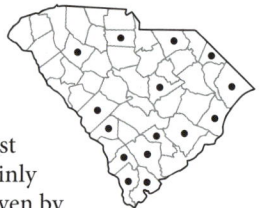

COMMENTS: Water Hyacinth was introduced into FL in 1884; through prolific growth, it has rapidly become a serious weed, clogging waterways in the warmer, frost-free coastal areas of the Southeast. The inflated leaf stalks consist of aerenchyma tissue that gives the plant great buoyancy. Reproduction is mainly by vegetative means, with sections breaking off and carried by currents or driven by wind. The genus name honors Johann Eichhorn of Berlin (1779–1856).

867. American Frog's-bit

Limnobium spongia (Bosc) Steudel
Lim-nó-bi-um spón-gi-a
Hydrocharitaceae (Frog's-bit Family)

DESCRIPTION: Perennial herb, generally free-floating in dense mats or becoming rooted in mud of a drying aquatic habitat; plantlets develop at the ends of runners; leaves with a petiole and blade; the blade reniform to suborbicular; flowers solitary on stems to 2.5″ long; flowers June–September.

RANGE-HABITAT: DE south to FL and west to IL and TX; in SC, common in the Coastal Plain in a variety of shallow, quiet-water habitats such as swamps, ponds, lakes, marshes, and drainage ditches.

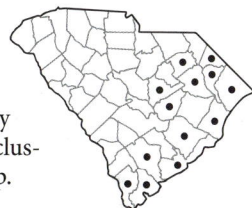

COMMENTS: Two leaf forms occur during American Frog's-bit's life cycle: floating leaves and emersed leaves. Vegetative, floating, basal clusters of nearly kidney-shaped leaves give rise to more robust plants. These plants consist of clusters of ascending and stalked, erect leaves on which flowers and fruits develop.

868. American Lotus; Water-chinquapin

Nelumbo lutea (Willdenow) Persoon
Ne-lúm-bo lù-te-a
Nelumbonaceae (Lotus-lily Family)

DESCRIPTION: Rhizomatous, perennial, emergent aquatic herb; leaves produced early in the season lie on the water surface, later in the season leaves extend up to 4′ above the surface; leaf blade orbicular, to 2′ wide, indented in the center; flowers solitary on a long flowering stalk; tepals pale yellow, numerous; acornlike fruits scattered in pits across a flat-topped, 4″ in diameter receptacle; flowers June–September.

RANGE-HABITAT: NY and s. Ontario to MN and IA, south to FL, and west to OK and TX; in SC, infrequent and scattered throughout the Coastal Plain in muddy areas of ponds, sluggish streams, and margins of natural and human-made lakes.

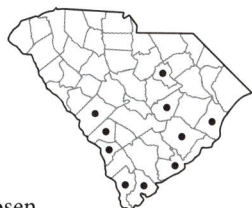

COMMENTS: American Lotus was a favorite Native American food. The tender, immature fruits were eaten raw or cooked. The ripe seeds were parched to loosen the shell, then husked and eaten dry, baked, boiled, or ground to make bread. The tuberous enlargements of the rootstocks become filled with starch in the fall and make a tasty food when baked or boiled and seasoned.

The pistil develops into a funnel-shaped fruit about 4″ in diameter, the apex of which contains several cavities, each containing a single seed that shakes out when ripe. Wildlife, especially ducks, prize the fruits.

869. Fragrant Waterlily; Sweet Waterlily

Nymphaea odorata Aiton
Nym-phaè-a o-do-rà-ta
Nymphaeaceae (Water-lily Family)

DESCRIPTION: Perennial floating aquatic herb, anchored from a thick rhizome; leaf blades almost orbicular, to 1′ wide, often purplish below; distinctive for its showy floating flowers, sweet-scented, white (rarely pinkish), to 7.5″ wide; flowers June–September.

RANGE-HABITAT: Newfoundland west to Manitoba and south to FL and TX; in SC, common in the Coastal Plain; scattered in the Piedmont and mountains, where it is rare; pools, ponds, and sluggish stream margins.

SIMILAR SPECIES: Yellow Waterlily (*N. mexicana* Zuccarini) is similar but has a yellow flower. It is found at widely scattered locations in the Coastal Plain and Piedmont, where it is probably introduced for waterfowl management.

COMMENTS: Morton (1974) indicates that rural people of the Coastal Plain used a root infusion of waterlily as a treatment for diarrhea. For itch of private parts, they bathed with and/or drank a decoction of the root. Waterlily is used as an ornamental in small ponds where it often develops dense stands that may interfere with boating and fishing.

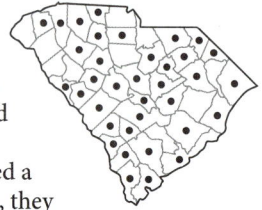

870. Bog-mat; Mud-midgets

Wolffiella gladiata (Hegelmann) Hegelmann
Wolf-fi-él-la flo-ri-dà-na
Lemnaceae (Duckweed Family)

SYNONYM: *Wolffiella floridana* (J. D. Smith) Thompson—RAB, PR

DESCRIPTION: Fronds floating near the surface of the water, submerged; elongate, curved fronds are rarely solitary, usually 2 or more attached by short, basal stalks forming a starlike colony; fronds 0.3–0.6″ long; no roots; reproduction is mostly vegetative by budding, but occasionally a frond produces one male and one female flower.

RANGE-HABITAT: From MA south to FL and west to IL and TX; in SC, common throughout the Sandhills, Coastal Plain, and maritime strand; slightly acidic and highly organic waters of ponds, streams, swamps, marshes, and roadside ditches.

COMMENTS: Bog-mat most often occurs intermixed with the other three genera of Lemnaceae.

871. Duckmeat; Greater Duckweed

Spirodela polyrhiza (L.)
Schleiden
Spi-ro-dè-la po-ly-rhì-za
Lemnaceae (Duckweed Family)

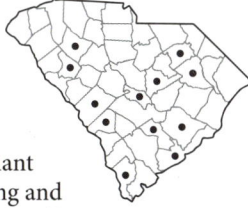

DESCRIPTION: Free-floating aquatic herb; plant body in this family ("frond")about 0.2″ long and has 7–21 roots; circular to obovate fronds usually in groups of 2–5, rarely solitary; flowers reduced to anthers and ovaries, surrounded by a membranelike scale.

RANGE-HABITAT: Widespread worldwide; in SC, common in the Sandhills, Coastal Plain, and maritime strand; pools, ponds, swamps, margins of sluggish streams, and ditches.

SIMILAR SPECIES: Spotted Duckmeat (*Landoltia punctata* (G. F. W. Meyer) Les & D. J. Crawford) is similar but has only 2–7 roots per frond. It is found nearly worldwide, but in SC has been introduced from aquarium and ornamental pond use. Six species of other duckweeds in the genus *Lemna* are also found in SC; they all have a single root per frond.

COMMENTS: *Spirodela* can cause problems in small ponds when it clogs irrigation pumps and covers the surface and interferes with livestock drinking. *S. polyrhiza* produces starch-filled turions (dormant buds) during adverse temperature and drought conditions. The turion sinks to the bottom until favorable conditions return, at which time it expels a small bubble of gas that carries it to the surface where it germinates.

Coastal Plain:
The Maritime Strand

872. Coastal beaches

873. Seashore-elder;
Dune Marsh-elder

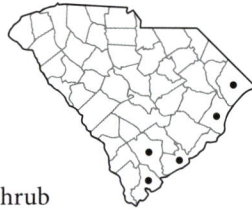

Iva imbricata Walter
Ì-va im-bri-cà-ta
Asteraceae (Aster or Sunflower
Family)

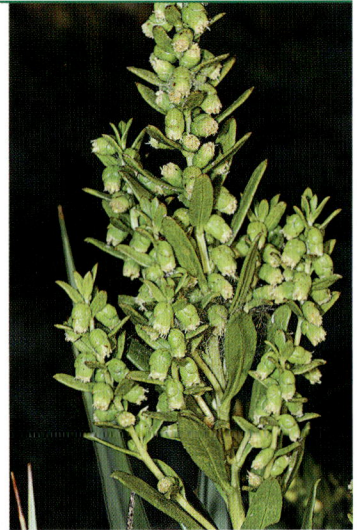

DESCRIPTION: Bushy-branched, perennial shrub
to 3′ tall; stems fleshy, smooth, commonly creeping
at the base; branches often reclining, tips often drying during the
winter; lower leaves opposite, midstem and upper leaves alternate;
fruit a large achene, yellow-brown when mature; flowers and fruits
late August–November.

RANGE-HABITAT: VA south to FL and west to TX; in SC, common
in the maritime strand on coastal dunes, the upper beach, and
saltwater overwash areas.

COMMENTS: Seashore-elder is often the most oceanward perennial plant, colonizing the upper beach or incipient dunes where it
grows mixed with Seabeach Amaranth, Sea Rocket, and Carolina
Saltwort.

874. Southeastern Sea Rocket

Cakile harperi Small
Ca-kì-le hár-per-i
Brassicaceae (Mustard Family)

DESCRIPTION: Smooth, succulent, branched annual, rarely woody at the base; up to 30″ tall;
leaves entire to shallowly crenate; flowers white,
produced in terminal racemes; flowers March–
October.

RANGE-HABITAT: NC south to FL; in SC, common in the maritime strand on sparsely vegetated coastal beaches above the high tide line and adjacent coastal dune communities.

COMMENTS: The name "rocket" refers to its use as a salad green or potherb, like true Rocket/Arugula (*Eruca vesicaria* (L.) Cavanilles). Young cooked plants are of good quality but without a distinctive taste. The fleshy young foliage and young fruits are palatable when mixed with milder leaves in a salad; eaten raw, they have the flavor of mild horseradish.

The fruit of Southeastern Sea Rocket is a modified silique that is indehiscent and divided into two segments, each 1-seeded. The terminal segment becomes dry and corky at maturity, breaking off from the basal segment and able to float great distances. The basal noncorky segment usually falls later and does not travel far. This allows both short-distance dispersal, in what is likely to be optimal habitat, and long-distance dispersal for colonization of new areas. The specific epithet honors Roland McMillan Harper (1878–1966), American botanist and geographer.

---

875. Carolina Saltwort; Southern Saltwort

Salsola kali L. var. *caroliniana* (Walter) Nuttall

Sál-so-la kà-li var. ca-ro-li-ni-à-na

Chenopodiaceae (Goosefoot Family)

SYNONYM: *Salsola caroliniana* Walter—PR

DESCRIPTION: Herbaceous, branched annual, 10–25″ tall; leaves fleshy, awn-shaped, and sharp-pointed; the flowers whitish or gray to yellowish gray in summer, pink in the fall; flowers June–frost.

RANGE-HABITAT: ME south to FL and west to TX; common along the SC coast on the upper beach above the high tide line.

SIMILAR SPECIES: The famous Tumbleweed (*S. tragus* L.) is introduced from Eurasia and has spread across much of the United States. It is rare in SC in disturbed areas inland. Tumbleweed has soft sepals with the midrib obscure and not excurrent. The SC (native) saltwort has a similar method of seed dispersal.

COMMENTS: Contact with fragments of the plant that fall on the sand can be painful to exposed skin, especially bare feet. The stems turn from green in the summer to red or pinkish purple in the fall. The dead plants are often blown loose and tumble down the beach, in the process disseminating the seeds.

---

876. Seabeach Amaranth

Amaranthus pumilus Rafinesque

A-ma-rán-thus pù-mi-lus

Amaranthaceae (Amaranth Family)

DESCRIPTION: Annual herb from a taproot; stems 4–24″ long, prostrate, erect, or somewhat reclining at their tips, abundantly branched from the base, forming mats; leaves fleshy, clustered near tips of the branches, broadly rounded, emarginate, 0.5–1.0″ in diameter; flowers in short axillary clusters; male and female flowers on the same plant; fruit an utricle, 1-seeded and indehiscent; flowers and fruits mature June–frost.

RANGE-HABITAT: Historically from MA south to upper Charleston County, SC; presently extant only in NC, SC, and Long Island, NY; its usual habitat is nearly pure silica sand substrate in front of the fore-dunes on coastal barrier islands.

COMMENTS: Coastal erosion during the twentieth and twenty-first centuries has significantly reduced the populations of Seabeach Amaranth along the coast of SC. One of the best sites to see this species is Huntington Beach State Park on the barrier beach north to Murrells Inlet.

Seeds of Seabeach Amaranth germinate in May; first they form a small, un-branched sprig and soon branch profusely. The plant then acts as a dune builder, trapping wind-blown sand to form a mound around it as the season progresses. As the mound builds, the plant continues to grow, earlier leaves are buried, and new leaves rise above the mound. Seeds are spread by wind or water in the fall.
CONSERVATION STATUS: Federally threatened

877. Coastal dunes and maritime grassland

878. Poison Ivy

Toxicodendron radicans (L.) Kuntze
var. *radicans*
Tox-i-co-dén-dron ra-dì-cans
Anacardiaceae (Cashew Family)

DESCRIPTION: Deciduous high-climbing, woody vine or trailing, sometimes shrubby woody vine; stems often with adventitious roots attaching the stem to tree trunks or other surfaces; leaves alternate, trifoliolate, margins entire to irregu-larly lobed or toothed, smooth above, sparsely pubescent below; flowers produced in panicles; flowers greenish, yellow, small; fruits are pubescent, whitish berries; flowers April–May; fruits mature August–October.
RANGE-HABITAT: Nova Scotia south to FL and west to TX and AR; other varie-ties range over most of temperate North America; in SC, common in rich soils of many forest systems, forest margins, dunes, and maritime grasslands.
TAXONOMY: Two varieties occur in SC. Midwestern Poison Ivy (*T. radicans* var. *negundo* (Greene) Reveal) is very rare in the mountains. It is confined to open wood-lands on high pH soils over amphibolite. It can be distinguished by the smooth fruits.
COMMENTS: Poison Ivy is infamous for the contact dermatitis caused by the chemical urushiol. Though it causes considerable discomfort to many, the fruits are important sources of food for birds and other wildlife. The hairlike adventitious roots of the stems often cause species such as Climbing Hydrangea (*Hydrangea barbara* (L.) B. Schulz, plate 32), which produce similar roots, to be confused for stems of Poison Ivy. Though it causes considerable discomfort to many, the fruits are important sources of food for birds and other wildlife.

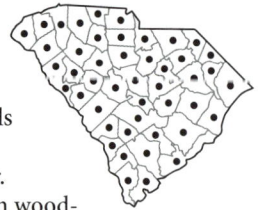

879. Dune Greenbrier

Smilax auriculata Walter
  Smì-lax au-ri-cu-là-ta
  Liliaceae (Lily Family)

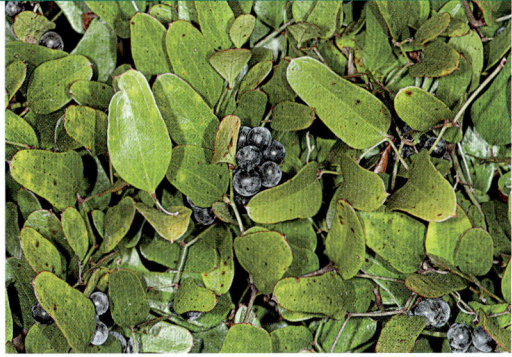

DESCRIPTION: Coarse-stemmed, evergreen vine; usually forming dense, low thickets; stems green, with scattered short, spinelike projections or stems without the spinelike projections; leaves oblong to oblong-lanceolate, sometimes hastate-lobed, leathery, smooth, and green on both sides; flowers May–July; fruits mature October–November.

RANGE-HABITAT: NC south to FL and west to LA and AR; in SC, common in the maritime strand on dunes, in maritime shrub thickets, and sandy openings in maritime forests.

COMMENTS: The extensive and tuberous-thickened rootstock helps Dune Greenbriar survive disturbance on the dunes.

880. Sea Oats

Uniola paniculata L.
  U-nì-o-la pa-ni-cu-là-ta
  Poaceae (Grass Family)

DESCRIPTION: Unmistakable when flowering or fruiting; a coarse rhizomatous perennial with stems 3–6′ tall; stems readily root at the nodes as the stem becomes covered with sand; reproduction is mainly by rhizomes; flowers June–November.

RANGE-HABITAT: VA south to FL and west to TX and Mexico; in SC, common in the maritime strand on coastal dunes and adjacent beaches and in interdune swales.

COMMENTS: Sea Oats is tolerant of strong winds, sand abrasion, and salt spray. The stems readily root at the nodes as the stem becomes covered with sand. These adaptations make it one of the most important maritime plants in dune formation and stabilization. On public property, a state law protects any and all parts of the plant from being removed or damaged.

881. Dune Sandspur

Cenchrus tribuloides L.
Cén-chrus tri-bu-loì-des
Poaceae (Grass Family)

DESCRIPTION: Sprawling perennial or annual rooting at the nodes; stem branches 4–28″ long; the small 2-flowered spikelets are enclosed within a spiny bur; fruit mature August–October, persisting much of the fall and winter.

RANGE-HABITAT: NY south to FL and west to TX; in SC, common in the maritime strand on coastal dunes along the coast and adjacent beaches; also common in sandy fields and woodlands.

COMMENTS: The spines forming the bur can inflict painful puncture wounds on exposed skin and are equally painful to remove because of the backward-pointing barbs. The spines protect the plant from herbivory and provide an effective mechanism for fruit dispersal. The burs turn from green to reddish with age and may remain on the dead stems throughout the winter. Several species of sandspurs occur in SC. Dune Sandspur is the large, common species familiar to almost everyone who has visited SC beaches.

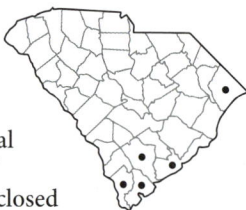

882. Bitter Seaside Panic-grass

Panicum amarum Elliott var. *amarum*
Pá-ni-cum a-mà-rum var. a-mà-rum
Poaceae (Grass Family)

DESCRIPTION: Perennial grass with long rhizomes, usually rooting at lower nodes; stems usually solitary, 15–40″ tall; leaves and stems grayish-green; spikelets of inflorescence dense, forming a narrow panicle, to 14″ long and 2″ wide; flowers September–October.

RANGE-HABITAT: CT south to FL and west to TX; in SC, common in the maritime strand on coastal dunes and swales behind foredunes.

COMMENTS: Bitter Seaside Panic-grass contributes to dune building; however, since it becomes buried or uprooted when sand shifts, it is not as effective as other dune grasses in binding soil. Two varieties of this species are recognized. Southern Seaside Panic-grass (*P. amarum var. amarulum* (A. S. Hitchcock & Chase) P. G. Palmer) is similar and occurs in similar habitats and inland in Longleaf Pine sandhills. It has short rhizomes and forms clumps. The panicle is also less densely flowered than Bitter Seaside Panic-grass.

883. Sweet Grass

Muhlenbergia sericea (Michaux) P. M. Peterson
Muh-len-bérg-i-a se-rí-ce-a
Poaceae (Grass Family)

SYNONYM: *Muhlenbergia filipes* M. A. Curtis—PR

DESCRIPTION: Tufted perennial to 40″ tall; inflorescence a loose, limber panicle that turns bright pink when mature; lemma awns 0.5–1.1″ long; flowers and fruits October–December.

RANGE-HABITAT: NC south to FL and west to TX; primarily on barrier islands along the coast; occasional, but often locally abundant; flats between coastal dunes, salt shrub thickets, stable dunes, and edges of freshwater or brackish marshes.

COMMENTS: This is the locally famous Sweet Grass used by basket makers in Mt. Pleasant. The baskets use Sweet Grass in combination with Longleaf Pine needles, Black Needle Rush stems, and Cabbage Palmetto leaves. There is concern that, in the future, there will not be an adequate supply of Sweet Grass in the wild for this use. This is one of the most commonly cultivated ornamental grasses in the South. It is sold under the erroneous name of *M. capillaris,* with the common name of Pink Muhly. Examination of the extremely long lemma awns of cultivated material easily shows that all cultivars, including the white-flowered White Cloud, are representative of Sweet Grass, not the true *M. capillaris.* Sweet Grass is a beautiful, drought-resistant perennial grass. Plants tend to require cutting back or burning in the spring for rejuvenation. The genus name honors Henry Ernst Muhlenberg (1753–1815), a Pennsylvanian botanist.

884. Beach Blanket-flower; Fire-wheel

Gaillardia pulchella Fougeroux
     var. *drummondii* (Hooker)
B. L. Turner
     Gail-lárd-i-a pul-chél-la var.
     drum-mónd-i-i
Asteraceae (Aster Family)

DESCRIPTION: Short-lived, pubescent, perennial herb, 6–28″ tall, creeping to erect; flowers arranged into an involucrate head with ray and disk flowers; ray flowers lobed at their tips, reddish, tipped with yellow, or occasionally all yellow; flowers April–frost.

RANGE-HABITAT: NC south to FL and west to TX; in SC, common on the Outer Coastal Plain and maritime strand on beach dunes, along sandy roadsides, and in other sandy habitats.

COMMENTS: This is a coastal variety of a common, more western species. It may have been inadvertently introduced in SC and may not be native. The definitively nonnative var. pulchella is sometimes planted in wildflower mixes and temporarily naturalizes. Beach Blanket-flower is a native and important member of the SC dune community. Many color variants occur in the red-pink-yellow range. The genus honors Gaillard de Charentonneau, an eighteenth-century French amateur botanist and patron of botanists.

Beach Blanket-flower is one of the best native plants to include in the sunny, well-drained garden. It provides year-long flowering that is very attractive to pollinators, provides one of the favored foods of American Goldfinches, and reseeds readily.

885. Dune Pennywort

Hydrocotyle bonariensis Lamarck
     Hy-dro-có-ty-le bo-na-ri-én-sis
Apiaceae (Parsley Family)

DESCRIPTION: Smooth, fleshy perennial rooting from the nodes of slender, creeping stems; leaves simple and peltate (umbrellalike); inflorescence compound in umbels or verticillate with pedicellate flowers; flowers April–November.

RANGE-HABITAT: VA south to FL and west to TX; also found throughout tropical America; in SC, common along the coast on stable coastal dunes; swales; and moist, open, sandy areas.

COMMENTS: The flowers and fruits of Dune Pennywort are often present at the same time because the branching umbellate or verticillate inflorescence is indeterminate and continues to produce new sections, with new flowers, for an extended period during the growing season. Dune Pennywort is an iconic and abundant member of the dune habitat.

886. Beach Evening-primrose

Oenothera drummondii Hooker
Oe-no-thè-ra drum-mónd-i-i
Onagraceae (Evening-primrose Family)

DESCRIPTION: Spreading to creeping, densely hairy perennial, sometimes appearing shrubby in mild winters; flowers about 3″ wide, turning toward the sun; flowers April–October.

RANGE-HABITAT: Coastal NC south to FL and west to TX; in SC, common on sandy beaches and coastal dunes on the barrier islands.

SIMILAR SPECIES: Dune Evening-primrose (Oenothera humifusa Nuttall) also is common on the coastal dunes. It has much smaller flowers, about 1″ wide, that are yellow tinged with pink.

COMMENTS: According to some sources, this species is not native to SC. Herbarium collections from SC include specimens from the nineteenth century, and it is unlikely that this plant, which is common in the dune habitats today, was intentionally introduced. The authors conclude that it must be considered native unless its introduction to SC can be proven. The specific epithet honors Thomas Drummond (1780–1835), a Scottish botanist who collected plants in Texas.

887. Spoonleaf Yucca; Curlyleaf Yucca

Yucca filamentosa L.
Yúc-ca fi-la-men-tò-sa
Agavaceae (Agave Family)

DESCRIPTION: Evergreen clumping perennial, sometimes with a very short trunk; flowering stems to 8′ tall; leaves clustered at the base, stiff, not easily bending, to 2.4″ wide, linear, tapering gradually, or with tips wider than the base and abruptly pointed; margins of leaves fraying into "threads" more than 2″ long; branches of inflorescence smooth; flowers white, composed of 6 tepals, to 2″ long; fruit a capsule; flowering April–June; fruits ripen September–October.

RANGE-HABITAT: NJ south to GA and west to MS; in SC, common in the maritime strand, Coastal Plain, and Sandhills; on dunes, in maritime grasslands, Longleaf Pine sandhills, dry flatwoods and sandy roadsides; uncommon in the Piedmont where found in shallow soils of the margins of granitic outcrops and other dry, open habitats.

SIMILAR SPECIES: Mound-lily Yucca (*Y. gloriosa* L.) is similar but is generally a larger plant with leaves that have nonfilamentous entire margins and flowers from August to October. Weakleaf Yucca (*Y. flaccida* Haworth, plate 19) is very similar but has thinner leaves that are easily bent, narrower (to 2″ wide), marginal filaments that don't exceed 2″ in length, and pubescent branches in the inflorescence. Weakleaf Yucca is native in the Piedmont and mountains but is widely planted and has escaped elsewhere.

COMMENTS: Spoonleaf Yucca is a very commonly planted species. It has escaped well beyond its native range. It is sometimes encountered deep in a forest, where it has persisted growing in the shade, many years after the homestead where it was grown has disappeared.

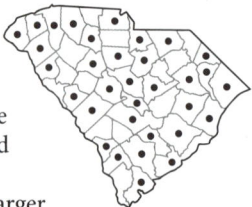

888. Creeping Frogfruit

Phyla nodiflora (L.) Greene var.
  nodiflora
    Phỳ-la no-di-flò-ra var. no-di-flò-ra
Verbenaceae (Vervain Family)

DESCRIPTION: Perennial herb with stems mostly prostrate and creeping; leaves opposite, obovate, with a blunt tip; 1–5 shallow teeth along the leaf margins; flowers produced in a dense head of small light pinkish to light purplish flowers held on stalks 1–4″ long; flowers May–December.
RANGE-HABITAT: VA south to FL and west to CA; also throughout the tropics (pantropical); in SC, common in the maritime strand and Outer Coastal Plain on dune swales, maritime grasslands, sandy roadsides, ditches, marshes, and other moist and sandy open areas.
COMMENTS: Creeping Frogfruit is a distinctive and characteristic coastal plant that flowers most of the year. The prostrate stems and small but attractive flowers that are heavily visited by small pollinators (such as skipper butterfly species) make this one of the best native groundcovers for SC gardens. It is ideal as a "living mulch," with larger seasonal interest plants planted among the dense Creeping Frogfruit groundcover. It is surprisingly hardy and tolerant of drought, flooding, and heavy soils.

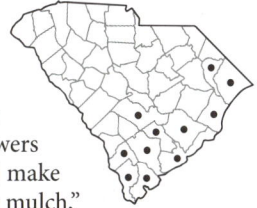

889. Dune Ground-cherry

Physalis walteri Nuttall
    Phý-sa-lis wál-ter-i
Solanaceae (Nightshade Family)

SYNONYM: *Physalis viscosa* L. ssp. *maritima* (M.A. Curtis) Waterfall—RAB

DESCRIPTION: Rhizomatous perennial herb, 8–24″ tall; leaves entire, ovate to elliptic; stems and leaves covered with stellate (star-shaped) pubescence; flowers produced from the axils; petals united into a funnel-like inverted bell; fruit a berry surrounded by an enlarged papery calyx; flowers May–September; fruits mature July–November.
RANGE-HABITAT: VA south to FL and west to MS; in SC, common in the maritime strand on dunes, maritime grasslands, and sandy roadsides.
COMMENTS: This is one of several native ground cherries found in SC. This species is characteristic of the maritime strand. The other species are found in a variety of habitats throughout SC. The berries, when cooked, are reported to make a very pleasant preserve. All other parts of the plant are potentially toxic. The specific epithet honors Thomas Walter (1740–89), British-born botanist, resident of Berkeley County, SC, and author of *Flora Caroliniana* (1788).

890. Northern Seaside Spurge;
  Northern Sandmat

Euphorbia polygonifolia L.
    Eu-phór-bi-a po-ly-go-ni-fò-li-a
Euphorbiaceae (Spurge Family)

SYNONYM: *Chamaesyce polygonifolia*
  (L.) Small—PR

DESCRIPTION: Smooth, creeping to ascending annual with stems radiating from a taproot;

individuals rarely overwinter as a weak perennial; sap milky white; flowers inconspicuous; flowers May–frost.

RANGE-HABITAT: Nova Scotia south to FL; also along the Great Lakes; in SC, common in the maritime strand on coastal dunes, maritime grasslands, the upper strand of beaches, dune blowouts, and areas of saltwater overwashes.

SIMILAR SPECIES: Northern Seaside Spurge is similar to Southern Seaside Spurge (*C. bombensis* Jacquin (=*Euphorbia ammannioides* Kunth)), which is rare. The two sometimes grow mixed. The two species may be separated by seed shape and size. The inflorescence of E. polygonifolia is terminal on the stems, while the inflorescence of E. bombensis is terminal and axillary. Northern Seaside Spurge tends to be a pioneer species on the upper beach and foredune front. E. bombensis prefers areas behind the foredune.

COMMENTS: The milky sap may cause skin irritation; when eaten, may cause severe poisoning. Many other sandmat species occur as weeds in inland areas.

891. Silver-leaf Croton

Croton punctatus Jacquin
Crò-ton punc-tà-tus
Euphorbiaceae (Spurge Family)

DESCRIPTION: Annual or short-lived perennial, to 3′ tall; entire plant, except upper leaf surface, covered with a dense layer of small scales and glands that give the plant a brownish gray appearance; inconspicuous male and female flowers produced in separate inflorescences on the same plant (monoecious); fruit a 3-lobed capsule, about 0.5″ wide; flowers late May–November.

RANGE-HABITAT: NC south to FL and west to TX; in SC, common and restricted to the maritime strand on coastal dunes and the upper strand of beaches.

COMMENTS: No other plant on the dunes resembles this plant. Plants can overwinter and produce new shoots from the old stems.

892. Southern Prickly-pear

Opuntia mesacantha Rafinesque ssp.
lata (Small) Majure
O-pún-ti-a me-sa-cán-tha ssp. là-ta
Cactaceae (Cactus Family)

SYNONYM: *Opuntia compressa* (Salisbury) MacBride—RAB, in part

DESCRIPTION: Perennial with leaves reduced to short-lived phyllodes and spines; stems succulent, photosynthetic, and separated into distinct segments (cladodes); cladodes elliptical to rounded in outline with scalloped margins; spines very thin and delicate; flowers large and yellow with many showy tepals; fruit an edible berry; flowers April–June; fruits mature July–February.

RANGE-HABITAT: SC south to FL and west to MS; in SC, occasional in the maritime strand and Coastal Plain in dry, sandy woodlands, maritime grasslands, and Longleaf Pine flatwoods.

SIMILAR SPECIES: This subspecies, a diploid (2 sets of chromosomes), has distinctive scalloped margined cladodes and is restricted to the Coastal Plain. It is similar to

Common Prickly-pear (*O. mesacantha* Rafinesque ssp. *mesacantha*), a tetraploid (4 sets of chromosomes), which lacks the scalloped margins, has stouter spines, and is widespread in SC. Two additional large cacti, *O. stricta* var. *stricta* (Haworth) Haworth and *O. stricta* var. *dillenii* (Ker-Gawler) L. D. Benson, both commonly called Shell Midden Prickly-pear, are known from the maritime strand in SC, where they are rare, generally on shell middens. Both are much larger plants, growing upright as shrubs.

893. Dune Devil-joint; Dune Prickly-pear

Opuntia drummondii Graham
  O-pún-ti-a drum-mónd-i-i
Cactaceae (Cactus Family)

SYNONYM: *Opuntia pusilla* (Haworth) Nuttall—PR

DESCRIPTION: Perennial with creeping succulent stems; stems photosynthetic and segmented into cylindrical or slightly flattened, loosely attached joints (cladodes) that separate readily; scattered over the stem are clusters of tiny hairlike spines (glochids), with or without 2–4″ long, stout, sharp spines; flowers May–June.

RANGE-HABITAT: NC south to FL and west to TX; in SC, common in the maritime strand on sandy dunes, maritime grasslands, and shell middens; rare in the Piedmont on granitic flatrocks.

COMMENTS: This plant is often inconspicuous and hidden in the dune vegetation; if stepped on with bare feet, its spines can inflict a painful wound. Because of the readiness with which the joints become detached, the spines may become embedded in shoes, after which they can work their way to skin (if the shoes are thin). The spines are retrorsely barbed, making them difficult to remove. Although typically a dune plant, it is also found at Forty Acre Rock Heritage Preserve in Lancaster County and near Clover, SC, where it is remarkably disjunct. Diploid, triploid, and tetraploid populations of this plant are present in SC. The specific epithet honors Thomas Drummond (1780–1835), a Scottish botanist who collected plants in Texas.

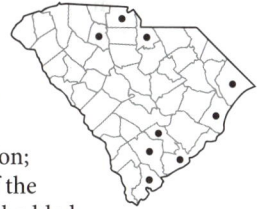

894. Beach Pea

Strophostyles helvola (L.) Elliott
  Stro-pho-stỳ-les hél-vo-la
Fabaceae (Bean Family)

DESCRIPTION: Trailing or twining annual herbaceous vine; leaves trifoliolate; leaflets ovate to rhombic-ovate or often 3-lobed; flowers produced in racemes; petals rose to purple in color, often turning green; fruit a legume, 2–4″ long, appressed pubescent; flowers June-September; fruits mature August–October.

RANGE-HABITAT: Quebec west to SD and south to FL and TX; in SC, common throughout; dunes and maritime grasslands, sandy roadsides, and other open areas.

COMMENTS: The pods and seeds are edible when cooked. Although this species is a common and abundant member of the dune community, it may be found in open, disturbed, often sandy soil throughout the state.

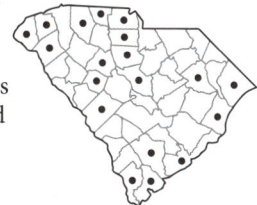

895. Large Sea-purslane

Sesuvium portulacastrum L.
Se-sù-vi-um por-tu-la-cás-trum
Aizoaceae (Carpetweed Family)

DESCRIPTION: Fleshy, smooth, perennial herb with elongate, creeping branches rooting at the nodes and forming mats; leaves opposite; pink flowers and fruits, solitary on distinct stalks from the leaf axils; flowers May–frost.

RANGE-HABITAT: NC south to FL and west to TX; also found in Central and South America, Europe, and Africa; in SC, common in the maritime strand in dune swales, coastal dunes, high salt marshes, dredged spoil disposal sites, and beaches.

SIMILAR SPECIES: Small Sea-purslane (*S. maritimum* (Walter) Britton, Sterns, & Poggenburg) grows in similar habitats. It differs in having sessile flowers and fruits, typically appearing as an erect to spreading annual, and not rooting at the nodes.

COMMENTS: When growing in sandy sites, it forms mounds where sand builds up around the plant. Large Sea-purslane is edible and was consumed by Native Americans. It can be eaten raw or cooked. The raw plants may have a bitter aftertaste that is removed by cooking.

896. Annual Sea-pink

Sabatia stellaris Pursh
Sa-bà-ti-a stel-là-ris
Gentianaceae (Gentian Family)

DESCRIPTION: Annual herb to 25″ tall; stems freely branched with opposite leaves; flowers with 5–7 petals; sepals less than ¾ the length of the petals; petals pink with yellowish, star-shaped centers edged with red; flowers July–October.

RANGE-HABITAT: MA south to FL and west into LA; also found in the Caribbean and Mexico; in SC, common in the maritime strand; brackish swales within dune systems, brackish and salt marshes, and marl spoil banks or flats.

SIMILAR SPECIES: Slender Marsh-pink (*S. campanulata* (L.) Torrey) of Longleaf Pine savannas and Pond Cypress savannas and other moist, open areas in the Coastal Plain and Sandhills, is similar but is a perennial with the sepals more than ¾ the length of the petals.

897. Camphorweed

Heterotheca subaxillaris (Lamarck) Britton & Rusby
He-te-ro-thè-ca sub-ax-il-là-ris
Asteraceae (Aster Family)

DESCRIPTION: Annual or short-lived perennial herb, pubescent throughout; stems erect to ascending, 1–5′ tall; leaves alternate, spreading to ascending, the lower with a petiole, the upper sessile; leaf blades ovate to lanceolate with serrate margins; flowers produced in involucrate heads

of yellow ray and yellow disk flowers; the heads in a corymblike arrangement; flowers July–October.

RANGE-HABITAT: Widespread through most of North America and south into the tropics; in SC, common in the maritime strand and Outer Coastal Plain on dunes, swales, maritime grasslands, and roadsides; common throughout the rest of SC in disturbed habitats such as roadsides and powerlines, often on sandy soils where assumed to be adventive.

COMMENTS: Two forms of this common seaside plant exist in SC. One form is much smaller and is a native component of the SC maritime strand; the other is a coarse and tall weedy form that has invaded or been accidentally introduced to disturbed sites inland. These two forms have variously been treated as varieties or even species. Currently, they are considered part of a single, highly variable species.

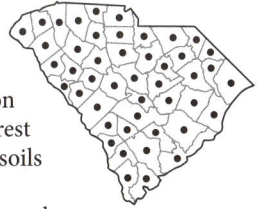

---

898. Beach Morning-glory;
Fiddle-leaf Morning-glory

Ipomoea imperati (Vahl) Grisebach
I-po-moè-a im-pé-ra-ti
Convolvulaceae (Morning-glory Family)

SYNONYM: *Ipomoea stolonifera* (Cyrillo) Poiret—RAB, PR

DESCRIPTION: Smooth, fleshy, trailing perennial, rooting at the nodes; most leaf blades lobed at the base, often deeply so; corolla funnelform, white with a yellowish center; August–October.

RANGE-HABITAT: NC south to FL and west to TX; in SC, occasional in the maritime strand; upper strand of the beach and on front dunes.

COMMENTS: The fleshy leaves and low-growth form (making it less exposed to wind and salt spray) are adaptations that allow it to grow in the hostile beach and dune environments. Another succulent morning-glory, Railroad Vine (*Ipomoea brasiliensis* (L.) Sweet), is common in the tropics and occasionally establishes along SC coastlines. It is a very robust species with deeply cleft leaves and ropelike stems, often with purplish flowers.

CONSERVATION STATUS: SC-Imperiled

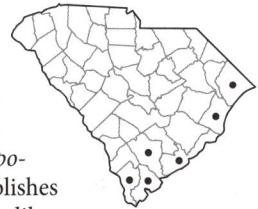

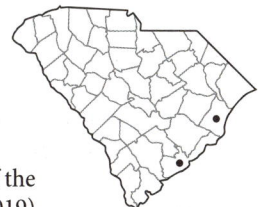

---

899. Dune Blue Curls

Trichostema nesophilum K. S. McClelland & Weakley
Tri-chos-tè-ma ne-só-phi-lum
Lamiaceae (Mint Family)

DESCRIPTION: Short, shrubby, perennial, often branched from near the base; stems often maroon colored and short pubescent; leaves small, up to 0.8″ long; oblong, rhombic, or obovate; flowers August–November.

RANGE: Endemic to the maritime strand of NC and SC; in SC, rare, known only from the maritime strand north of Cape Romain; maritime grasslands, back dunes, openings, and edges of maritime forests.

COMMENTS: "Blue curls" refers to the long, coiled, blue-purple stamens. The scientific name is very descriptive and translates as "hairlike stamens and island-loving." This distinctive species of blue curls is easily separated from the other two species found in SC by the perennial habit and bushy nature of the plants. This species was very recently described (McClelland and Weakley, 2019).

CONSERVATION STATUS: SC-Imperiled

900. Spanish Dagger

Yucca aloifolia L.
   Yúc-ca a-loi-fò-li-a
Agavaceae (Agave Family)

DESCRIPTION: Evergreen treelike plant, to about 15′ tall; trunk usually without branches; leaves bayonetlike, crowded, with spinulose-serrate margins, tapering to a sharp point; flowers white, bell-shaped, hanging, in a large terminal panicle; flowers June–July; fruits are fleshy capsules, mature October–December.
RANGE-HABITAT: VA south to FL and west to LA; also found in Mexico; in SC, common in the maritime strand as a native; less abundant inland, where persistent as escapes from cultivation; margins of maritime forests, shell hammocks, middens, dunes, and other open sandy areas.
SIMILAR SPECIES: Two similar species are found in the maritime strand. Mound-lily Yucca (*Yucca gloriosa* L.) is similar but is generally a smaller plant with leaves that have entire margins and flowers produced in August–October. Spoonleaf Yucca (*Y. filamentosa* L., plate 887) has leaf edges that fray into twisted fibers.
COMMENTS: Spanish Dagger makes an enormous but beautiful ornamental in open, sunny gardens. Since it can withstand salt spray, it is often planted along the coast as an evergreen landscaping choice. Like all our yuccas, portions of the plant are considered edible. The petals are used in salads, or the entire flowers are fried as fritters. The fruits can be cooked and eaten after the seeds are removed. Spanish Dagger is so potentially inhospitable that it was often used by Spanish troops to fortify strongholds during the colonial period–hence the common name. It is sometimes considered to have been introduced in SC, but it appears to be native; if it was introduced, it occurred so long ago that it cannot be corroborated.

901. Maritime forests

902. Southern Red Cedar

Juniperus silicicola (Small) Bailey
   Ju-ní-pe-rus si-li-cí-co-la
Cupressaceae (Cypress Family)

DESCRIPTION: Aromatic, evergreen tree, 40–60′ tall; mature trees generally with a flattened crown; leaves on saplings and seedlings linear and spreading from the twigs, giving the seedling a prickly feeling; leaves of mature tree are short scales in close, overlapping pairs; often both juvenile and mature leaves occur on the same plant; male and female cones on separate trees; mature

female cones bluish-black, berrylike, resinous, 0.10–0.16″ long; male cones shed pollen January–March; female cones mature October–November.
RANGE-HABITAT: From NC south to FL and west to MS; also found in TX; common in the Outer Coastal Plain and maritime strand; shell deposits, maritime forests, dunes, salt shrub thickets, and brackish marshes.
COMMENTS: The following comments refer to both Eastern Red Cedar and Southern Red Cedar. The wood has an essential oil that makes it durable—that is, not readily attacked by fungi or insects—hence, cedar has been used for shingles, fence posts, and storage chests. Before the larger trees with straight trunks were exhausted, its wood was used extensively for making lead pencils. Early Charleston furniture makers used cedar until mahogany became available. Today, it is more of a specialty wood and is used for cedar chests and interior trim. Many mammals and birds, including Cedar Waxwings, eat the mature fleshy cones. Cedar is now grown on plantations for Christmas trees.

There has been considerable debate concerning the distinctiveness of this species from the common and widespread Eastern Red Cedar (*Juniperus virginiana* L.). In SC, Southern Red Cedar is confined to the maritime strand and Outer Coastal Plain; Eastern Red Cedar is more abundant in the mountains and Piedmont. Southern Red Cedar has narrower terminal twigs, smaller leaves, smaller female cones, and a different architecture. Southern Red Cedar is confined to areas near salt marshes or the shore. The authors think they are best assigned to separate species. A very popular horticultural selection, "Brodie," is commonly planted in SC.

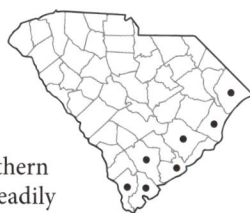

903. Live Oak

Quercus virginiana P. Miller
Quér-cus vir-gi-ni-à-na
Fagaceae (Beech Family)

DESCRIPTION: Large- to medium-size tree with a wide crown from low-spreading branches; leaves evergreen, margins not heavily curved downward and not shaped like an overturned boat, veins not deeply impressed; female flowers 1–2 on a short stalk, separate from the male flowers, which are in catkins; flowers April; acorns mature September–November.
RANGE-HABITAT: Endemic to the southeastern Coastal Plain from VA south to FL and west to TX; in SC, common as a native species in the maritime strand, Coastal Plain, and Sandhills and widely planted throughout the state; maritime forests, sandy oak-hickory forests, flatwoods, and open, sandy woods.
SIMILAR SPECIES: Sand Live Oak (*Q. geminata* Small, plate 519) is a very similar species but its leaves display deeply impressed veins on the upper surface and have a distinct shape, with the margins deeply curled under and resembling an overturned boat. It is a much smaller tree that is confined to scrub ridges, dry sandhills, and dry flatwoods.
COMMENTS: The deep-grooved bark makes Live Oak prime habitat for the establishment of epiphytes; occasionally Spanish Moss, Green-fly Orchid, and Resurrection Fern grow on the same branch. The acorns, which ripen in the fall, are the sweetest of all the oaks and contain so little tannin that they can be eaten off the tree. The acorns are a major food source for wildlife.

Live Oak was an important lumber tree in colonial times. Curved pieces cut from the junction of the limb and trunk were used for ribs in wooden ships. Expeditions were sent from the ship-building ports of the northeastern US to the barrier islands of the Atlantic and Gulf Coast states to harvest Live Oak.

Contrary to popular belief, Live Oak is a fast-growing tree in rich soils and has a normal life span of only around 350 years. By observing the Live Oaks that were planted in the 1700s and 1800s along plantation avenues, one can appreciate how fast Live Oak grows. Widely planted as an ornamental, it withstands high winds, generally suffering only pruning. It also tolerates some inundation of saltwater during hurricanes.

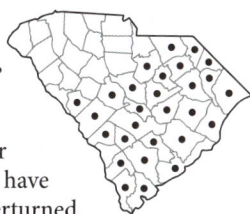

904. Hercules'-club; Southern Toothache-tree

Zanthoxylum clava-herculis L.
Zan-tho-xỳ-lum clà-va-hér-cu-lis
Rutaceae (Rue Family)

DESCRIPTION: Shrub or small tree 20–30′ tall; leaves alternate, odd-pinnately compound with spiny rachis; leaflets thick and leathery with oblique bases; trunk of tree covered with pyramid-shaped, corky, spine-tipped outgrowths; flowers April–May, after the new leaves appear; fruit a 2-locular capsule bearing a few seeds; typically one black seed hangs out of the capsule at maturity.

RANGE-HABITAT: VA south to FL and west to TX, AR, and OK; in SC, occasional in the maritime strand in shell hammock forests, shell rings and middens, dunes, and sandy, thin woods, almost always with high-calcium soil.

COMMENTS: Hercules'-club has a prominent place in American folklore. F. P. Porcher (1869) lists numerous medicinal uses of this tree. Another source states that an oil derived from leaves and bark was used as a drug to treat toothaches. Chewing the leaves initially gives a pleasant sensation, but later, it causes a numbing sensation. The alkaloid chelerythrine is the major active compound in *Zanthoxylum*. It has been studied extensively for its antibacterial and anticancer properties. A tincture of the dried bark is now available as an herbal supplement.

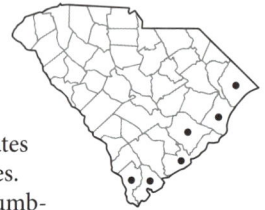

905. Southern Magnolia; Bull Bay

Magnolia grandiflora L.
Mag-nòl-i-a gran-di-flò-ra
Magnoliaceae (Magnolia Family)

DESCRIPTION: Large, fast-growing tree; leaves evergreen, persisting 2 years, shiny dark green above, often covered with reddish rust-colored hairs on the lower surface; flowers large, fragrant; flowers May–June.

RANGE-HABITAT: NC south to FL and west to TX and AR; in SC, common in the maritime strand and Coastal Plain, occasional in the Piedmont, often escaping from cultivation throughout the state; maritime forests, alluvial swamp forests, low woods, ravine slopes, and beech forests.

COMMENTS: In SC, as a native, it rarely grows in pure stands. It usually grows in association with other hardwoods and is not abundant at any one site. The wood turns brown after exposure to air, so it has never been a commercially important lumber tree. It is planted extensively as an ornamental tree, both within and beyond its natural range. It has become naturalized in many areas where it is not native.

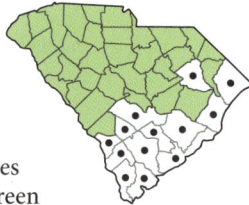

906. Red Bay

Persea borbonia (L.) Sprengel
Pér-se-a bor-bò-ni-a
Lauraceae (Laurel Family)

DESCRIPTION: Aromatic shrub or small tree to 60–70′ tall; leaves evergreen, simple, alternate, dark green, and smooth above; lower leaf surface, appressed pubescent, pubescence not spreading or ascending; leaf margins often with swollen knots due to galls; peduncles of inflorescence 0.4–1.2″ long; flowers May–June; fruit a dark blue or black drupe; fruits mature September–October.

RANGE-HABITAT: From NC south to FL and west to TX; in SC, common in the maritime strand on dunes, in maritime forests, and in dry, sandy soils of barrier islands.

SIMILAR SPECIES: Red Bay is similar to Swamp Bay (P. palustris (Rafinesque) Sargent, plate 721), which is common in wet, acidic soils of swamps, pocosins, and bay forests. The lower surface of the leaf of Red Bay has minute, silvery to shinning golden hairs appressed to the surface; the lower surface of the leaf of Swamp Bay has longer, rusty, often crooked hairs that are not appressed.

COMMENTS: The aromatic, spicy leaves of Red Bay and Swamp Bay have been used as a substitute for Bay Laurel leaves (*Laurus nobilis* L.), native to Asia Minor, and the laurel of history. Fresh or dried leaves can be used as a spice to flavor meats, soups, and other dishes. Because of its attractive evergreen leaves, Red Bay is often cultivated. The fruits are of limited importance to wildlife.

Both Swamp Bay and Red Bay have recently been devastated by the accidental introduction in the 1990s of an Asian Ambrosia Beetle (*Xyleborus glabratus*), which bore into the trees and infected them with a fungus of the genus Ophiostoma that causes the death of the stem. Swamp Bay, Red Bay, and (to a lesser extent) other members of the Lauraceae are the primary host for the caterpillars of the Palamedes Swallowtail, one of the most common swallowtail butterflies in the Coastal Plain.

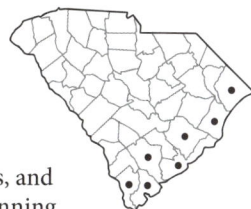

907. Cabbage Palmetto

Sabal palmetto (Walter)
Loddiges ex J. A. & J. H.
Schultes
Sà-bal pal-mét-to
Arecaceae (Palm Family)

DESCRIPTION: Branchless palm to 65′ tall; evergreen leaves fanlike at the top of the thick stems; leaves with the costa (midrib) extending to the tip of the leaf on the undersurface; hastula (midrib extension) 2–7″ long, acuminate to acute, on the upper leaf surface; leaf scars persist as shallow, incomplete rings on the trunk; flowers July; fruit a drupe; drupes mature October–November.

RANGE-HABITAT: Along the coast from NC south to FL; in SC, common in the Outer Coastal Plain and maritime strand; maritime forests on barrier islands, edges of ponds, salt and brackish marshes, and flatwoods; does not occur naturally more than 75 miles from the coast.

COMMENTS: Cabbage Palmetto is adapted to withstand high winds. Its leaves offer little wind resistance, and its soft "wood" allows it to bend with hurricane-force winds without being uprooted or broken. The trunks are used in wharf construction because they are not subject to injury from sea worms. The inner portion of the apical meristem is very tender and palatable,

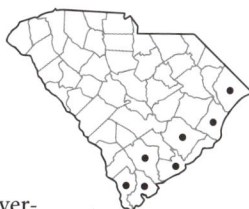

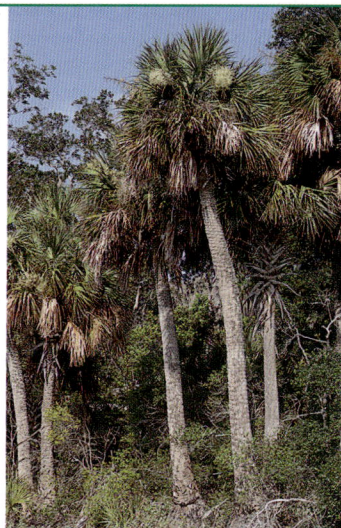

tasting like artichoke and cabbage (hence its common name). Unfortunately, removing the meristem kills the tree. Cabbage Palmetto can tolerate a variety of soils and can grow in either sun or shade.

Cabbage Palmetto is the state tree of SC but is a monocot without typical secondary growth;strictly speaking, it is not a tree. During the Revolutionary War, coastal forts were made of Cabbage Palmetto logs. The soft stems absorbed the force of cannon balls without shattering. It is prized today as an ornamental. Cabbage Palmetto was once common along the NC coast but was virtually extirpated for heart-of-palm. Now it grows naturally only on Bald Head Island in Brunswick County, NC.

Young Cabbage Palmetto can be told easily from Dwarf Palmetto (*Sabal minor* (Jacquin) Persoon, plate 802): Dwarf Palmetto has no filaments on the leaf segments and is without a midrib on the blade except at the very base; Cabbage Palmetto has filaments on the leaf segments and a prominent midrib curving downward at the leaf tip.

908. Yaupon

Ilex vomitoria Aiton
  Ì-lex vo-mi-tò-ri-a
Aquifoliaceae (Holly Family)

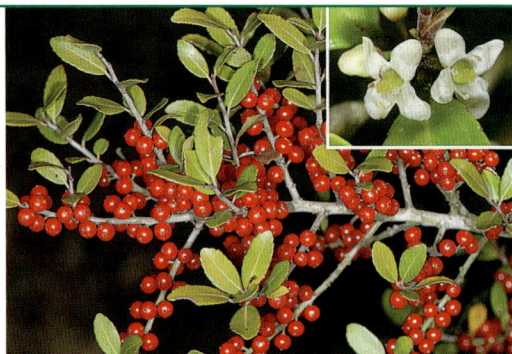

DESCRIPTION: Evergreen shrub or small tree to 25′ tall; male and female flowers on separate plants; leaves small, dark green leathery, shiny above, with crenate margins; fruit often persisting through the winter; flowers March–May; fruits October–November.

RANGE-HABITAT: From VA south to FL and west to TX; in SC, common in the maritime strand, Coastal Plain, and some areas of the Sandhills; uncommon in the lower Piedmont; salt shrub thickets, maritime forests, shell deposits, maritime shrub thickets, fencerows, pond margins, and swamps.

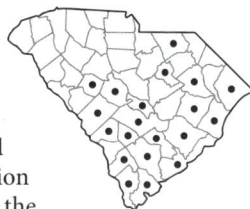

COMMENTS: The specific epithet, vomitoria, refers to the alleged emetic effect. Native Americans used a decoction of the dried old leaves (which were boiled down until the tea was very black and strong) to induce vomiting in purification rites. Hudson (1979), however, indicates that the emetic effect may have been the result of other herbs added to the drink. The young, dried leaves have been and are still used today for a tea. The authors have used the tea and can report it has good flavor and does not cause vomiting. The leaves are known to contain considerable amounts of caffeine, providing the lift people expect from tea. In the early 1900s, there were two attempts to grow Yaupon commercially in Mt. Pleasant, Charleston County, SC; both failed due to competition from oriental teas.

Duncan and Duncan (1988) report that Native Americans transplanted the species to new campsites; consequently, some inland populations are probably a result of these transplants. Yaupon does well in cultivation and is popular as an ornamental shrub for its evergreen leaves and red berries. Often the plants are trimmed into hedges.

909. Common Wax-myrtle; Southern Bayberry

Morella cerifera (L.) Small
  Mo-rél-la ce-rí-fe-ra
Myricaceae (Bayberry Family)

SYNONYM: *Myrica cerifera* L. var. *cerifera*—RAB, PR

DESCRIPTION: Aromatic shrub or small tree to 25′ tall, not stoloniferous; leaves 1.6–3.5″ long and 0.3–0.8″ wide; male and female flowers on separate plants (dioecious); leaves evergreen, although dropping in severe winters; leaves coated with orange, resinous glands on both surfaces; flowers April; fruits August–October, often persisting through winter.

RANGE-HABITAT: NJ south to FL and west to TX; in SC, common in the Piedmont, Sandhills, Coastal Plain, and maritime strand; found in a wide variety of habitats, including maritime shrub thickets, maritime forests, salt shrub thickets, shell deposits, pine-mixed hardwood forests, loblolly pine plantations, swamp forests, pocosins, and roadsides.

SIMILAR SPECIES: Three species of similar, evergreen species of Morella are found in SC. Dwarf Wax-myrtle (*M. pumila* (Michaux) Small, plate 562) is a low, stoloniferous shrub, seldom over 3′ tall, that has shorter leaves (0.6–2.0″ long) and a distinct odor of Eucalyptus when crushed. It is common in the Coastal Plain in Longleaf Pine flatwoods habitats. Pocosin Bayberry (*M. carolinensis* (P. Miller) Small) is also similar but has larger elliptic to oblanceolate leaves that are tardily deciduous–semi-evergreen and larger fruits. It is found in pocosins, wet flatwoods, and savannas in the Coastal Plain and Sandhills.

COMMENTS: Myrtle Beach, SC, gets its name from this plant. The berries are covered with a wax that is used to make fragrant candles. The wax may be irritating to some people. Common Wax Myrtle is commonly planted as an ornamental. The powdered root bark was an ingredient in "composition powder," once used as a folk remedy for chills and colds. The root bark was also used to make an astringent tea and emetic. Common Wax-myrtle is generally found in habitats that are not frequently burned, while its relative, Dwarf Wax Myrtle, is confined to such habitats. The fruits of all wax-myrtles are consumed by birds, especially Yellow-rumped Warblers (also called the Myrtle Warbler).

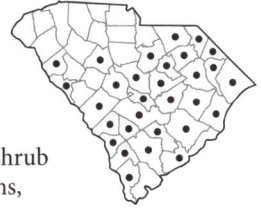

910. Slimleaf Pawpaw; Dog Apples

Asimina angustifolia Rafinesque
A-sí-mi-na an-gus-ti-fò-li-a
Annonaceae (Custard-apple Family)

DESCRIPTION: Shrub to 4′ tall; stems arching; leaves oblinear to spatulate, broadest toward the tip, thick and leathery; flowers solitary, produced in the leaf axils of the current year's growth; tepals white, often with streaks of pink or red toward the base; fruit a curved berry; flowers April–May; fruits mature July–August.

RANGE-HABITAT: SC south to FL; in SC, rare in the maritime strand in a single location in maritime forest at the edge of the salt marsh.

COMMENTS: Slimleaf Pawpaw was first documented for SC by Richard Porcher and Patrick McMillan (McMillan and Porcher, 2005). The habitat here is very different from that of GA or FL where it is known from Longleaf Pine/Slash Pine flatwoods and sandhill habitats. The site is also disjunct from the next nearest locations in Bryan County, GA. This distinctive species is easily recognized as a pawpaw species by the odor of the crushed leaves (like green bell peppers). It should be sought in additional locations along the coast as well as in flatwoods habitats. Additional populations should be reported to the SC Department of Natural Resources, Heritage Trust Program.

CONSERVATION STATUS: SC-Critically Imperiled

911. Creeping Bluet; Roundleaf Bluet

Houstonia procumbens (J. F. Gmelin)
Standley
Hous-tòn-i-a pro-cúm-bens
Rubiaceae (Madder Family)

DESCRIPTION: Prostrate or creeping, perennial herb; leaves opposite, ovate to nearly round with ciliate margins; flowers on short erect stalks, solitary, with 4 white petals; flowers October–April.
RANGE-HABITAT: SC south to FL and west to LA; common in the Outer Coastal Plain and maritime strand; open, sandy sites in maritime forests, beach dunes, pinelands, and sandy roadsides.
COMMENTS: This delicate little wildflower blooms mostly during the winter months. Though often listed as rare, it is actually common in the maritime strand in SC but flowers during a time of year when few botanists are looking for flowers.

912. Spanish Moss

Tillandsia usneoides L.
Til-lánds-i-a us-ne-oì-des
Bromeliaceae (Pineapple Family)

DESCRIPTION: Rootless epiphyte on trees; stems usually curled, wiry; leaves filiform; leaves and stems bear gray scales that absorb atmospheric moisture and minerals; reproduction mainly vegetative via movement of broken stems, but changes in distribution are caused mostly by dispersal of seeds; flowers April–June; spent fruit (capsules) remain throughout the winter.
RANGE-HABITAT: From VA south to FL and west to TX; also south through much of the Neotropics; in SC, common in the Coastal Plain and maritime strand where relative humidity is high; maritime forests, swamp forests, and upland forests.

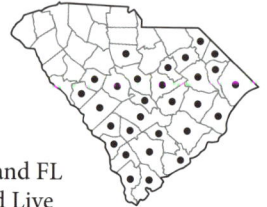

SIMILAR SPECIES: Ball-moss (*T. recurvata* (L.) L.) is similar but forms dense spherical balls on limbs and twigs. It has expanded its range north from GA and FL in recent years and is now commonly seen, particularly on Crepe Myrtles and Live Oaks, in many coastal cities in SC.
COMMENTS: The cultural and economic uses of Spanish Moss are legend (Martinez, 1959). Its durable fiber is resistant to insects and is highly resilient. These characteristics made it sought after as stuffing for mattresses and upholstery. It was also used as a binder in construction of mud and clay chimneys.

The common name Spanish Moss probably came from the French who settled in Louisiana. It may have reminded them of the long, gray beards of the Spanish explorers who had come before them and called it "Spanish Beard." It was later changed to "Spanish Moss." The specific epithet usneoides refers to its resemblance to the lichen Usnea, also common on trees in SC. This note on the genus name comes from Fernald (1950): "The genus honors Elias Tillandz (1640–1693), professor at Abo, who, as a student crossing directly from Stockholm, was so seasick that he returned to Stockholm by walking more than 1000 miles around the head of the Gulf of Bothnia and hence assumed his surname (meaning: by land); the genus erroneously supposed by L. to dislike water."

Spanish Moss is the most conspicuous epiphyte in the SC Coastal Plain. Much of the aesthetic appeal of the Coastal Plain comes from the moss-draped live oaks.

913. Coral Bean; Cherokee Bean

Erythrina herbacea L.
E-ry-thrì-na her-bà-ce-a
Fabaceae (Pea or Bean Family)

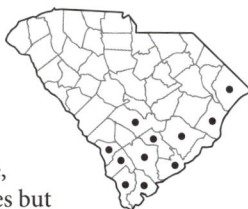

DESCRIPTION: Perennial herb, 2–5′ tall; branchlets usually prickly; leaves alternate, trifoliolate; flowers appear before the leaves but may continue well into the summer; fruit pod constricted between the seeds, and upon breaking open, brilliant scarlet seeds often hang from the open pod; flowers May–July; fruits mature July–September.

RANGE-HABITAT: NC south to FL and west to TX; in SC, common in the maritime strand and Outer Coastal Plain; maritime forests, open coastal dunes, and sandy, dry, open woods and clearings; often persisting around abandoned house sites.

COMMENTS: Coral Bean is often cultivated in gardens. The seeds and bark possess alkaloids that have a curarelike action and may cause death if taken internally. The crushed stems are sometimes employed as fish poisons. In Mexico, the seeds are used for poisoning rats and dogs. Coral Bean is a woody shrub in FL where exposed to fewer frosts but an herb in the rest of its mainly coastal range.

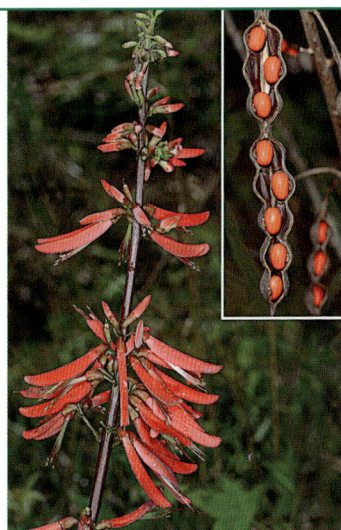

914. Florida Atamasco-lily

Zephyranthes simpsonii
Chapman
Ze-phy-rán-thes simp-són-i-i
Amaryllidaceae (Amaryllis Family)

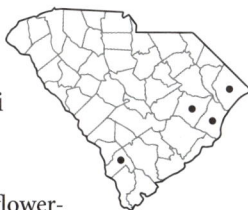

DESCRIPTION: Perennial herb from a bulb; flowering stalk to 1′ tall, generally solitary, terminated by a single large white flower; leaves basal, linear; perianth usually white, rarely pink or fading to pink with tepals ascending to erect at flowering; style and stigma shorter than to equaling the anthers; flowers April–May.

RANGE-HABITAT: NC south to FL; in SC, rare and restricted to the maritime strand in northeastern SC in maritime grasslands, edges of maritime forests, and maritime sandhills; also rare and local in the Coastal Plain in southeastern SC in Longleaf Pine flatwoods and ditches and roadsides through current or former Longleaf Pine flatwoods.

SIMILAR SPECIES: Common Atamasco-lily (*Z. atamasco* (L.) Greene, plate 806) is similar but has longer tepals that are recurved when in flower and has anthers that exceed the stigma and style. It is common throughout the Coastal Plain and Piedmont of SC in moist forests.

COMMENTS: Florida Atamasco-lily has an unusual range and occurs in two widely different habitats. It is more common in flatwoods in GA and FL. Plants from northeastern SC and southeastern NC may, in the future, be recognized as distinct. It often flowers a few days after rains, like many species in the genus, and hence sometimes is called Rain Lily. The genus name refers to the Greek god of the west wind (Zephyrus) and flower (anthos). The species honors American botanist Joseph Herman Simpson (1841–1918).

CONSERVATION STATUS: SC-Critically Imperiled

915. Resurrection Fern

Pleopeltis michauxiana (Weatherby)
 Hickey & Sprunt
  Ple-o-pél-tis mi-chaux-i-à-na
Polypodiaceae (Polypody Family)

SYNONYM: *Polypodium polypodioides* (L.)
 Watt—RAB, PR

DESCRIPTION: Small, evergreen, epiphytic or epilithic (rock-dwelling) fern from a scaly, creeping rhizome and forming large masses on branches or rocks; leaf stalks and underside of leaves covered with copious rusty scales; spores produced in sori that are arranged as spots in two lines on the backs of the pinnae; spores produced June–November.

RANGE-HABITAT: From MD south to FL and west to IL, MO, and KS; in SC, common throughout; limbs and crotches of large trees in any habitat with large, hardwood trees having a deep-grooved bark; grows on circumneutral rocks such as amphibolite and occasionally on granite in the mountains and Piedmont.

COMMENTS: The common name comes from the curling of its leaves during prolonged drought. The curled, dry leaves appear to be dead. However, after a few hours of rain, the leaf absorbs water and uncurls, and it is as alive and as green as ever. Evidence suggests that the scales on the underside of the leaf act as channels for water absorption, hastening the recovery process. Resurrection Fern spreads from tree to tree by wind-borne spores; once established on a limb, it spreads by its creeping rhizome. Though most often thought of as a Coastal Plain species on the limbs of large Live Oaks, it may be found high in the branches of trees in cove forests and on the trunks of Eastern Red Cedar in the mountains or Piedmont. The specific epithet honors Andre Michaux (1746–1802), French explorer and botanist.

916. Salt Marsh

917. Sea Ox-eye

Borrichia frutescens (L.) Augustin de
 Candolle
  Bor-rích-i-a fru-tés-cens
Asteraceae (Aster Family)

DESCRIPTION: Rhizomatous shrub, 6″–4′ tall and forming extensive colonies; little branched; leaves grayish, opposite, thick, somewhat fleshy; flowers produced in involucrate heads of yellow ray and yellow disk flowers; receptacle bracts hard and rigid, with sharp spine tips, remaining on plant through the winter; flowers May–September.

RANGE-HABITAT: VA south to the FL Keys and west to TX; also found in Mexico and Bermuda; common throughout its range in brackish marshes, salt shrub thickets, salt marshes, salt flats, and often in vacant lots or along roadsides near the ocean.

COMMENTS: This is one of the few woody members of the aster family in SC. The genus name honors Ole Borrich (1626–90), a Danish botanist.

918. Common Groundsel-tree; Consumption Weed

Baccharis halimifolia L.
Bác-cha-ris ha-li-mi-fò-li-a
Asteraceae (Aster or Sunflower Family)

DESCRIPTION: Freely branched shrub, 3–9′ tall; leaves tardily deciduous (with some hanging on through the winter), alternate, fleshy, glandular-punctate, margins with few to several teeth; male and female flowers produced on separate plants (dioecious); flowers produced in long-stalked, involucrate heads of white disk flowers; female plants have a satiny, white look in the fall from the mass of bristle-tipped achenes; male plants have a dull, yellow appearance; flowers and fruits September–December.

RANGE-HABITAT: MA south to FL and west to TX and OK; in SC, common in natural habitats in the maritime strand and Coastal Plain; also common in disturbed habitats in the Piedmont; salt shrub thickets, high salt marshes, brackish marshes, freshwater marshes, dune swales, fencerows, old fields, dredged disposal sites, pond margins, roadsides, and other disturbed sites.

SIMILAR SPECIES: Silverling (*B. glomuliflora* Persoon) is similar to B. halimifolia but is strictly a coastal species and has not spread inland. Silverling has sessile rather than pedunculate heads that are scattered along the leafy branches. In B. halimifolia, the flowering heads at the end of the branches are long-stalked. Their habitats overlap in the maritime strand and Outer Coastal Plain. Baccharis glomulerifera, however, is more frequently associated with sites with an elevated pH and is especially common along the margins of roads where coquina limestone (marl) is used as surfacing.

COMMENTS: F. P. Porcher (1869) gives numerous medicinal uses of Common Groundsel-tree and states: "This plant is of undoubted value, and of very general use in popular practice in SC, as a palliative and demulcent in consumption and cough." The common name Consumption Weed is based on this use.

In the fall, during windy days in the maritime strand, it is not unusual for the air to be filled with the whitish achenes of Common Groundsel-tree. It is thought that it was originally a coastal plant that spread throughout SC as disturbed areas increased. It appears that the establishment of the interstate road system, with large grassy margins, was critical in its spread inland. The late autumn flowers of all Baccharis species are attractive to pollinators, especially wasps.

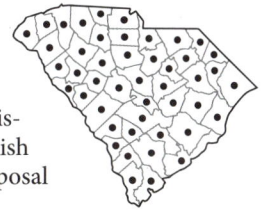

919. Smooth Cordgrass

Spartina alterniflora Loiseleur
Spar-tì-na al-ter-ni-flò-ra
Poaceae (Grass Family)

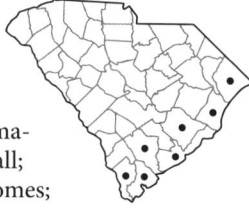

DESCRIPTION: Facultative halophyte; rhizomatous perennial grass; stems (culms) 1–8′ tall; reproduction primarily by spreading rhizomes; flowers August–October.

RANGE-HABITAT: Newfoundland south to FL and west to TX; in SC, the dominant plant (often a monoculture) of intertidal salt marshes along the coast.

COMMENTS: Smooth Cordgrass is a dominant salt marsh plant, covering vast areas and often excluding other species. Plants in the high salt marshes, especially along the edges of salt flats, may grow only 1′ tall (the short form), while in deep water (the low marsh) along the edges of tidal creeks they may grow 8′ tall. Smooth Cordgrass is a facultative halophyte that can grow in brackish or freshwater systems, but it rarely does because it cannot compete with other species. It dominates the intertidal area because it is the species that best responds to the daily inundation by saltwater.

Five major ecological roles have been assigned to salt marshes, in which Smooth Cordgrass plays the dominant role: formation of detritus, habitat for animals, stabilization of coastal substratum through spreading rhizomes, filtration of coastal runoff, and removal of organic waste.

920. Swallow-wort; Sand-vine

Pattalias palustre (Pursh) Fishbein
Pat-tá-li-as pa-lús-tre
Apocynaceae (Dogbane Family)

SYNONYM: *Cynanchum palustre* (Pursh) Heller—RAB

DESCRIPTION: Slender, perennial, twining herbaceous vines with milky sap; leaves linear, 1.5–3.0″ long; inflorescence an umbel of 7–12 flowers, similar to a miniature milkweed flower; petals yellowish or greenish-white; corona white; flowers June–July; fruits are slender follicles that mature July–October.

RANGE-HABITAT: NC south to FL and west to TX; also in the Caribbean and Central America; in SC, common in the maritime strand in salt shrub thickets, upper edges of salt marshes, shell hammock forests, and other moist and sunny conditions.

SIMILAR SPECIES: Leafless Swallow-wort (*Orthosia scoparia* (Nuttall) Liede & Meve) is similar but has mostly leafless green stems with linear leaves that fall off soon after expanding. This species is very rare in SC and is only known from a single extant location, a shell hammock at Daws Island in Beaufort County.

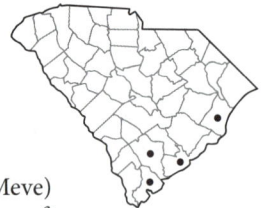

921. Thinleaf Orach

Atriplex prostrata Boucher ex A. P. de Candolle
Á-tri-plex pros-trà-ta
Chenopodiaceae (Goosefoot Family)

SYNONYM: *Atriplex patula* L., in part—RAB

DESCRIPTION: Annual herb with slightly angular, usually grooved stems; plant dark green, often purple-tinged throughout; lower leaves opposite, middle and upper leaves alternate; principal lower leaves ovate to triangular, mostly truncate at the base; separate male and female flowers on same plant (monoecious); fruits enclosed by 2 small, spongy, and thickened bracts; flowers July–frost.

RANGE-HABITAT: Widespread in eastern North America and in western North America and Eurasia; in SC, common in the maritime strand in the upper edge of salt marshes and in brackish flats.

SIMILAR SPECIES: Occurring in the same habitats in the maritime strand is A. arenaria Nuttall, which has principal lower leaves that are linear to lanceolate or elliptic, with wedge-shaped leaf bases.

COMMENTS: Throughout the summer, the young leaves of *Atriplex* species are tender and pleasant to taste, either raw or cooked. Thinleaf Orach is high in vitamins and minerals. Some authors consider this species to be introduced from Eurasia.

922. Saltmarsh Aster

Symphyotrichum tenuifolium (L.) G.L. Nesom
Sym-phy-o-trì-chum te-nu-i-fò-li-um
Asteraceae (Aster or Sunflower Family)

SYNONYM: *Aster tenuifolius* L.—RAB

DESCRIPTION: Perennial herb, 1–4′ tall from slender, creeping rhizomes; branches few to many, curved, slightly zigzag; stems and leaves fleshy, entire; flowers produced in involucrate heads of pinkish, lavender, or white ray flowers and yellow disk flowers; flowers late August–November.

RANGE-HABITAT: NH south to FL and west along the Gulf Coast to se. TX; in SC, common in the maritime strand; salt and brackish marshes, sand mud flats, salt shrub thickets, and dredged soil disposal sites.

COMMENTS: Saltmarsh Aster is never a major ecological component of the salt marsh. It is generally found in a high marsh dominated by the short form of Smooth Cordgrass and becomes conspicuous only when it blooms. In dredged soil disposal sites along the Carolina coast, it grows more robustly and forms extensive stands.

923. Carolina Sea Lavender

Limonium carolinianum (Walter) Britton

Li-mò-ni-um ca-ro-li-ni-à-num

Plumbaginaceae (Leadwort Family)

DESCRIPTION: Perennial, fleshy herb, 6″ to 2′ tall, with basal rosette of elliptic to oblanceolate leaves (2–10″ long); flowers about 0.13″ wide with white sepals and lavender to purple petals; flowers August–October.

RANGE-HABITAT: Newfoundland and Quebec south to FL and west to TX; in SC, common in the maritime strand in the upper edges of salt marsh, edges of salt flats, edges of salt marsh thickets, interdune swales, and saline ditches.

COMMENTS: Salt glands on leaves and stems allow Carolina Sea Lavender to excrete excess salt. It varies in size and vigor, depending on the habitat. It grows about 6–8″ tall in salt flats and to 2′ tall on the edges of salt shrub thickets.

924. Southern Seaside Goldenrod

Solidago mexicana L.

So-li-dà-go mex-i-cà-na

Asteraceae (Aster Family)

SYNONYM: *Solidago sempervirens* L.—RAB

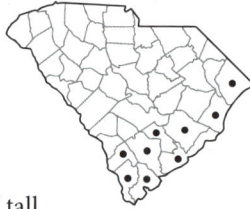

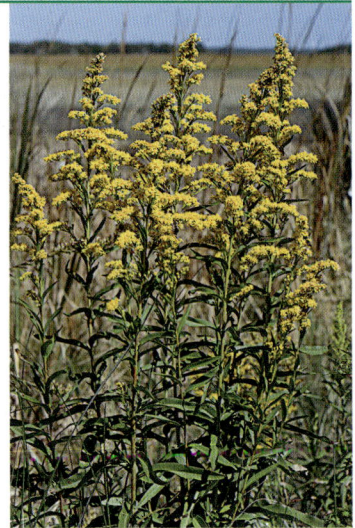

DESCRIPTION: Perennial, fleshy herb, 1.5–6′ tall, with basal rosette of ovate to oblanceolate leaves; stem leaves lanceolate with entire margins; flowers arranged in involucrate heads of yellow ray and yellow disk flowers that are held in a one-sided (secund) raceme, racemes in a paniculate arrangement; flowers August–December.

RANGE-HABITAT: MA south to FL and west to TX; also found in the Caribbean and Mexico; in SC, common in the maritime strand in the upper edges of salt marshes, maritime grasslands, moist openings in maritime forests, martime shrub thickets, ditches, and moist roadsides.

COMMENTS: This is the most common goldenrod species near the coastline. It has recently been recognized as distinct from the Northern Seaside Goldenrod (*S. sempervirens* L.), which does not occur in SC.

925. Salt flats

926. Perennial Glasswort; Samphire

Salicornia ambigua Michaux
Sa-li-cór-ni-a am-bí-gu-a
Chenopodiaceae (Goosefoot Family)

SYNONYM: *Salicornia virginica* L.—RAB, PR, misapplied

DESCRIPTION: Perennial halophyte with fleshy, smooth, somewhat woody stems; stems trailing or weakly arching to erect, rooting at the nodes and forming mats; this year's stem green, last year's stem tan; leaves opposite, reduced to scales; flowers inconspicuous, sunken in pits in the opposite leaf axils of each fleshy stem tip; flowers July–October.

RANGE-HABITAT: NH south to FL and west along the Gulf Coast to TX, the Yucatan, and s. CA; salt flats, high salt marshes, and brackish marshes.

COMMENTS: Perennial Glasswort stems are filled with salty water and make a pleasant salty salad. It also has been popular as a pickle, by first boiling the stems in their own salted water, then adding spiced oil or vinegar. Recently, wild-collected plants have started to show up in high-end grocery chains where they are sold by their Old World name, Samphire.

Perennial Glasswort is one of the few plants that can tolerate the high salinity of salt flats. This is possible because its fleshy stems maintain a sodium concentration equal to that of its environment. Although Perennial Glasswort has long been known as *S. virginica,* that name is properly applied to an annual species, Virginia Samphire, that is more closely related to Old World Samphire (*S. europaea* L.).

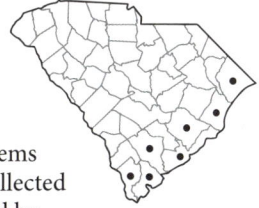

927. Saltwort

Batis maritima L.
Bá-tis ma-rí-ti-ma
Bataceae (Saltwort Family)

DESCRIPTION: Fleshy halophytic, perennial shrub; stems trailing and rooting at the nodes, forming dense colonies from which arise erect, flowering stems seldom greater than 20″ in height; leaves opposite, to 0.8″ long and 0.12″ wide, one surface flat, the other rounded; flowers unisexual, on separate plants (dioecious), small and obscure, crowded in fleshy spikes and solitary in leaf axils; flowers June–July.

RANGE-HABITAT: SC south to FL and west to TX; also found throughout the Neotropics; in SC, occasional in the maritime strand; salt flats and high salt marshes.

COMMENTS: Along with the glassworts, Saltwort is one of the few species that can tolerate the high salinity of salt flats. Most manuals list Saltwort as rare in SC; however, it is widespread along the coast, and a classification of occasional is justified.

928. Maritime shell forests

**929. Godfrey's Forestiera;
    Godfrey's Swamp-privet**

Forestiera godfreyi L. C. Anderson
    Fo-res-ti-èr-a gód-frey-i
    Oleaceae (Olive Family)

DESCRIPTION: Deciduous shrub or small tree 3–6′
tall; main stem arching or leaning; occasionally a
few branchlets develop enlarged bases to become
spinelike; leaves opposite, simple, pubescent,
elliptic, with acuminate to acute tips; male and
female flowers on separate plants (dioecious);
flowers without petals, in tight axillary clusters;
flowers mid-January–mid-February; fruit a drupe, dark blue, with a whitish
bloom; mature late April–early May.

RANGE-HABITAT: SC south to FL; in SC, rare in the maritime strand in mesic
shell hammock forests.

COMMENTS: Loran C. Anderson described this species in 1985. The specific
epithet honors Robert K. Godfrey (1911–2000), a distinguished botanist at Tall
Timbers Research Station, and Florida State University. In his paper, Anderson
cites J. H. Mellichamp's collections of Forestiera from an area near Bluffton in
Beaufort County in the 1800s, which Anderson used in describing this species. The location of Mellichamp's population of F. godfreyi near Bluffton is unknown. It seems likely, however, from Patrick D.
McMillan's research, that the original collection was made along the May River on shell associated
with Buzzard Island.

In May 1975, Richard Porcher collected a plant on a Native American shell mound on Pig
Island at the intersection of Townsend Creek and the North Edisto River in Charleston County,
but he did not recognize the specimen as a species of Forestiera. In April 1996, John F. Townsend
and Patrick D. McMillan accompanied Richard Porcher to the shell mound. The plant noted above
was identified as *F. godfreyi* by Townsend and McMillan. Additional populations of the rare shrub
have been located by botanists since 1996, but it remains one of the rarest shrubs in SC and is rare
throughout its entire range.

The genus is named in honor of André Robert Forestier (1736–1812), a physician in Saint-Quentin,
France, who was Poiret's first botany teacher. Poiret described the genus *Forestiera* in 1810.

CONSERVATION STATUS: SC-Critically Imperiled

930. Tough Bumelia

Sideroxylon tenax L.
    Si-de-róx-y-lon té-nax
Sapotaceae (Sapodilla Family)

SYNONYM: *Bumelia tenax* (L.)
    Willdenow—RAB, PR

DESCRIPTION: Small scrubby tree or shrub growing to 30′ tall; bark thick, fissured, reddish brown; branches armed with stout thorns; twigs tough, flexible, pubescent; leaves oblanceolate, with rounded tips; underside of leaves covered with appressed, silky, rust to coppery-gold hairs that are matted and shiny; fruit a drupelike berry, mature September–October.

RANGE-HABITAT: NC south to FL; in SC, common in the maritime strand and the Outer Coastal Plain; maritime forests, sand dunes, sandy pinelands, and shell deposits.

SIMILAR SPECIES: Eastern Gum-bumelia (*S. lanuginosum* Michaux ssp. lanuginosum) is found in the central and lower Savannah River drainage on bluffs with high pH soils. It can be distinguished by the undersides of the leaves having woolly and unmatted hairs.

COMMENTS: Some birds and mammals eat the fruits and seeds. Though it occurs in habitats other than shell midden areas, it is most common where soils have an elevated pH.

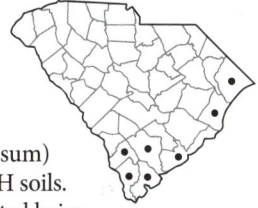

931. Carolina Buckthorn

Frangula caroliniana (Walter) Gray
    Frán-gu-la ca-ro-li-ni-à-na
Rhamnaceae (Buckthorn Family)

SYNONYM: *Rhamnus caroliniana*
    Walter—RAB, PR

DESCRIPTION: Short-lived shrub or small tree, to 30–40′ tall; leaves deciduous, but some often remaining through the winter; parallel veins adjacent to the midrib of the lower leaf surface are prominent, evenly spaced; leaves with skunklike odor when crushed; fruit a drupe with 3 stones (pits), turning red, then black, maturing September–October.

RANGE-HABITAT: VA south to FL and west to OH, MO, and TX; in SC, common in the Piedmont in association with basic rock in moist deciduous woods; rare in the Coastal Plain and maritime strand on limestone bluffs, shell hammock forests, and middens.

COMMENTS: Many birds eat the sweet fruits. The attractive dark foliage and pleasing form make it a desirable small tree in the partly shaded landscape.

**932. Small-flowered Buckthorn;
Shell-midden Buckthorn**

Sageretia minutiflora (Michaux) Mohr
    Sa-ge-rèt-i-a mi-nu-ti-flò-ra
Rhamnaceae (Buckthorn Family)

DESCRIPTION: Sprawling, weak-stemmed shrub to 10′ tall, with many short, thornlike branches; leaves opposite or nearly so; flowers very fragrant, flowering in August–October; fruit drupelike, purplish black when mature, persistent over winter and maturing in the spring.

RANGE-HABITAT: From NC south to FL and west to MS; in SC, uncommon in the maritime strand; shell hammock forests and shell middens and rings.
SIMILAR SPECIES: Small-flowered Buckthorn can be mistaken for Yaupon (*Ilex vomitoria* Aiton, plate 908), with which it typically grows. The leaves of the Small-flowered Buckthorn are opposite or nearly so; the leaves of Yaupon are alternate.
COMMENTS: Small-flowered Buckthorn is probably more abundant than is indicated by manuals, but it is certainly not common. Botanists have observed it on numerous sites along the coast, especially on shell deposits. It has the habit of draping over other vegetation, but it is not a woody vine. Its weak stems apparently give it the flexibility to survive the wind and water surge of coastal hurricanes. The genus is named in honor of French botanist Augustin Sageret (1763–1851).
CONSERVATION STATUS: SC-Vulnerable

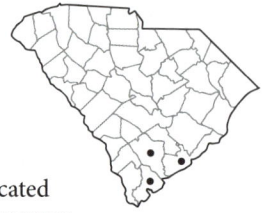

933. Shell-midden Morning-glory

Ipomoea macrorhiza Michaux
I-po-moè-a ma-cro-rhì-za
Convolvulaceae (Morning-glory Family)

DESCRIPTION: Robust perennial vine from large tubers; leaves pubescent, crinkled; older leaves often 3-lobed; calyx hairy; corolla tubular, to 3″ in diameter, white to pale pink or bluish, throat purplish within; flowers June–July
RANGE-HABITAT: From NC south to FL and west to AL; also in tropical America; in SC, local and rare in the maritime strand in shell hammock forests and on shell middens; also rare inland in the Coastal Plain, where possibly escaped from cultivation.
SIMILAR SPECIES: Shell-midden Morning-glory is distinguished from Man-of-the-earth (*I. pandurata* (L.) G. F. W. Meyer, plate 988) by its hairy, crinkled leaves and hairy calyx. The leaves and calyx of I. pandurata are hairless.
COMMENTS: The species only blooms at night and early in the morning. According to Small (1927), Native Americans, who cultivated the plant for its huge, starchy tubers, may have introduced it into the Carolinas from further south. The seeds are held in a cottony mass that aids in floatation on the tides. Though this species is sometimes considered introduced from tropical America, it certainly appears native and may have been introduced by Native Americans as a food source.
CONSERVATION STATUS: SC-Critically Imperiled

934. Mellichamp's Skullcap

Scutellaria mellichampii Small
Scu-tel-là-ri-a mel-li-chámp-i-i
Lamiaceae (Mint Family)

DESCRIPTION: Herbaceous perennial; stems 1–3′ tall, pubescent with the second node below the inflorescence not glandular or only very sparsely glandular; leaves ovate with a cuneate to truncate base and bluntly serrate to crenate margins; flowers produced in racemes; flowers with a pubescent, light purple tubular, bilabiate corolla to 1.0″ long; flowers June–July.
RANGE-HABITAT: SC, GA, and AL; in SC, rare in the Coastal Plain and near the fall-line; shell hammocks and middens, calcareous bluff forests, and other forest systems with high-calcium soils.

SIMILAR SPECIES: Ocmulgee Skullcap (*Scutellaria ocmulgee* Small) is similar but has stems that are densely glandular at the second node below the inflorescence and leaves with truncate to cordate bases. It is found on calcareous bluff forests and other high-calcium areas along the fall-line in Aiken and Edgefield Counties.

COMMENTS: Mellichamp's Skullcap is a beautiful and rare native that makes a showy garden plant. It should never be disturbed in the wild; it is available from native plant mail-order nurseries and the South Carolina Botanical Garden. The specific epithet honors Joseph Hinson Mellichamp (1829–1903), physician, botanist, and prolific collector of plant specimens from Bluffton, SC.

CONSERVATION STATUS: SC-Imperiled

---

935. Catesby's Virgin's-bower

Clematis catesbyana Pursh
    Clé-ma-tis cates-by-à-na
Ranunculaceae (Buttercup
    Family)

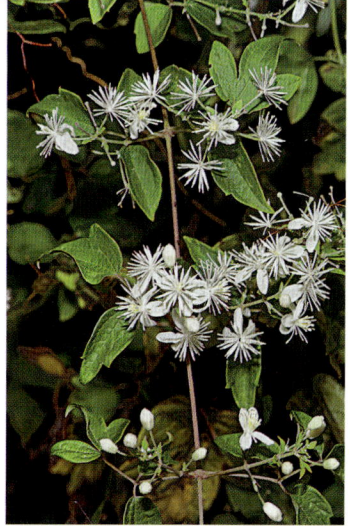

DESCRIPTION: Perennial vine with clambering stems to 20′ long; leaves ternately compound, with 5–9 leaflets; leaflets ovate to lanceolate, with toothed margins and conspicuous venation; flowers produced in simple to compound panicles; petals lacking; sepals white; fruit an aggregate of achenes with long, plumose, and persistent styles; flowers June–September; fruits mature July–October.

RANGE-HABITAT: VA south to FL and west to LA; also in scattered inland populations in NC, TN, KY, AR, and MO; in SC, uncommon and restricted to shell hammock forests and calcareous maritime forests in the maritime strand.

SIMILAR SPECIES: Common Virgin's-bower (*C. virginiana* L., plate 217) is a widespread and common species that in SC is restricted to the mountains and Piedmont, where it is found along forest margins and wet meadows. It can be distinguished from Catesby's Virgin's-bower by the leaves having only 3 leaflets. Sweet Autumn Clematis (*C. terniflora* A. P. de Candolle) is a common invasive exotic throughout SC that has leaves without toothed margins and 3–5 pinnately arranged leaflets.

COMMENTS: The species is named in honor of Mark Catesby (1683–1749), British explorer and naturalist, who wrote and illustrated *The Natural History of Carolina, Florida and the Bahama Islands*. Catesby's Virgin's-bower makes an exceptional flowering vine for trellises in the home landscape.

CONSERVATION STATUS: SC-Vulnerable

The Ruderal Communities

936. Chickasaw Plum

Prunus angustifolia Marshall
Prù-nus an-gus-ti-fò-li-a
Rosaceae (Rose Family)

DESCRIPTION: Shrub or small tree to 15′ tall, forming thickets by means of root suckers; stems slightly zigzag, with short lateral twigs often ending as a thorn; leaves deciduous, developing after the flowers, with marginal teeth tipped with red glands that may fall with age, leaving a scar; flowers white, solitary, or in clusters from twigs of previous year; flowers February–April; fruit a drupe, red or yellow, mature in May–early July.

RANGE-HABITAT: Throughout the southeastern and mid-southern US; in SC, common throughout; fencerows, roadsides, pastures, old fields, woodland borders, beach dunes, and around abandoned house sites.

SIMILAR SPECIES: Chickasaw Plum can be differentiated from Hog Plum (*P. umbellata* Elliott, plate 534) by its gland-tipped teeth on the margin of the leaves; Hog Plum has glands at the base of the leaf blade but no glands on the leaf margins.

COMMENTS: The native distribution of Chickasaw Plum is not clear. It is probably native west of SC and introduced in prehistoric times by Native Americans, who cultivated it for the fruits. The ripe fruits are generally sweet and can be eaten fresh or used for sauces, pies, preserves, jams, and jellies. The fruits are an important food source for wildlife, including deer, bears, raccoons, squirrels, other mammals, and birds. The thicket-forming habit of this plum makes it useful for erosion control and wildlife cover. In addition, the profuse flowering in the spring attracts a myriad of wasp, fly, and bee species and is one of the best choices for attracting pollinators.

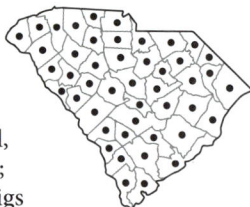

937. Callery Pear; Bradford Pear

Pyrus calleryana Decaisne
Pỳ-rus cal-le-ry-à-na
Rosaceae (Rose Family)

DESCRIPTION: Small- to medium-sized tree to 60′ tall; branches reddish brown, becoming grayish with age; twigs and young branches frequently with stout thorns; leaves alternate, smooth; leaf blades 1.5–3.5″ long, ovate to broadly ovate, deep glossy green above, lighter below, margins serrate or entire; flowers February–April; fruits are brown pomes, mature in July–September.

RANGE-HABITAT: Native of Asia, now thoroughly naturalized and a problematic invasive throughout the southeastern United States; abundant throughout SC on roadsides, open woodlands, fields, and virtually all other early successional habitats.

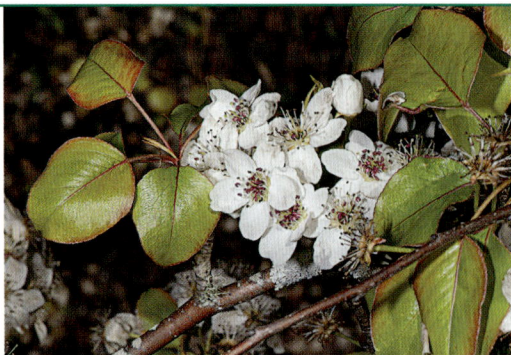

COMMENTS: Callery Pear has become one of the most noxious invasive exotics in SC. It was introduced as a thornless variety that was believed to be a sterile hybrid "Bradford Pear." The species quickly reverted and today has spread to become a serious pasture weed and a threat to SC's endangered Piedmont prairie and forest margin species. Mile after mile of roadside habitats are now white in the early spring with the flowers of this invasive exotic. Bradford Pear should never be planted in a landscape; doing so only further burdens those attempting to control this tree.

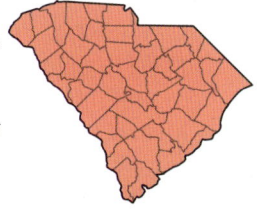

938. White Mulberry

Morus alba L.
Mò-rus ál-ba
Moraceae (Mulberry Family)

DESCRIPTION: Shrubby, small, deciduous tree that may reach 50′ tall; leaves variously lobed or unlobed, smooth and shiny above, hairless below; flowers unisexual, in spikes on the same tree or on different trees; flowers March–May; fruit compound, berrylike; white, pink, or dark purple; mature May–June.

RANGE-HABITAT: Native to east Asia; widely naturalized in eastern and southern North America; in SC, widely scattered throughout; along streams and fencerows, around dwellings, in pastures and vacant lots, and forming dense stands in the maritime strand in dredged soil disposal sites.

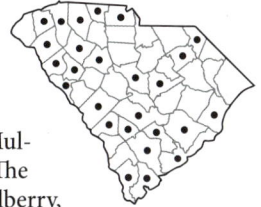

SIMILAR SPECIES: White Mulberry is similar in appearance to the native Red Mulberry (Morus rubra L.), which occurs throughout the eastern United States. The leaves of Red Mulberry are dull green and usually rough above. In White Mulberry, the leaves are shiny green, smooth above and hairless beneath except on the main veins.

COMMENTS: White Mulberry was introduced into the United States from China in the 1600s in an attempt to establish a silk industry. The leaves are the chief food of the silkworm. The industry failed, but the trees survived. They quickly spread widely because birds eat the fruits and disperse the seeds. The wood has little commercial value. Although not as sweet and tasty as the fruits of Red Mulberry, the ripe fruits of White Mulberry can be used in a variety of ways.

939. Mimosa; Silk Tree

Albizzia julibrissin Durazzini
Al-bíz-zi-a ju-li-brís-sin
Fabaceae (Bean Family)

DESCRIPTION: Small, flat-topped tree to 40′ tall; leaves large, bipinnately compound, and feather-like; leaves and leaflets droop as light diminishes in the evening; flowers clustered in fluffy, pink heads; flowers April–May.

RANGE-HABITAT: A native of Asia; now found essentially throughout the southeastern states; common throughout SC in a variety of disturbed sites such as roadsides, woodland borders, and around abandoned home sites.

COMMENTS: Mimosa is not a true mimosa, which is in the genus Mimosa L. Mimosa is cultivated for its wide-spreading crown, showy flowers, and graceful leaves. It was introduced by French botanist André Michaux. It has become a serious weed in the Southeast and invades open habitats. In the Piedmont and mountains it is considered an invasive exotic. The genus honors eighteenth century Florentine botanist, Filippo degli Albizzi, who introduced this genus into European cultivation. Fernald (1950) states the correct spelling should be Albizzia.

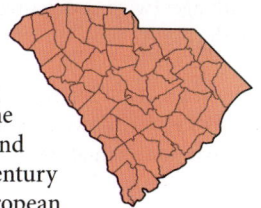

940. China-berry; Pride-of-India

Melia azedarach L.
Mé-li-a a-zé-da-rach
Meliaceae (Mahogany Family)

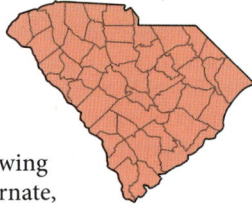

DESCRIPTION: Small- to medium-size tree, producing abundant root suckers; fast growing but short-lived; twigs aromatic; leaves alternate, bipinnately compound; inflorescence a panicle; flowers April–May; fruit a drupe, yellow, matures September–October.

RANGE-HABITAT: Native to Asia but naturalized throughout the Southeast; common throughout SC, except at high altitudes; woodland borders and fencerows, around abandoned homes, and in vacant lots and old fields.

COMMENTS: French botanist André Michaux introduced China-berry into North America, planting it in his Charleston garden. From this and other gardens, it escaped and became naturalized. China-berry quickly came to have many uses. It was used as a vermifuge to expel worms from the body, and broken branches were placed in a house to keep out fleas. Its use as a fleabane is based on chemicals in the wood that repel insects. Both the green and ripe fruits of China-berry are poisonous, although poisoning is rare because of the bitter taste of the fruits. Birds eat the over-ripe fruits, causing a mild intoxication (from alcohol in the fermented fruits) and temporary paralysis if too many are consumed. A complete account of the poisonous nature of China-berry can be found in Kingsbury (1964).

941. Princess Tree; Paulownia

Paulownia tomentosa
(Thunberg) Siebold &
Zuccarini ex Steudel
Pau-lòw-ni-a to-men-tò-sa
Paulowniaceae (Paulownia Family)

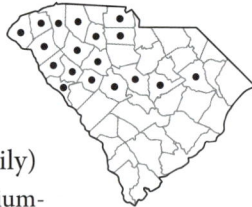

DESCRIPTION: Fast-growing, small- to medium-size tree; leaves deciduous, opposite, ovate, entire or slightly lobed, 6–12″ long; flowers large, blue-purple in an open panicle, produced before the leaves, in April–May; fruit is a two-locular capsule, matures September–October.

RANGE-HABITAT: Native to China but now naturalized throughout the eastern US as far north as Boston, MA, and west to TX; widely scattered throughout SC in open woodlands, along roadsides and fencerows, and in other disturbed areas.

COMMENTS: Princess Tree is native to central China and was introduced into North America in 1834 as an ornamental tree because of its large, upright clusters of purple flowers. It has spread readily from cultivation along roadsides and stream sides because its tiny, winged seeds are blown a considerable distance. It is considered an invasive exotic. Its ability to invade native woodlands is a serious concern. It has been cultivated in Japan for several hundred years for use in making wooden shoes and expensive dower chests. The genus honors Anna Pavlovna (1795–1865), a princess of the Netherlands.

942. Canary Island Salt Cedar; Canary Island Tamarisk

Tamarix canariensis Willdenow
Tá-ma-rix ca-na-ri-én-sis
Tamaricaceae (Tamarisk Family)

DESCRIPTION: Shrub or small tree to 25′ tall with evergreen, awn-shaped, gray-green leaves; the young leaves are papillose (covered with bumps); branches flexible; flowers 0.1–0.2″ wide, produced in slender racemes; flowers April–July.

RANGE-HABITAT: A native of southern Europe, Africa, and the Canary Islands; naturalized throughout much of the eastern US and to a lesser extent into the western states; in SC, generally confined to the maritime strand; occasional and locally abundant; sandy roadsides, dredged soil disposal sites, old fields, and margins of salt and brackish marshes.

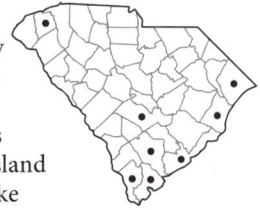

COMMENTS: The common name, Salt Cedar, refers to its tolerance of salty soils and its small, leafy twigs resembling forms of Southern Red Cedar. Canary Island Salt Cedar is a flowering plant (angiosperm), not a conifer (gymnosperm), like Southern Red Cedar. It is cultivated as an ornamental and is used for windscreens and erosion control. Interestingly, the nectar from some species in the desert regions of the Middle East can form thin wafers as it dries on the sand. These wafers are edible and are still referred to by locals as "manna." At least three species of Tamarix are known from SC; this species is by far the most common. Species can be very difficult to distinguish. It was erroneously included as French Tamarisk (*T. gallica* L.) in the *Guide to the Vascular Flora of the Carolinas* (Radford, Ahles, and Bell, 1968), the first edition of the *Guide to the Wildflowers of South Carolina* (Porcher and Rayner, 2001), and most other regional floras.

943. Popcorn Tree; Chinese Tallow Tree

Triadica sebifera (L.) Small
Tri-á-di-ca se-bí-fe-ra
Euphorbiaceae (Spurge Family)

SYNONYM: *Sapium sebiferum* (L.) Roxburgh—RAB, PR

DESCRIPTION: Small- to medium-size tree to 50′ or taller with milky sap; leaves alternate, with a pair of glands near the base of the blade; flowers produced in long, slender spikes, with male flowers above and female flowers toward the base; fruits are capsules, with the capsule walls falling away at maturity and exposing the white seeds; flowers May–June; fruits mature August–November.

RANGE-HABITAT: Native to China; naturalized from NC to FL and west to TX and OK; in SC, primarily in the Outer Coastal Plain; common in moist areas in maritime forests, field margins, barnyards, impoundments, ditch banks, and disturbed areas in coastal cities.

COMMENTS: Popcorn Tree was introduced into the colonies as an ornamental or shade tree as early as the 1700s. It quickly became naturalized and is a serious weed in many areas because it has adapted to a wide range of soil types. It is unfortunately still cultivated because of the brilliant yellow to red leaves in the fall. The Chinese used the waxy coating on the seeds to make soap and candles. All parts contain a poisonous milky juice. The common name, Popcorn Tree, alludes to the cluster of white seeds, exposed when the capsule opens, that look like popcorn. In the SC Coastal Plain, the seed clusters are used in Christmas decorations.

944. Chinese Privet

Ligustrum sinense Loureiro
Li-gús-trum si-nén-se
Oleaceae (Olive Family)

DESCRIPTION: Shrub or small tree; leaves evergreen or somewhat deciduous in severe winters or in colder parts of its range; leaves opposite, simple, short-stalked; blades mostly elliptic; flowers white, with a disagreeable odor; flowers May–June; fruits are bluish or black drupes, mature September–November.

RANGE-HABITAT: Native to China; widely naturalized throughout the Southeast; common throughout SC; moist forests, especially alluvial bottomlands, fencerows, wet thickets, well-drained and poorly drained places, and around abandoned house sites.

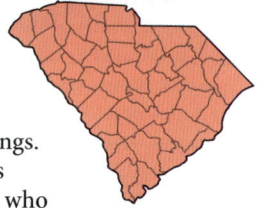

COMMENTS: Chinese Privet has presumably spread as a result of birds feeding on the fruits of ornamental plantings and dispersing the seeds in their droppings. It is one of SC's most invasive exotics, choking out native species in thousands of square miles. How quickly it has spread can be inferred from Small (1933), who stated that it "occurs as an escape in S La." Once established in a site, it is difficult to eradicate by mechanical or biological methods; chemical treatment is the only effective method.

945. Tree-of-heaven

Ailanthus altissima (Miller) Swingle
Ai-lán-thus al-tís-si-ma
Simaroubaceae (Quassia Family)

DESCRIPTION: Fast-growing tree (up to 10′ in a season), short-lived, to 80′ tall; often colonizing by root sprouts; leaves alternate, odd-pinnately compound; leaflets 15–27, entire except for 1–5 rounded, basal teeth, each with a prominent dark green gland; male and female flowers on separate trees, on the same tree, or with perfect flowers; flowers late May–early June; fruits mature July–October.

RANGE-HABITAT: Native to east Asia; naturalized from MA and Ontario south through the eastern US; to a lesser degree from the southern Rocky Mountains to CA; common throughout SC; in wind-throw gaps in forests, along railroad embankments and woodland borders, and in a variety of other disturbed sites with high light intensities.

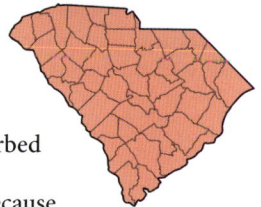

COMMENTS: Tree-of-heaven was introduced from East Asia as an urban tree because of its rapid growth and ability to withstand the stresses of cities. However, it is difficult to eradicate because it spreads by copious winged seeds and root sprouts. It is a serious invasive exotic because it often colonizes windthrow gaps in forests, where it replaces the native vegetation. Another unpopular feature is the unpleasant odor produced by the male flowers and crushed foliage. The tree is of little value to wildlife and has no commercial value. It is easily downed by snow and heavy winds.

946. Southern Catalpa

Catalpa bignonioides Walter
Ca-tál-pa big-no-nio-oì-des
Bignoniaceae (Bignonia Family)

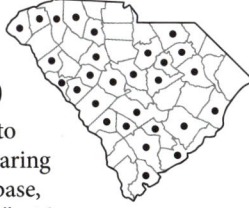

DESCRIPTION: Small- to medium-sized tree to
65′ tall; leaves deciduous, opposite, or appearing
whorled, simple, very broad, widest at the base,
abruptly acuminate at the tip; corolla to 1.6″ wide,
the lower lip petal undulate but not notched with purple spots and
lines; flowers May–June; capsule cylindrical, 12″ or more long, fruits
mature July–August.

RANGE-HABITAT: Native from SC south to FL and west to LA; widely
introduced and escaping elsewhere; in SC, as a native along the
margins of larger rivers such as the Santee and Savannah in the
Coastal Plain and lower Piedmont; escaped and naturalized elsewhere in moist, disturbed habitats or persisting from cultivation.

SIMILAR SPECIES: A similar species is Northern Catalpa (*C. speciosa*
(Walter) Warder ex Engelmann). It is native to the midwest-
ern United States but is now naturalized throughout the eastern states. It is widely scattered
throughout SC in a variety of disturbed habitats and occasionally cultivated. The two species
differ in the following details: Southern Catalpa has flower clusters with many flowers, each
flower 1.2–1.6″ wide, lower lip of the petals without a notch, leaves short-pointed at the tip,
fruits .23–0.4″ wide, and foliage with a fetid odor. Northern Catalpa has flower clusters with few
flowers, each flower 2.4–2.8″ wide, lower lip of the petals with a notch, leaves long-pointed at
the tip, fruits 0.4–0.6″ wide, and fresh foliage that is mainly odorless.

COMMENTS: Southern Catalpa has long been considered an escaped species native to the Gulf
Coastal Plain. Research by Patrick McMillan, Amy Blackwell, and Chris Blackwell revealed
that the species was collected by Mark Catesby in 1723–24 along the Savannah River, according to Catesby "a long distance from the settlements" (McMillan et al., 2013). This information, as well as the fact that a specimen from SC was used by SC resident Thomas Walter
to write the species description, has led these authors to conclude that the plant should be
considered native. The big black caterpillars (catalpa "worms") that feed on the leaves are a
nuisance on cultivated trees. However, fisherman prize them as bait.

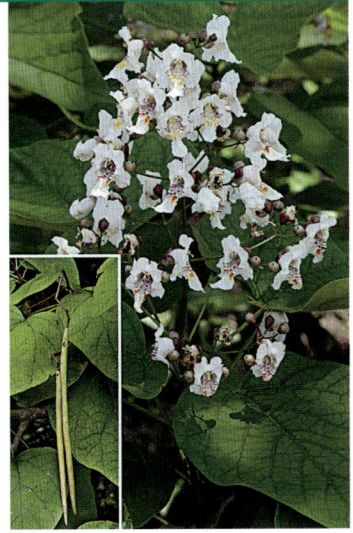

947. Smooth Sumac

Rhus glabra L.
Rhùs glà-bra
Anacardiaceae (Cashew Family)

DESCRIPTION: Rhizomatous shrubs or small trees to
20′ tall; stems smooth and often glaucous; leaves
deciduous, pinnately compound, lower leaf surface
glaucous, rachis not winged; flowers greenish white,
in dense erect, terminal panicles; flowers late May–
July; fruits are red drupes, mature June–October.

RANGE-HABITAT: Widespread in eastern North
America; in SC, common in the mountains, Pied-
mont, Sandhills, and Inner Coastal Plain; along roadsides and in pastures,
meadows, thickets, woodland borders, and other disturbed sites.

COMMENTS: Because sumacs rapidly form dense colonies, they are well suited
as a landscape plant to halt erosion on sloping road cuts and fills. They are used as
a fall ornamental because of their bright red leaves. Smooth Sumac has countless
uses as folk remedies, both to Native Americans and early European colonists. Native
Americans used the berries to stop bed-wetting, and in Appalachia the leaves are rolled
and smoked as a treatment for asthma. The fruits of Smooth Sumac are sticky and sour and have been
used as a gargle for sore throats. See Winged Sumac (plate 948) below for additional comments.

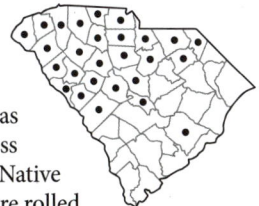

948. Winged Sumac

Rhus copallina L.
Rhùs co-pal-lì-na
Anacardiaceae (Cashew Family)

DESCRIPTION: Rhizomatous shrub or small tree, 20–25′ tall; stems densely short-pubescent; leaves deciduous, alternate, pinnately compound; rachis winged; leaves turn bright red in fall; male and female flowers in clusters on different plants; flowers July–September; fruits are dark red drupes, ripening in the fall.

RANGE-HABITAT: NY south to FL, west to TX, and north to KS and WI; common throughout SC in oak-hickory and pine-mixed hardwood forests; Longleaf Pine flatwoods; and disturbed sites such as fencerows, roadsides, thickets, pastures, and old fields.

COMMENTS: These comments apply to both Smooth and Winged Sumac. The hairs on the surface of the drupes contain malic acid, a pleasant tasting acid. Native Americans, then European settlers, used the fruit as the source of a cool, summer drink. It is prepared by bruising the fruits in water to free the acid, then straining the water to remove the hairs. Sugar can be added. The resulting drink is similar to pink lemonade. The fruits are rich in vitamin A and are a valuable food source for birds and other wildlife in winter when other fruits are scarce. Both sumacs are native species that exploit disturbed habitats.

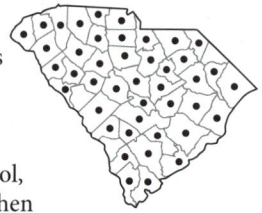

949. Cherokee Rose

Rosa laevigata Michaux
Rò-sa lae-vi-gà-ta
Rosaceae (Rose Family)

DESCRIPTION: Robust, high-climbing, evergreen vine; stems smooth but with prickles that are curved and flattened and have a broad base; leaves trifoliolate; leaflets with prickles on larger veins; flowers late March–April; fruit a red hip, matures September–October.

RANGE-HABITAT: Native to Asia; SC south to FL and west to MS; in SC, common in the Coastal Plain and maritime strand and occasional elsewhere; low woods, roadsides, and abandoned homesites.

COMMENTS: Cherokee Rose is native to China and Japan. It was naturalized very early in the colonial era in the South. Though native to Asia, it was first described by Michaux from plants collected in America. It is the state flower of GA.

950. Multiflora Rose

Rosa multiflora Thurnberg
Rò-sa mul-ti-flò-ra
Rosaceae (Rose Family)

DESCRIPTION: Thorny shrub, often climbing over other vegetation, to 6–15′ tall; branches vigorous, 10′ or so long, forming nearly impenetrable thickets; prickles curved, flattened; leaves pinnately compound; leaflets usually 7–9; flowers white; flowers May–June; fruit a red hip, matures September–October.

RANGE-HABITAT: Native to Asia; naturalized from NY south to FL and west to TX; in SC, common throughout in forests, along fencerows and roadsides, and in pastures.

COMMENTS: Multiflora Rose was once planted as a living hedge and provides excellent wildlife cover. It is, however, a pest in many areas, spreading into fields and pastures and rich bottomland forests as an invasive exotic.

951. Japanese Honeysuckle

Lonicera japonica Thunberg
Lo-níc-er-a ja-pó-ni-ca
Caprifoliaceae (Honeysuckle Family)

DESCRIPTION: Left to right twining, woody vine with opposite, evergreen leaves; often climbing to 30′ or more; flowers sweet-smelling, in pairs from the leaf axils; flowers April–June and sporadically into September; fruit a black berry, matures in August–October.

RANGE-HABITAT: A native of Asia; widely naturalized and abundant throughout SC in almost any disturbed habitat including woodlands, fields, fencerows, thickets, abandoned buildings, and along railroad banks; it also invades intact, rich forest systems, often as a low, creeping groundcover.

COMMENTS: Japanese Honeysuckle quickly invades any disturbed opening in native woodlands, sometimes replacing native flora. Birds carry the seeds, which allows it to spread rapidly. It is one of SC's most problematic invasive exotics. The species can persist and even expand in relatively undisturbed woodlands and is a serious threat to SC native forest ecosystems.

When pulled free of the flower, a sweet nectar can be sucked from the base of the stigma–style.

952. Cow-itch; Trumpet Vine

Campsis radicans (L.) Seemann ex Bureau
Cámp-sis ra-dì-cans
Bignoniaceae (Trumpet-creeper Family)

DESCRIPTION: Deciduous, woody vine, trailing or high-climbing by means of 2 short rows of aerial roots from the nodes; sometimes climbing over 100′; stem with yellowish and shreddy bark; leaves opposite, pinnately compound with 7–15 leaflets; flowers trumpet-shaped, red to orange with a yellow throat, produced in terminal cymes of 4–12; flowers June–July.

RANGE-HABITAT: NJ south to FL and west to OH and TX; common throughout SC; swamps, bottomlands, and woodlands, along fencerows and roadsides, and in vacant lots and yards.

COMMENTS: Cow-itch is a native vine that has exploited disturbed habitats throughout its range. It is often cultivated for its attractive flowers, which attract and are pollinated by Ruby-throated Hummingbirds. Contact with Cow-itch may cause skin inflammation and blisters in sensitive people–hence, the common name.

953.   Kudzu

Pueraria montana (Loureiro) Merrill
   var. *lobata* (Willdenow) van der
   Maesen & S. Almeida
   Pu-e-rà-ri-a lo-bà-ta
   Fabaceae (Bean Family)

SYNONYM: *Pueraria lobata* (Willdenow)
   Ohwi—RAB, PR

DESCRIPTION: Trailing or climbing, robust, pubescent, semi-woody vine up to 90′ long, from a
large rootstock; leaves pinnately 3-foliolate; flowers reddish purple, grape scented; flowers July–October.

RANGE-HABITAT: PA to FL and west to TX and AR; common throughout SC;
along roadsides; around abandoned home sites; and in forest margins, fields,
and other disturbed areas.

COMMENTS: Kudzu was introduced from east Asia into the southeastern US to
stabilize eroded areas. It can quickly dominate disturbed areas and kill off competing vegetation by blocking sunlight. Since it can climb to the tops of tall trees, entire
forests can be destroyed. Fortunately, it does not invade very deeply into natural areas
as readily, because of the high light requirement. The leaves and younger stems are highly sensitive to
frost; older stems tend to be resistant to the coldest winters in the South. The medicinal uses of Kudzu
are too numerous to list. The rhizome is a good source of starch and is used as food in Asia. The genus
is named in honor of M. W. Puerari (1765–1845), a Swiss botanist.

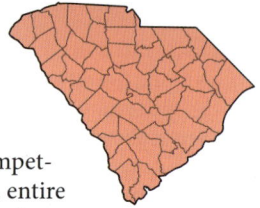

954.   Common Chickweed

Stellaria media (L.) Villars
   Stel-là-ri-a mè-di-a
   Caryophyllaceae (Pink Family)

DESCRIPTION: Herbaceous, prostrate or creeping
annual, 3–8″ tall; usually dying completely by
June; stems with hairs in lines; leaves opposite;
petals 5, white, each deeply cleft and appearing
to be 10; flowers January–May, but occasionally
throughout the year.

RANGE-HABITAT: Native to Eurasia; naturalized
throughout the world; common throughout SC in
lawns, fields, gardens, and meadows and along roadsides.

COMMENTS: Chickweed often invades natural communities but not to the
exclusion of native vegetation. It has long been used as a potherb; the young,
tender, growing tips can be eaten. It is a favorite food of chickens and wild
birds. Duke (1997) recommends chickweed as an herbal remedy for the treatment of obesity.

955. Henbit

Lamium amplexicaule L.
   Là-mi-um am-plex-i-caù-le
Lamiaceae (Mint Family)

DESCRIPTION: Winter annual to 14″ tall; stems soft, 4-angled, freely branched from the decumbent base; flowering stems erect; leaves roundish to ovate, opposite, often shallowly 3-lobed; flower clusters in the axils of opposite, sessile, leaflike bracts; flowers mid-winter to May.

RANGE-HABITAT: Native of Eurasia; naturalized in eastern North America and the Pacific Coast; throughout SC; pastures, abandoned fields, lawns, gardens, and almost any other disturbed site.

SIMILAR SPECIES: Purple Dead-nettle (*L. purpureum*) is similar and often co-occurs with Henbit. It can be separated by the fact that all the leaves are petiolate. It is also a native of Eurasia and naturalized throughout SC.

COMMENTS: The specific epithet, amplexicaule, refers to the sessile leaflike bracts that subtend the flowers, appearing to grasp the stem. The young plants have been used as a potherb in the US and Japan. Kingsbury (1964) reports that it caused "staggers" in sheep, cattle, and horses. Eating large quantities is not recommended.

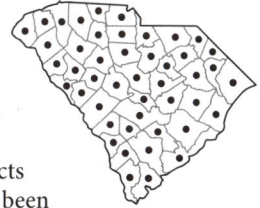

956. Common Shepherd's Purse

Capsella bursa-pastoris (L.)
   Medikus
   Cap-sél-la búr-sa-pas-tò-ris
Brassicaceae (Mustard Family)

DESCRIPTION: Winter annual with one main stem and ascending branches; leaves basal and on the stem, reduced upward; flowers white, in a raceme to 12″ long in fruit; fruits said to be in the shape of the purse European shepherds once hung from their belts; seeds reddish brown; flowers February–June or all year long if weather permits.

RANGE-HABITAT: Introduced from Europe and now a weed throughout the world; common throughout SC in fields, lawns, along roadsides, and in many other disturbed habitats.

COMMENTS: James and Patricia Pietrepaolo (1986) describe shepherd's purse seeds as carnivorous. The seeds are covered in a thin layer that attracts protozoans, nematodes, and motile bacteria and then kills them. The seed covering then releases enzymes that breaks down the dead organisms, providing nutrients for the germinating seed.

Numerous sources refer to the plant's ability to stop bleeding. F. P. Porcher (1869) reports the juice of the plant, when placed on a cotton ball, was used to plug nostrils to help stop nosebleeds. Dried or fresh herb tea, made from the seeds and leaves, was used to allay profuse menstrual bleeding. The dried plant was also a useful styptic against hemorrhage. The dried, ground seeds can be used as a substitute for pepper, and the young leaves can be cooked like spinach.

957. Mock Strawberry;
Indian Strawberry

Potentilla indica (Andrews) T. Wolf
Po-ten-tíl-la ín-di-ca
Rosaceae (Rose Family)

SYNONYM: *Duchesnea indica* (Andrews)
Focke—RAB, PR

DESCRIPTION: Low, trailing, perennial herb with
stolons; leaves trifoliolate; flowers yellow; flowers
and fruits February–frost; fruit an aggregate of
achenes embedded in a red, fleshy receptacle.

RANGE-HABITAT: Native of India; naturalized from
CT south to northern FL and west to OK; common throughout SC; found in
lawns, pastures, and open woods and along roadsides.

COMMENTS: Mock Strawberry is often mistaken for Wild Strawberry when in
fruit. Although the fruits appear edible, they are flat and tasteless. It is not a true
strawberry, which belongs to the genus Fragaria.

958. Common Dandelion

Taraxacum officinale G. H. Weber ex
Wiggers
Ta-ráx-a-cum of-fi-ci-nà-le
Asteraceae (Aster Family)

DESCRIPTION: Perennial herb with milky sap; root
thick, deep, bitter tasting; leaves basal, deeply and
irregularly lobed and toothed; flowering stalks
hollow, erect, 2–18″ tall; flowers produced in
solitary involucrate heads of yellow ray flowers;
flowers December–June.

RANGE-HABITAT: Native of Eurasia; naturalized
nearly throughout the US and southern Canada; common throughout SC in
lawns, pastures, vacant lots, and fallow fields and along roadsides.

COMMENTS: Common Dandelion is a native of Eurasia. In the Old and New
World, the leaves are used as a potherb, and the ground roots can be used to
make a palatable, bitter drink. The roots can be cooked for food, and a strong
wine can be made from the flowers and leaves.

The specific epithet officinale, meaning "of the shops," indicates the medicinal
uses of this versatile plant. For examples, it is used as a laxative and to alleviate symptoms of
arthritis.

959. Early Winter-cress; Creasy

> *Barbarea verna* (P. Miller)
>   Ascherson
>   Bar-ba-rè-a vér-na
>   Brassicaceae (Mustard Family)

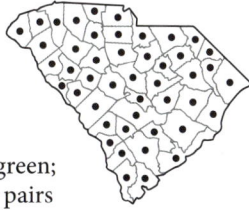

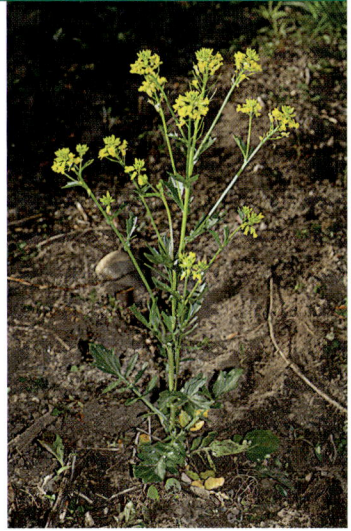

DESCRIPTION: Erect herb to 32″ tall; stems green; leaves pinnately dissected, each with 4–10 pairs of lateral lobes; leaf stalks ciliate; flowers March–June.

RANGE-HABITAT: Native of Eurasia; naturalized practically throughout the US; in SC, common throughout; along roadsides and in fields and disturbed areas.

SIMILAR SPECIES: Early Winter-cress is similar to Common Winter-cress (*B. vulgaris* R. Brown). Early Winter-cress has basal leaves with 4–10 pairs of lateral lobes, whereas Common Winter-cress has basal leaves with 1–4 pairs of lateral lobes.

COMMENTS: The young foliage and new young stems, while still tender, are a good potherb. They should be boiled twice or more with water changes. The first water removes the strongest bitters. People of the mountains and upper Piedmont who eat the plant call it "creasy greens."

960. Common Yellow Thistle; Southern Bull Thistle

> *Cirsium horridulum*
>   Michaux var. *horridulum*
>   Cír-si-um hor-rí-du-lum var.
>   hor-rí-du-lum
>   Asteraceae (Aster Family)

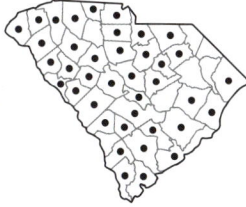

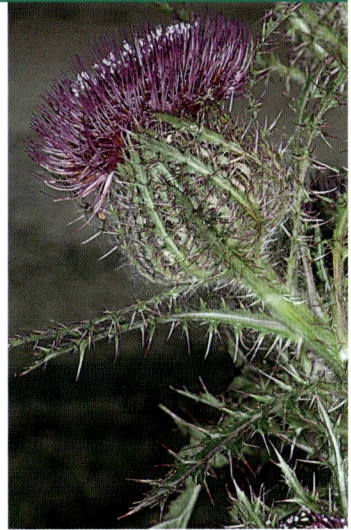

> **SYNONYM:** *Carduus spinosissimus*
>   Walter—RAB

DESCRIPTION: Biennial herb to 5′ tall; stems covered with cobweb-like hairs; leaves spiny on the margin, pinnately lobed, stalkless, and clasping the stem; flowers arranged in involucrate heads of purple to light yellow disk flowers; flowering heads subtended by an involucre of narrow, spiny-toothed bracts; flowers late March–early June.

RANGE-HABITAT: ME south to FL and west to TX; in SC, common throughout (except the mountains); found in Longleaf Pine flatwoods, forest margins, along roadsides and in fields, meadows, and other disturbed sites.

TAXONOMY: Two forms occur, yellow-flowered and purple-flowered; the former is restricted to the Coastal Plain, and the latter grows throughout SC.

COMMENTS: This is a native species that has successfully exploited disturbed sites.

961. Fumitory; Earth-smoke

Fumaria officinalis L.
  Fu-mà-ri-a of-fi-ci-nà-lis
Fumariaceae (Fumitory Family)

DESCRIPTION: Erect, branching, annual herb; stem and branches 8–40″ long; leaves finely dissected; flowers in racemes; corolla purplish and crimson at the tip; one petal with a spur at base; flowers March–May.

RANGE-HABITAT: Naturalized from Europe and widespread in the eastern US; in SC, occasional in the maritime strand and Coastal Plain in fields and along roadsides and other open, sunny, disturbed sites.

COMMENTS: The specific epithet officinalis means "of the shops," in reference to its early repute in medicine. The genus name comes from the Latin words *fumus*, "smoke," and *terrae*, "of the earth." The smoke name apparently has many explanations. One alludes to the gray-green color of the plant that from a distance has a smoky appearance. Another explanation is in reference to the nitrous odor of the roots when they are first pulled from the ground.

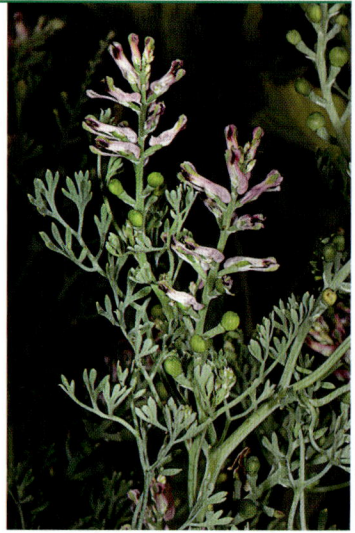

962. Common Toadflax

Linaria canadensis (L.) Dumont de
  Corset
  Li-nà-ri-a ca-na-dén-sis
Plantaginaceae (Plantain Family)

SYNONYM: *Nuttallanthus canadensis* (L.)
  D. A. Sutton—PR

DESCRIPTION: Winter annual or biennial herb with numerous prostrate stems; flowering stems erect, to 30″ tall, slender, with linear, alternate leaves; opposite leaves, radiate from the base of the upright stem; conspicuous spur projects down from the corolla; flowers March–May.

RANGE-HABITAT: Common throughout the US; throughout SC; in a wide variety of natural and disturbed habitats such as fallow fields, roadsides, lawns, pastures, and vacant lots.

SIMILAR SPECIES: Texas Toadflax (*L. texana* Scheele) is very similar but has larger flowers and is also found throughout the state. The two species are very difficult to tell apart if not growing together. The seeds of Texas Toadflax are tuberculate rather than having raised lines, but this requires microscopic examination.

COMMENTS: This species is probably native to the thin soils of rock outcrops but has expanded its range into many disturbed habitats. Toadflax and Sourgrass (*Rumex hastatulus* Baldwin, plate 965) often grow together in fallow fields where they form a colorful mix of red and blue in early spring. Both species are native but weedy.

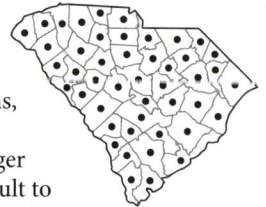

963. Running Five-fingers

Potentilla canadensis L.
    Po-ten-tíl-la ca-na-dén-sis
Rosaceae (Rose Family)

DESCRIPTION: Low, stoloniferous, herbaceous perennial with short, usually erect, rhizomes; stolons elongating to 20″ or more; leaves palmately compound with 5 leaflets; flowers solitary on axillary stalks from the stolons; first flower is in the axil of first fully developed leaf or of an undeveloped leaf below it; flowers March–May.

RANGE-HABITAT: Nova Scotia south to GA and west to TN, MO, and OH; in SC, common throughout in forest margins, pastures, lawns, and disturbed sites and along roadsides.

SIMILAR SPECIES: Old Field Five-fingers (*P. simplex* Michaux) is similar but has the first flower in the axil of the second fully developed leaf. It occurs throughout SC in similar habitats and flowers April–June.

COMMENTS: Native Americans used a tea of the pounded roots as an astringent to treat diarrhea.

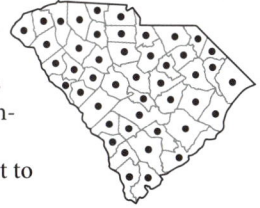

964. Wild Radish; Jointed Charlock

Raphanus raphanistrum L.
    Rá-pha-nus ra-pha-nís-trum
Brassicaceae (Mustard Family)

DESCRIPTION: Coarse winter annual from a taproot; stems erect, to 2′ tall, freely branched, with bristlelike hairs; leaves reduced upward, basal ones deeply dissected; petals sulfur yellow, distinctly veined, fading white (rarely purple); flowers March–June and sporadically throughout rest of the year; fruit a silique, constricted between seeds.

RANGE-HABITAT: Native to Europe; naturalized from Newfoundland south to GA and west to British Columbia and CA; common nearly throughout SC but less so in the mountains; cultivated fields, roadsides, and most other disturbed areas.

COMMENTS: Wild Radish is naturalized from Mediterranean Europe. Kingsbury (1964) states that Wild Radish has been considered dangerous to livestock in Europe and the United States.

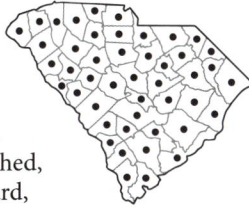

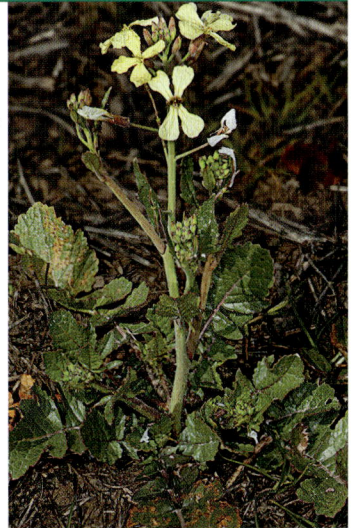

965. Sourgrass; Wild Sorrel

Rumex hastatulus Baldwin ex Elliott
    Rù-mex has-tá-tu-lus
Polygonaceae (Buckwheat Family)

DESCRIPTION: Annual or short-lived perennial with a taproot, to 4′ tall, but typically less than 2′ tall; leaves arrow-shaped with spreading, basal lobes; stems single or in large clumps; male and female flowers on separate plants; flowers March–May.

RANGE-HABITAT: MA south to FL and west to TX and KS; a native species that has become weedy; common throughout SC; sandy fields, roadsides, vacant lots, lawns, and sandy, open woods.

SIMILAR SPECIES: A similar species, Sheep-sorrel (*R. acetosella L.*), is a native of Eurasia and occurs throughout SC in similar habitats. Sheep-sorrel is a perennial with rhizomes, while Sourgrass lacks rhizomes.

COMMENTS: Sourgrass often grows in association with Common Toadflax (*Linaria canadensis* (L.) Dumont de Corset, plate 962) in fallow fields, where these two species form a colorful mix of red and blue. The stem of Sourgrass has an acid taste, caused by oxalic acid, and is often chewed as a trail nibble. Eating large amounts, however, can result in poisoning because of the oxalic acid.

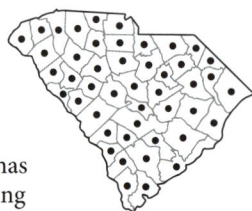

966. Stiff Verbena; Veiny Vervain

Verbena rigida Sprengel
Ver-bè-na rí-gi-da
Verbenaceae (Vervain Family)

DESCRIPTION: Erect perennial 4–28″ tall, with elongate, underground stolons; often forming large patches; stems and leaves rough-hairy; leaves opposite, simple; spikes stiffly erect; flowers late March–July.

RANGE-HABITAT: Native to South America; naturalized from VA south to FL and west to TX; in SC, common in the lower Piedmont, Coastal Plain, and maritime strand; along roadsides and in fields, pastures, and other open, sunny, disturbed sites.

COMMENTS: Stiff Verbena is drought resistant and makes an attractive ornamental.

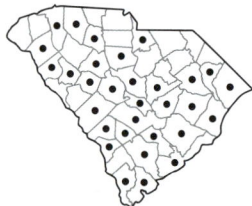

967. Narrowleaf Vetch

Vicia sativa L. ssp. *nigra* (L.) Ehrhart
Ví-ci-a sa-tì-va ssp. nì-gra
Fabaceae (Bean Family)

SYNONYM: *Vicia angustifolia* Richard—RAB, PR

DESCRIPTION: Annual herb with decumbent to ascending stems, more or less climbing; leaves pinnately compound with the terminal leaflet usually modified into a branched tendril; inflorescence is nearly sessile with 1–4 flowers produced from the axil of the leaves; flowers March–June.

RANGE-HABITAT: A native of Europe, naturalized nearly throughout the eastern US; common throughout SC along roadsides and fencerows and in lawns, fields, and other waste places.

COMMENTS: The seeds of vetches, like many members of the legume family, are an important source of food for animals. It is occasionally used for forage. Vetch species are often utilized as cover crops to increase soil nitrogen as they are capable of fixing atmospheric nitrogen. Species of vetches, many cultivated, have occasionally been reported as producing disease or loss of life in livestock and humans (Kingsbury, 1964). The seeds have been reported to contain cyanide.

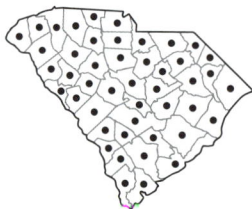

968. Wild-pansy; Johnny-jump-up

Viola bicolor Pursh
Vì-o-la bì-co-lor
Violaceae (Violet Family)

SYNONYM: *Viola rafinesquii* Greene—RAB, PR

DESCRIPTION: Winter annual, about 4–12″ tall; stems simple or branched above; leaves taper gradually to the stem, with large stipules cleft into narrow lobes (like a cock's-comb); lower 3 petals with purple veins; petals greatly exceed the sepals; flowers March–April.

RANGE-HABITAT: Native and widely distributed in the Southeast; in SC, common throughout along roads, in lawns, pastures, and other disturbed sites.

SIMILAR SPECIES: European Field-pansy (*V. arvensis* Murray) is similar; it is also a winter annual but has small cream-colored petals that are shorter than the sepals. It is a native of Europe but is naturalized throughout SC and often co-occurs with Wild-pansy.

COMMENTS: Wild-pansy often forms dense but ephemeral populations, especially along roadsides. The root contains methyl salicylate, which imparts a wintergreen smell and taste. Methyl salicylate is poisonous if ingested. Methyl salicylate is the active ingredient in some commercial muscle rubs.

969. Alligator-weed

Alternanthera philoxeroides (Martius) Grisebach
Al-ter-nán-the-ra phi-lox-e-roì-des
Amaranthaceae (Amaranth Family)

DESCRIPTION: Emergent, perennial, aquatic herb with creeping stems, rooting at the nodes or in free floating mats; stems to 3′ long; leaves opposite; flowers April–October.

RANGE-HABITAT: Native to tropical America; introduced and invasive from VA to TX; in SC, abundant in the Coastal Plain, maritime strand, Sandhills, and parts of the Piedmont; various freshwater habitats, such as tidal freshwater marshes, ditches, ponds, and swamps.

COMMENTS: Alligator-weed was introduced into the United States in the early 1950s, probably from ballast water or cargo from tropical America. One source states that viable seeds have not been found in the United States and that reproduction occurs only by vegetative means. It grows in a wide range of water and soil conditions. Mats of the plant can quickly block canals and ditches, reducing water flow and boat movement.

970. White Prickly-poppy

Argemone albiflora Hornemann var. *albiflora*
Ar-ge-mò-ne al-bi-flò-ra var. al-bi-flò-ra
Papaveraceae (Poppy Family)

DESCRIPTION: Annual, prickly herb, 1–3′ tall; sap milky white, but quickly turning yellow as it dries; leaves irregularly and coarsely pinnately lobed; flowers large and showy, petals white; flowers April-May.

RANGE-HABITAT: Presumptive native from NC south to FL, and west to LA; naturalized elsewhere in North America; in SC, common in the Coastal Plain and maritime strand in a variety of disturbed habitats such as roadsides, railroad right-of-ways, abandoned lots, and open, sandy fields.

COMMENTS: This showy, weedy species is apparently native in our region. A yellow-flowered species, Yellow Prickly-poppy (*A. mexicana* L.), is also known from similar habitats in our area. It is introduced from farther south and has become naturalized.

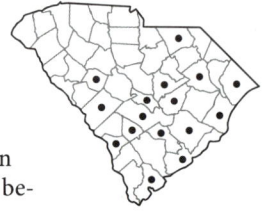

---

971. Ox-eye Daisy

Leucanthemum vulgare Lamarck
    Leu-cán-the-mum vul-gà-re
Asteraceae (Aster or Sunflower
    Family)

SYNONYM: *Chrysanthemum
    leucanthemum* L.—RAB, PR

DESCRIPTION: Perennial herb to 3′ tall, with short rhizomes; leaves alternate, the numerous basal leaves usually pinnately lobed or cleft; flower heads one or a few per stalk; flowers April–August.

RANGE-HABITAT: A native of Eurasia that is naturalized throughout the US; throughout SC, although less frequent in the southern part of SC; along roadsides, around buildings, in lawns, old fields, pastures, meadows, and other disturbed sites.

COMMENTS: It is used as an ornamental but can become a serious invasive exotic, especially in pastures and meadows. It can dominate fields and, if eaten by cattle, can impart an unwanted flavor to milk. Deer, rabbits, and other herbivores generally avoid eating Ox-eye Daisy.

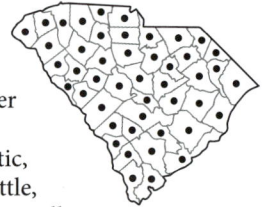

---

972. Clasping Heliotrope

Heliotropium amplexicaule M. Vahl
    He-li-o-trò-pi-um am-plex-i-caù-le
Boraginaceae (Borage Family)

DESCRIPTION: Perennial glandular-hairy herb from a strong rootstalk; stems several, spreading, creeping or ascending to 20″ tall; plant glandular-hairy; leaves alternate; flowers produced in a helicoid cyme (expanding from a coil, like a snail's shell); flowers April–September.

RANGE-HABITAT: A native of South America that is naturalized from NC south to FL and west to TX; in SC, common in the Piedmont, Coastal Plain, and maritime strand along city roadsides, in cultivated fields, roadsides, abandoned lots, and other disturbed sites.

COMMENTS: The genus name comes from the Greek *helios* (the sun) and *trope* (a turn). Ancient writers believed the flowers turn to follow the sun. All parts of the plant are toxic and cause problems in livestock if eaten in quantity.

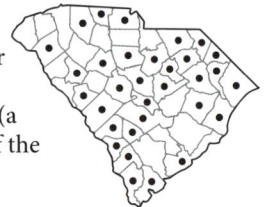

973. Spotted Cat's-ear

Hypochaeris radicata L.
     Hy-po-chaè-ris ra-di-cà-ta
Asteraceae (Aster Family)

DESCRIPTION: Perennial herb from stout roots and with milky sap; leaves all basal, coarsely pubescent, few-toothed or pinnately cut to near the midrib; flowering stems 12–24″ tall; flowers produced in involucrate heads of yellow ray flowers; flowers April–July and sporadically later.

RANGE-HABITAT: Native to Europe but widely naturalized throughout the US; in SC, common throughout along roadsides, in fields, lawns, and disturbed areas.

SIMILAR SPECIES: Smooth Cat's-ear (*H. glabra* L.) is similar and is naturalized throughout the Piedmont, Sandhills, Coastal Plain, and maritime strand in similar disturbed places. *H. radicata* is conspicuously hairy, while *H. glabra* is glabrous or apparently so.

COMMENTS: Cat's-ears are edible and can be used in salads or as a potherb.

974. Annual Phlox

Phlox drummondii Hooker var.
     peregrina Shinners
          Phlóx drum-mónd-i-i var. pe-re-
          grì-na
Polemoniaceae (Phlox Family)

DESCRIPTION: Erect, herbaceous annual, 4–28″ tall; stems glandular-hairy; lowermost leaves opposite, others alternate; flowers rose-red, pink, white, or variegated depending on the cultivar; flowers April–July.

RANGE-HABITAT: A native of TX and OK that has escaped eastward; in SC, common in the Coastal Plain and maritime strand where found around abandoned home sites; in lawns, stable dune areas, and meadows; and along sandy roadsides.

COMMENTS: There are numerous cultivated forms of Annual Phlox. The different forms are often found growing together, especially around abandoned home sites. Plants in SC are the progeny of various cultivars derived from hybrids and selections of the species.

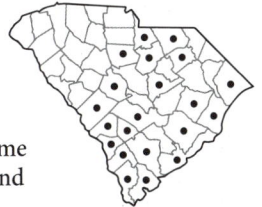

975. English Plantain

Plantago lanceolata L.
     Plan-tà-go lan-ce-o-là-ta
Plantaginaceae (Plantain Family)

DESCRIPTION: Perennial herb with a basal cluster of long, narrow, elliptic to lanceolate, strongly ribbed leaves; flowering stalks solid, 5-angled, 4–20″ tall; flowers in dense, cylindrical heads; flowers April–frost.

RANGE-HABITAT: Native to Europe; naturalized throughout the US; common throughout SC in disturbed areas, including lawns, vacant lots, pastures, and fields and along roadsides and railroad beds.

COMMENTS: English Plantain is often abundant in lawns. American Plantain (*P. rugelii* Decaisne) has elongate, more loosely flowered, cylindrical spikes and broad, elliptic leaves. It is also common throughout and is a native of the US that often also occurs in lawns. Though weedy, plantains are beneficial for wildlife. Birds eat the seeds, the leaves are food for rabbits, and the plants serve as a host for the native Buckeye butterfly. Duke (1997) recommends all plantain species as an herbal remedy for obesity and hemorrhoids. The crushed leaves of plantains are also used as a remedy for rashes caused by Poison Ivy.

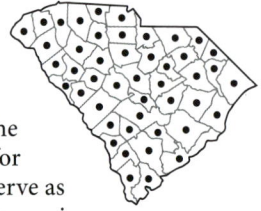

976. American Heal-all

Prunella vulgaris var. *lanceolata*
(W. Barton) Fernald
Pru-nél-la vul-gà-ris var. lan-ce-o-là-ta
Lamiaceae (Mint Family)

DESCRIPTION: Perennial herb, 6–12″ tall, with short branches below the central, terminal inflorescence; stems 4-angled, leaves opposite; flowers in globose spikes, but spikes becoming cylindrical as fruits mature; flowers April–frost.

RANGE-HABITAT: Newfoundland west to AR and south to NC, SC, TN, MO, KS, NM, AZ, and CA; common throughout SC in moist forests, roadsides, lawns, fields, and meadows.

SIMILAR SPECIES: Eurasian Heal-all (*P. vulgaris* L. var. vulgaris) was naturalized early on from Europe and is widespread in the US and in SC. They differ as follows: var. lanceolata has the main or median stem leaves lanceolate to oblong and wedge-shaped at the base; var. vulgaris has the principal or median stem leaves ovate to ovate-oblong and broadly rounded at the base.

COMMENTS: As the common names suggest, this plant has been used for a variety of folk remedies in the belief it could cure all ailments. For example, an infusion of the flowers and leaves has been used as a gargle for sore throat. Duke (1997) recommends its use as an herbal remedy for hyperthyroidism. With repeated mowing or grazing, the plants become stunted and matted, flowering when only 2″ tall.

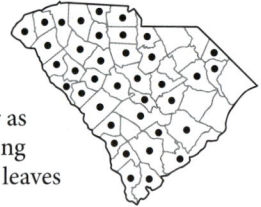

977. False-dandelion; Carolina Desert-chicory

Pyrrhopappus carolinianus (Walter)
Augustin de Candolle
Pyr-rho-páp-pus ca-ro-li-ni-à-nus
Asteraceae (Aster Family)

DESCRIPTION: Annual or short-lived perennial, with milky juice, from a well-developed taproot, 1 to 4′ tall; basal leaves pinnately lobed or dissected to merely toothed; upper leaves reduced; flowers arranged in an involucrate head of light-yellow ray flowers; heads often solitary on long stalks; flowers April–June.

RANGE-HABITAT: DE south to FL and west to KS and TX; common throughout its range and throughout SC (but less common in the mountains); along roadsides and in pastures, fallow and cultivated fields, lawns, and meadows.

COMMENTS: This attractive, native weedy species is particularly abundant along roadsides in the Coastal Plain. It can be confused with the non-native Spotted

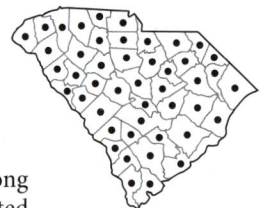

Cat's-ear (*Hypochaeris radicata* L., plate 973) but is a less pubescent plant, has lighter green foliage, much lighter colored ray flowers, and larger heads.

978. Bulbous Buttercup

Ranunculus bulbosus L.
Ra-nùn-cu-lus bul-bò-sus
Ranunculaceae (Buttercup Family)

DESCRIPTION: Erect perennial, 1–2′ tall, from a cormlike base; basal leaves 1–4″ wide, stalked, cut into 3-lobed or cleft parts; stem leaves smaller, fewer; flowers with tightly reflexed sepals; fruits are aggregates of achenes with beaks that are recurved or hooked; flowers April–June.

RANGE-HABITAT: A native of Europe; naturalized throughout North America; in SC, common in the mountains and Piedmont along moist roadsides, fields, gardens, lawns, and meadows.

COMMENTS: Kingsbury (1964) reports that several species of buttercups, including Bulbous Buttercup, are poisonous to animals such as hogs or cattle.

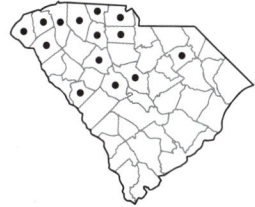

979. Crimson Clover

Trifolium incarnatum L.
Tri-fò-li-um in-car-nà-tum
Fabaceae (Bean Family)

DESCRIPTION: Erect, annual herb, 8–16″ tall; plant softly downy pubescent; leaves palmately 3-foliolate; petals scarlet, deep red, or rarely white; flowers April–June.

RANGE-HABITAT: Native to Eurasia; widely naturalized in the southeastern US; common throughout SC along roadsides and in fields.

COMMENTS: Crimson Clover is planted as a cover crop, for fodder, along roads as a soil binder, and for its attractive flowers. The fibrous calyx from overripe flowers may be dangerous to horses because it may become impacted in their digestive tract.

980. Venus' Looking-glass

Triodanis perfoliata (L.)
Nieuwland
Tri-o-dà-nis per-fo-li-à-ta
Campanulaceae (Bellwort Family)

SYNONYM: *Specularia perfoliata* (L.)
Alphonse de Candolle—RAB

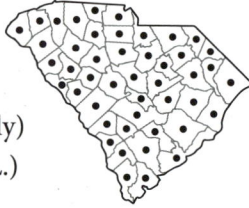

DESCRIPTION: Herbaceous, winter annual, 8–40″ tall; stems erect, freely branched at base, unbranched above; leaves alternate, sessile and clasping; flowers sessile in leaf axils; lower flowers do not open and are self-pollinating; upper several flowers open and are openly pollinated; flowers April–June.

RANGE-HABITAT: From southern Ontario and Quebec, south to FL, and west to TX and the Dakotas; also in tropical America; common throughout SC; in pastures, fields, along roadsides, in abandoned lots, and lawns; also in less disturbed habitats such as glades, open woodlands, and margins of rock outcrops.

SIMILAR SPECIES: Southern Venus' Looking-glass (*T. biflora* (R. & P.) Greene) is similar to and usually occurs mixed with *T. perfoliata*. Both species are unusual in dispersing their seeds through holes (pores) in the sides of the fruit (capsule). In *T. perfoliata* the pores of the capsule are at or below the middle, while in *T. biflora* the pores are near the top. Southern Venus' Looking-glass also typically produces only a single, open, sexually reproducing flower at the tip of the stem.

COMMENTS: Venus' Looking-glass is an attractive, weedy native. The native habitat is obscure but is presumed to be margins of rock outcrops or open woodlands.

981. Smooth Vetch

Vicia villosa Roth var. *varia* (Host)
Corbière
Ví-ci-a vil-lò-sa var. và-ri-a
Fabaceae (Bean Family)

DESCRIPTION: Annual or rarely perennial herb, trailing or climbing by tendrils; stems, flower stalks, and leaves smooth with age or with appressed pubescence; leaves alternate, pinnately compound with 10–20 leaflets; terminal leaflet modified into a branched tendril; racemes with flowers borne on one side; calyx bulging at base, the flower stalk appearing lateral; flowers May–September.

RANGE-HABITAT: Naturalized throughout the US; common throughout SC; fallow fields, along roadsides and fencerows, and in other open, disturbed sites.

SIMILAR SPECIES: A second variety, Hairy Vetch (*V. villosa* Roth var. *villosa*), also occurs naturalized throughout SC in similar habitats. The stem of Hairy Vetch is pubescent with spreading hairs; otherwise it is similar.

COMMENTS: Vetch can be poisonous to livestock and animals grazing on them develop granulomatous disease that can affect many organs.

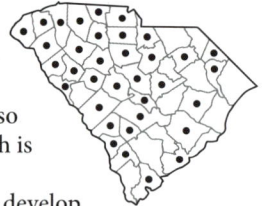

982. Florida Betony; Florida Hedge-nettle

Stachys floridana Shuttleworth ex
  Bentham
    Stá-chys flo-ri-dà-na
  Lamiaceae (Mint Family)

DESCRIPTION: Rhizomatous perennial, 8–20″ tall,
extensively colonial; rhizomes slender, becoming
tuberous and thickened at intervals; stems square;
leaves opposite, with long stalks; corolla whitish
to pale pink, with purple spots and lines; flowers
April–July.

RANGE-HABITAT: VA south to FL and west to TX;
in SC, common in the Coastal Plain and scattered locations in the Piedmont;
in a variety of habitats where open or semi-open moist areas exist, including
gardens and lawns, along roadsides, shrub thickets, and other moist, disturbed areas.

COMMENTS: Florida Betony's native range is obscure, and it was probably introduced from farther south. It can become a weed in moist lawns and gardens and is
very difficult to eradicate. The extremely rare Carolina Hedge-nettle (*S. caroliniana*
J. B. Nelson & D. A. Rayner, plate 850) is endemic to the northeastern Coastal Plain of SC
and is easily distinguished by the sessile leaves.

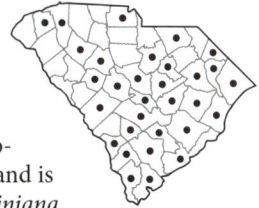

983. Wild Leek; Elephant Garlic

Allium ampeloprasum L.
    Ál-li-um am-pe-ló-pra-sum
  Alliaceae (Onion Family)

DESCRIPTION: Perennial, bulbous, scapose,
herb with odor of garlic; stems 3–4′ tall; leaves
whitish-green, flat, free portion 12–24″ long,
sheathing portion 0.5–2.5″ long, creating a leafy
lower stem; leaves die back at flowering, leaving
a leafless flowering stem terminated by a large,
many-flowered umbel; perianth lavender; flowers
May–early July.

RANGE-HABITAT: Native to Eurasia; naturalized from OH and VA and south to
FL and TX; uncommon throughout SC; abandoned home sites, roadsides,
and other disturbed places.

COMMENTS: Wild Leek is naturalized from Eurasia. Several vegetables have been
selected and cultivated from this species, including Elephant Garlic, Pearl Onion,
Kurrat, and Persian Leek.

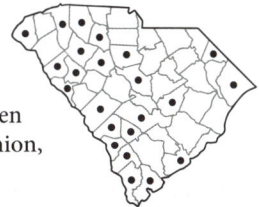

984. Indian-hemp; Marion's Weed

Apocynum cannabinum L.
    A-pó-cy-num can-ná-bi-num
  Apocynaceae (Dogbane Family)

DESCRIPTION: Perennial herb with milky sap, to
4′ tall, and branched only at the top; leaves opposite, entire, spreading or ascending; corolla
white to greenish; flowers May–July; fruit consists
of paired follicles which hang down, maturing
September–October.

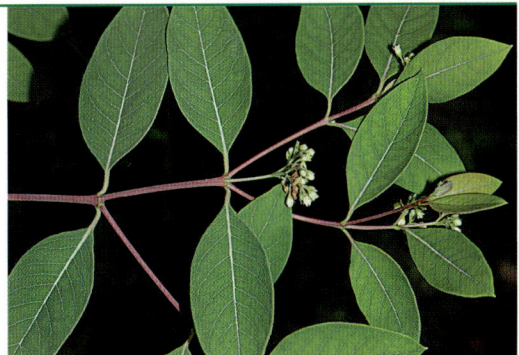

RANGE-HABITAT: Widely distributed in the US and Canada; common throughout SC in woodland margins, along roadsides, and in dry, disturbed sites.

COMMENTS: The name "Indian hemp" comes from Native Americans' use of the stem fiber for cordage, fishing nets, and coarse cloth. In St. John's Parish, Berkeley County, it was called Marion's weed because it was a favorite remedial agent in General Francis Marion's camp during the Revolutionary War. F. P. Porcher (1869) describes the many and varied historical uses of the plant, such as a "substitute for quinine, purgative, cure for dropsy, diuretic, and as an agent to remove ascarides." The milky sap was a folk remedy for venereal warts. The plant contains toxic cardioactive glycosides.

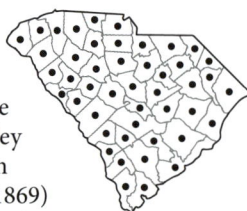

985. Butterfly-weed; Pleurisy-root

> *Asclepias tuberosa* L. ssp. *tuberosa*
> As-clè-pi-as tu-be-rò-sa ssp. tu-be-rò-sa
> Apocynaceae (Dogbane Family)

DESCRIPTION: Perennial herb from a thick root; stems 1–2.5′ tall, one to several, erect, ascending or creeping; sap clear, not milky; plants rough-hairy throughout; leaves abundant, alternate, obovate to oblanceolate, the margins flat; flowers bright orange, yellow or reddish; flowers May–August.

RANGE-HABITAT: Native wide-ranging species, from NH west to OH and south to FL and TX; common throughout SC in sandy oak-hickory forests; sandy, dry, open woods; dry fields; along roadsides; and in forest margins.

TAXONOMY: Another subspecies of *A. tuberosa* found in SC is Sandhills Butterfly-weed (*A. tuberosa* ssp. *rolfsii* (Britton ex Vail) Woodson). The leaves of ssp. rolfsii are hastate, and the margins usually revolute. It is uncommon; found in dry sandy habitats in the Inner Coastal Plain and Sandhills of SC.

COMMENTS: The common name, Butterfly-weed, comes from the attractiveness of its brilliant flowers to butterflies, particularly Monarchs, whose larvae also use this plant, like all milkweed species, for food. Pleurisy-root comes from Native Americans chewing the root as a cure for pleurisy and other pulmonary ailments. A tea made of fresh leaves has been used in Appalachia to induce vomiting. The roots of Butterfly-weed are poisonous.

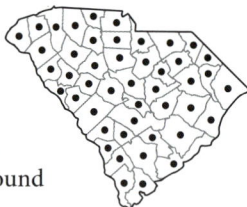

986. Queen Anne's Lace; Wild Carrot

> *Daucus carota* L.
> Daù-cus ca-rò-ta
> Apiaceae (Carrot Family)

DESCRIPTION: Freely branched biennial herb, 1–3′ tall, with a long, slender taproot; leaves bipinnately compound; flowers in umbels with the centralmost flower often maroon; fruits with bristles that are not barbed; flowers May–September.

RANGE-HABITAT: Native to Europe; naturalized throughout North America; common throughout SC, although only occasional in the Outer Coastal Plain and maritime strand; most often found along roadsides, but also in fallow fields, fencerows, abandoned lots, and other disturbed areas.

SIMILAR SPECIES: Queen Anne's Lace closely resembles the native American Queen Anne's Lace (*D. pusillus* Michaux). This native species is widespread in the southeastern states and occurs

in the Piedmont, Coastal Plain, and maritime strand of SC. It differs from wild carrot in being unbranched (or rarely few-branched), the central flower of each umbel is white, and the bristles of the fruits are barbed.

COMMENTS: For centuries, the root was eaten in Europe and America; however, the root is bitter and stringy and needs a long period of cooking. As is often the case, a wild plant is recognized as having important qualities (here, vitamin A), and a cultivar is developed from it, the commercial carrot.

---

987. Bitterweed

Helenium amarum (Rafinesque)
H. Rock
He-lén-i-um a-mà-rum
Asteraceae (Aster or Sunflower
Family)

DESCRIPTION: Annual herb to 40″ tall from a taproot; crushed stems/leaves have a harsh, bitter odor; freely branched above; leaves linear, often with smaller leaves on very short branches in the leaf axils; basal leaves soon deciduous; flowers arranged in involucrate heads of yellow ray and yellow disk flowers; flowers May–frost.

RANGE-HABITAT: Widespread in eastern North America, appears to have spread east from its prairie habitats; common throughout SC in a variety of disturbed areas with poor soils, including margins of open woods, pastures, roadsides, abandoned lots, and fields.

COMMENTS: Helenium is named for Helen of Troy, daughter of Zeus and Leda, and cause of the Trojan War. Bitterweed can be a serious weed in pastures where cattle graze. When other forage is scarce, cattle will graze on Bitterweed, giving their milk a bitter taste. Bitterweed is suspected of human poisonings.

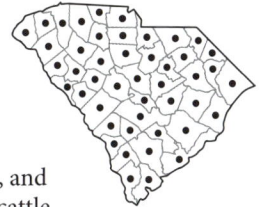

---

988. Man-of-the-earth; Man-root

Ipomoea pandurata (L.) G. F. W. Meyer
I-po-moè-a pan-du-rà-ta
Convolvulaceae (Morning-glory
Family)

DESCRIPTION: Perennial, trailing vine from a deep, vertical, enlarged root; root growing to 30 pounds; corolla white, always with a purple center; flowers May–July.

RANGE-HABITAT: From CT, NY, and s. Ontario, south to FL, and west to OH, KS, and TX; common throughout SC along roadsides and fencerows, in lawns, sandy and open woods, and fallow fields.

COMMENTS: Several sources state that Native Americans used the root for food, but only after long roasting. Caution should be taken since the fresh root is reported to be purgative. A great quantity of starch can be extracted from the roots, which is probably safe. It is interesting to note that man-of-the-earth is related to sweet potato (*Ipomoea batatas* Lamarck).

989. Pink-ladies; White Evening-primrose

Oenothera speciosa Nuttall
Oe-no-thè-ra spe-ci-ò-sa
Onagraceae (Evening-primrose
Family)

DESCRIPTION: Stoloniferous, erect, usually branched perennial herb, 8–24″ tall; leaves linear to oblong-lanceolate, the wider ones irregularly dentate or narrowly lobed; flower buds nodding, opening into showy pink or white flowers; flowers May–August.

RANGE-HABITAT: Native west of the Mississippi River; naturalized and widespread throughout the US; in SC, common in the Piedmont and Coastal Plain; along roadsides and in meadows, lawns, fields, roadsides, and various dry, disturbed places.

COMMENTS: According to several sources, Pink-ladies is native to the west but has spread widely from plants escaped from cultivation. It is drought resistant, making it ideal for cultivation in dry sites. The aggressively rhizomatous nature of the plant allows it to outcompete nearly all other plants with which it is interplanted, and it can spread far beyond the desired locations in the garden. The specific epithet, speciosa, means beautiful.

990. Maypops; Passion-flower

Passiflora incarnata L.
Pas-si-flò-ra in-car-nà-ta
Passifloraceae (Passion-flower Family)

DESCRIPTION: Tendril-bearing, perennial, herbaceous vine, to 6′ long; either creeping and rooting at the nodes or clambering over vegetation; leaves deeply 3-lobed; flowers May–September; fruit fleshy, yellow when ripe; fruits mature July–October.

RANGE-HABITAT: FL to TX and north to MD, MO, and OK; very common throughout SC along roadsides and fences, in fallow fields, hedge rows, vacant lots, thickets, and other dry, disturbed sites.

SIMILAR SPECIES: Yellow Passion-flower (*P. lutea* L., plate 755) has similar shaped flowers, but they are much smaller and greenish yellow. It flowers in June–September and is common throughout SC in mixed deciduous woodlands and thickets.

COMMENTS: Maypops is a native vine that is weedy but often cultivated because it is the primary host plant for the caterpillars of several beautiful butterflies such as the Zebra Longwing, Gulf Fritillary, and Variegated Fritillary. The common name, Maypops, comes from the fact that the unripe fruit may make a loud popping sound when stepped on. The common name, Passion-flower, comes from the resemblance of the floral parts to the story of Christ's Passion: the styles resemble nails; the 5 stamens, the wounds Jesus received; the purplish corona, the bloody crown; the 10 perianth parts, the 10 disciples (Peter and Judas being absent); the coiled tendrils, the whips for scourging; the pistil, the column where Christ was scourged; and the flower in the background of dull, green leaves represents Christ in the hands of his enemies. Interestingly, the flower's life is generally 3 days.

The juicy pulp surrounding the seeds is edible raw but is more esteemed when made into jelly.

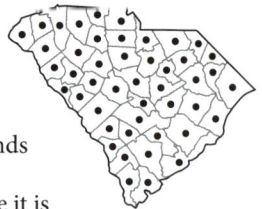

991. Southern Beard-tongue

Penstemon australis Small
Pen-stè-mon aus-trà-lis
Plantaginaceae (Plantain
Family)

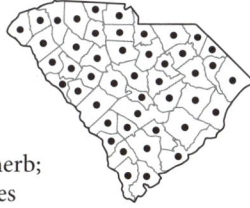

DESCRIPTION: Opposite-leaved, perennial herb; stems one to several, 8–28″ tall; basal leaves shaped differently from the stem leaves; stems below the inflorescence pubescent with a mix of glandular and nonglandular hairs; the corolla with a nearly closed throat; lower corolla lobes with purplish lines; flowers May–July.

RANGE-HABITAT: VA south to FL and west to AL; in SC, common throughout the Coastal Plain, Sandhills, and Piedmont; a native species more common as a weedy roadside species than in native communities but also occurring in Longleaf Pine sandhills; glades and margins of granitic outcrops; sandy, dry, open woods; dry, fallow fields; and burned-over thickets.

COMMENTS: The common name comes from the beardlike appearance of the sterile stamens in the corolla throat of some species.

992. Pokeweed

Phytolacca americana L.
Phy-to-lác-ca a-me-ri-cà-na
Phytolaccaceae (Pokeweed Family)

DESCRIPTION: Robust, perennial herb, 3–10′ tall, from a thick root; leaves alternate, entire, smooth; plant unpleasantly scented; flowers in narrow, long, drooping racemes produced opposite the upper leaves; flowers May–frost; berries in long racemes, dark purple when ripe in the fall.

RANGE-HABITAT: A native weed widespread in North America and southeastern Canada; common throughout SC in a wide variety of natural and disturbed sites, including fields, pastures, vacant lots, railroad embankments, abandoned house sites, newly created openings in forests and pastures, and barnyards.

SIMILAR SPECIES: Maritime Pokeweed (*Phytolacca rigida* Small) has shorter, erect racemes and is a more rigid plant. As the name would suggest, it is found in the maritime strand in disturbed areas, maritime grasslands, shrub thickets, and shell middens.

COMMENTS: All parts of the plant are poisonous, and poisons include a virtual cornucopia of toxic compounds, including its own distinctive toxins—phytollaccatoxin and pytolaccin. The young shoots are eaten as a potherb before the red color appears and after boiling in multiple changes of water. Tinctures obtained from various parts of the plant are available as herbal remedies, and the plant is even available as a canned vegetable. Native Americans and European colonists used the juice of the berry as a dye. Consumption of the fruits by birds, especially by Mourning Doves, allows for the rapid spread of the seeds. It quickly colonizes bare soil in natural communities, especially mounds created by windthrows.

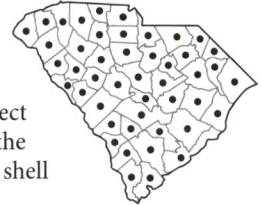

993. Sulphur Five-fingers

Potentilla recta L.
Po-ten-tíl-la réc-ta
Rosaceae (Rose Family)

DESCRIPTION: Perennial, hairy herb; stems several, 1–2′ tall; leaves palmately compound, with 7–9 coarsely serrate leaflets; inflorescence terminal, flat-topped with pale yellow flowers with notched petals; flowers in May–August.

RANGE-HABITAT: Native to Europe; naturalized from Newfoundland south to FL and west to MN and TX; common throughout SC in pastures and meadows and along roadsides and railroad beds.

994. Black-eyed Susan

Rudbeckia hirta L.
Rud-béck-i-a hír-ta
Asteraceae (Aster Family)

DESCRIPTION: Tap-rooted annual or more often biennial or fibrous-rooted perennial, 12–40″ tall; leaves and stems very rough and bristly-pubescent; as a biennial it forms a rosette of leaves the first year, followed by a flowering stem with alternate leaves the second year; flowers arranged in involucrate heads of blackish-brown disk flowers and yellow ray flowers; flowers May–June.

RANGE-HABITAT: Nearly throughout North America; common throughout SC along roadsides and in pastures, open woods, forest margins, Longleaf Pine flatwoods, fields, and meadows.

COMMENTS: Black-eyed Susan exists as a native species but large-flowered, showy, and often bicolored forms from farther west have escaped from cultivation. It has been used as a dye source and to treat skin infections. It is a highly polymorphic species, particularly in relation to duration and leaf size and shape. Several varieties are recognized.

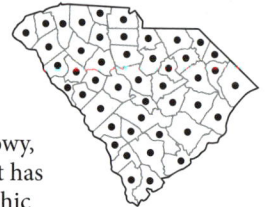

995. Soapwort; Bouncing Bet

Saponaria officinalis L.
Sa-po-nà-ri-a of-fi-ci-nà-lis
Caryophyllaceae (Pink Family)

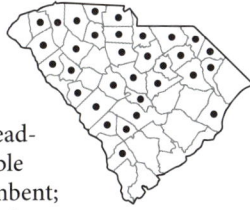

DESCRIPTION: Smooth, perennial herb, spreading from seeds and rhizomes to form sizable colonies; stems 20–60″ tall, erect or decumbent; leaves opposite; petals white to light pink; flowers May–October.

RANGE-HABITAT: Native to Europe; naturalized essentially throughout the US; common throughout SC in fields and around abandoned home sites and other disturbed places.

COMMENTS: Soapwort has had a long history of folk use. When mixed with water, the crushed leaves and roots make lather— hence, the common name. Native Americans used a poultice of leaves for boils. The specific epithet, officinalis, indicates its former official or accepted use in medicine. The second common name, Bouncing Bet, is an old-fashioned nickname for a washerwoman, again in reference to its use as soap. The seeds contain saponins, the agents that cause the foaming action and are poisonous when ingested.

996. Small's Ragwort

Packera anonyma (Wood)
W. A. Weber & Á. Löve
Páck-er-a a-nón-y-ma
Asteraceae (Aster Family)

SYNONYM: *Senecio smallii*
Britton—RAB, PR

DESCRIPTION: Clump-forming annual, 12–28″ tall; stem densely woolly at base; basal leaves wedge-shaped, lanceolate, serrate, and pinnately dissected; flowers produced in involucrate heads of yellow ray and yellow disk flowers; heads in a corymblike arrangement; flowers May–early June.

RANGE-HABITAT: NY south to FL and west to IN and LA; common throughout SC along dry roadsides and in meadows, fallow fields, pastures, and open woodlands; in the Piedmont and mountains it also grows on the margins of granitic flatrocks and granitic domes.

COMMENTS: This is the most common ragwort species in SC. It frequently lines roadways with splashes of yellow color. This species, while native, has apparently recently invaded natural granitic dome communities in the mountains, where it was formerly absent. It now poses a threat to the genetic integrity of the rare and endemic Yarrowleaf Ragwort (*P. millefolium* (Torrey & Gray) W. A. Weber & Á. Löve, plate 13), with which it readily hybridizes.

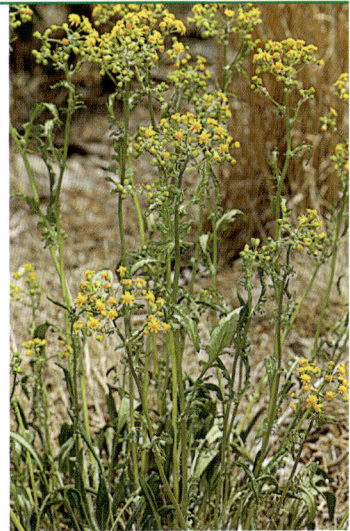

997. Carolina Horse Nettle

Solanum carolinense L.
So-là-num ca-ro-li-nén-se
Solanaceae (Nightshade Family)

DESCRIPTION: Erect, weakly branched, perennial herb, 1–3′ tall; stems and underside of leaves with sharp prickles; leaves coarsely lobed, both surfaces with star-shaped hairs; corolla light purple to white, with yellow center; flowers May–July; berry yellow, matures August–September.

RANGE-HABITAT: From southern Ontario to New England and NY, south to FL, west to TX, and north to NE; common throughout SC; along roadsides, in fallow and cultivated fields, around farm lots and abandoned house sites, and in lawns.

COMMENTS: The berries are poisonous. Kingsbury (1964) reports one case of a child's dying from eating the berries. Poisoning in cattle and deer have been reported. This is another native weedy species that has become more abundant as disturbed habitats have increased.

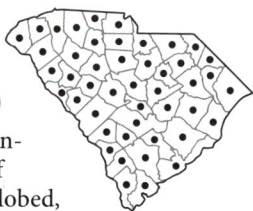

998. Sand Blackberry

Rubus cuneifolius Pursh
Rù-bus cu-nei-fò-li-us
Rosaceae (Rose Family)

DESCRIPTION: Semi-woody plant with erect to arching stems, to 5′ tall; prickles broad-based, flat and curved; leaves palmately compound, densely light pubescent below; leaflets usually wider toward the tips; fruit an aggregate of drupes; fruits mature June–July.

RANGE-HABITAT: Primarily Coastal Plain from CT and NY, south to FL and AL; throughout SC (except in the mountains); sandy and rocky woodland borders, fields, disturbed areas, forests, and along roadsides.

COMMENTS: Humans throughout history have used blackberries (and dewberries) for more than just food. F. P. Porcher (1869) relates that the root was a valuable astringent and that a decoction easily checks diarrhea. A laxative was made from the fruits, which worked because of the mechanical irritation of the seeds. Wine, syrup, cordials, jelly, and jam are made from the fruits. Blackberries and related dewberries are among the most important summer foods for songbirds and many mammals. There are currently thought to be 12 blackberry species in SC, and many are very difficult to distinguish.

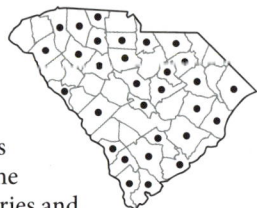

999. Standing-cypress

Ipomopsis rubra (L.) Wherry
I-po-móp-sis rù-bra
Polemoniaceae (Phlox Family)

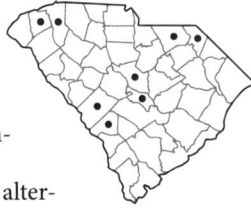

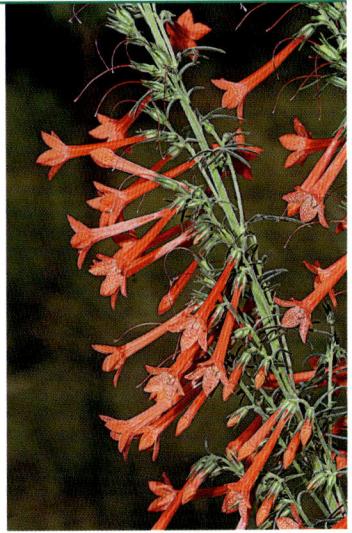

DESCRIPTION: Erect, hairy, herbaceous biennial, to 3′ tall (exceptionally 5′); first-year leaves in a large, basal rosette; stem leaves alternate, divided into numerous, narrow segments; flowers in racemes, often one-fourth to one-third of the total plant height; flowers bright red, or sometimes pale yellow; flowers June–August.

RANGE-HABITAT: Ranging as a native from NC south to FL and west to TX and OK; scattered and rare in SC; Piedmont, Fall-line Sandhills, and Coastal Plain; Longleaf Pine sandhills, riverbanks, roadsides, sand ridges of Carolina bays, openings in dry woodlands, and margins of granitic flatrocks and domes.

COMMENTS: Standing-cypress is a native that has adapted to ruderal sites as fire suppression has reduced woodland openings. In gardens, it is found sometimes under the synonym *Gilia coronopifolia* Persoon. It is closely related to the large western genus *Gilia* and thus represents another example of the affinity between the Sandhills flora of the Southeast and the flora of the dry southwestern US. It is often cultivated. It grows in FL orange groves, where it is often called "Spanish-larkspur." The long, tubular flowers are adapted to pollination by hummingbirds.

Standing-cypress can be found at Glassy Mountain Heritage Preserve in Pickens County.

1000. Perennial Sand Bean

Strophostyles umbellata (Muhlenberg ex Willdenow) Britton
Stro-pho-stỳ-les um-bel-là-ta
Fabaceae (Pea or Bean Family)

DESCRIPTION: Trailing or climbing, twining, herbaceous perennial vine; stems to about 5′ long; leaf blades pinnately 3-foliolate; petals pink or pale purple, often fading yellowish; flowers few to several in headlike clusters on stalks to 12″ long; flowers mature in June–September; fruit a legume; fruits mature August–October.

RANGE-HABITAT: From NY south to FL and west to IL and TX; in SC, common in the Coastal Plain and Sandhills, uncommon in the Piedmont; Longleaf Pine sandhills and flatwoods, fields, thickets, dry, sandy woods, and clearings.

COMMENTS: Perennial Sand Bean is a native species that is more abundant in disturbed sites than in native, wooded habitats. There is no reference to this species having any food value for humans. A variety of birds eat the seeds, including Northern Bobwhite and Wild Turkey.

1001. Woolly Mullein

Verbascum thapsus L.
ssp. *thapsus*
Ver-bás-cum tháp-sus ssp.
tháp-sus
Scrophulariaceae (Figwort Family)

DESCRIPTION: Densely woolly biennial, 2–6′ tall; only a basal cluster of thick, velvety leaves present in the first year; flowering stem develops the second year, with leaves gradually reduced upward; flowers fragrant, yellow, in a long, dense spike; flowers June–September.

RANGE-HABITAT: Native of Eurasia; naturalized throughout the US; common throughout SC along roads and in pastures, fallow fields, sandy, open woods, vacant lots, and coastal shell deposits.

COMMENTS: Woolly Mullein is one of our most widely distributed naturalized plants. Few plants have had as many uses in the past as Woolly Mullein, both in the Old and New Worlds. Roman soldiers dipped the stalks into grease for torches. Native Americans, enslaved people, and colonists lined their shoes with the velvety leaves as cushions and to keep out the cold. Quaker women, not allowed to use cosmetics, rubbed their cheeks with the leaves to achieve a rosy look. A tea from the plant was used to treat colds. Native Americans smoked the dried leaves, flowers, or roots for pulmonary ailments, and enslaved people on southern plantations used a decoction of the plant to relieve the pain of hemorrhoids or to induce sleep. Mullein does contain a narcotic principle, and Duke (1997) recommends it as an herbal remedy for bronchitis and laryngitis. The leaves and flowers contain saponins, which are poisonous when ingested in quantity. Tinctures of various parts of the plant are readily available as herbal remedies.

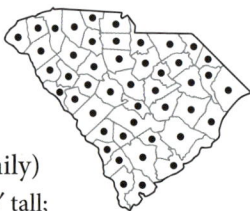

1002. Sicklepod

Senna obtusifolia (L.) H. S. Irwin &
Barneby
Sén-na ob-tu-si-fò-li-a
Fabaceae (Bean Family)

SYNONYM: *Cassia obtusifolia* L.—RAB, PR

DESCRIPTION: Erect annual herb to 5′ tall; leaves evenly once-pinnately compound; leaflets 4–6, obovate, terminal pair largest; elongate gland on the rachis just above the attachment of the stalks of the lowest pair of leaflets; flowers July–September.

RANGE-HABITAT: Common weedy species from FL to TX and northward to PA and KS; in SC, common throughout in a variety of weedy habitats, such as abandoned agricultural fields, pastures, and other disturbed sites.

COMMENTS: Sicklepod is a tropical species that has invaded the southern US. It is reported to be poisonous when eaten in large quantities. It is known to cause severe muscle and liver problems in cattle when consumed in large amounts.

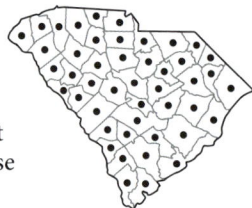

1003. Mexican-tea; Epazote; Worm-seed

Dysphania ambrosioides (L.)
   Mosyakin & Clemants
   Dys-phá-ni-a am-bro-si-oì-des
   Chenopodiaceae (Goosefoot Family)

SYNONYM: *Chenopodium ambrosioides*
   L.—RAB, PR

DESCRIPTION: Annual or sometimes perennial herb from a stout; stems 2–4′ tall, covered with glandular-resin dots; entire plant with a strongly pungent odor; leaves alternate; flowers in dense clusters on short to long spikes; flowers July–frost.

RANGE-HABITAT: Native to tropical America; naturalized throughout much of eastern North America; common throughout SC in cultivated and fallow fields, pastures, vacant lots, fencerows, and along railroads and roads.

COMMENTS: In earlier times, rural people in the coastal area used it as a vermifuge–hence, the common name Worm-seed. Oil of Chenopodium is still widely used for this purpose, especially for hookworm infections and especially in Latino cultures.

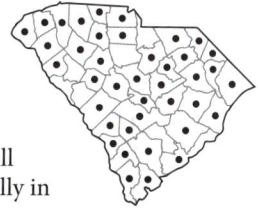

1004. Jimson-weed; Jamestown-weed

Datura stramonium L.
   Da-tù-ra stra-mò-ni-um
   Solanaceae (Nightshade Family)

DESCRIPTION: Rank-smelling, coarse, annual weed, 1–5′ tall; leaves alternate, irregularly lobed; flowers white, funnel-shaped, very large; flowers July–September; fruit an erect, spiny, many-seeded capsule, matures August–October.

RANGE-HABITAT: Common throughout the US; common throughout SC along roadsides, in barnyards, fields, vacant lots, overgrazed pastures, and other disturbed areas.

COMMENTS: Jimson-weed is a corruption of Jamestown-weed, the latter name derived from its growing around Jamestown, VA, where it was reported that soldiers who were sent there in 1676 to quell Bacon's Rebellion became intoxicated from smoking the plant. Inadvertently, leaves of Datura were included in their tobacco, and, as reported in Robert Beverly's (1705) *The History and Present State of Virginia*, "the effect of which was a very pleasant Comedy; for they turn'd natural Fools upon it for several days . . . and after Eleven Days, returned to themselves again, not remembering any thing that pass'd." Jimson-weed is the only hallucinogenic flowering plant in SC. All parts of the plant are poisonous. Cattle and sheep have died from grazing on it. Children have been poisoned and recreational drug users have died from ingesting just a few seeds. Touching the leaves or flowers may cause dermatitis in sensitive people.

There seems to be a question in the literature about where Jimson-weed is native. One source says it comes from tropical America, another source says Europe. Yet another source questions both of these by pointing out that it was reported growing around Jamestown in 1676, suggesting it is native to North America.

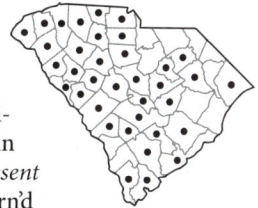

1005. Beaked Hawkweed

Hieracium gronovii L.
Hi-e-rà-ci-um gro-nòv-i-i
Asteraceae (Aster Family)

DESCRIPTION: Perennial herb with milky juice; stems pubescent, to 3′ tall; leaves mostly basal or on the lower part of the stem and reduced above; leaf blade with long, ascending to erect, light-colored hairs on the upper surface; inflorescence a panicle with numerous involucrate heads composed of yellow ray flowers only; flowers July–frost.

RANGE-HABITAT: MA to KS and south to FL and TX; common throughout SC in Longleaf Pine flatwoods, sandhills, open, sandy or rocky woods; old fields; roadsides;, and meadows.

COMMENTS: The term "beaked" in the common name refers to the plants' achenes (fruits), which are distinctly narrowed toward the apex. All other hawkweeds in our area have achenes that are truncate at the apex.

The specific epithet honors Jan Fredick Gronovius (1690–1762), professor at Leyden, teacher of Linnaeus, and author of *Flora Virginica* (1739).

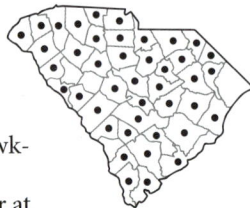

1006. Common Morning-glory

Ipomoea purpurea (L.) Roth
I-po-moè-a pur-pú-re-a
Convolvulaceae (Morning-glory Family)

DESCRIPTION: Twining, annual, herbaceous vine with one main stem from a taproot; leaves simple, heart-shaped; corolla purple, bluish, white or variegated; flowers July–September.

RANGE-HABITAT: Throughout the southeastern US; common throughout SC in fallow and cultivated fields, along roadsides and fencerows, and in abandoned lots.

COMMENTS: Common Morning-glory is native to tropical America and was introduced into North America as an ornamental. It escaped from gardens and is now widely naturalized. Its broad, heart-shaped leaves are distinctive.

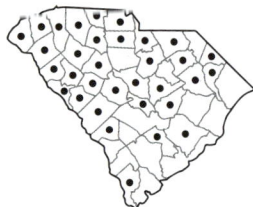

1007. Sericea

Lespedeza cuneata (Dumont
de Courset) D. Don
Les-pe-dèz-a cu-ne-à-ta
Fabaceae (Bean Family)

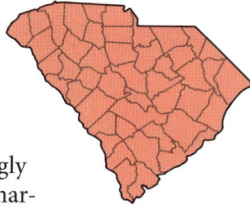

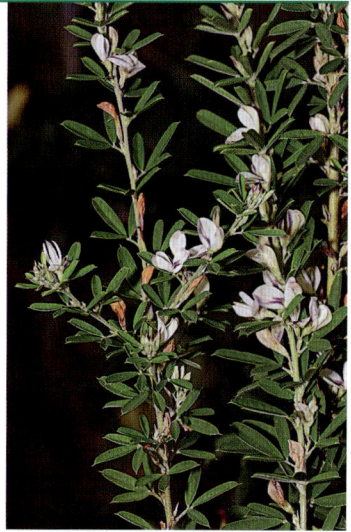

DESCRIPTION: Perennial with erect or strongly
ascending, strigose stems to 5′ tall; leaves narrowly oblong, wedge-shaped to linear-oblanceolate,
with a bristle at the very tip (mucro); petals creamy-white, with
violet-purple along the veins of the standard petal; flowers July–
September; legume matures October–November.

RANGE-HABITAT: Naturalized and common throughout the southeast; common throughout SC; along roadsides, in fields, and in
other disturbed areas.

COMMENTS: Sericea was introduced into the Southeast from Asia
as a hay plant and as a soil-binder along roads. It has become a
serious invasive exotic, invading pastureland and natural habitats.
The genus honors Vincente Manuel de Céspedes, governor of
Florida in 1790. A note from *Gray's Manual of Botany* (Fernald, 1950) explains the difference
in the spelling of the genus Lespedeza and de Céspedes: "Dedicated to Vincente Manuel de
Céspedes, Spanish Governor of East FL during the explorations there of Michaux late in the
18th Century; the name later misspelled, probably by Michaux's editor, as de Lespedez."

1008. Northern Obedient-plant

Physostegia virginiana (L.)
Bentham ssp. *virginiana*
Phy-sos-té-gi-a vir-gi-ni-
à-na ssp. vir-gi-ni-à-na
Lamiaceae (Mint Family)

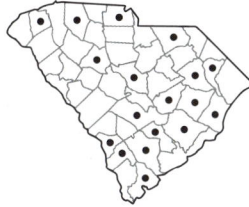

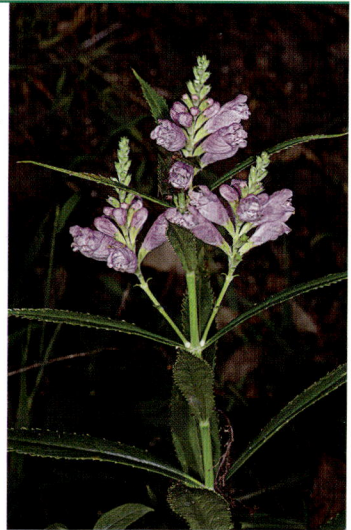

SYNONYM: *Dracocephalum
virginianum* L.—RAB

DESCRIPTION: Upright, nearly glabrous perennial herb with
rhizomes; stems with 15–22 nodes, up to 5′ tall; leaves opposite,
simple, sessile, elliptic to lanceolate, long-tapering, with toothed
margins; corolla white to deep pink, spotted with reddish purple;
flowers July–October.

RANGE-HABITAT: Native from Quebec west to Manitoba and south
to VA, TN, and KS; in SC, naturalized and common throughout
on stream sides, in seepages and marshes, on roadsides, and in
abandoned home sites and other moist, disturbed sites.

COMMENTS: Several references state that it is probably not native in the eastern part of its
range but that it escaped from cultivation. The common name, Obedient-plant, relates to the
flowers' tendency to stay in a new position for a while after being twisted to one side.

1009. Bladderpod; Bagpod

Sesbania vesicaria (Jacquin) Elliott
Ses-bàn-i-a ve-si-cá-ri-a
Fabaceae (Bean Family)

SYNONYM: *Glottidium vesicarium*
(Jacquin) Harper

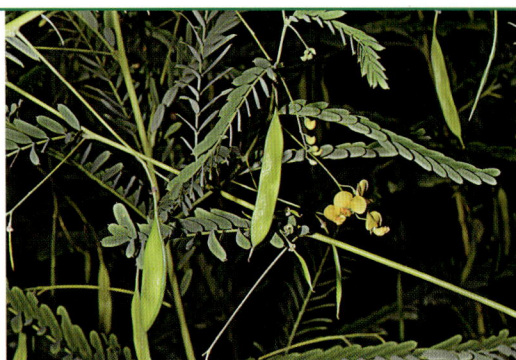

DESCRIPTION: Robust annual, to 12′ tall; leaves even-pinnate, with 12–40 leaflets; petals yellow, tinged with red; legume 2-seeded, at maturity the firm outer layer separating from a thin, soft layer enclosing the seeds; flowers July–September; fruits mature August–November.

RANGE-HABITAT: Coastal plain from NC south to FL and west to TX; in SC, common in the Coastal Plain and maritime strand; a weedy plant, found in a variety of disturbed habitats such as ditches, freshwater marshes, low fields, and other wet, disturbed areas.

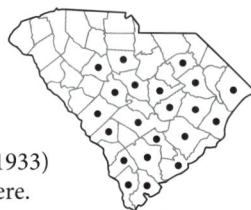

COMMENTS: Bladderpod was probably introduced from farther South. Small (1933) states that "Bladder Pod" was evidently introduced but does not say from where. One source reported that it is probably native to the West Indies.

Bladderpod is a vigorous annual, and the firm, woody stems, with the dangling pods, may persist throughout the winter. Kingsbury (1964) reports that the seeds and fruits of bladderpod are poisonous to a wide variety of animals, including sheep, cattle, and fowl. Hundreds of cattle from a single herd have been killed. Cattle poisoning occurs in the fall and winter when the pods (still carried by the erect, dead stems) are available to cattle after other forage has become scarce.

1010. New York Ironweed

Vernonia noveboracensis
(L.) Michaux
Ver-nòn-i-a no-ve-bo-ra-cén-sis
Asteraceae (Aster Family)

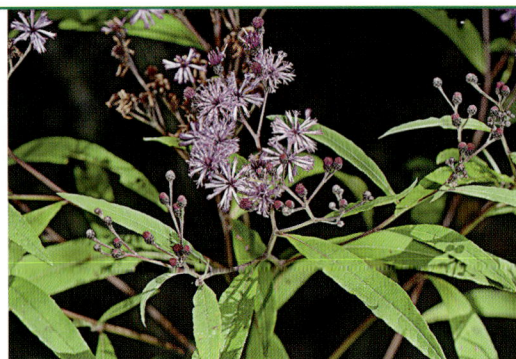

DESCRIPTION: Robust, perennial herb, 3–7′ tall; stem green to dark purple; basal leaves absent; stem leaves alternate, numerous, 4–6 times as long as wide; flowers arranged into involucrate heads of purple disk flowers; pappus brown to purple; flowers July–September.

RANGE-HABITAT: MA and NY, south to FL, and west to MS; inland to WV and OH; common throughout SC; wet meadows, stream sides, low, open woodlands, thickets, and swales.

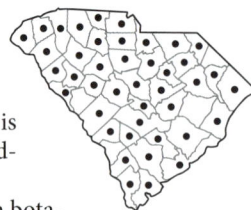

SIMILAR SPECIES: Appalachian Ironweed (*V. glauca* (L.) Willdenow) is similar but has leaves 2.5–3 times as long as wide and a white to yellowish pappus. It is found in open areas mostly on rich soils derived from amphibolite in the Piedmont and mountains.

COMMENTS: The genus honors William Vernon (ca. 1666/67–1711), an English botanist who traveled in North America.

1011. Cocklebur

Xanthium chinense P. Miller
Xán-thi-um chi-nén-se
Asteraceae (Aster Family)

SYNONYM: *Xanthium strumarium*
L.—RAB, PR

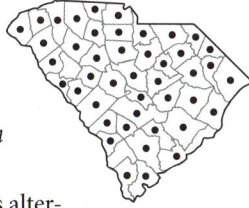

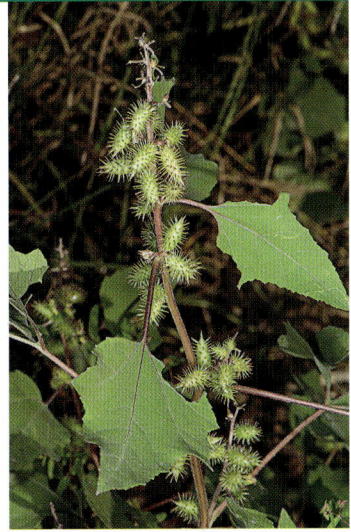

DESCRIPTION: Annual herb, to 6′ tall; leaves alternate, triangular-ovate; male and female flowers in separate heads; female flowers two each in burs (involucres) beset with strong, hooked bristles; flowering and fruiting July–frost.

RANGE-HABITAT: Throughout the US and southern provinces in Canada; common throughout SC in a wide variety of weedy habitats, including cultivated and fallow fields, pastures, floodplain forests and their clearings, ditches, pond shores, and along roadsides.

COMMENTS: Cocklebur poisoning has occurred in all classes of domestic livestock. Kingsbury (1964) states: "It is always associated with ingestion of seedlings in the cotyledonary stage of growth. Cocklebur seeds sprout readily when present in soil that has been under water but is drying out. Such conditions are found along streams, or about the shores of shallow ponds as summer progresses. Frequently, as the shore area is extended by a receding water margin, there is continual germination and sprouting of cocklebur seedlings which, as the water withdraws, provide potentially toxic forage over an extended period of time. Although cocklebur is an annual, its seeds may persist for several years before germinating. This must be taken into consideration when planning control measures for its eradication." The burs of cocklebur can cause puncture wounds to animals if they become tangled in their fur.

Despite the specific epithet, *chinense,* this species is native to North America. Apparently, Miller was mistaken about where the plant had been collected.

1012. Common Ragweed

Ambrosia artemisiifolia L.
Am-brò-si-a ar-te-mi-si-i-fò-li-a
Asteraceae (Aster Family)

DESCRIPTION: Annual herb to 6′ tall, from a taproot; stems freely branched; leaves deeply bipinnately dissected, lower opposite, upper alternate; flowers inconspicuous; male and female flowers on same plant; male flower heads in terminal racemes; female flowers heads in upper leaf axils; flowers and fruits August–frost.

RANGE-HABITAT: A native species growing virtually throughout the US; throughout SC on margins of rock outcrops and dry, open sandy woodlands but invading almost any open, disturbed site, including fields, roadsides, pastures, and vacant lots.

SIMILAR SPECIES: Giant Ragweed (*A. trifida* L., var. trifida) grows in the mountains and Piedmont in waste places and alluvial fields. Its pollen, spread by wind, is also a significant cause of hay fever. It differs from Common Ragweed in that it is about twice as tall and has 3–5 lobed leaves.

COMMENTS: Wind-borne pollen from the genus Ambrosia is the cause of approximately 90% of pollen-induced allergies in the United States. Today's way of life creates an abundance of disturbed sites, so there is little relief for hay fever sufferers. The best control is to allow perennials to crowd out this annual.

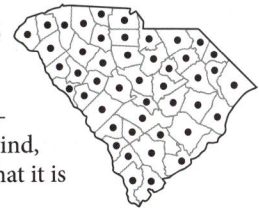

1013. Rattlebox

Crotalaria spectabilis Roth
    Cro-ta-là-ri-a spec-tá-bi-lis
Fabaceae (Bean Family)

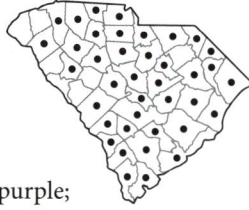

DESCRIPTION: Annual herb, 2–4′ tall, from a woody taproot; leaves simple, alternate, distinctly obovate; stems commonly dark purple; flowers yellow, in terminal racemes; flowers July–September; fruit an inflated legume, many seeded, black at maturity; fruits mature August–October.

RANGE-HABITAT: Native to tropical Asia, naturalized from VA south to FL and west to TX and MO; naturalized and common in the Piedmont, Sandhills, Coastal Plain, and maritime strand; roadsides, cultivated and fallow fields, and other disturbed areas.

COMMENTS: Rattlebox was introduced as a soil-building green manure. It quickly was naturalized and became a severe pest in agricultural crops. It is poisonous and has caused substantial losses in fowl, cattle, horses, and swine. It is also poisonous to humans. The common name comes from the legume, the seeds of which, when mature, break loose inside and rattle when shaken.

1014. Boneset

Eupatorium perfoliatum L.
    Eu-pa-tò-ri-um per-fo-li-à-tum
Asteraceae (Aster or Sunflower Family)

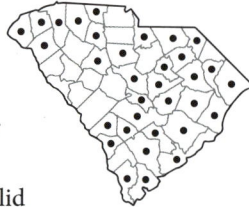

DESCRIPTION: Perennial herb with erect, solid stems, 2–4′ tall, often growing in clumps; leaves opposite, perfoliate; plant conspicuously hairy; flowers produced in involucrate heads of white disk flowers; flowers August–October.

RANGE-HABITAT: Widespread in eastern North America; common throughout SC in damp, moist areas such as wet meadows, alluvial woods, marshes, ditches, thickets, clearings, and pastures.

COMMENTS: The genus name, Eupatorium, is named for Mithridates Eupator, king of Pontus. (Pontus is in what is now the eastern Black Sea region of Turkey.) Boneset is a native species that has become weedy as disturbed habitats increased.

Boneset was a common home remedy of nineteenth-century America, extensively used by Native Americans and European colonists. The common name comes from the belief that the plant could cause rapid union of broken bones, as suggested by the basal union of the pairs of opposite leaves, an example of the Doctrine of Signatures. F. P. Porcher (1869) reported that enslaved people on coastal plantations employed it as a febrifuge in fevers and typhoid pneumonia. In addition, a tea from the leaves was a substitute for quinine to treat malaria and (in Appalachia) as a laxative and as a treatment for coughs and chest illnesses. Not surprisingly, extracts are readily available today as herbal remedies.

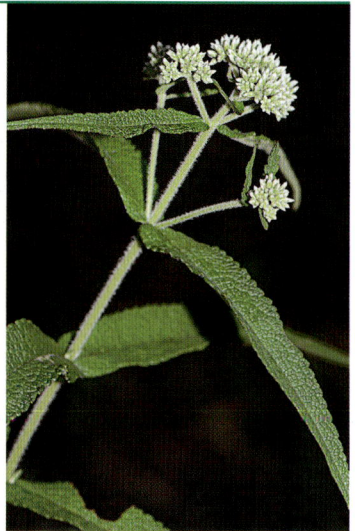

1015. Rabbit Tobacco; Life Everlasting

Pseudognaphalium obtusifolium (L.) Hilliard & Burtt
Pseu-do-gna-phà-li-um ob-tu-si-fò-li-um

Asteraceae (Aster Family)

SYNONYM: *Gnaphalium obtusifolium* L.—RAB, PR

DESCRIPTION: Erect, fragrant, winter or summer annual herb; stems erect, 1–3′ tall, densely whitish woolly; dead stems tend to remain throughout the winter; stem leaves alternate, green above, whitish woolly beneath; flowers August–October.

RANGE-HABITAT: Quebec to Manitoba and south to FL and TX; common native plant found throughout SC along roadsides, in fallow fields, pastures, open and sandy woodlands, and other disturbed sites.

COMMENTS: Rabbit Tobacco was once the most popular native cold remedy in coastal SC (Morton, 1974). "Flower ladies" in the Market Place in Charleston still sell it. The tea is drunk as a febrifuge. The tea is bitter and lemon juice is added to make it palatable for children. Rural people smoke the plant for asthma. F. P. Porcher (1869) said that the plant was an astringent; the leaves and flowers were chewed, and the juice swallowed to relieve ulcerations in the mouth and throat. Foster and Duke (1990) give a long list of the past medicinal uses of Rabbit Tobacco.

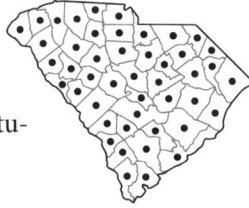

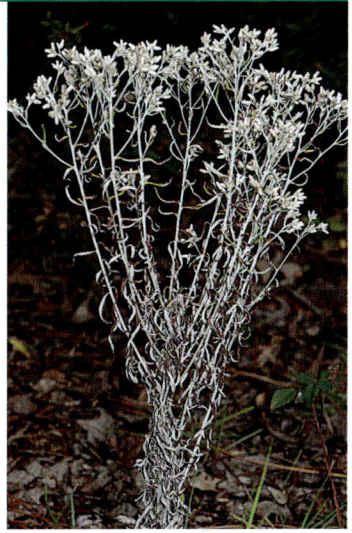

1016. Rattle-bush; Purple Sesban

Sesbania punicea (Cavanilles) Bentham
Ses-bàn-i-a pu-ní-ce-a

Fabaceae (Bean Family)

SYNONYM: *Daubentonia punicea* (Cavanilles) Augustin de Candolle—RAB

DESCRIPTION: Shrub 3–10′ tall; leaves even-pinnately compound, with 12–40 entire leaflets; flowers brilliant reddish-orange; fruit a legume 4-winged, with a sharp tip; flowers June-September; fruits mature August–November.

RANGE-HABITAT: Presumably native to South America, now naturalized from NC south to FL and west to TX; common weed in the Coastal Plain and maritime strand of SC; along roadsides and in ditches and sandy wet places.

SIMILAR SPECIES: Drummond's Rattle-bush (*Sesbania drummondii* (Rydberg) Cory) is similar but has yellow flowers and a blunt-tipped legume. It was first recorded in SC from a spoil island near the mouth of the Savannah River by Richard Porcher, John Townsend, and Patrick McMillan and has since been found at additional nearby sites. Drummond's Rattle-bush is a native of the Gulf Coast and is thought to be introduced in SC.

COMMENTS: Rattle-bush was introduced into FL as an ornamental. It escaped and spread throughout the Southeast. The leaves, and especially the seeds, are poisonous to humans and a wide variety of animal species.

1017. Small-flowered Morning-glory

Ipomoea lacunosa L.
  I-po-moè-a la-cu-nò-sa
Convolvulaceae (Morning-glory Family)

DESCRIPTION: Twining, annual, herbaceous vine with sparsely hairy stems; leaves ovate, entire, or with two basal lobes, heart-shaped, usually with maroon margins; corolla small, white, bell-shaped, 0.25–0.75″ long; flowers September–frost.

RANGE-HABITAT: Native weedy species from NJ south to FL and west to IL and TX; in SC, common throughout in moist fields and thickets, along roadsides, and in waste places.

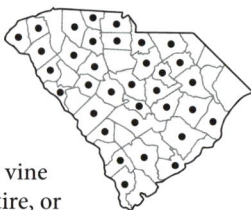

1018. Cypress-vine

Ipomoea quamoclit L.
  I-po-moè-a quá-mo-clit
Convolvulaceae (Morning-glory Family)

DESCRIPTION: Twining, herbaceous, annual vine; leaves pinnately divided into narrow, linear segments; corolla crimson, with a long tube; flowers September–frost:

RANGE-HABITAT: Native to tropical America; now widely naturalized from VA south to FL and west to KS and TX; in SC, common throughout on roadsides, along fences, in cultivated fields, and in other disturbed areas.

COMMENTS: The common name refers to its leaves, which resemble the terminal, needle-bearing twigs of cypress trees. This species is heavily visited by Ruby-throated Hummingbirds. It is treated as a weed in most gardens, but leaving a few of these vines where they are unobtrusive will certainly increase sightings of hummingbirds.

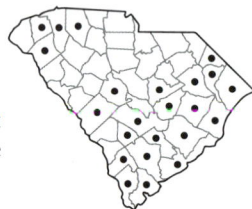

1019. Frost Aster (A)

Symphyotrichum pilosum
  (Willdenow) G. L. Nesom
  var. *pilosum*
    Sym-phy-o-trì-chum
    pi-lò-sum var. pi-lò-sum
Asteraceae (Aster Family)

SYNONYM: *Aster pilosus* Willdenow—RAB

Long-stalked Aster (B)

Symphyotrichum dumosum (L.) G. L. Nesom var. *dumosum*
　　Sym-phy-o-trì-chum du-mò-sum var. du-mò-sum
Asteraceae (Aster Family)

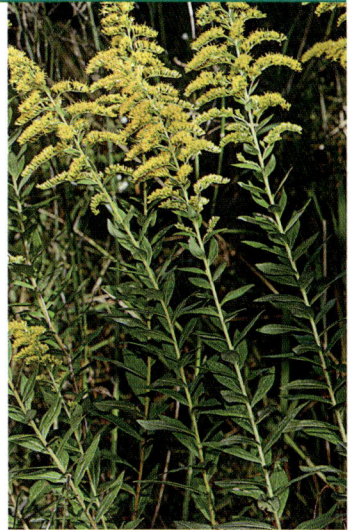

SYNONYM: *Aster dumosus* L.—RAB

DESCRIPTION: Perennial herbs, 2–5′ tall, sometimes with short rhizomes, forming clumps; stems of *S. pilosum* pubescent; stems of *S. dumosum* smooth; leaves linear to elliptic, pubescent in *S. pilosum*, smooth in *S. dumosum*: flowers produced in involucrate heads of white ray flowers and yellow disk flowers; heads numerous, arranged in a paniclelike structure; flowers September–frost, and sporadically at other times of the year.

RANGE-HABITAT: Both species are widespread and common throughout SC in fields, roadsides, and other open, disturbed habitats; *S. dumosum* is also common in Longleaf Pine flatwoods and savannas, where it assumes a much smaller stature.

COMMENTS: Frost Aster and Long-stalked Aster are among the most common white "asters" visible in great masses in open, disturbed habitats and roadsides throughout SC. Frost Aster is much shorter-lived and often transient, invading an area early in the successional process. Both are very important food sources for native pollinators and attract huge numbers of bees, flies, and butterflies, including migrating monarchs. Homeowners are encouraged to leave some areas of their landscape feral to provide flowering of these species to support SC's native biodiversity.

1020. Common Goldenrod;
　　Field Goldenrod

Solidago altissima L. var. *altissima*
　　So-li-dà-go al-tís-si-ma var. al-tís-si-ma
Asteraceae (Aster Family)

DESCRIPTION: Perennial herb, 2–7′ tall, with long and slender rhizomes; stems with appressed hairs; lower leaves narrowly lanceolate, 3-nerved; flowers produced in involucrate heads of a few small yellow ray flowers and yellow disk flowers; heads all directed in one direction (secund) in a pyramidal, paniclelike arrangement; flowers September–October.

RANGE-HABITAT: Native to the eastern US and common throughout SC in dry soils in old fields, pastures, meadows, and along roadsides.

COMMENTS: Common Goldenrod is one of the most abundant and widely distributed species of Solidago. Identification is facilitated by the usual presence of stem ball galls produced by the fly Eurosta solidaginis, which is usually on the mid to upper stems. Common Goldenrod is the state wildflower of SC. The species aggressively spreads via stolons to form large patches. In prairie restoration projects and prairie gardens, it often becomes a nuisance by crowding out less-aggressive species. Though often blamed for allergies, goldenrods simply flower at the time of year that ragweeds (*Ambrosia* spp.) flower. Goldenrods have heavy, insect-pollinated flowers and ragweeds, the true culprit, have light, wind-dispersed pollen that causes the allergies.

1021. Huguenot Fern;
Spider Brake Fern (A)

Pteris multifida Poiret
Pté-ris mul-tí-fi-da
Pteridaceae (Maidenhair Fern
Family)

Ladder Brake (B)

Pteris vittata L.
Pté-ris vit-tà-ta
Pteridaceae (Maidenhair Fern
Family)

A

B

DESCRIPTION: Small evergreen ferns; P. *multi-fida*: rhizome slender, with dark brown scales, sending out close-set spreading leaves (fronds); rachis winged; blade divided pedately below and pinnately above into 5–7 well-spaced, long, narrow segments; sori marginal; P. *vittata*: rootstock stout, to 0.4″ wide, with orangish scales; rachis not winged; fronds with blade pinnately divided, lanceolate in outline; pinnae simple, broadly linear with serrate margins.

RANGE-HABITAT: P. *multifida*: native to the Old World tropics; naturalized from New Hanover County, NC, south to FL and west to TX; in SC, locally common in the Coastal Plain in circumneutral soils of wooded slopes and ravines and in damp crevices of masonry around towns and cemeteries; P. *vittata*: native to Asia; naturalized from SC south to FL and west to Texas; in SC, locally common on masonry in larger, older, coastal towns.

COMMENTS: Huguenot Fern and Ladder Brakes escaped from cultivation. Their presence in the crevices of masonry is due to the calcium content from the limestone that was used to make the masonry. Over the years, as the masonry weathers, crevices are created that allow for the establishment of new plants from spores. Repointing of brick foundations with modern cement is reducing the habitat for Huguenot Fern, and its numbers are decreasing.

The common name, Huguenot Fern, refers to its 1868 discovery on the masonry in the cemetery of the Huguenot Church on Church Street in Charleston. It still grows there and at the adjacent St. Philip's Episcopal Church. Huguenot Fern and Ladder Brake are readily grown in gardens.

1022. Moonwort; Winter Grapefern

Sceptridium lunarioides
(Michaux) Holub
Scep-trí-di-um lu-na-ri-oì-des
Ophioglossaceae (Adder's-tongue
Family)

SYNONYM: *Botrychium lunarioides*
(Michaux) Swartz—RAB

DESCRIPTION: Small fern, typically less than 4″ tall; leaf (frond) blades 2–3 pinnate, less than 3″ long and 5″ wide; held horizontally; pinnules (ultimate segments) fan-shaped with a broadly rounded tip, the margins minutely toothed (denticulate); pinnules lack a midrib but have multiple veins arranged like the ribs of a fan; spores produced January-April.

RANGE-HABITAT: NC south to FL and west to OK and TX; in SC, very rare, or rarely observed and scattered in the Piedmont and

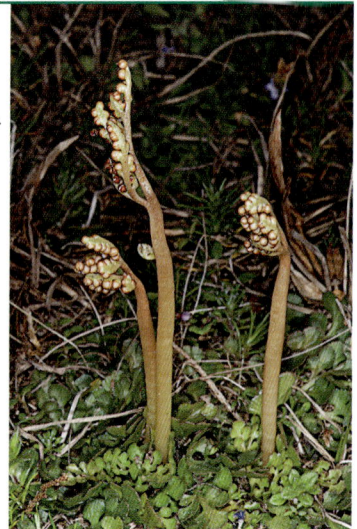

Coastal Plain; most frequently encountered in old graveyards, often associated with Little Bluestem (*Schizachyrium scoparium* (Michaux) Nash), also on the margins of rock outcrops in shallow soil and in old fields.

COMMENTS: This very strange, small fern is very rarely noticed. It produces fronds in the late fall and is green through the winter. The broad, fan-shaped pinnules are distinctive among SC grapeferns. The common name Moonwort comes from the shape of the pinnules, which resemble some phases of the moon. Throughout its range it is associated with old graveyards. The tendency to grow in this unlikely habitat has given rise to much folklore concerning this peculiar fern.

CONSERVATION STATUS: SC-Critically Imperiled

GLOSSARY

The terms marked with an asterisk (*) are illustrated in figure 8 in the Appendix.

ACE Basin. Lower part of the watershed of the Ashepoo, Combahee, and Edisto Rivers.

Achene. A small, dry, one-loculate, one-seeded, indehiscent fruit (e.g., in the Asteraceae—sunflower family).

Acuminate. Tapering to the apex, the sides more or less pinched in before reaching the tip.

Acute. Ending in a point less than 90° (a right angle); applied to tips and bases of structures.

Aerenchyma. Aerating tissue in aquatic plants, characterized by large intercellular spaces filled with gasses; functions in flotation of the plant and storage and diffusion of gasses.

Alkaloid. Organic compound that is alkaline and produces a marked physiological effect in animals, including man (e.g., nicotine, morphine, and cocaine).

Alluvial soil. Soil developing from recent alluvium (material deposited by running water); exhibits no horizon development; typical of floodplains.

*Alternate leaves. Only one leaf at a node.

Amphibolite. Rock, usually metamorphic, with large amounts of amphibole, complex silicate minerals that weather to produce soils high in calcium, magnesium, and iron.

Angiosperm. A flowering plant.

Annual. Plant growing from seed to fruit in one year, then dying.

*Anther. The pollen producing part of the stamen.

Aphrodisiac. Stimulator of sexual desire.

Apical. At the tip or summit.

Appressed. Lying flat against something.

Aromatic. Having a fragrant, sweet-smelling, or spicy aroma.

Artesian. Water capable of rising to the surface by internal hydrostatic pressure.

Ascending. Growing obliquely upward at about a 40–60° angle from the horizontal.

Asexual reproduction. Reproduction that does not involve fusion of gametes; for example, fragmentation of a rhizome, each fragment growing into a new plant.

Asphyxia. Unconsciousness or death resulting from lack of oxygen.

Astringent. Agent that causes contraction of tissues, thereby lessening secretion.

Auricle. Any earlike lobe or appendage.

Axil. The angle formed by the upper side of a leaf and the stem from which it grows.

Axillary. In an axil.

Barrier island. Narrow islands of sand that run parallel to the shoreline; a river, marsh, or lagoon separates them from the mainland. Otter Island and Capers Island are examples of barrier islands.

Basal leaves. Leaves at the base of the stem.

Basic soil. Soil with an alkaline pH, that is, a pH greater than 7.0.

Berry. A simple, fleshy, usually indehiscent fruit with one or more seeds (e.g., a tomato or grape).

Biennial. Living for two years, then dying naturally.

**Bipinnately compound leaf.* Twice pinnate, the primary leaflets once again pinnate. (*See* pinnate)

Bisexual. Having both stamens and pistils, usually used in reference to a flower.

**Blade.* Flattened and expanded part of a leaf.

Blackwater river. River that originates in the coastal plain, receiving its water from local rain, with a narrow floodplain, its black color caused by organic acids from decaying leaves (e.g., the Black and Edisto Rivers).

Bract. Modified leaf, usually smaller than a foliage leaf, often situated at the base of a flower or inflorescence.

Bracteate. Having bracts.

Bractlet. A very small bract.

Branchlet. A small branch.

Brownwater river. River that originates in the mountains or Piedmont, has a wide, alluvial floodplain, with brown water colored by suspended silt and clay that originates from erosion of the Piedmont and mountains (e.g., the Santee and Savannah Rivers).

Bryophyte. Terrestrial, nonflowering, nonvascular plant comprising mosses, liverworts, and hornworts.

Bulb. An underground, fleshy enlargement of stem and leaves, as in the onion.

Buttress. Additional, often flattened, supporting tissue at the base of the trunk, as in Bald Cypress.

Calcareous. Consisting, having, or typical of calcium carbonate, calcium, or limestone (e.g., calcareous soil).

Calcicole. A plant that thrives in soil abundantly supplied with calcium ions.

*Calyx (pl. *calyces*).The collective term used to describe all the sepals of a flower.

Canopy. The top layer of leaf growth within most woody communities.

Capsule. A dry dehiscent fruit with more than one chamber; it may open by pores, by splitting vertically, or by the top coming off like a lid.

**Catkin.* A spikelike inflorescence bearing either male or female flowers, as in willows and oaks.

Cespitose. Tufted, growing in clumps.

Chasmogamy. Having flowers that open to expose reproductive organs, allowing cross-pollination.

Chlorophyll. The green pigment of plants that traps light energy and makes photosynthesis possible.

Ciliate. Beset with a marginal fringe of hairs.

Circumboreal. Around northern regions.

Circumneutral soils. Soils that are only slightly acidic or slightly alkaline; such soils are usually relatively high in calcium and magnesium.

Clasping. Partly wrapped around another structure, such as a leaf whose base wholly or partly wraps around or surrounds the stem.

Cleistogamy. Self-pollination within a perfect flower that does not open.

Climax community. A stable natural community culminating succession that is capable of self-perpetuation under prevailing environmental conditions.

Coadaptation. The adaptive response of two organisms to one another (e.g., flower structure and pollinator).

Colonial. Adjective of colony.

Colony. Growing in clumps produced asexually from underground structures such as rhizomes, rootstocks, stolons, or roots.

Column. In orchids, a structure formed by the union of stamens, style, and stigma; the supporting structure of the hood in pitcherplants.

Community. Group of interacting plants and animals inhabiting a given area.

Composite. Any member of the Asteraceae (sunflower family).

*Compound leaf. Leaf in which the blade is subdivided into two or more leaflets or pinnae.

Cone-bearing plants. Technically the gymnosperms, plants that produce seeds not enclosed by an ovary, as in pine trees.

Conifer. Any of the cone-bearing gymnosperms, such as Carolina hemlock (*Tsuga caroliniana*).

Connective. Filament extension between thecae.

Consumption. Colloquial term for tuberculosis.

Cordate. With a sinus and rounded lobes at the base of the leaf.

Corm. A thickened, vertical, underground stem with thin, scalelike leaves.

*Corolla. All the petals of a flower, separate or united; the inner whorl of the perianth.

Corymb. A flat-topped or rounded inflorescence with the outer flowers on longest stalks and opening first.

Cross-pollination. Pollination between two different plants of the same species.

Culm. The flowering stems of grasses and sedges.

Cultivated plant. A purposely grown plant; it may be a native plant moved from local woodlands into a garden or yard or a plant introduced from another country.

Cusp. A strong sharp point.

Cuticle. Waxy, noncellular layer on outer surface of epidermal wall of plant organs (mostly leaves) that prevents water loss.

Cyme. A broad, flattish inflorescence, the central flowers maturing first.

Deciduous. Falling away, not persistent or evergreen (e.g., leaves falling from oak trees in the autumn).

Decoction. An extraction of a plant made by boiling a plant part in water.

Decomposition. Breakdown of complex organic substances from dead organisms into simpler, inorganic substances.

Decumbent. Prostrate at or near the base, the upper parts erect or ascending.

Decurrent. Fused to the stem or leaf stalk and extending beyond the point of attachment, as in the leaf base of blackroot.

Dehiscent. Opening by pores or slits to discharge the contents.

Demulcent. A soothing usually mucilaginous or oily substance used for relieving pain in irritated mucous surfaces.

Dentate. Toothed with the teeth directed outward.

Dermatitis. Inflammation of the skin.

Detritus. Accumulated mass of partially decomposed remains of animals and plants that form in aquatic systems.

Diabase. Dark-colored, igneous rocks that are granitic and fine-grained in texture and weather to produce soils high in calcium and magnesium; diabase rocks are fine-textured variants of gabbro rocks.

Dichotomous. Two-forked, the branches equal or nearly so.

Dioecious. Having male and female flowers on different plants of the same species.

Disjunct. A population of plants growing far from its main range.

*Disk flower. The small tubular flowers in the central part of a floral head, as in most members of the Asteraceae (sunflower family).

*Dissected leaf. The blade cut into more or less fine divisions.

*Divided leaf. Any blade cut into divisions that reaches three-fourths or more of the distance from the margin to the midvein or to the base.

Doctrine of Signatures. Medieval belief that the key to human use of plants was hidden in the form of the plant itself (e.g., the red juice of bloodroot to treat blood disorders).

Dredged soil disposal site. Site used to dispose of soil dredged to maintain harbor or river depths; in the coastal area most often a section of banked marsh.

Drupe. A stone fruit; fleshy fruit with the single seed covered by a hard covering (stone).

Ecosystem. The biotic community and its abiotic environment functioning as a system.

Ecotone. Transitional area between two different communities, having characteristics of both, yet having a unique character of its own.

*Elliptic. Widest in the center and narrowed to two equal ends.

Emarginate. Having a shallow notch at the tip.

Emergent. Aquatic plant with its lower part submerged and its upper part extended above water.

Emergent, nonpersistent. Emergent marsh species that fall to the surface of the water at the end of the growing season.

Emergent, persistent. Emergent marsh species that usually remain standing at least until the beginning of the next growing season.

Emersed. Rising above the surface of the water; applies to leaves.

Emetic. An agent that induces vomiting.

Endemic. Restricted to a small area or region.

Endosperm. The nutritive tissue of most seeds.

*Entire leaf. Leaf margin without teeth, lobes, or divisions.

Ephemeral. Lasting only a short time.

Epiphyte. A plant growing on another plant but obtaining no nutrition from it; often referred to as an air plant (e.g., Spanish moss).

Ericaceous. Having characteristics similar to those of the heath family (Ericaceae).

Ericad. A member of the Ericaceae (heath family).

Escarpment. A long cliff or steep slope separating two comparatively level or more gently sloping surfaces and resulting from erosion or faulting.

Essential oil. Volatile oils with characteristic odor, composed of various constituents and contained in plant organs.

Estuary. An area where freshwater and seawater meet and mix.

*Even-pinnately compound. Compound leaves that have an even number of leaflets; this is easily determined by the terminal pair of leaflets.

Evergreen. Bearing green leaves throughout the year; holding live leaves over one or more winters until new ones appear.

Exfoliating. To come off or separate as scales, flakes, sheets, or layers.

Facultative. Ability to adjust optionally to different environmental conditions.

Falcate. Sickle-shaped.

Fascicle. A small bundle or tuft, as of leaves.

Febrifuge. Agent that relieves or reduces fever.

Female flower. With pistils and without fertile stamens.

Fibrous roots. Root system composed of a mass of fiberlike roots with no main root predominating.

*Filament. Part of the stamen that bears the anther.

Filiform. Threadlike; long and very slender.

Flatwoods. Poorly drained, low-lying, nearly level timberland.

Fleshy. A plant having tissue that serves to store moisture, such as a cactus.

Floodplain. A level, flat area bordering a river, subject to frequent flooding.

Flora. Collective term to refer to all plants of an area; a book dealing with the plants of an area.

Flowering plants. Technically the Angiosperms, plants that produce seeds enclosed in an ovary.

*Flower stalk. The stalk of a single flower.

Fluvial. Caused by the action of flowing water.

FMNF. Francis Marion National Forest.

Follicle. A dry fruit produced from a single ovary, opening along one suture.

Food. An organic compound from which an organism can derive a source of energy and contributes materially to growth and repair of tissues.

Forb. Herbaceous plant other than a grass, sedge, or rush.

Forest. Vegetation dominated by trees with their crowns overlapping, generally forming 60–100% cover.

Frond. In Lemnaceae (duckweed family), the expanded leaflike stem that functions as a leaf; the leaf of ferns.

Fruit. The seed-bearing structure of the plant; a matured ovary with its contents, often with attached parts.

Fusiform. Cylindrical except thick near the middle and tapering to both ends.

Gabbro. Dark-colored, dense, igneous rocks that are medium to coarse-grained in texture, very tough, and weather to produce soils high in calcium and magnesium.

Glade. An open space in a forest or woodland.

Gland. A secreting surface or structure, or an appendage having the general appearance of such an organ.

Glandular. Bearing glands.

Glaucous. Whitened with a bloom.

Globose. Globular or spherical.

Gneiss. A metamorphic rock corresponding in composition to granite.

Granite. A coarse-grained igneous rock.

Gymnosperm. A seed plant in which the seeds are not enclosed in an ovary; examples include pines, firs, spruces, and cedars.

Habitat. Place where a plant or animal lives.

Halophyte. Plant able to survive and complete its life cycle in high salinity.

Hardpan. A cemented or compacted and often clayey layer in soil that is impermeable to water.

Hastate. Like an arrowhead but diverging at the base.

Haustorium. A bridge of xylem between host and parasite in plants through which water, minerals, and limited amounts of food pass from host to parasite.

*Head. A dense inflorescence of sessile or subsessile flowers on a short or broadened axis, as in the Asteraceae (sunflower family).

Hemiparasite. Dependent on the host for water and minerals; contains chlorophyll and can make its own food, but may receive some food from the host, as in chaff-seed (*Schwalbea americana*).

Herb. Having no persistent woody stem above ground; also, a plant used in seasoning.

Herbaceous. Having the characteristic of an herb.

Herbarium. A collection of dried plants; an institution that houses a collection of dried plants.

Heterotrophic. Requiring a supply of organic matter (food) from the environment.

Hip. The fleshy to leathery hollow fruit of roses.

Holoparasite. Parasite completely dependent on the host for water, minerals, and food, as in beech-drops (*Epifagus virginiana*).

Hummock. A low mound or ridge of earth; like the small hummocks that rise above the salt marsh along the coast.

Humus. Organic material derived from partial decay of plant and animal matter.

Hydric. Habitats characterized by an abundant water supply.

Hydrophyte. A plant that grows in water.

Hypanthium. A cup-shaped or tubular organ below, around, or adhering to the side of the ovary.

Hyphae. The threadlike filaments that make up the mycelium, or major part of the body of a fungus.

Indehiscent. Remaining persistently closed; not opening by definite pores or sutures.

Indigenous. Native to an area.

Inflorescence. A flower cluster on a plant, or (especially) the arrangement of flowers on a plant.

Infusion. An extraction of a plant made by soaking the plant part in water.

Internode. The portion of a stem between two nodes.

Introduced. Plant brought intentionally from another area; such a plant may escape and become naturalized. For example, *Crotalaria spectabilis* was brought as a green manure from India.

Involucre: A dense cluster of bracts that subtend an inflorescence, as in the Asteraceae.

*Irregular flower. A flower with petals that are not uniform in shape but are usually grouped to form upper and lower "lips."

Knee. Vertical outgrowth from the lateral roots of trees growing on soil subjected to long periods of inundation; the function is unknown (e.g., Bald Cypress).

*Lanceolate. Lance-shaped, much longer than wide and broadest near the base.

*Leaflet. One of the leaflike parts of a compound leaf.

*Leaf stalk. The basal stalk of a leaf.

Legume. A dry fruit from a single ovary usually dehiscent along two sutures.

Lenticels. Small openings in the bark of roots and stems of flowering plants used for gas exchange.

Levee. Soil deposits along channels of large rivers that form a natural embankment.

Lichen. Unique composite organism formed by a symbiotic relationship between some sac fungi (and, to a lesser extent, club fungi) and a photosynthetic partner, either a blue-green or green algae.

Limestone. Sedimentary rock composed mostly of calcium carbonate.

*Linear. Narrow and elongate with essentially parallel sides.

*Lip. The lower petal of some irregular flowers, often showy, as in the orchids.

Litter. Accumulated mass of partially decomposed remains of plants (and animals) that collects on the forest floor.

*Lobed leaf. Blade divided into parts separated by rounded sinuses extending one-third to one-half the distance between the margin and the midrib.

Locular. Having one or more locules.

Locule. Compartment of an ovary or an anther.

Lowcountry. Physiographically the Outer Coastal Plain; traditionally the tidewater and rice growing area of South Carolina.

Mafic. Dark-colored minerals high in magnesium and iron.

Male flower. Having stamens and no functional pistil.

Malodorous. Having a foul odor.

Manual. Book that provides an inventory of the flora of a specific region and provides keys to identifying plants.

Many-ranked. Leaves that are arranged in many rows along the stem.

Maritime. Located on or close to the sea.

Marl. Sedimentary rock formation composed of unconsolidated mixture of 35–65% calcium carbonate and 65–35% clay.

Marsh. Wetland dominated by emergent, herbaceous vegetation.

Marsh, brackish. Marsh flooded regularly or irregularly by water of low salt content.

Marsh, freshwater. Marsh saturated or flooded with freshwater.

Marsh, tidal freshwater. The zone of marsh along a coastal river above the influence of salt water (the salt point), and to the point where the tidal amplitude vanishes.

Mesic. Moist but well-drained soils.

Mesophyte. Plant adapted to a mesic environment.

Monadnock. A hill or mountain of resistant rock surmounting a peneplain.

Monoecious. Having both male and female flowers on the same plant.

Monotypic. A genus with only a single species in it.

Montane. Related to mountains.

Mucilage. A substance of varying composition produced in cell walls of plants; hard when dry, swelling and slimy when moist.

Mycelium. A large, entangled network of filaments that forms the body of a fungus.

Mycorrhiza (pl. *mycorrhizae*). A special compound structure formed between a fungal mycelium and certain underground parts of a vascular plant, particularly a root.

Mycotrophy. The nutrition of most vascular plants (and bryophytes) directly tied up with the nutrition of higher fungi.

Native plant. One that originated in the area where it grows.

Naturalized. Plant from another area that is thoroughly established in a new area because it is able to naturally and successfully reproduce; for example, white clover introduced from Europe is now naturalized throughout the southeastern United States.

Natural selection. Natural process that results in survival of the best-adapted individual of a species and elimination of individuals less well adapted to their environment.

Naval stores. Crude turpentine, and its products (e.g., pitch, resin), from southern yellow pines used chiefly in connection with wooden sailing ships.

Nectar. Sweet substance secreted by special glands (nectaries) in flowers and in certain leaves.

Needle. A stiff, narrow leaf, as in pine trees.

Nitrogen fixation. Conversion of atmospheric nitrogen to forms usable by plants.

Node. Point on a stem where one or more leaves are borne or attached.

Nomenclature. The assignment of scientific names to plants and animals.

Nut. Indehiscent, one-seeded fruit having a hard outer wall, as in oaks and hickories.

Nutlet. A small nut or nutlike fruit.

Oblanceolate. Lanceolate, and attached at the narrow end.

Obligate. Limited to one mode of life or action, as an obligate parasite.

Oblique. Sides unequal, especially the base of a leaf.

Oblong. Elongate and with parallel (or nearly parallel) sides.

Obovate. Ovate and attached at the narrow end.

Obtuse. Blunt or rounded at end, the angle at the end exceeds 90° but is less than 180°.

Odd-pinnately compound leaf. Refers to compound leaves having an odd number of leaflets; this is usually easily determined because there is a single, terminal leaflet.

Opposite leaves. Two leaves inserted at the node opposite each other on the stem.

Ornamental. A plant cultivated for its beauty.

Outcrop. A stratum or formation, as of limestone or marl or granite, that protrudes above the soil.

Ovary. The basal, enlarged part of the pistil that contains the ovules or seeds.

Ovate. Egg-shaped and attached at the broad end.

Ovule. The structure that develops into the seed.

Palmate. Having three or more divisions or lobes, looking like the outspread fingers of a hand.

Palmately compound leaf. The leaflets diverge from a common point at the end of the leaf stalk.

Panicle. A compound inflorescence in which the main axis is branched one or more times and supports racemes.

Parasitic. A plant (or animal) deriving food or mineral nutrition, or both, from another living organism.

Peat. The partially decomposed remains of plants and animals.

Peduncle. The main flower stalk, supporting either a cluster of flowers or the only flower.

Peltate. Leaf attached to its stalk inside the margin, as in species of *Hydrocotyle.*

Peneplain. A nearly flat land surface representing an advanced stage of erosion.

Perennating. Surviving from one year to the next.

Perennial. A plant lasting for three or more years.

Perfect flower. A flower having both female (pistil) and male (stamen) parts.

Perfoliate. Describes those stalkless leaves whose base surrounds the stem, the stem thus apparently passing through it.

Perianth. The calyx and corolla collectively; the calyx alone if the corolla is absent.

Persistent. Remaining attached; not falling off.

Petal. One of the individual parts, separate or united, of the corolla.

pH. A measure of the acidity or alkalinity of a solution (e.g., the soil solution).

Pharmacopoeia. Book containing an official list of drugs along with recommended procedures for their preparation and use.

Photodermatitis. Dermatitis in the form of a sunburnlike rash resulting from contact with plants and containing compounds that sensitize the skin to subsequent exposure to ultraviolet light.

Photosynthesis. Synthesis of carbohydrates from carbon dioxide and water by chlorophyll-containing plants using light as energy and releasing molecular oxygen as a by-product.

Pinna (pl. *pinnae*). One of the first or primary divisions of a pinnately compound leaf; applied especially in ferns.

Pinnate. Arranged along the sides of a common axis.

*Pinnately compound leaf. Leaf with leaflets placed on either side of the rachis; featherlike.

Pinnatifid. Pinnately lobed, cleft, or parted, usually halfway or more to the midrib.

*Pistil. The central, seed-bearing organ of a flower, usually composed of stigma, style, and ovary; the female part of a flower.

Pith. The central portion of a dicot stem.

Plantlet. Literally a small or young plant; used in this book to refer to vegetative offshoots produced by some specialized horizontal stems or from buds in the inflorescences of some plants that can develop into new plants.

Pollen. Spores formed in the anthers that produce the sperm.

Pollination. The transfer of pollen from an anther to a stigma.

Pollinium (pl. *pollinia*). An agglutinated pollen mass in orchids and other plants.

Polygamous. Having perfect, male, and female flowers all on the same plant.

Polymorphic. With three or more forms, such as the entire, two-lobed, or three-lobed leaves in sassafras.

Pome. A simple, fleshy fruit like an apple in which the fleshy part is derived from nonovary parts.

Poultice. A moist, soft mass (usually heated) of an adhesive substance, such as meal or clay, spread on cloth and applied to warm, moisten, or stimulate a sore or inflamed part of the body.

Prairie. An extensive tract of flat or rolling grassland.

Prescribed burn. An intentionally set fire, usually by professional foresters or land managers, under the right conditions of wind, humidity, and temperature to remove ground litter, reduce wildfire threat, promote growth of desirable plants, and control underbrush buildup.

Prickle. A small, usually slender outgrowth of the epidermis.

Primary succession. Vegetational development starting on a site never before colonized by plant communities.

Propagule. Any of the various structures of plants capable of developing into a new individual.

Prostrate. Lying flat on the ground; if a stem, may or may not root at the nodes.

Pubescent. Covered with short, soft hairs.

Punctate. With translucent or colored dots, depressions, or pits scattered over the surface.

Pungent. Affecting the organs of smell or taste with a strong, acrid sensation.

Purgative. Tending to cleanse or purge, especially tending to cause evacuation of the bowels.

Quartzite. Hard and tough rock that consists mostly of quartz; usually recrystallized sandstone.

Quinine. Main drug for treatment of malaria, derived from bark of species of *Cinchona* native to South America.

*Raceme. A simple, indeterminate inflorescence of stalked flowers borne on a single more or less elongated axis.

*Rachis. The central elongated axis of an inflorescence or a compound leaf.

Rank. A vertical row.

*Ray flower. The regular flowers around the edge of the head in many members of the Asteraceae (sunflower family); each ray flower resembles a single petal.

*Receptacle. The base of the flower, where flower parts are attached.

Recurved. Curved outward, downward, or backward.

Reflexed. Abruptly recurved or bent downward or backward.

*Regular flower. With petals and/or sepals arranged around the center, like the spokes of a wheel; always radically symmetrical.

Resin. Any of numerous clear or translucent yellow or brown solid or semi-solid viscous substances of plant origin, such as amber.

Resinous. With the appearance of resin; glandular-dotted.

Revolute. Rolled backward, with margins rolled toward the lower side.

Rhizomatous. Bearing rhizomes.

Rhizome. Underground stem, usually horizontally orientated; sometimes functions in food storage.

Rhizosphere. The soil immediately surrounding the root system.

Rootstock. An erect, rootlike stem or branch under or sometimes on the ground.

Rosette. Arrangement of leaves radiating from a crown or center, usually at or close to the ground.

Ruderal. Growing in waste places or among rubbish.

Runcinate. Margins that are coarsely serrate to sharply incised with the segments pointing toward the base, as in Common Dandelion (*Taraxacum officinale,* plate 615).

Sagittate. Like an arrowhead, with the basal lobes pointing downward or inward toward the leaf stalk.

Saline. Of, relating to, or containing salt; salty.

Samara. A dry, indehiscent, winged fruit, as in red maple (plate 52).

Saprophyte. An organism that derives its nourishment from dead or decaying organic matter.

Savanna. A flat area with widely spaced trees, usually dominated by grasses.

Scape. A leafless stem bearing flowers and rising from the ground or near it.

Scapose. Having a scape.

SCDNR. South Carolina Department of Natural Resources.

SCDPRT. South Carolina Department of Parks, Recreation and Tourism.

Schist. Highly metamorphosed, crystalline rocks, usually splitting along parallel planes, and named for their most prominent mineral (e.g., mica schist).

Secondary forest. The forest occupying a site where the original forest was removed.

Secondary succession. Plant succession taking place on sites that previously supported vegetation.

*Seed. The matured ovule consisting of an embryo, seed coat, and stored food.

Seepage. Water that has passed slowly through porous soils.

Self-fertilization. Fertilization of an egg by sperm from the same flower or by sperm from another flower of the same plant.

Self-pollination. The transfer of pollen from an anther to a stigma on the same plant.

*Sepals. One of the parts of the calyx, either separate or united.

Serotinous cones. Cones that remain on the tree several years and require the heat of fire to open them to release the seeds, as in Pond Pine (*Pinus serotina*).

Serrate. Having sharp teeth pointed terminally.

Sessile. Without a stalk of any kind, as a sessile leaf.

Sexual reproduction. Reproduction resulting from the fusion of gametes (sex cells).

Shade tolerance. Capacity of a tree to develop and grow to maturity in the shade.

Sheath. A tubular envelope, usually used for that part of the leaf of a sedge or grass that envelops the stem.

Shoal. A shallow area in a body of water; if covered with rocks, a rocky shoal.

Shrub. A woody plant that remains low to the ground and produces several shoots or trunks from the base.

*Simple leaf. A leaf with a blade in a single part, although it may be variously divided.

Sinus. The depression or recess between two adjoining lobes.

Slate. Metamorphosed shale (clays hardened into rocks by heat and pressure).

Sp. (pl. *spp.*). Species.

*Spadix. A spike with a fleshy axis in which the flowers are embedded, as in jack-in-the-pulpits.

*Spathe. A large bract enclosing an inflorescence.

Sphagnum. A large genus of distinctive mosses that grow in wet, acidic sites; often called peat mosses because of their importance in the formation of peat.

*Spike. An elongated, indeterminate inflorescence of sessile or subsessile flowers.

Spine. A sharp-pointed, rigid, deep-seated outgrowth from the stem, not pulling off with the bark, as in the Cactaceae (cactus family).

Spur. A tubular or saclike projection from a petal or sepal.

*Stamen. The pollen-producing organ of a flower; the male part of a flower.

Staminode. A sterile stamen.

Staminodium (pl. *staminodia*). A sterile stamen or any structure lacking an anther but corresponding to a stamen.

*Standard. Uppermost petal in a pea flower; also called the banner.

*Stigma. The part of the pistil that receives the pollen; usually hairy or sticky.

Stipitate. Provided with a stipe or with a slender stalklike base.

*Stipule. Small appendages, often leaflike, on either side of a leaf stalk at the base.

Stolon. A slender stem that runs along the surface of the ground, or just below, and produces a new plant at the tip.

Stoma (pl. *stomata*). Small opening on the surface of a leaf through which gas exchange takes place with the atmosphere.

Strigose. With sharp, stiff, straight appressed hairs that are often basally swollen.

Style. The elongated portion of the pistil that connects the stigma and ovary.

Styptic. Agent that stops bleeding.

Subcanopy. The layer of leaves just below the canopy in a forest.

Subclimax community. A long persisting vegetational stage immediately preceding the climax community.

Submerged. Growing entirely under water.

Subsessile. Nearly sessile; having almost no stalk.

Subtending. Situated closely beneath something, often enclosing or embracing it.

Subterranean. Below the surface of the ground.

Subulate. Narrowly triangular and tapering to a sharp point.

Sucker. Lateral underground shoot that breaks from the roots or rhizomes and roots itself, forming an independent, individual plant.

Swamp. Forested, freshwater wetland often with saturated soil or standing water.

Symbiosis. Two or more individuals of different species living together in intimate association, as in the fungal-algal symbiosis in the lichens.

Synonym. An invalid or illegitimate scientific name no longer in currant taxonomic use.

Taproot. A large, elongated root, usually vertical.

Tendril. A slender twining or clasping structure that enables plants to climb.

Ternate. Arranged in threes.

Terrestrial. Growing or living on land.

Theca (pl. *thecae*). One half of an anther containing two pollen sacs.

Topography. Physical structure of the landscape.

Trailing. Prostrate but not rooting.

Transpiration. Loss of water vapor from aerial parts of land plants.

Tree. A perennial woody plant of considerable stature at maturity and with one or few main trunks.

Trifoliolate leaf. A compound leaf with three leaflets.

Truncate. Base or apex essentially straight across.

Tuber. Fleshy, thickened, short, usually subterranean stem having numerous buds called "eyes," such as the potato.

Tuft. A dense clump, especially of bushes or trees.

Twining. Ascending by coiling around a support.

Two-ranked. Leaves that are in two rows along a stem.

Umbel. A flat-topped or rounded inflorescence having flowers on stalks of nearly equal length and attached to the summit of the peduncle, the characteristic order of blooming being from the outside toward the center.

Unisexual. Of one sex, either male or female.

USFS. The United States Forest Service.

Utricle. A bladderlike, one-seeded, usually indehiscent fruit.

Valve. One of the parts or segments into which a dehiscent fruit splits, as in a legume.

Vascular plant. Any of various plants typified by a conducting and supporting system of xylem and phloem; includes the ferns and cone-bearing and flowering plants.

Velamen. A specialized moisture-absorbing tissue, especially in roots of epiphytic orchids.

Vermifuge. An agent that expels worms from the intestine.

Vine. A plant that climbs by tendrils or other means, or creeps or trails on the ground.

Water, brackish. Water that has salt concentration greater than freshwater and less than seawater.

WCL. Wildflowers of the Carolina Lowcountry and Lower Pee Dee (Porcher, 1995).

Weed. In the broadest sense, a plant growing in a place it is not wanted. Usually weeds are aggressive colonizers of disturbed areas and are frequently nonnative; they are generally noxious and of no economic value and compete with agricultural crops.

*Whorled leaves. Three or more leaves inserted at one node.

*Wing. A thin, flat extension found at the margins of plant parts; the lateral petal in a pea flower.

Witches broom. An abnormal broomlike growth of weak, closely clustered shoots or branches on a woody plant caused by fungi or viruses.

Woodland. Open stands of trees with crowns not usually touching, generally forming 25–60% cover.

Xeric. Dry soils and sites, or adapted to dry conditions.

Xerophyte. Plant adapted to a xeric environment

Illustrations of Plant Structures

This section presents the shapes and arrangements of basic flower and leaf parts (figure 8). The illustrations will help clarify the terminology in the text. Each term is defined in the glossary.

Illustrations of Plant Structures.

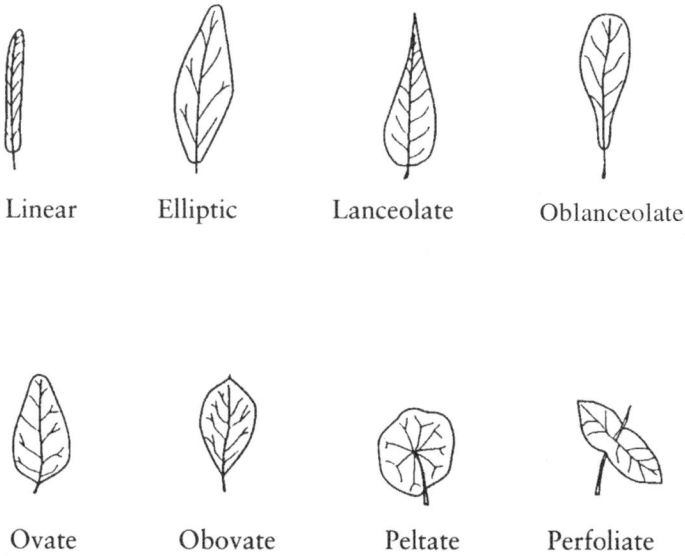

Leaf Shapes

Linear Elliptic Lanceolate Oblanceolate

Ovate Obovate Peltate Perfoliate

Leaf Parts

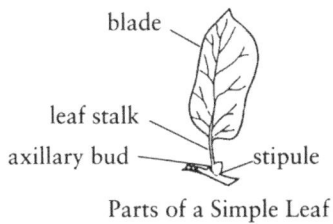

blade

leaf stalk

axillary bud ———— stipule

Parts of a Simple Leaf

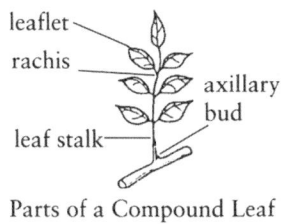

leaflet

rachis

axillary bud

leaf stalk

Parts of a Compound Leaf

Leaf Arrangement

Alternate

Opposite

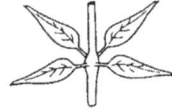

Whorled

Types of Compound Leaves

Palmately Compound

Trifoliolate

Odd-pinnate

Even-Pinnate

Bipinnate

Leaf Margins

Entire   Divided

Dissected   Lobed

Parts of a Generalized Flower

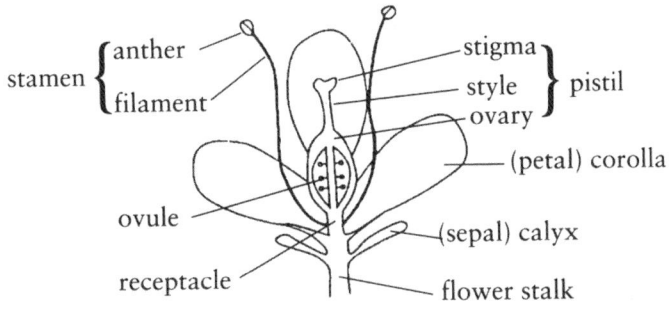

stamen { anther / filament }   stigma / style / ovary } pistil

(petal) corolla

ovule

(sepal) calyx

receptacle

flower stalk

Irregular Flowers

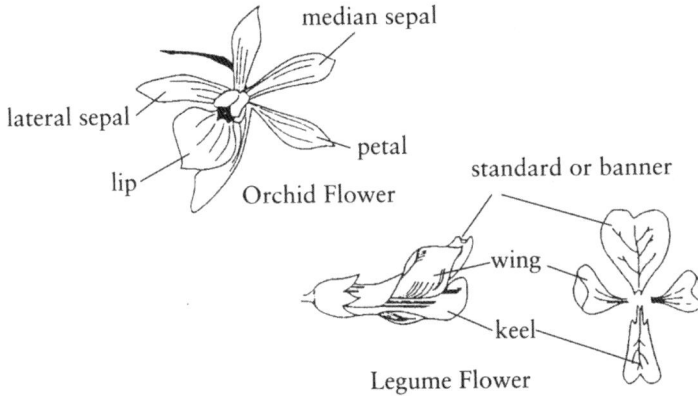

median sepal

lateral sepal

petal

lip

Orchid Flower

standard or banner

wing

keel

Legume Flower

Types of Inflorescences

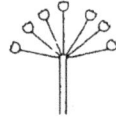

Raceme   Spike   Umbel

Panicle   Catkin

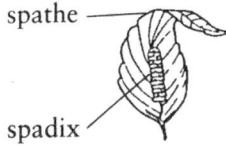

Spadix

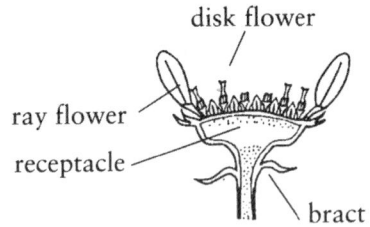

Head

GENERAL REFERENCES

Popular Wildflower Books for the Southeastern States

Ajilvsgi, Geyata. 1979. *Wild Flowers of the Big Thicket: East Texas, and Western Louisiana.* College Station: Texas A&M University Press.

Batson, Wade T. 1987. *Wildflowers in the Carolinas.* Columbia: University of South Carolina Press.

Bell, C. Ritchie, and Anne H. Lindsey. 1990. *Fall Color and Woodland Harvests: A Guide to the More Colorful Fall Leaves and Fruits of the Eastern Forests.* Chapel Hill, NC: Laurel Hill Press.

Bell, C. Ritchie, and Bryan J. Taylor. 1982. *Florida Wild Flowers.* Chapel Hill, NC: Laurel Hill Press.

Brown, Clair A. 1972. *Wildflowers of Louisiana and Adjoining States.* Baton Rouge: Louisiana State University Press.

Case, Frederick W., Jr., and Roberta B. Case. 1997. *Trilliums.* Portland, OR: Timber Press.

Coffey, Timothy. 1993. *The History and Folklore of North American Wildflowers.* New York: Houghton Mifflin.

Cotterman, L., D. Waitt, and A. Weakley. 2019. *Wildflowers of the Atlantic Southeast: A Timber Press Guide.* Portland, OR: Timber Press.

Dean, Blanche E., Amy Mason, and Joab L. Thomas. 1973. *Wildflowers of Alabama and Adjoining States.* Tuscaloosa: University of Alabama Press.

Duncan, Wilbur H., and Marion B. Duncan. 1999. *Wildflowers of the Eastern United States.* Athens: University of Georgia Press.

Duncan, Wilbur H., and Leonard E. Foote. 1975. *Wildflowers of the Southeastern United States.* Athens: University of Georgia Press.

Gaddy, L. L. 2000. *A Naturalist's Guide to the Southern Blue Ridge Front.* Columbia: University of South Carolina Press.

Gupton, Oscar W., and Fred C. Swope. 1979. *Wildflowers of the Shenandoah Valley and Blue Ridge Mountains.* Charlottesville: University Press of Virginia.

Gupton, Oscar W., and Fred C. Swope. 1982. *Wildflowers of Tidewater Virginia.* Charlottesville: University Press of Virginia.

Hemmerly, Thomas E. 2000. *Appalachian Wildflowers.* Athens: University of Georgia Press.

Hemmerly, Thomas E. 1990. *Wildflowers of the Central South.* Nashville, TN: Vanderbilt University Press.

Hunter, Carl G. 1984. *Wildflowers of Arkansas.* Little Rock, AR: Ozark Society Foundation.

Justice, William S., and C. Ritchie Bell. 1968. *Wild Flowers of North Carolina.* Chapel Hill: University of North Carolina Press.

Kingsbury, John M. 1964. *Poisonous Plants of the United States and Canada.* Englewood Cliffs, NJ: Prentice-Hall.

Martin, Laura C. 1989. *Southern Wildflowers.* Atlanta, GA: Longstreet Press.

Midgley, Jan W. 1999. *All About Mississippi Wildflowers.* Birmingham, AL: Sweetwater Press.

Murdy, William H., and Eloise Brown Carter. 2000. *Guide to the Plants of Granite Outcrops.* Athens: University of Georgia Press.

Niering, William A., and Nancy C. Olmstead. 1979. *The Audubon Society Field Guide to North American Wildflowers (Eastern Region).* New York: Alfred A. Knopf.

Nourse, Hugh, and Carol Nourse. 2000. *Wildflowers of Georgia.* Athens: University of Georgia Press.

Rickett, Harold W., and the New York Botanical Garden. 1967. *Wild Flowers of the United States.* Volume 2, *The Southeastern States.* New York: New York Botanical Garden and McGraw-Hill.

Schmidt, J. M. & J. A. Barnwell. 2002. "A Flora of the Rockhill Blackjacks Heritage Preserve." *Castanea* 67 (3): 247–79.

Smith, Richard M. 1998. *Wildflowers of the Southern Mountains.* Knoxville: University of Tennessee Press.

Sorrie, B. A. 2011. *A Field Guide to the Wildflowers of the Sandhills Region: A Southern Gateway Guide.* Chapel Hill: University of North Carolina Press.

Taylor, Walter K. 1992. *The Guide to Florida Wildflowers.* Dallas, TX: Taylor Publishing.

Timme, S. Lee. 1989. *Wildflowers of Mississippi.* Jackson: University Press of Mississippi.

Wharton, Mary E., and Roger W. Barbour. 1979. *Wildflowers and Ferns of Kentucky.* Lexington: University of Kentucky Press.

Bibliographies of Botanists and Naturalists

Berkeley, Edmund, and Dorothy S. Berkeley. 1969. *Dr. Alexander Garden of Charles Town.* Chapel Hill: University of North Carolina Press.

Earnest, Ernest. 1940. *John and William Bartram: Botanists and Explorers.* Philadelphia: University of Pennsylvania Press.

Frick, George, and Raymond Stearns. 1961. *Mark Catesby.* Urbana: University of Illinois Press.

Gee, Wilson. 1918. "South Carolina Botanists: Biography and Bibliography." *Bulletin of the University of South Carolina* 72: 1–52.

Haygood, Tamara M. 1987. *Henry William Ravenel, 1814–1887.* Tuscaloosa: University of Alabama Press.

Herr, J. M., Jr. 1984. "A Brief Sketch of the Life and Botanical Work of A. C. Moore." *Abstracts. Botanical Society of America,* in *American Journal of Botany* 71 (5, pt. 2): 106–7.

Kastner, Joseph. 1977. *A Species of Eternity.* New York: Alfred A. Knopf.

Reveal, James L. 1996. *America's Botanical Beauty.* Golden, CO: Fulcrum.

Sanders, Albert E., and William D. Anderson Jr. 1999. *Natural History Investigations in South Carolina from Colonial Times to the Present.* Columbia: University of South Carolina Press.

Savage, Henry, Jr. 1970. *Lost Heritage.* New York: William Morrow.

Savage, Henry, Jr., and Elizabeth J. Savage. 1986. *André and François André Michaux.* Charlottesville: University Press of Virginia.

Shuler, Jay. 1995. *Had the Wings: The Friendship of Bachman and Audubon.* Athens: University of Georgia Press.

Slaughter, Thomas P. 1996. *The Nature of John and William Bartram.* New York: Alfred A. Knopf.

Taylor, David. 1998. *South Carolina Naturalists.* Columbia: University of South Carolina Press.

Waring, Joseph I. 1967. *A History of Medicine in South Carolina, 1825–1900.* Columbia: South Carolina Medical Association.

Wilson, Scofield David. 1978. *In the Presence of Nature.* Amherst: University of Massachusetts Press.

Carnivorous Plants

Lloyd, Francis E. 1976. *The Carnivorous Plants.* New York: Dover.

Pietropaolo, James, and Patricia Pietropaolo. 1986. *Carnivorous Plants of the World.* Portland, OR: Timber Press.

Schnell, Donald C. 1976. *Carnivorous Plants of the United States and Canada.* Winston-Salem, NC: John F. Blair.

Slack, Adrian. 1979. *Carnivorous Plants.* Cambridge, MA: MIT Press.

Carolina Bays

Bennett, Stephen H., and John B. Nelson. 1991. *Distribution and Status of Carolina Bays in South Carolina.* Columbia: Nongame and Heritage Trust Section, South Carolina Wildlife and Marine Resources Department.

Johnson, Douglas. 1942. *The Origin of the Carolina Bays.* New York: Columbia University Press.

Savage, Henry, Jr. 1982. *The Mysterious Carolina Bays.* Columbia: University of South Carolina Press.

Cultivated and Ornamental Plants

Bailey, L. H. 1949. *Manual of Cultivated Plants.* 2d ed. New York: MacMillan.

Batson, Wade T. 1984. *Landscape Plants for the Southeast.* Columbia: University of South Carolina Press.

Briggs, Loutrel W. 1951. *Charleston Gardens.* Columbia: University of South Carolina Press.

Foote, Leonard E., and Samuel B. Jones Jr. 1989. *Native Shrubs and Woody Vines of the Southeast.* Portland, OR: Timber Press.

Halfacre, Gordon R., and Anne Shawcroft. 1989. *Landscape Plants of the Southeast.* Raleigh, NC: Sparks Press.

Shaffer, Edward T. H. 1963. *Carolina Gardens.* 3d ed. New York: Devin-Adair.

Economic and Cultural

Butler, Carroll B. 1998. *Treasurers of the Longleaf Pine.* Shalimar, FL: Tarkel.

Hill, Albert F. 1952. *Economic Botany.* New York: McGraw-Hill.

Rosengarten, Dale. 1986. *Row upon Row: Sea Grass Baskets of the South Carolina Lowcountry.* Columbia: McKissick Museum, University of South Carolina.

Schery, Robert W. 1972. *Plants for Man.* Englewood Cliffs, NJ: Prentice-Hall.

Simpson, Beryl B., and Molly C. Ogorzaly. 1986. *Economic Botany: Plants in Our World.* New York: McGraw-Hill.

Wood, Virginia S. 1981. *Live Oaking: Southern Timber for Tall Ships.* Boston: Northeastern University Press.

Edible Wild Plants

Angier, Bradford. 1974. *Field Guide to Edible Wild Plants.* Harrisburg, PA: Stackpole Books.

Berglund, Berndt, and Clare E. Bolsby. 1971. *The Edible Wild*. New York: Charles Scribner's Sons.

Brown, Tom, Jr. 1985. *Tom Brown's Guide to Wild Edible and Medicinal Plants*. New York: Berkley Publishing Group.

Elias, Thomas S., and Peter A. Dykeman. 1982. *Field Guide to North American Edible Wild Plants*. New York: Outdoor Life Books.

Fernald, Merritt L., and Alfred C. Kinsey. 1958. *Edible Wild Plants of Eastern North America*. New York: Harper and Row.

Gibbons, Euell. 1966. *Stalking the Healthful Herbs*. New York: David McKay.

Gibbons, Euell. 1962. *Stalking the Wild Asparagus*. New York: David McKay.

Hall, Alan. 1976. *The Wild Food Trailguide*. New York: Holt, Rinehart and Winston.

Harris, Ben C. 1968. *Eat the Weeds*. Barre, MA: Barre.

Peterson, Lee Allen. 1977. *Edible Wild Plants, A Peterson Field Guide*. Boston: Houghton Mifflin.

Exhibit Books

Blagden, Tom, Jr. 1992. *South Carolina's Wetland Wilderness: The Ace Basin*. Englewood, CO: Westcliffe.

Blagden, Tom, Jr., and Barry Beasley. 1999. *The Rivers of South Carolina*. Englewood, CO: Westcliffe.

Blagden, Tom, Jr., Jane Lareau, and Richard D. Porcher. 1988. *Lowcountry: The Natural Landscape*. Greensboro, NC: Legacy.

Blagden, Tom, Jr., and Thomas Wyche. 1994. *South Carolina's Mountain Wilderness: The Blue Ridge Escarpment*. Englewood, CO: Westcliffe.

Ferns

Dunbar, Lin. 1989. *Ferns of the Coastal Plain*. Columbia: University of South Carolina Press.

Wherry, Edgar T. 1964. *The Southern Fern Guide*. New York: Doubleday.

Folk Remedies and Medicinal Plants

Duke, James A. 1997. *The Green Pharmacy*. New York: St. Martin's Paperbacks.

Erichsen-Brown, Charlotte. 1979. *Medicinal and Other Uses of North American Plants*. New York: Dover.

Foster, Steven, and James A. Duke. 1990. *A Field Guide to Medicinal Plants*. Boston: Houghton Mifflin.

Hamel, Paul B., and Mary U. Chiltoskey. 1975. *Cherokee Plants: Their Uses—A 400-Year History*. Sylva, NC: Herald.

Hudson, Charles M., ed. 1979. *Black Drink: A Native American Tea*. Athens: University of Georgia Press.

Hutchens, Alma R. 1992. *A Handbook of Native American Herbs*. Boston: Shambala.

Hutchens, Alma R. 1973. *Indian Herbology of North America*. Boston: Shambala.

Krochmal, Arnold, and Connie Krochmal. 1973. *A Guide to the Medicinal Plants of the United States*. New York: Quadrangle/New York Times Book Company.

Lewis, W. H., and M. P. F. Elvin-Lewis. 1977. *Medical Botany: Plants Affecting Man's Health*. New York: John Wiley and Sons.

Moerman, Daniel E. 1998. *Native American Ethnobotany*. Portland, OR: Timber Press.

Morton, Julia F. 1974. *Folk Remedies of the Low Country*. Miami, FL: E. A. Seemann.

Moss, Kay K. 1999. *Southern Folk Medicine: 1750–1820*. Columbia: University of South Carolina Press.

Porcher, Francis P. 1869. *Resources of the Southern Fields and Forests*. Charleston, SC: Walker, Evans and Cogswell.

Tyler, Varro E. 1994. *Herbs of Choice: Therapeutic Use of Phytomedicinals*. Binghamton, NY: Haworth Press.

Hiking Guides in South Carolina

Dehart, Allen. 1994. *Hiking South Carolina Trails*. 3d ed. Old Saybrook, CT.: Globe Pequot Press.

Gaddy, L. L. 2000. *A Naturalist's Guide to the Southern Blue Ridge Front*. Columbia: University of South Carolina Press.

Giffen, Morrison. 1997. *South Carolina: A Guide to Backcountry Travel and Adventure*. Asheville, NC: Out There Press.

Manning, Phillip. 1995. *Palmetto Journal: Walks in the Natural Areas of South Carolina*. Winston-Salem, NC: John F. Blair.

Historical Books

Drayton, John M. 1802. *A View of South Carolina*. Charleston, SC: W. P. Young. Reprint, Spartanburg, SC: The Reprint Company 1972.

Ramsey, David. 1858. *History of South Carolina*. Newberry, SC: W. J. Duffie. Reprint, Spartanburg, SC: The Reprint Company, 1960.

Native Orchids

Correll, Donovan S. 1978. *Native Orchids of North America.* Stanford, CA: Stanford University Press.

Gupton, Oscar W., and Fred C. Swope. 1986. *Wild Orchids of the Middle Atlantic States.* Knoxville: University of Tennessee Press.

Luer, Carlyle A. 1972. *The Native Orchids of Florida.* New York: New York Botanical Garden.

Luer, Carlyle A. 1975. *The Native Orchids of the United States and Canada Excluding Florida.* New York: New York Botanical Garden.

Natural History Guides and References

Barry, John M. 1980. *Natural Vegetation of South Carolina.* Columbia: University of South Carolina Press.

Braun, E. Lucy. 1950. *Deciduous Forests of Eastern North America.* Philadelphia: Blakistin.

Conner, R. C. 1998. "South Carolina's Forests, 1993." *Resource Bulletin SRS-25.* Asheville, NC: USDA Forest Service, Southern Research Station.

Cowdry, A. E. 1996. *This Land, This South: An Environmental History.* Lexington: University of Kentucky Press.

Dennis, John V. 1988. *The Great Cypress Swamps.* Baton Rouge: Louisiana State University Press.

Duncan, Wilbur H., and Marion B. Duncan. 1987. *Seaside Plants of the Gulf and Atlantic Coasts.* Washington, DC: Smithsonian Institution Press.

Fenneman, N. M. 1938. *Physiography of the Eastern United States.* New York: McGraw-Hill.

Godfrey, Michael A. 1980. *The Piedmont: A Sierra Club Naturalist's Guide.* San Francisco: Sierra Club Books.

Hackney, C. T., S. M. Adams, and W. H. Martin. 1992. *Biodiversity of the Southeastern United States: Aquatic Communities.* New York: John Wiley and Sons.

Horton, J. W., Jr., and V. A. Zullo. 1991. *The Geology of the Carolinas.* Knoxville: University of Tennessee Press.

Jones, S. M. 1988. "Old-growth, Steady-state Forests within the Piedmont of South Carolina." Ph.D. dissertation, Clemson University.

Kovasik, Charles F., and John J. Winberry. 1987. *South Carolina: The Making of a Landscape.* Columbia: University of South Carolina Press.

Lyons, Janet, and Sandra Jordan. 1989. *Walking the Wetlands.* New York: John Wiley and Sons.

Nelson, John B. 1986. *The Natural Communities of South Carolina.* Technical Report. Columbia: South Carolina Wildlife and Marine Resources Department.

Pirkle, E. C., and W. H. Yahi. 1977. *Natural Regions of the United States.* Dubuque, IA: Kendall Hunt.

Porcher, Richard D. 1985. *A Teacher's Field Guide to the Natural History of The Bluff Plantation Wildlife Sanctuary.* New Orleans, LA: Kathleen O'Brien Foundation.

Radford, A. E., D. K. S. Otte, L. J. Otte, J. R. Massey, and P. D. Whitson. 1981. *Natural Heritage Classification, Inventory, and Information.* Chapel Hill: University of North Carolina Press.

Shafale, Michael P., and Alan S. Weakley. 1990. *Classification of the Natural Communities of North Carolina. Third Approximation.* Raleigh: North Carolina Department of Environment, Health, Natural Resources, Division of Parks and Recreation, Natural Heritage Program.

Smith, Richard M. 1989. *Wild Plants in America.* New York: John Wiley and Sons.

Tansey, J. B. 1986. "Forest Statistics for the Piedmont of South Carolina, 1986." *Resource Bulletin SE-89.* Asheville, NC: USDA Forest Service, Southeastern Forest Experiment Station.

Wells, B. W. 1932. *The Natural Gardens of North Carolina.* Chapel Hill: University of North Carolina Press.

Parasitic Vascular Plants

Kuijt, Job. 1969. *The Biology of Parasitic Flowering Plants.* Berkeley: University of California Press.

Photography of Wildflowers

Adams, Kevin, and Mary Casstevens. 1996. *Wildflowers of the Southern Appalachians: How to Photograph and Identify Them.* Winston-Salem, NC: John F. Blair.

Shaw, John. 1987. *Closeups in Nature.* New York: AMPHOTO.

Poisonous Plants

Kingsbury, John M. 1964. *Poisonous Plants of the United States and Canada.* Englewood Cliffs, NJ: Prentice-Hall.

Westbrooks, Randy G., and James W. Preacher. 1986. *Poisonous Plants of Eastern North America.* Columbia: University of South Carolina Press.

Rice Culture

Doar, David. 1970. *Rice and Rice Planting in the South Carolina Low Country.* Charleston, SC: Charleston Museum.

Heyward, Duncan C. 1937. *Seed from Madagascar.* Chapel Hill: University of North Carolina Press.

Littlefield, Daniel C. 1991. *Rice and Slaves.* Chicago: University of Illinois Press.

Wood, Peter H. 1974. *Black Majority: Negroes in Colonial South Carolina from 1670 through the Stono Rebellion.* New York: Alfred A. Knopf.

Technical Manuals and Floras

Aulbach-Smith, C. A., and S. J. deKozlowski. 1996. *Aquatic and Wetland Plants of South Carolina.* 2d ed. Columbia: South Carolina Department of Natural Resources and the South Carolina Aquatic Plant Management Council.

Batson, Wade T. 1984. *Genera of Eastern Plants.* 3d ed., rev. Columbia: University of South Carolina Press.

Conquist, A. J. 1980. *Vascular Flora of the Southeastern United States.* Volume 1, *Asteraceae.* Chapel Hill: University of North Carolina Press.

Godfrey, Robert K., and Jean W. Wooten. 1981. *Aquatic and Wetland Plants of the Southeastern United States: Dicotyledons.* Athens: University of Georgia Press.

Godfrey, Robert K., and Jean W. Wooten. 1979. *Aquatic and Wetland Plants of the Southeastern United States: Monocotyledons.* Athens: University of Georgia Press.

Radford, Albert E., Harry E. Ahles, and C. Ritchie Bell. 1968. *Manual of the Vascular Flora of the Carolinas.* Chapel Hill: University of North Carolina Press.

Small, John K. 1933. *Manual of the Southeastern Flora.* Chapel Hill: University of North Carolina Press.

Strausbaugh, P. D., and Earl L. Core. 1977. *Flora of West Virginia.* Grantsville, WV: Seneca Books.

Tobe, John D., et al. 1988. *Florida Wetland Plants: An Identification Manual.* Tallahassee: Florida Department of Environmental Protection.

Weakley, A. S. 2020. *Flora of South Carolina, A Derivative of the Flora of Southeastern North America.* Available online at http://herbarium.unc.edu/flora.htm.

Trees and Shrubs

Brown, Clair A., and Glen N. Montz. 1986. *Baldcypress: The Tree Unique, the Wood Eternal.* Baton Rouge, LA: Claitor's Publishing Division.

Brown, Claud L., and L. Katherine Kirkman. 1990. *Trees of Georgia and Adjacent States.* Portland, OR: Timber Press.

Duncan, Wilbur H., and Marion B. Duncan. 1988. *Trees of the Southeastern United States.* Athens: University of Georgia Press.

Elias, Thomas S. 1980. *The Complete Trees of North America.* New York: Times Mirror Magazines.

Godfrey, Robert K. 1988. *Trees, Shrubs, and Woody Vines of Northern Florida and Adjacent Georgia and Alabama.* Athens: University of Georgia Press.

Peattie, Donald C. 1966. *A Natural History of Trees of Eastern and Central North America.* New York: Bonanza Books.

Walker, Lawrence C. 1990. *Forests: A Naturalist's Guide to Trees and Forest Ecology.* New York: John Wiley and Sons.

Walker, Lawrence C. 1991. *The Southern Forest: A Chronicle.* Austin: University of Texas Press.

Woody Vines

Duncan, Wilbur H. 1975. *Woody Vines of the Southeastern United States.* Athens: University of Georgia Press.

LITERATURE CITED

Ajilvsgi, G. 1984. *Wildflowers of Texas*. Fredericksburg, TX: Shearer.

Anderson, Loran C. 1985. "*Forestiera godfreyi* (Oleaceae), A New Species from Florida and South Carolina." *Sida* 11 (1): 1–5.

Azuma, H., M. Toyota, Y. Asakawa, R. Yamaoka, J. Garcia-Franco, G. Dieringer, L. B. Thien, and S. Kawano. 2005. "Chemical Divergence in Floral Scents of *Magnolia* and Allied Genera (Magnoliaceae)." *Plant Species Biology* 12 (2–3): 69–83.

Bailey, L. H. 1949. *Manual of Cultivated Plants*. 2d ed. New York: MacMillan.

Baird, W. V., & J. L. Riopel. 1986. Life History Studies of *Conopholis americana* (Orobanchaceae)." *The American Midland Naturalist* 116 (1): 140–51.

Barden, L. S. 1997. "Historic Prairies in the Piedmont of North and South Carolina, USA." *Natural Areas Journal* 17: 149–52.

Barry, John M. 1980. *Natural Vegetation of South Carolina*. Columbia: University of South Carolina Press.

Bartram, William. 1791. *Travels through North & South Carolina, Georgia, East & West Florida, the Cherokee Country, the Extensive Territories of the Muscogulges, or Creek Confederacy, and the Country of the Choctaws*. Reprint, Mark Van Doren, ed. New York: Dover Publications, 1955.

Bennett, Stephen H., and John B. Nelson. 1991. *Distribution and Status of Carolina Bays in South Carolina*. Columbia: Nongame and Heritage Trust Section, South Carolina Wildlife and Marine Resources Department.

Berkeley, Edmund, and Dorothy S. Berkeley. 1969. *Dr. Alexander Garden of Charles Town*. Chapel Hill: University of North Carolina Press.

Beverley, Robert. 1705. *The History and Present State of Virginia*.

Blackwell, Amy H., and Patrick D. McMillan. 2013. "Collected in South Carolina 1704–1707: The Plants of Joseph Lord." *Phytoneuron* 59: 1–15.

Blomquist, H. L. 1948. "*Asplenium mananthes* in South Carolina." *American Fern Journal* 38 (4): 171–76.

Brown, D. S. 1953. *A City without Cobwebs: Rock Hill, South Carolina*. Columbia: University of South Carolina Press.

Butterweck, V., T. Bockers, B. Korte, W. Wittkowski, and H. Winterhoff. 2002. "Long-term Effects of St. John's Wort and Hypericin on Monoamine Levels in Rat Hypothalmus and Hippocampus." *Brain Research* 930 (1–2): 21–29.

Case, Frederick W., Jr., and Roberta B. Case. 1997. *Trilliums*. Portland, OR: Timber Press.

Catesby, Mark, and George Edwards. *The natural history of Carolina, Florida, and the Bahama Islands: containing the figures of birds, beasts, fishes, serpents, insects, and plants: particularly the forest-trees, shrubs, and other plants, not hitherto described, or very incorrectly figured by authors. Together with their descriptions in English and French. To which are added, observations on the air, soil, and waters: with remarks upon agriculture, grain, pulse, roots, &c. To the whole is prefixed a new and correct map of the countries treated of*. [London: Printed for C. Marsh etc, 1754]. Retrieved from the Library of Congress, https://lccn.loc.gov/11000175.

Chapman, A. W. 1860. *Flora of the Southern United States*. New York: Ivison, Phinney.

Childs, Arney Robinson, ed. 1947. *The Private Journal of Henry William Ravenel, 1859–1887*. Columbia: University of South Carolina Press.

Coker, William C., and Henry R. Totten. 1945. *Trees of the Southeastern States*. Chapel Hill: University of North Carolina Press.

Cross, J. K. 1973. "Tar Burning, Forgotten Art?" *Forests and People* 23 (2): 21–23.

Cushman, Laary J., Vincent P. Richardsm and Patrick D. McMillan. 2020. "*Micranthes petiolaris* variety *shealyi*: A New Variety of *Micranthes* (Section *Stellares*, Saxifragaceae) from South Carolina." *Phytotaxa* 452 (2).

Dauncey, Elizabeth A., and Melanie-Jayne R. Howes. 2020. *Plants That Cure: Plants as a Source for Medicines, from Pharmaceuticals to Herbal Remedies*. Princeton, NJ: Princeton University Press.

Duke, James A. 1997. The Green Pharmacy. New York: St. Martin's Paperbacks.

Dunbar, Lin. 1989. *Ferns of the Coastal Plain*. Columbia: University of South Carolina Press.

Duncan, Wilbur H., and Marion B. Duncan. 1987. *Seaside Plants of the Gulf and Atlantic Coasts*. Washington, DC: Smithsonian Institution Press.

Duncan, Wilbur H., and Marion B. Duncan. 1988. *Trees of the Southeastern United States*. Athens: University of Georgia Press.

Fernald, Merritt L. 1950. *Gray's Manual of Botany*. 8th ed. New York: American Book.

Fernald, Merritt L., and Alfred C. Kinsey. 1958. *Edible Wild Plants of Eastern North America.* New York: Harper and Row.

Foote, Leonard E., and Samuel B. Jones Jr. 1989. *Native Shrubs and Woody Vines of the Southeast.* Portland, OR: Timber Press.

Foster, Steven, and James A. Duke. 1990. *A Field Guide to Medicinal Plants.* Boston: Houghton Mifflin.

Gaddy, L. L. 1986. "A New Heartleaf (*Hexastylis*) from Transylvania County, North Carolina." *Brittonia* 38: 82–85.

Gaddy, L. L. 2008. "A New Sessile-flowered *Trillium* (Liliaceae: subgenus *Phyllantherum*) from South Carolina." *Phytologia* 90 (3): 374–82.

Gaddy, L. L., T. H. Carter, B. Ely, S. Sakaguchi, A. Matsuo, and Y. Suyama. 2020. "*Shortia brevistyla comb. et stat. nov.,* (Diapensiaceae), a Narrow Endemic from the Headwaters of the Catawba River in North Carolina, USA." *Phytologia* 101 (2): 113–19.

Gee, Wilson. 1918. "South Carolina Botanists: Biography and Bibliography." *Bulletin of the University of South Carolina* 72: 1–52.

Gibbes, Lewis R. 1859. "Botany of Edings Bay." *Proceedings of the Elliott Society of Natural History* 1 (4): 241–48.

Godfrey, Robert K., and Jean W. Wooten. 1981. *Aquatic and Wetland Plants of the Southeastern United States: Dicotyledons.* Athens: University of Georgia Press.

Godfrey, Robert K., and Jean W. Wooten. 1979. *Aquatic and Wetland Plants of the Southeastern United States: Monocotyledons.* Athens: University of Georgia Press.

Hale, Peter M., and Moses A. Curtis. 1883. *The Woods and Timbers of North Carolina.* Raleigh, NC: P. M. Hale Publisher.

Hardin, James W., and Committee. 1977. Vascular Plants. In *Endangered and Threatened Plants and Animals of North Carolina,* edited by J. E. Cooper, S. S. Robinson, and J. B. Funderburg. Raleigh: North Carolina State Museum of Natural History.

Haslam, S. M. 1971. "Community Regulation in *Phragmites communis* Trinius I. Monodominant Stands." *Journal of Ecology* 59: 65–73.

Haygood, Tamara M. 1987. *Henry William Ravenel, 1814–1887.* Tuscaloosa: University of Alabama Press.

Haynes, Jake E., Whitney D. Phillips, Alexander Krings, Nathan P. Lynch, and Thomas G. Ranney. 2020. "Revision of *Fothergilla* (Hamamelidaceae), Including Resurrection of *F. parviflora* and a New Species, *F. milleri.*" *PhytoKeys* 144: 57–80.

Heafner, Kerry D. 2001. "*Pellaea wrightiana* Hooker (Pteridaceae) in North Carolina Revisited with a New Record for Eastern North America and a Key to *Pellaea* Species in the Carolinas." *Castanea* 68: 319–26.

Hill, Steven R., and Charles N. Horn. 1997. "Additions to the Flora of South Carolina." *Castanea* 62 (3): 194–208.

Hudson, Charles M., ed. 1979. *Black Drink: A Native American Tea.* Athens: University of Georgia Press.

Hunt, Kenneth W. 1947. "The Charleston Woody Flora." *American Midland Naturalist* 37: 670–756.

Hunt, Kenneth W. 1942. "Ferns of the Vicinity of Charleston." *Charleston Museum Leaflet* no. 17.

Hunt, Kenneth W. 1943. "Floating Mats on a Southeastern Coastal Plain Reservoir." *Bulletin of the Torrey Botanical Club* 70 (5): 481–88.

Justice, William S., and C. Ritchie Bell. 1968. *Wild Flowers of North Carolina.* Chapel Hill: University of North Carolina Press.

Kelly, Howard A. 1914. *Some American Medical Botanists.* Troy, NY: Southworth.

Kingsbury, John M. 1964. *Poisonous Plants of the United States and Canada.* Englewood Cliffs, NJ: Prentice-Hall.

Kirkman, W. B., and J. R. Ballington. 1990. "Creeping Blueberries (Ericaceae: *Vaccinium* sect. *Herpothamnus*)—a New Look at. *V. crassifolium* including *V. sempevirens.*" *Systemic Botany* 15 (4): 679–99.

Kral, R., and G. L. Nesom. 2003. "Two New Species of *Liatris* Series Graminifoliae (Asteraceae: Eupatorieae) from the Southeastern United States." *Sida, Contributions to Botany* 20 (4): 1573–83.

Krochmal, Arnold, and Connie Krochmal. 1973. *A Guide to the Medicinal Plants of the United States.* New York: Quadrangle/New York Times Book Company.

Kron, Kathleen A., and Mike Creel. 1999. "A New Species of Deciduous Azalea (*Rhododendron* section *Pentanthera;* Ericaceae) from South Carolina." *Novon* 9: 337–80.

Lance, R. 2014. *Haws, A Guide to Hawthorns of the Southeastern United States.* Mills River, NC: Author.

Lawson, John. 1709. *A New Voyage to Carolina.* Reprint edition, Readex Microprint.

LeBlond, Richard J., Edward E. Schilling, Richard D. Porcher, and Bruce A. Sorrie. 2007. *Eupatorium paludicola*, sp. nov. (Asteraceae): A new species form the coastal plain of North and South Carolina." *Rhodora* 109: 137–77.

Martin, C. E., and S. K. Francke. 2015. "Root Aeration and Function of Bald Cypress Knees (*Taxodium distichum*)." *International Journal of Plant Sciences* 176 (2): 170–73.

Martinez, R. J. 1959. *The Story of Spanish Moss and Its Relatives.* New Orleans, LA: Home Publications.

McMillan, Patrick D. 2003. "Noteworthy collections: South Carolina." *Castanea* 68 (4): 345–47.

McMillan, Patrick D. 2007. "Andre Michaux, Botanist and Explorer of the South Carolina Upstate." In *An Oconee Bell Celebration,* edited by B. Sanders. Athens, GA: Fevertree Press.

McMillan, Patrick D., A. H. Blackwell, C. Blackwell, and M. Spencer. 2013. "The Vascular Plants in the Mark Catesby Collection at the Sloane Herbarium, with Notes on Their Taxonomic and Ecological Significance." *Phytoneuron* 2013 (7): 1–37.

McMillan, Patrick D., and Richard D. Porcher. 2002. "Noteworthy collections: South Carolina." *Castanea* 70 (3): 237–40.

McMillan, Patrick D., and Richard D. Porcher. 2005. "Noteworthy collections: South Carolina." *Castanea* 70 (3): 237–40.

McMillan, P. D., R. K. Peet, R. D. Porcher, and B. A. Sorrie. 2002. "Noteworthy Botanical Collections from the Fire-maintained Pineland and Wetland Communities of the Coastal Plain of the Carolinas and Georgia." *Castanea* 67 (1): 61–83.

McMillan, P. D., E. B. Pivorun, R. D. Porcher, C. Davis, D. Whitten, and K. Wade. 2018. "Three Remarkably Disjunct Fern Species Discovered in Pickens County, South Carolina." *Phytoneuron* 2018 (21): 1–5.

Melton, Frank A., and William Schriever. 1933. "The Carolina Bays: Are They Meteorite Scars?" *Journal of Geology* 41: 52–66.

Mills, Robert. 1826. *Statistics of South Carolina.* Charleston: privately printed.

Morris, Karen, ed. 1983. *Guide to the Foothills Trail.* Greenville, SC: Foothills Trail Conference.

Morton, Julia F. 1974. *Folk Remedies of the Low Country.* Miami, FL: E. A. Seemann.

Naturaland Trust. 1994. *Guide to the Mountain Bridge Trails.* Greenville, SC: Author.

Nesom, G. L. 2015. "Taxonomy of *Galactia* (Fabaceae) in the USA." *Phytoneuron* 2015 (42): 1–54.

Peet, R. K. 1993. "A Taxonomic Study of *Aristida stricta* and *A. beyrichiana*." *Rhodora* 9: 209–10.

Pietropaolo, James, and Patricia Pietropaolo. 1986. *Carnivorous Plants of the World.* Portland, OR: Timber Press.

Platt, S. G., and J. F. Townsend. 1996. "Noteworthy Collections: *Pellaea wrightiana* in Pickens County, South Carolina." *Castanea* 61 (4): 397–98.

Platt, S. G., and C. G. Brantley. 1997. "Canebrakes: An Ecological and Historical Perspective." *Castanea* 62 (1): 8–21.

Porcher, Francis P. 1869. *Resources of the Southern Fields and Forests.* Charleston, SC: Walker, Evans and Cogswell.

Porcher, Richard D. 1987. "Rice Culture in South Carolina: A Brief History, the Role of the Huguenots, and Preservation of its Legacy." *Transactions of the Huguenot Society* 92: 1–22.

Porcher, Richard D. 1995. *Wildflowers of the Carolina Lowcountry and Lower Pee Dee.* Columbia: University of South Carolina Press.

Porcher, Richard D., and Douglas A. Rayner. 2001. *A Guide to the Wildflowers of South Carolina.* Columbia: University of South Carolina Press.

Radford, Albert E. 1959. "A Relic Plant Community in South Carolina." *Journal of the Elisha Mitchell Scientific Society* 75: 33–34.

Radford, Albert E., Harry E. Ahles, and C. Ritchie Bell. 1968. *Manual of the Vascular Flora of the Carolinas.* Chapel Hill: University of North Carolina Press.

Radford, Albert E., et al. 1974. *Vascular Plant Systematics.* New York: Harper and Row.

Rayner, Douglas A., and J. Henderson. 1980. "*Vaccinium semperevirens* (Ericaceae), a New Species from Atlantic White-Cedar Bogs in the Sandhills of South Carolina." *Rhodora* 82: 503–7.

Rembert, David H. 1979. "The Carolina Plants of André Michaux." *Castanea* 44: 65–80.

Rembert, David H. 1980. *Thomas Walter: Carolina Botanist.* Museum Commission Bulletin, no. 5. Columbia: South Carolina Museum Commission.

Reveal, James L. 1996. *America's Botanical Beauty.* Golden, CO: Fulcrum.

Rickett, Harold W., and the New York Botanical Garden. 1967. *Wild Flowers of the United States: Volume 2, The Southeastern States.* New York: New York Botanical Garden and McGraw-Hill.

Salley, A. S., Jr. 1919. "The Introduction of Rice into South Carolina." *Bulletin of the Historical Commission of South Carolina,* no. 6. Columbia, SC: State Company.

Sanders, Albert E., and William D. Anderson Jr. 1999. *Natural History Investigations in South Carolina from Colonial Times to the Present.* Columbia: University of South Carolina Press.

Savage, Henry, Jr. 1982. *The Mysterious Carolina Bays.* Columbia: University of South Carolina Press.

Savage, Henry, Jr., and Elizabeth J. Savage. 1986. *André and François André Michaux.* Charlottesville: University Press of Virginia.

Schilling, Edward E., Aaron Floden, Jayne Lampley, Thomas S. Patrick, and Susan B. Farmer. 2017. A new species in *Trillium* subgen. *Delostylium* (Melanthiaceae, Paridae). *Phytotaxa* 296: 287–91.

Semple, John C., T. Shea, M. El-Swesi, H. Rahman, and Y. Ma. 2016. "A Multivariate Study of the *Solidago stricta* Complex (Asteraceae: Astereae: *S.* subsect. *Maritimae*)." *Phytoneuron* 2016 (86): 1–34.

Skinner, M. W., and B. A. Sorrie. 2002. "Conservation and Ecology of *Lilium pyrophilum,* a New Species of Liliaceae from the Sandhills Region of the Carolinas and Virginia, USA." *Novon* 12: 94–105.

Slack, Adrian. 1979. *Carnivorous Plants.* Cambridge, MA: MIT Press.

Small, John K. 1927. "Among Floral Aborigines." *Journal of the New York Botanical Garden* 28 (325): 1–20.

Small, John K. 1917. "Cactus Hunting on the Coast of South Carolina." *Journal of the New York Botanical Garden* 18: 237–46.

Small, John K. 1933. *Manual of the Southeastern Flora.* Chapel Hill: University of North Carolina Press.

Sorrow, James A., John F. Townsend, and Richard W. Christie. 1999. *South Carolina Plant and Fish Atlas: Helixatl.* Beta CD-ROM. Columbia: South Carolina Department of Natural Resources, and Clemson, SC: Clemson University.

Stearn, William T. 1992. *Botanical Latin.* 4th ed. Brunel House, England: David & Charles.

Strausbaugh, P. D., and Earl L. Core. 1977. *Flora of West Virginia.* Grantsville, WV: Seneca Books.

Taylor, David. 1998. *South Carolina Naturalists.* Columbia: University of South Carolina Press.

Taylor, M. S. 1938. "Filmy Ferns in South Carolina." *Journal of the Elisha Mitchell Scientific Society,* 54 (2): 345–48.

Thomas, D. T. A. R. Ahedor, C. F. Williams, C. de Pamphilis, D. J. Crawford, and Q. Xiang. 2008. "Genetic Analysis of a Broad Hybrid Zone in *Aesculus* (Sapindaceae): Is There Evidence of Long-distance Pollen Dispersal? *International Journal of Plant Sciences* 169 (5): 647–57.

Trimble, S. W. 1972. *Man-induced Erosion on the Southern Piedmont 1700–1970.* Ankeny, IA: Soil Conservation Society of America.

Tyler, V. E. 1994. *Herbs of Choice: Therapeutic Use of Phytomedicinals.* Binghamton, NY: Haworth Press.

Weakley, Alan S. 2001. *Flora of the Carolinas and Virginia.* Chapel Hill, NC: Nature Conservancy, Southeast Regional Office.

Weakley, Alan S. 2020. *Flora of the Southeastern United States.* Chapel Hill: University of North Carolina Herbarium, North Carolina Botanical Garden.

Weakley, Alan S., R. Kevan Schoonover McClelland, Richard J. LeBlond, Keith A. Bradley, James F. Matthews, Chad Anderson, Alan R. Franck, and James Lange. 2019. "Studies in the Vascular Flora of the Southeastern United States: V". *Journal of the Botanical Research Institute of Texas* 13 (1): 107–29. https://journals .brit.org/jbrit/article/view/832.

Wharton, C. H. 1978. *The Natural Environments of Georgia.* Atlanta: Geologic and Water Resources Division and Resource Planning Section, Office of Planning and Resource, Georgia Department of Natural Resources.

Wieboldt, T. F., and J. C. Semple. 2003. "*Solidago faucibus* (Asteraceae: Astereae), a new mesic forest goldenrod from the Appalachian mountains." *Sida* (4): 1595–1603.

Wooten, J. W. 1973. "Taxonomy of Seven Species of *Sagittaria* from Eastern North America." *Brittonia* 25: 64–74.

INDEX

Boldface indicates page number of a species or habitat photograph.

PHOTOGRAPHY CREDITS

All photographs by Richard D. Porcher unless otherwise indicated.

Keith A. Bradley: 0056 (right), 0133, 0293 (2 images), 0295, 0303 (2 images), 0412, 0413 (left), 0517

Alan Cressler: 0043, 0135 (top), 0455 (2 images), 0485 (left)

Cecelia N. Dailey: 0928

Jim Fowler: 0071, 0153, 0173, 0194B, 0241 (2 images), 0685, 0734

Patrick D. McMillan: 0007, 0019, 0020 (inset), 0021, 0024B, 0031, 0036, 0039 (main), 0040 (2 images), 0042, 0048A (inset), 0048B (2 images), 0049, 0060, 0061 (left), 0062 (2 images), 0063 (2 images), 0068, 0069 (2 images), 0076, 0077 (2 images), 0087 (2 images), 0093, 0094 (2 images), 0096 (bottom), 0125 (right), 0126 (right), 0130, 0135 (bottom), 0136 (2 images), 0141, 0144 (right), 0151 (2 images), 0158, 0159 (2 images), 0165 (right), 0175 (2 images), 0181 (left), 0185 (2 images), 0193 (2 images), 0195 (main), 0198, 0199A, 0199B, 0209, 0216 (inset), 0217 (main), 0230 (right), 0242, 0243, 0250, 0260 (inset), 0261 (2 images), 0263A (main), 0266, 0270, 0304 (2 images), 0309 (2 images), 0332 (main), 0339, 0346 (right), 0371A, 0371B, 0372, 0374, 0376, 0377 (4 images), 0384, 0388, 0392 (2 images), 0399, 0408 (2 images), 0413 (right), 0414 (2 images), 0419 (main), 0422 (2 images), 0426 (left), 0468, 0474, 0497 (inset), 0500 (2 images), 0510 (right), 0511, 0512, 0588B (inset), 0593 (center), 0633 (2 images), 0634 (3 images), 0662 (2 images), 0691, 0744 (main), 0753, 0784, 0892 (left)

Waynna McMillan: 0179 (left)

Edward Pivorun: 0006, 0008, 0018 (inset), 0023, 0024A (inset), 0025 (inset), 0027, 0028, 0032, 0033, 0034, 0035 (2 images), 0039 (inset), 0041, 0044 (inset), 0047, 0055 (2 images), 0056 (left), 0057, 0059 (2 images), 0073 (inset), 0075 (right), 0084 (inset), 0084 (inset), 0096 (top), 0098 (bottom), 0102, 0104 (2 images), 0106, 0110 (left), 0111, 0114, 0116 (inset), 0118 (left), 0120 (2 images), 0122 (2 images), 0123 (2 images), 0138 (right), 0144 (left), 0146 (top), 0147 (right), 0148 (2 images), 0149 (2 images), 0160, 0161 (inset), 0162, 0164, 0165 (left), 0166 (right), 0167 (right), 0170 (top), 0180, 0188, 0190 (inset), 0194A, 0195 (inset), 0196 (inset), 0218, 0219, 0226, 0233 (lower right), 0238 (right), 0245, 0247 (inset), 0252 (2 images), 0254 (2 images), 0256, 0258B (2 images), 0263A (inset), 0263B (inset), 0275, 0279A, 0282, 0284, 0286, 0288 (inset), 0300 (inset), 0305, 0306 (2 images), 0313 (left), 0315, 0321 (inset), 0323A, 0327, 0336 (left), 0337 (inset), 0338 (inset), 0343 (inset), 0350, 0364 (right), 0379, 0407 (right), 0421 (2 images), 0425, 0484 (right), 0485 (right), 0572 (upper right), 0648, 0662 (top), 0804, 0812A (2 images), 0812B (2 images), 0816 (top), 0973 (right), 0996, 1001, 1019B (2 images)

Bruce A. Sorrie: 0469

Will Stuart: 0508

David B. White: 0269 (3 images), 0299 (right), 0386, 0593 (2 images), 0655 (center)